Geometry

Explorations and Applications

Authors

Douglas B. Aichele

Patrick W Hopfensperger

Miriam A. Leiva

Marguerite M. Mason

Stuart J. Murphy

Vicki J. Schell

Matthias C. Vheru

McDougal Littell
A HOUGHTON MIFFLIN COMPANY

Evanston, Illinois • Boston • Dallas

Authors

Douglas B. Aichele

Regents Professor of Mathematics,
Oklahoma State University, Stillwater, Oklahoma

Patrick W. Hopfensperger

Mathematics Teacher, Homestead High School,
Mequon, Wisconsin

Miriam A. Leiva

Cone Distinguished Professor for Teaching and Professor
of Mathematics, University of North Carolina at Charlotte

Marguerite M. Mason

Assistant Professor of Mathematics Education,
University of Virginia, Charlottesville, Virginia

Stuart J. Murphy

Visual Learning Specialist, Evanston, Illinois

Vicki J. Schell

Professor of Mathematics and Mathematics Education,
Lenoir-Rhyne College, Hickory, North Carolina

Matthias C. Vheru

Mathematics Teacher, Locke High School,
Los Angeles, California

2001 Impression

ISBN: 0-395-72285-3   8 9 10 11 12   VH   05 04 03 02

Internet Web Site: http://www.hmco.com

Contents

Patterns, Lines, and Planes

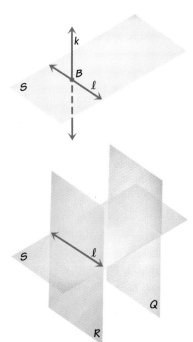

1.4 *Representing planes* 23

Applications

Interview:
Latecia Leavy
Tumbler
12, 17, 37

——— Connection ———

Astronomy	7	**Manufacturing**	31
Communication	13	**Science**	39
History	19	**Crafts**	45
Art	26		

Applying Algebra 16, 32, 41, 43, 44, 46

Additional applications: chemistry, travel, social studies, history, skiing, racing, boating

Triangles and Polygons

2

2.2 *Sorting Triangles*   64

Applications

Interview:
Virgil Lueth
Mineralogist
68, 82, 85, 98

――――――― Connection ―――――――

Biology	61	**Crafts**	77
Communications	62	**Art**	83
History	70	**Optics**	91

Applying Algebra 66, 67, 68, 80, 84, 86, 88, 104

Additional applications: architecture, bicycling, sports, computer-aided design, games

Reasoning in Geometry

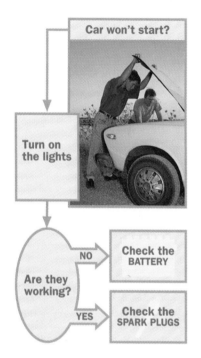

Car won't start?

Turn on the lights

NO → Check the BATTERY

Are they working?

YES → Check the SPARK PLUGS

3.7 *Auto Maintenance*   151

Applications

Interview: *Mary-Jacque Mann*
Forensic Scientist 128, 133

———— Connection ————

Zoology	115	**Language**	139
Navigation	121	**History**	146
Measurement	122	**Auto Maintenance**	151
Literature	127		

Applying Algebra 112, 115, 142, 144

Additional applications: personal finance, marketing, carpentry, archeology, architecture, design, travel

Coordinates in Geometry

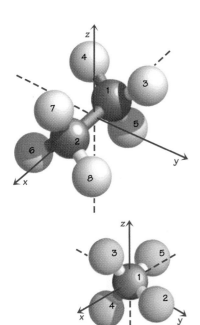

4.6 *Ethane and methane molecules*
201, 205

Applications

Interview: *Marc Hannah*
Computer-system Designer 170, 198, 206

─────── Connection ───────

Puzzles   171   **Quilt Patterns**   185
Engineering   177   **Architecture**   192
City Planning   178

Applying Algebra   170, 177, 185, 193, 196, 200

Additional applications: sports, recreation, accessibility, city planning, carpentry, printing, chemistry

Parallel Lines

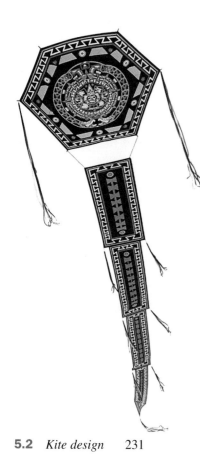

5.1   **Parallel Lines and Transversals**   219
      Exploration: *Angles and Transversals*   221

5.2   **Properties of Parallel Lines**   226

5.3   **Types of Proofs**   234

5.4   **Conditions for Parallel Lines**   242
      Exploration: *Drafting Parallel Lines*   242

5.5   **Proving Theorems About Parallels**   249

5.6   **Parallels in Space**   256
      Exploration:
      Investigating Planes and Their Intersections   256

5.7   **Constructing Parallels and Perpendiculars**   263

Review   **Chapter Review**   272

Assessment   **Ongoing Assessment**   225, 233, 241, 248,
                                                    254, 262, 269

             Assess Your Progress   233, 255, 269

             Journal writing   233, 255, 269

             Portfolio Project:
             Creating Technical Drawings   270

             Chapter Assessment   274

5.2 *Kite design*   231

Applications

Conjectures About Triangles

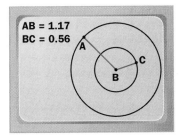

AB = 1.17
BC = 0.56

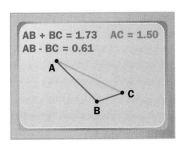

AB + BC = 1.73   AC = 1.50
AB - BC = 0.61

6.1 *Exploring using technology*   279

                                             312, 318, 324

               Assess Your Progress   291, 312, 325

               Journal writing   291, 312, 325

               Portfolio Project: *Building a Mobile*   326

               Chapter Assessment   330

               Cumulative Assessment *Ch. 4–6*   334

Applications

Interview:
Madeleine Fleming
Optical Physicist
290, 295, 317

───── Connection ─────

Probability	283	**Aerial**
Interior Design	284	**Photography** 310
Rescue Safety	297	**Physics** 323
Astronomy	305	

Applying Algebra   286, 289, 310, 315, 317, 322,
323, 325, 330

Additional applications: art, pottery, biology, cars,
rock climbing, architecture

Quadrilaterals, Areas, and Volumes

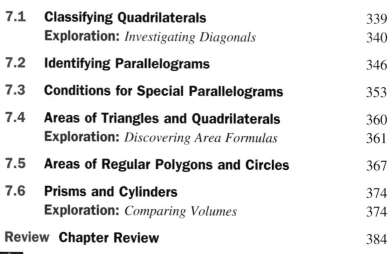

7.5 *Grass for the Silverdome*   367

Applications

Interview:
Walt Stone
Alderman
366, 380

———— Connection ————

Engineering	344	**Biology**	371
Machinery	351	**Consumer**	
Social Studies	357	**Economics**	379
Gardening	365		

Applying Algebra 341, 351, 354, 357, 365, 380

Additional applications: auto repair, carpentry, drafting, crafts, quilting, history, clothing, sports, architecture, package design, nutrition

Using Transformations

8.4 *Hawaiian quilts*   417

Applications

Interview: ***Terri Johnson***
               Architect   395, 409, 418

——— Connection ———

Audio-Visual		**Hawaiian Quilts**	417
Presentation	396	**Computer**	
Miniature Golf	403	**Simulation**	424
Architecture	410	**Literature**	431

Applying Algebra 395, 402

Additional applications: periscopes, optics, computers, furniture design, fabric design, animal tracks, manufacturing

Similar Polygons

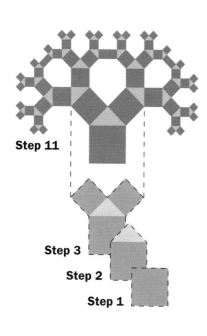

Step 11

Step 3

Step 2

Step 1

Applications

Interview:
Loy Arcenas
Set Designer
465, 468, 472

———— Connection ————

Astronomy	450	**Consumer**	
Surveying	457	**Economics**	473
Electronics	459	**Zoology**	479

Applying Algebra 448, 449, 462, 464, 466, 470, 474

Additional applications: graphic design, technical
drawing, movies, geography, fractals, sculpture,
ballooning, advertising

Applying Right Triangles

Angles of grain piles

	wheat	corn	oat
$m \angle A$	27.0	27.5	28.0
tan A	0.5095	0.5206	0.5317

10.3 *Angles of grain piles*   507, 513

Applications

Interview:
Debby Tewa
Solar Engineer
503, 519

Additional applications: home repair, music, sports,
history, forestry, agriculture, navigation, paleontology,
theater, aviation, orienteering, archaeology, packaging

Circles and Spheres

11.4 *Global positioning* 575

Applications

Interview:

Ron Courson
*Athletic Trainer/
Physical Therapist*
557, 589

—————— **Connection** ——————

Biology	564	**Biology**	595
Highway Safety	571	**Cooking**	596
Earth Science	577	**Geography**	601
Rock Climbing	584	**Hyperbolic**	
Sports	590	**Geometry**	602

Applying Algebra 558, 562, 568, 574, 576, 587, 595

Additional applications: global positioning, sports, irrigation, catering, architecture, astronomy, physics

12

Coordinates for Transformations

12.3   *Morphing*   626

Applications

Student Resources

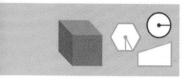

About *the* Interviews

Using Mathematics in Careers

Each chapter of this book starts with a personal interview with someone who uses mathematics in his or her life. You may be surprised by the wide range of careers that are included. These are the people you will be reading about:

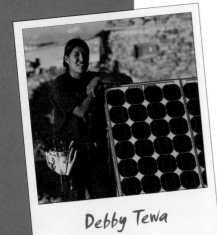

Debby Tewa

- **Tumbler**
  Latecia Leavy

- **Mineralogist**
  Virgil Lueth

- **Forensic Scientist**
  Mary-Jacque Mann

- **Computer-system Designer**
  Marc Hannah

- **Kite Designer**
  José Saínz

- **Optical Physicist**
  Madeleine Fleming

- **Alderman**
  Walt Stone

- **Architect**
  Terri Johnson

- **Set Designer**
  Loy Arcenas

- **Solar Electric Technician**
  Debby Tewa

- **Athletic Trainer/Physical Therapist**
  Ron Courson

- **Planetary Geologist**
  Adriana Ocampo

After each interview, the *Explore and Connect* questions guide you in learning more about the topic being discussed. In each chapter there are *Related Examples and Exercises* that show how the mathematics you are learning is used by the person highlighted in the interview.

José Saínz

Marc Hannah

Welcome *to* Geometry

Explorations and Applications

GOALS OF THE COURSE

This book will help you use mathematics in your daily life and prepare you for success in future courses and careers.

In this course you will:
- Study the geometry concepts that are most important for today's students
- Apply these concepts to solve many different types of problems
- Learn how calculators and computers can help you solve problems

You will have a chance to develop your skills in:
- Reasoning and problem solving strategies
- Using geometric properties to solve real-world problems
- Communicating orally and in writing
- Studying and learning independently and as a team member

Pyramid Arena, Memphis, Tennessee

MATHEMATICAL CONTENT

This contemporary geometry course gives you a strong background in the types of mathematical reasoning and problem solving that will be important in your future.

The book emphasizes:

- Visualizing and analyzing geometric relationships in two and three dimensions
- Developing inductive and deductive reasoning skills
- Investigating connections of geometry to algebra, probability, trigonometry, and discrete mathematics

ACTIVE LEARNING

To learn geometry successfully, you need to get involved!

There will be many opportunities in this course for you to participate in:

- Explorations of mathematical concepts
- Cooperative learning activities
- Small-group and whole-class discussions

So don't sit back and be a spectator. If you join in and share your ideas, everyone will learn more.

Course Overview

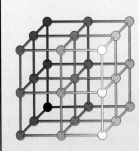

To get an overview of your course, turn to pages xx–xxxi to see some of the types of problems you will solve and topics you will explore.

"What does geometry have to do with me?"

Applications and Connections

Geometry is about you and the world around you.

In this course you'll learn how geometry can help answer many different types of questions in daily life and in careers.

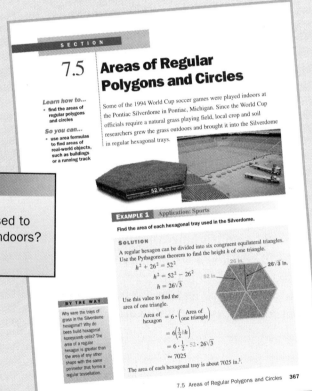

SECTION

7.5 Areas of Regular Polygons and Circles

Learn how to...
- find the areas of regular polygons and circles

So you can...
- use area formulas to find areas of real-world objects, such as buildings or a running track

Some of the 1994 World Cup soccer games were played indoors at the Pontiac Silverdome in Pontiac, Michigan. Since the World Cup officials require a natural grass playing field, local crop and soil researchers grew the grass outdoors and brought it into the Silverdome in regular hexagonal trays.

EXAMPLE 1 Application: Sports

Find the area of each hexagonal tray used in the Silverdome.

SOLUTION

A regular hexagon can be divided into six congruent equilateral triangles. Use the Pythagorean theorem to find the height h of one triangle.

$$h^2 + 26^2 = 52^2$$
$$h^2 = 52^2 - 26^2$$
$$h = 26\sqrt{3}$$

Use this value to find the area of one triangle.

$$\text{Area of hexagon} = 6 \cdot \left(\text{Area of one triangle} \right)$$
$$= 6\left(\tfrac{1}{2}bh\right)$$
$$= 6 \cdot \tfrac{1}{2} \cdot 52 \cdot 26\sqrt{3}$$
$$\approx 7025$$

The area of each hexagonal tray is about 7025 in.^{2}.

BY THE WAY

Why were the trays of grass in the Silverdome hexagonal? Why do bees build hexagonal honeycomb cells? The area of a regular hexagon is greater than the area of any other shape with the same perimeter that forms a regular tessellation.

7.5 Areas of Regular Polygons and Circles **367**

Engineering

Why were hexagonal trays used to install a grass playing field indoors?

(Chapter 7, page 367)

Auto Repair

What kind of quadrilateral can help you change a tire?

(Chapter 7, page 345)

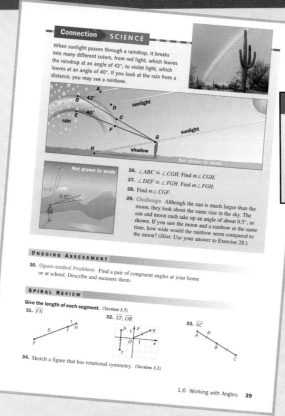

Connection SCIENCE

When sunlight passes through a raindrop, it breaks into many different colors, from red light, which leaves the raindrop at an angle of 42°, to violet light, which leaves at an angle of 40°. If you look at the rain from a distance, you may see a rainbow.

26. $\angle ABC \cong \angle CGH$. Find $m\angle CGH$.

27. $\angle DEF \cong \angle FGH$. Find $m\angle FGH$.

28. Find $m\angle CGF$.

29. **Challenge** Although the sun is much larger than the moon, they look about the same size in the sky. The sun and moon each take up an angle of about 0.5°, as shown. If you saw the moon and a rainbow at the same time, how wide would the rainbow seem compared to the moon? (*Hint:* Use your answer to Exercise 28.)

ONGOING ASSESSMENT

30. **Open-ended Problem** Find a pair of congruent angles at your home or at school. Describe and measure them.

SPIRAL REVIEW

Give the length of each segment. (*Section 1.5*)

31. $\overline{FN}$

32. $\overline{ST}$; $\overline{OR}$

33. $\overline{AC}$

34. Sketch a figure that has rotational symmetry. (*Section 1.2*)

1.6 Working with Angles **39**

Connections Exercises

These clusters of exercises, which appear throughout each chapter, focus on the connections of geometry to other subjects, careers, and branches of mathematics.

Biology

What does the angle of the crest of a Stellar's jay tell you?

(*Chapter 2, page 61*)

Crafts

How can you fold a piece of paper to cut out a perfect five-pointed star?

(*Chapter 2, page 77*)

"Do we just sit back and listen?"

Explorations and Cooperative Learning

In this course you'll be an active learner.

Working individually and in groups, you'll investigate questions and then present and discuss your results. Here are some of the topics you'll explore.

Combining Congruent Triangles

What kinds of shapes can you make by combining two congruent triangles?

(Chapter 6, page 287)

SECTION

5.6 **Parallels in Space**

There are two possibilities for how two planes can be related: the planes may be parallel or they may intersect in a line. But what happens if there are three planes?

Learn how to...
• recognize relationships among parallel and intersecting planes

So you can...
• identify parallel lines and planes in space
• analyze real-world examples of parallel and perpendicular lines in space

EXPLORATION
COOPERATIVE LEARNING

Investigating Planes and Their Intersections

Work with another student. You will need:
• three index cards
• scissors
• two toothpicks

1 Cut and label the cards as shown.

2 Hold cards A and B so that they model parallel planes. Can you position plane C so it is parallel to both A and B? Can plane C intersect one plane but not the other? Can C intersect both planes? If so, how are the lines of intersection related?

3 Use cards A and B to model intersecting planes. Can plane C intersect one plane but not the other? Can C intersect both planes?

4 Can three planes form exactly one line of intersection? two lines? three lines? Explain.

5 Draw a pair of intersecting lines on one card. Place one end of a toothpick on the point of intersection and position the toothpick so it is perpendicular to both of the lines. A line that is perpendicular to both lines is perpendicular to the plane of the card.

6 Hold two toothpicks to model two different lines that are each perpendicular to the same card. Describe the relationship between the lines.

256 Chapter 5 *Parallel Lines*

Investigating Intersecting Planes

How many lines of intersection can three planes form?

(Chapter 5, page 256)

Portfolio Projects

These open-ended projects give you a chance to explore applications of the topics you have studied.

Tessellating the Plane

A tessellation is a pattern formed by repeating a shape to cover a plane without any gaps or overlapping. A brick walkway is an example of a tessellation that uses rectangles to cover the plane. A tiled wall is another example. So is a honeycomb—not all tessellations are made with quadrilaterals.

PROJECT GOAL Use transformations with different polygons to make tessellations.

Making Quadrilateral Tilings

1. Work with a partner. Cut several congruent rectangles from cardboard to represent bricks, and use transformations to make three different tessellations with them. Discuss with your partner the transformations you used.

2. Cut a quadrilateral that is not a rectangle from a piece of cardboard. Trace the quadrilateral several times to make a tessellation. Discuss the transformations that you used to make your pattern.

3. Choose any vertex on your sketch and measure the angles at that vertex. What do you notice about the sum of the measures of these angles? How might you have predicted this result without measuring?

434   Chapter 8 *Using Transformations*

Investigating Symmetry

What kind of symmetry can you find in objects around you?

(Chapter 1, page 48)

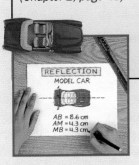

Constructing a Perpendicular Bisector

How can you draw the perpendicular bisector of a line without measuring the angle or the length of the line?

(Chapter 5, page 265)

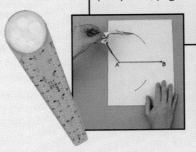 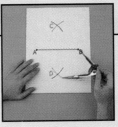 

"How is geometry different from algebra?"

Logical Reasoning and Visualization

In this course you'll learn new ways to see and understand mathematics.

You will study the properties of two- and three-dimensional shapes and learn new ways to reason mathematically.

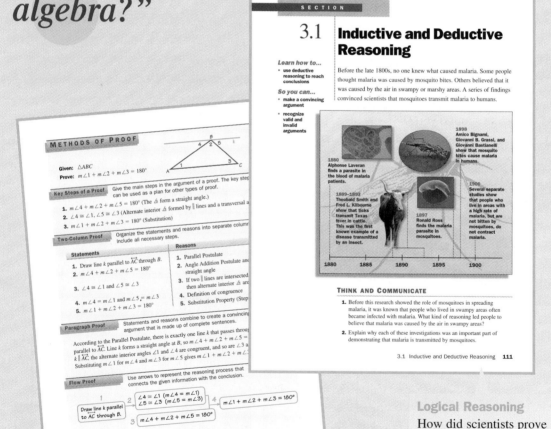

SECTION

3.1 Inductive and Deductive Reasoning

Learn how to...
* use deductive reasoning to reach conclusions

So you can...
* make a convincing argument
* recognize valid and invalid arguments

Before the late 1800s, no one knew what caused malaria. Some people thought malaria was caused by mosquito bites. Others believed that it was caused by the air in swampy or marshy areas. A series of findings convinced scientists that mosquitoes transmit malaria to humans.

1898 Amico Bignami, Giovanni B. Grassi, and Giovanni Bastianelli show that mosquito bites cause malaria in humans.

1880 Alphonse Laveran finds a parasite in the blood of malaria patients.

1889–1893 Theobald Smith and Fred L. Kilbourne show that ticks transmit Texas fever in cattle. This was the first known example of a disease transmitted by an insect.

1897 Ronald Ross finds the malaria parasite in mosquitoes.

1900 Several separate studies show that people who live in areas with a high rate of malaria, but are not bitten by mosquitoes, do not contract malaria.

1880   1885   1890   1895   1900

THINK AND COMMUNICATE

1. Before this research showed the role of mosquitoes in spreading malaria, it was known that people who lived in swampy areas often became infected with malaria. What kind of reasoning led people to believe that malaria was caused by the air in swampy areas?

2. Explain why each of these investigations was an important part of demonstrating that malaria is transmitted by mosquitoes.

3.1 Inductive and Deductive Reasoning **111**

METHODS OF PROOF

Given: △ABC
Prove: $m\angle 1 + m\angle 2 + m\angle 3 = 180°$

Key Steps of a Proof Give the main steps in the argument of a proof. The key step can be used as a plan for other types of proof.

1. $m\angle 4 + m\angle 2 + m\angle 5 = 180°$ (The △ form a straight angle.)
2. $\angle 4 \cong \angle 1, \angle 5 \cong \angle 3$ (Alternate interior △ formed by ∥ lines and a transversal ar
3. $m\angle 1 + m\angle 2 + m\angle 3 = 180°$ (Substitution)

Two-Column Proof Organize the statements and reasons into separate column Include all necessary steps.

Statements	Reasons
1. Draw line k parallel to $\overleftrightarrow{AC}$ through B.	1. Parallel Postulate
2. $m\angle 4 + m\angle 2 + m\angle 5 = 180°$	2. Angle Addition Postulate an straight angle
3. $\angle 4 \cong \angle 1$ and $\angle 5 \cong \angle 3$	3. If two ∥ lines are intersected then alternate interior △ are
4. $m\angle 4 = m\angle 1$ and $m\angle 5 = m\angle 3$	4. Definition of congruence
5. $m\angle 1 + m\angle 2 + m\angle 3 = 180°$	5. Substitution Property (Step

Paragraph Proof Statements and reasons combine to create a convincing argument that is made up of complete sentences.

According to the Parallel Postulate, there is exactly one line k that passes throug parallel to $\overleftrightarrow{AC}$. Line k forms a straight angle at B, so $m\angle 4 + m\angle 2 + m\angle 5 =$ $k \parallel \overleftrightarrow{AC}$, the alternate interior angles $\angle 1$ and $\angle 4$ are congruent, and so are $\angle 3$ a Substituting $m\angle 1$ for $m\angle 4$ and $m\angle 3$ for $m\angle 5$ gives $m\angle 1 + m\angle 2 + m\angle$

Flow Proof Use arrows to represent the reasoning process that connects the given information with the conclusion.

1   Draw line k parallel to AC through B. → 2 $\angle 4 \cong \angle 1$ ($m\angle 4 = m\angle 1$) $\angle 5 \cong \angle 3$ ($m\angle 5 = m\angle 3$) → 4 $m\angle 1 + m\angle 2 + m\angle 3 = 180°$

3   $m\angle 4 + m\angle 2 + m\angle 5 = 180°$

Reasons
1. Parallel Postulate
2. If two ∥ lines are intersected by a transversal, then alternate interior △ are ≅.
3. Angle Addition Postulate and definition of straight angle
4. Substitution Property

Proof You will learn different ways to organize and write proofs.

Logical Reasoning
How did scientists prove that mosquitoes transmit malaria?

Visualizing 3-Dimensional Objects

What does the surface of a three-dimensional shape look like when you unfold it?

(Chapter 2, page 95)

Unfolding a Prism

Imagine cutting this prism along some of its edges, then opening and unfolding it. The resulting plane figure is called a **net**.

You can construct a prism by folding up its net and taping the edges together.

EXAMPLE 2

Sketch a net for this prism.

SOLUTION

Method 1

Sketch all the lateral faces attached to one of the bases of the prism.
  Sketch the other base on the opposite edge of one of the lateral faces.

Method 2

Sketch all the lateral faces attached to each other.
  Place the bases at opposite ends of two of the lateral faces.

THINK AND COMMUNICATE

3. Describe another way to sketch a net for the prism above. How many different methods do you think there are?

4. Describe one possible net for a cube.

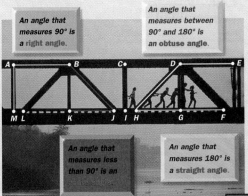

An angle that measures 90° is **a right angle.**

An angle that measures between 90° and 180° is **an obtuse angle.**

An angle that measures less than 90° is an

An angle that measures 180° is **a straight angle.**

Analyzing Geometric Relationships

Geometry plays an important role in objects you see every day.

(Chapter 2, page 57)

Visual Thinking

How can you show that the measures of the angles of a triangle add up to 180°? *(Chapter 2, page 65)*

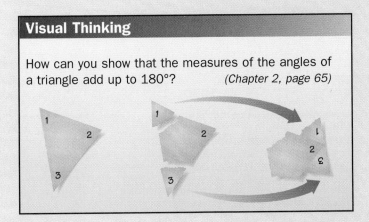

"How can I visualize the problem?"

Using Technology

Calculators and computers can help you see mathematical relationships.

In this course there are many opportunities to use technology to model problem situations, identify patterns, and find solutions.

Graphing Technology

How are the slopes of parallel and perpendicular lines related?

(Chapter 4, page 180)

Matrix Operations

How are matrix operations used to change images on a computer screen?

(Chapter 12, page 614)

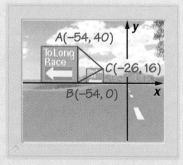

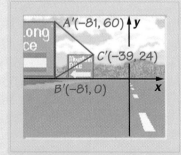

45. a. DESIGN The Danish artist Per Arnoldi designed a chair made of triangles and squares formed by one rectangle of metal. Find the value of *d*.

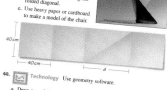

  b. Draw a net for the chair. Give all dimensions, including the folded diagonal.

  c. Use heavy paper or cardboard to make a model of the chair.

46. **Technology** Use geometry software.

  a. Draw a quadrilateral. Draw the diagonals of the quadrilateral. Measure the length of each segment.

  b. Repeat part (a) with at least five different quadrilaterals. Record the lengths in a table.

  c. For what quadrilaterals will the equation $a^2 + b^2 + c^2 + d^2 = e^2 + f^2$ be true? Explain your reasoning.

ONGOING ASSESSMENT

47. a. Visual Thinking Explain how this diagram illustrates a proof of the Pythagorean theorem. Write the algebraic steps for the proof.

  b. How is this diagram different from the one that you used in the Exploration on page 141? Explain.

SPIRAL REVIEW

Write each statement as a conditional. Circle the hypothesis and underline the conclusion. *(Section 1.3)*

48. Angles in a linear pair are supplementary angles.

49. Perpendicular lines intersect at right angles.

50. Write the converse of each statement in Exercises 48 and 49. *(Section 3.5)*

3.6 The Pythagorean Theorem **147**

Technology Exercises

In these exercises you will be using geometry software or spreadsheets to practice, apply, and extend what you have learned.

Geometry Software

How can you use a computer to investigate a relationship between parallel lines and similar triangles?

(Chapter 9, page 461)

	A	B	C	D	E
1		Diameter at Equator (km)	Mean distance from sun (million km)	Scale diameter (cm)	Scale distance (cm)
2	Sun	1400000	0		
3	Mercury	4870	57.9		
4	Venus				
5	Earth				

Spreadsheet

What is an appropriate scale to use to make a scale model of the solar system?

(Chapter 9, page 482)

"Can I solve this problem with geometry?"

Integrating Math Topics

Sometimes you need to combine geometry with other math topics in order to find a solution.

In this course you'll see how you can solve problems by integrating geometry with algebra, probability, trigonometry, and discrete mathematics.

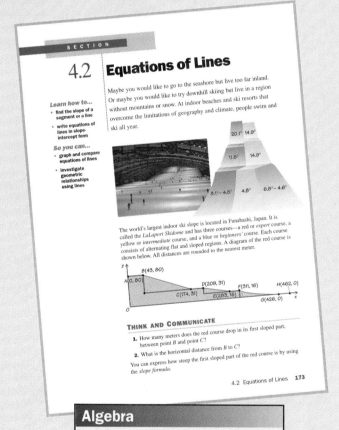

SECTION

4.2 Equations of Lines

Maybe you would like to go to the seashore but live too far inland. Or maybe you would like to try downhill skiing but live in a region without mountains or snow. At indoor beaches and ski resorts that overcome the limitations of geography and climate, people swim and ski all year.

Learn how to...
- find the slope of a segment or a line
- write equations of lines in slope-intercept form

So you can...
- graph and compare equations of lines
- investigate geometric relationships using lines

The world's largest indoor ski slope is located in Funabashi, Japan. It is called the *LaLaport Skidome* and has three courses—a red or *expert* course, a yellow or *intermediate* course, and a blue or *beginners'* course. Each course consists of alternating flat and sloped regions. A diagram of the red course is shown below. All distances are rounded to the nearest meter.

THINK AND COMMUNICATE

1. How many meters does the red course drop in its first sloped part, between point *B* and point *C*?
2. What is the horizontal distance from *B* to *C*?

You can express how steep the first sloped part of the red course is by using the *slope formula*.

4.2 Equations of Lines **173**

Trigonometry

What is the horizontal distance a jet travels in the first ten seconds after takeoff?

(Chapter 10, page 519)

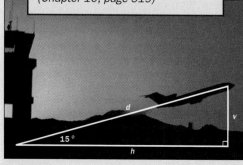

Algebra

How is the slope of an expert ski course different from the slope of a beginner course?

(Chapter 4, page 173)

Transformational Geometry

How can you model a reflection using transformations?

(Chapter 1, pages 8–9)

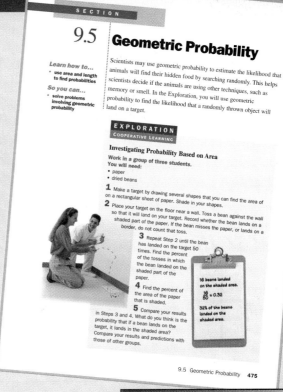

9.5 Geometric Probability **475**

Probability

What is the probability that an object thrown at random will land on a target?

(Chapter 9, page 475)

Discrete Mathematics

How many squares remain after the third step in creating Sierpinski's carpet?

(Chapter 9, page 480)

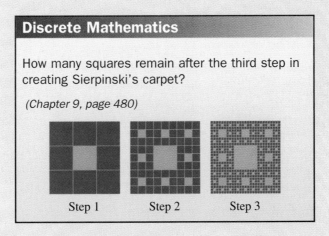

Step 1   Step 2   Step 3

"When will I ever use this?"

Building for the Future

The skills you'll learn in this course will form a strong foundation for the future.

They'll prepare you for more advanced courses and increase your career opportunities.

Problem Solving and Communication

The exercise sets help you develop your problem solving and communication skills.

22. Open-ended Problem Give an example of a scale drawing or a scale model you have seen that is an enlargement of a real-life object or group of objects.

23. a. TECHNICAL DRAWING If you drew the structures in the table below so that 1 in. represents 100 ft, what would the scale be? How tall would each structure be in your drawing?

CN Tower	TMG Offices	Washington Monument	Empire State Building
Toronto, Canada	Tokyo, Japan	Washington, DC	New York City, NY
1815 ft	793 ft	555 ft	1250 ft

b. Open-ended Problem On a sheet of $8\frac{1}{2}$ in. by 11 in. paper, sketch and label a scale drawing of the four structures. Explain how you chose the scale for your drawing.

24. TRANSFORMATIONS Explain why dilations are sometimes called *similarity transformations*.

25. Challenge What do you think must be true for two prisms to be similar? Write a definition for similar prisms.

26. MOVIES For most of the movie *King Kong*, the scale models of King Kong were built to appear about 18 ft tall. When King Kong climbs the Empire State Building, the models make him appear to be 24 ft tall.

a. Make a scale drawing that shows King Kong at both 18 ft and 24 ft tall, a 6 ft tall person, and the Empire State Building (1250 ft).

b. Writing Why do you think the movie makers chose to make King Kong appear taller when he climbs the Empire State Building?

27. SAT/ACT Preview Select the pair of words that *best* expresses a relationship similar to that expressed in the pair *gallon : liquid*.

A. ruler : paper   **B.** week : month   **C.** length : width
D. degree : temperature   **E.** foot : meter

28. GEOGRAPHY The Nile, in Africa, is the world's longest river. It would be about the length of a 24 in. shoelace (61 cm) if drawn so that 1 cm represents 110 km. What is the scale? About how long is the Nile?

451

SAT/ACT Preview
These exercises help you prepare for college entrance exams.

Like many real-world problems, **Open-ended Problems** have more than one solution.

Connection ASTRONOMY

During a solar eclipse, the shadow of the moon travels across the face of Earth. If you are in the path of the shadow, you see the sun disappear and reappear as the moon passes over you. The map shows the path of the moon's shadow for three eclipses that are all from the same series.

May 10, 1994
April 29, 1976
April 19, 1958

BY THE WAY

Each year there are two to five eclipses somewhere on Earth's surface. Each eclipse belongs to a different series based on the positions of Earth, the moon, and the sun.

14. Estimate the date of the next eclipse in the series shown.
15. Describe the path of the next eclipse in the series.
16. Open-ended Problem Describe something that happens regularly and explain how to predict when it will happen next. Will your method of prediction always work?

ONGOING ASSESSMENT

17. Writing Describe a situation in which you used inductive reasoning to help you make a decision. How many different examples did you use?

SPIRAL REVIEW

Graph each point on a coordinate plane. *(Toolbox, page 696)*
18. (2, 4)   **19.** (−1, 1)   **20.** (−3, −1)   **21.** (0, 2)

Evaluate each expression when $a = -2$ and $b = 3$. *(Toolbox, page 690)*
22. $4b + a$ **23.** $5b - a$ **24.** $b^2 + 4a$ **25.** $\frac{b - a^2}{ab}$

1.1 Patterns and Reasoning   **7**

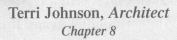

CHAPTER

8 Using Transformations

BLUEPRINTS FOR BALLGAMES

INTERVIEW Terri Johnson

Were it not for Terri Johnson's love of baseball, she might never have started designing sports stadiums. Johnson graduated from architecture school in Chicago while construction was starting at the new White Sox stadium, Comiskey Park. "I really wanted to get involved in that project," she says. She contacted HOK Sport, the Kansas City architectural firm handling the job but did not land a position with them for two years. By then, construction on Comiskey Park was completed, though she did get to work on the stadiums where the Cleveland Indians, the Colorado Rockies, and the Carolina Panthers play.

388

Terri Johnson, *Architect*
Chapter 8

Career Preview

In the Interviews and exercises, you will find problems that show how math is used in a variety of careers.

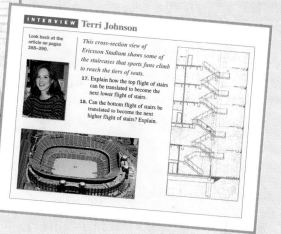

INTERVIEW Terri Johnson

Look back at the article on pages 388–390.

This cross-section view of Ericsson Stadium shows some of the staircases that sports fans climb to reach the tiers of seats.

17. Explain how the top flight of stairs can be translated to become the next lower flight of stairs.

18. Can the bottom flight of stairs be translated to become the next higher flight of stairs? Explain.

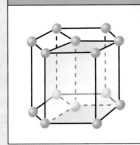

Lattice for Beryl

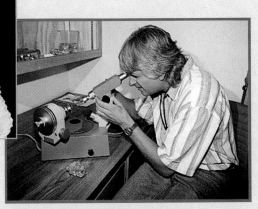

Virgil Lueth, *Mineralogist*
Chapter 2

1 | Patterns, Lines, and Planes

SPRINGING *Across Obstacles*

Latecia Leavy

Latecia Leavy studies mathematics by day at a Chicago high school and then, in the late afternoon, she takes the "crash course." In a typical session with the Jesse White Tumbling Team, the 15-year-old Leavy is just a blur as she whips across the ground in a dizzying series of backward half-flips ("flip-flops") and back somersaults. Leavy and her teammates also catapult from a mini-trampoline, soaring over people, cars, and other objects, before doing a roll and safely returning to the planet Earth.

"Everybody should find something they like to do as much as I enjoy tumbling."

For Leavy, geometry is more than an academic exercise. "You have to take off from the trampoline at the right angle, so you don't go off sideways and crash into somebody," she explains. Leavy began tumbling when she was eight years old and continues to practice several hours each week. It barely seems like work to her because she enjoys tumbling so much. "Although this is not for everyone," she says, "everybody should find something they like to do as much as I enjoy tumbling."

Entertaining Teamwork

Leavy and about 60 other youths—mostly residents of Chicago housing projects—make up the tumbling team founded and coached by Jesse White. The group has traveled around the world, displaying breathtaking feats. They have appeared in movies and TV shows and entertained audiences at professional sports events. "You've got to have some experience to make this team," says White, "but members also need a good attitude, and they have to understand teamwork."

"You have to spring from the trampoline at a precise angle for a vertical landing."

Angling for Success

In one trick, teammates form a "Pyramid," with three people on the base, two more on the second level, and another two on top. Then someone else springs from the trampoline after a running take-off and sails over the human pyramid. Tumblers have to run fast before hitting the trampoline or they won't make it over the top.

White sets the trampoline at about a 45 degree angle, as steep as it will go, to ensure maximum height and distance. The 45 degree setting is crucial for the most spectacular trick of the show, called "Superman." A tumbler springs from the trampoline and soars over 20 or more people—a feat Leavy is still training for. "It's an amazing thing to see…and do," she says.

Explore and Connect

Latecia speaks with a spectator before a performance.

1. Writing White sets the trampoline at the angle shown above so the tumbler will fly as far as possible. How do you think the tumbler's path would be different if the trampoline were horizontal? Explain your reasoning.

2. Project Make a model trampoline by taping a piece of a large balloon over the top of an open can or box. Throw a pebble at the trampoline while it is tilted at different angles. How does the path of the pebble vary? Make a poster to report your results to your class.

3. Research Some gymnasts use a *springboard* instead of a trampoline. Find out what a springboard is. Can you set it at different angles?

Mathematics
& Latecia Leavy

In this chapter, you will learn more about how ideas of geometry are related to tumbling.

Related Exercises

Section 1.2
• **Exercises 11–13**

Section 1.3
• **Exercises 3–5**

Section 1.6
• **Exercise 10**

1.1

Patterns and Reasoning

Learn how to...
* reason inductively

So you can...
* make predictions about patterns such as eclipses and the occurrence of the popularity of car colors

In this course, you will explore many kinds of patterns. For example, each branch of a sea fan looks like the fan itself. Each part of each branch looks like the fan too. The figure in the Exploration grows in a pattern like the sea fan.

EXPLORATION
COOPERATIVE LEARNING

Analyzing Patterns

Work with another student.

1 Draw a segment.

2 Add 2 shorter segments.

3 At each new point, draw 2 segments.

4 At each new point, draw two segments half as long as the segments from Step 3. Copy the table, including several more steps. Complete the row for Step 4.

5 Continue the pattern as long as you can. Fill in the table after each step.

Step	Number of new segments	Total number of segments
1	1	1
2	2	3
3	4	7
4	?	?

6 What patterns do you notice in the figure you drew? Circle three parts of your diagram that look like each other but are different sizes.

7 What patterns do you notice in your table?

8 Predict the total number of segments for the tenth step. Explain.

When you make a prediction based on several examples, you are using **inductive reasoning**. For example, if your eyes itch each time you play with a cat, you use inductive reasoning to generalize and decide that you are allergic to cats.

EXAMPLE

Predict the number of small triangles in the seventh figure of the pattern.

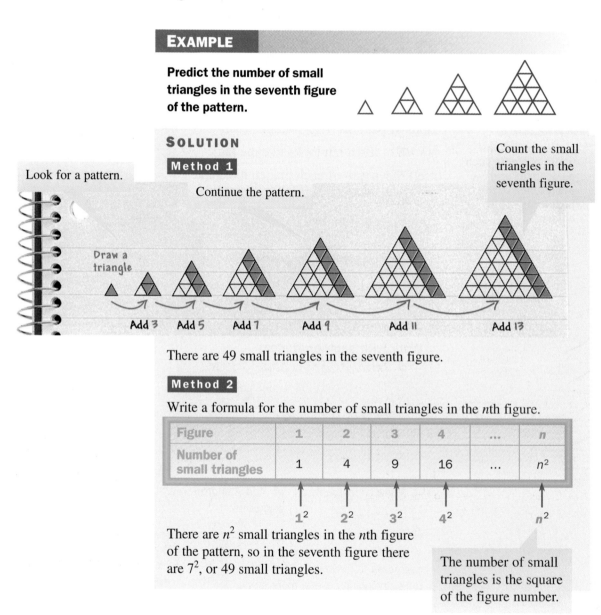

SOLUTION

Method 1

Look for a pattern.

Continue the pattern.

Count the small triangles in the seventh figure.

Draw a triangle

Add 3   Add 5   Add 7   Add 9   Add 11   Add 13

There are 49 small triangles in the seventh figure.

Method 2

Write a formula for the number of small triangles in the nth figure.

Figure	1	2	3	4	...	n
Number of small triangles	1	4	9	16	...	n^2

1^2 2^2 3^2 4^2 n^2

There are n^2 small triangles in the nth figure of the pattern, so in the seventh figure there are 7^2, or 49 small triangles.

The number of small triangles is the square of the figure number.

THINK AND COMMUNICATE

Describe each pattern and give the next two numbers or figures.

1. 5, 4, 3, 2, 1, 0, . . .

2.

3. Describe at least two ways to continue this pattern: 1, 2, 1, ...

✓ CHECKING KEY CONCEPTS

For Questions 1–4, use the shapes and table below.

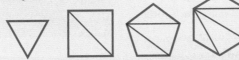

Number of sides	3	4	5	6
Number of triangles	1	2	3	4

1. Copy the table, leaving room for two more columns. What patterns do you notice?

2. Sketch the next two shapes in the sequence. Record the number of triangles in each shape.

3. Give a formula for the number of triangles in a shape with *n* sides.

4. Use your formula from Question 3 to find the number of triangles in a shape that has 100 sides.

1.1 Exercises and Applications

Extra Practice exercises on page 659

1. **a.** The table describes the number of points in the figure you drew in the Exploration on page 3. Copy the table and add three rows.

   b. What patterns do you notice in your table? What will the total number of points be for the tenth stage? Explain.

Step	Number of new points	Total number of points
1	2	2
2	2	4
3	4	8

Use inductive reasoning to find the next two numbers in each pattern.

2. $-1, -3, -5, -7, \underline{?}, \underline{?}$

3. $0, 3, 8, 15, \underline{?}, \underline{?}$

4. $1, 3, 7, 13, \underline{?}, \underline{?}$

5. $2, \dfrac{3}{2}, \dfrac{4}{3}, \underline{?}, \underline{?}$

Use inductive reasoning to sketch the next shape in each pattern.

6.
7.

8. **Open-ended Problem** Describe one way you have used inductive reasoning outside of your mathematics class. What did you predict? How much information did you base your prediction on? How accurate was your prediction?

9. Writing Each year, auto makers introduce new colors for their cars. One way auto makers choose colors for new cars is to find out which colors sold well in the past. The table shows color popularity for compact cars and sports cars from 1990 through 1995.

Popular Car Colors (percent of total sales for the year)						
Color	1990 sales	1991 sales	1992 sales	1993 sales	1994 sales	1995 sales
White	20.6%	21.0%	19.3%	15.3%	15.3%	14.4%
Bright red	22.0%	19.3%	12.4%	12.9%	11.2%	9.5%
Medium red	5.8%	7.7%	10.6%	11.4%	10.5%	11.3%
Bright blue	4.8%	7.5%	4.3%	3.8%	5.4%	2.3%
Silver	6.4%	6.8%	4.9%	5.2%	2.4%	6.3%
Green	0.4%	0.6%	6.3%	16.1%	12.8%	15.2%

a. Describe any trends that you observe in color popularity from 1990 through 1995.

b. Based on the trends, what do you think were the most popular colors in 1996? Explain.

Write a formula for the value of the nth term in each pattern.

10.

Term	1	2	3	4	5	6	...	n
Value	−3	−2	−1	0	1	2	...	_?_

11.

Term	1	2	3	4	5	6	...	n
Value	4	7	10	13	16	19	...	_?_

12. CHEMISTRY Give the molecular and structural formulas for the next two compounds in the alkane series. The first four compounds are shown below.

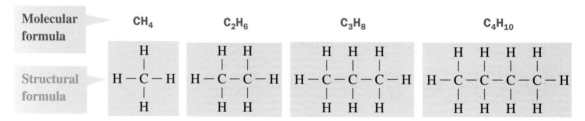

13. Challenge Find the next two numbers in the pattern below and explain your reasoning. −1, 0, 7, 26, 63, ...

During a solar eclipse, the shadow of the moon travels across the face of Earth. If you are in the path of the shadow, you see the sun disappear and reappear as the moon passes over you. The map shows the path of the moon's shadow for three eclipses that are all from the same series.

━━━━ May 10, 1994
━━━━ April 29, 1976
━━━━ April 19, 1958

BY THE WAY

Each year there are two to five eclipses somewhere on Earth's surface. Each eclipse belongs to a different series based on the positions of Earth, the moon, and the sun.

14. Estimate the date of the next eclipse in the series shown.

15. Describe the path of the next eclipse in the series.

16. Open-ended Problem Describe something that happens regularly and explain how to predict when it will happen next. Will your method of prediction always work?

ONGOING ASSESSMENT

17. Writing Describe a situation in which you used inductive reasoning to help you make a decision. How many different examples did you use?

SPIRAL REVIEW

Graph each point on a coordinate plane. *(Toolbox, page 696)*

18. $(2, 4)$ **19.** $(-1, 1)$ **20.** $(-3, -1)$ **21.** $(0, 2)$

Evaluate each expression when $a = -2$ and $b = 3$. *(Toolbox, page 690)*

22. $4b + a$ **23.** $5b - a$ **24.** $b^2 + 4a$ **25.** $\dfrac{b - a^2}{ab}$

1.2 | Transformations and Symmetry

Learn how to...

- **identify and perform transformations**

So you can...

- **describe movement and patterns, such as in tumbling and Braille letters**

In the photos, the signs are **reflected** in the water. The potter makes pottery by sculpting clay as it **rotates** on the potter's wheel. When a goose runs in a straight line across snow, it leaves **translated** sets of tracks.

Rotation, reflection, and translation are examples of **transformations**. When you transform a figure, the new figure is called the **image**. In these transformations, the image is the same size and shape as the original figure.

The reflection in the water is the **image** of the sign.

The back of the bowl is an **image** of the front.

Each set of goose tracks is an **image** of the first set.

EXAMPLE 1

Perform each transformation.

a. Reflect the triangle over the red line.

b. Rotate the triangle by any amount around the green point.

c. Translate the triangle any distance to the right.

d. Rotate the triangle a complete turn around the green point.

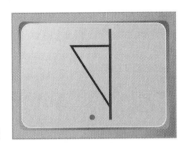

SOLUTION

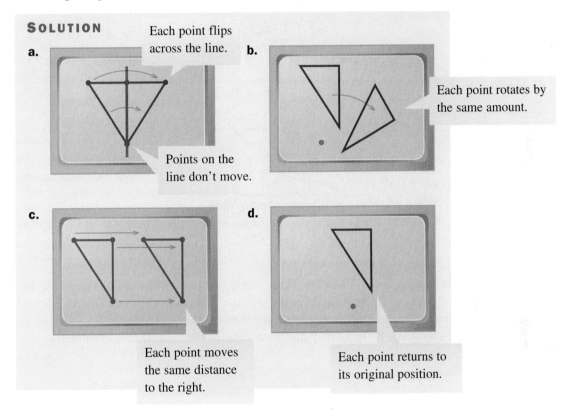

a. Each point flips across the line.

Points on the line don't move.

b. Each point rotates by the same amount.

c. Each point moves the same distance to the right.

d. Each point returns to its original position.

THINK AND COMMUNICATE

1. How are the two transformations below alike? How are they different?

2. The second transformation is called a *half-turn*. Explain what this term means.

3. Is there more than one way to rotate a figure around a point? Explain.

4. Is there more than one way to reflect a figure over a line? Explain.

EXAMPLE 2

How do the coordinates of the points on a triangle change when you translate the triangle to the right? Try many different examples.

SOLUTION

Translate a triangle 3 units to the right.

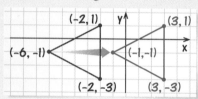

Each *x*-coordinate increases by 3.

Translate a triangle 1 unit to the right.

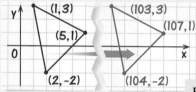

Each *x*-coordinate increases by 1.

Translate a triangle 5 units to the right.

Each *x*-coordinate increases by 5.

Translate a triangle 102 units to the right.

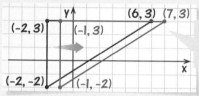

Each *x*-coordinate increases by 102.

When you translate a triangle to the right *n* units, the *x*-coordinate of each point increases by *n* units.

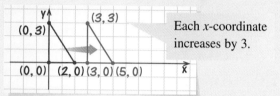

If a figure and its image coincide, the figure has **symmetry**. For example, if you reflect the butterfly over the red line, the image coincides exactly with the original. The butterfly has *reflection symmetry* and the line is a **line of symmetry**. A figure may also have rotational or translational symmetry.

EXAMPLE 3

Describe the symmetry of each object.

a.

b.

SOLUTION

a. The umbrella has rotational symmetry. It looks exactly the same after you spin it part way around the handle.

b. The staircase has both translational and reflection symmetry. If you translate the staircase up one step, it looks the same. Also, a plane of reflection runs down the center of the stairs.

plane of reflection

For Questions 1–4, copy the triangle at the right.

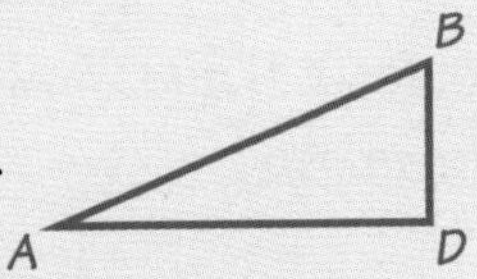

1. Reflect the triangle over side *AD*.

2. Rotate the triangle a half-turn around point *D*.

3. Translate the triangle 2 in. to the left.

4. The *vertices*, or corner points, of a triangle are $A(0, -2)$, $B(5, 4)$, and $C(3, -3)$. Predict the coordinates of the vertices of the image if you translate this triangle 7 units to the right.

Describe the symmetry of each object.

5.  **6.**   **7.**

1.2 **Exercises and Applications**

Extra Practice
exercises on
page 659

Open-ended Problems **Describe at least two real-world objects that have each type of symmetry.**

1. reflection symmetry

2. rotational symmetry

3. translational symmetry

Copy each diagram. Sketch the reflection of the triangle over the red line.

4. **5.** **6.**

Sketch the fabric. Include the blue line and green point in your sketch. Sketch the image of each transformation.

7. Reflect the figure over the blue line.

8. Rotate the figure a half-turn around the green point.

9. Translate the figure any distance to the right.

10. Research Find a crossword puzzle in a newspaper. Describe the symmetry of the puzzle.

Look back at the article on pages xxxii–2.

From the opening floorwalk through the closing somersault, each Jesse White show contains many tumbling transformations and patterns.

11. Which transformation best describes the floorwalk? Why?

12. What kind of symmetry does this pose have? Explain.

13. Which transformation best describes this flip? Why?

14. Cooperative Learning Work in a group of three people. Use the results from Example 2 on page 10. One person should complete part (a), another should complete part (b), and the third should complete part (c). Work together for parts (d) and (e).

a. Graph $D(-2, 0)$, $E(0, 2)$, and $F(-5, 2)$ and connect the points to draw a triangle. How do the coordinates change when you translate the triangle up 2 units? up 3 units? up n units?

b. Graph $D(-2, 0)$, $E(0, 2)$, and $F(-5, 2)$ and connect the points to draw a triangle. How do the coordinates change when you translate the triangle down 2 units? down 3 units? down n units?

c. Graph $D(-2, 0)$, $E(0, 2)$, and $F(-5, 2)$ and connect the points to draw a triangle. How do the coordinates change when you translate the triangle to the left 2 units? to the left 3 units? to the left n units?

d. Summarize your answers from parts (a)–(c) and the results in Example 2.

e. If you translate a triangle to the right a units and then down b units, how will the coordinates change? Explain.

Describe the symmetry of each object or group of objects.

15.

16.

17.

Braille is a code of raised dots that can be read by touch. It was developed by a 15-year-old blind French student named Louis Braille. He got the idea from a dot code used to send messages to soldiers at night. The Braille alphabet is based on a cell three dots high and two dots wide.

A sportswriter uses Braille while working at a game.

A	B	C	D	E	F	G	H	I	J

K	L	M	N	O	P	Q	R	S	T

U	V	X	Y	Z					W

BY THE WAY

Louis Braille did not include the letter W in his original alphabet because it is rarely used in French. W was added to the Braille alphabet for use in English.

18. Writing   Compare the first ten letters of the Braille alphabet with the second ten letters. What patterns do you see?

19. Which Braille letters are reflections of each other?

20. Which Braille letters are rotations of each other?

21. Writing   There are no Braille letters that are translations of each other. Why do you think this is so?

22. Challenge   Find a word that has rotational symmetry or reflection symmetry when written in Braille.

ONGOING ASSESSMENT

23. Open-ended Problem   A door rotates when you open it. What other objects do you rotate? translate? What can you use to reflect an object?

SPIRAL REVIEW

Solve each equation. *(Toolbox, pages 695 and 696)*

24. $-2x = 35$ **25.** $\frac{n}{3} = -12$ **26.** $5 - 3p = -13$

27. Anna buys a book about juggling for $9 and some juggling bags for $1.50 each. Write a variable expression for the amount she pays, based on the number of juggling bags she buys. *(Toolbox, page 692)*

1.3 Making Conjectures

Can you draw a house like this one without lifting your pencil or retracing an edge? The house is an example of a *network*, a figure made by connecting points. In the Exploration, you will discover how to tell whether a network is traceable.

EXPLORATION
COOPERATIVE LEARNING

Tracing Networks

Work with another student.

1 Copy each network.

vertex

edge

A   B   C   D   E

2 Try to trace each network without lifting your pencil or retracing an edge. Which networks are traceable?

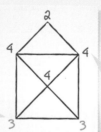

A vertex is **odd** if an odd number of edges meet at the vertex.

A vertex is **even** if an even number of edges meet at the vertex.

3 For each network, count how many edges meet at each vertex.

4 Copy and complete the table, including a row for each figure. (*Note:* Vertices is the plural of vertex.) Describe any patterns that you notice.

5 Sketch a network with four odd and zero even vertices. Is it traceable?

6 Make a conjecture about how you can tell if a network is traceable.

Network	Number of odd vertices	Traceable?
House	2	Yes
A	?	?
B	?	?
C	?	?

A prediction based on inductive reasoning, like the rule you wrote in the Exploration, is called a **conjecture**. Conjectures are usually *conditional statements*. A **conditional statement** can be written in the form "if *P*, then *Q*." A conditional statement may be true or false.

If all the vertices of a network are even, then **it is traceable**.

If today is July 4, then **it is Independence Day in the United States**.

The *if* part is called the **hypothesis** and the *then* part is called the **conclusion**. Sometimes the hypothesis comes after the conclusion.

A network is traceable if all its vertices are even.

It is Independence Day in the United States if today is July 4.

EXAMPLE 1

For each conditional statement, identify the hypothesis and the conclusion.

a. Cars with underinflated tires waste gasoline.

b. Every point on the *y*-axis has an *x*-coordinate of zero.

SOLUTION

a. If a car has underinflated tires, then it wastes gasoline.

> Hypothesis: a car has underinflated tires
> Conclusion: the car wastes gasoline

b. If a point is on the y-axis, then its x-coordinate is zero.

> Hypothesis: a point is on the y-axis
> Conclusion: the x-coordinate of the point is zero

First rewrite each statement using the words *if* and *then*.

The *if* part is the hypothesis and the *then* part is the conclusion.

THINK AND COMMUNICATE

Tell whether each conditional statement is *True* or *False*. If it is false, explain why.

1. If you translate a triangle three units to the right, then the *x*-coordinate of each point increases by three units.

2. If it's January, then it's cold out.

3. A number is divisible by 9 if it is divisible by 3.

Counterexamples

As you probably decided in Questions 1–3 on page 15, a conjecture is not true unless it is always true. If you can find one example for which the hypothesis is true but the conclusion is false, then the conjecture is false. Such an example is called a **counterexample**

EXAMPLE 2   Connection: Algebra

Kevin used a computer to test the integers from 1 to 35 in the formula $P = n^2 - n + 41$ and see if the result is a prime number. He found that P was a prime number each time. Kevin made a conjecture:

If n is a positive integer, then $n^2 - n + 41$ is a prime number.

Is this conjecture true?

	Formula for Primes?	
n	$n^2 - n + 41$	Prime?
1	41	Yes
2	43	Yes
3	47	Yes
4	53	Yes
33	1097	Yes
34	1163	Yes
35	1231	Yes

100%

SOLUTION

Test 41 in the formula: $41^2 - 41 + 41 = 41^2$, and 41^2 is not prime. Since there is one value of n that doesn't work in the formula, Kevin's conjecture is false.

> **WATCH OUT!**
>
> Even though Kevin's conjecture works for 35 numbers, one counterexample is enough to show that the conjecture is false.

THINK AND COMMUNICATE

4. Write a conjecture that has a counterexample.

5. Write a conjecture that does not have a counterexample.

☑ CHECKING KEY CONCEPTS

Identify the hypothesis and the conclusion of each statement.

1. If a figure is a square, then it has four lines of symmetry.

2. The expression $(a - b)^2$ represents a positive number if a does not equal b.

3. People who live in glass houses shouldn't throw stones.

Tell whether each statement is *True* or *False*. If it is false, give a counterexample.

4. If a person has seen a doctor, then the person has a broken arm.

5. If $n = 3$, then $2n - 2 = 4$.

6. If n is a whole number, then $n^2 + n + 11$ is an odd number.

1.3 | Exercises and Applications

Extra Practice
exercises on
page 660

1. **TRAVEL** The map shows part of the city of Kyoto, Japan. The bureau of tourism recommends a walking tour to appreciate the beauty of the path along the canal. Nancy Kim has planned the path shown. Can she follow the path without retracing her steps? How do you know?

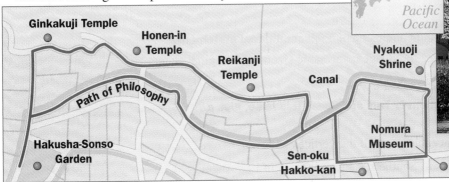

2. **SOCIAL STUDIES** Compare the two systems of counting marks shown below. Make a conjecture about the figures for 8 and 9 in the Japanese system.

American	|	||	|||	||||	⊞	⊞|	⊞||
Japanese	一	丁	下	疒	正	疋	疋丁

INTERVIEW Latecia Leavy

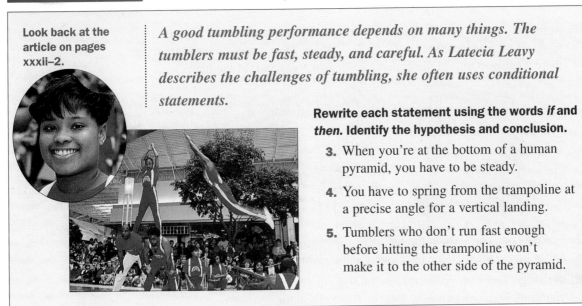

Look back at the article on pages xxxii–2.

A good tumbling performance depends on many things. The tumblers must be fast, steady, and careful. As Latecia Leavy describes the challenges of tumbling, she often uses conditional statements.

Rewrite each statement using the words *if* and *then*. Identify the hypothesis and conclusion.

3. When you're at the bottom of a human pyramid, you have to be steady.

4. You have to spring from the trampoline at a precise angle for a vertical landing.

5. Tumblers who don't run fast enough before hitting the trampoline won't make it to the other side of the pyramid.

Visual Thinking For each figure, write a conjecture in the form "If the figure has _?_ symmetry, then the hidden part of the figure looks like..." Complete the conjecture with a sketch of the hidden part.

6.

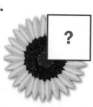

7.

8.

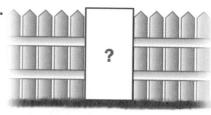

Identify the hypothesis and conclusion of each conditional statement.

9. If you reflect a shape, the image is the same size as the original shape.

10. I'll buy a book if I have enough money.

Tell whether each statement is *True* or *False*. If it is false, give a counterexample.

11. If a network is traceable, then all the vertices are even.

12. A person who lives in Southern California lives in Los Angeles.

13. If a and b are real numbers, then $a + b = b + a$.

14. Challenge If $x^2 - 5x + 6 = 0$, then $x = 2$.

Investigation For Exercises 15–18, you will need a piece of paper that is at least $8\frac{1}{2}$ in. by 11 in.

Fold the paper in half. Unfold it and notice that the crease divides the paper into two sections.

15. Refold the paper. Fold it in half again. How many sections are there when you unfold the paper?

16. Refold the paper. Fold it in half again. How many sections are there when you unfold the paper?

17. Make a conjecture about the number of sections if you fold the paper in half n times. Identify the hypothesis and conclusion of your conjecture.

18. Writing Test your conjecture by folding the paper two more times. Do you think your conjecture is *True* or *False*? Explain.

Natural events have had a great effect on history. For example, on the night of August 29, 1776, George Washington's army was able to escape the British because the East River was covered in a thick fog.

Identify the hypothesis and conclusion of each conditional statement. (*Hint:* Write the hypothesis and conclusion using "was not" instead of "had not been" and so on.)

19. If the East River had not been cloaked by fog, George Washington's army might have been trapped at the western end of Long Island.

20. If there had not been a drought in the Southern Great Plains in the 1930s, the area might not have been hit by the huge dust storms that drove thousands of families away.

21. If the winter of 1620 had not been unusually mild, then the Pilgrims might not have survived their first winter in Massachusetts.

22. Open-ended Problem A solar eclipse during a battle in ancient Mesopotamia frightened the armies so much that they made peace. Modern astronomers used what they know about eclipses to calculate the exact date of the battle. Use these two facts to write a conditional statement like those in Exercises 19–21. Identify your hypothesis and conclusion.

Use each hypothesis to write a true conditional statement.

23. Hypothesis: Today is January 1.

24. Hypothesis: $5x + 7 = -3x - 5$.

25. Hypothesis: A point is located on the x-axis.

26. Hypothesis: A shape has reflection symmetry.

27. Hypothesis: You translate a point up one unit on a coordinate plane.

28. SAT/ACT Preview If $A = x$ and $B = x^2$, then:

  A. $A \geq B$ **B.** $A \leq B$ **C.** $A = B$ **D.** relationship cannot be determined

ONGOING ASSESSMENT

29. Open-ended Problem Write a conditional statement about yourself. Identify the hypothesis and conclusion. Is there a counterexample?

30. Open-ended Problem Sketch a simple figure that has reflection symmetry. Draw the line of symmetry. *(Section 1.2)*

31. Graph the equation $y = 4 + \frac{1}{3}x$. *(Toolbox, page 697)*

ASSESS YOUR PROGRESS

VOCABULARY

inductive reasoning (p. 4)
reflect (p. 8)
rotate (p. 8)
translate (p. 8)
transformation (p. 8)
image (p. 8)
symmetry (p. 10)

line of symmetry (p. 10)
conjecture (p. 15)
conditional statement (p. 15)
hypothesis (p. 15)
conclusion (p. 15)
counterexample (p. 16)

For Exercises 1–4, use the toothpick figures below.

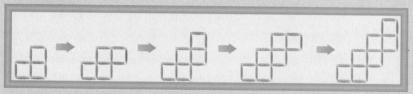

1. Copy and complete the table. What pattern do you notice? *(Section 1.1)*

Number of squares	3	4	5	6	7
Number of toothpicks	10	?	?	?	?

2. Find the number of toothpicks needed to make 10 squares. Describe how you found your answer. *(Section 1.1)*

3. Write a conjecture about the number of toothpicks needed to make *n* squares. Identify the hypothesis and conclusion. *(Section 1.3)*

4. Describe the symmetry of each figure. *(Section 1.2)*

5. Copy the diagram. Then sketch the image after each transformation. *(Section 1.2)*

   a. Reflect the letter over the red line.

   b. Rotate the letter around the green point by a half turn.

   c. Translate the letter two inches down.

6. Tell whether the statement "if *a* and *b* are integers, then $a \div b \le a$" is *True* or *False*. If it is false, give a counterexample. *(Section 1.3)*

7. **Journal** Describe how you look for a pattern in a series of numbers.

1.4 Modeling Points, Lines, and Planes

Learn how to...

- use points, lines, and planes

So you can...

- describe and sketch relationships between points, lines, and planes in real-world objects

Figures in geometry are made of points, lines, and planes. Instead of defining *point*, *line*, and *plane*, mathematicians describe the relationship between them. A point represents a location and has no size. A line is made of many points, and a plane contains many points and lines.

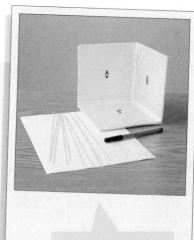

EXPLORATION
COOPERATIVE LEARNING

Representing Points, Lines, and Planes

Work with another student.
You will need:

- three foam trays or pieces of stiff paper
- tape
- scissors
- several pieces of uncooked spaghetti

Make a model like the one shown. The pieces of spaghetti represent lines and the trays represent planes.

Tape three trays together. Poke one hole in each tray. Label the holes as points A, B, and C.

1 On the tray that contains point A, draw a line through A. Label the line ℓ. Put a piece of spaghetti through the holes at A and B. Do line ℓ and the spaghetti have any points in common other than point A?

2 Draw a line on the same tray as line ℓ so that the two lines don't share any points, even though they continue forever. Label the new line m.

3 Add a line to your model that is not on the same tray as line ℓ and does not share any points with ℓ.

4 Is it possible for two different trays to contain the same line? If so, add such a line to your model and label it s.

5 Is it possible for a line and a plane to have exactly one point in common? exactly two points in common? no points in common? Give an example of each possibility, drawing lines on your model as necessary.

For any two points, there is exactly one line through the points. You can name a line by using a single letter or by naming two points on the line. Line ℓ can also be called $\overleftrightarrow{AB}$ or $\overleftrightarrow{BA}$.

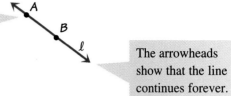

Points that are on the same line are **collinear**.

The arrowheads show that the line continues forever.

For any three noncollinear points, there is exactly one plane through the points. You can name a plane by using three noncollinear points in the plane. Plane *S* can also be called plane *ABP* or plane *PAB*, for example.

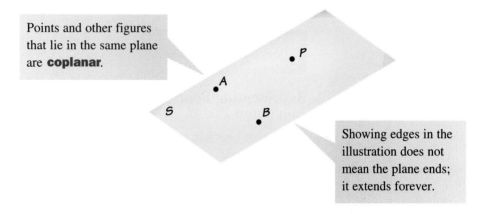

Points and other figures that lie in the same plane are **coplanar**.

Showing edges in the illustration does not mean the plane ends; it extends forever.

If two different figures share at least one point, then they **intersect**. The shared points are called the **intersection**. If two lines in the same plane do not intersect, then they are **parallel**. If two lines do not intersect and are not parallel, then they are **skew**.

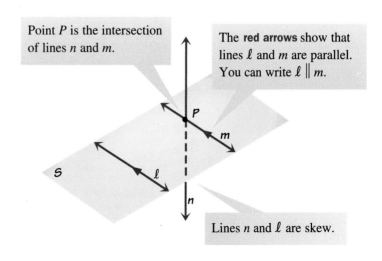

Point *P* is the intersection of lines *n* and *m*.

The **red arrows** show that lines ℓ and *m* are parallel. You can write $\ell \parallel m$.

Lines *n* and ℓ are skew.

Two planes can intersect in a line. Planes that do not intersect are parallel. If a line and a plane intersect, then their intersection is a point or a line.

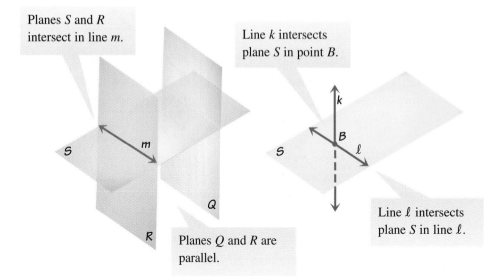

Planes *S* and *R* intersect in line *m*.

Line *k* intersects plane *S* in point *B*.

Planes *Q* and *R* are parallel.

Line *ℓ* intersects plane *S* in line *ℓ*.

EXAMPLE

Sketch each situation.

a. $\overleftrightarrow{HI}$ intersects plane *E* in point *Q*.   **b.** Line *m* and line *n* are skew.

SOLUTION

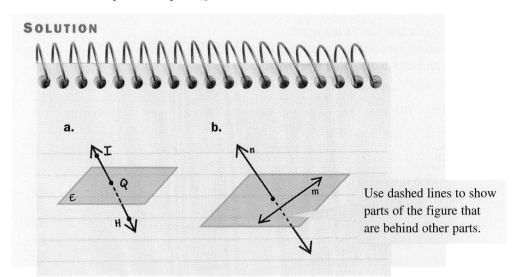

a.

b.

Use dashed lines to show parts of the figure that are behind other parts.

THINK AND COMMUNICATE

Tell whether each statement is *True* or *False*. Explain your answer.

  1. If two lines intersect, then they are coplanar.

  2. For any line, there are many planes that contain the line.

  3. It is possible for three points to be noncoplanar.

  4. It is possible for four points to be noncoplanar.

For Questions 1–7, use the figure at the right.

1. What is the intersection of $\overleftrightarrow{AB}$ and $\overleftrightarrow{EA}$?

2. Which lines are parallel to $\overleftrightarrow{EF}$?

3. Tell whether each group of points or lines is *collinear*, *coplanar but noncollinear*, or *noncoplanar*.

   a. *E, F,* and *H*   **b.** *D* and *B*   **c.** $\overleftrightarrow{AD}$ and *B*

   d. *A, D,* and *G*   **e.** *A, B, C,* and *G*   **f.** $\overleftrightarrow{AD}$ and $\overleftrightarrow{CG}$

4. Name all the lines in the diagram that are skew to $\overleftrightarrow{FD}$.

5. What is the intersection of $\overleftrightarrow{AB}$ and plane *HGC*?

6. What is the intersection of plane *EHB* and plane *GCB*?

7. Which plane is parallel to plane *EFH*?

8. Are points that are collinear also coplanar? Explain.

1.4 Exercises and Applications

Extra Practice exercises on page 660

HISTORY For Exercises 1–6, use the kite at the right. The kite was part of a survival kit for pilots in World War II. The kite string was an antenna for sending an S O S signal.

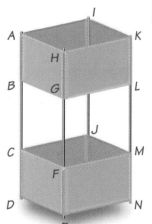

1. What is the intersection of $\overleftrightarrow{CJ}$ and $\overleftrightarrow{MJ}$?

2. Name four lines parallel to $\overleftrightarrow{BG}$.

3. Name a line through point *K* that is skew to $\overleftrightarrow{HE}$.

4. What is the intersection of plane *CDE* and plane *MNE*?

5. Name two planes that intersect in $\overleftrightarrow{AB}$.

6. Name a plane parallel to plane *ABG*.

Tell whether each statement is *True* or *False*. If the statement is false, sketch a counterexample or give a counterexample from the figure at the right.

7. Two points are always collinear.

8. If two lines are not skew, then they are coplanar.

9. If three points are collinear, then there is exactly one plane through the points.

10. If three lines intersect at one point, then the lines are coplanar.

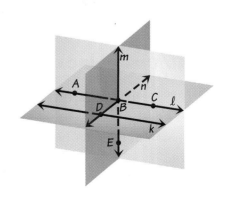

SKIING Describe the relationship between the two skis in each photo.

11.

12.

13.

Open-ended Problem Describe something in your school or classroom that fits each description.

14. parallel lines

15. skew lines

16. parallel planes

17. intersecting planes

18. a. What do the arrowheads on the ends of the lines *t* and *n* mean?

b. What do the red arrows in the middle of the lines mean?

19. Which name best describes the plane?

A. plane *n*

B. plane *ADC*

C. plane *S*

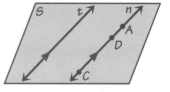

Sketch each situation.

20. Line *ℓ* contains points *A*, *B*, *X*, and *Y*.

21. $\overleftrightarrow{FG}$ and $\overleftrightarrow{DE}$ intersect in point *C*.

22. $s \parallel r$

23. Noncollinear points *D*, *E*, and *F* lie in plane *W*.

24. Collinear points *A*, *F*, *X*, and *Z* lie in plane *H*.

25. $\overleftrightarrow{RM}$ lies in plane *A*.

26. The intersection of $\overleftrightarrow{PQ}$ and plane *B* is point *D*.

Use the diagram at the right.

27. Tell which line is parallel to $\overleftrightarrow{BE}$.

28. Tell which plane is parallel to plane *BXE*.

29. Tell which lines are skew to $\overleftrightarrow{CF}$.

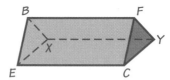

30. TRANSFORMATIONS Copy the triangle and translate it up about 2 in. Label the new triangle so that the image of *A* is *D*, the image of *B* is *E*, and the image of *C* is *F*.

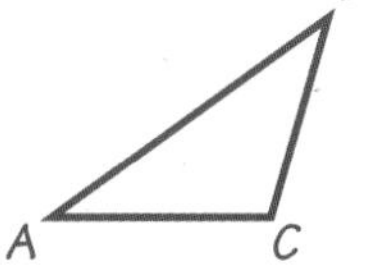

a. Draw $\overleftrightarrow{AD}$, $\overleftrightarrow{BE}$, and $\overleftrightarrow{CF}$. Tell whether the lines are *parallel*, *intersecting*, or *skew*.

b. Draw a simple figure and translate it in any direction. Draw lines through points on the figure and their images as in part (a). Tell whether the lines are *parallel*, *intersecting*, or *skew*.

c. Make a conjecture based on your answers to parts (a) and (b).

How do you make a flat surface appear to have depth? Ancient Egyptians used overlapping forms to suggest depth. Chinese artists used misty spaces to separate nearby forms from distant ones. In the fifteenth and sixteenth centuries, Italian artists explored painting objects as though seen through a window with one eye shut. This method is called *linear perspective*.

Above: Egypt, 1550–1295 B.C.

Right: China, fifteenth century

Italy, around 1470

31. a. Imagine that you are in the painting. Are $\overleftrightarrow{AB}$ and $\overleftrightarrow{CD}$ parallel?

b. Think of the painting as a flat canvas with paint on it. Are $\overleftrightarrow{AB}$ and $\overleftrightarrow{CD}$ parallel? Explain.

32. a. Imagine that you are in the painting. Name two skew lines.

b. Think of the painting as a flat canvas. Are the lines from part (a) still skew? Why or why not?

33. Name two lines that are parallel both when you look at the painting as a flat canvas and when you imagine that you are in the painting.

34. Investigation You will need tracing paper. Carefully trace the columns and the lines of the floor and roof.

a. Extend $\overleftrightarrow{AB}$ and $\overleftrightarrow{CD}$. Label the point of intersection P. Where is P?

b. Extend $\overleftrightarrow{EF}$ and $\overleftrightarrow{GH}$. Where do they intersect?

c. Open-ended Problem Make a conjecture based on your answers to parts (a) and (b). Can you find a counterexample?

d. Point P is called the *vanishing point*. Where is the vanishing point in the illustration at the left?

35. Challenge The etching at the right is by M. C. Escher, a Dutch artist.

   a. Look closely at the building. What is unusual about it?

   b. Describe two lines that are skew and intersect.

   c. Could you build a model of this building? Explain.

36.  **Technology** Use geometry software and these steps to draw a cube.

   a. Draw a point. A point is a *zero-dimensional* figure.

   b. Copy the point. Connect the old point and the new point by drawing a segment between them. Move the new point until the segment is horizontal. A segment is a *one-dimensional* figure.

   c. Copy the segment. Connect each endpoint of the segment with the same endpoint on the new segment. Move the new segment until you form a square. A square is a *two-dimensional* figure.

   d. Copy the square and connect the endpoints. Move the new square until the figure looks like a cube. A cube is a *three-dimensional* figure.

ONGOING ASSESSMENT

37. Writing How would you describe a point, a line, and a plane to someone who has not taken geometry?

SPIRAL REVIEW

Tell whether each statement is *True* or *False*. If it is false, give a counterexample. *(Section 1.3)*

38. If I turn the lights off, then I can't see.

39. If $x = 5$, then $x^2 = 25$.

Give the slope of each line. *(Toolbox, page 697)*

40. $y = 4x + 7$ **41.** $y = x$ **42.** $y = -\frac{1}{2}x + 3$

For each graph, give the slope and the *y*-intercept. *(Toolbox, page 697)*

43.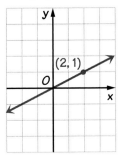

44.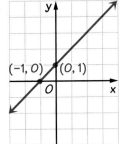

45.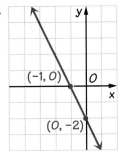

1.5 Segments and Their Measures

Learn how to...
- represent rays and segments
- interpret diagrams

So you can...
- find lengths and distances in various situations, such as manufacturing and astronomy

Tunnels beneath cities and towns are carefully constructed. One way to make sure a tunnel is straight is to use a *laser alignment tool*. The laser projects a straight ray of light in the direction of the tunnel, and guides the machine that digs the tunnel.

In geometry, a **ray** is part of a line with one endpoint. You can name a ray by its endpoint and another point on the ray. Always name the endpoint first.

endpoint

You can call this ray $\overrightarrow{AB}$ or $\overrightarrow{AC}$.

A **segment** is part of a line with two endpoints. Name a segment by using its endpoints. Segments that are equal in length are **congruent segments**.

Tick marks are used to show congruence.

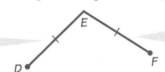

You can call this segment $\overline{EF}$ or $\overline{FE}$.

You can write "$\overline{DE}$ is congruent to $\overline{EF}$" by writing $\overline{DE} \cong \overline{EF}$. The length of the segment $\overline{DE}$, written DE, is the distance between D and E.

EXAMPLE 1

Give the length of each segment.

a. $\overline{QN}$

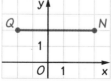

b. $\overline{EF}$

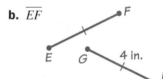

4 in.

SOLUTION

a. Count how many units long $\overline{QN}$ is. $QN = 5$.

b. The tick marks show that $\overline{EF} \cong \overline{GH}$, so $EF = GH$. $EF = 4$ in.

You might say that a store is *between* home and school even though the store is not on a straight line from home to school. But in geometry, a point is *between* two other points only if the three points are collinear.

Betweenness and Segment Addition

If $XY + YZ = XZ$, then point Y is between points X and Z on $\overline{XZ}$.

If point Y is between points X and Z, then $XY + YZ = XZ$.

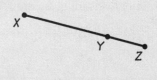

EXAMPLE 2

Find each length.

a. AC

b. CB

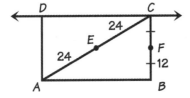

SOLUTION

a. Point E is between points A and C, so $AC = AE + EC$.

$$AC = AE + EC \qquad \text{Use Segment Addition.}$$
$$= 24 + 24$$
$$= 48$$

b. The red tick marks show that $\overline{CF} \cong \overline{FB}$, so $CF = FB$.

$$CF = 12$$

Point F is between C and B, so $CB = CF + FB$.

$$CB = CF + FB \qquad \text{Use Segment Addition.}$$
$$= 12 + 12$$
$$= 24$$

THINK AND COMMUNICATE

Use the figure in Example 2.

1. Which two segments are congruent to $\overline{AE}$? How do you know?

2. Describe how $\overrightarrow{DC}$ is different from $\overrightarrow{CD}$.

3. Make a sketch that shows that $MN = NO = 1$ in. In your sketch, how long is $\overline{MO}$? Must M, N, and O be collinear? Compare your answer with your classmates' answers.

Diagrams

In geometry, you should be careful about what you conclude from a sketch or diagram. For example, you can conclude that all the points shown in the diagram below are coplanar, that $\overline{AB}$, $\overline{AC}$, and $\overline{AD}$ all intersect at point A, and that E is between A and C. Notice that segments and rays that are contained in parallel lines are also parallel.

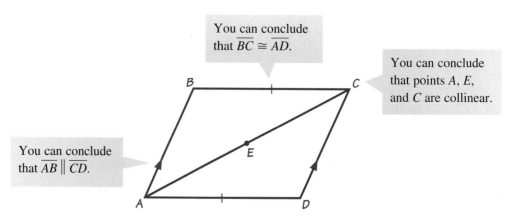

You can conclude that $\overline{BC} \cong \overline{AD}$.

You can conclude that points A, E, and C are collinear.

You can conclude that $\overline{AB} \parallel \overline{CD}$.

The diagram does *not* tell you that $\overline{AD} \parallel \overline{BC}$ or that $\overline{AB} \cong \overline{CD}$.

☑ CHECKING KEY CONCEPTS

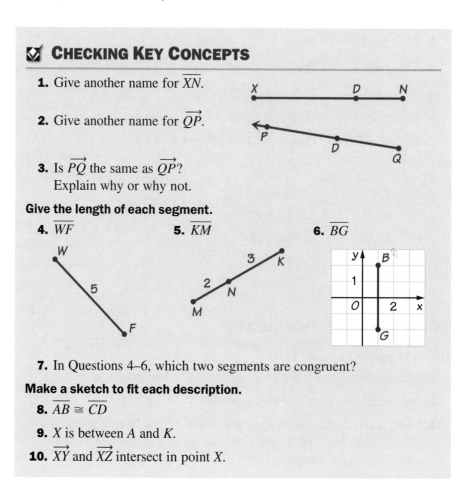

1. Give another name for $\overline{XN}$.

2. Give another name for $\overrightarrow{QP}$.

3. Is $\overrightarrow{PQ}$ the same as $\overrightarrow{QP}$? Explain why or why not.

Give the length of each segment.

4. $\overline{WF}$

5. $\overline{KM}$

6. $\overline{BG}$

7. In Questions 4–6, which two segments are congruent?

Make a sketch to fit each description.

8. $\overline{AB} \cong \overline{CD}$

9. X is between A and K.

10. $\overrightarrow{XY}$ and $\overrightarrow{XZ}$ intersect in point X.

Extra Practice
exercises on
page 660

1.5 **Exercises and Applications**

Give the length of each segment.

1. $\overline{AB}$

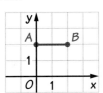

2. $\overline{BE}$

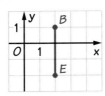

3. $\overline{DH}$

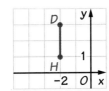

Give the length of each segment.

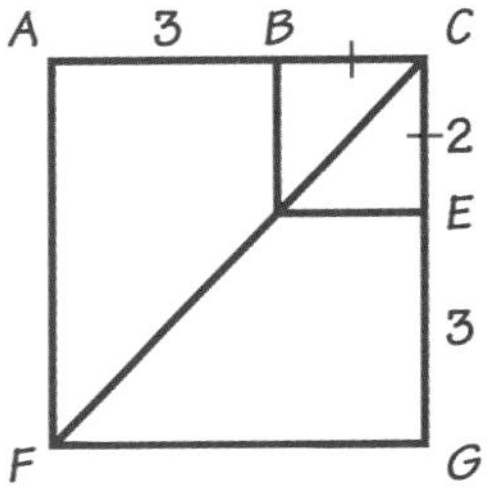

4. $\overline{CG}$

5. $\overline{BC}$

6. $\overline{AC}$

7. List three pairs of congruent segments in the figure.

Make a sketch to fit each description.

8. $\overrightarrow{AB}$

9. $\overline{CD}$

10. $ST = 2$ in.

11. $\overline{EF} \cong \overline{GE}$

12. Point M is between points N and P.

13. $\overrightarrow{JK}$ intersects $\overline{LQ}$ in point R.

Connection ▶ MANUFACTURING

When designers plan a new product, they must consider how the product will be transported or stored. Often, long narrow objects are designed to come apart into pieces.

About how long will each object be when it is assembled?

14. The pieces are all congruent.

20 in.

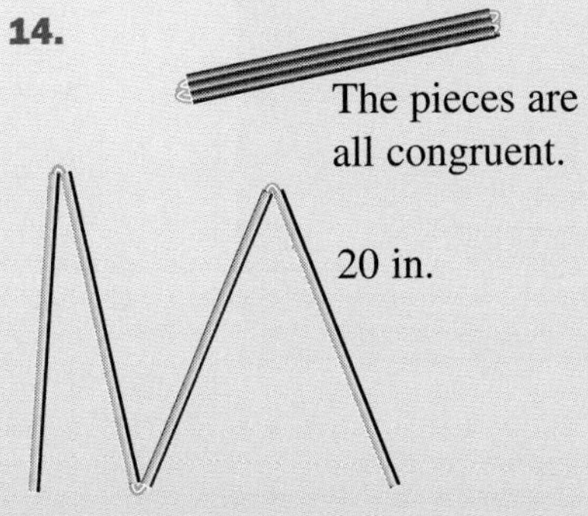

Tent poles would never fit in a car trunk at their full length. But folded up, the poles can fit into a bag with the tent.

15. $2\frac{1}{2}$

$5\frac{1}{2}$

$4\frac{3}{4}$

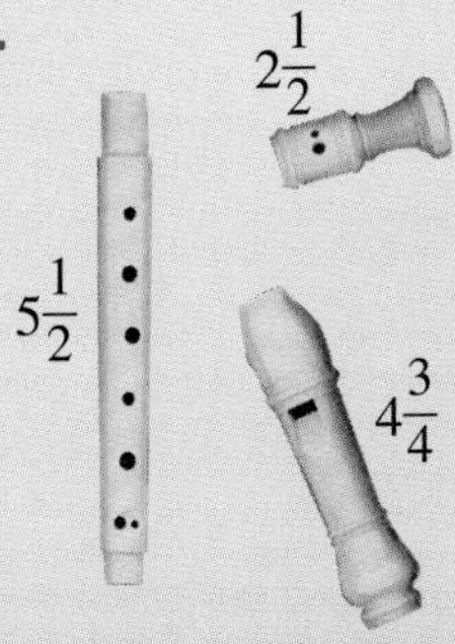

A recorder comes apart so it can be cleaned and stored easily. You can store a recorder in pieces.

16. 3 ft

3 ft

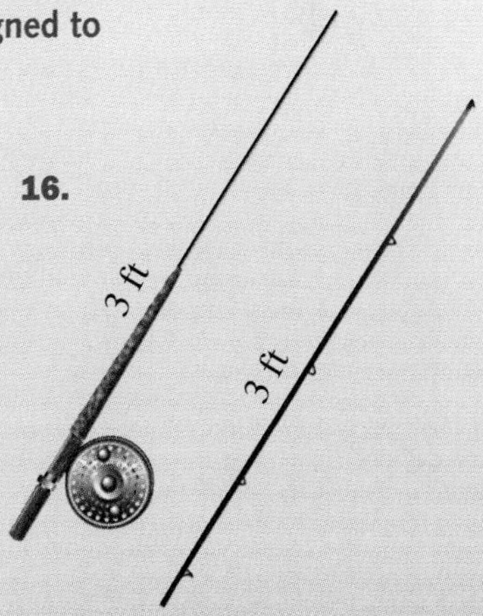

Portable fishing poles are popular with hikers who don't want to carry much food with them.

17. a. Writing In the picture at the right, Tyler is between Juna and Ian, Ian is between Tyler and Charelra, and Charelra is next to Nicole and Ian. Juna is wearing a light blue shirt. What is each of the other students wearing? Explain how you got your answer.

b. Is Ian between Juna and Nicole? Explain.

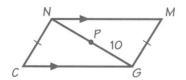

For Exercises 18–23, give the letter of the best description.

18. $\overline{AB}$

19. $\overleftrightarrow{AB}$

20. $\overrightarrow{AB}$

21. AB

22. $\cong$

23. $\parallel$

A. the line through points A and B

B. the distance between points A and B

C. is congruent to

D. the segment connecting points A and B

E. the ray from point A through point B

F. is parallel to

24. ALGEBRA $PM = 3x + 2$, $MN = x + 4$, and $PN = 18$. Find the values of x and of PM.

25. ALGEBRA $ST = 2y + 7$, $RT = 7y + 1$, and $RS = 9$. Find the values of y and of RT.

26. Open-ended Problem Make a sketch that has five points labeled, and includes one pair of congruent segments, one ray, and three collinear points. Name the ray and the congruent segments.

27. Challenge Earth and Saturn both orbit the sun. On average, Earth is 1.5×10^{11} m from the sun and Saturn is 1.4×10^{12} m from the sun. About how far can Saturn be from Earth? About how close can Saturn be to Earth? Explain your answer. Include a diagram.

O N G O I N G   A S S E S S M E N T

28. Writing What information does the diagram give you about the points and segments in the diagram? Explain how you know.

Sketch each situation. *(Section 1.4)*

29. $\overline{ST} \parallel \overline{NJ}$

30. Lines n and m intersect in point T.

31. Line ℓ intersects plane ABC in point B.

32. Points P, M, and R are noncollinear.

Sketch the next figure in each pattern. *(Section 1.1)*

33.

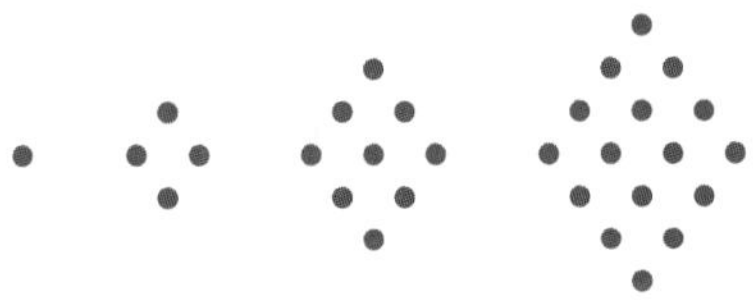

34. $\vee < \wedge > \vee < \wedge > \vee <$

ASSESS YOUR PROGRESS

VOCABULARY

collinear (p. 22)
coplanar (p. 22)
intersect (p. 22)
intersection (p. 22)
parallel (p. 22)

skew (p. 22)
ray (p. 28)
endpoint (p. 28)
segment (p. 28)
congruent segments (p. 28)

For Exercises 1–5, use the figure at the right. *(Section 1.4)*

1. Name two lines that intersect at point C.

2. What is the intersection of $\overleftrightarrow{CD}$ and plane AED?

3. What is the intersection of $\overleftrightarrow{AE}$ and plane BCD?

4. Which two planes intersect in $\overleftrightarrow{AD}$?

5. Name a line that is skew to $\overleftrightarrow{AB}$.

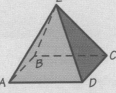

Sketch each situation. *(Section 1.4)*

6. $n \parallel m$

7. Points A, B, and C are collinear.

8. $\overleftrightarrow{QF}$ lies in plane Z.

9. Line ℓ intersects $\overleftrightarrow{RS}$ in point T.

Make a sketch to fit each description. *(Section 1.5)*

10. $AB = 7$

11. Point S is between T and U.

12. Open-ended Problem Draw a segment on a coordinate graph and give its length.

Give the length of each segment. *(Section 1.5)*

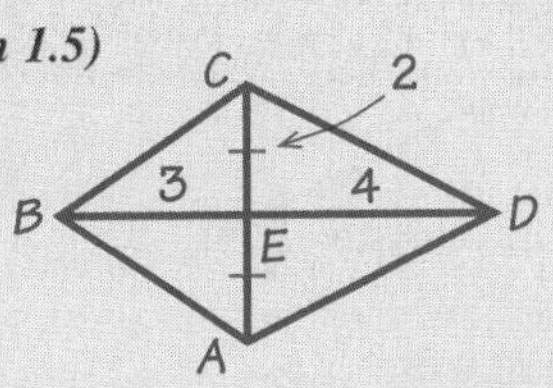

13. $\overline{BD}$

14. $\overline{AE}$

15. $\overline{AC}$

16. Name a segment that is congruent to $\overline{AE}$. *(Section 1.5)*

17. Journal How many lines contain a particular segment? How many rays contain a particular segment? Explain your answers.

SECTION

1.6 Working with Angles

Learn how to...

- **name and measure angles**

So you can...

- **describe angles in various situations, such as space flight, tumbling, and finding your latitude**

In 1970, *Apollo 13* blasted into space. The astronauts hoped to set foot on the moon. Just over two days into the trip, an oxygen tank exploded and the space capsule had to return to Earth. Getting back was not easy. The capsule had to maintain its course for the correct *reentry angle*.

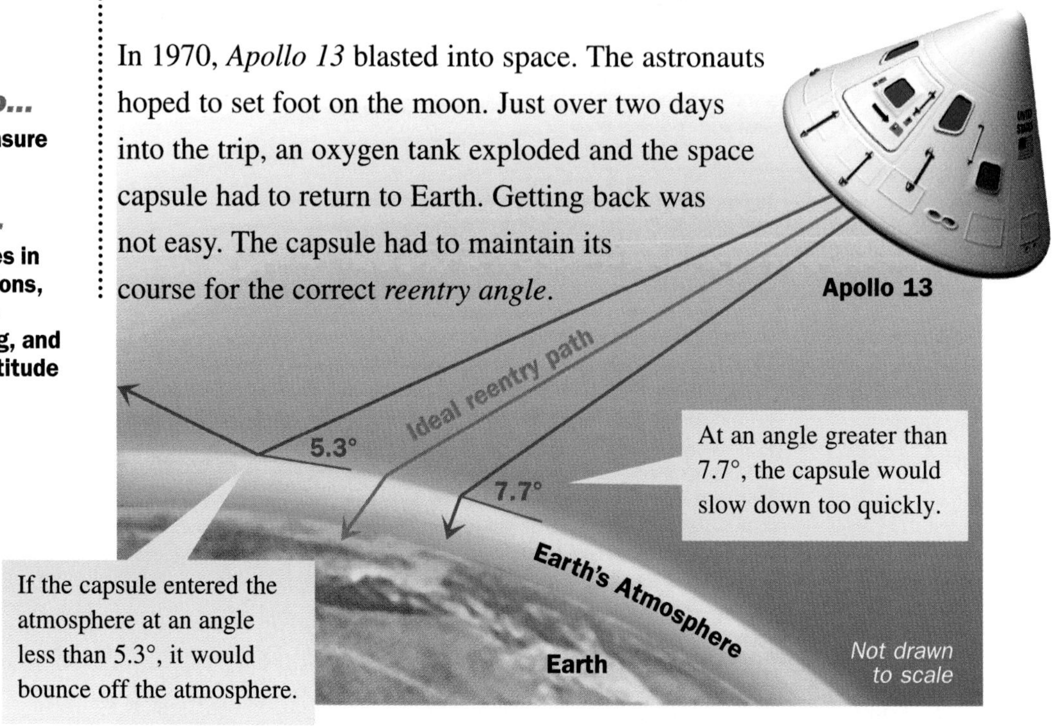

Apollo 13
Ideal reentry path
5.3°
7.7°
At an angle greater than 7.7°, the capsule would slow down too quickly.
If the capsule entered the atmosphere at an angle less than 5.3°, it would bounce off the atmosphere.
Earth's Atmosphere
Earth
Not drawn to scale

An **angle** is formed by two rays with a common endpoint. The endpoint is the **vertex** of the angle and the rays are the sides of the angle. You can name an angle by its vertex, by the vertex and a point on each ray, or by a number. When three points name an angle, the vertex is in the middle.

EXAMPLE 1

Give as many different names as you can for each angle.

a. $\angle ABC$

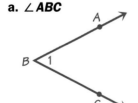

A
B
1
C

b. $\angle XWZ$

X
Y
W
Z

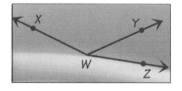

SOLUTION

a. $\angle B$, $\angle 1$, $\angle CBA$

b. $\angle ZWX$. You cannot use $\angle W$ as a name for $\angle ZWX$ because $\angle W$ could also refer to $\angle XWY$ and $\angle YWZ$.

You use a protractor to measure an angle in *degrees* (°). The measure of an angle is greater than 0° and less than or equal to 180°. Write the measure of ∠*TSV* as *m*∠*TSV*. Two angles that have the same measure are **congruent angles**.

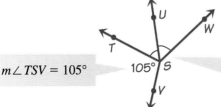

m∠*TSV* = 105°

The red marks show that the angles are congruent.
∠*TSU* ≅ ∠*USW*.

EXAMPLE 2

Use a protractor to measure each angle, to the nearest degree.

a. ∠*AVM*

b. ∠*AVK*

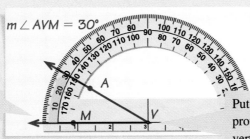

SOLUTION

a. Copy the angle first.

Line up one ray with 0°.

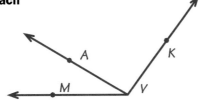

m∠*AVM* = 30°

Put the center of the protractor over the vertex of the angle.

b.

When you copy the angle, you may need to extend rays so that they cross the scale.

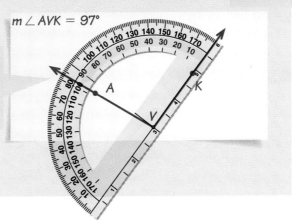

m∠*AVK* = 97°

THINK AND COMMUNICATE

1. Look at the numbers along the scale of the protractor. When should you use the numbers that are on the outside? When should you use the numbers that are on the inside?

2. Measure ∠*MVK*. How is *m*∠*MVK* related to *m*∠*AVM* and *m*∠*AVK*?

To reenter Earth's atmosphere safely, *Apollo 13* had to stay inside the safe reentry corridor. The inside of an angle is called the angle's *interior*.

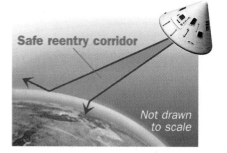

Safe reentry corridor

Not drawn to scale

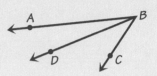

Angle Addition

If *D* is in the interior of $\angle ABC$, then $m\angle ABD + m\angle DBC = m\angle ABC$.

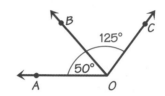

EXAMPLE 3

Find *m* ∠ *BOC*.

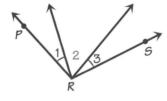

SOLUTION

Point *B* is in the interior of $\angle AOC$, so

Use Angle Addition.

$$m\angle AOB + m\angle BOC = m\angle AOC$$
$$50° + m\angle BOC = 125°$$
$$m\angle BOC = 75°$$

EXAMPLE 4

$m\angle 1 = 24°$
$m\angle 2 = 62°$
Find *m* ∠ *PRS*.

SOLUTION

The red marks mean that $\angle 1 \cong \angle 3$, so $m\angle 1 = m\angle 3$.

Substitute **24** for $m\angle 1$ and $m\angle 3$, and **62** for $m\angle 2$.

$$m\angle PRS = m\angle 1 + m\angle 2 + m\angle 3$$
$$= 24 + 62 + 24$$
$$m\angle PRS = 110°$$

Give another name for each angle.

1. $\angle 1$

2. $\angle 2$

3. $\angle XYZ$

4. $m\angle 1 = 62°$ and $m\angle 2 = 47°$. Find $m\angle XYZ$.

Find the measure of each angle to the
nearest degree.

5. $\angle ABG$ **6.** $\angle CBD$ **7.** $\angle CBG$

8. Name two angles that are congruent.

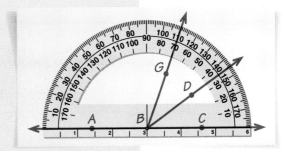

1.6 | **Exercises and Applications**

*Extra Practice
exercises on
page 660*

1. Writing Sketch three rays that share a vertex. Label the vertex A and
label another point on each ray. Explain why you can't name any of the
angles $\angle A$.

**Use a protractor to find the measure of each angle,
to the nearest degree.**

2. $\angle BAD$ **3.** $\angle CAB$

4. $\angle CAD$ **5.** $\angle D$

6. $\angle B$ **7.** $\angle BCD$

8. $\angle BCA$ **9.** $\angle ACD$

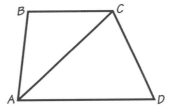

10. TUMBLING One of the Jesse White tumblers' most
exciting tricks is called *Superman*. A tumbler springs
from the trampoline and soars over 20 or more
people. If the trampoline is at about a 45° angle
to the floor, the tumbler will fly as far as possible.
Estimate the angle of each trampoline.

a.

b.

c.

d.

e. Which of these trampoline positions would be best for the *Superman* stunt?

Give the measure of each angle.

11. ∠MNO

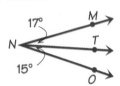

12. ∠PQR

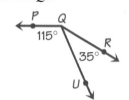

13. ∠YZA

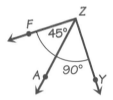

For Exercises 14–18, use the diagram at the right.

14. Find $m\angle CEG$.

15. Find $m\angle GED$.

16. Find $m\angle HED$.

17. Which angle is congruent to ∠CEG?

18. $m\angle HEV = 60°$. Find $m\angle CEV$.

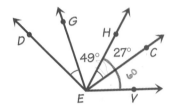

19. Open-ended Problem Look at the objects and surfaces around you and estimate the measures of any angles you see. Which angle measure(s) seems most common? Why do you think this is so?

20. Research Work with another person. Use an encyclopedia, star map, or astronomy guide to find the North Star.

   a. You can use the North Star to estimate your latitude. One of you should point one arm at the North Star and the other at the horizon. The other should estimate the angle between the arms. The measure of the angle is about the same as your latitude.

   b. Use a map or an almanac to find your latitude. How close is your estimate? What could cause your estimate to be far from the actual latitude?

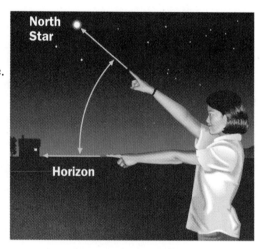

 Technology You will need geometry software.

Step 1 Construct $\overline{AB}$ and point O.

Step 2 Rotate $\overline{AB}$ around point O by 90°. Label the new endpoints C and D, as shown, so C is the image of A and D is the image of B.

Step 3 Construct $\overline{AO}$ and $\overline{CO}$.

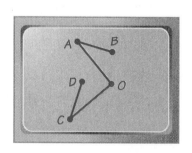

21. Measure ∠AOC.

22. Move $\overline{AB}$. Describe the effect on $m\angle AOC$.

23. Move point O and describe the effect on $m\angle AOC$.

24. Make a conjecture concerning the measure of ∠DOB.

25. Measure ∠DOB. Is your conjecture correct?

When sunlight passes through a raindrop, it breaks into many different colors, from red light, which leaves the raindrop at an angle of 42°, to violet light, which leaves at an angle of 40°. If you look at the rain from a distance, you may see a rainbow.

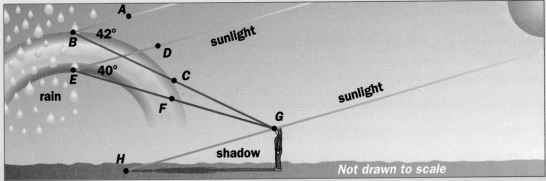

26. $\angle ABC \cong \angle CGH$. Find $m\angle CGH$.

27. $\angle DEF \cong \angle FGH$. Find $m\angle FGH$.

28. Find $m\angle CGF$.

29. **Challenge** Although the sun is much larger than the moon, they look about the same size in the sky. The sun and moon each take up an angle of about 0.5°, as shown. If you saw the moon and a rainbow at the same time, how wide would the rainbow seem compared to the moon? (*Hint:* Use your answer to Exercise 28.)

ONGOING ASSESSMENT

30. **Open-ended Problem** Find a pair of congruent angles at your home or at school. Describe and measure them.

SPIRAL REVIEW

Give the length of each segment. *(Section 1.5)*

31. $\overline{FN}$

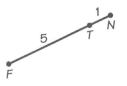

32. $\overline{ST}$; $\overline{OR}$

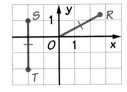

33. $\overline{AC}$

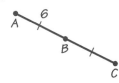

34. Sketch a figure that has rotational symmetry. *(Section 1.2)*

1.7 Bisecting Segments and Angles

Learn how to...

- identify bisectors of segments and angles

So you can...

- find real-world distances, such as the distance between two cities

- find angle measures, such as in boating

Origami, the art of paper folding, originated in China almost 2000 years ago. Although the word *origami* is Japanese, the art was also practiced in North Africa and Spain. The folded figures suggest not only the form of an object, but its motion or even its character. One of the most popular figures is the crane, sometimes used as a symbol of peace.

THINK AND COMMUNICATE

The first two folds for making an origami crane are shown.

fold and unfold

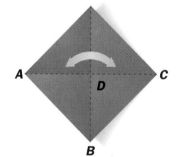

fold and unfold

1. Do you think that $\angle ABD \cong \angle CBD$? Why or why not?

2. Do you think that $\overline{AD} \cong \overline{DC}$? Why or why not?

3. What transformation is suggested by these folds?

When you reflect, translate, or rotate a figure, the image is congruent to the original figure. In the diagram on page 40, $\overline{DC}$ is the reflection of $\overline{AD}$ over $\overline{DB}$, and $\overline{AD} \cong \overline{DC}$. Point D is called the *midpoint* of $\overline{AC}$. The **midpoint** of a segment is the point that divides the segment into two congruent segments.

If M is the midpoint of $\overline{XY}$, then:

$$XM = MY \text{ and } \overline{XM} \cong \overline{MY}$$

$$XM = \tfrac{1}{2}XY \text{ and } MY = \tfrac{1}{2}XY$$

A **bisector of a segment** is a line, segment, ray, or plane that intersects the segment at its midpoint.

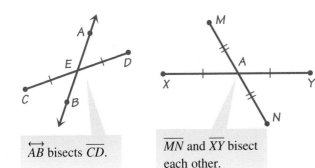

$\overleftrightarrow{AB}$ bisects $\overline{CD}$.

$\overline{MN}$ and $\overline{XY}$ bisect each other.

Plane E bisects $\overline{RT}$.

EXAMPLE 1 — Connection: Algebra

$\overline{RS}$ bisects $\overline{PQ}$.
Find the value of n.

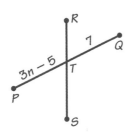

SOLUTION

Because $\overline{RS}$ bisects $\overline{PQ}$, T is the midpoint of $\overline{PQ}$.

$$PT = TQ$$
$$3n - 5 = 7 \qquad \text{Add 5 to both sides of the equation.}$$
$$3n = 12$$
$$n = 4 \qquad \text{Divide both sides by 3.}$$

> **Toolbox p. 696**
> *Solving Two-Step Equations*

THINK AND COMMUNICATE

4. Can a line have a midpoint? Explain why or why not.

5. In the diagram, $\overline{DE} \cong \overline{EF}$, but E is not the midpoint of $\overline{DF}$. Explain why not.

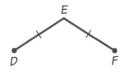

6. Explain how to find PQ in Example 1.

EXAMPLE 2

Find the coordinates of the midpoint of $\overline{GH}$.

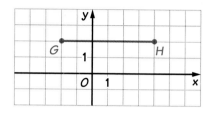

SOLUTION

$\overline{GH}$ is 6 units long. The midpoint of $\overline{GH}$ divides it into two segments, each 3 units long.

Count how far G is from H. $GH = 6$.

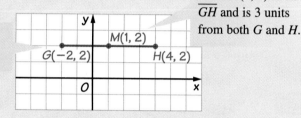

Point $M(1, 2)$ is on $\overline{GH}$ and is 3 units from both G and H.

THINK AND COMMUNICATE

7. In Example 2, explain how you could fold the paper to find the coordinates of the midpoint of $\overline{GH}$.

8. In Example 2, how is the x-coordinate of the midpoint related to the x-coordinates of the endpoints?

Angle Bisectors

If you fold one ray of an angle onto the other ray, and unfold the paper, the crease divides the angle into two congruent parts. The crease is an example of an *angle bisector*.

A **bisector of an angle** is a ray or line that divides the angle into two congruent angles.

If $\overrightarrow{QS}$ bisects $\angle PQR$, then:

$$m\angle PQS = m\angle SQR \text{ and } \angle PQS \cong \angle SQR$$

$$m\angle PQS = \frac{1}{2}m\angle PQR \text{ and } m\angle SQR = \frac{1}{2}m\angle PQR$$

EXAMPLE 3   Connection: Algebra

$\overrightarrow{SV}$ bisects $\angle RST$.
Find the value of
x and $m\angle VST$.

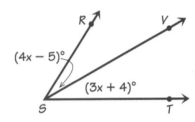

$(4x - 5)^\circ$

$(3x + 4)^\circ$

SOLUTION

Because $\overrightarrow{SV}$ bisects $\angle RST$, $m\angle RSV = m\angle VST$.

$$m\angle RSV = m\angle VST$$

$$(4x - 5)^\circ = (3x + 4)^\circ \quad \text{Add 5 to both sides.}$$

$$4x = 3x + 9$$

$$x = 9 \quad \begin{array}{l}\text{Subtract } 3x \text{ from} \\ \text{both sides.}\end{array}$$

$$m\angle VST = (3x + 4)^\circ$$

$$= 3(9) + 4 \quad \text{Substitute 9 for } x.$$

$$= 31^\circ$$

So $x = 9$ and $m\angle VST = 31^\circ$.

☑ CHECKING KEY CONCEPTS

The midpoint of $\overline{QR}$ is M, the midpoint of $\overline{PR}$ is Q, and $QM = 1.7$.
Find each length.

P   Q   1.7   M   R

1. QR **2.** MR **3.** PR **4.** PM

$\overrightarrow{DB}$ bisects $\angle ADC$.
Find each measure.

5. $m\angle BDC$

6. $m\angle ADC$

7. Sketch a segment with endpoints J and
L. Show how to use paper folding
to find its midpoint, K.

8. $\overrightarrow{CD}$ bisects $\angle BCE$. Find the value of x.

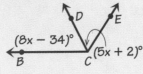

$(8x - 34)^\circ$ $(5x + 2)^\circ$

1.7 Exercises and Applications

**Extra Practice
exercises on
page 661**

In the diagram, $\overline{AB}$ and $\overline{EF}$ bisect each other. Find each length.

1. *MB*   **2.** *EF*   **3.** *MF*   **4.** *AB*

Tell whether each statement is *True* or *False*.

5. *Y* and *W* are both midpoints of $\overline{XZ}$.

6. *Y* is the midpoint of $\overline{XW}$.

7. If *YZ* = 6, then *YW* = 12.

8. If *XW* = 12, then *YZ* = 12.

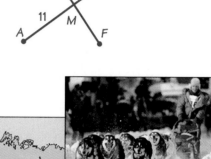

9. RACING The Iditarod Sled Dog Race
is held each spring. Racers (called
"mushers") are pulled by teams of sled
dogs from Anchorage to Nome, Alaska.

a. The route changes each year. In
odd-numbered years, the city of
Iditarod is roughly at the midpoint
of the 1161 mi race. About how far
from Nome is Iditarod?

b. Writing How is the midpoint of a
race like the midpoint of a segment?
How is it different?

Find the coordinates of the midpoint of each segment.

10. $\overline{AB}$ **11.** $\overline{CD}$

12. $\overline{EF}$ **13.** $\overline{GH}$

14. $\overline{IJ}$ **15.** $\overline{OA}$

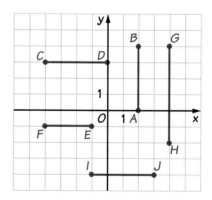

Make a sketch to illustrate each statement.

16. *M* is the midpoint of $\overline{SG}$.

17. *A* is the midpoint of $\overline{CD}$ and *B* is the midpoint of $\overline{CA}$.

18. $\overline{PQ}$ bisects $\overline{DE}$, but $\overline{DE}$ does not bisect $\overline{PQ}$.

19. $\overrightarrow{XY}$ bisects $\angle AXP$.

**ALGEBRA In each figure, *P* is the midpoint of the segment. Find the value
of each variable.**

20.

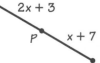

2x + 3

P   x + 7

21.

4x + 3

P

8x − 11

22.

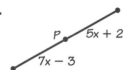

P   5x + 2

7x − 3

You can use origami to make a drinking cup that will actually hold water. The figures below show a square piece of paper with vertices *A, B, C,* and *D.* You will locate points *E, F, G,* and *H* by folding.

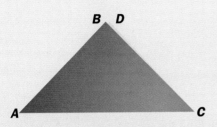

Step 1 Fold the bottom half of the paper up along $\overline{AC}$, bringing point *B* to *D*.

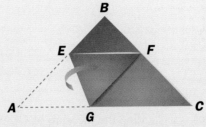

Step 2 Fold $\overline{AC}$ so point *A* meets $\overline{BC}$ at point *F* and $\overline{EF} \parallel \overline{GC}$.

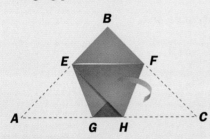

Step 3 Fold $\overline{BC}$ at *F* so point *C* meets $\overline{AB}$ at point *E*.

Step 4 Separate the two points, *B* and *D*, at the top. Fold the rear piece back. Fold the front piece forward and tuck it into the front pocket.

Step 5 Squeeze the sides together lightly and you have a drinking cup.

Unfold your cup. Make a sketch showing the creases you have made. Label the corners of the paper and the endpoints of each crease as shown above.

23. **Writing** Name three pairs of congruent segments and at least three pairs of congruent angles. Explain how you know that the parts are congruent.

24. **Open-ended Problem** Find four angles that are all congruent to each other. How do you know that they are congruent?

25. Do any of the creases represent angle bisectors? If so, tell which ones.

26. Do any of the creases represent segment bisectors? If so, tell which ones.

27. **Research** Find directions for another origami figure, such as a crane or a pajarita. Fold the figure. Describe any segment bisectors or angle bisectors in the completed figure.

28. **BOATING** A boat traveling through water forms two waves as part of its *wake*. If the boat is traveling in a straight line and there is no wind, then the trail of the boat bisects the angle formed by the waves. The measure of the angle depends on the speed of the boat and the shape of its hull.

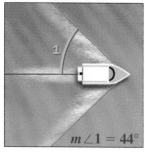

$m\angle 1 = 44°$

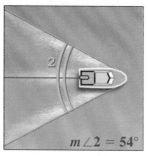

$m\angle 2 = 54°$

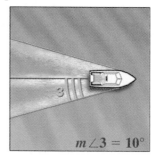

$m\angle 3 = 10°$

a. Find the measure of the angle between the waves of the wake.

b. Find the measure of the angle between the boat's trail and one wave of the wake.

c. Find the measure of the angle between the waves of the wake.

ALGEBRA In each figure, $\overrightarrow{PQ}$ is the bisector of the angle. Find the value of x.

29.

$(5x - 1)°$
Q
$(3x + 15)°$
P

30.

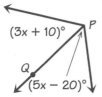

$(3x + 10)°$ P
Q
$(5x - 20)°$

31.

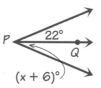

P $22°$
Q
$(x + 6)°$

32. $\overrightarrow{BD}$ bisects $\angle ABE$ and $\overrightarrow{BE}$ bisects $\angle ABC$.

a. Find the value of x.

b. Find $m\angle DBC$.

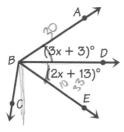

A
$(3x + 3)°$ D
B
$(2x + 13)°$
C
E

33. **Challenge** In the diagram, $MP = 24$. Is N the midpoint of $\overline{MP}$? Explain why or why not.

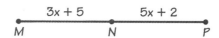

$3x + 5$ $5x + 2$
M N P

34. **Cooperative Learning** Work with another person.

a. Draw a triangle on a blank piece of paper. Fold each side of the triangle to find its midpoint. What do you notice about the three creases?

b. Draw another large triangle on a piece of paper. Fold each angle of the triangle to form its bisector. What do you notice about the three creases?

Find the measure of each angle. *(Section 1.6)*

35. ∠*DFG* **36.** ∠1 **37.** ∠*ZXQ*

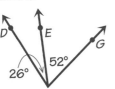

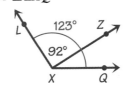

Use inductive reasoning to find the next two numbers in each pattern.
(Section 1.1)

38. 2, −3, −8, −13, _?_, _?_ **39.** 2, 5, 10, 17, _?_, _?_

ASSESS YOUR PROGRESS

VOCABULARY

angle (p. 34)   **midpoint** (p. 41)
vertex of an angle (p. 34)   **bisector of a segment** (p. 41)
congruent angles (p. 35)   **bisector of an angle** (p. 42)

Give as many names as you can for each angle. *(Section 1.6)*

1. ∠1

2. ∠*EHF*

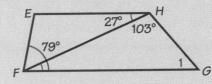

Find the measure of each angle.
(Section 1.6)

3. ∠*EHG*   **4.** ∠*GFH*   **5.** ∠*EFH*

6. *K* is the midpoint of $\overline{JL}$ and *JK* = 13. Find *KL* and *JL*. *(Section 1.7)*

7. $\overrightarrow{AB}$ bisects $\overline{PQ}$. Find the value of *x*. *(Section 1.7)*

8. Find the coordinates of the midpoint of $\overline{RS}$. *(Section 1.7)*

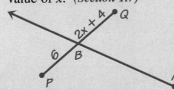

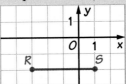

9. $\overrightarrow{NP}$ bisects ∠*MNO*. Find the value of *a* and *m*∠*MNO*. *(Section 1.7)*

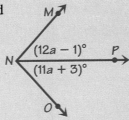

10. **Journal** Draw an angle dark enough so you can see it through the paper. Describe how you can fold the paper to form:
a. the angle bisector   **b.** a congruent angle

Investigating Symmetry

From hummingbirds to hubcaps, from leaves to ladders, symmetry can be found all around you. Countless natural and artificial objects have symmetry.

PROJECT GOAL Identify objects that have symmetry and analyze transformations that preserve their shapes and sizes.

Reflection, Translation, and Rotation

1. Find five or more objects that exhibit symmetry. At least one of them should be something that is found in nature, and at least one should be artificial. Make sure you have examples of each type of symmetry (reflection, translation, and rotation).

2. Identify a way to transform each object so that the image coincides with the object. Measure the transformation as described below. Then sketch the object. Label the angles and distances you measured.

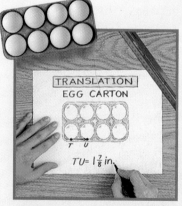

ROTATION Choose a point *R* on the object and locate point *S*, the image of *R* after a rotation. Find the point *O* around which the object is rotated and measure $\angle ROS$, the angle of rotation.

TRANSLATION Choose a point *T* on the object and locate point *U*, the image of *T* after a translation. Measure *TU*.

REFLECTION Choose a point *A* on the object and locate point *B*, the image of *A* after a reflection. Measure *AB* and find the midpoint *M* of $\overline{AB}$. Include the line (or plane) of symmetry in your sketch.

Presenting Your Results

Make a poster, write a report, or give a verbal presentation of your results. Include sketches of each object that you used. Label each sketch with the type of symmetry the object has. On your sketches, label the points that you used and the angles and distances that you measured.

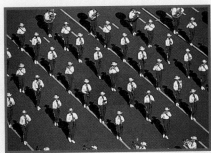

You may also want to consider the following ideas:

- Do any of the objects have more than one type of symmetry? Tell how many types of symmetry each has and explain how you know.

- Choose one of the objects that has rotational symmetry. Are there other angles that can be used to rotate the object so that the image coincides with the object? Explain your answer.

- Do any of the objects that you measured have features that make them not truly symmetrical? Explain.

Extending Your Project

You can extend your project by examining some of the ideas below:

- Describe how symmetry is used by a group that you know, such as a marching band or gymnastics team. Use sketches or photographs to help you explain what types of symmetry are used.

- Ask a graphic designer, scientist, architect, or other professional how symmetry is used in his or her field.

Self-Assessment

Describe how you took the measurements of your objects. Were any of the objects particularly difficult to measure? Why? How did measuring and sketching the objects improve your understanding of symmetry?

1 | Review

STUDY TECHNIQUE

What study techniques have you tried before? Write two brief paragraphs starting with these two phrases:

- To study for a mathematics test I usually . . .
- A study technique that has helped me in the past is . . .

VOCABULARY

inductive reasoning (p. 4)
reflect (p. 8)
rotate (p. 8)
translate (p. 8)
transformation (p. 8)
image (p. 8)
symmetry (p. 10)
line of symmetry (p. 10)
conjecture (p. 15)
conditional statement (p. 15)
hypothesis (p. 15)
conclusion (p. 15)
counterexample (p. 16)
collinear (p. 22)
coplanar (p. 22)

intersect (p. 22)
intersection (p. 22)
parallel (p. 22)
skew (p. 22)
ray (p. 28)
endpoint (p. 28)
segment (p. 28)
congruent segments (p. 28)
angle (p. 34)
vertex of an angle (p. 34)
congruent angles (p. 35)
midpoint (p. 41)
bisector of a segment (p. 41)
bisector of an angle (p. 42)

SECTIONS 1.1, 1.2, *and* 1.3

Inductive reasoning involves identifying patterns and making a prediction based on those patterns.

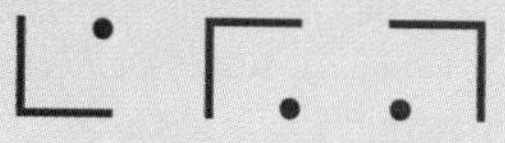

Prediction: next figure will be

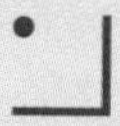

Inductive reasoning may lead to a **conjecture**, which is a **conditional statement** that says if one thing is true, then something else will be true.

Consider the prime numbers: 2, 3, 5, 7, 11, 13, 17, 19, 23, . . .

You might conjecture: If **a number ends in 3**, then **the number is prime**.

                  hypothesis   conclusion

This conjecture is false because it has a **counterexample**: 33 is not prime.

A **transformation** shifts a figure by **reflecting** it over a line or plane, **rotating** it around a given point, or **translating** it in a given direction. The new figure is called the **image**. A figure has **symmetry** if it coincides with its image after a transformation.

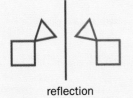

reflection

rotation

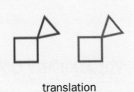

translation

SECTION 1.4

Points are **collinear** if they are on the same line.

Points or lines are **coplanar** if they lie on the same plane.

Two figures **intersect** if they have at least one point in common.

Parallel lines lie in the same plane but do not intersect.

Skew lines do not intersect and are not parallel.

The intersection of plane *R* and plane *S* is line *n*.

Lines *ℓ* and *m* are **skew**.

Points *A*, *C*, and *D* are **collinear**.

Lines *m* and *n* are **parallel**.

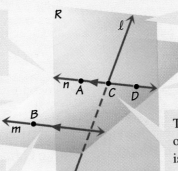

The intersection of lines *n* and *ℓ* is point *C*.

Points *A*, *B*, and *C* and lines *m* and *n* are **coplanar**.

SECTIONS 1.5, 1.6, *and* 1.7

A **ray** is a part of a line with one endpoint.

Two rays with a common endpoint, or **vertex**, form an angle. The **bisector of an angle** is a ray or line that divides an angle into two congruent angles.

If $\overrightarrow{XV}$ bisects $\angle YXW$ and $m\angle YXW = 40°$, then $m\angle YXV = m\angle VXW = 20°$, and $\angle YXV \cong \angle VXW$.

A **segment** is part of a line with two endpoints.

A **midpoint** divides a segment into two **congruent** segments.
If *S* is the midpoint of $\overline{RT}$, then $\overline{RS} \cong \overline{ST}$ and $RS = ST$.

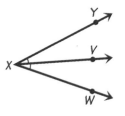

The segments are **congruent**.   The segments' lengths are **equal**.

If three points such as *R*, *S*, and *T* are collinear, then $RS + ST = RT$.

Assessment

VOCABULARY QUESTIONS

1. What kind of reasoning can help you identify patterns?

2. What are the two parts of a conditional statement?

3. After a transformation, what is the new figure called? What are three common transformations?

4. Writing Explain the difference between *congruent* and *equal* for two segments.

SECTIONS 1.1, 1.2, *and* 1.3

Give the next two numbers or figures and describe the pattern.

5. 4, 12, 36, 108, _?_ , _?_

6. • :: :::

7. a. Rewrite the statement "All tall people are good basketball players" in if-then form. Identify the hypothesis and conclusion.

b. Do you think the statement is *True* or *False*? If it is false, give a counterexample.

Copy the diagram at the left. Sketch each transformation. Give the coordinates of the vertices of the images.

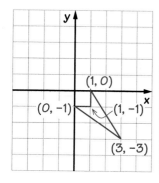

8. reflect over the *x*-axis

9. translate 6 units up

10. rotate a half-turn about the origin

SECTION 1.4

For Exercises 11–14, tell whether each statement is *True* or *False*. If a statement is false, explain why.

11. Two intersecting lines are coplanar.

12. Three noncollinear points are always coplanar.

13. Two skew lines can be coplanar.

14. Two planes can intersect in a point.

For Questions 15–19, use the diagram at the right.

15. Give at least three names for plane ABC.

16. What is the intersection of plane ABE and $\overleftrightarrow{EC}$?

17. What is the intersection of $\overleftrightarrow{AD}$ and $\overleftrightarrow{DE}$?

18. Name a pair of skew lines.

19. Name two lines that appear to be parallel.

Sketch each situation.

20. Line ℓ intersects $\overleftrightarrow{AB}$ at point B.

21. Plane ABC intersects line m at point B.

22. Noncollinear points A, B, C, and D lie in plane H and $\overleftrightarrow{AB} \parallel \overleftrightarrow{CD}$.

SECTIONS 1.5, 1.6, *and* 1.7

Use the diagram at the right.

23. List as many conclusions as you can about the diagram.

24. Describe a conclusion that looks correct but may not be.

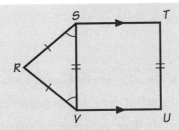

25. Point C is the midpoint of $\overline{XB}$ and $XB = XE$.

    a. Find XC.   **b.** Find XB.   **c.** Find XF.

26. $XD = 3y - 1$, $DH = 3$, and $XH = 4y + 1$. Find the value of y and the length XH.

27. $\overrightarrow{XE}$ bisects $\angle BXD$, $\overrightarrow{XD}$ bisects $\angle BXG$, and $m\angle BXF = 30°$. Find $m\angle BXD$ and $m\angle BXG$.

28. **Open-ended Problem** Draw two triangles. Measure the angles and sides of the triangles. Make a conjecture about the relationship between the shortest side and the smallest angle in any triangle. Draw another triangle and test your conjecture.

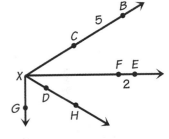

PERFORMANCE TASK

29. Design a border for wallpaper or fabric with a repeating pattern. Use as many of the following geometric relationships and objects as you can: angle bisector, parallel lines, midpoint, reflection symmetry, rotational symmetry, translational symmetry, vertex, congruent segments, congruent angles. Draw at least two repeats of your design. Label the geometric figures and relationships that you used.

2 | Triangles and Polygons

Exploring CRYSTALS

Virgil Lueth

"The crystals in the rocks were always intriguing."

"Crystals are everywhere," explains Virgil Lueth, state mineralogist of New Mexico and crystallographer. "All gems are crystals. Rocks are made up of crystals. Crystals are the basis of everything."

Each type of crystal has a predictable shape because the angles found in the molecules determine the angles formed by the faces of the crystal. "Ever notice when you look at snowflakes that they're always six-sided, or hexagonal?" Lueth asks. "When you freeze water into ice, the water molecules line up. They always line up in the same way and repeat these six-sided patterns."

Lueth got hooked on crystals when he was very young. "I always liked rocks, and where I grew up there were a lot of little cavities in the rocks that were lined with quartz crystal. The crystals in the rocks were always intriguing to me."

A Multifaceted Career

After studying geology in college and working as a professor, Lueth became state mineralogist for New Mexico. What does Lueth do now? His job is anything but dull. "I identify rocks and minerals for the public free of charge. I do the same thing for my scientific colleagues. I also maintain the museum at the Bureau of Mines. Here we exhibit classic minerals from around the world and especially New Mexico," Lueth says.

"**When I'm telling people about minerals and crystallography, that's pure enjoyment for me.**"

"I also do geological and geochemical research aimed at finding new ore deposits," he continues. "I do field work mostly in New Mexico, and I map the rocks, using geometry to determine the orientation of rocks in three dimensions. If there's a gold deposit down there, you want to know exactly where it is in order to drill for it and dig it out." A favorite part of Lueth's job is explaining the science he loves to visitors—everyone from young children and college students to senior citizens. "When I'm telling people about minerals and crystallography, that's pure enjoyment for me," Lueth says.

The Geometry of Crystals

Crystallographers study the organization of atoms and molecules in crystals. To unlock this hidden geometry, they measure the crystal's flat surfaces. "We can do it by measuring the angles between faces using what's called a contact goniometer," Lueth explains. The crystallographer rests the base of the goniometer on one crystal face and the movable arm on the adjoining face. The arm points at a scale that gives the angle between the two faces.

"By compiling these angles," Lueth says, "we decide which geometric system the crystal belongs to. The angles measured correspond to specific forms which are characteristic of particular crystal systems. For example, right angles are characteristic of the cubic system."

Using these facts, crystallographers can determine the exact internal structure of the crystal and what forms are possible—key information in nearly every area of science and technology.

Pyrite
Composition: Iron sulfide FeS$_2$

Cubic   Octahedral

Pyritohedral

Pyrite is the most common sulfide. Its symmetry is isometric, which means its axes are the same length and meet at right angles. Pyrite usually takes the form of a cube, though it can also be octahedral or pyritohedral.

Explore and Connect

The machine above is another kind of goniometer used to measure much smaller crystals.

1. Project Find out more about goniometers. Then make your own goniometer and use it to measure the angles of everyday objects, like the roof of a birdhouse.

2. Research What types of rocks and minerals are found in your area? Choose a mineral and learn about its structure. Summarize your research in a written report or an oral presentation.

3. Writing The pyrite crystal often takes the form of a cube. What type of angles does a cube have? What shape is each face of a cube? Write a possible definition of a cube.

Mathematics
& Virgil Lueth

In this chapter, you will learn more about how mathematics is related to crystallography.

Related Examples and Exercises

Section 2.2
• Exercises 12 and 13

Section 2.4
• Example 3
• Exercises 36–41

Section 2.6
• Exercises 25 and 26

2.1

Types of Angles

The beams of this bridge over the Kwai River in Thailand form many angles. When describing structures like this bridge, you will find it helps to be able to name and recognize different kinds of angles.

Learn how to...
- identify and classify angles

So you can...
- estimate and calculate the measures of angles

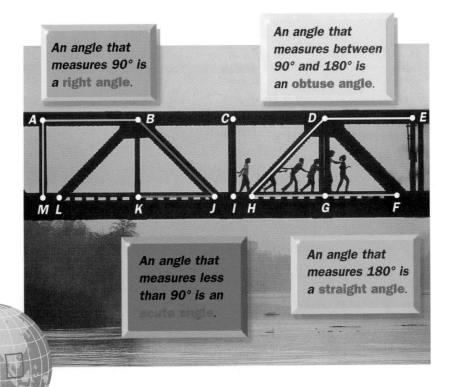

An angle that measures 90° is a right angle.

An angle that measures between 90° and 180° is an obtuse angle.

An angle that measures less than 90° is an acute angle.

An angle that measures 180° is a straight angle.

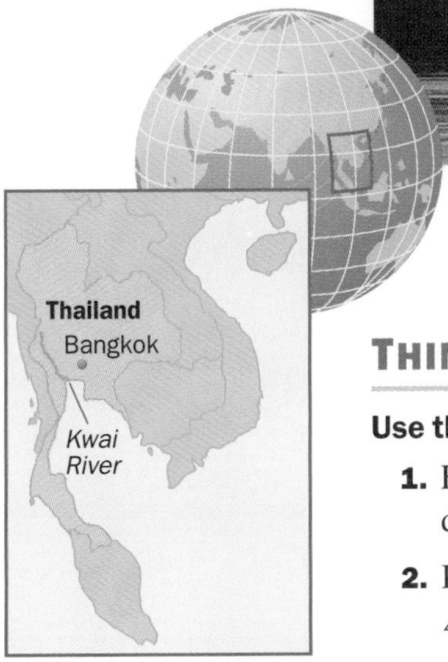

Thailand
Bangkok

Kwai River

THINK AND COMMUNICATE

Use the photograph of the bridge.

1. Find an acute angle, an obtuse angle, a right angle, and a straight angle different from those used in the definitions above.

2. In the diagram, ∠BCI and ∠ICD are both right angles. Explain why ∠BCD must be a straight angle.

3. In the diagram, ∠CDG is a right angle.
   a. Explain why ∠CDF must be an obtuse angle.
   b. Explain why ∠CDH and ∠HDG must be acute angles.

You can classify a pair of angles by looking at points and rays in common or by finding the sum of their measures.

Adjacent angles are two coplanar angles that share a vertex and a side but do not overlap.

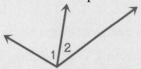

∠1 and ∠2 are adjacent angles.

Two adjacent angles form a **linear pair** if their nonshared rays form a straight angle.

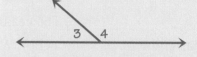

∠3 and ∠4 form a linear pair.

Two lines, segments, or rays are **perpendicular** if they intersect to form right angles.

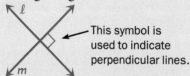

This symbol is used to indicate perpendicular lines.

The statement ℓ ⊥ m is read, "ℓ is perpendicular to m."

Vertical angles are non-adjacent, non-overlapping angles formed by two intersecting lines.

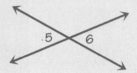

∠5 and ∠6 are vertical angles.

Two angles are **complements** of each other if their measures add up to 90°.

They are called *complementary angles*.

Two angles are **supplements** of each other if their measures add up to 180°.

They are called *supplementary angles*.

EXAMPLE 1

  a. Find the measures of a complement and a supplement of ∠ *SPT*.
  b. Sketch ∠ *SPT* and its complement adjacent to one another.

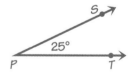

SOLUTION

  a. Complement: 90° − 25° = 65°
     Supplement: 180° − 25° = 155°

  b. Sketch ∠ *SPT*. Draw a ray perpendicular
     to $\overrightarrow{PT}$ at P. Because ∠ *RPT* is a right
     angle, ∠ *RPS* and ∠ *SPT* are complements.

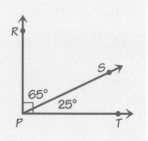

The position of some angles can tell you how their measures are related. The needle on this fuel gauge forms a linear pair with the base of the gauge. The two angle measures change when the needle moves, but their sum is always 180°.

$24° + 156° = \mathbf{180°}$ $110° + 70° = \mathbf{180°}$

Linear Pair Property

The angles that form a linear pair are supplementary.

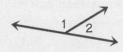

If $\angle 1$ and $\angle 2$ form a linear pair, then $m\angle 1 + m\angle 2 = 180°$.

These scissors form vertical angles. The angle between the blades stays congruent to the angle between the handles when you open and close the scissors.

Vertical Angles Property

Vertical angles are congruent.

If $\angle 1$ and $\angle 3$ are vertical angles, then $\angle 1 \cong \angle 3$.

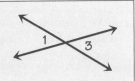

EXAMPLE 2

Find the value of each variable.
Give the properties that you use.

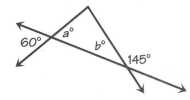

SOLUTION

Vertical angles are congruent.

$$a = 60$$

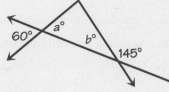

Angles in a linear pair are supplementary.

$$b = 180 - 145$$
$$b = 35$$

Complete each statement.

4. a. All right angles measure __?__ °.

  b. Angles with the same measure are __?__ to each other.

  c. All right angles are __?__ to each other.

5. Use the diagram at the right.

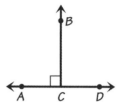

  a. Since $\overrightarrow{CB} \perp \overleftrightarrow{AD}$, $m \angle ACB = $ __?__ °.

  b. Since $\angle ACB$ and $\angle BCD$ form a linear pair, $m \angle ACB + m \angle BCD = $ __?__ °.

  c. Therefore, $m \angle BCD = $ __?__ °.

Notice that rays and segments can also be perpendicular, if they intersect to form right angles. By definition, perpendicular lines form right angles. Since all right angles are congruent, the angles formed by perpendicular lines are congruent.

Angles Formed by Perpendicular Lines

If two lines are perpendicular, then they form congruent adjacent angles.

If two lines form congruent adjacent angles, then the lines are perpendicular.

If $m \perp n$, then $\angle 1 \cong \angle 2$.
If $\angle 1 \cong \angle 2$, then $m \perp n$.

EXAMPLE 3

Tell whether each statement is *True* or *False*. Explain your reasoning.

a. $\overline{AJ} \perp \overline{IB}$

b. $\angle CGH \cong \angle HGK$

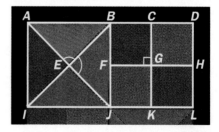

SOLUTION

a. True. Since $\angle AEB \cong \angle BEJ$ and they are adjacent angles, $\overline{AJ} \perp \overline{IB}$.

b. True. Since $\overline{CK} \perp \overline{FH}$, the segments form congruent adjacent angles.

For Questions 1–3, find the measures of a complementary angle, a supplementary angle, and a vertical angle for the given angle.

1. $m \angle Z = 40°$ **2.** $m \angle W = 82°$ **3.** $m \angle Y = x°$

4. The measure of $\angle ABC$ is 80°. $\angle ABC$ and $\angle CBD$ form a linear pair.

   a. Sketch $\angle ABC$ and $\angle CBD$.

   b. What is the measure of $\angle CBD$? How do you know?

5. Complete the following statement with *always*, *sometimes*, or *never*. Lines that form supplementary vertical angles are __?__ perpendicular.

2.1 | **Exercises and Applications**

Extra Practice
exercises on
page 661

Complete each statement.

1. $\angle AHB$ and __?__ are vertical angles.

2. $\angle BJG$ and __?__ are congruent adjacent angles.

3. $\angle FJE$ and __?__ are complementary angles.

4. $\angle GJE$ and __?__ are supplementary angles that do not form a linear pair.

Connection ▶ **BIOLOGY**

angle of crest

Scientists who study birds have noticed that the angle of the crest of the Stellar's jay can indicate the mood or activity of the bird. For Exercises 5–7, match the description of the crest with the corresponding picture.

A.

5. The angle of the crest is close to 0° during courtship.

6. While eating sunflower seeds, the angle of the crest is about 30°.

7. The crest is almost at a right angle during aggression.

B.

8. Writing What type of angles can the crest of a Stellar's jay form? Explain.

C.

9. Writing Explain why the angles in each pair are *not* adjacent.

a. ∠1 and ∠2   **b.** ∠3 and ∠4   **c.** ∠ABC and ∠ABD

   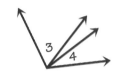

Find the measure of each angle in the diagram at the right. Tell which properties or definitions you use.

10. ∠EFB   **11.** ∠FEB   **12.** ∠ABC

13. ∠EBD   **14.** ∠GFE   **15.** ∠ABE

Use the diagram at the right. Tell whether each statement is *True* or *False*.

16. ∠FEB and ∠BED are complements.

17. ∠HFB ≅ ∠GFE

18. ∠ABG and ∠HFB are adjacent angles.

19. ∠ABC and ∠EBD are vertical angles.

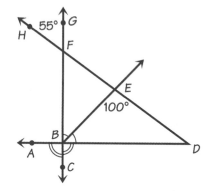

Connection **COMMUNICATION**

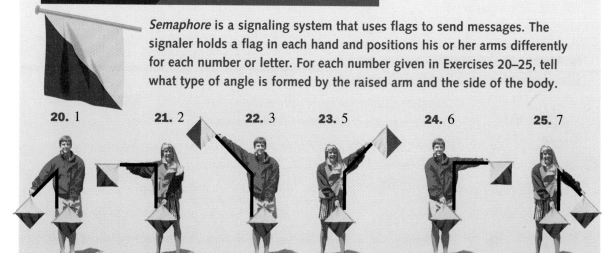

Semaphore is a signaling system that uses flags to send messages. The signaler holds a flag in each hand and positions his or her arms differently for each number or letter. For each number given in Exercises 20–25, tell what type of angle is formed by the raised arm and the side of the body.

20. 1   **21.** 2   **22.** 3   **23.** 5   **24.** 6   **25.** 7

1   **2**   **3**   **5**   **6**   **7**

26. a. Open-ended Problem Write instructions for signaling one of the numbers shown. Describe the position of each arm by the type of angle it makes with the side of the body.

b. Cooperative Learning Have a classmate follow the instructions you wrote in part (a) and guess which number you chose. If your partner finds the instructions unclear, work together to revise them.

27. What is the measure of a supplement of a complement of a 50° angle?

28. SAT/ACT Preview What is the measure of a complement of a supplement of a 140° angle?

    A. 40°   **B.** 90°   **C.** 140°   **D.** 50°   **E.** 180°

29. Writing The nonshared rays of two adjacent acute angles are perpendicular. How are their measures related? Explain your reasoning.

30. a. ∠BFC and ∠CFD are complementary angles. Find m∠CFD.

    b. ∠AFB and ∠BFC are complementary angles. Find m∠AFB.

    c. Make a conjecture about the measures of two angles that are complementary to the same angle. Explain your reasoning.

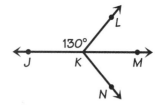

31. a. Find m∠LKM. How did you get your answer?

    b. ∠JKL and ∠MKN are supplementary angles. Find m∠MKN.

    c. Make a conjecture about the measures of two angles that are supplementary to the same angle. Explain your reasoning.

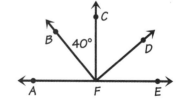

32. Challenge Show that $\overleftrightarrow{TU} \perp \overleftrightarrow{PW}$.

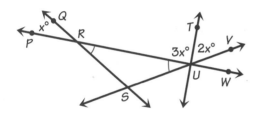

O N G O I N G   A S S E S S M E N T

33. Open-ended Problem Sketch a building, a bridge, or a stained glass window. Include and label examples of acute, obtuse, right, straight, vertical, supplementary, and complementary angles, perpendicular segments, and a linear pair. Mark all right angles and congruent angles in your sketch.

S P I R A L   R E V I E W

34. Make a sketch of plane R intersecting plane M in $\overleftrightarrow{XY}$. *(Section 1.4)*

35. Use a protractor to measure each angle. *(Section 1.6)*

    a. ∠A

    b. ∠B

    c. ∠C

2.2 | Classifying Triangles

A **triangle** is the figure formed by the segments whose endpoints are three noncollinear points. The shape of a triangle depends on the measures of its angles and the lengths of its sides.

Learn how to...

- name, label, and classify triangles
- sketch triangles and find the measures of their angles

So you can...

- use types of triangles to describe objects such as crystals

EXPLORATION
COOPERATIVE LEARNING

Sorting Triangles

Work in a group of four students. You will need:

- a set of triangles like the ones at the right

1 One student should write down a "secret rule" that a second student will use to sort the triangles.

2 The second student, the sorter, should choose triangles that fit the rule as the other students watch. At any time, another student can stop the sorter and guess the rule.

One rule might be "triangles with an obtuse angle."

Another possible rule is "triangles with three congruent sides."

3 Repeat Steps 1 and 2 until five rules have been used. Make sure each student has a chance to write a rule, sort the triangles, and guess a rule.

4 Make a list of the rules your group used. What other rules can you think of? What rules did the other groups in your class use?

The segments that form a triangle are called its **sides**. The endpoints of the sides are called **vertices** (plural of *vertex*). The triangle with vertices A, B, and C is called △ABC, read as "triangle ABC."

You can classify a triangle by considering its sides.

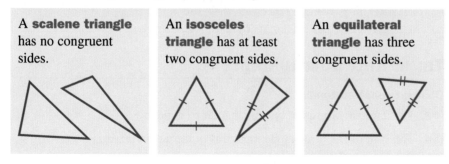

A **scalene triangle** has no congruent sides.

An **isosceles triangle** has at least two congruent sides.

An **equilateral triangle** has three congruent sides.

You can also classify a triangle by considering its angles.

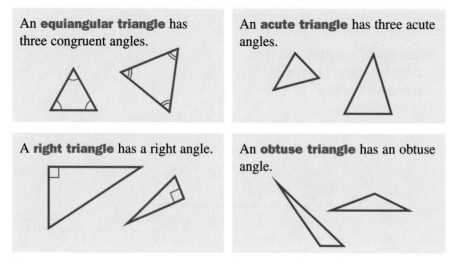

An **equiangular triangle** has three congruent angles.

An **acute triangle** has three acute angles.

A **right triangle** has a right angle.

An **obtuse triangle** has an obtuse angle.

THINK AND COMMUNICATE

1. Which triangles on page 64 are scalene? Which are acute? right? obtuse?

2. Are all equilateral triangles isosceles? Explain.

You may already know what the sum of the angle measures of a triangle is. Here is a quick way to recall this sum:

Cut any triangle out of a piece of paper.

Tear off two angles of the triangle.

Place the angles so that they are adjacent.

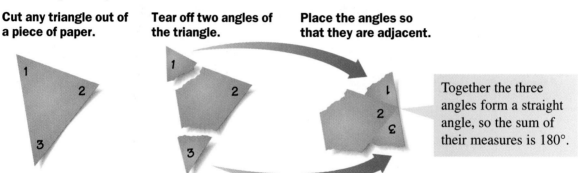

Together the three angles form a straight angle, so the sum of their measures is 180°.

The Angles of a Triangle

The sum of the angle measures of a triangle is 180°.

In $\triangle ABC$, $m \angle A + m \angle B + m \angle C = 180°$.

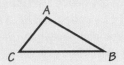

THINK AND COMMUNICATE

Complete each statement.

3. The angle measures of an equiangular triangle are ? .

4. The sum of the three angle measures of a triangle is ? .

5. The measure of each angle of an equiangular triangle is ? .

EXAMPLE 1	Connection: Algebra

In the figure, $m \angle BDC = 10°$ and $\angle BAD \cong \angle BDA$. $\triangle ACD$ is an equiangular triangle. Find $m \angle ABD$.

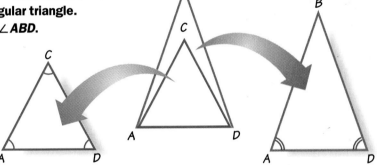

SOLUTION

First find $m \angle BDA$ and $m \angle BAD$.

Because $\angle BDC$ and $\angle CDA$ are adjacent:

$m \angle BDA = m \angle BDC + m \angle CDA$

$ = 10° + 60°$ Since $\triangle ACD$ is an equiangular triangle, $m \angle DAC = m \angle ACD = m \angle CDA = 60°$.

$ = 70$

Because $m \angle BAD = m \angle BDA$, $m \angle BAD = 70°$ as well.

Now find $m \angle ABD$.

$m \angle ABD = 180° - (m \angle BAD + m \angle BDA)$

$ = 180 - (70 + 70)$

$ = 180 - 140$

$ = 40$

Therefore, $m \angle ABD = 40°$.

Toolbox p. 690
Simplifying and Evaluating Expressions

Notice that Triangle B on page 64 is both scalene and obtuse. Not all combinations of types are possible, however.

EXAMPLE 2

Sketch each triangle, if possible. If it is not possible, explain why not.

a. an acute isosceles triangle   **b. an obtuse right triangle**

SOLUTION

a. Sketch an acute angle.

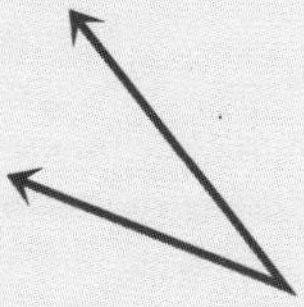

Make the sides into congruent segments.

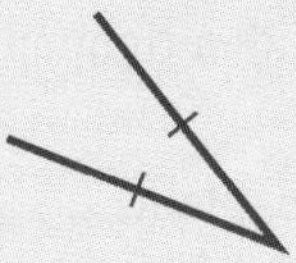

Sketch the third side. △*XYZ* is an acute isosceles triangle.

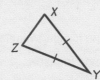

b. Sketch an obtuse angle.

Sketch a ray perpendicular to one of its sides.

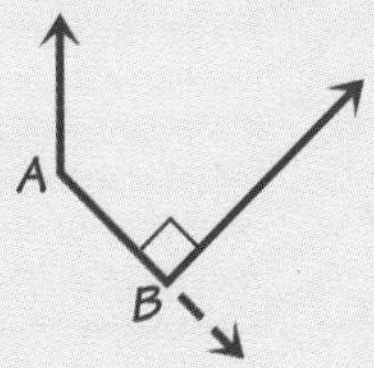

The sum of $m\angle A$ and $m\angle B$ is greater than 180°, so no such triangle can exist.

✓ CHECKING KEY CONCEPTS

Classify each triangle. Be as specific as possible.

1.

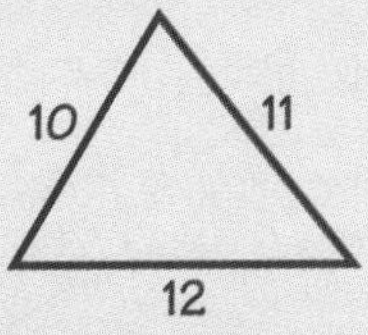

2.

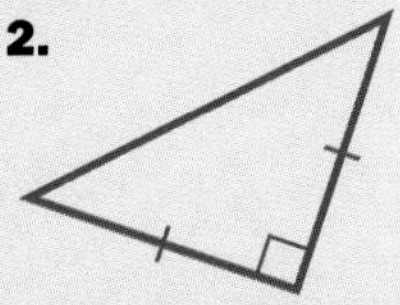

Sketch each triangle. Mark any right angles, congruent angles, and congruent sides.

3. a scalene right triangle   **4.** an isosceles obtuse triangle

ALGEBRA Find each unknown angle measure.

5.

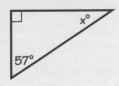

6.

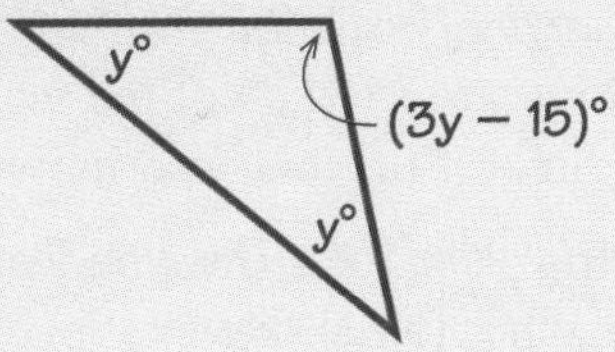

(3y − 15)°

2.2 Exercises and Applications

Extra Practice
exercises on
page 662

Sketch each triangle, if possible. If it is not possible, explain why not.

1. an isosceles right triangle

2. a scalene equilateral triangle

3. an obtuse equiangular triangle

4. an obtuse scalene triangle

5. Writing A given triangle has two congruent angles. Are the two angles acute, right, or obtuse? How do you know?

6. Open-ended Problem Draw an obtuse triangle with two congruent angles. Label each angle with its measure.

7. Logical Reasoning Sketch a right triangle, $\triangle PQR$, with a right angle at Q.

   a. $m \angle P + m \angle R = \underline{\ ?\ }$

   b. How are the acute angles in a right triangle related? Explain.

ALGEBRA Find each unknown angle measure.

8.

9.
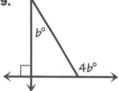

10.
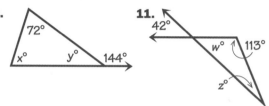

11.

INTERVIEW Virgil Lueth

Look back at the article on pages 54–56.

The geometric shape you see when you look at a crystal is called its form. *The different forms are often described in terms of their faces. Since many of the faces are triangular, identifying different types of triangles can help you learn the forms that crystals have.*

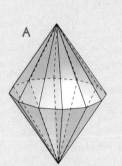

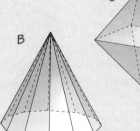

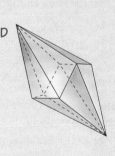

12. A *hexagonal scalenohedron* has twelve faces. When the crystal develops perfectly, all of the faces are scalene triangles. Which diagram best fits this description?

13. Writing Use the terms you learned in this section to describe each of the other forms shown.

EQUILATERAL TRIANGLE

Construct an equilateral triangle.

A geometric *construction* is a method of drawing figures using only a compass and a straightedge (a ruler with no marks). The steps below show how to construct an equilateral triangle.

1. Draw a ray. Label the endpoint *A*.

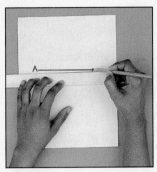

2. Place the point of the compass at *A*. Using any compass setting, draw an arc that intersects the ray at a point *B*.

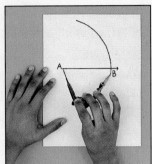

3. Without changing the setting, place the point of the compass at *B* and draw another arc. Label the intersection *C*.

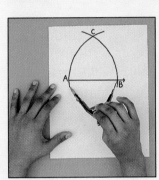

4. Draw $\overline{AC}$ and $\overline{BC}$ to form equilateral $\triangle ABC$.

14. How do you know that $\overline{AB} \cong \overline{AC}$ and that $\overline{AB} \cong \overline{BC}$?

15. Construct an equilateral triangle.

16. Construct an equilateral triangle with side length *JK*. (*Hint:* Open the compass so that the distance from the point to the pencil is *JK*.)

17. Writing Suppose you extended the arcs in Steps 2 and 3 on both sides of $\overline{AB}$. Explain why the result is two equilateral triangles that share a side.

18. Challenge Explain how to construct an isosceles triangle by drawing one arc with the point of the compass at a given point *X*.

Complete each statement with *always*, *sometimes*, or *never*.

19. A triangle ? has two right angles.

20. A triangle ? has exactly one acute angle.

21. A triangle ? has at least two acute angles.

22. An obtuse triangle ? has exactly two acute angles.

23. An isosceles triangle ? has three congruent sides.

24. An equilateral triangle is ? isosceles.

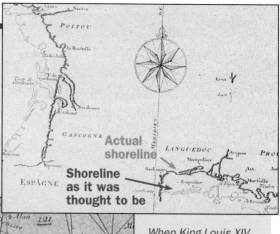

In the seventeenth and eighteenth centuries, a survey was conducted in France. The method used, *triangulation,* involved constructing a network of triangles throughout the country. By knowing some of the side lengths and angle measures of the triangles, the surveyors calculated the size of France, finding it much smaller than they had thought.

Actual shoreline

Shoreline as it was thought to be

Find each angle measure in the map at the right.

25. $m \angle CAB$

26. $m \angle ABC$

27. $m \angle DCB$

28. $m \angle CED$

29. $m \angle CDE$

When King Louis XIV saw the difference between the shoreline as they believed it to be and the actual shoreline, he remarked that the survey had cost him more land than a war.

The map at the left shows some of the triangles used for the survey.

30. Research Find out more about triangulation. What kind of mathematics is involved? What information do the surveyors need to determine distances? How do they measure the angles?

 Technology For Exercises 31 and 32, you will need geometry software or a ruler and a protractor.

31. a. Use the steps on page 69 or the software to construct an equilateral triangle. Measure each angle. What are the measures?

b. Draw an equiangular triangle. Measure each side of the triangle. How are the lengths related?

c. Make a conjecture about the relationship between equiangular triangles and equilateral triangles.

32. a. Draw two congruent segments that share an endpoint, as shown. Connect the other two endpoints. What kind of triangle is it?

b. Measure the two angles that are opposite the congruent sides. These are called the *base angles.* What do you notice?

c. Repeat parts (a) and (b) using a different side length and a different angle measure.

d. Make a conjecture about the base angles of an isosceles triangle.

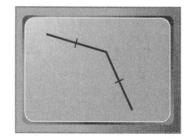

33. **Open-ended Problem** Sketch three or four triangles so that you have at least one of each type: scalene, isosceles, equilateral, equiangular, acute, right, and obtuse. Classify each triangle. Mark all congruent angles and sides, and give angle measures when possible.

SPIRAL REVIEW

For each figure, (a) state how many segments are shown, (b) list any congruent segments, and (c) list any congruent angles. *(Sections 1.5 and 1.6)*

34.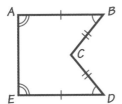

35.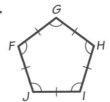

36.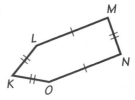

ASSESS YOUR PROGRESS

VOCABULARY

types of angles (p. 57)
adjacent angles (p. 58)
linear pair (p. 58)
perpendicular (p. 58)
vertical angles (p. 58)
complements (p. 58)

supplements (p. 58)
triangle (p. 64)
side of a triangle (p. 65)
vertex of a triangle (p. 65)
types of triangles (p. 65)

Find the value of each variable. Tell which property or definition you use. *(Sections 2.1 and 2.2)*

1. a
2. b
3. c
4. d

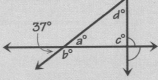

BICYCLING Imagine sitting on a bicycle. For each pedal position, (a) sketch the angle formed at your knee, (b) estimate the measure of the angle, and (c) tell what type of angle it is. *(Section 2.1)*

5. when the pedal is farthest from the ground

6. when the pedal is closest to the ground

7. when the pedal is in the middle of a stroke

8. Name and classify the four triangles shown in the diagram. Be as specific as possible. *(Section 2.2)*

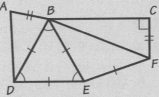

9. **Journal** Choose ten vocabulary words from Sections 2.1 and 2.2. Describe a way to remember the definition of each.

2.3 Types of Polygons

The *Generalife* is a thirteenth century palace overlooking the Alhambra in Granada, Spain. The plans below show part of the Generalife gardens which contain many different geometric designs and shapes.

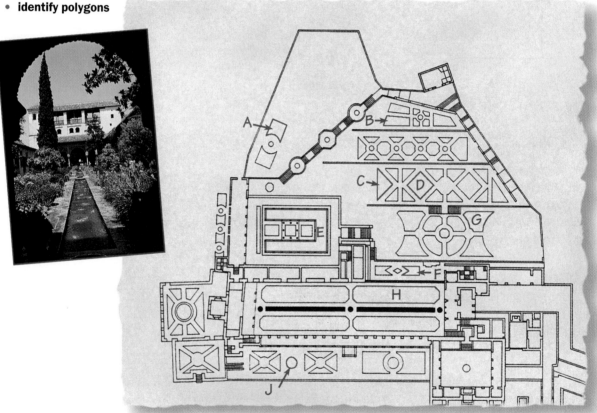

THINK AND COMMUNICATE

1. How are shapes E and H alike?

2. How are shapes E and F different from shapes B, C, D, and H?

3. a. Shapes B, C, D, E, F, and H are examples of *polygons.* Shapes A, G, and J are not polygons. How would you define polygon?

   b. The figures at the right are not polygons. Does this affect your definition from part (a)? If so, change it as needed.

2.3 **Exercises and Applications**

Extra Practice
*exercises on
page 662*

Classify each polygon. Be as specific as possible.

1.   **2.**   **3.**   **4.**

5. Logical Reasoning Copy and complete the table. Use any
patterns you see to predict what the next column will look like.

triangle	quadrilateral	pentagon	hexagon	heptagon	octagon
![triangle]	![quadrilateral]	?	?	?	?
3 sides	4 sides	?	?	?	?
0 diagonals	2 diagonals	?	?	?	?

6. Writing Can a regular polygon be concave? Explain your reasoning.

7. a. Use a circle to sketch a regular nonagon.

 b. Use a circle to sketch a nonagon that is *not* regular. How is the method
 you used different from your method in part (a)?

8. a. Imagine stretching a regular hexagon *ABCDEF* so that $\overline{AB}$ and $\overline{DE}$
 become longer than the other sides. Sketch the resulting figure.

 b. Is hexagon *ABCDEF* still regular? Is it still equilateral? equiangular?
 Explain.

9. a. When naming the polygon at the right, how many
 different vertices can you start with?

 b. If you start with vertex *G*, how many different
 vertices could come next?

 c. How many different names does this polygon have?

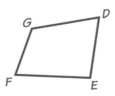

10. How many different names does each polygon have?

a. 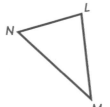 **b.** 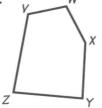 **c.**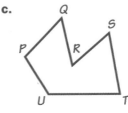

THINK AND COMMUNICATE

Use the polygon at the right.

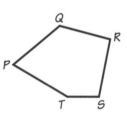

6. Name two consecutive angles of the polygon.

7. Name a pair of consecutive sides of the polygon.

8. Name a diagonal of the polygon.

9. Name this polygon two different ways.

10. Explain why *PRTSQ* is not a name for the polygon.

EXAMPLE 2

Sketch a hexagon that has two consecutive right angles.

SOLUTION

Sketch a segment with a right angle at each end.

Add three more sides to make a hexagon.

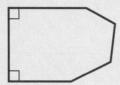

☑ CHECKING KEY CONCEPTS

Classify each polygon. Be as specific as possible. Then use the vertices to name the polygon.

1.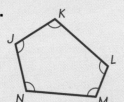

2.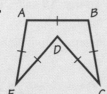

3.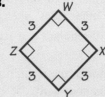

Look back at the plans for the gardens on page 72.

4. What kind of polygon is shape F?

5. Explain why shape G is not a polygon.

Tell whether each statement is *True* or *False*. Explain your reasoning.

6. Every regular polygon is equilateral.

7. Every equiangular polygon is regular.

8. A diagonal of a convex polygon can lie outside of the polygon.

9. Every equilateral polygon is convex.

EXAMPLE 1

How many lines of symmetry does a regular octagon have?

SOLUTION

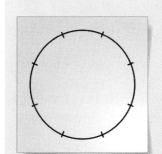

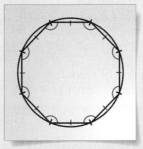

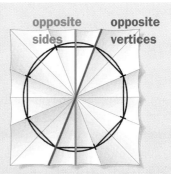

opposite sides   opposite vertices

Step 1 Sketch a circle. Mark eight points evenly spaced around the circle.

Step 2 Connect the points to form a regular octagon.

Step 3 Fold the paper or use a transparent mirror to find the lines of symmetry.

There are four lines of symmetry that pass through opposite vertices and four lines of symmetry that bisect opposite sides. A regular octagon has 8 lines of symmetry.

THINK AND COMMUNICATE

4. a. Sketch a regular 12-gon. How many lines of symmetry does it have?

b. Repeat part (a), using a different regular polygon. Make a conjecture about how many lines of symmetry a regular *n*-gon has.

5. Rachel says that all regular polygons have rotational symmetry as well as line symmetry. Do you agree or disagree? Explain.

Parts of Polygons

Two vertices of a polygon connected by a side are called *consecutive vertices*. Also, two angles that share a side are called **consecutive angles**, and two sides that share a vertex are called **consecutive sides**. A segment that connects nonconsecutive vertices is called a **diagonal**.

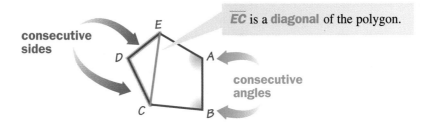

consecutive sides

$\overline{EC}$ is a **diagonal** of the polygon.

consecutive angles

To name a polygon, start at any vertex and write consecutive vertices in order. Two names for the pentagon above are *BAEDC* and *ABCDE*.

A **polygon** is a closed plane figure whose sides are segments that intersect only at their endpoints. No two sides with a common endpoint are collinear.

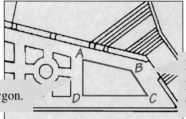

$\overline{AD}$ is a side of the polygon.

Point B is a vertex of the polygon. $\angle B$ is an angle of the polygon.

One way to classify a polygon is by the number of sides it has.

A **triangle** is a polygon with three sides.

A **quadrilateral** is a polygon with four sides.

A **pentagon** is a polygon with five sides.

A **hexagon** is a polygon with six sides.

Here are names for some other polygons:

Number of sides	7	8	9	10	n
Name	heptagon	octagon	nonagon	decagon	n-gon

A polygon can be either *concave* or *convex*. In a convex polygon, no segment can be drawn outside of the polygon to connect two vertices.

For example, on page 72, shape E is concave and shape B is convex.

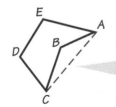

This polygon is concave because $\overline{AC}$ lies outside of the polygon.

You can describe a polygon by comparing the lengths of its sides or the measures of its angles.

If all the sides of a polygon are congruent, then it is an **equilateral** polygon.

If all the angles of a polygon are congruent, then it is an **equiangular** polygon.

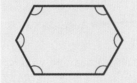

If a polygon is both equilateral and equiangular, then it is a **regular** polygon.

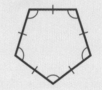

Paper-cutting is a craft found in many countries, including China, Japan, Mexico, and Poland. Use an $8\frac{1}{2}$ in. by 10 in. piece of paper for this paper-cutting activity.

Step 1 Fold the paper in half, as shown. Then crease the paper by folding and unfolding it in half horizontally and vertically.

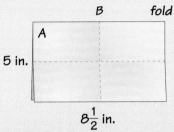

5 in.

$8\frac{1}{2}$ in.

Step 2 Folding at B, bring corner A right and down until it meets the horizontal crease.

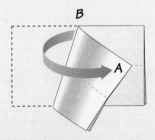

Step 3 Bring corner A left until the folded edges meet.

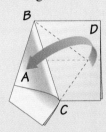

Step 4 Bring corner D left and fold on $\overline{BC}$.

Step 5 Bring corner D right until $\overline{BD}$ lines up with $\overline{BC}$.

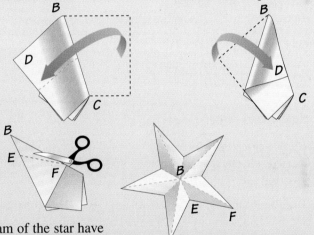

Step 6 Cut along a segment, $\overline{EF}$, so that $\triangle BEF$ is an obtuse triangle. When you unfold the paper, you should have a five-pointed star.

11. How many triangles in the diagram of the star have the same side lengths and angle measures as $\triangle BEF$?

12. What type of polygon is the resulting star?

13. How many lines of symmetry does the star have?

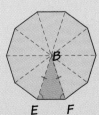

14. a. The polygon at the left can be made by changing the shape of $\triangle BEF$. What type of polygon is it? Is it equilateral? equiangular? Explain.

   b. What kind of triangle is $\triangle BEF$ in the polygon at the left?

   c. Describe how to cut $\triangle BEF$ in Step 6 to make the polygon.

15. Suppose you cut along $\overline{EF}$ so that $\triangle BEF$ is a right triangle, as shown at the right. What type of polygon is the resulting figure? Sketch the polygon, marking any congruent angles and sides.

16. Challenge Describe how to cut $\triangle BEF$ in Step 6 to make an equilateral convex decagon that is not regular.

How many lines of symmetry does each polygon have?

17.

18.

19.

Open-ended Problems **Sketch an example of each figure.**

20. a pentagon with three congruent obtuse angles

21. a concave hexagon with one line of symmetry

22. a quadrilateral with three right angles

23. a convex hexagon with two consecutive congruent sides

24. a polygon divided into two pentagons by a diagonal

25. Try to make or sketch each polygon described in the table. Copy and complete the table by sketching each polygon that is possible.

Polygon	Triangle	Quadrilateral	Pentagon
Concave	not possible	?	?
Convex	?	?	?
Regular	?	?	?
Equilateral only	?	?	?
Equiangular only	?		?

ONGOING ASSESSMENT

26. **Writing** Describe at least two ways to classify a polygon. What parts of the polygon should you examine?

SPIRAL REVIEW

27. Identify the hypothesis and conclusion of this conditional statement:
If a polygon is regular, then it is equiangular. *(Section 1.3)*

Find each unknown angle measure. *(Sections 2.1 and 2.2)*

28.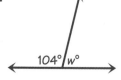

29.

30.

2.4 | Angles in Polygons

You can find the sum of the angle measures of any polygon just by knowing the number of sides it has. From now on in this book, unless stated otherwise, the term "polygon" refers to a convex polygon.

EXPLORATION
COOPERATIVE LEARNING

Finding Angle Measures in Polygons

Work in a group of four students.
You will need:

- dot paper or a geoboard and rubber bands
- a graphing calculator or graph paper

Repeat Steps 1 and 2 for each of the following polygons: triangle, quadrilateral, pentagon, hexagon, and heptagon.

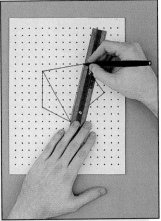

1 Sketch the polygon on dot paper or make it on the geoboard. Draw all of its diagonals from one vertex. The diagonals should divide the polygon into triangles.

2 Make a table like the one below. How are the numbers in each column related to the numbers in the previous column?

Type of polygon	Number of sides	Number of triangles formed	Sum of angle measures of triangles	Sum of angle measures of polygon
triangle	3	1	180°	180°
?	?	?	?	?

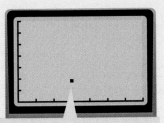

The point (3, 180) represents the sum of the measures of the angles of a triangle.

3 Graph the data as shown. Put the number of sides on the *x*-axis and the sum of the angle measures on the *y*-axis. Describe any patterns you see in the graph.

4 Extend your table to include an *n*-gon. Make a conjecture about the relationship between the number of sides of a polygon and the sum of its angle measures.

In the Exploration you saw how to divide an n-gon into $(n - 2)$ triangles so that the sum of the angle measures of the triangles equals the sum of the angle measures of the polygon. This leads to the following formula.

The Angles of a Polygon

In a polygon with n sides, the sum of the angle measures is $(n - 2)180°$.

EXAMPLE 1   Application: Sports

In baseball and softball, home plate is a pentagon with three right angles and two congruent obtuse angles. Find the measure of each obtuse angle in home plate.

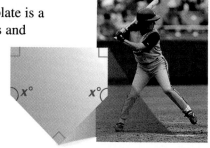

SOLUTION

Step 1 Find the sum of the angle measures of a pentagon.

$$\text{Sum of the angle measures} = (n - 2)180°$$
$$= (5 - 2)180$$
$$= 540$$

Step 2 Write and solve an equation.

$$90° + 90° + x° + 90° + x° = 540°$$

Solve for x.

$$270 + 2x = 540$$
$$2x = 270$$
$$x = 135$$

Each obtuse angle of home plate measures 135°.

EXAMPLE 2   Connection: Algebra

What polygon, if any, has angle measures that add up to 1080°?

SOLUTION

Use the formula above.

$$(n - 2)180° = \text{Sum of the angle measures of the polygon}$$
$$(n - 2)180 = 1080 \qquad \text{Divide both sides by 180.}$$
$$n - 2 = 6$$
$$n = 8$$

The polygon whose angle measures add up to 1080° is an octagon.

THINK AND COMMUNICATE

1. How could you use the graph that you made in the Exploration on page 79 to solve Example 2?

2. Can the sum of the angle measures of a polygon be 450°? Explain.

Exterior Angles of a Polygon

The angles that you looked at in the Explorations are *interior angles* of the polygons. If one side of a polygon is extended, the ray forms an *exterior angle* with an adjacent side of the polygon.

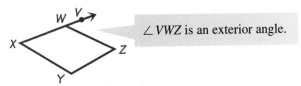

∠VWZ is an exterior angle.

At each vertex of a polygon, an interior angle and its adjacent exterior angle form a linear pair. You can use this fact to find the sum of the measures of the exterior angles.

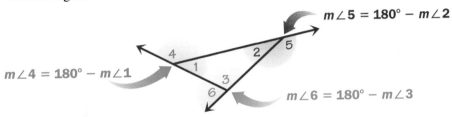

$m\angle 5 = 180° - m\angle 2$

$m\angle 4 = 180° - m\angle 1$

$m\angle 6 = 180° - m\angle 3$

$$m\angle 4 + m\angle 5 + m\angle 6 = (180° - m\angle 1) + (180° - m\angle 2) + (180° - m\angle 3)$$
$$= 3(180) - m\angle 1 - m\angle 2 - m\angle 3$$
$$= 540 - (m\angle 1 + m\angle 2 + m\angle 3)$$
$$= 540 - 180$$
$$= 360°$$

For an *n*-gon:

$$\text{Sum of the measures of } n \text{ exterior angles} = \text{Sum of the measures of } n \text{ linear pairs} - \text{Sum of the measures of } n \text{ interior angles}$$
$$= n(180°) - (n - 2)180°$$
$$= 180n - 180n + 360$$
$$= 360°$$

The Exterior Angles of a Polygon

The sum of the measures of the exterior angles of any polygon is 360°.

THINK AND COMMUNICATE

3. Repeat the calculations in the middle of page 81 for a hexagon. Is the result different? Explain.

4. What is the measure of each exterior angle of a regular quadrilateral?

EXAMPLE 3   Interview: Virgil Lueth

Some crystals take the form of a *dihexagonal prism*, whose cross-section is an equiangular 12-gon. Find the measure of each interior and each exterior angle of an equiangular 12-gon.

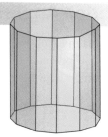

SOLUTION

Method 1

First find the sum of the measures of the interior angles.

$$(n - 2)180° = (12 - 2)180°$$
$$= 1800°$$

Then divide the sum by 12 to find the measure of each **interior angle**.

$$\frac{1800°}{12} = 150°$$

Divide 360° by 12 to find the measure of each **exterior angle**.

$$\frac{360°}{12} = 30°$$

Method 2

Start by finding the measure of one **exterior angle**.

$$\frac{360°}{12} = 30°$$

Each **interior angle** is the supplement of an exterior angle.

$$180° - 30° = 150°$$

☑ CHECKING KEY CONCEPTS

Find the sum of the measures of the interior angles of each polygon.

1. hexagon   **2.** 25-gon   **3.** 2*x*-gon

What polygon, if any, has angle measures that add up to the given sum?

4. 1440°   **5.** 4860°   **6.** 2700°

7. Find $m \angle 1$.

125° 1
118°
68°
112°

2.4 **Exercises and Applications**

Extra Practice exercises on page 662

Find the sum of the measures of the interior angles of each polygon.

1. quadrilateral

2. heptagon

3. 75-gon

Find the value of each variable.

4.

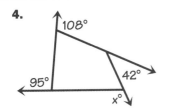

5.

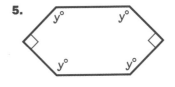

6.

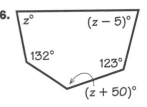

Find the measure of each interior and exterior angle of the polygon.

7. regular pentagon

8. regular 11-gon

9. regular 32-gon

Connection ▶ **ART**

A *tessellation* is a pattern of polygons that covers the plane without gaps or overlaps. Tessellations have been used by many cultures to cover walls, floors, and ceilings. A regular tessellation is formed by repeating a single regular polygon. The angle measures of a polygon determine whether it can form a regular tessellation.

Floor tiles from Mexico

10. Writing Which one of the tessellations at the right is regular? Explain.

11. What is the sum of the angle measures at each vertex of a tessellation? Explain your reasoning.

Wall tiles from Spain

Floor tiles from Italy

vertex

For Exercises 12–15, answer the following questions:
a. What is the measure of an interior angle of a regular *n*-gon?
b. How many regular *n*-gons can meet at a vertex without overlapping?
c. Can a regular *n*-gon form a tessellation?

12. $n = 3$ **13.** $n = 4$ **14.** $n = 5$ **15.** $n = 6$

16. a. How many regular heptagons can meet at a vertex without overlapping?
b. How many regular octagons can meet at a vertex without overlapping?
c. Explain why only two regular *n*-gons can meet at a vertex without overlapping for all $n \geq 7$.
d. Which regular polygons can form a regular tessellation?

20.4

Exercises 17 and 18 show equiangular windows from the wall of a Chinese garden. Find the measure of an interior and exterior angle of each.

17.

18.

What polygon, if any, has angle measures that add up to the given sum?

19. $3600°$

20. $3060°$

21. $2160°$

22. $(180y - 360)°$

23. Open-ended Problem Find a possible set of eight measures for the interior angles of an octagon that is not equiangular.

ALGEBRA In Exercises 24–26, the measures of the exterior angles of a polygon are given. Find the value of each variable.

24. $65°, 70°, 115°, w°$

25. $105°, z°, 4z°$

26. $90°, 90°, 40°, 40°, x°, x°$

ALGEBRA Find the measure of each interior angle.

27.

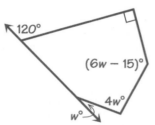

28.

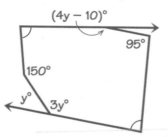

29.

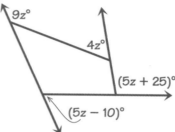

The measure of one interior angle of a regular n-gon is given. Find the value of n.

30. $144°$

31. $162°$

32. $171°$

33. Challenge Show that the interior angles of a convex polygon cannot include four angles that measure $90°, 75°, 130°,$ and $60°$.

34. SAT/ACT Preview Which of the following *cannot* be the sum of the measures of the interior angles of a polygon?

A. $1260°$ **B.** $360°$ **C.** $1500°$ **D.** $180°$ **E.** $2700°$

35. Investigation Draw a polygon. From a point inside the polygon, draw a segment to each vertex.

   a. How many sides does the polygon have? How many triangles are formed by the segments?

   b. What is the sum of the angle measures of the triangles?

   c. Which angles of the triangles do not share a vertex with the polygon? What is the sum of the measures of these angles?

   d. Show how these steps lead to the expression $(180n - 360)°$, or the same expression on page 80.

*Imagine two planes inside a crystal, each perpendicular to a face.
The angle that these planes form when they meet inside the crystal
is called the* **interfacial angle.** *When it is not possible to measure
this angle directly, a* **contact goniometer**
*is used to measure a congruent angle
outside the crystal.*

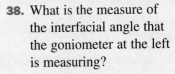

angle measured
by goniometer

interfacial angle

36. What is the sum of the measures
of the interior angles of
quadrilateral *ABCD*?

37. **Logical Reasoning** Complete
the statements in parts (a) and (b).
Explain your answers.

a. $m\angle ABC + m\angle ADC = \underline{\ ?\ }°$

b. $m\angle ABC + m\angle EBC = \underline{\ ?\ }°$

c. Use your completed statements from parts (a) and (b) to explain
why $\angle ADC \cong \angle EBC$.

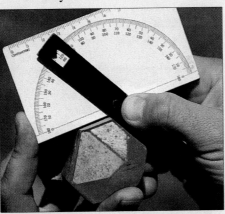

38. What is the measure of
the interfacial angle that
the goniometer at the left
is measuring?

**At the left is a cross section of an
aragonite crystal. Use the diagram for
Exercises 39–41.**

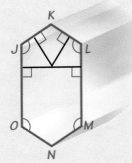

39. The interfacial angle of faces $\overline{JK}$ and $\overline{KL}$ is 64.8°.

a. Find $m\angle K$.

b. Find the measures of $\angle L$, $\angle M$, $\angle O$, and $\angle J$. How did you get
your answers?

c. What is the interfacial angle of faces $\overline{KL}$ and $\overline{LM}$?

40. Name the three pairs of faces that have an interfacial angle
congruent to the interfacial angle of faces $\overline{KL}$ and $\overline{LM}$. Explain
why these four angles must be congruent.

41. **Writing** Explain why $\angle K \cong \angle N$.

42. Writing Suppose you know that the angles of a regular polygon are acute. Explain how you can tell how many sides the polygon has. Can you do the same thing if you know that the angles are right? if they are obtuse? Explain why or why not.

SPIRAL REVIEW

43. Explain the difference between skew lines and parallel lines. *(Section 1.4)*

Sketch each figure. *(Section 2.3)*

44. a hexagon with three right angles

45. a concave octagon

ASSESS YOUR PROGRESS

VOCABULARY

polygon (p. 73)
triangle (p. 73)
quadrilateral (p. 73)
pentagon (p. 73)
hexagon (p. 73)
equilateral (p. 73)

equiangular (p. 73)
regular (p. 73)
consecutive angles (p. 74)
consecutive sides (p. 74)
diagonal (p. 74)

Classify each polygon. Be as specific as possible. *(Section 2.3)*

1. **2.** **3.** **4.**

5. a. Sketch a regular polygon *ABCDEF*.

   b. How many lines of symmetry does *ABCDEF* have?

   c. Name two consecutive sides of *ABCDEF*.

   d. Name two consecutive angles of *ABCDEF*. *(Section 2.3)*

6. ALGEBRA Three interior angles of a hexagon measure 70°, 120°, and 110°. The other three angles are congruent. Find the measure of each angle. *(Section 2.4)*

7. Find the measure of each interior and exterior angle of a regular 24-gon. *(Section 2.4)*

8. Journal Make a diagram of the different polygons you learned about in Section 2.3. Give angle measures whenever possible.

2.5 | Parallelograms

A *parallelogram* is a type of quadrilateral. In this section, you will explore the properties of parallelograms and learn about the special types of parallelograms.

Learn how to...
- **identify types of quadrilaterals**

So you can...
- **find the measures of angles and segments in quadrilaterals**

EXPLORATION
COOPERATIVE LEARNING

Analyzing Parallelograms

**Work in a group of four students.
You will need:**

- lined notebook paper
- patty paper or tracing paper
- a ruler
- a protractor

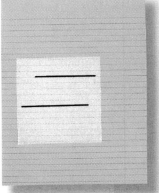

1 Place a piece of patty paper on the notebook paper. Trace two lines as shown.

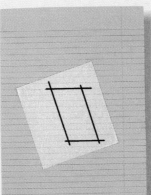

2 Turn the patty paper so that the lines on the notebook paper intersect the lines you drew in Step 1. Trace two more lines to form a quadrilateral. This figure is called a *parallelogram*.

3 Measure each side of the parallelogram. What do you notice?

4 Measure each angle. How do the measures of opposite angles compare?

5 Draw a diagonal of the parallelogram and find its midpoint. Draw the other diagonal. What do you notice?

6 Repeat Steps 1–5 at least two more times. Use different side lengths and angle measures for each parallelogram.

7 Make three conjectures about the parts of a parallelogram.

A **parallelogram** is a quadrilateral with both pairs of opposite sides parallel. The expression □*ABCD* is read "parallelogram *ABCD*." You may have discovered the following properties of parallelograms in the Exploration.

Opposite Parts of a Parallelogram

The opposite sides of a parallelogram are congruent.

The opposite angles of a parallelogram are congruent.

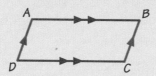

If *ABCD* is a parallelogram, then $\overline{AB} \cong \overline{DC}$ and $\overline{AD} \cong \overline{BC}$.

If *ABCD* is a parallelogram, then $\angle A \cong \angle C$ and $\angle B \cong \angle D$.

EXAMPLE 1 Connection: Algebra

Find the value of each variable in □*PQRS*. **Explain your reasoning.**

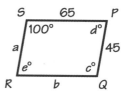

SOLUTION

Since opposite sides of a parallelogram are congruent, *SR = PQ* and *SP = RQ*. So *a* = 45 and *b* = 65.

Opposite angles of a parallelogram are congruent. So $\angle S \cong \angle Q$ and *c* = 100. Use the sum of the angle measures of a quadrilateral to find the other angle measures.

$c° + d° + e° + 100° = 360°$

$100 + d + d + 100 = 360$

$2d + 200 = 360$

$d = 80$

Since the opposite angles are congruent, you can substitute 100 for *c* and *d* for *e*.

Solve for *d*.

Therefore, *d* = 80 and *e* = 80.

There are three special kinds of parallelograms.

A **rectangle** is an equiangular parallelogram.

A **rhombus** is an equilateral parallelogram.

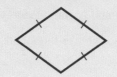

A **square** is a regular parallelogram.

This diagram shows how these quadrilaterals are related.

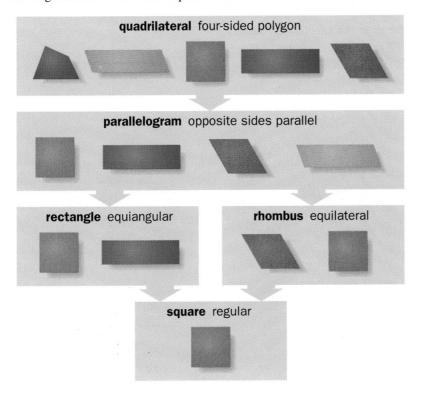

A quadrilateral may or may not have some of the properties of the quadrilaterals linked below it, but each quadrilateral in the diagram has all the properties of the quadrilaterals linked above it.

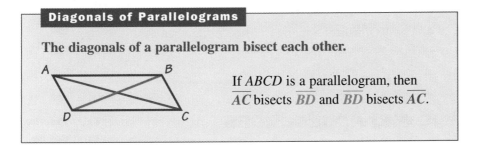

Diagonals of Parallelograms

The diagonals of a parallelogram bisect each other.

If *ABCD* is a parallelogram, then $\overline{AC}$ bisects $\overline{BD}$ and $\overline{BD}$ bisects $\overline{AC}$.

THINK AND COMMUNICATE

1. What is the measure of each angle of a rectangle? How do you know?

2. How does the diagram above show that a parallelogram is a special type of quadrilateral?

3. When is a rectangle a quadrilateral? a parallelogram? a square?

4. When is a rhombus a quadrilateral? a parallelogram? a square?

5. Is a square a quadrilateral? a parallelogram? a rectangle? a rhombus?

6. Do the diagonals bisect each other in a rectangle? in a rhombus? in a square? Explain.

EXAMPLE 2

In rhombus *WXYZ*, *XW* = 15 and *VY* = 9.
Find *XY* and *WY*.

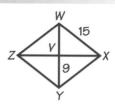

SOLUTION

Since *XW* = 15 and a rhombus is equilateral, *XY* =15.

Rhombus *WXYZ* is a parallelogram. The diagonals of a parallelogram bisect each other, so *V* is the midpoint of $\overline{WY}$.

$$WY = 2(VY) = 2(9) = 18$$

☑ CHECKING KEY CONCEPTS

1. Find each length or angle measure of ▱*ABCD*.

  a. $m \angle A$

  b. $m \angle B$

  c. *CD*

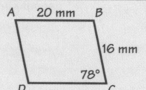

Tell whether each statement is *True* or *False*. If the statement is false, sketch a counterexample and justify your answer.

  2. Every square is a rhombus.

  3. Every rectangle is a square.

  4. The diagonals of a rhombus bisect each other.

2.5 Exercises and Applications

Extra Practice exercises on page 663

Find each length or angle measure.

  1. rectangle *UVWX*

    a. *UW*

    b. $m \angle VWX$

    c. *XW*

  2. rhombus *EFGH*

    a. *HG*

    b. *GF*

    c. $m \angle F$

  3. ▱*JKLM*

    a. $m \angle JLM$

    b. $m \angle KLM$

    c. $m \angle JMN$

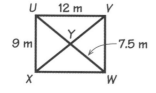

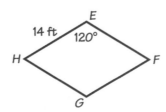

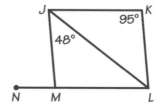

Find each length or angle measure of rectangle _JKLM_.

4. _NK_

5. _ML_

6. _m ∠ JML_

7. _m ∠ JNK_

8. _m ∠ KJN_

9. _m ∠ JKN_

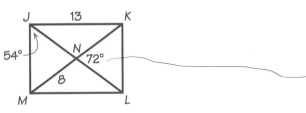

10. **Writing** Sarita says that a square is an equiangular rhombus. Bob says that a square is an equilateral rectangle. Who is correct? Explain.

Investigation **For Exercises 11 and 12, you will need scissors and a sheet of paper.**

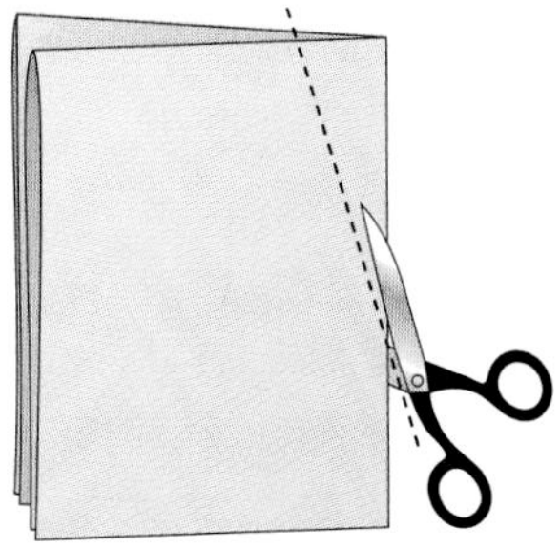

11. **a.** Fold a sheet of paper in quarters as shown. Cut off the folded corner. What shape do you think you will have when you unfold the paper?

   b. Unfold the paper. What shape did you get?

12. Tell whether it is possible to cut the folded paper to produce each quadrilateral. If it is possible, describe how. If not, explain why not.

   a. a parallelogram that is not a rhombus

   b. a rhombus

   c. a rectangle that is not a square

   d. a square

Connection OPTICS

When amateur astronomers or birdwatchers need to hold binoculars steady for a long period of time, they use special binocular mounts. This mount features a parallelogram that shifts to adjust to the height of the viewer.

13. In the diagram, $\overline{BE}$ stays vertical as ▱_BCDE_ shifts. Explain why $\overline{CD}$ stays vertical as well.

14. Suppose the mount is set so that $m ∠ ABE = 55°$.

   a. Find $m ∠ BCD$.

   b. Find $m ∠ FCB$.

   c. What is the angle of elevation when the line of sight, $\overline{GC}$, makes a 67° angle with the top piece of the mount, $\overline{AC}$?

15. **Challenge** Show that the angle of elevation of the binoculars does not change as ▱_BCDE_ shifts.

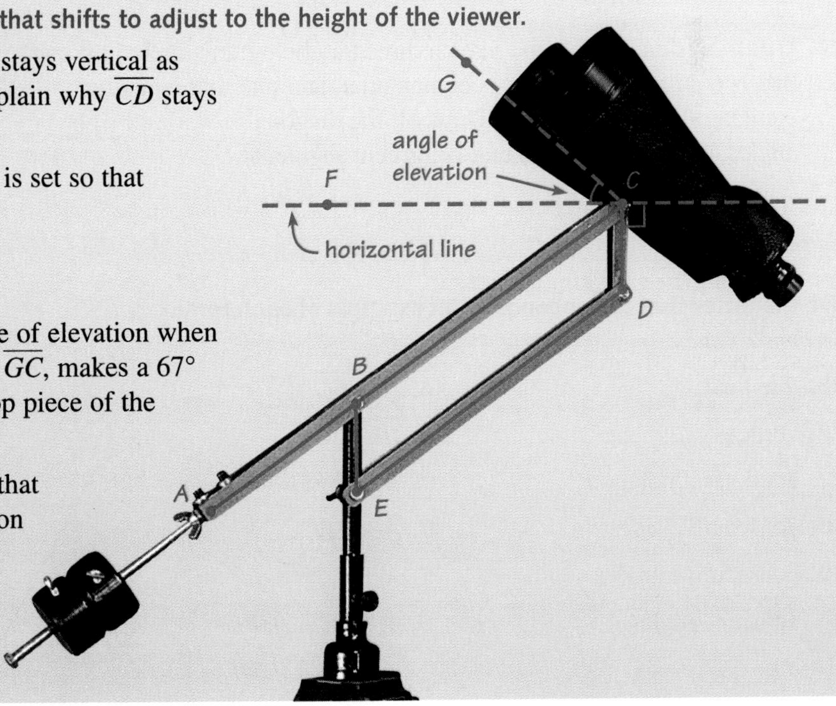

16. Logical Reasoning In □*ABCD*, ∠*D* is a right angle.

    a. Find *m* ∠*B*. How did you get your answer?

    b. Find *m* ∠*A* and *m* ∠*C*. Explain your reasoning.

    c. Complete: A parallelogram that has one right angle is a ? .

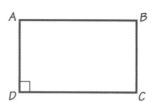

Copy the table. Use check marks to show which polygons have each property.

		quadrilateral	parallelogram	rectangle	rhombus	square
17.	Opposite sides are parallel.	?	?	?	?	?
18.	Opposite sides are congruent.	?	?	?	?	?
19.	Opposite angles are congruent.	?	?	?	?	?
20.	All angles are congruent.	?	?	?	?	?
21.	All angles are right angles.	?	?	?	?	?
22.	All sides are congruent.	?	?	?	?	?
23.	Diagonals bisect each other.	?	?	?	?	?

O N G O I N G   A S S E S S M E N T

24. Open-ended Problem Sketch three parallelograms including one that is equilateral, one that is equiangular, and one that is neither equilateral nor equiangular. Draw all the diagonals. Mark all right angles, congruent angles, and congruent segments.

S P I R A L   R E V I E W

Give the letter that corresponds to an example of each term.
(Sections 2.1 and 2.3)

25. diagonals	**A.** $\overline{CB}$, $\overline{BG}$
26. vertices	**B.** $\overline{BH}$, $\overline{EG}$
27. equilateral polygon	**C.** *BCDEFG*
28. consecutive sides	**D.** *ABGH*
29. consecutive angles	**E.** *A, F*
30. adjacent angles	**F.** ∠*ABH*, ∠*HBG*
	G. ∠*BCD*, ∠*CDE*

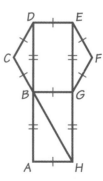

2.6 Building Prisms

Three-dimensional objects are often designed using a process called *computer-aided design.* For example, an architect might enter the plans for this house into a computer system in order to look at different aspects of the design.

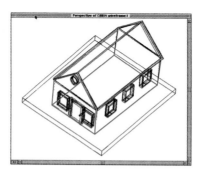

There are several ways to visualize a three-dimensional object. One way is to look through the *faces* of the object to the *edges,* as in the drawing at the left.

Another way to look at an object is from different positions. For instance, an architect might look at the front, side, and top views of the house, as shown.

Side View   **Top View**   **Front View**

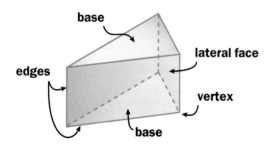

base

lateral face

edges

vertex

base

The house above and the figure at the left are *prisms.* A **prism** is a three-dimensional figure, with two congruent faces called the **bases**, that lie in parallel planes. The other faces of a prism, called the **lateral faces**, are formed by connecting the corresponding vertices of the bases. In this book, the lateral faces of prisms are rectangles. A prism's **vertices** are connected by segments called **edges**.

Prisms are classified by the shapes of their bases.

| The bases of this prism are triangles, so it is called a triangular prism. | A **cube** is a rectangular prism whose faces are all squares. |

WATCH OUT! ▶

Notice that the bases of a prism do not have to be horizontal.

THINK AND COMMUNICATE

1. How many faces does the triangular prism have? How many faces does the cube have?

2. a. Name three different possible pairs of bases for the cube.

   b. Are there different possible pairs of bases for the triangular prism? Explain why or why not.

EXAMPLE 1

Sketch each prism.

a. a rectangular prism

b. a pentagonal prism

SOLUTION

Start by sketching one base. Sketch the other base either above or below the first base, or next to it. Make sure that the corresponding edges of the bases are parallel.

Draw the bases. **Connect corresponding vertices.** **Make the edges that are hidden from view into dashed lines.**

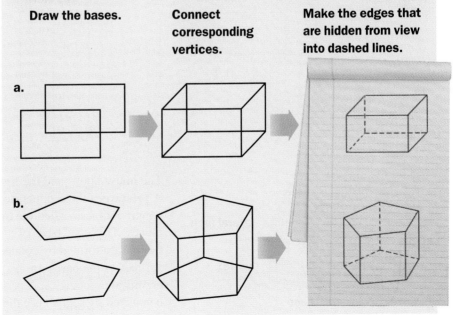

a.

b.

Unfolding a Prism

Imagine cutting this prism along some of its edges, then opening and unfolding it. The resulting plane figure is called a **net**.

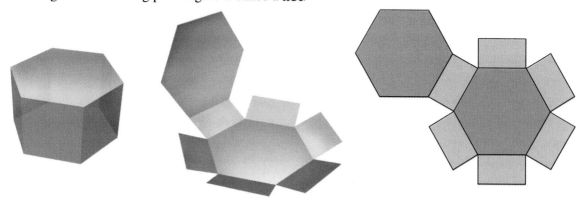

You can construct a prism by folding up its net and taping the edges together.

EXAMPLE 2

Sketch a net for this prism.

SOLUTION

Method 1

Sketch all the lateral faces attached to one of the bases of the prism.

Sketch the other base on the opposite edge of one of the lateral faces.

Method 2

Sketch all the lateral faces attached to each other.

Place the bases at opposite ends of two of the lateral faces.

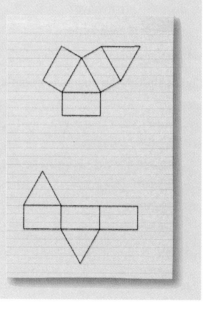

THINK AND COMMUNICATE

3. Describe another way to sketch a net for the prism above. How many different methods do you think there are?

4. Describe one possible net for a cube.

1. Sketch the front, top, and side views of a rectangular prism. Then sketch the prism.

2. Use the diagram of a cube on page 94. How many edges does it have? How many vertices?

Visual Thinking **Tell whether or not each figure is a net for a prism.**

3.

4.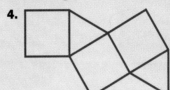

2.6 **Exercises and Applications**

Extra Practice exercises on page 663

Use the prism at the right.

1. Name the two bases of the prism. What type of prism is it?

2. Sketch the front, top, and side views of the prism.

3. How many lateral faces does the prism have?

4. How many edges meet at vertex C? Name them.

5. Name two skew edges of the prism.

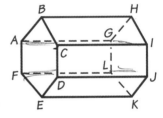

Use the prisms in this section or sketch the prisms to complete the chart.

	Type of Prism	Number of faces, F	Number of vertices, V	Number of edges, E
6.	triangular	?	?	?
7.	rectangular	?	?	?
8.	pentagonal	?	?	?
9.	hexagonal	?	?	?
10.	heptagonal	?	?	?

11. **Logical Reasoning** Describe any patterns that you see in the chart. Use these patterns to find F, V, and E for a prism whose base is an n-gon.

12. Use your answer to Exercise 11 to write an expression for E in terms of F and V. (This relationship is known as *Euler's formula.*)

13. Check that the relationship you found in Exercise 12 is true for another type of prism. Do you think that the relationship will work for any prism? Explain why or why not.

BY THE WAY

Leonhard Euler (pronounced "oiler") developed many important scientific as well as mathematical theories. Some of his greatest work was accomplished after he lost his sight at the age of 59.

Sketch each prism.

14. an octagonal prism

15. a prism whose bases are obtuse triangles

16. Open-ended Problem Sketch a prism that is not a cube but has two lateral faces that are squares.

Visual Thinking Tell whether or not each figure is a net for a prism.

17. **18.**

Sketch a net for each prism.

19. a rectangular prism

20. a pentagonal prism

21. a prism whose bases are nonregular hexagons

22. Research Most cardboard boxes are made from cutouts that are not actually nets. Examine an unfolded cardboard box. How is it different from a net? What are some possible reasons for the differences?

23. GAMES Many games use a die shaped like a cube, with the faces labeled 1 through 6. The sides are labeled so that when you add the numbers on any pair of opposite faces, the total is 7.

Since one dot is on the front face ...

... six dots must be on the back.

a. How many dots are on the left face of this die? How many dots are on the bottom face?

b. Sketch a net for the die, writing the numbers 1 through 6 inside the appropriate squares.

24. Writing All of the prisms that you have seen in this section and will see later in this book are *right prisms*. Prisms that are not right prisms are called *oblique prisms*. Here are some examples of both types of prisms.

Right prisms **Oblique prisms**

Suppose that your friend is going to take a quiz in which he or she is shown a prism and must tell whether it is right or oblique. Write a note that explains what your friend should know.

Look back at the article on pages 54–56.

In crystallography, a lattice is a set of points that shows the arrangement of molecules in a given crystal. Each point represents a molecule, and the length of each segment corresponds to the distance between the molecules. The diagrams below show two lattices that are shaped like prisms.

Lattice for Pyrite

Lattice for Beryl

Use the lattices above.

25. In the lattice for pyrite, the segments are all the same length and all meet at right angles. What type of prism do the segments form?

26. Each base of the lattice for beryl is a regular hexagon formed by three identical rhombuses. Find the measure of each angle of one of these rhombuses.

A *diagonal of a prism* connects two vertices that do not lie on the same face of the prism. Use this definition for Exercises 27 and 28.

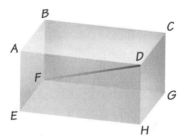

27. a. Name two other diagonals of this prism.

    b. Explain why $\overline{AH}$ is not a diagonal of the prism.

    c. How many diagonals does this prism have?

28. Challenge How many diagonals does a hexagonal prism have?

ONGOING ASSESSMENT

29. Cooperative Learning Work with another student.

    a. Sketch a prism. Without showing it to the other student, describe the prism. The other student should sketch the prism, based on the description.

    b. Compare the two sketches. How similar are they? Should the description have been different? If so, revise it so it is more accurate.

30. List all of the properties of a rhombus that you have learned. *(Section 2.5)*

31. Define *isosceles triangle* and sketch an example. *(Section 2.2)*

Use inductive reasoning to find the next two terms in each number pattern.
(Section 1.1)

32. 10, 7, 4, 1, _?_ , _?_   **33.** 1, 4, 9, 16, _?_ , _?_

ASSESS YOUR PROGRESS

VOCABULARY

parallelogram (p. 88)   **lateral face** (p. 93)
rectangle (p. 88)   **vertex of a prism** (p. 93)
rhombus (p. 88)   **edge** (p. 93)
square (p. 88)   **cube** (p. 94)
prism (p. 93)   **net** (p. 95)
base (p. 93)

Find each length or angle measure. *(Section 2.5)*

1. ▱*WXYZ*
  a. $m \angle X$
  b. $m \angle Y$
  c. *WX*

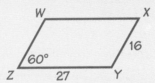

2. rectangle *ABCD*
  a. $m \angle ABC$
  b. *BC*
  c. *BD*

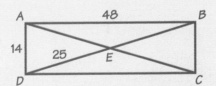

3. rhombus *LMNO*
  a. *MN*
  b. $m \angle M$
  c. $m \angle LOP$

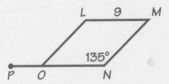

4. Sketch a hexagonal prism. Label one of its bases and one of its lateral faces. *(Section 2.6)*

5. Sketch two different nets for a rectangular prism. *(Section 2.6)*

6. Journal Sketch a rectangle, a rhombus, and a square. Draw all of the lines of symmetry for each shape. How are the lines of symmetry for the square related to the ones for the rectangle and the rhombus?

Building the Platonic Solids

A *regular solid* is made of only one type of regular polygon. Over 2000 years ago, the Greek mathematician Euclid proved that only five regular solids, called the *Platonic solids*, are possible. The subject of this activity is the last proposition in Euclid's book, *The Elements*.

PROJECT GOAL — Work with a partner to discover the regular solids and investigate how to make nets for them.

Discovering the Five Solids

1. Draw a regular triangle, a square, a regular pentagon, and a regular hexagon using what you know about the angle measures of regular polygons. Cut out six copies of each. Cut out additional copies as needed.

2. Tape three triangles together so that they share a common vertex, as shown below. Can they fold up to form a closed three-dimensional vertex? If so, use another piece of tape to hold the sides together. If not, explain why they don't make a closed vertex.

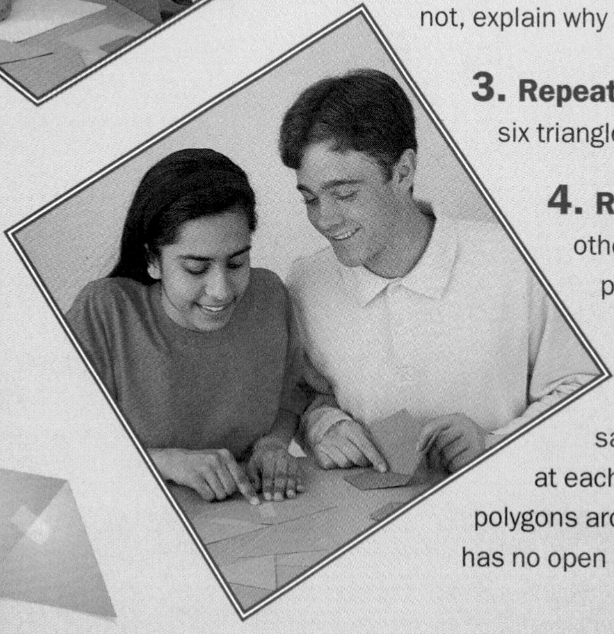

3. Repeat Step 2 for four, five, and six triangles.

4. Repeat Steps 2 and 3 for the other polygons. You will find five possible vertices.

5. Finish constructing the five solids so that the same number of polygons meet at each vertex. Continue adding polygons around each vertex until the solid has no open spaces.

Use the information at the right to identify each of your solids. Tell what kind of polygons were used for the faces of each solid, and how many polygons meet at each vertex.

tetrahedron ■	**4 faces**
hexahedron (cube) ■	**6 faces**
octahedron ■	**8 faces**
dodecahedron ■	**12 faces**
icosahedron ■	**20 faces**

Making Nets for the Solids

Cut or remove some of the tape from the icosahedron. Leave enough edges attached to keep it in one piece, but separate enough edges so that the pieces lie flat. Sketch the resulting figure. Use your sketch to make an accurate net for the solid from heavy paper. Do the same for the dodecahedron.

Try to make nets for the other solids without taking them apart. Use these nets to make solids out of heavy cardboard.

Extending Your Project

Continue your investigation of Platonic solids by exploring one of the following topics.

• Find all the different possible nets for the tetrahedron. Find at least three nets for each of the other solids. What makes one net easier to assemble than another?

• Count the number of vertices, faces, and edges of each solid. Use a table like the one on page 96 to record your data. What patterns do you see? Two pairs of Platonic solids are related by their vertices, faces, and edges. Find these pairs and describe the relationships.

• Use a protractor to make a contact goniometer like the kind crystallographers use (see page 56). Find the interfacial angles of each solid.

Self-Assessment
Describe the work you and your partner did for this project. What did you learn during your investigation? What advice would you give to someone who plans to do this project in the future?

2 Review

Make a sketch or give an example for each of the vocabulary words that you learned in this chapter.

VOCABULARY

types of angles (p. 57)
adjacent angles (p. 58)
linear pair (p. 58)
perpendicular (p. 58)
vertical angles (p. 58)
complements (p. 58)
supplements (p. 58)
triangle (p. 64)
side of a triangle (p. 65)
vertex of a triangle (p. 65)
types of triangles (p. 65)
polygon (p. 73)
triangle (p. 73)
quadrilateral (p. 73)
pentagon (p. 73)
hexagon (p. 73)
equilateral (p. 73)

equiangular (p. 73)
regular (p. 73)
consecutive angles (p. 74)
consecutive sides (p. 74)
diagonal (p. 74)
parallelogram (p. 88)
rectangle (p. 88)
rhombus (p. 88)
square (p. 88)
prism (p. 93)
base (p. 93)
lateral face (p. 93)
vertex of a prism (p. 93)
edge (p. 93)
cube (p. 94)
net (p. 95)

SECTIONS 2.1 *and* 2.2

Angles are classified by their measures. An angle can be **acute**, **right**, **obtuse**, or **straight**. Pairs of angles can be classified by their positions or by their measures.

Vertical angles are congruent. Angles that form a **linear pair** are supplementary.

If two lines are **perpendicular**, they form congruent adjacent angles. If two lines form congruent adjacent angles, the lines are perpendicular.

When classified by its sides, a triangle can be **scalene**, **isosceles**, or **equilateral**. When classified by its angles, a triangle can be **acute**, **obtuse**, **right**, or **equiangular**.

The sum of the measures of the interior angles of any triangle is 180°.

SECTIONS 2.3 *and* 2.4

A **polygon** is a closed plane figure whose sides are segments that meet only at their vertices.

A polygon is named by consecutive vertices. Two names for the polygon at the right are hexagon *BCDEFA* and hexagon *DCBAFE*.

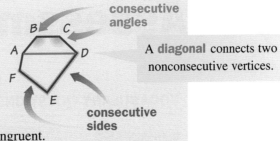

consecutive angles

A **diagonal** connects two nonconsecutive vertices.

consecutive sides

In an **equilateral** polygon, all sides are congruent.
In an **equiangular** polygon, all angles are congruent.
A **regular** polygon is both equilateral and equiangular.

Polygons are classified by the number of sides they have.

Number of sides	3	4	5	6	n
Name	triangle	quadrilateral	pentagon	hexagon	n-gon

The sum of the measures of the interior angles of an n-gon is $(n - 2)180°$.
The sum of the measures of the exterior angles of any polygon is $360°$.

SECTIONS 2.5 *and* 2.6

A **parallelogram** is a quadrilateral with both pairs of opposite sides parallel. The opposite sides of a parallelogram are congruent. The opposite angles of a parallelogram are congruent. The diagonals of a parallelogram bisect each other.

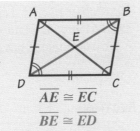

$$\overline{AE} \cong \overline{EC}$$
$$\overline{BE} \cong \overline{ED}$$

There are three special types of parallelograms.

rectangle
equiangular

rhombus
equilateral

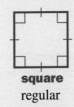

square
regular

A **prism** is a figure with two congruent parallel faces called **bases**. The **lateral faces** of the prism are rectangles. The line segments connecting the **vertices** are called **edges**.

Prisms are named by the shapes of their bases.
This is a triangular prism.

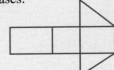

This **net** can be folded and taped to make the triangular prism.

Assessment

VOCABULARY QUESTIONS

Complete each statement.

1. The measure of a(n) _?_ angle is between 90° and 180°.

2. A(n) _?_ connects two nonconsecutive vertices of a polygon.

3. A regular quadrilateral is called a(n) _?_.

4. The sum of the measures of the _?_ angles of an octagon is 360°.

Use the diagram at the left. Give an example of each of the following.

5. vertical angles

6. congruent adjacent angles

7. Explain the difference between a prism and a net.

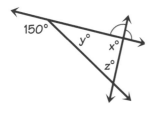

Classify each polygon. Be as specific as possible.

8.

9.

10.

11. **Writing** Explain why it is impossible to sketch an equilateral triangle that is not isosceles.

SECTIONS 2.1 *and* 2.2

12. Find the value of each variable in the diagram at the left.

    a. x **b.** y **c.** z

Find the measures of a complementary angle, a supplementary angle, and a vertical angle for an angle with the given measure.

13. 71°   14. 39°   15. 85°

ALGEBRA Find each unknown angle measure.

16.

17.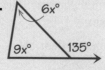

18.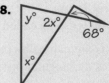

19. **Open-ended Problem** Find a possible set of three angle measures of a triangle. Use the measures to classify the triangle.

SECTIONS 2.3 *and* 2.4

For Questions 20–22, use the diagram at the right.

20. What type of polygon is the figure? Be as specific as possible.

21. Give three different names for the polygon.

22. Name all the diagonals with one endpoint at *A*.

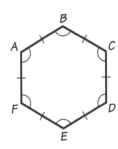

Find the value of each variable.

23.

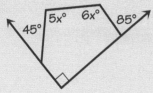

24.

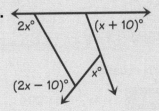

25. Each exterior angle of an equiangular *n*-gon measures 12°. Find the value of *n*.

26. Find the measure of each interior angle of a regular 24-gon.

SECTIONS 2.5 *and* 2.6

Find each length or angle measure.

27. rhombus *DEFG*

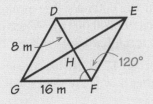

a. *FE*

b. *DF*

c. *m*∠*DEF*

28. parallelogram *STUV*

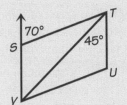

a. *m*∠*U*

b. *m*∠*TVU*

c. *m*∠*STV*

29. Sketch a pentagonal prism.

30. a. What type of prism is shown at the right?

b. How many lateral faces does the prism have?

c. Sketch two different nets for the prism.

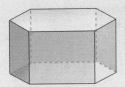

PERFORMANCE TASK

31. Make a poster that shows what you know about prisms. Identify the parts of a prism and give instructions for drawing a prism on paper. Show how a given prism can be made from different nets.

Algebra
Review/Preview

These exercises review algebra topics you will use in the next chapters.

SOLVING LINEAR EQUATIONS

Solve each equation.

Toolbox p. 694
Linear Equations

EXAMPLE
$$-4x + 5 = 29$$
$$-4x + 5 - 5 = 29 - 5 \quad \longleftarrow \text{Subtract 5 from both sides.}$$
$$-4x = 24$$
$$\frac{-4x}{-4} = \frac{24}{-4} \quad \longleftarrow \text{Divide both sides by } -4.$$
$$x = -6$$

1. $-81 = -37 + b$ **2.** $-3y - 4 = 8$ **3.** $5z + 16 = -34$

4. $27 = 9 + 8p$ **5.** $12 - 10q = 60$ **6.** $-200 = -32 - 7r$

7. $\dfrac{w - 6}{2} = 8$ **8.** $\dfrac{1 + k}{3} = -17$ **9.** $\dfrac{5}{4}n - 4 = 11$

Toolbox p. 692
Operations with Variable Expressions

10. ASTRONOMY The speed of light is about 186,000 mi/s. The mean distance between the sun and Earth is about 93,000,000 mi. About how long does it take for light from the sun to reach Earth?

11. Mario wants to buy a mountain bike that costs $780. He has already saved $130. If he saves $25 each week, after how many weeks will he have enough money to buy the bike?

SIMPLIFYING RADICALS

Simplify each expression.

Toolbox p. 700
Simplifying Radicals

EXAMPLES
$$\sqrt{48} = \sqrt{16 \cdot 3} \qquad\qquad \sqrt{50} + \sqrt{18} = \sqrt{25 \cdot 2} + \sqrt{9 \cdot 2}$$
$$= \sqrt{16} \cdot \sqrt{3} \qquad\qquad\qquad\quad = \sqrt{25} \cdot \sqrt{2} + \sqrt{9} \cdot \sqrt{2}$$
$$= 4\sqrt{3} \qquad\qquad\qquad\qquad\quad = 5\sqrt{2} + 3\sqrt{2}$$
$$\qquad\qquad\qquad\qquad\qquad\qquad\quad = 8\sqrt{2}$$

12. $\sqrt{12}$ **13.** $\sqrt{27}$ **14.** $\sqrt{45}$

15. $\sqrt{75}$ **16.** $5\sqrt{7} - 3\sqrt{7}$ **17.** $6\sqrt{32} + 9\sqrt{32}$

18. $\sqrt{20} + \sqrt{5}$ **19.** $\sqrt{72} - \sqrt{8}$ **20.** $\sqrt{24} + 7\sqrt{3}$

SOLVING QUADRATIC EQUATIONS

Solve each equation.

EXAMPLE

$$3x^2 + 9 = 84$$
$$3x^2 = 75 \quad \longleftarrow \text{ Subtract 9 from both sides.}$$
$$x^2 = 25 \quad \longleftarrow \text{ Divide both sides by 3.}$$
$$x = \pm\sqrt{25} \quad \longleftarrow \text{ Take the square root of both sides.}$$
$$x = \pm 5$$

> **Toolbox p. 701**
> *Quadratic Equations*

21. $x^2 = 81$ **22.** $134 = y^2 - 10$ **23.** $3r^2 = 27$

24. $-\frac{5}{2}t^2 = -160$ **25.** $5p^2 - 9 = 11$ **26.** $17 - 8n^2 = 15$

GRAPHING INEQUALITIES

Graph each inequality on a number line.

EXAMPLES $x \geq 1$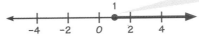

> A closed circle at 1 means that $x = 1$ *does* satisfy the inequality.

$-3 < p < 2$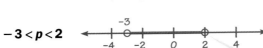

> Open circles at -3 and 2 mean that $p = -3$ and $p = 2$ *do not* satisfy the inequality.

$z \leq -2$ or $z > 0$

27. $x \geq -2$ **28.** $u < 1$ **29.** $-4 < v \leq 0$

30. $1 \leq y < 3$ **31.** $k < -2$ or $k > 2$ **32.** $m < 0$ or $m \geq 3$

GRAPHING EQUATIONS OF LINES

Graph each equation.

EXAMPLE $y = \frac{3}{2}x - 4$

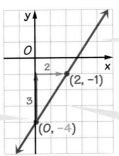

Step 3 Connect the two plotted points with a line.

Step 1 The y-intercept is -4. Graph the point $(0, -4)$.

Step 2 The slope is $\frac{3}{2}$. Start at $(0, -4)$. Count **3 units up** and **2 units right**. Graph this point.

33. $y = 2x + 1$ **34.** $y = 6x - 2$ **35.** $y = 4 - x$

36. $y = \frac{4}{3}x + 2$ **37.** $y = -\frac{2}{5}x - 3$ **38.** $3x + y = 5$

> **Toolbox p. 697**
> *Graphing Linear Equations*

3 | Reasoning in Geometry

Logical Crime ⟨SOLVER⟩

INTERVIEW **Mary-Jacque Mann**

Mary-Jacque Mann uses scientific tools and logical analysis to track down criminals. As a senior forensic scientist at the National Fish and Wildlife Forensics Laboratory in Ashland, Oregon, Mann investigates crimes against wildlife—hunting protected animals and importing outlawed products derived from these animals. Mann and her colleagues test the remains of animals from all over the world. They try to establish the cause of death and link a suspect to the crime.

Mann has been at the Oregon laboratory since it opened in 1989. In addition to completing an advanced degree in forensic science and working for six years in a crime lab, she was a volunteer at the center before she joined the staff. "This is the only full-service wildlife forensics lab in the world," she says. "I'm at the best possible place to pursue the kind of work I like to do."

" ... We use logic to solve actual crimes. We're looking for a higher standard ... "

Histogram of all Schreger angles (n = 260) of extinct and extant proboscidean ivory

The histogram shows the results of the research on mammoth and elephant ivory.

Making a Difference

"There's a lot of variety in this job, and the analysis is challenging," she says. "I also feel we are making a difference." Staff members at the lab identify samples of wild animal hair, feathers, fur, or blood that are found at a crime scene. Mann may try to identify the gun involved in a crime. One of the first projects she worked on at the lab was finding a way to identify different kinds of ivory.

In order to protect elephants worldwide, the United States banned the importation and sale of elephant ivory. Mammoth ivory and some other materials that look like elephant ivory are still legal. The lab wanted to find a way to tell whether an object is made of elephant ivory without damaging the object. Working with Ed Espinoza, Mann found a technique to distinguish between elephant ivory products and other forms of ivory. After doing thousands of measurements, they discovered that the faint etchings known as Schreger lines found in elephant ivory form different angles from those found in mammoth ivory. (Mammoths, relatives of the modern elephant, became extinct about 10,000 years ago.)

109

The Ivory Angle

Mann sums up this finding in two rules of thumb:

- **If the Schreger angles in an ivory object average 115° or more, then the object is made of elephant ivory.**

- **If the Schreger angles average 90° or less, then the object is made of mammoth ivory.**

If the object is neither elephant nor mammoth ivory, a set of rules formulated by Espinoza and Mann can be used to determine whether the object is made of whale, walrus, hippopotamus, or warthog ivory. "As in other sciences, if you follow a logical progression from beginning to end, going through a specified sequence of steps, you can get a predictable result," Mann says.

With this identification technique, officials are able to enforce a law banning the importation of elephant ivory, essentially shutting down sales of illegal ivory in the United States. Mann hopes to see more cases where the combination of logic, the scientific

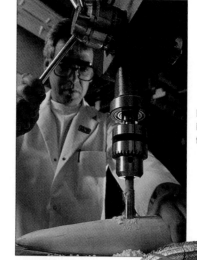

Ed Espinoza cuts ivory samples from a tusk.

Schreger lines on mammoth tusk

Schreger lines on elephant tusk

method, and international law can be applied to protect endangered animals like the African and Asian elephant.

EXPLORE AND CONNECT

Mary-Jacque Mann uses an electron microscope to determine the source of dust found on a piece of evidence.

1. Project   Make a poster about an endangered animal. Explain why the animal has become endangered. Have the measures taken to protect the animal affected local economy? Have the protection measures succeeded in helping the animal survive?

2. Writing   Find out what forensic scientists do in a crime lab. How do you think the work in a regular crime lab is different from the work done at the Wildlife Forensics Lab?

3. Research   Find out when the mammoths lived. In what parts of the world did they live? How did humans rely on mammoths?

Mathematics
& Mary-Jacque Mann

In this chapter, you will learn more about how mathematics is related to forensic science.

Related Exercises

Section 3.3
- Exercises 12–14

Section 3.4
- Exercises 5–8

3.1 Inductive and Deductive Reasoning

Learn how to...
- use deductive reasoning to reach conclusions

So you can...
- make a convincing argument
- recognize valid and invalid arguments

Before the late 1800s, no one knew what caused malaria. Some people thought malaria was caused by mosquito bites. Others believed that it was caused by the air in swampy or marshy areas. A series of findings convinced scientists that mosquitoes transmit malaria to humans.

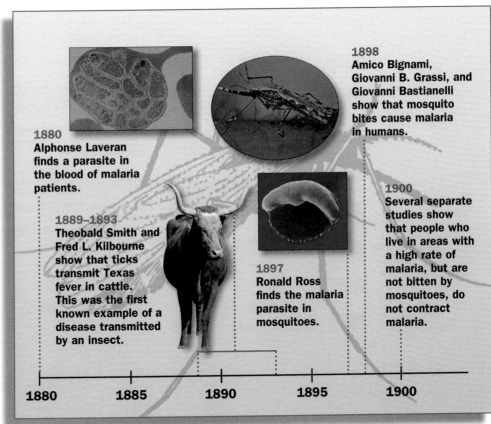

1880
Alphonse Laveran finds a parasite in the blood of malaria patients.

1889–1893
Theobald Smith and Fred L. Kilbourne show that ticks transmit Texas fever in cattle. This was the first known example of a disease transmitted by an insect.

1897
Ronald Ross finds the malaria parasite in mosquitoes.

1898
Amico Bignami, Giovanni B. Grassi, and Giovanni Bastianelli show that mosquito bites cause malaria in humans.

1900
Several separate studies show that people who live in areas with a high rate of malaria, but are not bitten by mosquitoes, do not contract malaria.

1880   1885   1890   1895   1900

THINK AND COMMUNICATE

1. Before this research showed the role of mosquitoes in spreading malaria, it was known that people who lived in swampy areas often became infected with malaria. What kind of reasoning led people to believe that malaria was caused by the air in swampy areas?

2. Explain why each of these investigations was an important part of demonstrating that malaria is transmitted by mosquitoes.

When you make conjectures based on patterns, such as noticing that people who live in swampy areas become infected with malaria, you use inductive reasoning. You use *deductive reasoning* to determine whether your conjectures are always true. **Deductive reasoning** involves using facts, definitions, and accepted properties in a logical order to reach a conclusion.

EXAMPLE 1   Connection: Algebra

Writing Is the product of an even integer and any other integer even or odd? Write a convincing argument that explains your conclusion.

SOLUTION

In order to find a pattern, you should try examples with several different types of numbers. You might want to make a table, or an organized list. Here is Ruby's solution.

I looked for a pattern by trying several examples.

$$4 \cdot 7 = 28 \qquad \text{(second number is odd)}$$
$$-8 \cdot 3 = -24 \qquad \text{(even number is negative)}$$
$$6 \cdot 6 = 36 \qquad \text{(second number is even)}$$
$$2 \cdot (-6) = -12 \qquad \text{(second number is negative)}$$

I think that the product of an even integer and any other integer will be even. I can use algebra to show that this is always true. I can write any even integer as $2n$ where n is an integer. I will let the second integer be m.

$$2n \cdot m = 2nm$$

Because $2nm$ is divisible by 2, the product of an even integer and any other integer is even.

THINK AND COMMUNICATE

3. What kind of reasoning did Ruby use when she looked for a pattern in the solution of Example 1? Using only this kind of reasoning, could she be certain of her conclusion?

4. What kind of reasoning did Ruby use to complete her solution? Which part of the argument do you find most convincing? Why?

5. Why do you think she used both kinds of reasoning?

EXAMPLE 2

Tell whether each argument uses *inductive* or *deductive* reasoning.

a. In the United States, you must be at least 18 years old to vote. Jamal is 19 years old. Jamal is old enough to vote in the United States.

b.

Find the next step in the pattern.

At each step, the black circle rotates clockwise to the next corner, so the next picture in the pattern is ⟶

SOLUTION

a. This argument uses deductive reasoning. The conclusion is based on two known facts.

b. This argument uses inductive reasoning. The conclusion is based on the observation of a pattern in the given steps.

THINK AND COMMUNICATE

6. In part (a) of Example 2, you cannot conclude that Jamal can vote in the United States. Give a reason why Jamal might not be able to vote in the United States.

7. Jonah is taller than Emily and Matt is shorter than Emily. Who is taller, Jonah or Matt? Did you use *inductive* or *deductive* reasoning?

☑ CHECKING KEY CONCEPTS

Tell whether each argument uses *inductive* or *deductive* reasoning.

1. By observing many individual cases, people concluded that malaria was caused by breathing air in swampy areas.

2. All students must study Algebra 1 before studying Geometry. Mia is studying Geometry. Therefore, Mia has studied Algebra 1.

3. Any quadrilateral with four congruent angles is a rectangle. A square has four congruent angles. A square is a rectangle.

4. PERSONAL FINANCE Martha Greene has been studying the profits of mutual funds for the last six months. She decides to invest in the fund that has had the highest profits during this time.

BY THE WAY

Malaria continues to be a health problem in many areas of the world. Scientists have found that one very effective way to prevent the disease is to use bed nets to prevent mosquito bites during the night.

3.1 | Exercises and Applications

Extra Practice
exercises on
page 663

Tell whether each argument uses *inductive* or *deductive* reasoning.

1. The sun has risen every day of my life, so the sun will rise tomorrow.

2.

Dilbert **reprinted by permission of United Feature Syndicate, Inc.**

3. The probability of being left-handed is about $\frac{1}{10}$. About 20% of the musicians that I know are left-handed. So left-handed people are more likely than right-handed people to be musicians.

4. All finches eat seeds. The Eurasian Hawfinch, Rosita's Bunting, and the Painted Bunting are all finches. These three birds all eat seeds.

Use inductive reasoning to predict the next two terms or figures in each pattern. Justify your prediction.

5.

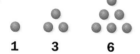

1　　3　　6

6.

3　　9　　18

7. 1, 2, 4, 8, . . .

8. 1, 8, 27, 64, 125, . . .

9. 1, 2, 6, 24, 120, . . .

10. 2, 9, 16, 23, . . .

For Exercises 11–13, select the conclusion that makes the statement true.

11. All equilateral triangles are isosceles. $\triangle ABC$ is equilateral. Therefore, $\triangle ABC$ (*is, is not,* or *may be*) isosceles.

12. Jean is a member of her school's outing club. Jean is a rock climber. So, the members of this club (*are, are not,* or *may be*) rock climbers.

13. If a triangle is isosceles, it has two congruent angles. $\triangle ABC$ has no congruent angles. Therefore, $\triangle ABC$ (*is, is not,* or *may be*) isosceles.

14. a. Give a convincing argument that the next two numbers in the sequence 1, 2, 3, . . . could be 4, 5.

　　b. Give a convincing argument that the next two numbers in the sequence 1, 2, 3, . . . could be 6, 12.

　　c. Give another pair of numbers that could be the next two numbers in the sequence 1, 2, 3, ... Explain your reasoning.

15. Open-ended Problem Describe a situation in your life when you used deductive reasoning.

BY THE WAY

The woodpecker finch of the Galapagos Islands uses cactus spines as a tool for finding grubs.

16. Visual Thinking This pattern is made by placing points on a circle and drawing each segment that connects a pair of points.

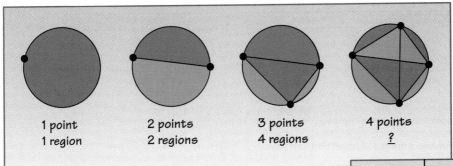

1 point
1 region

2 points
2 regions

3 points
4 regions

4 points
?

a. Complete the table at the right by sketching diagrams. Count the number of regions created.

b. Open-ended Problem Predict the number of regions formed by 6 points and their segments. Explain your prediction.

c. Sketch diagrams to check your prediction in part (b).

Number of points	Number of regions
2	2
3	4
4	?
5	?

17. Challenge Change the pattern in Exercise 16 so that you draw only segments connecting adjacent points. Write a formula for the number of regions as a function of the number of points. Give a convincing argument that your formula is correct.

18. ALGEBRA Suppose you multiply any two odd integers. Is the product odd or even? Explain your answer with a convincing argument. (*Hint:* An odd number can be written as $2n + 1$.)

Connection ▶ ZOOLOGY

Upside-down catfish live in water that may contain little oxygen except near the surface. Zoologist Lauren Chapman suspected that the fish swim upside down to reach this oxygen more easily. To test this theory, Chapman and colleagues studied these catfish in an aquarium. When they lowered the oxygen content of the water, the catfish swam upside down.

19. What kind of reasoning did Chapman use to test her theory? Do you think this experiment provides convincing evidence that her theory is correct? Explain why or why not.

20. The catfish eat surface plankton. Swimming upside down may allow the catfish to reach the surface plankton more easily. Describe an experiment you could use to test this theory.

BY THE WAY

Synodontis nigriventris is a species of catfish in Africa. These catfish, known as upside-down catfish, often swim upside down near the surface of the water.

21. What type of reasoning does Rosemary use in her note to her friend? Do you think her conclusion is reasonable?

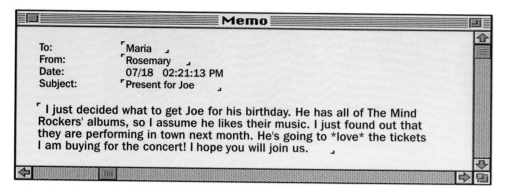

MARKETING Andre would like to work as a buyer for a music store. He made an appointment to talk to Karla Myers, the buyer at a music store.

22. Karla said that she makes her decisions based on what she observes, by analyzing her mistakes, and also by trying to think ahead. Explain how she uses both inductive reasoning and deductive reasoning.

23. Open-ended Problem Suppose that you are the buyer for a music store. A new band is releasing an album. What would you consider in deciding how many CDs and cassettes to order for your store?

24. Writing Explain the difference between inductive and deductive reasoning. Why do mathematicians and scientists use both deductive reasoning and inductive reasoning in their work?

ONGOING ASSESSMENT

25. Cooperative Learning Work in a group of three people. Find six advertisements in a magazine or newspaper. Each advertisement should make a claim about a service or product. Each of you should determine what kind of reasoning is used to reach the conclusions in each advertisement. Discuss whether the claims are true or false. Choose one advertisement and work together to give a deductive argument that could be used to support the claim made in it.

SPIRAL REVIEW

26. Sketch a triangular prism. *(Section 2.6)*

Tell whether each angle is *acute*, *right*, or *obtuse*. *(Section 2.1)*

27. ∠AFC

28. ∠CFE

29. ∠AFE

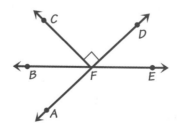

3.2 Postulates, Definitions, and Properties

Learn how to...

- **recognize and use postulates, definitions, and properties**

So you can...

- **justify statements about geometric figures**

Members of the Continental Congress signed the Declaration of Independence in August 1776. Thomas Jefferson wrote most of this document, which declared the colonies independent of British rule. In the document, Jefferson stated the assumptions that the congress made about the rights of men.

> "**...We** hold these truths to be self-evident, that all men are created equal, that they are endowed by their Creator with certain unalienable Rights, that among these are Life, Liberty, and the pursuit of Happiness ... "

Abigail Adams urged her husband John to include rights for women in the laws being written by the Continental Congress.

THINK AND COMMUNICATE

1. What does it mean for a statement to be self-evident? According to the Declaration of Independence, what truths are self-evident?

2. Why do you think Jefferson needed to state his assumptions?

3. Jefferson used the phrase "all men are created equal." What meanings does the word *men* have in English? How does the meaning of the word *men* affect the meaning of this statement? How do you think Jefferson defined *men*?

In mathematics, a statement that is accepted without proof is called a **postulate**. Jefferson's *self-evident truths* were his postulates for the Declaration of Independence. Postulates may be used as the starting points of our convincing arguments.

Some of the ideas that you learned in Chapters 1 and 2 are postulates. For example, the statement "If point *Y* is between points *X* and *Z*, then *X, Y,* and *Z* are collinear and *XY* + *YZ* = *XZ*" can now be called the Segment Addition *Postulate*. A list of the postulates that you can use in convincing arguments is given on page 712.

EXAMPLE 1

Sketch an illustration of the postulate:

If two lines intersect, they intersect in one and only one point.

SOLUTION

Draw any two lines that intersect. They will intersect at only one point.

Two lines that are parallel do not intersect at any point.

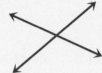

Definitions

Another important feature of an argument is the **definition**, or meaning, of each word that is used. In the Declaration of Independence, the definition of *men* that is used affects the meaning of Jefferson's self-evident truths. In mathematics, if a word is defined, then there is no question about its meaning.

For example, a polygon is a square *if and only if* it is a regular quadrilateral. This is an example of a *biconditional* statement.

WATCH OUT! ▶

Some words used in geometry are not defined. *Point, line,* and *plane* are undefined terms. In Chapter 1, these terms were described, but not defined.

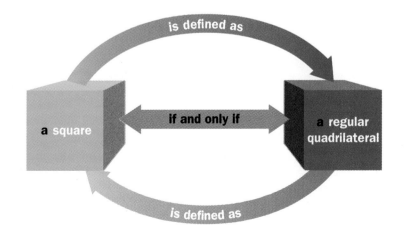

Properties

You can use properties of equality to reason about lengths of segments and measures of angles. Some of these properties are also true for congruence.

Equality and Congruence

Reflexive Property

$a = a$

$\overline{PQ} \cong \overline{PQ}$; $\angle 1 \cong \angle 1$

Symmetric Property

If $a = b$, then $b = a$.

If $\overline{PQ} \cong \overline{RS}$, then $\overline{RS} \cong \overline{PQ}$.

Transitive Property

If $a = b$ and $b = c$, then $a = c$.

If $\angle 1 \cong \angle 2$ and $\angle 2 \cong \angle 3$, then $\angle 1 \cong \angle 3$.

You can add or subtract the same value from both sides of an equation. In geometry, this can lead to new information about lengths and measures.

Properties of Equality

Addition Property

If $a = b$, then $a + c = b + c$.

If $PQ = RS$, then $PQ + QR = QR + RS$.

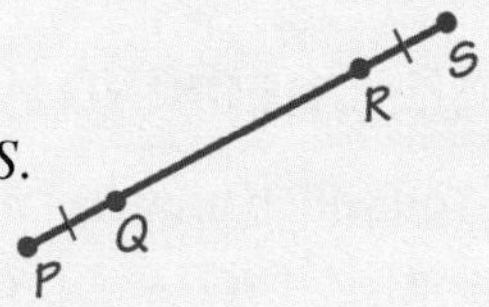

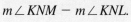

Subtraction Property

If $a = b$, then $a - c = b - c$.

If $m \angle JNL = m \angle KNM$, then $m \angle JNL - m \angle KNL = m \angle KNM - m \angle KNL$.

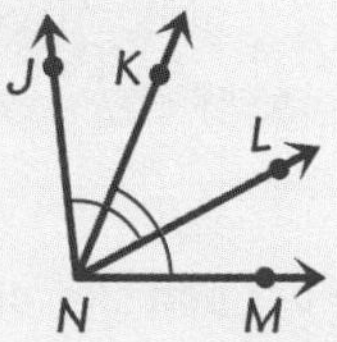

Substitution Property

If $a = b$, then a can be substituted for b (and b for a) in an expression.

If $m \angle 1 + m \angle 2 = 180°$ and $m \angle 1 = m \angle 3$, then $m \angle 3 + m \angle 2 = 180°$.

THINK AND COMMUNICATE

4. Use the diagram above. Explain how the Addition Property supports the statement "If $\overline{PQ} \cong \overline{RS}$, then $\overline{PR} \cong \overline{QS}$."

5. Use the diagram above. Explain how the Subtraction Property supports the statement "If $\angle JNL \cong \angle KNM$, then $\angle JNK \cong \angle LNM$."

You can use postulates, definitions, and properties as reasons in convincing arguments.

EXAMPLE 2

Give the postulate, definition, property, or previous statement that makes each statement true.

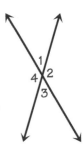

a. $\angle 1$ and $\angle 2$ are a linear pair; $\angle 2$ and $\angle 3$ are a linear pair.

b. $m\angle 1 + m\angle 2 = 180°$; $m\angle 2 + m\angle 3 = 180°$

c. $m\angle 1 + m\angle 2 = m\angle 2 + m\angle 3$

d. $m\angle 1 = m\angle 3$

SOLUTION

a. Definition of linear pair

b. Angles in a linear pair are supplementary. (Linear Pair Property)

c. Because both expressions in part (b) are equal to 180°, the Substitution Property says that they equal each other.

d. This follows from part (c) by the Subtraction Property.

☑ CHECKING KEY CONCEPTS

Tell which property makes each statement true.

1. $\angle ABC \cong \angle CBA$

2. COOKING If a recipe calls for $\frac{1}{2}$ cup buttermilk, you can use $\frac{1}{2}$ cup milk mixed with 1 teaspoon of lemon juice instead.

3. If $\overline{GH} \cong \overline{JK}$, then $\overline{JK} \cong \overline{GH}$.

4. If $m\angle 1 = m\angle 2$ and $m\angle 2 = m\angle 3$, then $m\angle 1 = m\angle 3$.

For each sketch, list the postulate(s) that it illustrates.

5.

6.
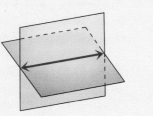

A. Through any two points there is exactly one line.

B. Two intersecting lines determine a plane.

C. Two planes can intersect in a line.

D. If A is between X and Y, then $XA + AY = XY$.

3.2 **Exercises and Applications**

Extra Practice exercises on page 664

Tell whether each statement is a _definition_ or a _postulate_.

1. If two lines intersect, then they intersect in exactly one point.

2. Two lines in the same plane that never intersect are parallel.

3. An isosceles triangle has at least two congruent sides.

Tell which property makes each statement true.

4. $m \angle ABC = m \angle ABC$

5. If $\overline{AB} \cong \overline{CD}$ and $\overline{CD} \cong \overline{EF}$, then $\overline{AB} \cong \overline{EF}$.

6. If $m \angle A + m \angle B = 180°$ and $m \angle A = 45°$, then $45° + m \angle B = 180°$.

For Exercises 7–14, give the postulate(s), definition(s), properties, or previous statement(s) that makes each statement about the diagram true.

7. $\overline{BC} \cong \overline{BC}$

8. $CE = AC$

9. $AB = DE$

10. $\overline{CE} \cong \overline{DF}$

11. $m \angle PTR = m \angle QTS$

12. $\angle PTQ \cong \angle QTP$

13. $m \angle PTQ = m \angle RTS$

14. $\angle QTR \cong \angle QTR$

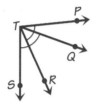

Connection ▶ NAVIGATION

Earth is divided into regions by imaginary circles called _meridians_. Because meridians are used as references for time and distance, it's important to have a common definition of the meridian that will be the "zero point." At a conference in 1884, representatives from twenty-five nations established the meridian through Greenwich, England, as the _prime meridian_, designated 0°.

15. The longitude of a location tells how many degrees east or west of the prime meridian it is. Before this agreement, many countries used different prime meridians. What information was needed in order to describe the longitude of a particular location?

16. Research Explain how degrees of longitude are related to degrees in an angle.

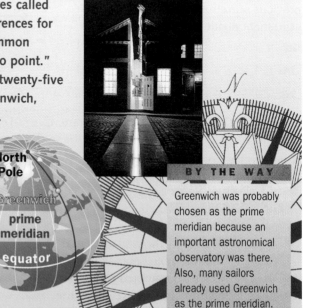

BY THE WAY

Greenwich was probably chosen as the prime meridian because an important astronomical observatory was there. Also, many sailors already used Greenwich as the prime meridian.

In order for scientific researchers around the world to communicate effectively with each other, they must agree on consistent definitions and standards for their procedures. They can calibrate a process by comparing their results to a standard.

17. **Writing** An ordinary bathroom scale is a measurement instrument. Explain how you can tell if your scale is calibrated correctly.

18. Explain why the thermostat on an oven must be calibrated correctly.

19. **Open-ended Problem** Name three other appliances or instruments that require calibration.

20. A researcher has a Standard Reference Material (SRM) and a list of the exact amounts of the chemicals in the SRM. Explain how the researcher can use this information to calibrate equipment.

BY THE WAY

The National Institute of Standards and Technology provides standard materials that researchers can use to check their measurements. For example, blubber from a beached whale can be used when studying chemical pollutants.

21. **CLOCKS** In the northern hemisphere, the shadows on a sundial rotate in the direction we call clockwise. In the southern hemisphere, shadows on a sundial rotate counterclockwise. If clocks had been invented in the southern hemisphere, what might the meanings of clockwise and counterclockwise have been?

Give the postulate, definition, property, or previous statement that makes the statement about the diagram true.

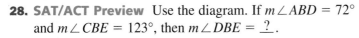

22. $\angle 1 \cong \angle 2$ 23. $\angle 2 \cong \angle 3$

24. $\angle 1 \cong \angle 3$ 25. $\angle 3 \cong \angle 4$

26. $\angle 1 \cong \angle 4$ 27. $m\angle 1 = m\angle 4$

28. **SAT/ACT Preview** Use the diagram. If $m\angle ABD = 72°$ and $m\angle CBE = 123°$, then $m\angle DBE = \underline{\ ?\ }$.

    A. 168°   **B.** 23°   **C.** 12°   **D.** 156°   **E.** 15°

29. **Challenge** Some anthropologists believe that the Chinese had contact with Central Americans before 100 A.D. Part of the evidence for this is the similarity in the designs of three-footed pots in both cultures. Explain why a three-footed pot is more stable than one with four feet. Use geometric postulates and properties to support your explanation.

Han dynasty jar, 206 B.C.–221 A.D.   *Teotihuacán jar, 100–200 A.D.*

30. Open-ended Problem Write three statements about the figure at the right. Give the postulate, definition, or property that makes each statement true.

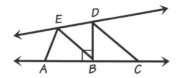

Find the measures of a complementary angle, a supplementary angle, and a vertical angle for the given angle. *(Section 2.1)*

31. $m\angle A = 63°$　　　**32.** $m\angle Y = 14°$　　　**33.** $m\angle H = 2y°$

34. Use a protractor to draw $\triangle ABC$ so that $m\angle A = 35°$, $m\angle B = 70°$, and $m\angle C = 75°$. *(Section 1.6)*

ASSESS YOUR PROGRESS

VOCABULARY

deductive reasoning (p. 112)　　　**definition** (p. 118)
postulate (p. 117)

Tell whether each argument uses *inductive* or *deductive* reasoning. *(Section 3.1)*

1. $\triangle JKL$ is a regular triangle. A regular *n*-gon has *n* lines of symmetry. $\triangle JKL$ has three lines of symmetry.

2. $7 \cdot 3 = 21$　　　$5 \cdot 9 = 45$　　　$-3 \cdot 11 = -33$
The product of two odd numbers is an odd number.

Select the conclusion that makes the statement true. *(Section 3.1)*

3. Quadrilateral *EFGH* is a parallelogram. It has four congruent sides. Therefore, it (*is*, *is not*, *might be*) a rhombus.

4. All of the Hawaiian Islands were formed by volcanoes. Little Diomede is not a Hawaiian Island. Therefore, Little Diomede (*was*, *was not*, *might have been*) formed by a volcano.

Give the postulate(s), definition(s), properties, or previous statement(s) that makes the statement about the diagram true. *(Section 3.2)*

5. $\overrightarrow{FC}$ bisects $\angle DFB$.

6. $\angle AFB \cong \angle EFD$

7. $m\angle AFC = m\angle EFC$

8. $\angle AFE \cong \angle DFB$

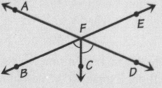

9. Journal Give an example of a postulate, a definition, and a property. Explain how the three are different. Explain how they are the same.

3.3 Paragraph Proof

How long do you think it would take to count out loud to one million? A day? A week? A year? You can use deductive reasoning to find out.

EXAMPLE 1

Writing In a paragraph, prove that you would need more than one week to count to one million without stopping. Assume that you could count one number per second.

SOLUTION

Here is one student's solution.

Given: You can count one number per second.
Prove: Counting to one million will take more than one week.

If you count one number each second, you need at least 1,000,000 s to count to one million. One million seconds is about 16,667 min because 1,000,000 s ÷ 60 s/min ≈ 16,667 min. In hours, this is 16,667 min ÷ 60 min/h ≈ 278 h. There are 24 • 7 = 168 h in one week, so it will take more than one week to count to one million.

THINK AND COMMUNICATE

1. In Example 1, is there another argument you could use to reach the same conclusion? If so, what is it?

2. Is it reasonable to assume that a person could count one number each second? Estimate the amount of time you think it would actually take to count to one million. Explain your reasoning.

To make a convincing argument, you should put your statements in a logical order, give a reason for every statement that you make, and use reasons that everyone agrees are true. A **proof** is a chain of deductive reasoning that follows these rules.

A **theorem** is a conjecture that can be proved to be true. A proof that is written in complete sentences is called a **paragraph proof**. When you write a proof, you make a series of **statements** and give the **reason** for each statement. Given information, postulates, definitions, and theorems can be used as reasons in a proof.

The **hypothesis** is the information you are **given**.

The **conclusion** is the statement you want to **prove**.

> Given: You can count one number per second.
> Prove: Counting to one million will take more than one week.
>
> If you count one number each second, you need at least 1,000,000 s to count to one million. One million seconds is about 16,667 min because 1,000,000 s ÷ 60 s/min ≈ 16,667 min. In hours, this is 16,667 min ÷ 60 min/h ≈ 278 h. There are 24 • 7 = 168 h in one week, so it will take more than one week to count to one million.

For each **statement** of a step in your reasoning, include the **reason** the statement is true.

In the Exploration, you will learn a theorem about the exterior angles of a triangle. You will see how to prove this theorem using facts and properties that you already know.

EXPLORATION
COOPERATIVE LEARNING

Measuring Exterior Angles

Work with another student.
You will need:
- geometry software
- a ruler and protractor

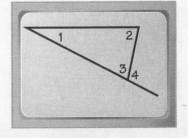

1 Draw a triangle with one side extended, as in the diagram. Find the measures of ∠1, ∠2, ∠3, and ∠4.

2 Round the angle measures to the nearest degree. Record the measures in a table.

m∠1	m∠2	m∠3	m∠4	m∠1 + m∠2
26°	78°	76°	104°	104°
?	?	?	?	?
?	?	?	?	?

3 Move one vertex, or draw new triangles, to form at least five different triangles. Record the angle measures for each triangle in your table.

4 Make a conjecture about the measures of the interior and exterior angles of any triangle. Compare your conjecture with other groups.

The measure of an exterior angle of a triangle is equal to the sum of the measures of the two interior angles that are not adjacent to it.

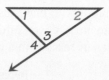

$$m\angle 4 = m\angle 1 + m\angle 2$$

EXAMPLE 2

Write a paragraph proof of the Exterior Angle Theorem.

SOLUTION

Step 1 Write the hypothesis and conclusion.

> **Given:** A triangle with interior angles $\angle 1$, $\angle 2$, $\angle 3$.
> $\angle 4$ exterior to $\angle 3$.
>
> **Prove:** $m\angle 4 = m\angle 1 + m\angle 2$

Step 2 Sketch a diagram based on the given information.

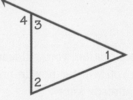

Step 3 Write the proof. Give a reason for each statement.

In any triangle, the sum of the measures of the angles is 180°, so $m\angle 1 + m\angle 2 + m\angle 3 = 180°$. Because $\angle 4$ and $\angle 3$ are a linear pair, $m\angle 4 + m\angle 3 = 180°$. Using the Substitution Property, substitute $m\angle 4 + m\angle 3$ for 180° in the first equation. This gives a new equation: $m\angle 1 + m\angle 2 + m\angle 3 = m\angle 4 + m\angle 3$. Using the Subtraction Property, subtract $m\angle 3$ from both sides to get $m\angle 1 + m\angle 2 = m\angle 4$.

☑ CHECKING KEY CONCEPTS

For Questions 1–3:

a. Sketch a diagram to illustrate the statement.

b. State the hypothesis and the conclusion.

1. If a quadrilateral is a square, then it has four right angles.

2. If a figure is a triangular prism, then it has five faces and nine edges.

3. If a quadrilateral is a rhombus, then its diagonals bisect each other.

Draw an obtuse triangle, an acute triangle, and a right triangle. Tell whether each statement is *True* or *False*.

4. All six exterior angles of a triangle may be obtuse.

5. All six exterior angles of a triangle can have different measures.

3.3 **Exercises and Applications**

Extra Practice
exercises on
page 664

Draw an obtuse triangle, an acute triangle, and a right triangle. Tell whether each statement is *True* or *False*.

1. All six exterior angles of a triangle can be right angles.

2. An exterior angle of a triangle can be smaller than one of the interior angles of the triangle.

3. An exterior angle of a triangle can be equal to one of the interior angles of the triangle.

Find the measure of each indicated angle.

4. $m \angle STV$

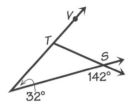

5. $m \angle PQR$

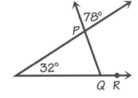

6. $m \angle BAC$

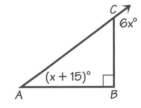

Connection ▶ LITERATURE

When you want to prove something, you must understand what you want to prove and state it clearly. In the book *Where Do You Stop?* by Eric Kraft, Peter Leroy and his classmates discover this when their science teacher reveals the topics for their term projects.

7. **Writing** Each of these questions can be stated in several different ways. Choose one of the questions. State clearly at least two different ways that you can interpret the question.

8. Explain why you should be able to interpret a theorem in only one way.

9. **Open-ended Problem** In *Where Do You Stop?*, the students have to include diagrams that illustrate an experiment to test their answers. Describe an experiment that might be included in one of the papers. Include a diagram for the experiment.

WHERE DOES THE LIGHT GO WHEN THE LIGHT GOES OUT?

WHEN IS NOW?

WHAT IS THE BIGGEST QUESTION OF ALL?

WHY ARE YOU YOU?

WHAT REALLY HAPPENS?

WHERE DO YOU STOP?

10. Complete the proof of the theorem: All right angles are congruent.

Given: ?

Prove: $\angle 1 \cong \angle 2$

From ? , $m\angle 1 = 90°$ and $m\angle 2 = 90°$. So $m\angle 1 = m\angle 2$ by ? . Therefore ? , by the definition of congruent angles.

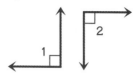
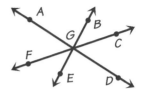

11. a. The measures of what angles are equal to $m\angle BGF - m\angle AGB$?

   b. The measure of which angle is equal to $m\angle BGC + m\angle CGD$? Why?

   c. What is the sum $m\angle CGD + m\angle DGE + m\angle EGF$? Why?

   d. If $\angle EGF \cong \angle AGF$, explain why $\angle BGC \cong \angle CGD$.

INTERVIEW # Mary-Jacque Mann

Look back at the article on pages 108–110.

At the Wildlife Forensics Lab, Mary-Jacque Mann's colleagues may need to identify an animal whose skin was used to create an object. At other times, they may need to identify a whole animal. They might use a key to identify the species of an alligator or a caiman.

1. **Iris of eye is dark brown, red, or orange.**
   - Yes – go to 2
   - No – go to 4

2. **Has fewer than 20 teeth on each side of the lower jaw.**
   - Yes – Chinese Alligator
   - No – go to 3

3. **Ear coverlets contrast with the lighter color of the head.**
   - Yes – Schneider's Caiman
   - No – Cuvier's Caiman

4. **Back is black with distinct yellow or white crossbands.**
   - Yes – go to 5
   - No – one of the other Caiman species

5. **Has three or more large dark spots on the sides of the jaws.**
   - Yes – Black Caiman
   - No – American Alligator

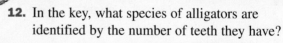

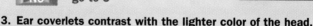

12. In the key, what species of alligators are identified by the number of teeth they have?

13. Identify the animal at the left. Write a statement and reason for each step in the process of identification.

14. Write a paragraph explaining how you would know that an animal is a Cuvier's Caiman.

For Exercises 15–18, tell whether each conclusion follows from the hypothesis. Explain why or why not.

15. △*VJS* is equilateral. Therefore, △*VJS* is isosceles.

16. ∠*H* is a supplement of ∠*M*. ∠*M* is a supplement of ∠*L*. Therefore, ∠*H* is a supplement of ∠*L*.

17. $m\angle 1 + m\angle 3 = 180°$ and $\angle 2 \cong \angle 3$. Therefore, $\angle 2 \cong \angle 1$.

18. In △*UAR*, ∠*U* is a right angle. Therefore, ∠*A* is a complement of ∠*R*.

19. **SAT/ACT Preview** In the figure, if $d = 97$, then $a = \underline{\ ?\ }$.

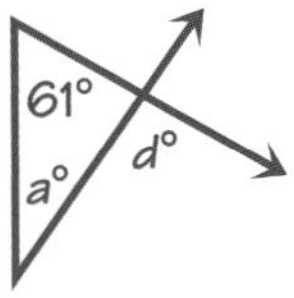

    A. 61   B. 36   C. 38

    D. 83   E. 45

20. **a.** Match each statement with the correct reason in the proof below.

    b. Challenge Arrange the statements and reasons in complete sentences to make a complete paragraph proof. Draw a diagram for the proof.

    Given: ∠1 is an exterior angle of △*XYZ* at *Z*.

    Prove: $m\angle Y = m\angle 1 - m\angle X$

    > Exterior Angle Theorem

    > ∠1 is an exterior angle of △XYZ at Z.

    > Subtraction Property

    > m∠y = m∠1 - m∠x

    > Given

    > m∠1 = m∠x + m∠y

ONGOING ASSESSMENT

21. **Writing** Write a paragraph proof of the theorem: The sum of the measures of the exterior angles of a triangle is 360°. Look at page 81 for an outline of an argument you can use for your proof. Be sure to include a diagram, given and prove statements, and reasons for your proof.

SPIRAL REVIEW

Tell which property makes each statement true. *(Section 3.2)*

22. $\angle ACE \cong \angle ACE$

23. If $\overline{AB} \cong \overline{BC}$ and $\overline{BC} \cong \overline{CD}$, then $\overline{AB} \cong \overline{CD}$.

24. If $m\angle 1 + m\angle 2 = 180°$ and $m\angle 2 = m\angle 5$, then $m\angle 1 + m\angle 5 = 180°$.

25. Describe at least two ways to continue the pattern 2, 4, 8, ... *(Section 1.1)*

26. ∠1 and ∠2 are vertical angles. If $m\angle 1 = 56°$, what is $m\angle 2$? How do you know? *(Section 2.1)*

3.4 Two-Column Proof

A proof shows that a statement is true. Often more than one correct proof can be written. Two proofs of the same theorem may present the same statements and reasons, but look very different. Compare Mary's and Carlotta's proofs of the Vertical Angles Theorem.

Learn how to...
- write a proof in two-column format

So you can...
- use and present convincing arguments in a variety of formats

Given: ∠1 and ∠3 are vertical angles.

Prove: ∠1 ≅ ∠3

$m\angle 1 + m\angle 2 = 180°$ and $m\angle 2 + m\angle 3 = 180°$, because ∠1 and ∠2 are a linear pair, and ∠2 and ∠3 are a linear pair. This means that $m\angle 1 + m\angle 2 = m\angle 2 + m\angle 3$, by the Substitution Property. Subtracting $m\angle 2$ from both sides gives $m\angle 1 = m\angle 3$, so ∠1 ≅ ∠3.

Given: ∠1 and ∠3 are vertical angles.

Prove: ∠1 ≅ ∠3

Statement	Reason
1. ∠1 and ∠3 are vertical angles.	1. Given
2. $m\angle 1 + m\angle 2 = 180°$ $m\angle 2 + m\angle 3 = 180°$	2. Angles in a linear pair are supplements.
3. $m\angle 1 + m\angle 2 = m\angle 2 + m\angle 3$	3. Substitution Property (Step 2)
4. $m\angle 1 = m\angle 3$	4. Subtraction Property
5. ∠1 ≅ ∠3	5. Definition of congruent angles

THINK AND COMMUNICATE

1. How are the two proofs different? How are they the same?

2. Mary and Carlotta proved the same theorem. Which proof do you prefer? Explain why.

Carlotta wrote a **two-column proof**. This format contains the statements and reasons of the proof arranged in two columns. Two-column proofs and paragraph proofs contain the same information, but they are organized differently.

In Chapter 2, you learned that vertical angles are congruent. You can use theorems like this one as reasons in proofs.

Vertical Angles Theorem

Vertical angles are congruent.

If $\angle 1$ and $\angle 2$ are vertical angles, then $\angle 1 \cong \angle 2$.

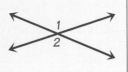

The proof of the Vertical Angles Theorem uses the fact that each of the vertical angles is supplementary to the same angle. You can prove the more general theorem that if two angles are supplementary to the same angle, the two angles are congruent.

EXAMPLE

Write a two-column proof of this theorem: If two angles are supplementary to the same angle, then the angles are congruent.

SOLUTION

Given: $\angle 1$ and $\angle 2$ are supplementary.
$\angle 1$ and $\angle 3$ are supplementary.

Prove: $\angle 2 \cong \angle 3$

Statements	Reasons
1. $\angle 1$ and $\angle 2$ are supplementary. $\angle 1$ and $\angle 3$ are supplementary.	**1.** Given
2. $m\angle 1 + m\angle 2 = 180°$ $m\angle 1 + m\angle 3 = 180°$	**2.** Definition of supplementary angles
3. $m\angle 1 + m\angle 2 = m\angle 1 + m\angle 3$	**3.** Substitution Property (Step 2)
4. $m\angle 2 = m\angle 3$	**4.** Subtraction Property
5. $\angle 2 \cong \angle 3$	**5.** Definition of congruent angles

When you use the Transitive Property or the Substitution Property as a reason, list each step you used.

You will use both equality and congruence in proofs about geometric relationships. Remember that equality refers to numbers and congruence refers to figures.

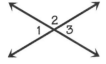

3. How could you use the theorem in the Example on page 131 to prove the Vertical Angles Theorem? Why can't you use the Vertical Angles Theorem to prove the theorem in the Example?

4. In the diagram at the left, ∠1 is supplementary to ∠2 and ∠2 is supplementary to ∠3, but you should not use this diagram for the proof in the Example. Explain why not.

5. In a two-column proof, the reasons are often stated very briefly. In the Example, explain how each reason supports the statement.

☑ CHECKING KEY CONCEPTS

1. Copy and complete the two-column proof.

Given: $m\angle 2 = m\angle 3$

Prove: $m\angle 1 = m\angle 4$

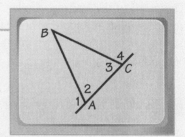

Statements	Reasons
1. $m\angle 1 + m\angle 2 = 180°$ $\quad m\angle 3 + m\angle 4 = 180°$	**1.** Angles in a linear pair are supplements.
2. $m\angle 1 + m\angle 2 = m\angle 3 + m\angle 4$	**2.** _?_
3. _?_	**3.** Given
4. _?_	**4.** Subtraction Property

2. State the theorem proved in Question 1.

State the conclusion that follows from each hypothesis. Give a reason for each conclusion.

3. The sum of the measures of the interior angles of a polygon is 3060°. Therefore, the number of sides is _?_ .

4. If $\angle ABC \cong \angle ABD$ and $\angle ABD \cong \angle CBE$, then _?_ .

3.4 Exercises and Applications

Extra Practice exercises on page 664

For Exercises 1–3, tell what is wrong with each statement. Then rewrite the statement so that it is always true.

1. Angles that are supplementary to the same angle are supplementary.

2. Angles that are congruent are vertical angles.

3. The measure of an exterior angle of a triangle is greater than the measure of each interior angle of a triangle.

4. Copy and complete this two-column proof.

Given: $\angle 2 \cong \angle 3$

Prove: $\angle 1 \cong \angle 4$

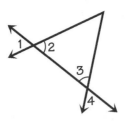

Statements	Reasons
1. ?	**1.** Given
2. ? ; ?	**2.** Vertical Angles Theorem
3. $\angle 1 \cong \angle 4$	**3.** ?

Look back at the article on pages 108–110.

INTERVIEW Mary-Jacque Mann

One of Mary-Jacque Mann's important projects involves finding a way to tell if an object is made of elephant ivory. For Exercises 5–9, use this part of the ivory identification chart that she helped develop.

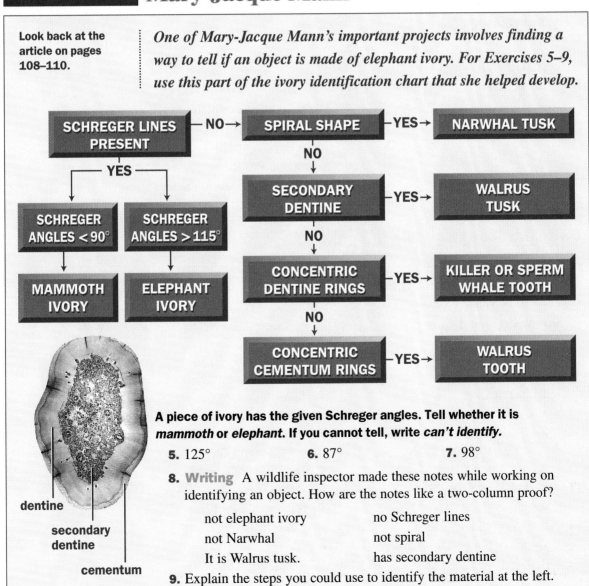

A piece of ivory has the given Schreger angles. Tell whether it is *mammoth* or *elephant*. If you cannot tell, write *can't identify*.

5. 125°   **6.** 87°   **7.** 98°

8. Writing A wildlife inspector made these notes while working on identifying an object. How are the notes like a two-column proof?

not elephant ivory	no Schreger lines
not Narwhal	not spiral
It is Walrus tusk.	has secondary dentine

9. Explain the steps you could use to identify the material at the left.

dentine

secondary dentine

cementum

Use the information in the diagram to find the measure of each angle.

10. $\angle FGE$ **11.** $\angle BGA$ **12.** $\angle BGC$ **13.** $\angle BGD$

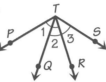

For Exercises 14–16:

a. Sketch and label a diagram for the proof of each theorem.

b. Write the Given and Prove statements for each theorem.

14. The angles in a linear pair are supplementary.

15. Two perpendicular lines form congruent adjacent angles.

16. Complements of the same angle are congruent.

17. Challenge Complete the proof.

    Given: $\angle PTR \cong \angle QTS$

    Prove: $\angle 1 \cong \angle 3$

Statements	Reasons
1. ?	1. Given
2. $m\angle PTR = m\angle QTS$	2. ?
3. $m\angle PTR = m\angle 1 + m\angle 2$	3. Angle Addition Postulate
4. ?	4. Angle Addition Postulate
5. ?	5. Substitution Property (Steps ?, ?, and ?)
6. ?	6. Subtraction Property
7. ?	7. Definition of congruent angles

O N G O I N G   A S S E S S M E N T

18. Open-ended Problem Arrange the statements and reasons to make a valid two-column proof. Can the steps be arranged in a different order and still be valid? Explain.

    Given: $\angle 1 \cong \angle 3$

    Prove: $\angle 2 \cong \angle 4$

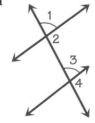

Statements	Reasons
1. $\angle 2 \cong \angle 4$	A. Substitution Property (Step(s) ?)
2. $\angle 1 \cong \angle 3$ ($m\angle 1 = m\angle 3$)	B. Angles in a linear pair are supplements.
3. $m\angle 1 + m\angle 2 = m\angle 1 + m\angle 4$	C. Definition of congruent angles
4. $m\angle 2 = m\angle 4$	D. Subtraction Property
5. $m\angle 1 + m\angle 2 = 180°$ $m\angle 3 + m\angle 4 = 180°$	E. Given
6. $m\angle 1 + m\angle 2 = m\angle 3 + m\angle 4$	F. Substitution Property (Step(s) ?)

For Exercises 19–21, rewrite each statement in conditional form. Underline the hypothesis and circle the conclusion of each statement. *(Section 1.3)*

19. Congruent angles have equal measures.

20. Two lines parallel to the same line are parallel to each other.

21. The diagonals of a parallelogram bisect each other.

Sketch each situation. *(Sections 1.4 and 1.5)*

22. Points *X*, *Y*, and *Z* lie on line *m*.

23. $\overline{PQ} \cong \overline{RS}$

ASSESS YOUR PROGRESS

VOCABULARY

proof (p. 124)	**conclusion of a proof** (p. 125)
theorem (p. 125)	**statement** (p. 125)
paragraph proof (p. 125)	**reason** (p. 125)
hypothesis of a proof (p. 125)	**two-column proof** (p. 131)

1. Write a paragraph proof of this theorem: If two angles are supplementary to the same angle, then the angles are congruent. Use the Example on page 131 as a guide. *(Section 3.3)*

2. Rewrite the proof in Example 2 on page 126 as a two-column proof. *(Section 3.4)*

3. Complete this proof. *(Section 3.4)*

Given: $\angle 1 \cong \angle 3$

Prove: $\angle 2 \cong \angle 4$

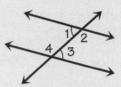

Statements	Reasons
1. $\angle 1 \cong \angle 3$	**1.** _?_
2. $m\angle 1 = m\angle 3$	**2.** _?_
3. $m\angle 1 + m\angle 2 = 180°$ $m\angle 3 + m\angle 4 = 180°$	**3.** _?_
4. $m\angle 1 + m\angle 2 = m\angle 3 + m\angle 4$	**4.** _?_
5. $m\angle 1 + m\angle 2 = m\angle 1 + m\angle 4$	**5.** _?_
6. $m\angle 2 = m\angle 4$	**6.** _?_
7. $\angle 2 \cong \angle 4$	**7.** _?_

4. **Journal** Is *a paragraph proof* or *a two-column proof* easier for you to read and understand? Which type of proof do you think is easier to write? Explain your answers.

3.5 Converses of Statements

Learn how to...
- write the converse of a conditional statement

So you can...
- recognize and use converses in logical arguments

It's true that if you're in Peru, then you're in South America. Is it also true that if you're in South America, then you must be in Peru? You will explore statements like this, called *converses*, in this lesson.

Angel Falls is the highest waterfall in the world.

South America's rain forests have more than 40,000 varieties of plants.

The Atacama Desert is one of the driest deserts in the world.

THINK AND COMMUNICATE

Tell whether each statement is *True* or *False*.
If it is false, give a counterexample.

1. a. If I am in the Atacama Desert, then I am in one of the driest places in the world.

   b. If I am in one of the driest places in the world, then I am in the Atacama Desert.

2. a. If I am in Paraguay, then I am not at the seashore.

   b. If I am not at the seashore, then I am in Paraguay.

3. What relationship do you see between the pairs of statements in Questions 1 and 2? If one statement in a pair is true, must the other statement also be true?

You can write the **converse** of a conditional statement by interchanging the
hypothesis and conclusion of the original statement.

Statement

If you are at Angel Falls, then you are at the tallest waterfall in the world.

Converse

If you are at the tallest waterfall in the world, then you are at Angel Falls.

EXAMPLE

**Use this theorem: If a triangle is a right triangle, then two of its angles
are complementary.**

a. Draw a diagram illustrating the theorem. State the hypothesis
and the conclusion of the theorem.

b. Draw a diagram illustrating the converse of the theorem. State
the hypothesis and the conclusion of the converse.

SOLUTION

a.

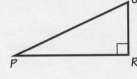

Given: $\triangle PQR$ is a right triangle, with
a right angle at R.

Prove: $\angle P$ and $\angle Q$ are complementary.

b.

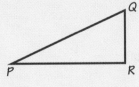

Given: In $\triangle PQR$, $\angle P$ and $\angle Q$ are
complementary.

Prove: $\triangle PQR$ is a right triangle.

The given information does
not tell you that the triangle
is a right triangle.

> **WATCH OUT!**
>
> Do not assume that
> the converse of a
> theorem is true. The
> converse of a theorem
> must be proved before
> you know that it is true.

THINK AND COMMUNICATE

**Use this statement and diagram: If two lines
are parallel, then they do not intersect.**

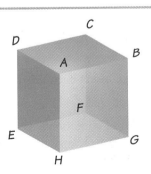

4. Name two lines that are parallel. Name
two lines that are skew.

5. Do skew lines ever intersect? Are skew
lines ever parallel?

6. Give the converse of the statement.
Is the converse true or false?

French Guiana
Suriname
Guyana
Venezuela
Colombia
Ecuador

Peru
Bolivia
Chile
Paraguay
Uruguay
Argentina

Brazil

- ■ Dutch
- □ English
- ▨ French
- ▨ Portuguese
- ▨ Spanish
- ■ Spanish, Guarani
- ■ Spanish, Quechua,
  Aymara

☑ CHECKING KEY CONCEPTS

Use the map of South America at the left. Write the converse of
each statement. Is the statement *True* or *False*? the converse?

1. If the official language is English, then you are in Guyana.

2. If you are in Peru, then an official language is Quechua.

3. If you are in Brazil, then the official language is
Portuguese.

4. Write a true conditional statement about South America.
Write the converse of your statement. Is it *True* or
False? (You may also use the map on page 136.)

For Questions 5–7:
a. Rewrite each statement as a conditional.
b. Write the converse of each statement.

5. A square is a rectangle.

6. I will buy a car if I earn enough money.

7. The sum of the measures of the angles in a triangle is 180°.

3.5 Exercises and Applications

*Extra Practice
exercises on
page 664*

Write the converse of each statement. Tell whether the converse is
True or *False*.

1. BASEBALL If the American League team won the first four games of
the World Series, then they are the champions.

2. If the opposite sides of a quadrilateral are parallel, then the quadrilateral
is a parallelogram.

3. If a rectangle is 2 ft long and 3 ft wide, then its area is 6 ft^2.

For Exercises 4–9:
a. Rewrite the statement as a conditional.
b. Write the converse of the statement.

4. The opposite angles of a parallelogram are congruent.

5. The supplements of congruent angles are congruent.

6. The supplement of an obtuse angle is an acute angle.

7. Isosceles triangles are not scalene.

8. The diagonals of a parallelogram bisect each other.

9. The diagonals of a square are congruent and perpendicular.

10. Research Find three examples of conditional statements in
newspapers or magazines. Give the converse of each statement.
Is the statement true? Is the converse true? Explain your reasoning.

11. Writing Match each statement with the correct reason. Then rewrite the statements in the correct order to prove the theorem in the Example.

Given: In $\triangle PQR$, $\angle R$ is a right angle.

Prove: $\angle P$ and $\angle Q$ are complementary angles.

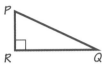

Statements	Reasons
1. $m\angle P + m\angle Q = 90°$	**A.** Definition of complementary angles
2. $m\angle P + m\angle Q + m\angle R = 180°$	**B.** Given
3. $m\angle R = 90°$	**C.** Definition of right angle
4. $\angle P$ and $\angle Q$ are complements.	**D.** Substitution Property (Step(s) ?)
5. In $\triangle PQR$, $\angle R$ is a right angle.	**E.** Subtraction Property
6. $m\angle P + m\angle Q + 90° = 180°$	**F.** The sum of the angle measures of a triangle is 180°.

12. Challenge Write a proof of the converse of the theorem in Exercise 11.

Connection ▶ **LANGUAGE**

Translating between languages presents several difficult problems. Words often have more than one meaning, so a translator must decide which definition is intended from the context.

For Exercises 13–16:

a. Tell whether the statement is *always*, *sometimes*, or *never* true.

b. Give the converse. Tell whether the converse is *True* or *False*.

13. If a Japanese sentence uses the word "sumu," an English translation would use the word "live."

14. If an English sentence uses the word "live," then a Japanese translation could use the word "ikiru."

Japanese	English
sumu	to live, reside **EXAMPLE:** I live on Maple Lane.
ikiru	to live, be alive, breathe **EXAMPLE:** I live in the twentieth century.

A. check (money)

B. check (investigate, look over, inspect)

15. If I want to sign the word "check," I would use the sign shown in diagram A.

16. If I use the sign shown in B, I want to sign the word "check."

17. Open-ended Problem Give an example of an English word that has several different meanings. Write a true conditional statement about using this word. Is the converse true?

18. Give the reasons for the following proof:

Given: $\angle FMR \cong \angle FRM$
$\angle 1 \cong \angle 3$

Prove: $\angle 2 \cong \angle 4$

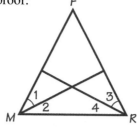

Statements	Reasons
1. $\angle FMR \cong \angle FRM$; $\angle 1 \cong \angle 3$	**1.** ?
2. $m\angle FMR = m\angle FRM$	**2.** ?
3. $m\angle FMR = m\angle 1 + m\angle 2$; $m\angle FRM = m\angle 3 + m\angle 4$	**3.** ?
4. $m\angle 1 + m\angle 2 = m\angle 3 + m\angle 4$	**4.** ?
5. $m\angle 1 = m\angle 3$	**5.** ?
6. $m\angle 1 + m\angle 2 = m\angle 1 + m\angle 4$	**6.** ?
7. $m\angle 2 = m\angle 4$	**7.** ?
8. $\angle 2 \cong \angle 4$	**8.** ?

ONGOING ASSESSMENT

19. Open-ended Problem Give an example of a true conditional statement with a true converse that you might hear at lunch or at a sports event. Give an example of a true conditional statement with a false converse.

SPIRAL REVIEW

20. Use this theorem: If two exterior angles at different vertices of a triangle are congruent, then two interior angles of the triangle are congruent. *(Section 3.4)*

   a. Sketch and label a diagram to illustrate a proof of this theorem.

   b. Write the Given and Prove statements for this theorem.

Find each length or angle measure in rectangle ABCD. *(Section 2.5)*

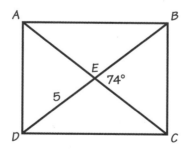

21. $m\angle BCD$ **22.** $m\angle CED$

23. BD **24.** EB

Sketch each triangle, if possible. If it is not possible, explain why not. *(Section 2.2)*

25. an acute scalene triangle

26. a right equiangular triangle

27. an isosceles obtuse triangle

28. an obtuse equilateral triangle

3.6 The Pythagorean Theorem

The *Pythagorean theorem* states a special relationship among the lengths of the sides of a right triangle. This theorem is named after the Greek mathematician Pythagoras (around 560–480 B.C.), although the relationship was known even earlier by other cultures.

EXPLORATION
COOPERATIVE LEARNING

Proving the Pythagorean Theorem

Work in a group of three students.
You will need:
- paper • scissors • ruler

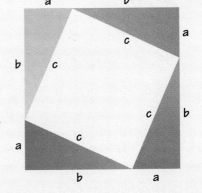

1 Cut out four identical right triangles. In a right triangle, each of the two shorter sides is called a **leg**. The side opposite the right angle is called the **hypotenuse**.

2 Arrange your triangles in a square. Measure each length and find the area of the square. Find the sum of the areas of the right triangles. The area of a triangle is $\frac{1}{2} \times$ height $\times$ base.

3 Subtract the areas in Step 2 to show that the area of the central square equals the square of the length of the hypotenuse of the right triangles.

$$\begin{array}{c} \text{Area of the} \\ \text{central square} \end{array} = \begin{array}{c} \text{Area of the} \\ \text{large square} \end{array} - \begin{array}{c} \text{Area of the} \\ \text{4 right triangles} \end{array}$$

4 Compare your results with those of other groups.

5 Use *a, b,* and *c* for the lengths of the sides of the triangles, as shown. Use algebra to show that in any right triangle, the sum of the squares of the lengths of the legs equals the square of the length of the hypotenuse.

The Pythagorean Theorem

In a right triangle, the sum of the squares of the lengths of the legs is equal to the square of the length of the hypotenuse.

$$a^2 + b^2 = c^2$$

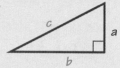

EXAMPLE 1

The legs of a right triangle are *a* and *b* units long. The hypotenuse is *c* units long. Find the unknown length for each right triangle.

a. $a = 12$, $b = 9$

b. $a = 3$, $c = 2\sqrt{6}$

SOLUTION

a.
$$a^2 + b^2 = c^2$$
$$12^2 + 9^2 = c^2$$
$$144 + 81 = c^2$$
$$\sqrt{225} = \sqrt{c^2}$$
$$\pm 15 = c$$

The length is positive, so $c = 15$.

> Find the square roots of both sides of the equation. The **radical** symbol, $\sqrt{\ }$, indicates the square root of a number.

b.
$$a^2 + b^2 = c^2$$
$$3^2 + b^2 = \left(2\sqrt{6}\right)^2$$
$$9 + b^2 = 24$$
$$b^2 = 15$$
$$\sqrt{b^2} = \sqrt{15}$$
$$b = \pm\sqrt{15}$$
$$b \approx \pm 3.87$$

The length is positive, so $b = \sqrt{15}$, or about 3.9.

In part (b) of Example 1, you can choose to give the answer as a radical or as a decimal approximation. In a real-world situation, the approximation that a calculator gives is usually accurate enough. In a mathematical situation, using radicals may make a calculation easier. Radicals in a solution are usually *simplified*.

EXAMPLE 2   Connection: Algebra

Simplify each expression.

a. $\pm\sqrt{48}$

b. $-\sqrt{7}$

SOLUTION

a.
$$\pm\sqrt{48} = \pm\sqrt{16 \cdot 3}$$
$$= \pm\sqrt{16} \cdot \sqrt{3}$$
$$= \pm 4\sqrt{3}$$

b. Because 7 is a prime number, $-\sqrt{7}$ is already simplified.

> **Toolbox p. 700**
> *Simplifying Radicals*

THINK AND COMMUNICATE

Tell whether each expression is in simplified form.

1. $2\sqrt{12}$ **2.** $\sqrt{51}$ **3.** $3\sqrt{19}$ **4.** $\sqrt{27}$

5. Two sides of a right triangle are 4 and 5 units long. Explain why the third side could be either 3 or $\sqrt{41}$ units long.

The Converse of the Pythagorean Theorem

The converse of the Pythagorean theorem is true. You can use the converse to determine whether a triangle is a right triangle.

> **Converse of the Pythagorean Theorem**
>
> If a, b, and c are the lengths of the sides of a triangle, and $a^2 + b^2 = c^2$, then the triangle is a right triangle.

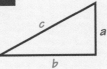

EXAMPLE 3   Application: Carpentry

Richard Mason is making a birdhouse for a Great Crested Flycatcher. The peak of the roof should be a right angle. He put the roof on the feeder and found that it does not fit correctly. What should he do to fix this problem?

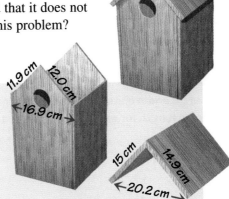

SOLUTION

Use the converse of the Pythagorean theorem to find out if the dimensions of the front of the birdhouse and the peak of the roof form a right triangle.

Front

$$a^2 + b^2 \stackrel{?}{=} c^2$$
$$11.9^2 + 12.0^2 \stackrel{?}{=} 16.9^2$$
$$141.61 + 144 \stackrel{?}{=} 285.61$$
$$285.61 = 285.61$$

This is a right triangle.

Roof peak

$$a^2 + b^2 \stackrel{?}{=} c^2$$
$$14.9^2 + 15^2 \stackrel{?}{=} 20.2^2$$
$$222.01 + 225 \stackrel{?}{=} 408.04$$
$$447.01 \neq 408.04$$

This is not a right triangle.

The angle at the top of the front is a right angle. The pieces of the roof do not meet at a right angle. Richard should change the angle of the roof to match the front of the birdhouse.

Use a calculator to approximate each square root to the nearest hundredth. Then simplify each expression.

1. $\sqrt{49}$ **2.** $\pm\sqrt{54}$ **3.** $-\sqrt{24}$

Find each unknown length.

4. 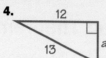   **5.** 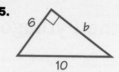   **6.**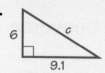

Can the given numbers be the lengths of the sides of a right triangle?

7. 20 ft, 21 ft, 29 ft   **8.** 8 m, 10 m, 12 m   **9.** 10 cm, 24 cm, 36 cm

---

3.6 Exercises and Applications

Extra Practice
exercises on
page 665

ALGEBRA Use a calculator to approximate each square root to the nearest hundredth. Then simplify each expression.

1. $\sqrt{121}$ **2.** $-\sqrt{80}$ **3.** $\pm\sqrt{27}$

4. $\pm\sqrt{47}$ **5.** $-\sqrt{405}$ **6.** $\sqrt{196}$

Can the given numbers be the lengths of the sides of a right triangle?

7. 1.2 m, 1.6 m, 2 m   **8.** 8 in., 9 in., 12 in.   **9.** 5.6 cm, 2.8 cm, 6.4 cm

10. 7, 25, 24   **11.** $6\frac{2}{3}$, 4, $5\frac{1}{3}$ **12.** $\sqrt{2}$, $\sqrt{3}$, $\sqrt{5}$

The legs of a right triangle are a and b units long. The hypotenuse is c units long. Find the unknown length for each right triangle, to the nearest hundredth.

13. $a = 5$, $c = 12$ **14.** $a = 6$, $b = 8$ **15.** $b = 9$, $c = 15$

16. $a = 21$, $b = 28$ **17.** $b = 5$, $c = 6$ **18.** $a = 273$, $b = 136$

Find each unknown length.

19.   **20.** 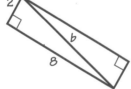   **21.**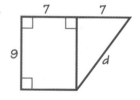

22. 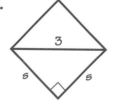   **23.**   **24.**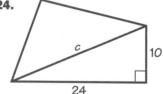

Tell whether or not each parallelogram can be a rectangle.

25.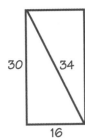

30   34

16

26.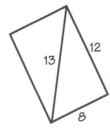

13   12

8

27.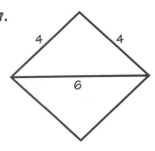

4   4

6

28. Open-ended Problem Each diagonal of a rectangle is 25 cm long. Find three possible pairs of lengths for the sides of the rectangle.

29. ARCHEOLOGY One of the two temples at Tikal in Guatemala has 49 steps leading up to the top of the pyramid. Each step is approximately 0.31 m deep and 0.36 m high. Estimate the height and depth of the entire flight of stairs. Estimate *d*, the length of the flight of stairs. Explain your reasoning.

30. a. Find the length of the diagonal, *d*, of the rectangular prism. (*Hint:* First find the value of *e*.)

b. Use the result of part (a) to find the length of the longest object that will fit inside a rectangular box with the given dimensions.

c. Find the length of the longest object that will fit inside a cube with edges that are 4 ft long.

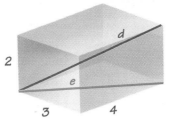

2

3   4

31. Writing Use rectangle *ABCD*. Find the lengths of the diagonals, $\overline{AC}$ and $\overline{BD}$. What do you notice? Do you think this will be true for all rectangles? Explain your reasoning.

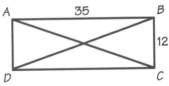

A   35   B

12

D   C

Plot each group of points on graph paper. Connect the points with segments. Find the length of each segment.

32. $A(3, 4)$, $B(6, 4)$, $C(6, 8)$

33. $C(2, 7)$, $D(2, -4)$, $E(10, -4)$

34. $M(-3, 9)$, $N(5, -6)$

35. $P(12, 7)$, $Q(5, 4)$

36. Writing Explain the difference between the Pythagorean theorem and its converse. Give an example of when to use each one.

ARCHITECTURE In the blueprint, the *pitch triangle* gives the ratio of the length of the roof to the height of the roof. Sketch each triangle and give its dimensions.

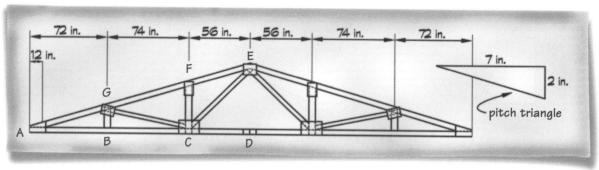

pitch triangle

37. △ADE   **38.** △ACF   **39.** △BGC   **40.** △CED

Connection ▶ HISTORY

Three integers that can be the lengths of the sides of a right triangle are called a *Pythagorean triple*. A Babylonian tablet made between 1800 and 1650 B.C. suggests that the Babylonians knew the Pythagorean theorem.

Caspian Sea

Mediterranean Sea

Babylonian Empire in 1686 B.C.

Persian Gulf

Red Sea

41. Spreadsheets The tablet gives two of the three numbers in 15 Pythagorean triples. Some of these pairs are given in the table. Find the missing number in each triple.

leg	hypotenuse
119	169
3367	4825
4601	6649
65	97
319	481
2291	3541
481	769
4961	8161

42. One way to find a Pythagorean triple is to choose two integers p and q, then find the values of $2pq$, $p^2 - q^2$, and $p^2 + q^2$. Choose three different pairs of values for p and q. Find the values above and show that they are Pythagorean triples.

43. Challenge The three numbers in a *reduced* Pythagorean triple have no common factors except 1. You can find a reduced Pythagorean triple by using values of p and q that have no common factors and are not both odd in the expressions given in Exercise 42.

a. Choose values of p and q that meet the conditions above. Show that they produce a reduced Pythagorean triple.

b. Use algebra to show that a Pythagorean triple found by using values of p and q with a common factor, n, where n is a positive integer greater than 1, will not be reduced.

44. ALGEBRA Show that $(2pq)^2 + (p^2 - q^2)^2 = (p^2 + q^2)^2$ when $q = 1$.

45. a. DESIGN The Danish artist Per Arnoldi designed a chair made of triangles and squares formed by one rectangle of metal. Find the value of d.

b. Draw a net for the chair. Give all dimensions, including the folded diagonal.

c. Use heavy paper or cardboard to make a model of the chair.

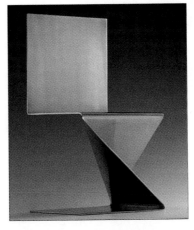

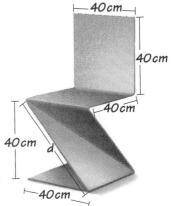

46. 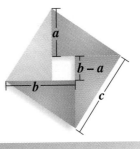 **Technology** Use geometry software.

a. Draw a quadrilateral. Draw the diagonals of the quadrilateral. Measure the length of each segment.

b. Repeat part (a) with at least five different quadrilaterals. Record the lengths in a table.

c. For what quadrilaterals will the equation $a^2 + b^2 + c^2 + d^2 = e^2 + f^2$ be true? Explain your reasoning.

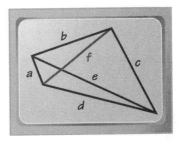

ONGOING ASSESSMENT

47. a. Visual Thinking Explain how this diagram illustrates a proof of the Pythagorean theorem. Write the algebraic steps for the proof.

b. How is this diagram different from the one that you used in the Exploration on page 141? Explain.

SPIRAL REVIEW

Write each statement as a conditional. Circle the hypothesis and underline the conclusion. *(Section 1.3)*

48. Angles in a linear pair are supplementary angles.

49. Perpendicular lines intersect at right angles.

50. Write the converse of each statement in Exercises 48 and 49. *(Section 3.5)*

3.7 Negations and Contrapositives

Learn how to...

- **determine whether a triangle is acute or obtuse from the lengths of its sides**

- **write inverses and contrapositives of statements**

So you can...

- **recognize inverses and contrapositives in logical arguments**

You can use the converse of the Pythagorean theorem to find out whether a triangle is a right triangle. Can the squares of the lengths of the sides tell you anything more about a triangle that is *not* a right triangle?

EXPLORATION
COOPERATIVE LEARNING

Analyzing Triangles

Work with another student.
You will need:
- graph paper
- scissors

1 Cut ten squares out of graph paper. Use several different side lengths.

2 Choose three squares. If you can, make a triangle with the three squares. Let c be the length of the longest side of the triangle. Let a and b be the lengths of the two shorter sides.

Lengths of sides	$a^2 + b^2$	c^2	Type of triangle
3, 4, 5	25	25	right
4, 4, 5	32	25	acute
?	?	?	?

3 Make at least ten different triangles. Copy and complete the table.

4 What conjectures can you make about the relationship between the squares of the side lengths and the type of triangle?

5 Use your conjectures to predict whether a triangle with the given side lengths is *right*, *obtuse*, or *acute*.

   a. 11, 11, 15

   b. 11, 13, 18

   c. 5, 11, 12

One conjecture you might have made in the Exploration is "If $\triangle ABC$ is **not** a right triangle, then $a^2 + b^2 \neq c^2$." The hypothesis of this statement is the **negation** of the statement "$\triangle ABC$ is a right triangle." You can use a letter to indicate a statement. For example, the negation of "P" is "**not** P." You can use negations to change a conditional into a different statement.

CONDITIONAL

If P, then Q.   If a serving of a food has less than 140 mg of sodium, then it may be labeled *low sodium*.

INVERSE

If *not P*, then *not Q*.   If a serving of a food does not have less than 140 mg of sodium, then it may *not* be labeled *low sodium*.

CONVERSE

If Q, then P.   If a food may be labeled *low sodium*, then it has less than 140 mg of sodium per serving.

CONTRAPOSITIVE

If *not Q*, then *not P*.   If a food may *not* be labeled *low sodium*, then it does *not* have less than 140 mg of sodium per serving.

THINK AND COMMUNICATE

1. The conditional statement above is true. Is the inverse *True* or *False*? Is the converse? Is the contrapositive?

2. Use the statement "If $AB = CD$, then $\overline{AB} \cong \overline{CD}$." Is the inverse *True* or *False*? Is the converse? Is the contrapositive?

EXAMPLE

Rewrite this statement as a conditional. Give the inverse, converse, and contrapositive of the statement.

An equilateral triangle is an isosceles triangle.

SOLUTION

Conditional: If a triangle is equilateral, then it is isosceles.

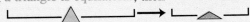

Inverse: If a triangle is *not* equilateral, then it is *not* isosceles.

Converse: If a triangle is isosceles, then it is equilateral.

Contrapositive: If a triangle is *not* isosceles, then it is *not* equilateral.

Tell whether the triangle with sides of the given lengths is *right*, *obtuse*, or *acute*.

   1. 10, 14, 17   **2.** 11, 12, 17   **3.** $\sqrt{3}$, 5, 6

Tell whether each statement from the Example on page 149 is *True* or *False*. If a statement is false, give a counterexample.

    4. the conditional   **5.** the inverse   **6.** the contrapositive

3.7 | **Exercises and Applications**

Extra Practice exercises on page 665

For Exercises 1–8, tell whether the triangle with sides of the given lengths is *right*, *obtuse*, or *acute*.

  1. 2, 3, 4   **2.** 3, 4, 5   **3.** 4, 5, 6   **4.** 1.5, 2.0, 2.5

  5. 0.6, 0.8, 0.9   **6.** 1, 1, $\sqrt{2}$ **7.** 1, $\sqrt{2}$, 2   **8.** $\sqrt{2}$, 2, $\sqrt{5}$

9. Open-ended Problem Using 3 and 7 as the lengths of two sides each time, draw an acute, a right, and an obtuse triangle. Estimate the length of the third side of each triangle, then measure the diagrams to check your estimates.

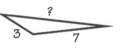

For Exercises 10–13, tell whether each statement is *True* or *False*.

10. An obtuse triangle may have sides of lengths 4, 7, and 10.

11. If the lengths of the sides of a triangle are 2 m, 7 m, and 8 m, then the triangle is an acute triangle.

12. In any triangle, the square of the length of one side is equal to the sum of the squares of the lengths of the other two sides.

13. If two sides of a triangle are 5 cm and 12 cm long, then the third side must be 13 cm long.

14. SAT/ACT Preview If A = $m \angle CED$ and B = 90°, then:
   A. A < B   **B.** A > B   **C.** A = B
   D. relationship cannot be determined

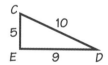

15. a. TRAVEL The highways Route 14, Route 92, and Route 175 form a triangle. Do these highways form a right triangle? Explain your reasoning.

   b. Writing Interstate highways that run north-south have odd numbers. Interstate highways that run east-west have even numbers. Do these three highways follow this numbering rule? Why do you think Route 14 has an even number?

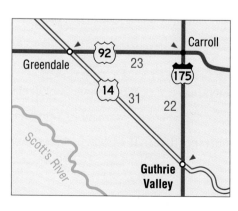

Write the inverse and the contrapositive of each statement.

16. If it is raining, then the streets are wet.

17. If it snows, then classes are canceled.

18. If a triangle is an acute triangle, then it has no obtuse angles.

19. The opposite sides of a parallelogram are congruent.

20. Perpendicular lines form adjacent supplementary angles.

For Exercises 21–23, write the inverse of each statement. Is the inverse *True* or *False*?

21. If two lines are perpendicular, then they form congruent adjacent angles.

22. The diagonals of a square are perpendicular.

23. Vertical angles are congruent.

24. Which of these statements has the same meaning as "The diagonals of a rhombus are perpendicular"?

A. If $ABCD$ is a rhombus, then $\overline{AC}$ is perpendicular to $\overline{BD}$.
B. If $\overline{AC}$ is perpendicular to $\overline{BD}$, then $ABCD$ is a rhombus.
C. If $ABCD$ is not a rhombus, then $\overline{AC}$ is not perpendicular to $\overline{BD}$.
D. If $\overline{AC}$ is not perpendicular to $\overline{BD}$, then $ABCD$ is not a rhombus.
E. If $ABCD$ is a parallelogram, then $\overline{AC}$ is not perpendicular to $\overline{BD}$.
F. If $\overline{AC}$ is not perpendicular to $\overline{BD}$, then $ABCD$ is a parallelogram.

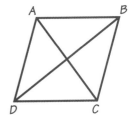

Connection ▶ AUTO MAINTENANCE

If you drive a car, chances are good that one day it will not start. The first thing to check is whether you have left the lights turned on and run down the battery. If the engine cranks when you turn the key, but the car does not start, then you can check several possible sources of the problem.

For Exercises 25–27, write the inverse and the contrapositive of the statement.

25. If the battery needs to be recharged, then the lights will dim when the key is turned.

26. If the lights are bright when the key is turned, then the spark plugs may be bad.

27. If the battery and spark plugs are working correctly, then fuel may not be getting to the engine.

28. Open-ended Problem Write another if-then statement based on auto maintenance. Is its converse true?

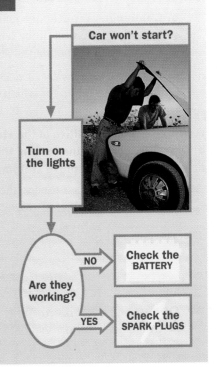

Without measuring, tell whether the side lengths of each triangle appear to be accurate. Explain your reasoning.

29.

30.

31.

32.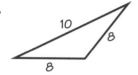

33.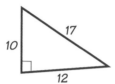

34.

35. **Cooperative Learning** Work with another person. One of you should do part (a) and the other should do part (b). Work together on parts (c) and (d). You will need a ruler and a protractor.

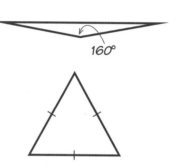

  a. Use the obtuse triangle. Find the ratio of the sum of the lengths of the two shorter sides to the length of the longest side. Experiment with other obtuse triangles. Write an inequality to describe the lower limit of the ratio.

  b. Use the equilateral triangle. What is the ratio of the sum of the lengths of two sides to the length of the third side?

  c. **Challenge** Write an inequality to describe all the possible values of the ratio of the sum of the lengths of the two shorter sides of a triangle to the length of the longest side.

  d. Did you use *inductive* or *deductive* reasoning to answer part (c)? Can you find a triangle for which your inequality does not work?

For Exercises 36–38:
a. Write the contrapositive of the statement.
b. Draw a diagram for the proof of the contrapositive.
c. Write the given and prove statements for the proof.

36. If a quadrilateral is a square, then it is a parallelogram.

37. If two angles are right angles, then they are congruent.

38. The diagonals of a parallelogram bisect each other.

ONGOING ASSESSMENT

39. **Open-ended Problem** Write two conditional statements. One should be from a real-world situation, and one should be mathematical. Write the inverse, converse, and contrapositive of each statement. Tell whether each of these statements is *True* or *False*. Give a counterexample for each false statement.

Find the mean of each pair of numbers. *(Toolbox, page 708)*

40. 7, 3   **41.** $-4, 8$ **42.** 2.7, 6.5

Tell whether the given lengths can be the sides of a right triangle.
(Section 3.6)

43. 15, 20, 25   **44.** 6, 9, 11   **45.** 1.6, 3, 3.4

ASSESS YOUR PROGRESS

VOCABULARY

converse (p. 137)   **negation** (p. 149)

leg of a right triangle (p. 141)   **inverse** (p. 149)

hypotenuse (p. 141)   **contrapositive** (p. 149)

radical (p. 142)

Find each unknown length. *(Section 3.6)*

1.   **2.** 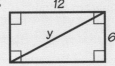   **3.**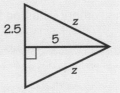

**Tell whether the triangle with sides of the given lengths is *acute, obtuse,*
or *right*.** *(Sections 3.6 and 3.7)*

4. 20, 29, 21   **5.** 6, 12, 8   **6.** 3.6, 7.2, 9.8

7. 4.3, 6.5, 5.8   **8.** $\sqrt{16}, \sqrt{12}, \sqrt{7}$ **9.** $4\frac{1}{2}, 3\frac{1}{3}, 7\frac{1}{2}$

**For Exercises 10 and 11, write the inverse, the converse, and the
contrapositive of each statement.** *(Sections 3.5 and 3.7)*

10. The diagonals of a rectangle are congruent.

11. If the exterior rays of two adjacent angles are perpendicular, then the
angles are complements.

12. Use the statement: If *PQRS* is a parallelogram, then $\angle P \cong \angle R$.

   a. Write the contrapositive of the statement.

   b. Draw a diagram for a proof of the contrapositive.

   c. Write the Given and Prove statements for a proof of the
contrapositive. *(Section 3.7)*

13. Journal Make a chart with examples of the four types of conditional
statements that you have learned. Assume that the conditional is true.
Indicate on the chart whether each of the others is *always, never,* or
sometimes true.

Classifying Information

How does a music store owner decide how to display instruments? Instruments could be grouped by the type of music performed on them or they could be grouped into types of instrument. The owner might use a classification system. We use systems like this to classify just about everything to help keep track of what we have or know.

PROJECT GOAL   Make and use a classification system for something of interest to you.

Using a Classification System

1. Classify a guitar as a *stringed*, *wind*, *percussion*, or *keyboard* instrument by using the diagram below.

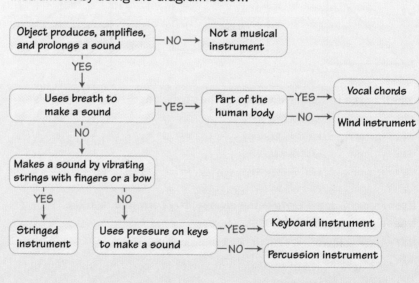

2. Choose some other instruments to classify using the diagram above. Does the diagram classify all of the instruments the way you expect them to be classified? How could you deal with instruments that fit into more than one category?

3. Discuss other ways the diagram could be arranged. Would it make sense to have more than two options in some places? What changes would you make to help classify instruments correctly?

Where Does it Belong?

Develop a system for classifying a group of objects that you collect. If you do not have a collection, choose something that you are interested in, such as types of music or Internet sites.

Draw a diagram for your classification system. Show how to classify at least five different items in the group using your system. If you need to, adjust your classification system.

Presenting Your Project

Make a display of your classification system and how it works. Explain what group of items your system classifies.

Look at some of the other displays. Analyze how well you think one other classification system works. Give a copy of your analysis to the group that created that system.

You may want to extend your project and explore one of the ideas below:

- Interview someone who works in marketing. Describe how they classify customers for various markets.

- Expand the musical instrument diagram. Add branches after the existing instrument categories.

- Research how astronomers classify stars. What type of criteria do they use to categorize a star?

Self-Assessment

What are some important points to remember when developing a classification system? What items are the most difficult to classify? Why?

3

Review

STUDY TECHNIQUE

Reread each section of the chapter. Pay special attention to the section titles and the description, below each section number, of what you should learn in that section. Review any topics that you found especially difficult. Write a short summary of each section to use as a study guide.

VOCABULARY

deductive reasoning (p. 112)
postulate (p. 117)
definition (p. 118)
proof (p. 124)
theorem (p. 125)
paragraph proof (p. 125)
statement (p. 125)
reason (p. 125)
hypothesis of a proof (p. 125)

conclusion of a proof (p. 125)
two-column proof (p. 131)
converse (p. 137)
leg of a right triangle (p. 141)
hypotenuse (p. 141)
radical (p. 142)
negation (p. 149)
inverse (p. 149)
contrapositive (p. 149)

SECTIONS 3.1 and 3.2

You can use inductive reasoning to make conjectures based on examples or patterns. **Deductive reasoning** is used to give a convincing argument that a conjecture is true. Use given information, definitions, and accepted assumptions, **properties** and **postulates**, in a logical chain of reasoning.

Some important postulates and properties include the following:

Reflexive Property	Addition Property
Substitution Property	Subtraction Property
Symmetric Property	Segment Addition Postulate
Transitive Property	Angle Addition Postulate

SECTIONS 3.3 *and* 3.4

A **proof** is a convincing argument that can be used to show that an important conjecture, called a **theorem**, is true.

Given: $\angle 1$ and $\angle 2$ are right angles.

Prove: $\angle 1 \cong \angle 2$

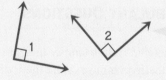

Use a diagram to represent the given information.

$\angle 1$ and $\angle 2$ are right angles, so by the definition of a right angle, $m\angle 1 = 90°$ and $m\angle 2 = 90°$. By the Substitution Property, $m\angle 1 = m\angle 2$. By the definition of congruent angles, $\angle 1 \cong \angle 2$.

In a **two-column proof**, the statements and reasons are arranged in columns.

SECTIONS 3.5, 3.6, *and* 3.7

You can use the Pythagorean theorem to find the lengths of the sides of a right triangle. The converse of the Pythagorean theorem tells you whether a triangle is a right triangle.

- If $\angle C$ is a right angle, then $a^2 + b^2 = c^2$.
- If $a^2 + b^2 = c^2$, then $\angle C$ is a right angle.

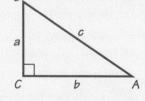

You can also use inequalities related to the converse of the Pythagorean theorem to tell if a triangle is acute or obtuse. Suppose the lengths of the sides of a triangle are a, b, and c, and c is the length of the longest side.

- If $a^2 + b^2 > c^2$, then the triangle is an acute triangle.
- If $a^2 + b^2 < c^2$, then the triangle is an obtuse triangle.

You can rewrite a conditional statement several different ways.

Conditional	If P, then Q.	If $x = 3$, then $x^2 = 9$.
Converse	If Q, then P.	If $x^2 = 9$, then $x = 3$.
Inverse	If *not P*, then *not Q*.	If $x \neq 3$, then $x^2 \neq 9$.
Contrapositive	If *not Q*, then *not P*.	If $x^2 \neq 9$, then $x \neq 3$.

If a conditional statement is true, then its contrapositive is also true. The converse and inverse of the statement are either both true or both false.

Assessment

VOCABULARY QUESTIONS

1. Sketch a right triangle. Indicate which sides are the legs and which is the hypotenuse. State the Pythagorean theorem in terms of your triangle.

2. Complete each statement. To show that a theorem is true, write a _?_. The given information is the _?_ and the statement that you want to prove is the _?_.

3. Write the inverse of this conditional: If point Y is between points X and Z, then $XY + YZ = XZ$.

SECTIONS 3.1 *and* 3.2

For Questions 4 and 5, select the conclusion that makes the statement true.

4. Deanna used deductive reasoning to show that two angles are congruent. The angles (*must, may,* or *cannot*) be congruent.

5. If a parallelogram is a square, then it has four right angles. Parallelogram *ABCD* is not a square. Therefore, *ABCD* (*has, does not have,* or *may have*) four right angles.

6. Tell whether the argument uses *inductive* or *deductive* reasoning.

$$1 = 1^2$$
$$1 + 3 = 4 = 2^2$$
$$1 + 3 + 5 = 9 = 3^2$$
$$1 + 3 + 5 + 7 = 16 = 4^2$$
$$\text{So } 1 + 3 + 5 + 7 + 9 \quad = 5^2$$

Give the postulate, definition, property, or previous statement that makes the statement about the diagram true.

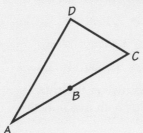

7. There is only one line that contains points A and B.

8. If $AB = BC$, then B is the midpoint of $\overline{AC}$.

9. If $\overline{AB} \cong \overline{BC}$ and $\overline{BC} \cong \overline{CD}$, then $\overline{AB} \cong \overline{CD}$.

10. $m \angle A + m \angle C + m \angle D = 180°$

SECTIONS 3.3 *and* 3.4

11. Copy and complete the proof.

Given: $\angle 2 \cong \angle 3$

Prove: $\angle 1 \cong \angle 4$

Statements	Reasons
1. $\angle 2 \cong \angle 3$	**1.** ?
2. $\angle 1 \cong \angle 2, \angle 3 \cong \angle 4$	**2.** ?
3. $\angle 1 \cong \angle 3$	**3.** ?
4. ?	**4.** ?

12. a. Sketch and label a diagram for the proof of the theorem: The diagonals of a rhombus are perpendicular.

b. Write the Given and Prove statements for the theorem.

SECTIONS 3.5, 3.6, *and* 3.7

13. Use the theorem in Question 12.

a. Write the converse, the inverse, and the contrapositive of the theorem.

b. Choose one of your statements from part (a). Draw a diagram for the proof of the statement you chose. Write the Given and Prove statements for the proof.

14. Open-ended Problem Give a conditional that is true, but whose converse is false.

15. A rectangle is 10 cm long and 8 cm wide. Find the length of each diagonal.

16. Find the length h in the diagram at the right.

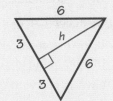

Tell whether a triangle with sides of the given lengths is *right*, *obtuse*, or *acute*.

17. 2.1, 2.8, 3.6   **18.** $\sqrt{7}, 3, 4$ **19.** 2, 9, 9

PERFORMANCE TASK

20. Use this diagram. Write at least two true conditionals and two false conditionals about the diagram. For each conditional, write the conditional, the converse, the inverse, and the contrapositive, and tell whether each is *True* or *False*. Do you notice a pattern in your results?

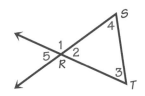

Cumulative Assessment

CHAPTERS 1–3

CHAPTER 1

For Questions 1–3, use the table below. Suppose that each person in a group of people shakes hands with every other person once.

Number of people	2	3	4	5	6	7
Number of handshakes	1	3	6	10	?	?

1. **LOGICAL REASONING** Use inductive reasoning to predict the total number of handshakes for 6 people and for 7 people.

2. Make a conjecture about the number of handshakes added when one more person joins a group of n people. Identify the hypothesis and conclusion of your conjecture.

3. Writing How could you test the conjecture that you made in Question 2? Would your method allow you to be sure that your conjecture is correct? Why or why not?

4. Open-ended Problem Draw a figure that has one line of symmetry.

5. Open-ended Problem Write a conditional statement that is false. Give a counterexample to show that the statement is false.

For Questions 6–9, use the diagram.

6. If $\overline{BD}$ bisects $\overline{AC}$, $AE = 3x + 4$, and $AC = 38$, find the value of x.

7. Give as many different names as you can for $\angle 1$.

8. Find the measures of $\angle ABC$ and $\angle ACF$.

9. Name an angle bisector and two congruent angles.

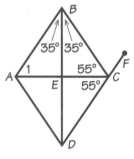

CHAPTER 2

Complete each statement with *always*, *sometimes*, or *never*.

10. Two angles formed by perpendicular lines are $\underline{\ ?\ }$ complementary.

11. An equilateral triangle is $\underline{\ ?\ }$ isosceles.

12. An isosceles triangle is $\underline{\ ?\ }$ an acute triangle.

13. The diagonals of a rhombus $\underline{\ ?\ }$ bisect each other.

14. Writing Suppose *ABCD* is a parallelogram. Explain how you know that ∠*A* and ∠*B* are supplementary angles.

For each figure, find the values of *x* and *y*.

15.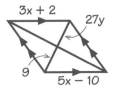
3x + 2
27y
9
5x − 10

16.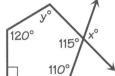
120°
y°
115° x°
110°

17.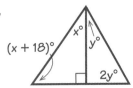
x°
(x + 18)°
y°
2y°

18. Sketch an equilateral pentagon and a regular quadrilateral. Describe the lines of symmetry of each figure and mark any congruent sides or angles.

19. Open-ended Problem Sketch a prism whose bases are isosceles triangles and sketch a net for the prism.

CHAPTER 3

20. Writing Write a convincing argument that uses deductive reasoning to explain why the opposite sides of a rectangle are parallel.

Give the postulate, definition, property, or theorem that makes the statement about the diagram true.

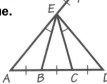

21. $m \angle A + m \angle D = m \angle DEF$

22. $m \angle AEC = m \angle DEB$

23. $AB + BC = AC$

24. Draw and label a diagram for the proof of the statement: The bisectors of the angles in a linear pair form a right angle. Write the Given and Prove statements for the proof.

25. Write a paragraph proof of the theorem: The measure of each angle of an equiangular triangle is 60°.

26. Write the inverse, the converse, and the contrapositive of the following statement: The diagonals of a rhombus are perpendicular.

Find the value of *x*.

27.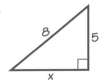
8
5
x

28.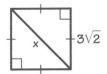
x
3√2

29.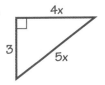
4x
3
5x

30. Open-ended Problem Use the theorem in Question 26. Draw a rhombus and label possible lengths for the sides and the diagonals of the rhombus.

4 Coordinates in Geometry

Making the Virtual Look Real

Marc Hannah

> "The most important thing, in terms of choosing a career, is that you enjoy what you do."

Marc Hannah has liked math and science ever since he was young. "I always found those courses easier and more interesting," he says. Hannah majored in electrical engineering in college and then studied computer graphics. Teaming up with university professor James Clark, Hannah began working on the "geometry engine," a computer system that performs calculations to describe objects and display them realistically on a computer screen from any perspective. This collaboration eventually led to the co-founding of Silicon Graphics, a world-leading company that makes computer workstations.

Marc Hannah and a co-worker discuss a design for a computer chip.

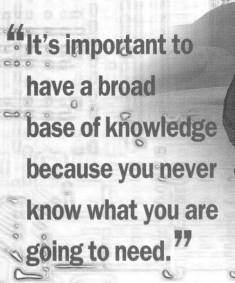

> **"It's important to have a broad base of knowledge because you never know what you are going to need."**

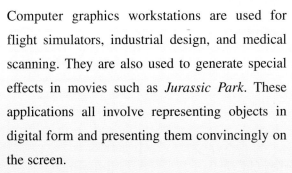

Visualizing Data in Three Dimensions

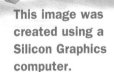

Computer graphics workstations are used for flight simulators, industrial design, and medical scanning. They are also used to generate special effects in movies such as *Jurassic Park*. These applications all involve representing objects in digital form and presenting them convincingly on the screen.

"Whether we're trying to depict a brain in a CAT scan or the view a pilot might see flying into Chicago, it's all done by choosing coordinates in three dimensions for every point in the image," Hannah explains. "These complicated pictures are broken down into geometric shapes, such as triangles and other polygons, which are illuminated and manipulated on the screen."

Building a Faster Machine

As a Vice President and Chief Scientist for Silicon Graphics, Hannah tries to predict what computer features will be needed in the future and the best way to build them. He designs computer chips that will perform calculations quickly. "I try to figure out how our customers use our systems, what they'd like to do, and how to build the fastest machine possible."

Hannah spends a lot of time working with various design groups in the company. "We put together complex systems, which means that no single individual can do it alone," he says. "It's important to be able to work with a team of people and express your ideas clearly."

163

Creating an Image

Coordinates are used in Marc Hannah's work to display images on a computer screen. To draw the interior of a triangle, the computer must color all the dots on the screen (called *pixels*) that lie inside the triangle.

"The way I define a triangle is by specifying the *x*- and *y*-coordinates of its vertices," says Hannah.

"Next I have to find out which pixels are within the boundaries of the triangle."

One way to do this is to find equations for the lines that pass through the vertices. Then the computer can calculate whether any particular pixel is above or below each of the three lines.

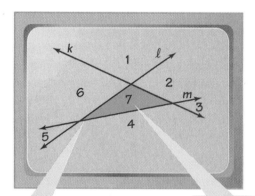

Each line passes through two vertices of the triangle.

Region 7, the interior of the triangle, is all the points that are *below* line *k*, *below* line *ℓ*, and *above* line *m*.

Explore and Connect

Marc Hannah near one of his company's workstations

1. Writing In the diagram above, each region can be described by referring to its place *above* or *below* each of the three lines. Describe the positions of Regions 1–6 this way.

2. Project Draw a polygon on a coordinate plane. Show how you can describe all the points in its interior by referring to the lines that pass through its vertices.

3. Research Find a movie in which computer graphics were used to create special effects. Rent or borrow a video of it and present selections to your class. Describe how the computer-generated special effects are combined with the real actors and settings.

Mathematics & Marc Hannah

In this chapter, you will learn more about how mathematics is related to computer graphics.

Related Exercises

Section 4.1
• Exercises 18–20

Section 4.5
• Exercises 4–7

Section 4.6
• Exercises 25–28

4.1 Finding Distances and Midpoints

Learn how to...

- **find the distance between two points**
- **find the coordinates of the midpoint of a segment**

So you can...

- **determine distances, such as at a swimming competition**
- **classify polygons by the lengths of their sides**

At hand-timed swimming competitions, timers at the end of each lane start their watches when they hear the starter's signal. If the starter is closer to the swimmers than to the timers, each swimmer will hear the signal before his or her timer does. This gives each swimmer a different advantage, which must be calculated to adjust the race times.

The lanes begin 1 m from the edge. Each lane is 2.4 m wide.

Each lane has a swimmer and a timer at opposite ends.

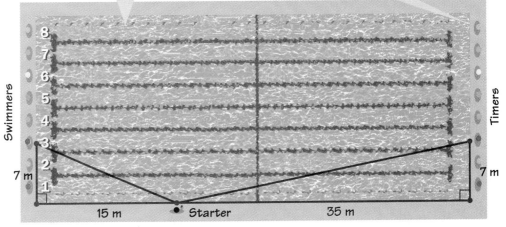

THINK AND COMMUNICATE

1. Use the Pythagorean theorem to find the distance, to the nearest hundredth of a meter, between the starter and the swimmer in lane 3.

2. Use the Pythagorean theorem to find the distance, to the nearest hundredth of a meter, between the starter and the timer in lane 3.

3. Use 346 m/s for the speed of the starting signal. How long does it take the sound, to the nearest thousandth of a second, to reach the swimmer in lane 3? to reach the timer in lane 3?

4. How much sooner, to the nearest thousandth of a second, does the swimmer in lane 3 hear the signal than the timer? How should you adjust the swimmer's race time to account for this advantage?

To find distances on a coordinate plane, the Pythagorean theorem is expressed as the Distance Formula. *Subscripts* are used to name the coordinates of points. You say "*x* sub 1" and "*y* sub 1" for a point with coordinates (x_1, y_1).

The Distance Formula

The distance between the points $A(x_1, y_1)$ and $B(x_2, y_2)$ is:

$$AB = \sqrt{(x_2 - x_1)^2 + (y_2 - y_1)^2}$$

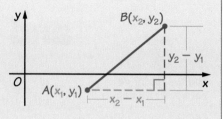

EXAMPLE 1　Application: Sports

Find the distance, to the nearest hundredth of a meter, between:

a. the starter and the swimmer in lane 8.

b. the starter and the timer in lane 8.

SOLUTION

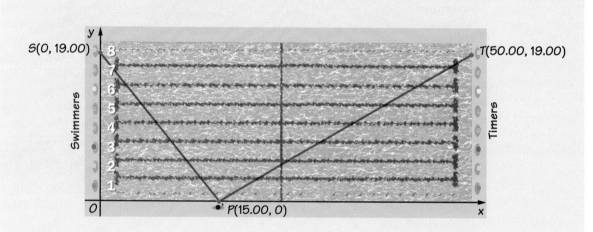

Place a coordinate grid with (0, 0) at one corner of the pool.

a. The starter is at $P(15.00, 0)$ and the swimmer in lane 8 is at $S(0, 19.00)$.

$$PS = \sqrt{(0 - 15.00)^2 + (19.00 - 0)^2}$$
$$= \sqrt{(-15.00)^2 + 19.00^2}$$
$$= \sqrt{225 + 361}$$
$$\approx 24.21$$

The swimmer in lane 8 is about 24.21 m from the starter.

b. The starter is at $P(15.00, 0)$ and the timer in lane 8 is at $T(50.00, 19.00)$.

$$PT = \sqrt{(50.00 - 15.00)^2 + (19.00 - 0)^2}$$
$$= \sqrt{35.00^2 + 19.00^2}$$
$$= \sqrt{1225 + 361}$$
$$\approx 39.82$$

The timer in lane 8 is about 39.82 m from the starter.

THINK AND COMMUNICATE

5. a. How long does it take the starting signal to reach the swimmer in lane 8? the timer in lane 8? (Use 346 m/s for the speed of sound.)

   b. How much sooner does the swimmer in lane 8 hear the signal than the timer in lane 8? How should you adjust the swimmer's race time to account for this advantage?

6. Refer to *Think and Communicate* Questions 4 and 5. Compare the time advantages of the swimmer in lane 3 and the swimmer in lane 8.

You can also use coordinates to find midpoints easily. The coordinates of the midpoint of a segment are the means of the coordinates of the endpoints.

The Midpoint Formula

The midpoint of the segment joining the points $A(x_1, y_1)$ and $B(x_2, y_2)$ has these coordinates:

$$\text{coordinates of midpoint of } \overline{AB} = \left(\frac{x_1 + x_2}{2}, \frac{y_1 + y_2}{2} \right)$$

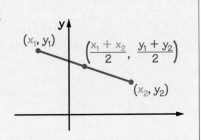

EXAMPLE 2

Use quadrilateral OBCD. Find the coordinates of the midpoint of $\overline{BD}$.

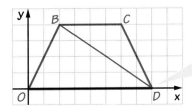

Unless marked otherwise, each grid square represents one unit.

SOLUTION

Use the Midpoint Formula with $B(2, 4)$ and $D(8, 0)$.

$$\text{coordinates of midpoint of } \overline{BD} = \left(\frac{x_1 + x_2}{2}, \frac{y_1 + y_2}{2} \right)$$

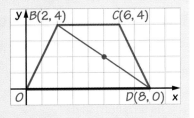

$$= \left(\frac{2 + 8}{2}, \frac{4 + 0}{2} \right)$$

$$= (5, 2)$$

The coordinates of the midpoint of $\overline{BD}$ are $(5, 2)$.

You can use the Distance Formula to classify triangles and other polygons based on the lengths of their sides.

EXAMPLE 3

The vertices of $\triangle ABC$ are $A(-3, 1)$, $B(2, 6)$, and $C(3, 0)$. Give the most specific name for $\triangle ABC$.

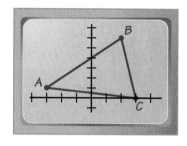

SOLUTION

Use the Distance Formula to find the lengths of the sides of $\triangle ABC$.

$$AB = \sqrt{(2 - (-3))^2 + (6 - 1)^2}$$
$$= \sqrt{5^2 + 5^2}$$
$$= \sqrt{50}$$
$$= 5\sqrt{2}$$

To find AB, use $A(-3, 1)$ as (x_1, y_1) and $B(2, 6)$ as (x_2, y_2).

$$BC = \sqrt{(3 - 2)^2 + (0 - 6)^2}$$
$$= \sqrt{1^2 + (-6)^2}$$
$$= \sqrt{37}$$

To find BC, use $B(2, 6)$ as (x_1, y_1) and $C(3, 0)$ as (x_2, y_2).

$$AC = \sqrt{(3 - (-3))^2 + (0 - 1)^2}$$
$$= \sqrt{6^2 + (-1)^2}$$
$$= \sqrt{37}$$

To find AC, use $A(-3, 1)$ as (x_1, y_1) and $C(3, 0)$ as (x_2, y_2).

Since $BC = AC$, $\triangle ABC$ is isosceles. Since $AC \neq AB$, $\triangle ABC$ is not equilateral.

THINK AND COMMUNICATE

7. Find the coordinates of the midpoint of $\overline{OC}$, the other diagonal of quadrilateral $OBCD$ in Example 2.

8. How can you check whether the isosceles triangle in Example 3 is a right triangle?

9. Draw the isosceles triangle in Example 3 on graph paper and draw a line of symmetry for it. Where does the line of symmetry you drew intersect $\overline{AB}$?

Use the diagram. Find each length.

1. AB **2.** CD

3. EF **4.** GH

Find the coordinates of the midpoint of each segment.

5. $\overline{AB}$ **6.** $\overline{CD}$

7. $\overline{EF}$ **8.** $\overline{GH}$

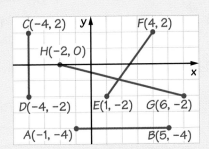

4.1 Exercises and Applications

Extra Practice exercises on page 665

For Exercises 1–3, refer to the swimmers in Example 1.

1. Cooperative Learning Work in a group of three people. Assume the starter is at $P(15.00, 0)$.

   a. Copy and complete the swimmers' time advantage table below. Each person should complete two rows. Two rows are already completed.

   b. Discuss your results. Which swimmers have greater time advantages? Which have smaller time advantages?

Lane	Swimmer position	Timer position	PS	PT	$\dfrac{PS}{346}$	$\dfrac{PT}{346}$	Advantage
8	$S(0, 19.00)$	$T(50.00, 19.00)$	24.21	39.82	0.070	0.115	0.045
7	$S(0, 16.60)$	?	?	?	?	?	?
6	$S(0, 14.20)$	?	?	?	?	?	?
5	$S(0, 11.80)$	?	?	?	?	?	?
4	$S(0, 9.40)$	?	?	?	?	?	?
3	$S(0, 7.00)$	$T(50.00, 7.00)$	16.55	35.69	0.048	0.103	0.055
2	$S(0, 4.60)$	?	?	?	?	?	?
1	$S(0, 2.20)$	?	?	?	?	?	?

2. Spreadsheets Use a spreadsheet to create a table like the one above. Include cells where you can specify the coordinates of P, the position of the starter. How does changing the position of the starter along the edge of the pool affect the results in your table?

3. Open-ended Problem Describe where the starter P could be positioned so that no swimmer has a time advantage.

Find the lengths of the sides of each polygon whose vertices are given. Give the most specific name for each polygon.

4. $A(2, -2)$, $B(-1, 1)$, $C(11, 7)$

5. $D(2, -1)$, $E(4, 2)$, $F(7, 0)$

6. $G(-5, 0)$, $H(1, -3)$, $K(4, 3)$

7. $J(-3, 0)$, $K(0, 3)$, $L(2, 1)$, $M(-1, -2)$

8. $P(2, -4)$, $Q(4, 0)$, $R(2, 2)$, $S(0, 0)$

9. $W(-3, 3)$, $X(2, 2)$, $Y(3, -3)$, $Z(-2, -2)$

Find the coordinates of the midpoint of the segment with each given pair of endpoints.

10. $A(0, 0)$, $B(8, 4)$

11. $C(-5, 7)$, $D(-5, 9)$

12. $E(2, -7)$, $F(-2, 7)$

13. $G(2, 5)$, $H(5, 2)$

14. $N(1, 1)$, $J(-3, -3)$

15. $K(7, 4)$, $L(-3, 1)$

Investigation For Exercises 16 and 17, use paper folding to locate midpoints.

Step 1 Draw a pair of axes on graph paper. Graph any two points, A and B.

Step 2 Carefully fold A onto B. Make a crease.

Step 3 Label the intersection of the crease and $\overline{AB}$ as midpoint M.

16. Open-ended Problem Choose two points, A and B, that have integer coordinates. Fold to locate the midpoint. Label the midpoint M and give its coordinates.

17. Writing Sally says that if a point P lies on the crease shown above, then $PA = PB$. Do you agree or disagree? Use the crease you made in Exercise 16 and the Distance Formula to support your answer.

INTERVIEW Marc Hannah

Look back at the article on pages 162–164.

Computer graphics programmers sometimes use this variation of the Midpoint Formula:

$$\left(x_1 + \frac{1}{2}(x_2 - x_1), y_1 + \frac{1}{2}(y_2 - y_1)\right)$$

18. Use the formula to find the midpoint of the segment with endpoints $(2, 6)$ and $(6, 14)$. Does your result agree with what you expect?

19. ALGEBRA Show that the formula above is the same as the Midpoint Formula shown on page 167.

20. Check that the formulas below give the coordinates of the points that divide the segment with endpoints $(0, 9)$ and $(3, 15)$ into three congruent parts.

$$\left(x_1 + \frac{1}{3}(x_2 - x_1), y_1 + \frac{1}{3}(y_2 - y_1)\right) \text{ and } \left(x_1 + \frac{2}{3}(x_2 - x_1), y_1 + \frac{2}{3}(y_2 - y_1)\right)$$

The Chinese puzzle game called *ch'i ch'ae pan* has seven pieces with straight edges. You may know this puzzle by the name *tangram*. The pieces can be arranged into a square and many other shapes.

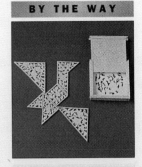

This set of Chinese tangram pieces is made of ivory.

Use the diagram of the tangram pieces in the form of a square.

21. Find the coordinates of all 10 points at the vertices of the tangram pieces.

22. The small square has edge length 1. Find the edge lengths of the other six pieces.

23. Use cardboard or heavy paper to make a set of tangram pieces. Use your answers to Exercise 22 to mark the lengths of the sides on your pieces.

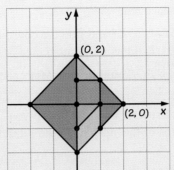

Use the diagram of the tangram pieces in the form of a running figure.

24. Each vertex marked in red on the diagram is a midpoint of a tangram edge. Find the coordinates of the unlabeled midpoints.

25. **Challenge** Find coordinates for all the vertices of the running figure.

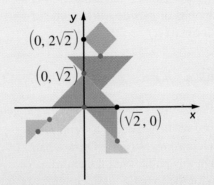

Use your tangram pieces to build each figure. Then sketch it on a coordinate plane.

26.

27.

28.

29. **Cooperative Learning** Work with another person. Create your own tangram figure and draw its outline on graph paper. Label the coordinates of some of the vertices. Exchange papers with the other person and try to build the figure you are given.

30. **Open-ended Problem** Sketch three different segments that have $(0, 0)$ as a midpoint. Write the coordinates of the endpoints of each segment. What do you notice about the coordinates?

31. **Writing** Suppose points A and B have integer coordinates. Under what circumstances will the midpoint of $\overline{AB}$ have integer coordinates?

Technology **For Exercises 32–34, use geometry software.**

Step 1 Draw a quadrilateral that has no parallel or congruent sides and no right angles. Label the vertices *P*, *Q*, *R*, and *S*.

Step 2 Find the length of each side of *PQRS*.

Step 3 Draw the diagonals of *PQRS* and find the coordinates of their midpoints. Label the midpoints *L* and *M*.

Step 4 Move the vertices of *PQRS* until *L* and *M* are the same point.

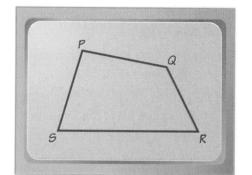

32. Repeat the steps above several times. What do you notice about the lengths of the sides of *PQRS* after Step 4?

33. What type of quadrilateral is *PQRS* after Step 4? Write a conjecture that describes what you have discovered.

34. Investigation Draw another quadrilateral. Locate and connect the midpoints of its sides. What type of quadrilateral is formed?

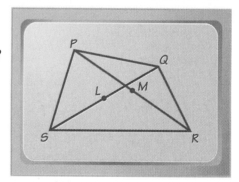

ONGOING ASSESSMENT

35. Open-ended Problem Use the triangle on the geoboard.

a. Choose one peg of the geoboard to represent the point with coordinates (0, 0). Draw a pair of axes through this point. Find the coordinates of each vertex of the triangle.

b. Find the lengths of the sides of the triangle.

c. Tell whether the triangle is *equilateral*, *isosceles*, or *scalene*.

d. Suppose you choose a different peg to represent the point with coordinates (0, 0). Which of the results above would change? Which would not?

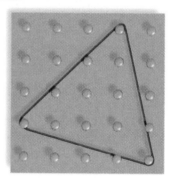

SPIRAL REVIEW

Tell whether the triangle with sides of the given lengths is *right*, *obtuse*, or *acute*. *(Section 3.7)*

36. $3, 7, \sqrt{58}$ **37.** $2, \sqrt{7}, 2\sqrt{3}$ **38.** $0.5, 1, 1$

Open-ended Problems **Sketch an example of each figure.** *(Section 2.3)*

39. a pentagon with exactly one line of symmetry   **40.** an equilateral parallelogram

Find the slope of each line. *(Toolbox, p. 697)*

41. $y = -4x + 7$ **42.** $2y = 3x + 8$ **43.** $x + y = 5$

4.2 Equations of Lines

Learn how to...
- **find the slope of a segment or a line**
- **write equations of lines in slope-intercept form**

So you can...
- **graph and compare equations of lines**
- **investigate geometric relationships using lines**

Maybe you would like to go to the seashore but live too far inland. Or maybe you would like to try downhill skiing but live in a region without mountains or snow. At indoor beaches and ski resorts that overcome the limitations of geography and climate, people swim and ski all year.

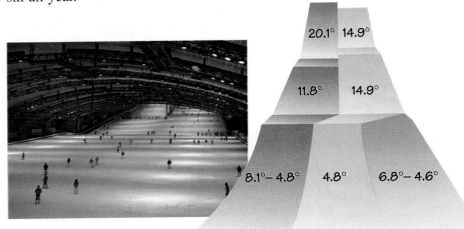

The world's largest indoor ski slope is located in Funabashi, Japan. It is called the *LaLaport Skidome* and has three courses—a red or *expert* course, a yellow or *intermediate* course, and a blue or *beginners'* course. Each course consists of alternating flat and sloped regions. A diagram of the red course is shown below. All distances are rounded to the nearest meter.

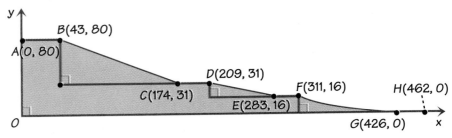

THINK AND COMMUNICATE

1. How many meters does the red course drop in its first sloped part, between point B and point C?

2. What is the horizontal distance from B to C?

You can express how steep the first sloped part of the red course is by using the *slope formula*.

The Slope Formula

The **slope** m of a line containing the points (x_1, y_1) and (x_2, y_2) is:

$$m = \frac{y_2 - y_1}{x_2 - x_1}$$

The slope of a **horizontal** line is 0.

The slope of a **vertical** line is **undefined**.

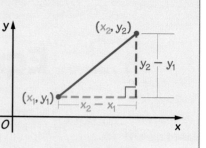

EXAMPLE 1   Application: Recreation

Find the slope of the first sloped part of the red course of the LaLaport Skidome.

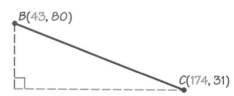

SOLUTION

The coordinates of point B are $(43, 80)$. The coordinates of point C are $(174, 31)$.

Use the slope formula.

$$m = \frac{y_2 - y_1}{x_2 - x_1}$$

$$= \frac{31 - 80}{174 - 43}$$

Substitute **43** for x_1, **80** for y_1, **174** for x_2, and **31** for y_2.

$$= -\frac{49}{131}$$

$$= -0.374$$

The slope of the first sloped part of the red course is about -0.37.

THINK AND COMMUNICATE

3. What does the negative sign mean in the value of the slope in Example 1?

4. In Example 1, Meg used $(174, 31)$ for (x_1, y_1) and $(43, 80)$ for (x_2, y_2). Will this affect the value of the slope? Explain why or why not.

5. Use the diagram on page 173. Calculate the slope of the second sloped part of the red course, between points D and E.

The y-coordinate of the point where a graph crosses the y-axis is called a **y-intercept**. If the y-intercept of a line is b, then $(0, b)$ is on the line. Substitute $(0, b)$ for (x_1, y_1) in the definition of slope and solve for y:

$$m = \frac{y - b}{x - 0} \rightarrow y = m(x - 0) + b$$

This leads to the *slope-intercept form* of an equation of a line.

Slope-Intercept Form of an Equation of a Line

The equation of a line with slope m and y-intercept b can be expressed in **slope-intercept form**:

$$y = mx + b$$

slope ↗   ↖ y-intercept

An equation of a **horizontal** line that contains $(0, b)$ is $y = b$.

An equation of a **vertical** line that contains $(a, 0)$ is $x = a$.

EXAMPLE 2   Application: Recreation

Write an equation for the line that represents the first sloped part of the red course in the LaLaport Skidome.

SOLUTION

Point C with coordinates $(174, 31)$ is on the line. From Example 1, the slope of this part of the course is about -0.37. Use the slope-intercept form. Find the value of b, the y-intercept:

$$y = mx + b$$
$$31 = (-0.37)(174) + b$$
$$b \approx 95$$

Substitute 174 for x, 31 for y, and -0.37 for m.

An equation of the line that represents the first sloped part of the red course is $y = -0.37x + 95$.

THINK AND COMMUNICATE

6. Use the equation from Example 2 to estimate the height of a point 100 m from the beginning of the red course.

7. Suppose that the first sloped part of the red course were extended to the left until it met the y-axis. How high would the top of the course be? How does this compare with its actual height?

1. Find the slope of the line that contains the points with coordinates (5, 1) and (7, 1).

2. Write an equation of the line that contains the point with coordinates (3, 4) and has slope 2.

3. Write an equation of the line that contains the point with coordinates (0, 2) and has slope −1.

4. Use the diagram on page 173. Suppose the LaLaport Skidome had a single course that started at a height of 80 m and continued at a constant slope to point G. What would be the slope of the course?

4.2 **Exercises and Applications**

Extra Practice
exercises on
page 666

Find the slope of the line that contains each pair of points given.

1. (0, 1) and (2, 9)
2. (2, 1) and (4, − 3)
3. (1, 3) and (5, − 2)
4. (− 4, 7) and (1, − 1)
5. (3, 2) and (− 3, − 1)
6. (2, − 5) and (− 1, − 5)

7. **ACCESSIBILITY** Ramps are cut into curbs to allow people in wheelchairs to enter and exit sidewalks. In the United States, ramps have sloped sides whenever the sidewalk at the top of the ramp is narrower than 4 ft. The slopes may not be greater than $\frac{1}{12}$.

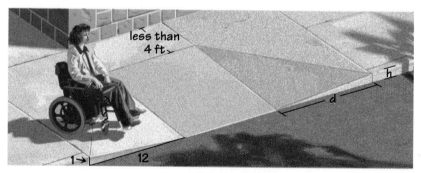

a. Suppose h is 3 in. Find the minimum length d.

b. Suppose d is 42 in. Find the maximum height h.

c. **Research** Find a wheelchair ramp at your school or another public building. If possible, find its dimensions and calculate its slope.

8. **Writing** Is $m = \dfrac{y_1 - y_2}{x_1 - x_2}$ another formula for the slope of a line? Explain why or why not.

Write an equation for each line described.

9. contains $(-3, -1)$; has slope 2

10. contains $(1, 1)$; has slope -2

11. contains $(4, -2)$; has slope $-\dfrac{3}{4}$

12. contains $(-3, 2)$; has slope 3.5

13. the vertical line that contains $(13, 5)$

14. the horizontal line that contains $(7, -9)$

15. contains $(-2, 1)$ and $(2, 2)$

16. contains $(-1, 3)$ and $(3, -1)$

Connection   ENGINEERING

To supply cities with water when the source is a long distance away, artificial channels called *aqueducts* may be built. More than 2000 years after it was built, a Roman aqueduct still stands in southern France. It brought water from a source in Uzès to the city of Nîmes. The water traveled downhill, dropping only 17 m across a distance of 50 km. The steepest part of the aqueduct is between Uzès and the Pont du Gard, a bridge across the Gardon River.

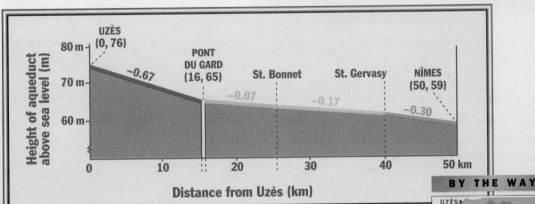

17. To approximate the slope of the steepest part of the aqueduct, what fraction could you use? (*Note:* Use "m/km" as the unit of measure for the slope.)

18. Suppose the aqueduct started at Uzès and ended at Nîmes, but had a constant slope. Write a linear equation to describe its course.

19. a. The Pont du Gard is located 16 km from Uzès. Using the equation you wrote in Exercise 18, find the height of the aqueduct at the bridge. How much lower than this is the actual bridge?

b. Writing Why do you think the Romans made the first part of the aqueduct steeper than the rest?

The sculptor Alexander Calder was born in Philadelphia, Pennsylvania. Both his father and his father's father were also named Alexander and were also sculptors. Artworks by three generations of Alexander Calders are placed along the line defined by the Benjamin Franklin Parkway in downtown Philadelphia.

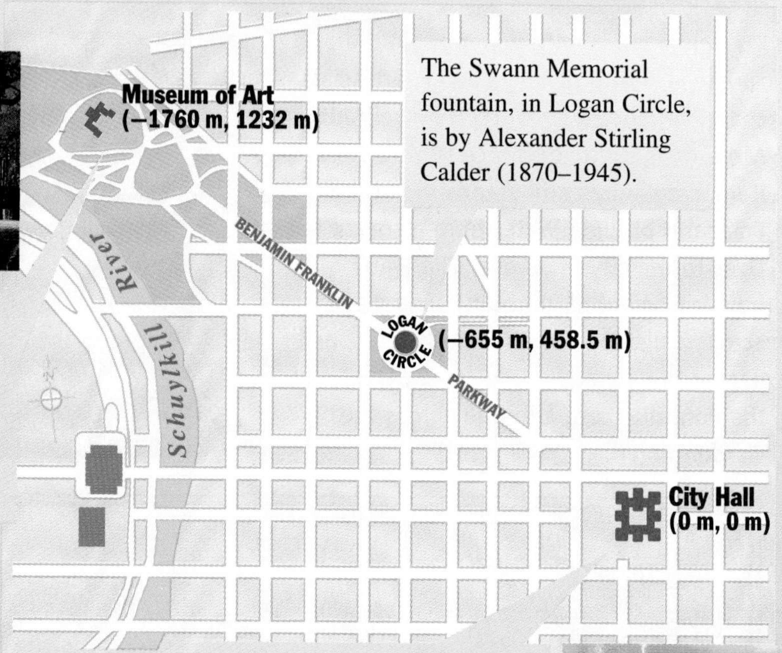

The Swann Memorial fountain, in Logan Circle, is by Alexander Stirling Calder (1870–1945).

Museum of Art (−1760 m, 1232 m)

LOGAN CIRCLE (−655 m, 458.5 m)

City Hall (0 m, 0 m)

Looking northwest from the fountain you can see the Museum of Art. Inside it is *Ghost*, a mobile by Alexander "Sandy" Calder (1898–1976).

Looking southeast from the fountain, you can see the William Penn statue atop City Hall. The statue is by Alexander Milne Calder (1846–1923).

Use the map of Philadelphia. Look at the line that passes through the William Penn statue and the *Ghost* mobile.

20. Use the coordinates of the William Penn statue and *Ghost* to find the slope of the line that connects them.

21. What is an equation for the line?

22. Show that the Swann Memorial fountain lies on the same line.

Find the slope and *y*-intercept of each line.

23. $y = 3x + 5$

24. $y = -2x + 4$

25. $y = -5(x + 5)$

26. $y = 4$

27. $y = -\frac{2}{3}x$

28. $y + 6 = \frac{1}{2}(x - 4)$

29. $y - 1 = 4(x - 1)$

30. $3y = 6x + 9$

Find the slopes of the sides of each quadrilateral.

31.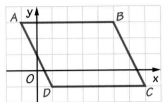

32.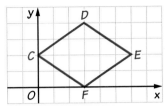

33.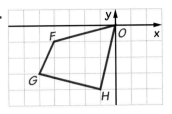

34. Open-ended Problem Draw several lines that have the same *y*-intercept but different slopes. What do you notice about them?

35. Challenge A line contains the points (2, 3) and (6, 1). Find the slope of the line. Then substitute the coordinates of each point in the definition of slope to create two equations for the line. Use algebra to show that the two equations are equivalent.

ONGOING ASSESSMENT

36. Writing A line contains the points $(a, 0)$ and $(0, b)$.
a. Find the slope of the line.
b. Write an equation for the line.
c. In parts (a) and (b) you used variables to find the slope and equation of a line. How is this different from using numeric coordinates?

SPIRAL REVIEW

Find the coordinates of the midpoint of the segment with each given pair of endpoints. *(Section 4.1)*

37. (0, 0) and (12, 6)

38. (−2, 5) and (2, −5)

39. (−7, 6) and (−7, −3)

40. (3, 5) and (6, −3)

Use the diagram at the right. Name all the lines in the diagram that fit each condition. *(Section 1.4)*

41. parallel to $\overleftrightarrow{AB}$

42. skew to $\overleftrightarrow{DH}$

43. parallel to $\overleftrightarrow{BC}$

44. skew to $\overleftrightarrow{EF}$

45. intersect $\overleftrightarrow{CG}$

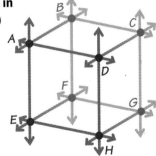

Tell whether each statement is *True* or *False*. If the statement is false, sketch a counterexample and justify your answer. *(Section 2.5)*

46. The opposite sides of a parallelogram are parallel.

47. All angles of a rhombus are right angles.

48. The diagonals of a square bisect each other.

4.3

Exploring Parallels and Perpendiculars

Learn how to...

- **find the slopes of parallel lines and perpendicular lines**

- **identify properties of perpendicular bisectors**

So you can...

- **write equations for parallel and perpendicular lines**

Do you think the long segments in this pattern are parallel? You can't always tell whether lines are parallel or perpendicular simply by looking. In the Exploration you will investigate a method for telling whether lines are parallel or perpendicular.

EXPLORATION

COOPERATIVE LEARNING

Comparing Slopes of Lines

Work with another student.
You will need:

- geometry software or graph paper

- a straightedge

1 Lines *j* and *k* are parallel. Find the slope of each line.

2 Graph a different pair of parallel lines. If you are graphing by hand, use the sides of your straightedge to draw the lines. Find the slopes of your lines.

3 Make a conjecture about the slopes of parallel lines.

4 Lines *ℓ* and *m* are perpendicular. Find the slope of each line.

5 Graph a different pair of perpendicular lines. If you are graphing by hand, use a corner of a piece of paper to help you. Find the slopes of the lines.

6 Make a conjecture about the slopes of perpendicular lines.

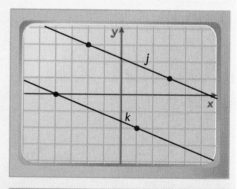

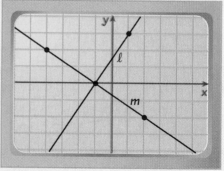

Slopes of Parallel Lines and Perpendicular Lines

Two nonvertical lines are parallel if and only if their slopes are equal.

If $m_1 = m_2$, then $\ell_1 \parallel \ell_2$.

If $\ell_1 \parallel \ell_2$, then $m_1 = m_2$.

Two nonvertical lines are perpendicular if and only if the product of their slopes is -1.

If $m_1 \cdot m_2 = -1$, then $\ell_1 \perp \ell_2$.

If $\ell_1 \perp \ell_2$, then $m_1 \cdot m_2 = -1$.

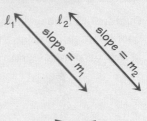

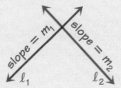

A **perpendicular bisector** of a segment is a line, ray, or segment that bisects the segment and is perpendicular to it. You can use what you know about perpendicular lines to find out if a line is a perpendicular bisector.

EXAMPLE 1

Is $\overleftrightarrow{CD}$ a perpendicular bisector of $\overline{AB}$? Explain.

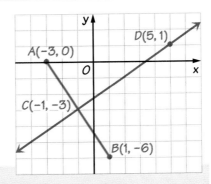

SOLUTION

The midpoint of the segment with endpoints $A(-3, 0)$ and $B(1, -6)$ is $C(-1, -3)$. So $\overleftrightarrow{CD}$ bisects $\overline{AB}$.

Next find the slopes of $\overline{AB}$ and $\overleftrightarrow{CD}$.

$$\text{slope of } \overline{AB} = \frac{-6 - 0}{1 - (-3)} \qquad \text{slope of } \overleftrightarrow{CD} = \frac{1 - (-3)}{5 - (-1)}$$

$$= \frac{-6}{4} \qquad\qquad\qquad = \frac{4}{6}$$

$$= -\frac{3}{2} \qquad\qquad\qquad = \frac{2}{3}$$

The product of the slopes is $-\frac{3}{2} \cdot \frac{2}{3} = -1$, so $\overleftrightarrow{CD} \perp \overline{AB}$.

Since $\overleftrightarrow{CD}$ is perpendicular to $\overline{AB}$, and $\overleftrightarrow{CD}$ bisects $\overline{AB}$, $\overleftrightarrow{CD}$ is a perpendicular bisector of $\overline{AB}$.

EXAMPLE 2   Application: Quilt Patterns

Quilters often copy patterns onto a coordinate grid to plan their work. Is *ABCD* a parallelogram?

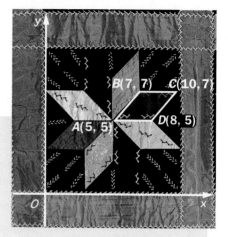

SOLUTION

Find the slope of each side of *ABCD*.

slope of $\overline{AB} = \dfrac{7-5}{7-5} = 1$

slope of $\overline{CD} = \dfrac{5-7}{8-10} = 1$

slope of $\overline{BC} = \dfrac{7-7}{10-7} = 0$ slope of $\overline{AD} = \dfrac{5-5}{8-5} = 0$

The slope of $\overline{AB}$ = the slope of $\overline{CD}$, so $\overline{AB} \parallel \overline{CD}$.

The slope of $\overline{BC}$ = the slope of $\overline{AD}$, so $\overline{BC} \parallel \overline{AD}$.

By the definition of a parallelogram, *ABCD* is a parallelogram.

THINK AND COMMUNICATE

1. State each theorem on page 181 as a conditional and its converse.

2. Why do the theorems on page 181 apply to nonvertical lines only?

3. Which formula can you use to find out if *ABCD* in Example 2 is a rhombus? Explain.

4. Explain how you know that *ABCD* in Example 2 is not a square.

EXAMPLE 3

Write an equation of the line that passes through (3, −3) and is perpendicular to the line *y* = −2*x* + 3.

SOLUTION

The line $y = -2x + 3$ has slope -2. Any line perpendicular to this line has slope $\dfrac{1}{2}$.

Use the slope-intercept form to find *b*.

$$y = mx + b$$
$$-3 = \dfrac{1}{2} \cdot 3 + b$$
$$b = -4.5$$

Substitute (3, −3) for (*x, y*) and $\dfrac{1}{2}$ for *m*.

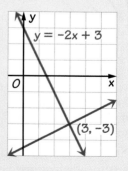

An equation of the line is $y = \dfrac{1}{2}x - 4.5$.

THINK AND COMMUNICATE

5. What is the slope of any line that is perpendicular to a vertical line?

6. What is the slope of any line that is parallel to a vertical line?

7. Give an equation for a line that is parallel to the line $x = 4$.

8. Give an equation for a line that is perpendicular to the line $x = 4$.

☑ CHECKING KEY CONCEPTS

Tell whether each pair of lines is *parallel*, *perpendicular*, or *neither*.

1. $y = -3x + 6$
$y = -\frac{1}{3}x$

2. $y = -5x - 11$
$y = \frac{1}{5}x - 11$

3. $x = 1$
$x = 5$

4. For the slope of a line that is perpendicular to the line $y = mx + b$, Emily uses $\frac{-1}{m}$. Will this method always work? Explain.

Find the slope of each line.

5. a line parallel to the line $y = 7x + 1$

6. a line perpendicular to the line $y = -2x - 3$

7. a line parallel to the x-axis

8. a line perpendicular to the x-axis

4.3 | Exercises and Applications

Extra Practice exercises on page 666

Tell whether each pair of lines is *parallel*, *perpendicular*, or *neither*.

1. $y = -3$
$x = -3$

2. $y = 2x - 7$
$y = 7x - 2$

3. $y = -\frac{4}{5}x + 1$
$y = -0.8x - 5$

Find the slope of each line.

4. a line parallel to the line $y = -x + 3$

5. a line parallel to the line $y = 5$

6. a line perpendicular to the line $y = 7x - 4$

7. a line perpendicular to the line $y = -3x$

8. SAT/ACT Preview Which equation represents a line that is perpendicular to the line with equation $y = -\frac{1}{2}x - 7$?

A. $y = -\frac{1}{2}x + 7$
B. $y = -2x - 7$
C. $y = 2x + 7$

D. $y = -7x - \frac{1}{2}$
E. $y = \frac{1}{2}x + 7$

Use slopes to tell whether each triangle with the given vertices is a right triangle.

9. $A(-4, 1), B(2, 3), C(5, -6)$

10. $D(-5, 0), E(0, 5), F(5, 0)$

11. $G(1, 1), H(4, -1), J(-1, 3)$

12. $K(-4, -4), L(0, 0), M(-6, 6)$

13. CITY PLANNING Parking lots may use different designs in different locations.

A

B

C

a. Which of these plans use perpendicular lines? Which use parallel lines?

b. **Writing** What are some advantages and disadvantages of each parking plan?

Give the most specific name for each quadrilateral with the given vertices.

14. $A(-4, -2)$, $B(5, 4)$, $C(7, 1)$, $D(-2, -5)$ **15.** $E(0, -3)$, $F(1, 1)$, $G(7, 0)$, $H(5, -4)$

16. $J(-3, 2)$, $K(-1, 6)$, $L(1, 0)$, $M(-1, -4)$ **17.** $P(-2, 3)$, $Q(3, 4)$, $R(4, -1)$, $S(-1, -2)$

18. Open-ended Problem Find the vertices of a quadrilateral with perpendicular diagonals that are neither horizontal nor vertical.

For each set of points, tell whether $\overleftrightarrow{CD}$ is a perpendicular bisector of $\overline{AB}$. If not, tell why not.

19. $A(0, 0)$, $B(4, -4)$, $C(2, -2)$, $D(5, 1)$ **20.** $A(-1, 1)$, $B(3, 3)$, $C(1, 2)$, $D(3, -3)$

21. $A(-2, 4)$, $B(2, -2)$, $C(1, 1)$, $D(4, 3)$ **22.** $A(1, 3)$, $B(3, -3)$, $C(-1, -1)$, $D(5, 1)$

23. CARPENTRY A *carpenter's square* can be used to find slopes and draw parallel and perpendicular lines. It looks like two perpendicular rulers that are joined at the 0 point.

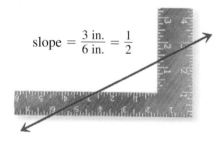

$$\text{slope} = \frac{3 \text{ in.}}{6 \text{ in.}} = \frac{1}{2}$$

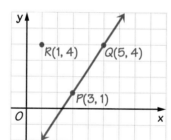

a. Describe how to use a carpenter's square to find the slope of $\overleftrightarrow{PQ}$. Write an equation for $\overleftrightarrow{PQ}$.

b. Describe how to use a carpenter's square to draw a line perpendicular to $\overleftrightarrow{PQ}$ that passes through R. Write an equation of the line.

c. Describe how to use a carpenter's square and the line from part (b) to draw a line parallel to $\overleftrightarrow{PQ}$ that passes through R. Write an equation of the line.

In order to transfer a quilt pattern, quilters may use a square that is 12 in. on a side, as shown below. By using slopes, they can check that they have copied a pattern correctly. Also, they can tell what kinds of quadrilaterals are in the pattern.

For each quadrilateral marked in the quilt patterns shown:

a. Give the coordinates of the vertices.

b. Find the slopes of the sides.

c. Tell whether it is a *parallelogram*, a *rhombus*, or *neither*.

24. guiding star

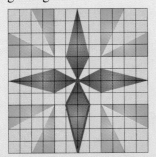

25. tumbling maple leaf

On each square, the origin is indicated by a red dot.

26. desert rose basket

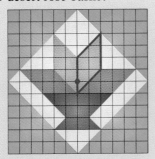

27. Arkansas traveler

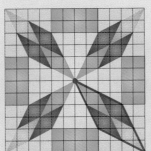

This Arapaho woman is preparing a quilt for the Sundance River Indian Reservation in Wyoming.

Write an equation for each line.

28. The line passes through $P(-4, 2)$ and is parallel to the line $y = -\frac{4}{3}x + 5$.

29. The line passes through $Q(3, 7)$ and is parallel to the line $y = 2$.

30. The line passes through $R(0, -2)$ and is perpendicular to the line $y = x - 4$.

31. The line passes through $S(2, 6)$ and is perpendicular to the *x*-axis.

32. Challenge Use the points $T(a, b)$, $U(c, d)$, $V(b, c)$, and $W(d, a)$, where $a \neq c$ and $b \neq d$.

   a. Show that $\overleftrightarrow{TU} \perp \overleftrightarrow{VW}$.

   b. Why is it necessary to state that $a \neq c$ and $b \neq d$?

33. Cooperative Learning Work with another person.

    a. Working with your partner, choose a slope. Each of you should write an equation for a line with that slope. Graph the lines on the same coordinate plane and label them *l* and *m*.

    b. Write an equation of a line that is perpendicular to *l*. Graph it and label it *q*.

    c. What is the relationship between lines *m* and *q*? Complete this conjecture: "If a line is perpendicular to one of two parallel lines, then ___?___."

Find the slope and *y*-intercept of each line. *(Section 4.2)*

34. $y = -3x - 2$ **35.** $y = x$ **36.** $y = 9$

37. Sketch a pentagonal prism. *(Section 2.6)*

The legs of a right triangle are *a* and *b* units long. The hypotenuse is *c* units long. Find the unknown length for each right triangle. *(Section 3.6)*

38. $a = 9.5, b = 16.8$ **39.** $a = 46.8, c = 49.3$ **40.** $b = 89.9, c = 90.1$

ASSESS YOUR PROGRESS

VOCABULARY

slope (p. 174)   **slope-intercept form** (p. 175)
y-intercept (p. 175)   **perpendicular bisector** (p. 181)

    1. Use $\triangle ABC$ at the right. *(Section 4.1)*

        a. Find the perimeter of $\triangle ABC$.

        b. Label the midpoints of the sides *M, N,* and *P.* Find their coordinates.

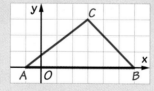

        c. Find the perimeter of $\triangle MNP$. What do you notice?

Find the slope of the line that contains the given points. *(Section 4.2)*

    2. $(0, 0)$ and $(5, -3)$ **3.** $(-2, 5)$ and $(7, 5)$ **4.** $(-3, -3)$ and $(1, 7)$

Give the slope of each line. *(Section 4.3)*

    5. a line parallel to the line $y = 4x + 3$

    6. a line perpendicular to the line $y = -5x - 2$

    7. Journal Describe how to find an equation for the perpendicular bisector of a segment if you know the coordinates of the segment's endpoints. Draw several examples, showing your method. Include at least one horizontal segment and one vertical segment.

4.4 Equations of Circles

Learn how to...
- **identify circles and parts of circles**
- **write equations for circles**

So you can...
- **describe circular shapes, such as architectural forms**

Architects have often used domes to create impressive interior spaces for public buildings. One of the best examples is the rotunda at the University of Virginia, designed by Thomas Jefferson. A rotunda is a circular building with a domed roof. A cross section of the building shows that a circle drawn to touch the dome also touches the center of the base.

$\overline{CD}$ is a *radius*.

$\overline{AB}$ is a *diameter*.

The center of a circle is not part of the circle.

A **circle** is the set of all points in a plane that are an equal distance from a given point, the **center**, which is also in the plane. A **diameter** is a segment that passes through the center of a circle and whose endpoints lie on the circle. A **radius** is a segment whose endpoints are the center and a point on the circle. In the diagram above, the circle with center C is called $\odot C$ (read "circle C").

THINK AND COMMUNICATE

In the cross section of the University of Virginia rotunda, *AB* is about 77 ft.

1. $\overline{DF}$ is a diameter of $\odot C$ that connects the dome of the building with its base. Find the length DF.

2. Find the lengths AC, CB, DC, and CF.

3. How do you know that all the *radii* (plural of *radius*) of a circle are congruent?

The Pantheon, a Roman building dating from the first century, is the inspiration for many later buildings, including the University of Virginia rotunda.

You can write an equation for a circle by placing it on a coordinate plane and using the Distance Formula.

EXAMPLE 1

Write an equation for the inner wall of the University of Virginia rotunda.

The inner wall of the rotunda is a circle with radius about 72 ft.

The center of the rotunda is at (0, 0).

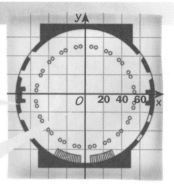

SOLUTION

Distance from (x, y) to $(0, 0) = 72$

$$\sqrt{(x - 0)^2 + (y - 0)^2} = 72 \qquad r \approx 72 \text{ ft}$$

$$x^2 + y^2 = (72)^2$$

$$x^2 + y^2 = 5184$$

In general, every point (x, y) on a circle with center $(0, 0)$ and radius r is r units from the center. An equation for such a circle is:

The distance between $(0, 0)$ and (x, y) is r.

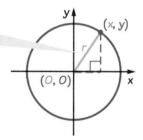

$$\sqrt{(x - 0)^2 + (y - 0)^2} = r$$

Squaring both sides of the above equation eliminates the square root symbol.

> **Equation of a Circle with Center (0, 0)**
>
> An equation of the circle with center $(0, 0)$ and radius r is:
> $$x^2 + y^2 = r^2$$

THINK AND COMMUNICATE

4. The outer wall of the University of Virginia rotunda has a radius of about 77 ft. What is an equation for this circle?

5. The round skylight in the dome of the University of Virginia rotunda has a diameter of about 16 ft. What is an equation for this circle?

The Distance Formula also leads to an equation for a circle that is not centered at (0, 0). The point (h, k) is usually used to represent the center of such a circle.

$$\sqrt{(x - h)^2 + (y - k)^2} = r$$

The center of the circle is (h, k).

Squaring both sides of the equation above eliminates the square root symbol.

Equation of a Circle with Center (h, k)

An equation of the circle with center (h, k) and radius r is:
$$(x - h)^2 + (y - k)^2 = r^2$$

EXAMPLE 2

Write an equation of the circle with center (2, 4) and radius 5.

SOLUTION

Use the equation of a circle.

$(x - h)^2 + (y - k)^2 = r^2$
$(x - 2)^2 + (y - 4)^2 = 5^2$
$(x - 2)^2 + (y - 4)^2 = 25$

Substitute **(2, 4)** for (h, k) and **5** for r.

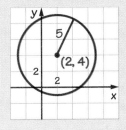

THINK AND COMMUNICATE

6. Suppose the center of the circle in Example 2 were $(-2, -4)$ instead of (2, 4). How would the equation of the circle change?

7. What happens to the equation of a circle centered at (h, k) if you use (0, 0) as the center?

8. In a plane, two or more circles that have the same center are called **concentric circles**. How are the equations for two concentric circles alike? How are they different? Explain.

EXAMPLE 3

Sketch the circle with equation $(x + 2)^2 + (y - 3)^2 = 16$.

SOLUTION

Rewrite the equation of the circle in the form $(x - h)^2 + (y - k)^2 = r^2$.

$$(x + 2)^2 + (y - 3)^2 = 16$$

Rewrite $x + 2$ as $x - (-2)$.   Rewrite 16 as 4^2.

$$(x - (-2))^2 + (y - 3)^2 = 4^2$$

The center is $(-2, 3)$.   The radius is **4**.

Now sketch the circle using the center $(-2, 3)$ and radius **4**.

Step 1 Graph the center, $(-2, 3)$.

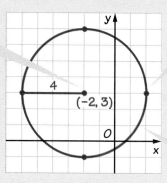

Step 2 Graph some points that are **4** units from the center.

Step 3 Sketch the circle by hand or with a compass.

THINK AND COMMUNICATE

9. What radius should you use to sketch a circle with equation $(x + 2)^2 + (y - 3)^2 = 25$?

10. What radius should you use to sketch a circle with equation $(x + 2)^2 + (y - 3)^2 = 7$?

☑ CHECKING KEY CONCEPTS

1. What is the radius of a circle whose diameter is 10?

2. Write an equation of a circle with center $(0, 0)$ and radius 7.

3. Write an equation of a circle with center $(2, -3)$ and radius 4.

4. Sketch the circle with equation $(x - 4)^2 + (y + 1)^2 = 9$.

4.4 Exercises and Applications

Extra Practice
*exercises on
page 666*

Write an equation of each circle.

1.
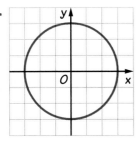

2.
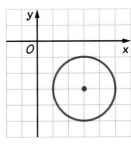

3.
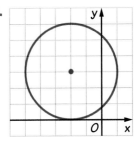

Write an equation of the circle with each given center and radius.

4. center $(0, 0)$
radius 1.5

5. center $(5, 2)$
radius 7

6. center $(3, 4)$
radius 5

7. center $(-2, 1)$
radius 1

8. center $(-6, -8)$
radius 3.5

9. center $(10, -20)$
radius 20

10. PRINTING Letters in a type design, or *typeface,* are carefully drawn, often with thick and thin areas. Typeface designers often work on a coordinate grid. In the design for the letter O shown below, three circles were used to draw the curves.

  a. The outside of the letter is $\odot A$, with center $(4.5, 4.5)$ and radius 4.5. Write an equation for $\odot A$.

  b. The inside of the letter is shaped using $\odot B$ and $\odot C$. Write equations for these circles.

  c. Open-ended Problem How do you think the curves shown in blue were drawn?

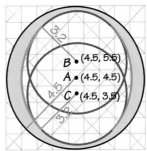

Sketch each circle. Label the coordinates of the center. Draw a radius and label it with its length.

11. $x^2 + y^2 = 64$

12. $(x - 1)^2 + (y + 5)^2 = 9$

13. $(x - 3)^2 + (y - 1)^2 = 10$

14. $(x + 5)^2 + (y - 3)^2 = 121$

15. $(x + 2)^2 + y^2 = 25$

16. $(y - 3)^2 + (x - 4)^2 = 25$

Make a sketch to fit each description.

17. $\odot A$ with diameter $\overline{CD}$ and radius $\overline{AB}$

18. two concentric circles with center Q

19. two circles that intersect in two points

20. two circles that intersect in exactly one point

The Inuit peoples of the Arctic use domes in the construction of their traditional homes, called *igloos*. To construct an igloo, the builder carves a circle in the snow, then arranges snow blocks cut from the interior of the circle to form a dome. Igloos are sometimes built in clusters, with arched passageways connecting the domes.

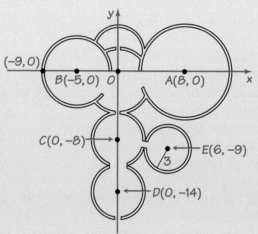

21. The Iglulik people of Hudson Bay, Canada, built this igloo cluster. Dimensions are marked in feet. Write equations for ⊙*B* and ⊙*E*.

22. Use the Distance Formula to find the distance between the center of igloo *C* and the center of igloo *E*.

A diameter of each circle is marked. Find the coordinates of the center of the circle, the length of the radius, and an equation of each circle.

23.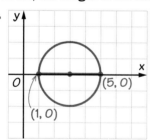

24.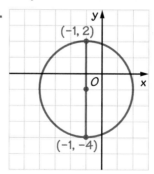

25.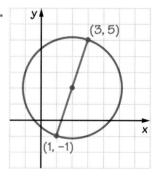

26. Writing Explain how to write an equation of a circle if you are given only the coordinates of the endpoints of a diameter of the circle, as in Exercise 25.

27. Challenge Sketch the circle with equation $x^2 - 4x + 4 + y^2 - 25 = 0$.

ALGEBRA Tell whether the given point is on the circle with the given equation. Explain your answer.

28. $(9, 0)$; $x^2 + y^2 = 18$

29. $\left(2\sqrt{2}, 2\right)$; $x^2 + y^2 = 12$

30. $(3, 5)$; $(x - 3)^2 + (y + 1)^2 = 36$

31. $\left(7, 5\sqrt{3}\right)$; $(x - 2)^2 + y^2 = 100$

32. SAT/ACT Preview Which point does *not* lie on the circle with equation $(x - 2)^2 + (y + 5)^2 = 36$?

    A. $(2, 1)$ **B.** $(8, -5)$ **C.** $(-4, -5)$ **D.** $(2, -5)$ **E.** $(2, -11)$

ONGOING ASSESSMENT

33. Cooperative Learning Work with another person. You will each need a circular object and graph paper.

    a. Each of you should draw a set of axes on graph paper. Without showing the other person your object, place it on your paper and trace around it with a pencil.

    b. On a different piece of paper, write an equation of the circle you traced.

    c. Exchange equations with the other person. Sketch the graph of the equation you receive.

    d. Compare your sketch with the circular object the other person traced. Explain any differences.

SPIRAL REVIEW

Give the most specific name for each quadrilateral with the given vertices.
(Section 4.3)

34. $A(0, 0)$, $B(5, 0)$, $C(7, 4)$, $D(2, 4)$

35. $W(7, 0)$, $X(0, 3)$, $Y(-7, 0)$, $Z(0, -3)$

36. $P(1, 0)$, $Q(0, 2)$, $R(4, 4)$, $S(5, 2)$

37. $J(-1, -1)$, $K(-1, 2)$, $L(2, 2)$, $M(2, -1)$

In the diagram, $\overline{AB}$ and $\overline{CD}$ bisect each other at point *E*. Find each length.
(Section 1.7)

38. *EB*

39. *CD*

40. *AB*

41. *CE*

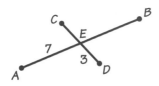

Tell whether each statement is *True* or *False*. If a statement is false, sketch a counterexample and justify your answer. *(Section 2.5)*

42. Every rhombus is a square.

43. Every rhombus is a parallelogram.

44. Every equiangular parallelogram is a square.

4.5 Coordinates and Proof

Learn how to...
- place figures on coordinate axes and label their vertices
- plan and write coordinate geometry proofs

So you can...
- use algebraic methods to verify conjectures about triangles and quadrilaterals
- improve your reasoning skills

Variables and equations can help you study geometric figures and prove theorems about them. The first step is to give coordinates for points on the figure. The Exploration will help you become familiar with this process.

EXPLORATION
COOPERATIVE LEARNING

Placing a Parallelogram on a Coordinate Plane

Work in a group of four students.
You will need: graph paper, scissors, straightedge

1 On graph paper, each student should draw a quadrilateral with vertices $E(0, 0)$, $F(6, 8)$, $G(11, 8)$, and $H(5, 0)$. Cut out your quadrilateral and label the vertices on both sides. Color it if you wish.

2 On a separate piece of graph paper, draw coordinate axes. Place the cut-out polygon on the coordinate plane so that all four vertices have integer coordinates. Each student should use a different placement.

3 Trace your cut-out polygon onto your graph paper. Label the vertices E, F, G, and H and write their coordinates.

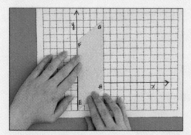 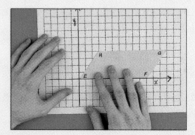

4 Use the Slope Formula to show that *EFGH* is a parallelogram.

5 Use the Midpoint Formula to show that the midpoint of $\overline{EG}$ is the same as the midpoint of $\overline{FH}$. Which theorem from Section 2.5 does this illustrate?

6 Compare your answers with those of others in your group. How are your answers different? How are they the same?

7 Does the position of a parallelogram on a coordinate plane affect whether or not the diagonals bisect each other? Explain.

In the Exploration, you showed that the diagonals of a given parallelogram bisect each other. To prove that this is true for *all* parallelograms, you must label a parallelogram in a general way, using variables to represent coordinates. Use only as many variables as necessary.

EXAMPLE 1

Prove that the diagonals of a parallelogram bisect each other.

SOLUTION

Plan ahead: Use variables for the coordinates of the vertices. Use the Midpoint Formula to show that the midpoint of $\overline{EG}$ is the same as the midpoint of $\overline{FH}$.

Proof

You need only three variables to label points E, F, G, and H. Use the definition of a parallelogram: the slopes of opposite sides are equal.

Step 1 Place the parallelogram with vertex E at (0, 0) and one side along the x-axis.

Step 3 Use (b, c) for point H.

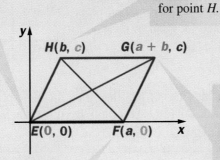

Step 4 Point G must have coordinates ($a + b$, c), so that the slopes of $\overline{EH}$ and $\overline{FG}$ will be equal.

Step 2 Use (a, 0) for point F because it is on the x-axis.

Now use the Midpoint Formula to find the midpoints of diagonals $\overline{EG}$ and $\overline{FH}$.

$$\text{midpoint of } \overline{EG} = \left(\frac{x_1 + x_2}{2}, \frac{y_1 + y_2}{2}\right) \qquad \text{midpoint of } \overline{FH} = \left(\frac{x_1 + x_2}{2}, \frac{y_1 + y_2}{2}\right)$$

$$= \left(\frac{0 + (a + b)}{2}, \frac{0 + c}{2}\right) \qquad\qquad = \left(\frac{a + b}{2}, \frac{0 + c}{2}\right)$$

$$= \left(\frac{a + b}{2}, \frac{c}{2}\right) \qquad\qquad\qquad = \left(\frac{a + b}{2}, \frac{c}{2}\right)$$

The midpoints of the diagonals are the same point because they have identical coordinates. Therefore, the diagonals bisect each other.

THINK AND COMMUNICATE

1. Show that the slopes of $\overline{EF}$ and $\overline{HG}$ are equal in the parallelogram in Example 1. Then show that the slopes of $\overline{EH}$ and $\overline{FG}$ are equal.

2. Vivian says that the result of Example 1 is true for a rhombus. Explain her reasoning.

The proof in Example 1 is a **coordinate geometry proof** that uses the Midpoint Formula. In Example 2, the Distance Formula and the Slope Formula are used in another proof.

EXAMPLE 2 | **Connection: Algebra**

Prove that the diagonals of a square are congruent and perpendicular.

SOLUTION

Plan ahead: Place a square on a coordinate plane. Use the Distance Formula to show that the diagonals are congruent. Use the Slope Formula to show that the product of the slopes of the diagonals is -1.

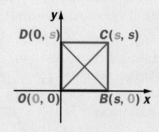

Proof

If the square has sides of length s, then it can be placed on a coordinate plane as shown. B and D are s units from O. C is s units from B and D.

Use the Distance Formula to show that the diagonals are congruent.

$$OC = \sqrt{(x_2 - x_1)^2 + (y_2 - y_1)^2} \qquad BD = \sqrt{(x_2 - x_1)^2 + (y_2 - y_1)^2}$$

$$= \sqrt{(s - 0)^2 + (s - 0)^2} \qquad\qquad = \sqrt{(0 - s)^2 + (s - 0)^2}$$

$$= \sqrt{s^2 + s^2} \qquad\qquad\qquad = \sqrt{(-s)^2 + s^2}$$

$$= \sqrt{2s^2} \qquad\qquad\qquad\quad = \sqrt{2s^2}$$

$$= s\sqrt{2} \qquad\qquad\qquad\quad = s\sqrt{2}$$

Because $OC = BD$, the *diagonals are congruent.*

Use the Slope Formula to show that the diagonals are perpendicular.

$$\text{slope of } \overline{OC} = \frac{y_2 - y_1}{x_2 - x_1} \qquad\qquad \text{slope of } \overline{BD} = \frac{y_2 - y_1}{x_2 - x_1}$$

$$= \frac{s - 0}{s - 0} \qquad\qquad\qquad\qquad = \frac{s - 0}{0 - s}$$

$$= 1 \qquad\qquad\qquad\qquad\qquad = -1$$

The product of the slopes is -1, so the *diagonals are perpendicular.*

Therefore, the diagonals of any square are congruent and perpendicular.

3. Adam and Clint each place squares on a coordinate plane, as shown. What properties of a square did Adam use in placing his square? What properties did Clint use?

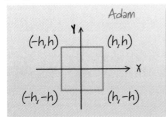

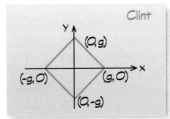

4. Can Adam use his diagram to prove the theorem in Example 2? Explain.

5. Can Clint use his diagram to prove the theorem in Example 2? Explain.

6. Example 1 shows that the diagonals of a parallelogram bisect each other. A square is a parallelogram. Can you conclude that the diagonals of a square bisect each other? Explain.

7. Example 2 shows that the diagonals of a square are congruent and perpendicular. A square is a rectangle. Can you conclude that the diagonals of a rectangle are congruent and perpendicular? Explain.

☑ CHECKING KEY CONCEPTS

1. Three vertices of a rectangle are $(0, 0)$, $(b, 0)$, and $(0, h)$. Sketch the rectangle and find the coordinates of the fourth vertex.

2. Show how to use the coordinates a, b, and c to label the vertices of a triangle on a coordinate plane.

3. Describe the steps needed to prove that the diagonals of a rectangle are congruent.

4.5 | **Exercises and Applications**

Extra Practice exercises on page 667

Find the missing coordinates without using any new variables.

1. rectangle

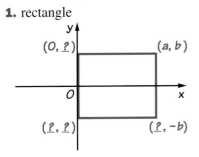

2. parallelogram

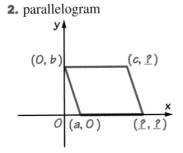

3. isosceles triangle

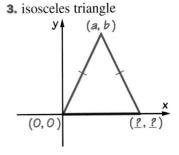

Look back at the article on pages 162–164.

On computer chips, like this one that Marc Hannah designed, components may be used in several locations. In designing a chip, coordinates define the placements of components. A grid has been placed on the chip.

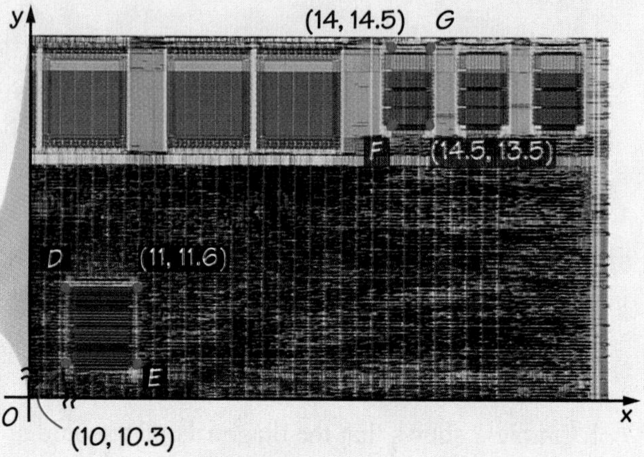

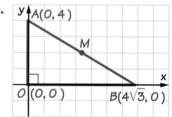

(14, 14.5) G

F   (14.5, 13.5)

D   (11, 11.6)

E

O

(10, 10.3)

4. The three Random Access Memory (RAM) components are congruent rectangles. Supply the missing coordinates for points *A*, *B*, and *C* in the diagram below.

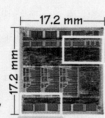

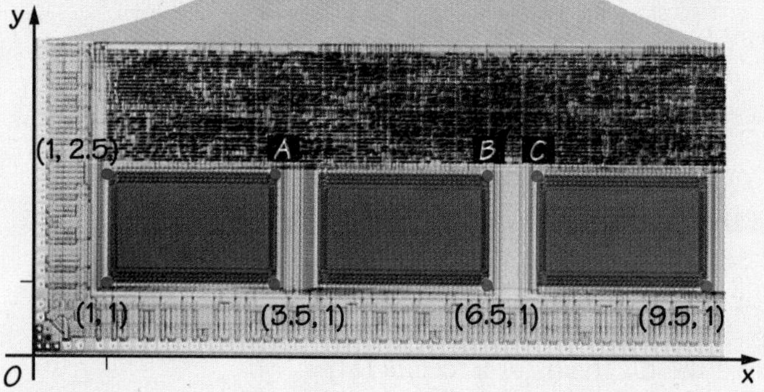

├─17.2 mm─┤

17.2 mm

5. The three register files are rectangles. Supply the missing coordinates for points *D* and *E* in the diagram above.

6. The three Address Units are rectangles. Supply the missing coordinates for points *F* and *G* in the diagram above.

7. TRANSFORMATIONS What type of transformation moves each of the components to a congruent component? Explain.

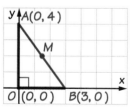

(1, 2.5)   A   B C

(1, 1)   (3.5, 1)   (6.5, 1)   (9.5, 1)

For each triangle *ABO*, find the midpoint *M* of the hypotenuse. Then find the lengths *MA*, *MB*, and *MO*.

8.

y A(0, 4)

M

O (0, 0)   B(3, 0)

9.

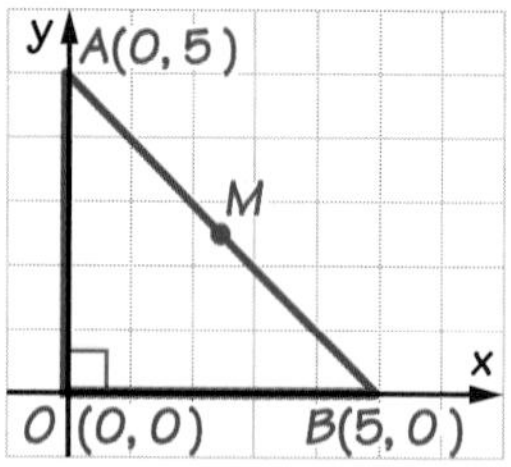

y A(0, 5)

M

O (0, 0)   B(5, 0)

10.

y A(0, 4)

M

O (0, 0)   B($4\sqrt{3}$, 0)

11. Draw a right triangle like the ones in Exercises 8–10, using variables for coordinates. Use your figure to prove that the midpoint of the hypotenuse of a right triangle is the same distance from all three vertices of the triangle.

The midpoints of two sides of a triangle are endpoints of a segment. Show that the segment is parallel to the third side and half as long.

12.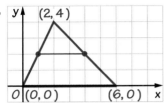

13.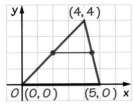

14.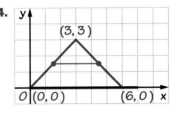

15. Draw a triangle like the ones in Exercises 12–14, using variables for coordinates. Use your figure to prove the result suggested by those exercises.

16. Use the diagram of rectangle *PQRS*.

    a. Find the missing coordinates.

    b. Find the lengths of the diagonals.

    c. Describe your results in part (b) as a theorem.

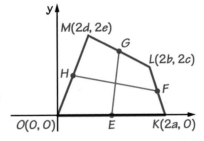

Use the diagram of quadrilateral *OKLM*. The midpoints of the sides are *E, F, G,* and *H*.

17. Find the coordinates of *E, F, G,* and *H*.

18. Find the midpoints of $\overline{EG}$ and $\overline{FH}$.

19. Describe your results from Exercise 18 as a theorem.

20. **Writing** Why do you think the coordinates 2*a*, 2*b*, 2*c*, 2*d*, and 2*e* are used in the diagram instead of *a*, *b*, *c*, *d*, and *e*?

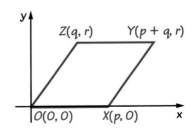

Open-ended Problems **For each type of quadrilateral, draw several examples on graph paper. Find the midpoints of the sides and join them to form a quadrilateral. Tell what kind of quadrilateral is formed.**

21. rectangle   **22.** rhombus   **23.** parallelogram

24. Prove that the midpoints of the sides of any quadrilateral form a parallelogram. Use a diagram similar to the one used in Exercises 17–20.

25. **Challenge** Use the coordinates given in the diagram below. Complete the proof.

    Given: The diagonals of parallelogram *OXYZ* are perpendicular.

    Prove: *OXYZ* is a rhombus.

    a. Find the slope of each diagonal.

    b. The diagonals are perpendicular, so the product of their slopes is -1. Use this fact and your results from part (a) to complete:

$$\frac{?}{?} \cdot \frac{?}{?} = -1$$

    c. Use algebra to solve the equation in part (b) for *p*.

    d. Show that all sides of *OXYZ* have length *p*. Therefore, *OXYZ* is a rhombus.

26. Cooperative Learning Work with another person. Each of you should draw a polygon on a coordinate plane, labeling some coordinates with variables but leaving other coordinates unlabeled. Write what kind of polygon it is. Exchange papers with your partner and supply the missing coordinates, using as few new variables as possible.

27. 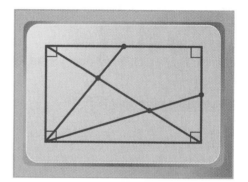 **Technology** Use geometry software to draw segments from one vertex of a rectangle to the midpoints of two other sides, as shown. Draw the diagonal of the rectangle that crosses these segments.

    a. Show that the segments divide the diagonal into three equal parts.

    b. ALGEBRA Prove the result from part (a) using a coordinate geometry proof. (*Hint:* Write equations for the lines that contain the segments and diagonal. Solve systems of equations to find the intersection points.)

ONGOING ASSESSMENT

28. Use the figure at the right. Assume $a > b$.

    a. Prove that the diagonals of *PQRS* are congruent.

    b. Label the midpoints of the diagonals *M* and *N* and find their coordinates.

    c. Show that $MN = \frac{1}{2}(PS - QR)$.

    d. Writing Write a paragraph that summarizes the main steps of your proof.

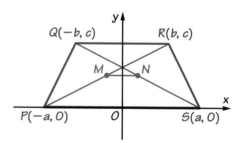

SPIRAL REVIEW

Write an equation of the circle with each given center and radius.
(Section 4.4)

29. center $(7, 0)$
    radius 6

30. center $(2, -5)$
    radius 5

31. center $(-1, 4)$
    radius 7

Find the sum of the measures of the interior angles of each polygon.
(Section 2.4)

32. pentagon

33. hexagon

34. 12-gon

Find each unknown length. *(Section 3.6)*

35.

36.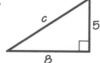

37.

4.6 Coordinates in Three Dimensions

Learn how to...

- **find coordinates in three dimensions**
- **use the Distance Formula and the Midpoint Formula in three dimensions**

So you can...

- **identify and describe relationships between geometric figures in three dimensions**

The element carbon is the basis of all life on Earth and is one of the key components of coal, oil, and natural gas. Atoms of carbon combine with other atoms to form molecules that have a three-dimensional structure.

The molecules of the natural gas *ethane* contain two carbon atoms and six hydrogen atoms. To describe the positions of the atoms within the molecule, you can use a **three-dimensional coordinate system**. The table shows the coordinates of the center of each atom.

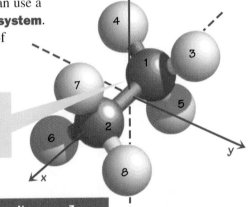

The **z-axis** is perpendicular to the x- and y-axes.

Atom	Type	x	y	z
1	carbon	−0.76	0.00	0.00
2	carbon	0.76	0.00	0.00
3	hydrogen	−1.12	0.89	0.51
4	hydrogen	−1.12	−0.89	0.51
5	hydrogen	−1.12	0.00	−1.03
6	hydrogen	1.12	−0.89	−0.51
7	hydrogen	1.12	0.00	1.03
8	hydrogen	1.12	0.89	−0.51

Coordinates are given to the nearest hundredth of an angstrom (Å). 10^{10} Å = 1 meter.

THINK AND COMMUNICATE

1. Name three atoms that have the same *x*-coordinate.

2. a. What do you notice about the coordinates of the two carbon atoms?

   b. What point is halfway between the centers of these two atoms?

The position of each point in space is given by an **ordered triple (x, y, z)**. For example, in the graph below, P has coordinates (2, 3, 4). To locate P, start at the origin. Move 2 units along the x-axis, 3 units parallel to the y-axis, and 4 units parallel to the z-axis.

The axes intersect at the *origin*, the point (0, 0, 0).

The **x-axis** appears to extend out from the plane of the paper.

The dashed part of each axis represents the negative direction.

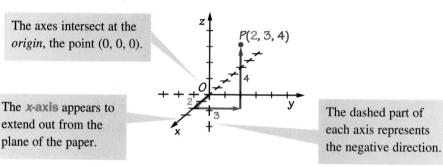

Each pair of axes determines a *coordinate plane*. At least one coordinate of every point that lies in a coordinate plane is 0.

Every point in the **yz-plane** has x-coordinate 0.

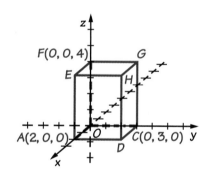

EXAMPLE 1

Find the coordinates of vertices D and E of the rectangular prism.

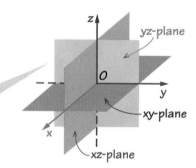

SOLUTION

D has the same x-coordinate as A and the same y-coordinate as C. Its z-coordinate is 0. D has coordinates (2, 3, 0).

E has the same x-coordinate as A and the same z-coordinate as F. Its y-coordinate is 0. E has coordinates (2, 0, 4).

THINK AND COMMUNICATE

3. In Example 1, what are the coordinates of O? of G? of H?

4. What is the z-coordinate of every point in the xy-plane?

In two dimensions, each coordinate of the midpoint of a segment is the mean of the corresponding coordinates of the endpoints. This is true in three dimensions as well.

The Midpoint Formula in Three Dimensions

The midpoint of the segment that joins any two points $A(x_1, y_1, z_1)$ and $B(x_2, y_2, z_2)$ has these coordinates:

$$\text{coordinates of the midpoint of } \overline{AB} = \left(\frac{x_1 + x_2}{2}, \frac{y_1 + y_2}{2}, \frac{z_1 + z_2}{2} \right)$$

EXAMPLE 2

The coordinates of E are (2, 0, 2), and the coordinates of C are (-1, 3, -2). Find the coordinates of the midpoint of $\overline{EC}$.

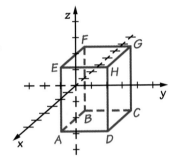

SOLUTION

Find the mean of each pair of corresponding coordinates:

$$\text{coordinates of midpoint of } \overline{EC} = \left(\frac{2 + (-1)}{2}, \frac{0 + 3}{2}, \frac{2 + (-2)}{2} \right)$$

$$= \left(\frac{1}{2}, \frac{3}{2}, \frac{0}{2} \right)$$

$$= \left(\frac{1}{2}, 1\frac{1}{2}, 0 \right)$$

The distance between two points in three dimensions can be found using a formula similar to the Distance Formula for two dimensions.

The Distance Formula in Three Dimensions

The distance AB between two points in space $A(x_1, y_1, z_1)$ and $B(x_2, y_2, z_2)$ is:

$$AB = \sqrt{(x_2 - x_1)^2 + (y_2 - y_1)^2 + (z_2 - z_1)^2}$$

EXAMPLE 3   Application: Chemistry

Use the data for the ethane molecule shown on page 201. Find the distance between the centers of atoms 1 and 3 to the nearest hundredth of an angstrom.

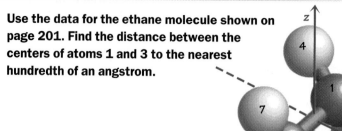

SOLUTION

Atom 1 has coordinates $(-0.76, 0, 0)$ and atom 3 has coordinates $(-1.12, 0.89, 0.51)$.

$$d = \sqrt{(-1.12 - (-0.76))^2 + (0.89 - 0)^2 + (0.51 - 0)^2}$$
$$= \sqrt{(-0.36)^2 + (0.89)^2 + (0.51)^2}$$
$$= \sqrt{1.1818}$$
$$\approx 1.09$$

The distance between atoms 1 and 3 is about 1.09 Å.

☑ **CHECKING KEY CONCEPTS**

Use the triangular prism. Give the coordinates of each vertex.

1. P

2. L

3. M

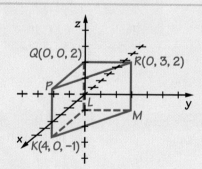

Find the coordinates of the midpoint of the segment with the given endpoints.

4. $(7, 0, 6)$ and $(5, -8, 1)$ **5.** $(2, 6, -4)$ and $(-4, 2, -4)$

6. $\left(1, \frac{1}{2}, -2\right)$ and $\left(-1, \frac{1}{2}, 3\right)$ **7.** $(0, 0, 0)$ and $(4, -6, -1)$

Find the length of the segment with the given endpoints.

8. $(2, 4, 3)$ and $(5, 0, -3)$ **9.** $(1, 9, -3)$ and $(0, 12, -5)$

4.6 Exercises and Applications

Extra Practice
exercises on
page 667

For Exercises 1–9, use the rectangular prism.
Give the coordinates of each vertex.

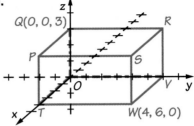

1. *P*
2. *R*
3. *S*
4. *T*
5. *O*
6. *V*

Tell which coordinate plane is parallel to each face of the prism.

7. *PQRS*
8. *SRVW*
9. *PSWT*

Tell which coordinate plane contains each point.

10. $A(2, 4, 0)$
11. $B(1, 0, 2)$
12. $C(0, -1, 4)$

Find the coordinates of the midpoint of the segment with the given endpoints.

13. $(0, 0, 6)$ and $(10, -4, 3)$
14. $(1, 3, -4)$ and $(8, 0, 4)$
15. $(6, -3.5, -9)$ and $(2, 2.5, 7)$

Find the length of the segment with the given endpoints.

16. $(1, 0, 2)$ and $(-2, 3, 5)$
17. $(1, 0, 0.5)$ and $(5, -2, 1.5)$
18. $(15, 3, -7)$ and $(10, -9, 10)$

The coordinates of the vertices of a triangle are given. Tell whether the triangle is *isosceles*, *equilateral*, or *scalene*.

19. $R(-2, 4, -3)$, $S(0, 1, 3)$, $T(-2, -3, -3)$

20. $A(5, 3, 4)$, $B(8, 8, 11)$, $C(4, -1, 8)$

21. $X(-1, -3, -1)$, $Y(2, 3, 6)$, $Z(3, -1, 6)$

22. **Writing** Use your answers to Exercises 19–21. Are any of the triangles right triangles? Explain how you know.

CHEMISTRY The table gives the coordinates of the atoms of a methane molecule to the nearest hundredth of an angstrom.

Atom	Type	x	y	z
1	carbon	0.26	−0.36	0.00
2	hydrogen	0.26	0.73	0.00
3	hydrogen	0.77	−0.73	0.89
4	hydrogen	0.77	−0.73	−0.89
5	hydrogen	−0.77	−0.73	0.00

23. **Open-ended Problem** Show that the triangle formed by the centers of the carbon atom and any two hydrogen atoms is isosceles.

24. **Open-ended Problem** Show that the triangle formed by any three hydrogen atoms is equilateral.

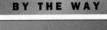

Methane gas can be captured at landfills. It is then converted into usable energy.

Look back at the article on pages 162–164.

The colors you see when you look at a color computer screen are all created by combining the three primary colors of light—red, blue, and green—in different proportions. The intensity of the red, blue, and green light can vary from 0% to 100%. The colors created this way can be displayed in a color cube.

25. What are the coordinates of the three vertices of the color cube that are not labeled? What colors are at those locations?

26. Open-ended Problem Choose two edges of the color cube. Find the coordinates of the midpoint of each edge that you chose. What color is at each midpoint?

27. What are the coordinates of the point at the center of the color cube? What color is at this location?

28. Open-ended Problem Find the coordinates of three points on the color cube that form an equilateral triangle.

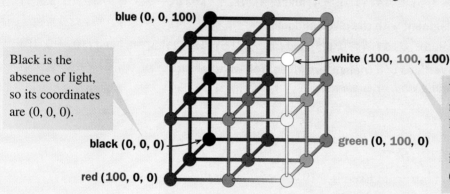

blue (0, 0, 100)

Black is the absence of light, so its coordinates are (0, 0, 0).

white (100, 100, 100)

White is made by mixing the three light primaries at 100% intensity, so its coordinates are (100, 100, 100).

black (0, 0, 0)

green (0, 100, 0)

red (100, 0, 0)

29. Challenge Given two points P and Q in space, a rectangular prism can be constructed so that P and Q are the endpoints of a diagonal of the prism. Each of the edges is parallel to an axis.

a. Use the Pythagorean theorem to find the length of the dashed diagonal.

b. Use the Pythagorean theorem to find PQ.

c. Writing Explain how finding the length of the diagonal of a prism is related to the Distance Formula in three dimensions.

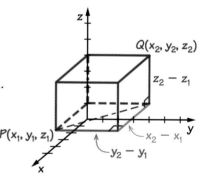

ONGOING ASSESSMENT

30. Open-ended Problem Choose any number. Imagine all the points in a three-dimensional coordinate system that have that number for a z-coordinate. What figure do you think they form? Describe the figure and sketch it. Make a conjecture about points that have the same z-coordinate.

Write the converse of each statement. *(Section 3.5)*

31. If a triangle is equilateral, then all its sides are congruent.

32. If a quadrilateral is a parallelogram, then its opposite sides are congruent.

Sketch each situation. *(Section 1.4)*

33. $\overleftrightarrow{AB} \parallel \overleftrightarrow{CD}$

34. $\overleftrightarrow{AB}$ intersects plane S at point N.

ASSESS YOUR PROGRESS

VOCABULARY

circle (p. 187)
center of a circle (p. 187)
diameter (p. 187)
radius, radii (p. 187)
concentric circles (p. 189)

coordinate geometry
   proof (p. 196)
three-dimensional coordinate
   system (p. 201)
z-axis (p. 201)
ordered triple (x, y, z) (p. 202)

Sketch each circle. Give the coordinates of the center. Draw a radius and label it with its length. *(Section 4.4)*

1. $x^2 + (y - 3)^2 = 4$

2. $(x - 4)^2 + y^2 = 16$

Answer Exercises 3–6 to complete the proof. Use the coordinates given in the diagram. *(Section 4.5)*

Given: The vertices of isosceles $\triangle ABC$ are $A(-k, 0)$, $B(0, 3k)$, and $C(k, 0)$; D and E are the midpoints of $\overline{AB}$ and $\overline{CB}$.

Prove: $\overline{AE} \perp \overline{CD}$

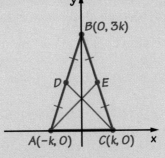

3. Find the coordinates of D and E.

4. Find the slopes of $\overline{AE}$ and $\overline{CD}$.

5. Find the product of the slopes of $\overline{AE}$ and $\overline{CD}$.

6. Explain why you can conclude that $\overline{AE} \perp \overline{CD}$.

The coordinates of the vertices of a triangle are given. Tell whether the triangle is *isosceles, equilateral,* or *scalene.* *(Section 4.6)*

7. $D(5, 0, 0)$, $E(0, 5, 0)$, $F(0, 0, 5)$

8. $U(0, 7, 1)$, $V(2, 5, 3)$, $W(4, 3, 1)$

9. Journal Create a coordinate geometry problem that requires the Distance Formula, the Midpoint Formula, or the Slope Formula. Your problem can be an exercise about a figure or a proof of a conjecture. Write your problem and a complete solution. Discuss any difficulties you had in creating and solving your problem.

Exploring Taxicab Geometry

You probably think that the shortest distance from A to B is along a straight line. This is true in the Euclidean geometry of the coordinate plane. What happens if you're in a city at point A and you want to go to point B? You can't go straight through buildings! You have to go along the streets. **Taxicab geometry** is the non-Euclidean geometry that a taxicab and a pedestrian must obey.

New York City

PROJECT GOAL	Compare Euclidean geometry to taxicab geometry and calculate traveling distances along a city grid.

Getting from Here to There

In taxicab geometry, there is more than one "shortest path" from A to B.

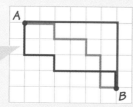

The rules in taxicab geometry are simple: You may move horizontally and vertically but never diagonally. The diagram below shows several ways of moving from A to B. Each path covers 10 blocks.

The **taxicab distance** between $A(x_1, y_1)$ and $B(x_2, y_2)$ is the sum of the horizontal and vertical distances along a shortest path from A to B. Pythagorean distance is less than or equal to taxicab distance.

The steps below show how to find the taxicab distance between $A(1, 3)$ and $B(7, -1)$.

Taxicab Distance
$$AB = |x_2 - x_1| + |y_2 - y_1|$$

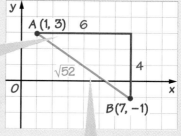

Absolute value is used because distances are always positive.

$$
\begin{aligned}
AB &= |x_2 - x_1| + |y_2 - y_1| \\
&= |7 - 1| + |-1 - 3| \\
&= |6| + |-4| \\
&= 6 + 4 \\
&= 10
\end{aligned}
$$

Pythagorean Distance
$$AB = \sqrt{(x_2 - x_1)^2 + (y_2 - y_1)^2}$$

Making a Taxicab Map

1. Make a map showing the grid of a real or invented city. Choose an origin. Plot two points at street intersections representing a school and a park entrance. The points should not lie on a horizontal or a vertical line. Label their coordinates.

The **points between A and B** form a rectangular region.

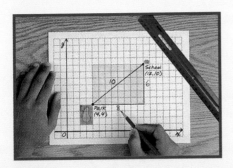

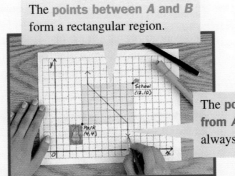

The **points equidistant from A and B** do not always form a straight line.

2. Find the Euclidean and taxicab distances between the school and the park. How do they compare? Draw the taxicab region that represents all points between the school and the park.

3. Choose a location for an apartment that is the same taxicab distance from the school as it is from the park. Find the coordinates of several possible locations for the apartment.

Making a Presentation

Suppose you walk your dog a round-trip distance of 10 blocks each day. Show on your map every place you can reach. This is the interior of a taxicab "circle" centered at the apartment with a radius of 5 blocks. Present your labeled and completed map to the class.

Extending Your Project

Find the taxicab circumference and taxicab diameter of the dog-walk "circle". If π is the ratio of circumference to diameter, what is the value of π in taxicab geometry?

Self-Assessment

What advice would you give to someone who plans to do the project? What are some other problems that you might think differently about after working with taxicab geometry?

4 | Review

STUDY TECHNIQUE

One way to prepare for a test is to write one. Write a test for this chapter, then exchange it with a classmate. After taking the tests, correct and discuss them. Review any topics that gave you difficulty.

VOCABULARY

slope (p. 174)
y-intercept (p. 175)
slope-intercept form (p. 175)
perpendicular bisector (p. 181)
circle (p. 187)
center of a circle (p. 187)
diameter (p. 187)
radius, radii (p. 187)

concentric circles (p. 189)
coordinate geometry proof (p. 196)
three-dimensional coordinate system (p. 201)
z-axis (p. 201)
ordered triple (x, y, z) (p. 202)

SECTIONS 4.1, 4.2, *and* 4.3

Use the *Distance Formula* to find the length of a segment.

$$AB = \sqrt{(x_2 - x_1)^2 + (y_2 - y_1)^2}$$

Use the *Midpoint Formula* to find the midpoint of a segment.

$$\text{midpoint of } \overline{AB} = \left(\frac{x_1 + x_2}{2}, \frac{y_1 + y_2}{2}\right)$$

An equation of $\overleftrightarrow{AB}$ is in **slope-intercept form** $y = mx + b$, where m is the slope and b is the y-intercept.

Use the slope formula to find the **slope** m of $\overline{AB}$.

$$m = \frac{y_2 - y_1}{x_2 - x_1}$$

A line parallel to $\overleftrightarrow{AB}$ has the same slope but a different y-intercept. The slope of a line perpendicular to $\overleftrightarrow{AB}$ is the negative reciprocal of the slope of $\overleftrightarrow{AB}$.

SECTIONS 4.4, 4.5, *and* 4.6

A **circle** is the set of all points in a plane that are the same distance from a fixed point, the **center** of the circle. The distance from the center of a circle to any point on the circle is the **radius**.

An equation of $\odot O$ with center at the origin $(0, 0)$ and radius $r = 3$ is:

$x^2 + y^2 = 3^2$ Use $x^2 + y^2 = r^2$.

An equation of $\odot P$ with center $(h, k) = (4, 3)$ and radius $r = 2$ is:

$(x - 4)^2 + (y - 3)^2 = 4$

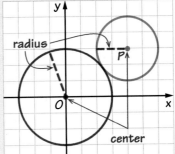

Use $(x - h)^2 + (y - k)^2 = r^2$.

Start a **coordinate geometry proof** by placing the figure on a coordinate plane. Use the properties of the figure to label its coordinates using as few variables as possible. Then use the Distance Formula, the Midpoint Formula, and the properties of slopes of parallel and perpendicular lines to prove theorems.

For example, in quadrilateral $RSTU$, you can use the Distance Formula to show $RT = \sqrt{(a + b)^2 + c^2} = SU$, so you can prove that the diagonals of $RSTU$ are congruent.

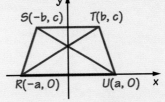

To locate a point in space, three coordinates are needed. Point P has x-coordinate 2, y-coordinate 0, and z-coordinate -3. It lies in the xz-plane. The coordinates of point Q are given by the ordered triple $(5, 4, 6)$.

Find the midpoint of $\overline{PQ}$ by using the Midpoint Formula.

$$\text{midpoint of } \overline{PQ} = \left(\frac{x_1 + x_2}{2}, \frac{y_1 + y_2}{2}, \frac{z_1 + z_2}{2}\right)$$

$$= \left(\frac{2 + 5}{2}, \frac{0 + 4}{2}, \frac{-3 + 6}{2}\right)$$

$$= \left(\frac{7}{2}, 2, \frac{3}{2}\right)$$

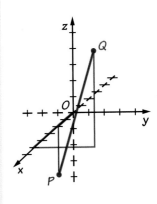

Find the length of $\overline{PQ}$ by using the Distance Formula.

$$PQ = \sqrt{(x_2 - x_1)^2 + (y_2 - y_1)^2 + (z_2 - z_1)^2}$$

$$= \sqrt{(5 - 2)^2 + (4 - 0)^2 + (6 - (-3))^2}$$

$$= \sqrt{3^2 + 4^2 + 9^2}$$

$$\approx 10.3$$

Assessment

VOCABULARY QUESTIONS

For each question, match a word to the given value.

The line with equation $y = -x + 1.8$

1. -1 **2.** 1.8

The circle with equation $x^2 + (y + 7)^2 = 100$

3. $(0, -7)$ **4.** 10 **5.** 20

A. diameter

B. slope

C. center

D. radius

E. y-intercept

SECTIONS 4.1, 4.2, *and* 4.3

For Exercises 6 and 7, (a) find the lengths of each side of the polygon, and (b) give the most specific name for the polygon.

6.

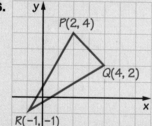

7.

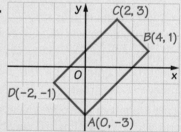

Find the slope of the line that contains each pair of points.

8. $(3, 7)$ and $(9, 4)$ **9.** $(-2, 9)$ and $(0, 1)$ **10.** $(0, 5)$ and $(0, 12)$

Give the slope and *y*-intercept of each line.

11. $y = 3x - 12$ **12.** $y = \frac{3}{4}x + 1$ **13.** $y = 4 - x$

For Exercises 14–17, write an equation for each line.

14. contains $(7, 3)$; has slope 2   **15.** contains $(0, 0)$; has slope $-\frac{3}{2}$

16. the vertical line that contains $(1, -1)$

17. contains $(9, -1)$; has slope 0

18. The vertices of $\triangle JKL$ are $J(3, 4)$, $K(4, 1)$, and $L(-2, -1)$. Use slopes to tell if $\triangle JKL$ is a right triangle.

19. Writing Suppose you know the coordinates of points P and M, and M is the midpoint of $\overline{PQ}$. Explain how to find the coordinates of Q.

Write an equation of the circle with the given center and radius.

20. center $(0, 0)$ and radius 3   **21.** center $(0, 0)$ and radius $\sqrt{5}$

22. center $(4, 2)$ and radius 9   **23.** center $(6, -2)$ and radius 2.5

24. a. Sketch the circle with equation $(x + 3)^2 + (y - 1)^2 = 36$.

    b. Give the coordinates of the center.

    c. Draw a radius and label it with its length.

Answer Exercises 25–27 to complete the proof.

Given: $\triangle PQR$ is isosceles.

    S, T, and O are the midpoints
    of $\overline{PQ}$, $\overline{QR}$, and $\overline{PR}$.

Prove: $\triangle STO$ is isosceles.

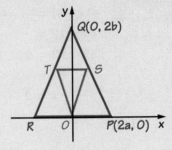

25. Find the coordinates of point R without using any new variables.

26. Find the coordinates of points S, T, and O.

27. a. Find the lengths TS, TO, and SO.

    b. Compare the lengths. Explain why $\triangle STO$ is isosceles.

28. Writing Give an example of an ordered triple, and explain how to locate that point in a three-dimensional coordinate system.

For Exercises 29 and 30, use points $M(-3, 5, 0)$ and $N(1, 7, -6)$.

29. Find the coordinates of the midpoint of $\overline{MN}$.

30. Find the length of $\overline{MN}$.

31. Open-ended Problem Sketch a rectangular prism that has point $(1, 3, 4)$ as a vertex. Label all the vertices of the prism.

PERFORMANCE TASK

32. Open-ended Problem Choose one of the following quadrilaterals: parallelogram, rectangle, rhombus, or square.

    a. Draw two examples of the quadrilateral on graph paper. One example should have vertices that are labeled with numbers as coordinates and the other should have vertices that are labeled with variables.

    b. State a fact about the quadrilateral and use coordinate geometry to prove the fact.

Algebra
Review/Preview

These exercises review algebra topics you will use in the next chapters.

SOLVING LINEAR EQUATIONS

Solve each equation.

Toolbox p. 695
Linear Equations

EXAMPLE

$$4(x + 3) = -x + 22$$
$$4x + 12 = -x + 22 \quad \longleftarrow \text{Use the distributive property.}$$
$$5x + 12 = 22 \quad \longleftarrow \text{Add } x \text{ to both sides.}$$
$$5x = 10 \quad \longleftarrow \text{Subtract 12 from both sides.}$$
$$x = 2 \quad \longleftarrow \text{Divide both sides by 5.}$$

1. $7x - 1 = 4x + 5$ **2.** $5(2u - 1) = 4(u - 1)$

3. $8s + 18 = 6(2s + 3)$ **4.** $2(r + 1) = 4r + 1$

Solve each equation.

EXAMPLE

$$\frac{a}{2} + \frac{a}{3} = 10$$
$$6\left(\frac{a}{2} + \frac{a}{3}\right) = 6 \cdot 10 \quad \longleftarrow \begin{array}{l}\text{Multiply both sides by the least}\\\text{common denominator, 6.}\end{array}$$
$$\frac{6a}{2} + \frac{6a}{3} = 60$$
$$3a + 2a = 60$$
$$5a = 60$$
$$a = 12 \quad \longleftarrow \text{Divide both sides by 5.}$$

5. $\frac{x}{2} + \frac{1}{7} = 1$ **6.** $\frac{y}{3} + \frac{y}{5} = 16$ **7.** $\frac{1}{12}w - \frac{1}{3}w = 2$

8. $\frac{5}{6}c + \frac{2}{15} = \frac{7}{10}c$ **9.** $\frac{t}{2} = \frac{3}{4} - \frac{6t}{24}$ **10.** $\frac{n + 3}{3} - \frac{n}{4} = \frac{n}{3}$

Solve each equation.

EXAMPLE

$$-1.6 + 3.68y = 4y$$
$$100(-1.6 + 3.68y) = 100(4y) \quad \longleftarrow \begin{array}{l}\text{Multiply both sides by 100 to}\\\text{eliminate the decimals.}\end{array}$$
$$-160 + 368y = 400y$$
$$-160 = 32y \quad \longleftarrow \text{Subtract 368}y \text{ from both sides.}$$
$$-5 = y \quad \longleftarrow \text{Divide both sides by 32.}$$

11. $1.3x + 7 = 10.9$ **12.** $8 - 0.7p = 2.18p + 3.5$

13. $2.2y - 12.8 = 6.6$ **14.** $3.12u - 5 = 4.37u$

Solve each system of equations.

> **EXAMPLE**
>
> $$x + y = 3$$
> $$5x - 2y = -13$$
>
> Solve one equation for y.
>
> $$y = 3 - x$$
>
> Substitute $3 - x$ for y in the other equation and solve for x.
>
> $$5x - 2(3 - x) = -13$$
> $$5x - 6 + 2x = -13$$
> $$7x - 6 = -13$$
> $$7x = -7$$
> $$x = -1$$
>
> To find y, substitute -1 for x in either of the original equations.
>
> $$x + y = 3$$
> $$-1 + y = 3$$
> $$y = 4$$
>
> The solution (x, y) is $(-1, 4)$.

> **Toolbox p. 694**
> *Evaluating Equations for Given Values*

15. $y = 2x$
$5x - y = 30$

16. $x + y = 15$
$4x + 3y = 38$

17. $2a + 3b = 0$
$a - 6b = -5$

18. $2p - q = 17$
$3p + 4q = -13$

19. $-r + 4s = -8$
$2r - 8s = 6$

20. $\frac{v}{2} = 2 - u$
$6u + 3v = 12$

21. Tickets to an amusement park cost \$20 for adults and \$12 for children. On one day, 4250 tickets were sold, and the total cost was \$63,400. How many adults and how many children visited the park that day?

Solve each inequality. Graph the solution on a number line.

> **EXAMPLES**
>
> $$4x - 7 \geq 1$$
> $$4x \geq 8$$
> $$\frac{4x}{4} \geq \frac{8}{4}$$
> $$x \geq 2$$
>
> $$-7 - 3m < 5$$
> $$-3m < 12$$
> $$\frac{-3m}{-3} > \frac{12}{-3}$$
> $$m > -4$$
>
> When multiplying or dividing by a negative number, reverse the direction of the inequality.

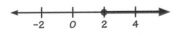

> **Toolbox p. 705**
> *Solving Inequalities*

22. $x - 2 \geq 3$

23. $\frac{2}{3}z < 12$

24. $-\frac{3}{5}y \geq 6$

25. $12p + 3 \geq 15$

26. $-8d + 19 < -29$

27. $5(1 - k) \leq 4(3 - k)$

5 | Parallel Lines

Watching the Colors Come Alive

José Saínz

José Saínz had no idea what he was getting into when a friend at work asked him to check out a dual-line stunt kite during lunch. But he was so taken with the stunt kite, which could perform tricks in the air, that by the end of the week he had bought one of his own. That was in 1989, and his life hasn't been the same since.

Now one of the top kite designers in the country, the Mexican-born Saínz spends two to three hours every night in his home workshop in San Diego, planning and building his latest creations. It's a hobby that takes him to kite exhibitions and contests all over the world, in which he often captures the top prizes. "Ironically, the friend who introduced me to all this dropped out of kiting soon afterwards," he says. "But I got stuck."

It's All in the Design

Saínz uses a variety of geometric shapes in his kite designs. His first kite, a bird design called "Ave," was square. His award-winning Aztec Calendar kite is a giant hexagon with a tail that is 55 ft long. Other designs are rectangular, triangular, hexagonal, or diamond-shaped.

In competition, kites are judged for aesthetics, structural integrity, craftsmanship, and, most importantly, flight characteristics. "The kite should be stable in the air and fly at a good angle, the higher the better," Saínz explains. The angle at which a conventional single-line kite will fly (and therefore the height of the kite), depends on its shape and the wind speed. "If you have two kites of the exact same shape, they will fly at the same angle and their lines will be parallel," he adds. The two kiters can stand very close together, yet the lines will not get tangled.

> **"The kite should be stable in the air and fly at a good angle, the higher the better."**

217

Synchronized Precision Maneuvers

Dual-line stunt kites are guided by two lines that run parallel from the kiter's hands to the *bridle points*, where they attach to the kite. One of the most exciting events in an exhibition takes place when members of a four-person team stand side by side, each flying a dual-line, controllable kite, performing synchronized precision maneuvers in the sky. "It's like an air show," Saínz explains. "All these kites are doing flips and loops at the same time which, of course, requires maintaining parallel lines."

In events where a three-person team stands side by side, the teammates on the left and right try to keep their lines parallel with the lines of the center person. As a result, the left and right teammates' lines are parallel to each other. This observation leads to the *Dual Parallels Theorem*, which you will learn more about in this chapter.

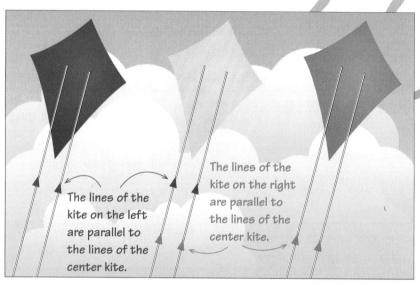

The lines of the kite on the left are parallel to the lines of the center kite.

The lines of the kite on the right are parallel to the lines of the center kite.

You can use the Dual Parallels Theorem to conclude that the lines of the left and right kites are parallel.

Explore and Connect

José Saínz with a *rokkaku* (Japanese for "six-sided") kite.

1. Writing In the team formation shown above, if the lines of the right-end person were not parallel to the lines of the center person, how would the lines of the right-end person be related to the lines of the left-end person?

2. Research Kites are used for practical as well as recreational uses. Write a short paragraph about one practical use for a kite. Explain the function of the kite and describe its structure.

3. Project Design a kite. It must have a balanced design in order to fly. First make a sketch of your kite. Then use paper, straws, string, and other materials to make the kite.

Mathematics
& José Saínz

In this chapter, you will learn more about how mathematics is related to kite design.

Related Exercises

Section 5.2
• Exercises 18–21

Section 5.5
• Exercises 4–6

Section 5.6
• Exercises 29 and 30

5.1 Parallel Lines and Transversals

Learn how to...
- identify pairs of angles formed by transversals and lines

So you can...
- find the measures of these angles
- analyze real-world examples of intersecting lines

What produces the oxygen that you breathe? Plants feeding themselves! The leaves of a plant produce food by combining water with carbon dioxide that people exhale. In the process, plants make the oxygen that people and animals depend on. The veins of a leaf are pipelines that make it all possible by carrying water and food through the leaf. In many leaves, the veins resemble intersecting lines.

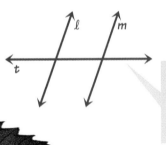

A **transversal** is a line that intersects two or more other lines in the same plane at different points.

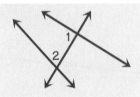

A transversal forms several angles with the two lines it intersects. Special names are given to pairs of these angles.

Same-side interior angles lie on the same side of a transversal between the two lines that it intersects.

∠1 and ∠2 are same-side interior angles.

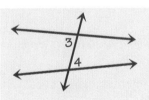

Alternate interior angles lie on opposite sides of a transversal between the two lines that it intersects.

∠3 and ∠4 are alternate interior angles.

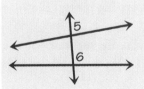

Corresponding angles lie on the same side of a transversal, in corresponding positions with respect to the two lines that it intersects.

∠5 and ∠6 are corresponding angles.

EXAMPLE 1

Name all pairs of same-side interior angles, alternate interior angles, and corresponding angles in this figure.

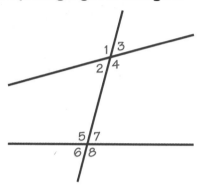

SOLUTION

Same-side interior angles	Alternate interior angles	Corresponding angles
∠2 and ∠5	∠2 and ∠7	∠1 and ∠5
∠4 and ∠7	∠4 and ∠5	∠2 and ∠6
		∠3 and ∠7
		∠4 and ∠8

THINK AND COMMUNICATE

Use the diagram.

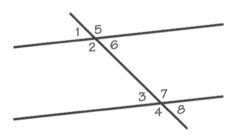

1. Name a pair of same-side interior angles and a pair of alternate interior angles.

2. Why is the word *interior* used in the terms *same-side interior angles* and *alternate interior angles*?

3. Why are same-side interior angles called *same-side*? Why are alternate interior angles called *alternate*?

4. a. Estimate and compare the measures of two corresponding angles in the diagram above.

   b. Estimate and compare the measures of two corresponding angles from Example 1.

EXPLORATION
COOPERATIVE LEARNING

Angles and Transversals

**Work in a group of three students.
You will need:**

• lined paper or geometry software

• a ruler or straightedge

• a protractor

1 Working individually, each member of your group should draw two parallel lines, using a ruler and paper or geometry software. Discuss how you know that the lines you drew are parallel.

2 Draw a transversal through the lines and number the angles formed.

3 Measure each angle in your diagram and record the results in a table.

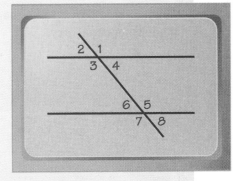

Angle	Measure
1	
2	
3	
4	
5	
6	
7	

4 Compare the measures of two corresponding angles. What do you notice? Compare the measures of a different pair of corresponding angles. Make a conjecture about corresponding angles.

5 Compare the measures of two same-side interior angles. What do you notice? Compare the measures of a different pair of same-side interior angles, and make a conjecture about same-side interior angles.

6 Compare the measures of two alternate interior angles. What do you notice? Measure a different pair of alternate interior angles. Make a conjecture about alternate interior angles.

7 Discuss the conjectures you made in Steps 4–6 with the other members of your group. Are your conjectures true for their diagrams?

8 If the lines you drew in Step 1 were *not* parallel, do you think your conjectures would still be true? Explain why or why not.

In the Exploration, you probably discovered several facts about the angles formed when parallel lines are intersected by a transversal. This postulate describes one of them.

Corresponding Angles Postulate

If two parallel lines are intersected by a transversal, then corresponding angles are congruent.

If $k \parallel \ell$, then $\angle 1 \cong \angle 2$.

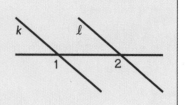

EXAMPLE 2   Connection: Algebra

Find the value of *x* and the measure of each numbered angle.

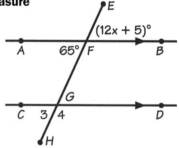

SOLUTION

$\angle EFB \cong \angle AFG$ because they are vertical angles.

$$(12x + 5)^\circ = 65^\circ$$
$$12x = 60$$
$$x = 5$$

$\angle 3 \cong \angle AFG$ because they are corresponding angles formed by a transversal and two parallel lines.

$$m\angle 3 = m\angle AFG$$
$$= 65^\circ$$

$\angle 4$ and $\angle 3$ are supplementary because they form a linear pair.

$$m\angle 4 = 180^\circ - m\angle 3$$
$$= 180 - 65$$
$$= 115^\circ$$

So $x = 5$, $m\angle 3 = 65^\circ$, and $m\angle 4 = 115^\circ$.

Use the diagram.

1. Which line is the transversal?

2. Name two pairs of same-side interior angles.

3. Name two pairs of alternate interior angles.

4. Name two pairs of corresponding angles.

5. Given that $m\angle 1 = 120°$, find:

    a. $m\angle 3$ **b.** $m\angle 6$ **c.** the value of x

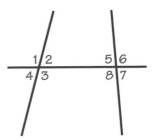

5.1   **Exercises and Applications**

Extra Practice exercises on page 667

Use the diagram for Exercises 1–4.

1. Name all pairs of same-side interior angles.

2. Name all pairs of alternate interior angles.

3. Name all pairs of corresponding angles.

4. **Writing** Find two pairs of angles that could be called *same-side exterior angles*, and explain why that term could be used.

Classify each pair of angles as *corresponding angles*, *alternate interior angles*, or *same-side interior angles*.

5. $\angle 5$ and $\angle 7$ 6. $\angle 7$ and $\angle 3$

7. $\angle 3$ and $\angle 1$ 8. $\angle 6$ and $\angle 2$

9. $\angle 2$ and $\angle 3$ 10. $\angle 8$ and $\angle 6$

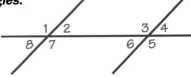

AUTOMOBILES The windshield wipers of this car move in such a way that they are always parallel to each other.

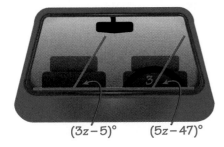

11. **ALGEBRA** Find the value of z and $m\angle 3$.

12. If you know that $\angle 4 \cong \angle 7$, what can you say about the windshield wipers? Explain.

Beech trees are found throughout the Northern Hemisphere. The secondary veins of the European beech, which all intersect the central vein, are often parallel to each other.

BY THE WAY

13. Is k a transversal of m and ℓ? Explain why or why not.

14. Given that $m \angle 1 = 142°$, find $m \angle 2$ and $m \angle 3$.

15. If you know $m \angle 2$, can you use the Corresponding Angles Postulate to find $m \angle 4$? Explain why or why not.

16. Research Find out about another type of leaf that has veins like this one. What makes this pattern of veins efficient?

Many insects, known as leaf insects, are naturally camouflaged because their wings mimic the veined pattern of a leaf.

ALGEBRA In Exercises 17 and 18, find each value specified.

17. Find the values of x, y, and z.

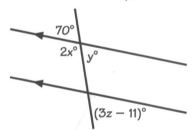

18. Find the value of w and $m \angle 1$.

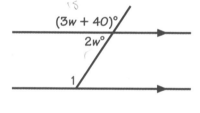

Classify each pair of angles as *corresponding angles, alternate interior angles, same-side interior angles,* or *none of these.*

19. $\angle 2$ and $\angle 3$

20. $\angle 7$ and $\angle 12$

21. $\angle 3$ and $\angle 16$

22. $\angle 10$ and $\angle 2$

23. $\angle 4$ and $\angle 10$

24. $\angle 6$ and $\angle 3$

25. $\angle 7$ and $\angle 11$

26. $\angle 14$ and $\angle 16$

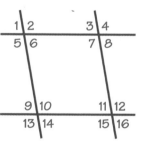

27. Open-ended Problem Sketch an example of lines intersected by a transversal in a piece of furniture, in the architecture of your school building, or in some other object. Label a pair of corresponding angles and a pair of same-side interior angles.

28. Copy and complete the proof.

Given: $m \parallel n; j \parallel k$
Prove: $\angle 1 \cong \angle 4$

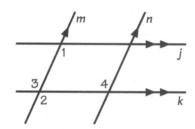

Statements	Reasons
1. $m \parallel n, j \parallel k$	**1.** ?
2. $\angle 1 \cong$?	**2.** If two $\parallel$ lines are intersected by a transversal, then corresponding angles are $\cong$.
3. $\angle 2 \cong \angle 3$	**3.** ?
4. $\angle 3 \cong \angle 4$	**4.** ?
5. ?	**5.** Transitive Property (Steps 2, 3, and 4)

Use this regular hexagon for Exercises 29–31.

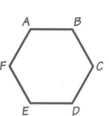

29. Name the three pairs of parallel segments in the hexagon.

30. Visual Thinking If $\overline{AB}$, $\overline{BC}$, and $\overline{ED}$ are extended into lines, which one is the transversal?

31. Challenge If all six sides of the hexagon are extended into lines, which of the lines are transversals of $\overleftrightarrow{AB}$ and $\overleftrightarrow{BC}$? Explain.

ONGOING ASSESSMENT

32. Open-ended Problem Sketch two pairs of parallel lines that intersect to form a parallelogram.

a. On your diagram, label a pair of same-side interior angles, a pair of alternate interior angles, and a pair of corresponding angles.

b. Choose one of the pairs from part (a) in which the two angles are congruent. Explain how you know that they are congruent.

SPIRAL REVIEW

33. The endpoints of $\overline{AB}$ are $A(3, -7, 4)$ and $B(2, 0, 6)$. Find the midpoint and length of $\overline{AB}$. *(Section 4.6)*

Tell whether each statement is *True* or *False*. If the statement is false, sketch a counterexample. *(Section 2.5)*

34. Every rectangle is a parallelogram.

35. If two sides of a quadrilateral are parallel, then it is a parallelogram.

5.2 Properties of Parallel Lines

Learn how to...

- **find the measures of alternate interior angles and same-side interior angles**

- **identify trapezoids**

So you can...

- **prove statements about special angles**

- **find congruent angles in kites, windows, and other real-world objects**

Cloth embroidered by the Shoowa people of central Zaire is admired for the beauty and complexity of its geometric designs. In the piece shown, parallel lines appear to be woven, suggesting basketwork.

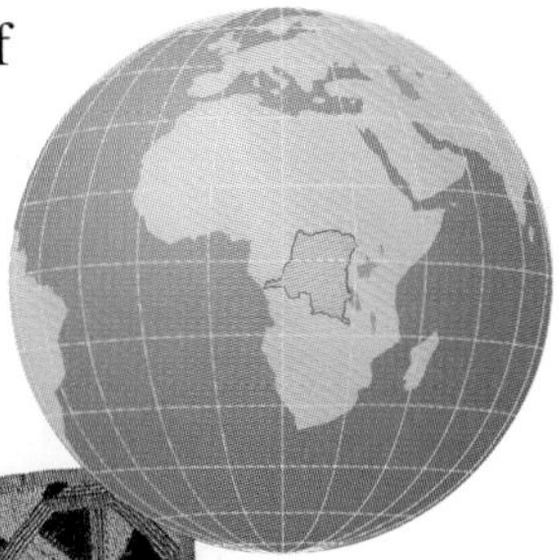

You know that the corresponding angles ∠1 and ∠2 are congruent. Also, ∠2 and ∠3 are congruent. Since ∠1 ≅ ∠2 and ∠2 ≅ ∠3, you can conclude that the alternate interior angles ∠1 and ∠3 are congruent.

THINK AND COMMUNICATE

Use the section of Shoowa cloth above.

1. How do you know that ∠2 ≅ ∠3?

2. If ∠1 ≅ ∠2 and ∠2 ≅ ∠3, how can you justify the conclusion that ∠1 ≅ ∠3?

Alternate Interior Angles Theorem

If two parallel lines are intersected by a transversal, then alternate interior angles are congruent.

If $k \parallel \ell$, then $\angle 1 \cong \angle 3$.

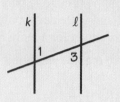

EXAMPLE 1   Connection: Algebra

Find the value of *y*.

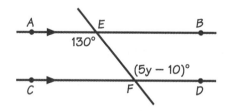

SOLUTION

Because $\angle EFD$ and $\angle AEF$ are alternate interior angles formed by a transversal and two parallel lines, they are congruent.

$$m \angle EFD = m \angle AEF$$
$$(5y - 10)° = 130°$$
$$5y = 140 \qquad \text{Solve for } y.$$
$$y = 28$$

THINK AND COMMUNICATE

Use the diagram in Example 1.

3. What is $m \angle EFD$?

4. What is $m \angle EFC$? How do you know?

5. What do you notice about $m \angle AEF$ and $m \angle EFC$?

Same-Side Interior Angles Theorem

If two parallel lines are intersected by a transversal, then same-side interior angles are supplementary.

If $j \parallel k$, then $m \angle 4 + m \angle 5 = 180°$.

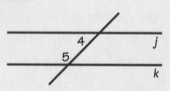

You already know how to write proofs of theorems in two different ways: two-column proof and paragraph proof. Sometimes, you may want to be less formal and write down only the *key steps of a proof*. The key steps can also serve as a plan for a more formal proof.

Here are the key steps of a proof of the Same-Side Interior Angles Theorem.

Given: $j \parallel k$
Prove: $m \angle 4 + m \angle 5 = 180°$

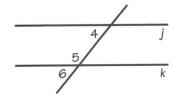

1. $m \angle 6 + m \angle 5 = 180°$ (The ∠s form a linear pair.)

2. $\angle 4 \cong \angle 6$ (Corresponding ∠s formed by $\parallel$ lines and a transversal are $\cong$.)

3. $m \angle 4 + m \angle 5 = 180°$ (Substitution)

A quadrilateral with exactly one pair of parallel sides is called a **trapezoid**. Each of the quadrilaterals highlighted in red below is a trapezoid.

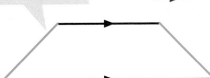
The two parallel sides of a trapezoid are called the **bases**.

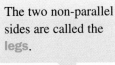
The two non-parallel sides are called the **legs**.

THINK AND COMMUNICATE

Use trapezoid WXYZ.

6. Which sides are the bases?

7. Which sides are the legs?

8. Name a pair of same-side interior angles.

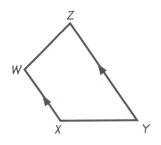

You can prove statements about trapezoids using the postulates and theorems that you know about parallel lines.

EXAMPLE 2

In trapezoid *ABCD*, $\overline{AB} \parallel \overline{DC}$. Write a paragraph proof showing that $m \angle A + m \angle D = 180°$.

SOLUTION

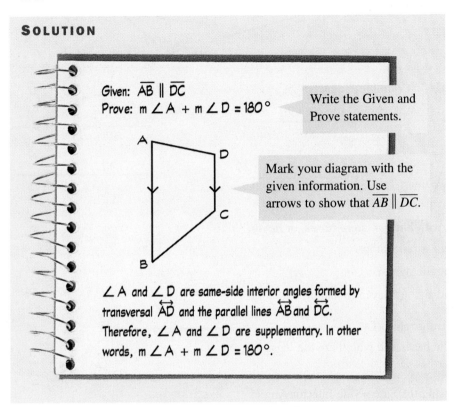

Given: $\overline{AB} \parallel \overline{DC}$
Prove: $m \angle A + m \angle D = 180°$

Write the Given and Prove statements.

Mark your diagram with the given information. Use arrows to show that $\overline{AB} \parallel \overline{DC}$.

$\angle A$ and $\angle D$ are same-side interior angles formed by transversal $\overleftrightarrow{AD}$ and the parallel lines $\overleftrightarrow{AB}$ and $\overleftrightarrow{DC}$. Therefore, $\angle A$ and $\angle D$ are supplementary. In other words, $m \angle A + m \angle D = 180°$.

☑ CHECKING KEY CONCEPTS

Find the values indicated.

1. Find the value of z and $m \angle 1$.

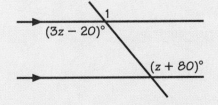

$(3z - 20)°$

$(z + 80)°$

2. Find the value of x and $m \angle 2$.

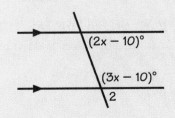

$(2x - 10)°$

$(3x - 10)°$

3. Sketch a trapezoid in which one base is twice as long as the other base.

5.2 **Exercises and Applications**

Extra Practice
exercises on
page 668

Find the measure of each numbered angle.

1.

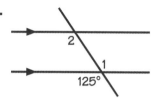

2.

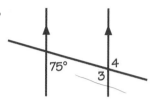

3.

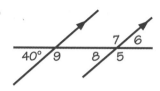

ALGEBRA Find the value of each variable.

4.

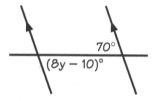

5.

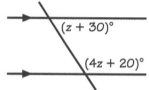

6.

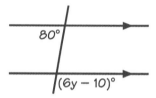

Complete each statement with *always*, *sometimes*, or *never*.

7. A trapezoid ? has two parallel sides.

8. A trapezoid is ? a parallelogram.

9. The legs of a trapezoid are ? the same length.

For Exercises 10–12, use trapezoid *JKLM*.

10. Tell which sides are the bases and which are the legs.

11. Find $m \angle M$.

12. Can you find $m \angle K$ using the Same-Side Interior Angles Theorem? If so, find it. If not, explain why not.

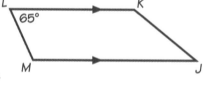

13. a. Sketch a copy of the diagram and mark it with the given information.

Given: $\overleftrightarrow{WX} \parallel \overleftrightarrow{YZ}$
Prove: $\angle 1 \cong \angle 2$

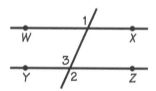

b. Copy and complete these key steps of the proof.

1. ? (Corresponding $\angle$s formed by $\parallel$ lines and a transversal are $\cong$.)

2. $\angle 3 \cong \angle 2$ (?)

3. ? (Transitive Property)

Sketch each trapezoid, if possible. If it is not possible, explain why not.

14. Each of the bases is longer than each leg.

15. The trapezoid has two right angles.

16. Each of the bases is shorter than each leg.

17. The trapezoid has three acute angles.

Look back at the article on pages 216–218.

Without its 55 ft long tail, the Aztec Calendar kite designed by José Saínz is a regular hexagon. The cloth is supported by six hollow sticks that are joined at the center of the kite and extend to the six vertices of the hexagon.

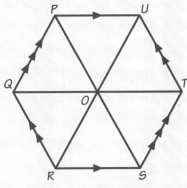

BY THE WAY

The Aztec Calendar kite won first prize in its class, the Grand Champion Award, and the People's Choice Award at the American Kitefliers Association's annual convention.

18. $\overleftrightarrow{PS}$ is a transversal of $\overleftrightarrow{UP}$ and $\overleftrightarrow{SR}$. Name another pair of parallel lines that have $\overleftrightarrow{PS}$ for a transversal.

19. Open-ended Problem Name a pair of angles on the kite that the Alternate Interior Angles Theorem guarantees are congruent.

20. Writing Explain why all of the acute angles are congruent in the diagram of the kite above.

21. Copy the diagram of the kite and extend $\overline{QR}$ and $\overline{UT}$ into lines. Label a point on each new ray. Name two angles that are supplementary by the Same-Side Interior Angles Theorem.

For Exercises 22 and 23, use the diagram.

22. SAT/ACT Preview Which angles are supplementary to $\angle 1$?

A. $\angle 2$ and $\angle 4$ only   **B.** $\angle 4$, $\angle 5$, and $\angle 8$

C. $\angle 2$, $\angle 5$, and $\angle 6$ **D.** $\angle 2$, $\angle 4$, $\angle 6$, and $\angle 8$

23. Tell which angles are congruent to $\angle 7$.

Cooperative Learning **For Exercises 24–27, work with another person.**

24. Each of you should use a ruler to draw a large diagram of two lines that are not parallel. Draw three non-parallel transversals of the lines and label three pairs of same-side interior angles *on the same side of the transversals*, as shown.

25. Measure the angles. Find $m \angle 1 + m \angle 2$ and $m \angle 3 + m \angle 4$.

26. Predict the value of $m \angle 5 + m \angle 6$. Then check your prediction by measuring.

27. Writing Compare your results with your partner's results. What can you conclude?

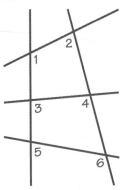

This stained glass window was designed by Frank Lloyd Wright for the Bradley house in Kankakee, Illinois. In the part of the window shown, $\overleftrightarrow{AC} \parallel \overleftrightarrow{HD}$ and $\overleftrightarrow{CE} \parallel \overleftrightarrow{BF}$.

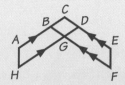

BY THE WAY

Frank Lloyd Wright (1867–1959) was one of the most famous and productive architects of his time.

28. Use the Corresponding Angles Postulate and the theorems that you learned in this lesson.

  a. Name four angles that are supplementary to $\angle CBG$.

  b. Name two angles that are congruent to $\angle CBG$.

29. Challenge Name four trapezoids in the part of the window shown above.

30. a. In parallelogram $WXYZ$, $m \angle X = 60°$. Sketch $WXYZ$ and find $m \angle W$ and $m \angle Y$.

  b. Sketch a diagram and write the key steps of a proof of this theorem: Consecutive angles of a parallelogram are supplementary.

ALGEBRA Find the measure of each numbered angle.

31.

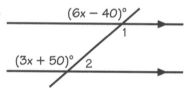

$(6x - 40)°$
1
$(3x + 50)°$ 2

32.

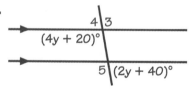

4 3
$(4y + 20)°$
5 $(2y + 40)°$

33. Write a two-column proof of the Alternate Interior Angles Theorem.

34. Each sentence of this paragraph proof has something wrong with it. Rewrite the proof correctly.

Given: $j \parallel k$; $m \perp j$
Prove: $m \perp k$

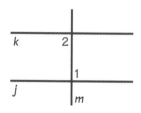

Since $j \parallel k$, $\angle 1$ is a right angle. Because $\angle 1$ and $\angle 2$ are corresponding angles formed by two parallel lines and a transversal, $\angle 1 \cong \angle 2$. Therefore, $\angle 2$ is an acute angle. So by the definition of perpendicular lines, $m \perp j$.

35. Open-ended Problem Sketch two parallel lines and two transversals that form a trapezoid. Number all of the angles formed.

a. List three pairs of supplementary angles. Explain how you know that they are supplementary.

b. Use your results from part (a) to write a theorem about the angles of a trapezoid.

SPIRAL REVIEW

ALGEBRA Find each unknown angle measure. *(Section 2.2)*

36.

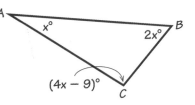

37.

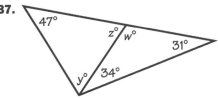

ASSESS YOUR PROGRESS

VOCABULARY

transversal (p. 219)
same-side interior angles (p. 219)
alternate interior angles (p. 219)
corresponding angles (p. 219)

trapezoid (p. 228)
bases of a trapezoid (p. 228)
legs of a trapezoid (p. 228)

Use the diagram. *(Section 5.1)*

1. Name all pairs of same-side interior angles, alternate interior angles, and corresponding angles.

ALGEBRA Find the value of each variable. *(Sections 5.1 and 5.2)*

2.

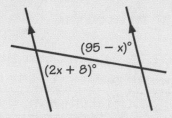

3.

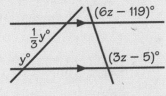

4. Sketch and label a trapezoid. Tell which sides are the bases and which are the legs. *(Section 5.2)*

5. Journal Sketch two parallel lines and a transversal and label the angles formed. Name all of the angles that are supplementary to one of the angles, and explain how you know they are supplementary.

5.3 | Types of Proofs

Learn how to...
- **write a flow proof**
- **introduce auxiliary lines in a diagram**

So you can...
- **choose from a variety of methods for proving statements**

Have you ever gotten stuck while trying to prove something? There are many different strategies you can use. One proof format that may help you to organize your ideas is called **flow proof**. As you can see in the proof below, this method displays the relationships between the statements in a proof.

Given: $n \parallel p$
Prove: $m \angle 1 + m \angle 2 = 180°$

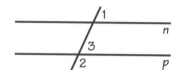

One way to start a flow proof is to write the given information on the left side of your paper and the statement that will be proved on the right side. You need to show how these statements can be logically connected.

① $n \parallel p$

$m \angle 1 + m \angle 2 = 180°$

<u>Reasons</u>
1. Given

You can try out different ideas with a flow proof. For example, even if you are not sure how to show that $\angle 1$ is supplementary to $\angle 2$, you might think it is related to the fact that $\angle 3$ is supplementary to $\angle 2$.

① $n \parallel p$

② $m \angle 3 + m \angle 2 = 180°$

? $m \angle 1 + m \angle 2 = 180°$

<u>Reasons</u>
1. Given
2. Angles in a linear pair are supplements.

If you can substitute $m \angle 1$ for $m \angle 3$, the proof will be complete. You can, because $\angle 1$ and $\angle 3$ are congruent and therefore $m \angle 1 = m \angle 3$.

The complete flow proof shows how the given information leads logically to the conclusion.

① $n \parallel p$ ③ $\angle 1 \cong \angle 3$ ($m \angle 1 = m \angle 3$) ④ $m \angle 1 + m \angle 2 = 180°$

② $m \angle 3 + m \angle 2 = 180°$

<u>Reasons</u>
1. Given
2. Angles in a linear pair are supplements.
3. If two ‖ lines are intersected by a transversal, then corresponding $\angle$s are $\cong$.
4. Substitution Property

Using a line or segment that is not part of the given information can sometimes help you prove a statement. An *auxiliary line* is a line that is added to a diagram to help complete a proof.

These two postulates are often used to justify including auxiliary lines in diagrams. You can use them in proofs about parallels and perpendiculars.

Parallel Postulate

Through a point not on a given line, there is exactly one line parallel to the given line.

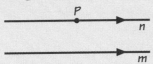

Through *P*, there is exactly one line *n* that is parallel to *m*.

Perpendicular Postulate

Through a point not on a given line, there is exactly one line perpendicular to the given line.

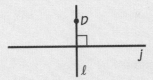

Through *D*, there is exactly one line *ℓ* that is perpendicular to *j*.

THINK AND COMMUNICATE

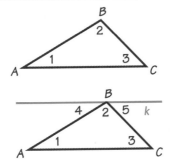

1. Does either of the theorems in Section 5.2 apply to this diagram? Why or why not?

2. In order to apply the theorems in Section 5.2 to the second diagram, what do you need to know about lines *k* and $\overleftrightarrow{AC}$?

3. What can you conclude about the angles in the diagram if $k \parallel \overleftrightarrow{AC}$?

In Questions 1–3 above, you saw how to introduce an auxiliary line to help prove that the sum of the angles of a triangle is 180°. On the next two pages, this familiar result is proved as the *Triangle Sum Theorem*. The proof is shown in four of the formats that you have learned.

METHODS OF PROOF

In each of these proofs of the Triangle Sum Theorem, the same chain of reasoning is used to connect the given information with the conclusion. You can choose which method of proof to use based on how you want to organize the information and how many details you want to include.

Given: $\triangle ABC$

Prove: $m\angle 1 + m\angle 2 + m\angle 3 = 180°$

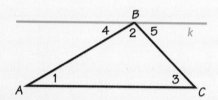

Key Steps of a Proof

Give the main steps in the argument of a proof. The key steps can be used as a plan for any of the other types of proof.

1. $m\angle 4 + m\angle 2 + m\angle 5 = 180°$ (The $\angle$s form a straight angle.)
2. $\angle 4 \cong \angle 1$, $\angle 5 \cong \angle 3$ (Alternate interior $\angle$s formed by $\parallel$ lines and a transversal are $\cong$.)
3. $m\angle 1 + m\angle 2 + m\angle 3 = 180°$ (Substitution)

Two-Column Proof

Organize the statements and reasons in separate columns. Include all necessary steps.

Statements	Reasons
1. Draw line k parallel to $\overleftrightarrow{AC}$ through B.	1. Parallel Postulate
2. $m\angle 4 + m\angle 2 + m\angle 5 = 180°$	2. Angle Addition Postulate and definition of a straight angle
3. $\angle 4 \cong \angle 1$ and $\angle 5 \cong \angle 3$	3. If two $\parallel$ lines are intersected by a transversal, then alternate interior angles are $\cong$.
4. $m\angle 4 = m\angle 1$ and $m\angle 5 = m\angle 3$	4. Definition of congruent angles
5. $m\angle 1 + m\angle 2 + m\angle 3 = 180°$	5. Substitution Property (Steps 2 and 4)

Statements and reasons combine to create a convincing argument that is made up of complete sentences.

According to the Parallel Postulate, there is exactly one line k that passes through B and is parallel to $\overleftrightarrow{AC}$. Line k forms a straight angle at B, so $m\angle 4 + m\angle 2 + m\angle 5 = 180°$. Because $k \parallel \overleftrightarrow{AC}$, the alternate interior angles $\angle 1$ and $\angle 4$ are congruent. For the same reason, $\angle 3$ and $\angle 5$ are congruent. Substituting $m\angle 1$ for $m\angle 4$ and $m\angle 3$ for $m\angle 5$ gives $m\angle 1 + m\angle 2 + m\angle 3 = 180°$.

Flow Proof

Use arrows to represent the reasoning process that connects the given information with the conclusion.

1 | Draw line k parallel to $\overleftrightarrow{AC}$ through B.

2 ↓

$\angle 4 \cong \angle 1$ $(m\angle 4 = m\angle 1)$
$\angle 5 \cong \angle 3$ $(m\angle 5 = m\angle 3)$

3 | $m\angle 4 + m\angle 2 + m\angle 5 = 180°$

4 ↓

$m\angle 1 + m\angle 2 + m\angle 3 = 180°$

Reasons

1. Parallel Postulate

2. If two $\parallel$ lines are intersected by a transversal, then alternate interior $\angle$s are $\cong$.

3. Angle Addition Postulate and definition of a straight angle

4. Substitution Property

THINK AND COMMUNICATE

4. Which of the methods of proof do you think is easiest to understand? Explain your choice.

5. Which method do you think is the most organized? Explain your choice.

6. Which of the methods of proof do you think shows the chain of logical reasoning the best? Explain your choice.

7. What other method of proof have you learned? How is it different from these methods? How is it like them?

Write a flow proof of this statement:

The sum of the measures of the angles of a quadrilateral is 360°.

SOLUTION

Plan Ahead: Sketch and label a diagram. To prove the statement, draw an auxiliary segment that splits the quadrilateral into two triangles.

Given: *ABCD* is a quadrilateral.

Prove: $m\angle A + m\angle B + m\angle C + m\angle D = 360°$

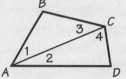

1 ┃ Draw $\overline{AC}$. → 2 ┃ $m\angle 1 + m\angle B + m\angle 3 = 180°$
　　　　　　　　　　　　　　　$m\angle 2 + m\angle 4 + m\angle D = 180°$

3 ↓

$m\angle 1 + m\angle 2 + m\angle B + m\angle 3 + m\angle 4 + m\angle D = 360°$

4 ┃ $m\angle 1 + m\angle 2 = m\angle A$
　　$m\angle 3 + m\angle 4 = m\angle C$

5 ↓

$m\angle A + m\angle B + m\angle C + m\angle D = 360°$

Reasons

1. Through any two points there is exactly one line.

2. The sum of the angle measures of a triangle is 180°.

3. Addition Property

4. Angle Addition Postulate

5. Substitution Property

☑ CHECKING KEY CONCEPTS

Questions 1 and 2 refer to this piece of a flow proof.

1. What statement should be in the blank in the proof labeled "2"?

2. **Open-ended Problem** Sketch a triangle that might accompany this flow proof.

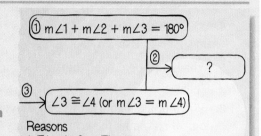

① $m\angle 1 + m\angle 2 + m\angle 3 = 180°$

② → ?

③ → $\angle 3 \cong \angle 4$ (or $m\angle 3 = m\angle 4$)

Reasons
1. Triangle Sum Theorem
2. Substitution Property

5.3 Exercises and Applications

Extra Practice
*exercises on
page 668*

For Exercises 1 and 2, copy and complete each proof.

1. **Given:** In $\triangle ABC$, $\angle A$ and $\angle B$ are complements.
   Prove: $m\angle C = 90°$

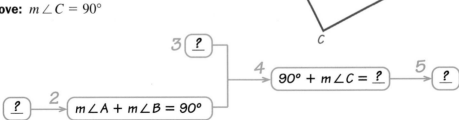

Reasons

1. Given

2. Definition of complementary angles

3. The sum of the angle measures of a triangle is 180°.

4. ?

5. ?

2. **Given:** $j \parallel k$; $m \parallel n$
   Prove: $\angle 1 \cong \angle 3$

   Since $\underline{\ ?\ } \parallel \underline{\ ?\ }$, it follows that $\angle 1 \cong \underline{\ ?\ }$
   because they are corresponding angles. Since $\underline{\ ?\ }$,
   it follows that $\underline{\ ?\ } \cong \angle 3$ because $\underline{\ ?\ }$.
   By the Transitive Property, $\underline{\ ?\ }$.

3. **Open-ended Problem** One advantage of using the flow proof method
   is that you can start at the beginning, middle, or end of the proof. How else
   might you have started writing the flow proof on page 234?

4. Rewrite the flow proof on page 234 as a two-column proof.

5. **Writing** In the piece of a
   proof to the right, Ashley
   introduces two auxiliary
   lines. Write a note to Ashley
   that explains what is wrong
   with her reasoning in each of
   the two sentences shown.

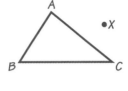

   Draw the line through A
   that is perpendicular to $\overline{BC}$
   at its midpoint. Then draw
   the line through A and X
   that is parallel to $\overline{BC}$.

**A *corollary* of a theorem is a statement that can be
easily proved using the theorem. Explain how each
corollary of the Triangle Sum Theorem follows from
the theorem.**

6. Each angle of an equiangular triangle measures 60°.

7. In a triangle, there can be at most one obtuse or right angle.

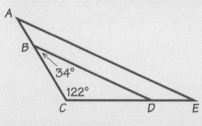

Spanish architect and engineer Santiago Calatrava designed the Alamillo Bridge for Exposicion Universal de 1992, in Seville, Spain. It is one of eight new bridges over the Guadalquivir River that were built for the Expo. The cables of the bridge are all parallel.

8. Sketch a copy of the diagram of the bridge at the left, including the two labeled angle measures. Find *m* ∠ *CAE*. What postulate or theorem did you use?

9. Find *m* ∠ *AEC*. What theorem or postulate did you use?

10. Find *m* ∠ *BDC*. What theorem or postulate did you use?

BY THE WAY

The 142 m tall pylon in the Alamillo Bridge is made of steel filled with concrete. It is heavy enough to counterbalance the deck of the bridge, so no cables are needed on the other side of the pylon.

11. Use this theorem: If two angles of one triangle are congruent to two angles of another triangle, then the third angles are congruent.

a. Does the theorem apply to the triangles that you explored in Exercises 8–10? Explain why or why not.

b. Write a two-column proof of the theorem using the diagrams below.

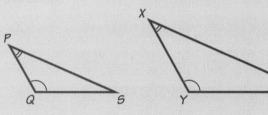

Use the method of proof indicated. First copy the diagram and mark it with the given information.

12. Flow proof
Given: $\overleftrightarrow{HV} \parallel \overleftrightarrow{TU}$; ∠6 ≅ ∠7
Prove: ∠7 ≅ ∠8

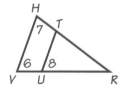

13. Key steps of a proof
Given: ∠1 ≅ ∠5
Prove: ∠2 ≅ ∠4

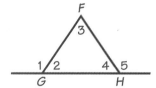

Use the method of proof indicated. First copy the diagram and mark it with the given information.

14. Flow proof

Given: $\overline{YU} \parallel \overline{ST}$; $\angle YXS \cong \angle UXT$

Prove: $\angle S \cong \angle T$

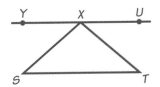

15. Two-column proof

Given: $\angle 2 \cong \angle 4$

Prove: $\angle 1 \cong \angle 3$

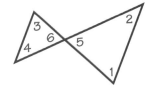

16. Paragraph proof

Given: $JKLM$ is a parallelogram.

Prove: $\angle 7 \cong \angle 9$

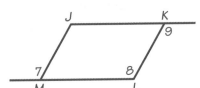

17. Use any method.

Given: $\overline{AB} \parallel \overline{CD}$; $\overline{BC} \parallel \overline{DE}$

Prove: $\angle ABC \cong \angle CDE$

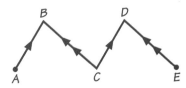

18. Challenge Sketch and label a diagram and write a two-column proof.

Given: In trapezoid $PQRS$, $\overline{PQ} \parallel \overline{RS}$ and $\angle P \cong \angle Q$.

Prove: $\angle R \cong \angle S$

ONGOING ASSESSMENT

19. Open-ended Problem Choose one of the proofs that you did in Exercises 12–18, and rewrite it using a different method of proof. Explain which method you like better and why.

SPIRAL REVIEW

20. ALGEBRA Find the measures of $\angle 1$ and $\angle 2$. *(Section 5.2)*

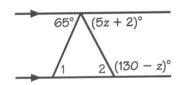

Write the converse of each statement. Tell whether the converse is *True* or *False*. *(Section 3.5)*

21. If a triangle is equilateral, then the triangle is isosceles.

22. If a bird has wings, then the bird can fly.

5.4

Conditions for Parallel Lines

Learn how to...
- apply the converses of theorems about parallel lines and transversals

So you can...
- use facts about angles to prove that two lines are parallel

Architects, engineers, and other designers may use drafting tools called T-squares and triangles to make precise technical drawings. In the Exploration, you will make your own triangle and use it to explore lines and transversals.

EXPLORATION
COOPERATIVE LEARNING

Drafting Parallel Lines

Work with another student.
You will need:
- a straightedge or ruler
- an index card
- scissors

1 Using a straightedge, draw one diagonal of the index card. Cut along the diagonal to form two right triangles, one for each student. Label the two triangles differently, as shown.

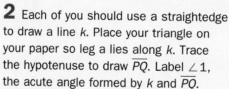

2 Each of you should use a straightedge to draw a line *k*. Place your triangle on your paper so leg a lies along *k*. Trace the hypotenuse to draw $\overline{PQ}$. Label ∠1, the acute angle formed by *k* and $\overline{PQ}$.

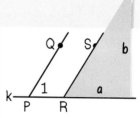

3 Slide the triangle along line *k* to a new position. Trace the hypotenuse to draw $\overline{RS}$. Label ∠2, the acute angle formed by *k* and $\overline{RS}$.

4 Use *k* as a transversal. What is the name for the pair of angles you labeled?

5 How do lines $\overleftrightarrow{PQ}$ and $\overleftrightarrow{RS}$ appear to be related? Compare your results with your partner's results. Make a conjecture based on your results.

Using the triangles in the Exploration, you probably discovered the following postulate. Notice that it is the converse of a postulate that you learned earlier.

Converse of the Corresponding Angles Postulate

If two lines are intersected by a transversal and corresponding angles are congruent, then the lines are parallel.

If $\angle 1 \cong \angle 2$, then $k \parallel l$.

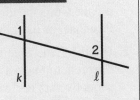

EXAMPLE 1 Connection: Algebra

Find the value of x that will allow you to prove that $\overleftrightarrow{AB} \parallel \overleftrightarrow{CD}$.

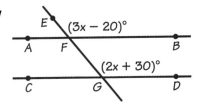

SOLUTION

$\angle EFB$ and $\angle FGD$ are corresponding angles. The postulate above states that if $\angle EFB \cong \angle FGD$, then $\overleftrightarrow{AB} \parallel \overleftrightarrow{CD}$.

$$m \angle EFB = m \angle FGD$$
$$(3x - 20)° = (2x + 30)°$$

> Subtract $2x$ from both sides.

$$3x - 2x - 20 = 30$$
$$x - 20 = 30$$
$$x = 50$$

If $x = 50$, then $m \angle EFB = 130°$ and $m \angle FGD = 130°$. So, $\overleftrightarrow{AB}$ will be parallel to $\overleftrightarrow{CD}$ if $x = 50$.

In the Exploration you saw that the converse of the Corresponding Angles Postulate is true. In the exercises, you will show that the converse of the Alternate Interior Angles Theorem is also true.

Converse of the Alternate Interior Angles Theorem

If two lines are intersected by a transversal and alternate interior angles are congruent, then the lines are parallel.

If $\angle 3 \cong \angle 4$, then $m \parallel n$.

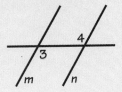

THINK AND COMMUNICATE

1. Suppose you know that $\overline{AB} \parallel \overline{CD}$ and you want to prove that $\angle BAC \cong \angle DCE$. Should you use the Corresponding Angles Postulate, or its converse?

2. Use the Same-Side Interior Angles Theorem to make a conjecture about two angles formed by $\overleftrightarrow{CD}$, $\overleftrightarrow{EF}$, and $\overleftrightarrow{AE}$.

3. What is the converse of your answer to Question 2?

Converse of the Same-Side Interior Angles Theorem

If two lines are intersected by a transversal and same-side interior angles are supplementary, then the lines are parallel.

If $m \angle 1 + m \angle 2 = 180°$, then $j \parallel k$.

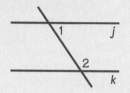

EXAMPLE 2

Prove the Converse of the Same-Side Interior Angles Theorem.

SOLUTION

Given: $m \angle 1 + m \angle 2 = 180°$
Prove: $j \parallel k$

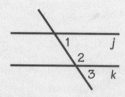

Statements	Reasons
1. $m \angle 1 + m \angle 2 = 180°$	**1.** Given
2. $m \angle 3 + m \angle 2 = 180°$	**2.** The $\angle\!\!s$ in a linear pair are supplementary.
3. $m \angle 1 + m \angle 2 = m \angle 3 + m \angle 2$	**3.** Transitive Property (Steps 1 and 2)
4. $m \angle 1 = m \angle 3$	**4.** Subtraction Property
5. $j \parallel k$	**5.** If two lines are intersected by a transversal and corresponding $\angle\!\!s$ are $\cong$, then the lines are $\parallel$.

Extra Practice
exercises on
page 669

☑ CHECKING KEY CONCEPTS

For each diagram, state the theorem or postulate that justifies why $\ell \parallel m$.

1.
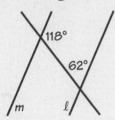

2.
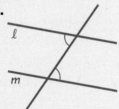

5.4 Exercises and Applications

ALGEBRA For each diagram, find the value of the variable that will allow you to prove that two lines are parallel.

1.
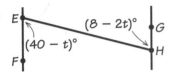

2.
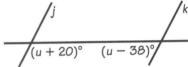

3.
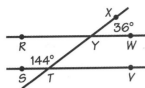

4.
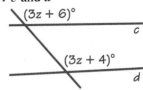

For each pair of lines indicated, tell whether they are *parallel* or *not parallel*. Explain your reasoning.

5. $\overleftrightarrow{RW}$ and $\overleftrightarrow{SV}$

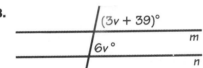

6. $\overleftrightarrow{EF}$ and $\overleftrightarrow{GH}$

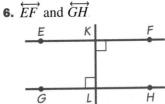

7. c and d

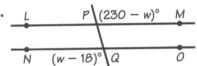

8. Sketch and label a diagram and write the key steps of a proof of the Converse of the Alternate Interior Angles Theorem.

For Exercises 9–11, use the diagram.

9. If $\angle DAC \cong \angle ACB$, which segments are parallel?

10. If $\angle DCA \cong \angle BAC$, which segments are parallel?

11. If $m \angle DCB + m \angle ABC = 180°$, which segments are parallel?

An *espalier* is a tree or shrub that is trained to grow in a flat plane, often in a regular pattern. Espalier gardening is often used when available space is limited. Also, the trellises on which espaliers are grown are often near walls, which shelters the plants from cold temperatures and frost.

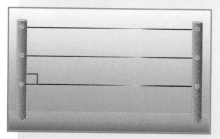

12. Open-ended Problem On this espalier trellis, the lowest wire is placed at a 90° angle to the vertical endposts. Describe how to position the remaining wires to make them parallel to the lowest one.

13. In the *palmette* style, shown in the middle diagram, all of the permanent branches on each side of the main trunk are parallel. Wooden stakes tied to the trellis wires train branches to grow in the desired direction.

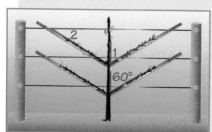

 a. With what measure of ∠1 should the stake on the right be tied to grow the upper right branch parallel to the lower right branch?

 b. For this espalier to be symmetrical about the main trunk, with what measure of ∠2 should the stake on the left be tied?

14. In the *single oblique* cordon style, single trunks (cordons) are grown in parallel rows.

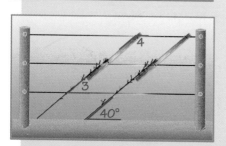

 a. If the right cordon is planted at an angle of 40° with the horizontal, describe how the other cordon should be positioned.

 b. Find $m\angle 3$ and $m\angle 4$. Explain your reasoning.

BY THE WAY

Commercial growers often grow fruit trees in oblique cordons to get them to produce fruit early. The closer to horizontal a fruit tree is, the more likely it is to produce fruit early.

15. a. Copy the diagram and mark it with the given information.

 b. Open-ended Problem Write a paragraph proof that uses either the Converse of the Corresponding Angles Postulate or the Converse of the Alternate Interior Angles Theorem.

 c. Write the theorem that you proved in part (b).

Given: $\angle JKF \cong \angle GLM$
Prove: $\overleftrightarrow{EF} \parallel \overleftrightarrow{GH}$

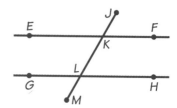

5.5 | Exercises and Applications

Extra Practice exercises on page 669

Use the diagram for Exercises 1–3.

1. What do you know about $\overleftrightarrow{LT}$ and $\overleftrightarrow{NU}$? Explain how you know.

2. Use your answer to Exercise 1. What can you conclude using the Dual Parallels Theorem?

3. **ALGEBRA** Find the value of x and $m\angle LPQ$.

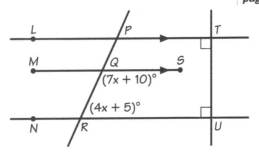

INTERVIEW **José Saínz**

Look back at the article on pages 216–218.

José Saínz has found inspiration for many of his designs in his own heritage. His Aztec Warrior kite displays the warrior in a black rectangle, surrounded by colorful flaps. There are eight supports across the back of the kite, and six vertical supports.

 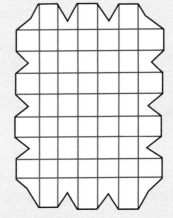

4. If all of the cross-supports are perpendicular to one vertical support, what do you know about all of the cross-supports? Explain.

5. In quadrilateral $ABCD$, $m\angle ABC = 60°$ and $m\angle BCD = 120°$.
   a. What do you know about $\overline{AB}$ and $\overline{CD}$? Explain how you know.
   b. What type of quadrilateral is $ABCD$?

6. If you know that $\overline{AB}$ is parallel to one of the cross-supports, what can you conclude about $\overline{CD}$? Explain.

EXAMPLE 2   Application: City Planning

In this map of part of Chicago, North Avenue is parallel to Chicago Avenue, and Orleans Street is perpendicular to both Division Street and Chicago Avenue. What can you conclude about ∠1 and ∠2?

SOLUTION

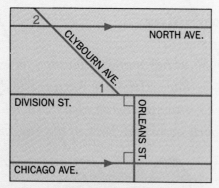

Division is parallel to Chicago, because they are both perpendicular to Orleans.

Because North is parallel to Chicago and Division is parallel to Chicago, the Dual Parallels Theorem allows you to conclude that North is parallel to Division.

∠1 ≅ ∠2, because they are corresponding angles formed by a transversal and two parallel lines.

☑ CHECKING KEY CONCEPTS

For Questions 1–3, use the diagram.

1. What can you conclude using the Dual Perpendiculars Theorem?

2. What do you know about $\overleftrightarrow{AB}$ and $\overrightarrow{FD}$? Explain how you know.

3. What can you conclude from your answers to Questions 1 and 2? Explain your reasoning.

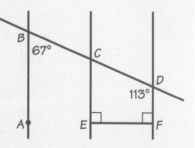

4. Sketch two coplanar lines and a transversal that is perpendicular to both of them. Describe all the ways that you can prove that the lines are parallel.

EXAMPLE 1

Write a paragraph proof of the Dual Perpendiculars Theorem.

SOLUTION

Write the given information and the statement you want to prove, and sketch a diagram showing the given information.

Given: Lines *m*, *n*, and *t*
are coplanar.
$m \perp t$; $n \perp t$

Prove: $m \parallel n$

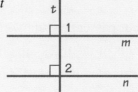

Using your diagram as a guide, write a proof of the theorem:

Since *m* and *n* are perpendicular to *t*, $\angle 1$ and $\angle 2$ are both right angles and are congruent. Since the transversal *t* forms congruent corresponding angles with the two lines *m* and *n*, $m \parallel n$.

Suppose you know that the first rung of a ladder is parallel to the second rung, the second rung is parallel to the third rung, and so on. The Dual Parallels Theorem will allow you to show that any two rungs of the ladder are parallel.

Dual Parallels Theorem

If two lines are both parallel to a third line, then the two lines are parallel.

If $j \parallel s$ and $k \parallel s$, then $j \parallel k$.

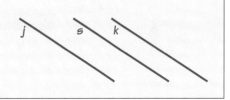

In the exercises, you will explain how to prove the Dual Parallels Theorem for lines that are coplanar.

THINK AND COMMUNICATE

1. If you used this diagram instead of the one shown in Example 1, would the proof for Example 1 be any different? Explain why or why not.

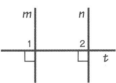

2. Are there other ways you could draw a diagram of three lines *j*, *k*, and *s* for the Dual Parallels Theorem? If there are, describe them.

5.5 Proving Theorems About Parallels

Learn how to...
- apply conditions to prove that lines are parallel

So you can...
- analyze real-world objects such as kites and desk lamps

For passenger safety, railway engineers need to make sure that the rails of the track are always the same distance apart. To measure the distance, they use a *gauge bar* that fits between the rails. The gauge bar must be perpendicular to both rails so that each measurement is accurate.

If the gauge bar fits snugly, then it is perpendicular to both rails.

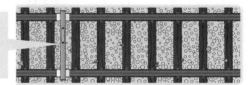

The following theorem guarantees that the rails are parallel if the gauge bar fits snugly.

Dual Perpendiculars Theorem

In a plane, if two lines are both perpendicular to a third line, then the two lines are parallel.

If *m, n,* and *t* are coplanar, $m \perp t$ and $n \perp t$, then $m \parallel n$.

The *distance between two parallel lines* is the length of a segment that connects the lines and is perpendicular to both of them.

The uniform distance between the rails of a track is called the *gauge.*

The **distance from a point to a line** is the length of the perpendicular segment from the point to the line. The distance from *Q* to *k* is 5 units.

21. In this two-column proof, each statement is paired with the correct reason, but the statements are in the wrong order. Write a correct proof by rearranging the lines in the proper order.

Given: $\angle 1 \cong \angle 3$

Prove: $\angle 4$ and $\angle 5$ are supplements.

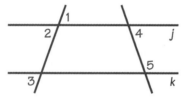

Statements	Reasons
1. $j \parallel k$	**1.** If two lines are intersected by a transversal and corresponding $\underline{\angle}$s are $\cong$, then the lines are $\parallel$.
2. $\angle 2 \cong \angle 3$	**2.** Transitive Property
3. $\angle 2 \cong \angle 1$	**3.** Vertical angles are congruent.
4. $\angle 4$ and $\angle 5$ are supplements.	**4.** If two $\parallel$ lines are intersected by a transversal, then same-side interior $\underline{\angle}$s are supplementary.
5. $\angle 1 \cong \angle 3$	**5.** Given

22. Open-ended Problem Sketch three lines m, n, and p, and a transversal that intersects them. Label all the angles formed. For each pair of lines, write two statements that could be used to prove that the lines are parallel.

23. Writing From this diagram, Ted concludes that $x = 60$. Explain how you know that he is wrong.

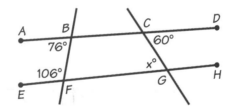

ONGOING ASSESSMENT

24. Open-ended Problem Explain how you can use the Converse of the Alternate Interior Angles Theorem and the triangle that you used in the Exploration on page 242 to draw parallel lines.

SPIRAL REVIEW

25. Write a flow proof. *(Section 5.3)*

Given: $\overleftrightarrow{LM} \parallel \overleftrightarrow{QW}$
$m\angle LMW + m\angle Q = 90°$

Prove: $\triangle WQD$ is a right triangle.

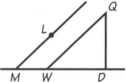

Write an equation for each line described. *(Section 4.3)*

26. The line passes through $P(-2, 3)$ and is parallel to $y = x - 4$.

27. The line passes through $R(0, -6)$ and is perpendicular to $y = -3x + 7$.

16. Copy and complete this flow proof.

Given: $\angle 3$ and $\angle 4$ are supplementary angles.
Prove: $h \parallel \ell$

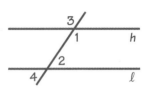

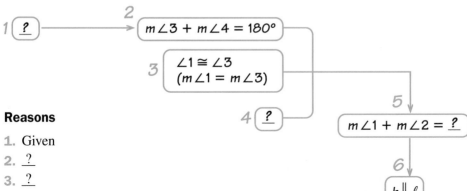

Reasons

1. Given
2. ?
3. ?
4. Vertical angles are congruent.
5. Substitution Property (Steps ?, ?, and ?)
6. ?

17. **Challenge** Find the value of each variable that will allow you to prove that *ABCD* is a parallelogram. Then find the measure of each angle of the parallelogram.

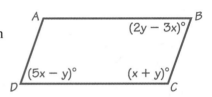

In dance, precise positioning of the arms and legs is crucial to give a pose its visual impact. In this move, each dancer must be sure that his leg is at the correct angle, so that it is parallel to the legs of the other dancers.

18. Given: $a \parallel b$; $\angle 1 \cong \angle 4$
   Prove: $c \parallel d$

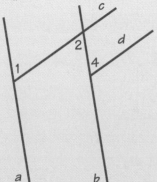

19. **Visual Thinking** Carefully redraw the diagram so that $\angle 1 \cong \angle 4$, but lines *a* and *b* are *not* parallel.

20. If the given information in Exercise 18 did not include "$a \parallel b$," would you still be able to prove that $c \parallel d$? Explain why or why not.

CITY PLANNING **For Exercises 7–10, use this street map.**

7. What can you conclude from the fact that Cleveland Avenue is perpendicular to both Webster Avenue and Grant Place? Explain your reasoning.

8. What does the Dual Perpendiculars Theorem tell you about Grant Place and Belden Avenue?

9. Based on your answers to Exercises 7 and 8, what do you know? Explain your reasoning.

10. Name each of the numbered angles that is congruent to ∠1, and explain how you know.

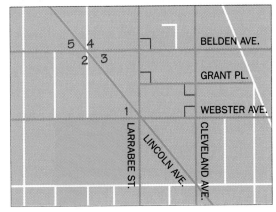

11. Find the distance from P to $\overleftrightarrow{AB}$.

12. Find the distance from O to $\overleftrightarrow{EF}$.

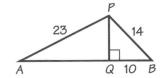

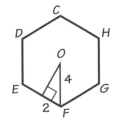

13. **Writing** The definition of the *distance from a point to a line* refers to the length of the perpendicular segment from the point to the line. Is it possible that there might be more than one such segment? Explain why or why not.

14. **SAT/ACT Preview** Which is greater, a or b?

    $a = LM$
    $b =$ the distance from L to $\overleftrightarrow{MN}$

    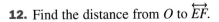

    A. a is greater.   **B.** b is greater.   **C.** a and b are equal.

15. **Open-ended Problem** Describe all the different ways that you can use to prove that two lines are parallel.

16. **Visual Thinking** Explain why the words *in a plane* are used in the statement of the Dual Perpendiculars Theorem. Include a sketch in your answer.

17. Carmela thinks that the distance between lines j and k is 5 units.

    a. **Writing** Do you agree or disagree with Carmela? Explain your answer.

    b. Find the distance between j and k.

18. **Writing** Explain why you can find the distance between two lines only if the lines are parallel.

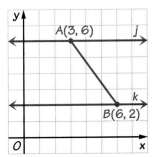

This desk lamp is designed with an adjustable arm so that different parts of the desk can be lit. The two quadrilaterals in the arm are flexible, making it possible to adjust the position of the lamp.

19. If $\overline{AD}$ and $\overline{BC}$ are both perpendicular to $\overline{AB}$, what do you know about $\overline{AD}$ and $\overline{BC}$? Explain your reasoning.

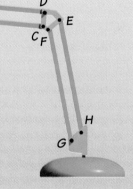

20. Logical Reasoning Is the information given in Exercise 19 enough for you to prove that *ABCD* is a rectangle? If so, explain how you would prove it. If not, give one additional piece of information that would allow you to prove it, and explain how.

21. Write a two-column proof.

Given: $\overline{FG} \parallel \overline{EH}$; $\angle FEH \cong \angle HGF$

Prove: *EFGH* is a parallelogram.

22. Challenge Find the exact distance from the point $P(6, 4)$ to the line $y = -\frac{1}{3}x - 4$. Describe your method.

23. Copy and complete this flow proof. First copy the diagram and mark it with the given information.

Given: $\angle 1 \cong \angle 2$; $\angle 3 \cong \angle 4$

Prove: $n \parallel p$

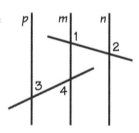

Reasons

1. ?
2. Given
3. ?
4. If two lines are intersected by a transversal and alternate interior angles are $\cong$, then the lines are $\parallel$.
5. ?

$$1 \boxed{\angle 1 \cong \angle 2} \xrightarrow{3} \boxed{n \parallel m}$$

$$2 \boxed{?} \xrightarrow{4} \boxed{?} \xrightarrow{5} \boxed{?}$$

ONGOING ASSESSMENT

24. Writing Use the diagram. Explain how drawing the transversal, *t*, allows you to prove the Dual Parallels Theorem for coplanar lines.

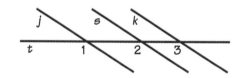

Sketch each situation. *(Section 1.4)*

25. Line *j* and line *k* are skew.

26. $\overline{PQ}$ lies in plane *W*.

27. The intersection of $\overleftrightarrow{RS}$ and plane *X* is point *T*.

Write an equation for each line described. *(Section 4.3)*

28. The line passes through $M(2, -5)$ and is parallel to the line $y = 3x - 18$.

29. The line passes through $N(3, 0)$ and is perpendicular to the line $y = 7x$.

ASSESS YOUR PROGRESS

VOCABULARY

flow proof (p. 234)
distance from a point to a line (p. 249)

Write a proof, using any method. *(Sections 5.3–5.5)*

1. Given: $m \angle J + m \angle L = m \angle K$
   Prove: $\triangle JKL$ is a right triangle.

2. Given: $\angle 1 \cong \angle 2; \overleftrightarrow{DC} \parallel \overleftrightarrow{BA}$
   Prove: $\overleftrightarrow{FE} \parallel \overleftrightarrow{BA}$

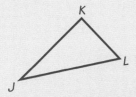

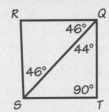

Tell which segments are parallel. Explain your reasoning. *(Section 5.4)*

3.

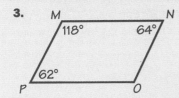

4.

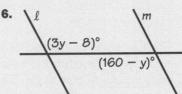

ALGEBRA **For each diagram, find the value of the variable that will allow you to prove that two lines are parallel.** *(Section 5.4)*

5.

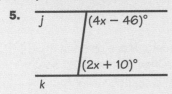

6.

7. Journal This chapter includes four different methods of proof: *paragraph proof, two-column proof, flow proof,* and *key steps of a proof.* Describe the advantages and disadvantages of each method.

5.6

Parallels in Space

There are two possibilities for how two planes can be related: the planes may be parallel or they may intersect in a line. But what happens if there are three planes?

EXPLORATION
COOPERATIVE LEARNING

Investigating Planes and Their Intersections

Work with another student. You will need:
- three index cards
- scissors
- two toothpicks

1 Cut and label the cards as shown.

2 Hold cards *A* and *B* so that they model parallel planes. Can you position plane *C* so it is parallel to both *A* and *B*? Can plane *C* intersect one plane but not the other? Can *C* intersect both planes? If so, how are the lines of intersection related?

3 Use cards *A* and *B* to model intersecting planes. Can plane *C* intersect one plane but not the other? Can *C* intersect both planes?

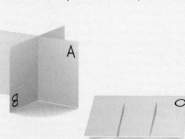

4 Can three planes form exactly one line of intersection? two lines? three lines? Explain.

5 Draw a pair of intersecting lines on one card. Place one end of a toothpick on the point of intersection and position the toothpick so it is perpendicular to both of the lines. A line that is perpendicular to both lines is perpendicular to the plane of the card.

6 Hold two toothpicks to model two different lines that are each perpendicular to the same card. Describe the relationship between the lines.

In the Exploration, you may have discovered this property of intersecting planes. The proof is not included.

> ### Intersecting Planes Theorem
>
> If two parallel planes are intersected by a third plane, then the lines of intersection are parallel.
>
> If plane $X \parallel$ plane Y, then $l \parallel n$.

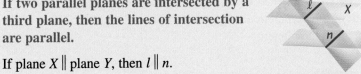

EXAMPLE 1

In this file box, faces *ABCD* and *EFGH* lie in parallel planes. How are edges $\overline{EF}$ and $\overline{AB}$ related?

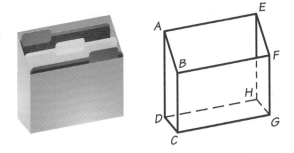

SOLUTION

$\overleftrightarrow{EF}$ and $\overleftrightarrow{AB}$ are the lines of intersection formed when the parallel planes *ABCD* and *EFGH* are intersected by plane *EFBA*. By the Intersecting Planes Theorem, the two lines are parallel. So $\overline{EF} \parallel \overline{AB}$.

THINK AND COMMUNICATE

Use the file box shown in Example 1.

   1. What does the Intersecting Planes Theorem allow you to conclude about $\overline{HG}$ and $\overline{DC}$? Explain.

   2. Noah believes he can use the Intersecting Planes Theorem to show that $\overline{AB} \parallel \overline{DC}$. Explain why this theorem *cannot* be used in this case.

You have learned that two lines parallel to a third line are parallel to each other. This theorem, which is not proved, shows that planes are related in the same way.

> ### Parallel Planes Theorem
>
> If two planes are both parallel to a third plane, then the two planes are parallel.
>
> If plane $P \parallel$ plane Z and plane $Q \parallel$ plane Z, then plane $P \parallel$ plane Q.

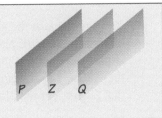

A **line is perpendicular to a plane** if it intersects the plane and is perpendicular to every line in the plane that passes through the point of intersection. In this figure, line $m \perp$ plane P.

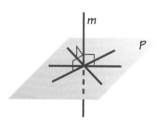

The **distance from a point to a plane** is the length of the perpendicular segment from the point to the plane.

EXAMPLE 2

Sketch and label a diagram for this theorem. Then write the Given and Prove statements.

If two lines are both perpendicular to the same plane, then the lines are parallel.

SOLUTION

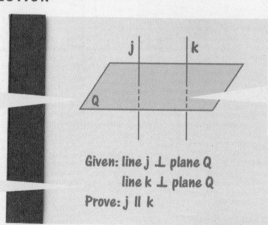

Sketch and label a diagram showing a plane and two lines perpendicular to the plane.

Use dashed lines for the parts of the lines that are hidden from view.

Use the labels in your sketch to write the Given and Prove statements.

Given: line $j \perp$ plane Q
line $k \perp$ plane Q
Prove: $j \parallel k$

☑ CHECKING KEY CONCEPTS

1. Use the cube at the right.
   a. $\overline{AB}$ is perpendicular to plane EAD. Name three segments that are perpendicular to $\overline{AB}$.
   b. Explain how to show that $\overline{AB} \parallel \overline{HG}$.

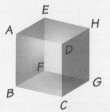

2. You can think of the slats of a vertical window blind as a group of planes. When the blind is partly open, each plane is parallel to the one next to it. Explain how you know that any two of the planes are parallel.

5.6 Exercises and Applications

Extra Practice exercises on page 670

Sketch each situation.

1. Plane $E \parallel$ plane F.

2. Plane A and plane B intersect in line p.

3. Lines j and k lie in plane Z and intersect at point D; line m is perpendicular to plane Z and also passes through D.

4. **Open-ended Problem** Tell which of the theorems and definitions that you learned in this section are illustrated by the planes formed by the triangles in the roof and the lines formed by the cross-supports. Explain your choices.

In this diagram, plane $X \parallel$ plane Y.

5. **a.** How are $\overleftrightarrow{AB}$ and $\overleftrightarrow{FC}$ related? Explain how you know.
   b. What type of quadrilateral is $ABDF$?

6. Suppose $m \angle ABD = 65°$. Find $m \angle FDB$.

7. Name two angles that are congruent to $\angle ABD$.

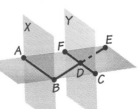

Connection ELECTRONICS

A *capacitor* is a basic component of televisions, radios, electronic toys, computers, and almost all electronic devices. Many old radios use an adjustable capacitor to tune in different stations.

8. What do you need to know about the front upper plate and the front lower plate in order to show that any upper plate and any lower plate lie in parallel planes? Explain your reasoning.

The planes of all of the plates in the upper group are parallel.

The planes of all of the plates in the lower group are parallel.

Tuning knob attaches here

9. **Writing** When the tuning knob of the radio is turned, the top group rotates to fit between the plates of the bottom group. Explain how the Parallel Planes Theorem shows that none of the plates from the upper group will touch any of the plates from the lower group.

In the diagram at the right, faces _MPTQ_ and _NOSR_ lie in parallel planes.

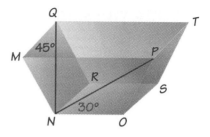

10. How are edges $\overline{OS}$ and $\overline{PT}$ related? Explain your reasoning.

11. Find $m \angle QNR$ and $m \angle MPN$, if possible. If it is not possible, explain why not.

For Exercises 12 and 13, sketch and label a diagram for the theorem. Then write the Given and Prove statements.

12. If a plane is perpendicular to one of two parallel lines, then the plane is also perpendicular to the other line.

13. If two planes are both perpendicular to the same line, then the planes are parallel.

14. a. Sketch a cube and label all eight vertices.

  b. Writing Use your diagram and the theorem in Exercise 13 to explain why two opposite faces of a cube lie in parallel planes.

AVIATION For Exercises 15–17, use this photograph of a Wright brothers' biplane. Although airplane wings are curved in order to produce lift, these two wings can be represented by parallel planes.

15. Each of the struts (vertical supports connecting the wings) is perpendicular to the plane of the upper wing. Explain how you know that all of the struts are parallel.

16. a. How do the struts appear to be related to the lower wing?

  b. Complete: If a line is perpendicular to one of two parallel planes, then _?_.

17. a. Sketch and label a diagram for the theorem that you completed in Exercise 16(b).

  b. Write the Given and Prove statements for the theorem.

In this diagram, $\overleftrightarrow{AD} \perp$ plane _CDE_.

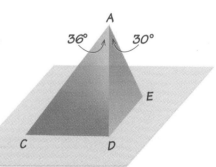

18. a. Find $m \angle ADC$. Explain your reasoning.

  b. Find $m \angle ACD$.

19. a. Find $m \angle AED$.

  b. Is it possible that $\overleftrightarrow{AE} \perp$ plane _CDE_? Explain your reasoning.

There are many ways to represent a three-dimensional object on a flat piece of paper. This cube is drawn using three different methods. *Oblique* drawing is the method you learned in Section 2.6.

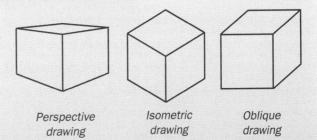

Perspective drawing   *Isometric drawing*   *Oblique drawing*

20. **Writing** Are segments that are parallel in a cube represented by parallel segments in each drawing?
Are segments that are congruent in a cube represented by congruent segments in each drawing? Write a comparison of the three methods based on your answers to these questions.

21. Follow the instructions at the right to make an isometric drawing of a prism that is 3 cm high with a square base that measures 5 cm by 5 cm.

22. Make an isometric drawing of a rectangular prism with edges of lengths 4 cm, 6 cm, and 8 cm.

23. **Research** Look in magazines, books, or newspapers to find a sketch or drawing of a building. Are any of the methods above used in the sketch? If so, why do you think that method was used?

Step 1 Draw three rays from a common endpoint. One should be vertical and the other two should be 30° from a horizontal line.

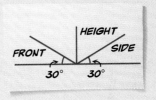

Step 2 Use a ruler to mark off segments 5 cm, 3 cm, and 5 cm long, respectively, on the rays.

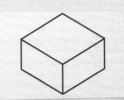

Step 3 Finish the diagram by drawing parallel and congruent segments as shown.

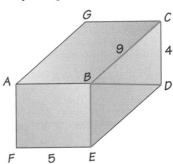

24. **Challenge** Sketch three planes that intersect in exactly one point.

In this rectangular prism, $\overleftrightarrow{AB} \perp$ plane *BCDE*.

25. **SAT/ACT Preview** Which of the green segments is (are) perpendicular to $\overleftrightarrow{AB}$?
   A. $\overline{BE}$ only
   B. $\overline{BC}$ only
   C. $\overline{BE}$ and $\overline{BC}$ only
   D. $\overline{BE}$, $\overline{BC}$, and $\overline{BD}$

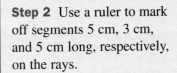

26. **a.** Find the distance from *F* to plane *BCD*.
   b. Find the distance from *A* to plane *BCD*.

27. Find the distance between points *D* and *B*.

28. Find the distance from *D* to plane *EFA*.

Look back at the article on pages 216–218.

Many kite enthusiasts enjoy flying dual-line stunt kites because the kites can perform flips and loops. Kite designers like José Saínz build stunt kites to be attractive as well as functional. This stunt kite is a stack of eight kites that are connected with string.

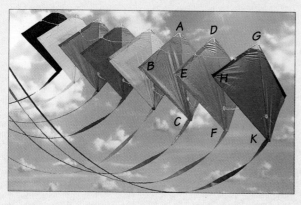

29. Plane *ABC* is parallel to plane *DEF*, and plane *DEF* is parallel to plane *GHK*. What does the Parallel Planes Theorem allow you to conclude? Explain your reasoning.

30. Edges $\overline{AB}$, $\overline{DE}$, and $\overline{GH}$ are coplanar. Explain what you can conclude using the Intersecting Planes Theorem.

ONGOING ASSESSMENT

31. Visual Thinking Complete each statement with *always*, *sometimes*, or *never*. If the statement is *always* true, sketch an example. If it is *never* true, sketch a counterexample. If it is *sometimes* true, sketch an example and a counterexample.

a. Two lines that lie in parallel planes are ? parallel.

b. Two lines that lie in intersecting planes are ? parallel.

c. Two lines that lie in parallel planes ? intersect.

d. Two lines that lie in intersecting planes ? intersect.

e. Two lines that intersect are ? coplanar.

f. Two lines that are parallel are ? coplanar.

SPIRAL REVIEW

32. ALGEBRA Find the value of each variable that will allow you to prove that *EFGH* is a parallelogram. *(Section 5.4)*

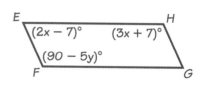

For each set of points, tell whether $\overleftrightarrow{CD}$ is a perpendicular bisector of $\overline{AB}$. If not, tell why not. *(Section 4.3)*

33. $A(-5, 1)$, $B(3, 5)$, $C(-2, 5)$, $D(0, 1)$

34. $A(-5, -1)$, $B(1, -9)$, $C(-2, -5)$, $D(3, -2)$

5.7 Constructing Parallels and Perpendiculars

Learn how to...
- **complete compass-and-straightedge constructions**

So you can...
- **construct lines parallel or perpendicular to given lines**

Measuring tools such as rulers and protractors can be used to draw precise geometric figures. Geometric figures can also be *constructed* using a compass and a straightedge (a ruler with no marks). A compass is used to construct circles and *arcs*, unbroken parts of circles. A straightedge is used to construct lines, rays, and segments.

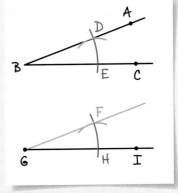

This arc has center O and radius OP.

CONSTRUCTION

CONGRUENT ANGLES

Given an angle, construct an angle congruent to it.

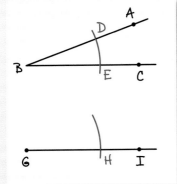

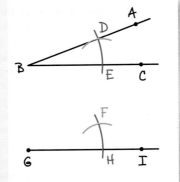

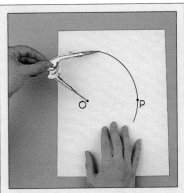

1. Using any radius and center *B*, draw an arc that intersects $\overrightarrow{BA}$ at *D* and $\overrightarrow{BC}$ at *E*. Draw a ray $\overrightarrow{GI}$. Using the same radius, draw an arc with center *G*. Label the intersection *H*.

2. Open the compass to distance *DE*. Draw an arc with this radius and center *H* that intersects the arc from Step 1. Label the intersection *F*.

3. Use your straightedge to draw $\overrightarrow{GF}$.
$\angle FGH \cong \angle ABC$

Given ∠1 and ∠2, construct an angle whose measure is equal to m∠1 + m∠2.

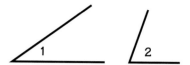

SOLUTION

Step 1 Draw a ray $\overrightarrow{QP}$. Using the Congruent Angles Construction, construct an angle with side $\overrightarrow{QP}$ that is congruent to ∠1. Label it ∠LQM.

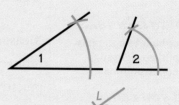

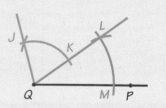

Step 2 Use the Congruent Angles Construction to construct an angle that is adjacent to ∠LQM and congruent to ∠2. Label it ∠JQK.

$$m\angle JQM = m\angle LQM + m\angle JQK$$

$$= m\angle 1 + m\angle 2$$

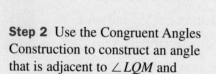

In the Example, the Congruent Angles Construction was used as part of a different construction. This is also true in the following construction.

CONSTRUCTION

PARALLEL LINES

Given a line, construct a line parallel to it.

1. Label any two points A and B on ℓ. Draw $\overleftrightarrow{AJ}$.

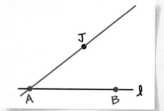

2. Using the Congruent Angles Construction, construct ∠CJD so that ∠CJD and ∠JAB are congruent corresponding angles. $\overleftrightarrow{JD} \parallel \ell$.

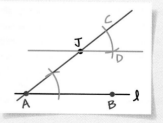

PERPENDICULAR BISECTOR

Given a segment, construct its perpendicular bisector.

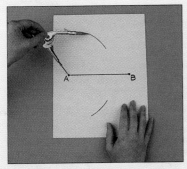
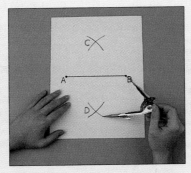
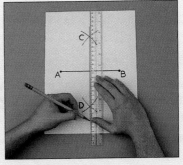

1. Using any radius greater than $\frac{1}{2}AB$, draw **two arcs with center A,** one on each side of $\overline{AB}$.

2. Using the same radius as in Step 1, draw **two arcs with center B** that intersect the arcs you drew in Step 1. Label the points C and D.

3. Draw $\overleftrightarrow{CD}$. $\overleftrightarrow{CD}$ is the perpendicular bisector of $\overline{AB}$.

THINK AND COMMUNICATE

1. In the Parallel Lines Construction, $\overleftrightarrow{JD}$ is constructed so that it is parallel to ℓ. Are there any other lines through J that are parallel to ℓ? Explain why or why not.

2. In the Perpendicular Bisector Construction, are there any other lines that contain the midpoint of $\overline{AB}$ and are perpendicular to $\overline{AB}$? Explain.

☑ CHECKING KEY CONCEPTS

Draw a figure like the one shown. Then complete the construction.

1. Construct an angle with side $\overrightarrow{TU}$ that is congruent to $\angle 1$.

2. Construct the line through R that is parallel to m.

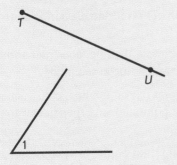

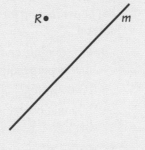

3. Draw any segment $\overline{GH}$. Construct its perpendicular bisector.

5.7 Exercises and Applications

Extra Practice exercises on page 670

1. Draw any obtuse ∠*RST*. Construct an angle congruent to ∠*RST*.

2. Draw any line *k* and a point *Q* not on *k*. Construct the line through *Q* that is parallel to *k*.

3. Draw a vertical segment and construct its perpendicular bisector.

Draw two angles like the ones shown. In Exercises 4–6, construct an angle with the indicated measure.

4. $(x + y)°$

5. $(x - y)°$

6. $(180 - 2y)°$

7. **a.** Explain how you know that $\overline{BD}$, $\overline{BE}$, and $\overline{GH}$ are all congruent in the Congruent Angles Construction on page 263.

   b. Writing Suppose you are given a segment, $\overline{XY}$. Explain how you could construct a segment congruent to $\overline{XY}$.

Connection ▶ PAPER FOLDING

Many constructions that can be done with a compass and straightedge can also be done using paper folding. Use paper you can see through. Every time you fold the paper, crease it and draw a dashed line along the crease.

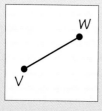
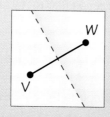

8. To construct the perpendicular bisector of $\overline{VW}$, fold the paper so that *W* meets *V*. The crease is the perpendicular bisector of $\overline{VW}$. How do you know that the crease is perpendicular to $\overline{VW}$? How can you tell that the crease intersects $\overline{VW}$ at its midpoint?

Use paper folding for each construction. Describe your method.

9. Given a line and a point on the line, construct a line that is perpendicular to the line at the point.

10. Given a line and a point not on the line, construct a line that contains the point and is perpendicular to the line.

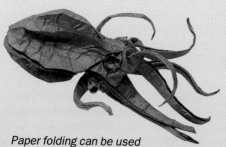

Paper folding can be used to make complex paper sculptures, such as this squid.

11. Given a line and a point not on the line, construct a line that contains the point and is parallel to the line.

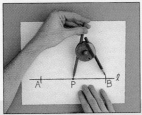

PERPENDICULAR LINES

In the construction on page 265, you learned how to construct the perpendicular bisector of a segment. Here are two related constructions. Only the first step of each construction is shown.

Given a line ℓ and a point P on ℓ, construct the line through P that is perpendicular to ℓ.

1. Using any radius, draw two arcs with center P that intersect ℓ at points A and B.

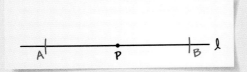

2. ?

Given a line g and a point Q not on g, construct the line through Q that is perpendicular to g.

1. Using any radius larger than the distance from Q to g, draw two arcs with center Q that intersect g at points M and N.

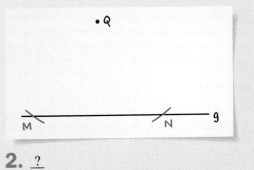

2. ?

12. a. Complete the construction on the left above by completing the missing step. (*Hint: P* is the midpoint of $\overline{AB}$.)

b. Draw any line *j* and a point *X* on *j*. Construct the line through *X* that is perpendicular to *j*.

13. a. Complete the construction on the right above.

b. Draw any line *k* and a point *H* not on *k*. Construct the line through *H* that is perpendicular to *k*.

14. Writing Describe a method for constructing parallel lines that is based on the Dual Perpendiculars Theorem. Use your method to construct two parallel lines. Do you prefer your method or the one on page 264?

15. **Technology** Here is a different method for the construction on the right above.

Step 1 Label any two points *X* and *Y* on *g*. Draw a circle with center *X* and radius *XQ*.

Step 2 Draw a circle with center *Y* and radius *YQ*. Label *Z*, the other point of intersection of the two circles.

Step 3 Draw $\overleftrightarrow{QZ}$.

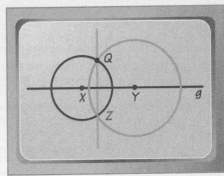

a. Repeat Exercise 13(b) using this method.

b. Do you prefer this method or the method above? Explain.

16. a. In the Parallel Lines Construction on page 264, which postulate or theorem guarantees that $\overleftrightarrow{JD} \parallel \ell$? Explain.

b. Describe a method for constructing parallel lines that uses the Converse of the Alternate Interior Angles Theorem. Then draw any line *n* and a point *S* not on *n*. Use your method to construct the line through *S* parallel to *n*.

17. Challenge Construct a 45° angle and a 135° angle. Explain why your methods work.

18. Follow these steps to construct a daisy pattern using only a compass.

Step 1 Draw a circle with any radius and choose a point *A* on the circle. Using the same radius, draw an arc with center *A* that intersects the circle at *B* and *F*.

Step 2 Draw an arc with center *B* and the same radius, and label the new intersection point *C*. Continue until you have completed the pattern.

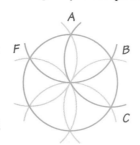

19. ART The pattern at the right is made up of nine square pieces. Given a square, describe how to use a straightedge and compass to construct one of the pieces.

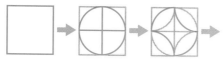

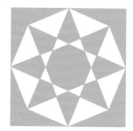

ART Many star patterns are based on a regular octagon.

20. Follow these directions to make a regular octagon. Start with a square. Draw the diagonals and locate their intersection. Using the intersection as the center, draw a circle that touches the square in four points. The eight intersection points shown are the vertices of a regular octagon.

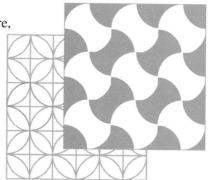

21. Explain how you know that all the intersection points in the diagram above are the same distance from the point where the diagonals intersect.

22. Open-ended Problem Starting with a regular octagon like the one you made in Exercise 20, design your own star pattern by connecting vertices and coloring in regions.

23. Cooperative Learning Work with another person. Draw an angle and a segment on a piece of paper. Then exchange papers with your partner.

a. Construct an angle congruent to the angle you received and construct the perpendicular bisector of the segment you received.

b. Exchange papers again. Check your partner's work by using a protractor and a ruler to measure the angles and segments.

SPIRAL REVIEW

Graph each system of inequalities. *(Toolbox, page 706)*

24. $x \leq 4$
$y \geq -2$

25. $y > x$
$y > 0$

26. $x > 5$
$x > -2$

ASSESS YOUR PROGRESS

VOCABULARY

line perpendicular to a plane (p. 258)
distance from a point to a plane (p. 258)

In the diagram, plane $P \parallel$ plane R and plane $R \parallel$ plane Q. *(Section 5.6)*

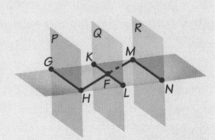

1. What does the Parallel Planes Theorem allow you to conclude?

2. How are $\overline{MN}$ and $\overline{KL}$ related? Explain your reasoning.

3. Suppose $m \angle NMF = 58°$. Find $m \angle MFL$ and $m \angle GHF$.

For Exercises 4–6, draw a figure like the one shown. Then complete the construction. *(Section 5.7)*

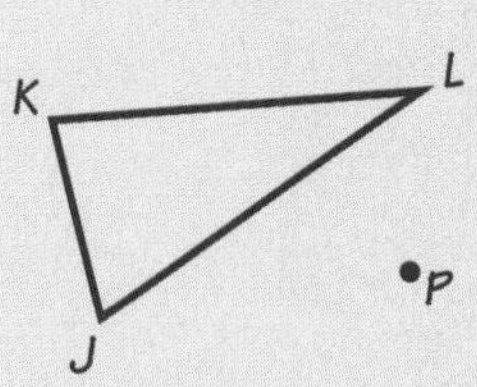

4. Construct an angle congruent to $\angle L$.

5. Construct the perpendicular bisector of $\overline{JL}$.

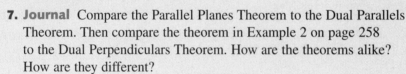

6. Construct the line through P that is parallel to $\overleftrightarrow{JL}$.

7. Journal Compare the Parallel Planes Theorem to the Dual Parallels Theorem. Then compare the theorem in Example 2 on page 258 to the Dual Perpendiculars Theorem. How are the theorems alike? How are they different?

Creating Technical Drawings

Did you know that many of the things you use every day began as technical drawings? Product design engineers use technical drawing to design and manufacture products for industry and for consumers.

PROJECT GOAL Make orthographic projections and isometric drawings of objects.

Visualizing Three-Dimensional Objects

Orthographic projection and *isometric drawing* are two ways that are often used to represent three-dimensional objects.

ORTHOGRAPHIC PROJECTION

This orthographic projection shows three perpendicular views of an object. Each view shows a single side of the object.

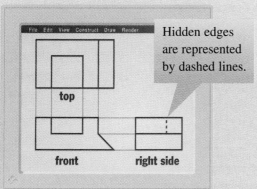

Hidden edges are represented by dashed lines.

top

front   right side

Each isometric line is drawn parallel to an isometric axis.

isometric axes

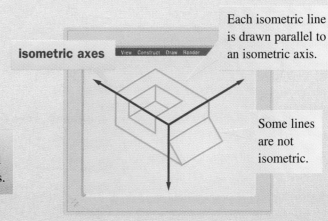

Some lines are not isometric.

ISOMETRIC DRAWING This isometric drawing shows three sides of the object in one diagram. At one corner of the object, three edges meet at 90° angles. These edges are drawn on the *isometric axes*, which meet at 120°. A line on the object that is parallel to one of these edges is called an *isometric line*.

Making the Drawings

1. Choose two objects at your school or where you live that have simple shapes and have most of their edges on isometric lines. Use objects that are already made, such as a bookshelf or table, or create your own, by stacking simple pieces, such as wooden blocks.

2. Make an orthographic projection and an isometric drawing of each object as shown. Measurements along isometric lines should be approximately proportional to the actual distances they represent.

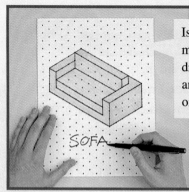

Isometric dot paper makes it easy to draw isometric lines and measure distances on them.

Displaying Your Drawings

Make a poster displaying the drawings you made of each object. Include a brief description or a photograph of each object. You can extend your project by exploring one of the following ideas.

- Draw an orthographic projection of a different object. Exchange drawings with another person and make isometric drawings from each others' orthographic projections.

- Use a computer and CAD (Computer-Aided Design) software to make orthographic projections and isometric drawings.

Self-Assessment

What did you find challenging about each type of drawing? Which type of drawing was easier for you to make? Explain your answers.

5 | Review

Without looking at the book, make a list of all the important postulates and theorems in this chapter. Make a sketch to illustrate each one. Then check each section to see if you missed any postulates or theorems.

VOCABULARY

transversal (p. 219)

same-side interior angles (p. 219)

alternate interior angles (p. 219)

corresponding angles (p. 219)

trapezoid (p. 228)

bases of a trapezoid (p. 228)

legs of a trapezoid (p. 228)

flow proof (p. 234)

distance from a point to a line (p. 249)

line perpendicular to a plane (p. 258)

distance from a point to a plane (p. 258)

SECTIONS 5.1 *and* 5.2

If two parallel lines are intersected by a **transversal**, then:

- corresponding angles are congruent.
  Example: If $r \parallel s$, then $\angle 1 \cong \angle 2$.

- Same-side interior angles are supplementary.
  Example: If $r \parallel s$, then $m\angle 2 + m\angle 3 = 180°$.

- Alternate interior angles are congruent.
  Example: If $r \parallel s$, then $\angle 3 \cong \angle 4$.

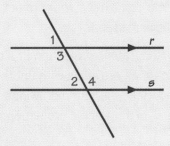

A **trapezoid** is a quadrilateral with exactly one pair of parallel sides called **bases**. The other two sides are the **legs**.

SECTIONS 5.3, 5.4, and 5.5

When you write a proof, you can choose from several different formats.

A *paragraph proof* combines the most important statements and reasons into sentences that form a convincing argument.

A *two-column proof* includes all the necessary steps with the statements and reasons listed in two separate columns.

A *flow proof* uses arrows to connect numbered statements in a logical order.

A *key-steps proof* includes the most important parts of a proof in an informal format.

Two lines must be parallel if they are intersected by a transversal and:

- corresponding angles are congruent.
  Example: If $\angle 5 \cong \angle 6$, then $j \parallel k$.

- alternate interior angles are congruent.
  Example: If $\angle 5 \cong \angle 7$, then $j \parallel k$.

- same-side interior angles are supplementary.
  Example: If $m\angle 7 + m\angle 8 = 180°$, then $j \parallel k$.

Also, two lines must be parallel if:

- they are coplanar and both are perpendicular to a third line.

- both lines are parallel to a third line.

The **distance from a point to a line** is length of the perpendicular segment from the point to the line.

SECTIONS 5.6 and 5.7

Two planes can either intersect or be parallel.

A **line is perpendicular to a plane** if the line intersects the plane and is perpendicular to every line in the plane that contains the point of intersection.

The **distance from a point to a plane** is the length of the perpendicular segment from the point to the plane.

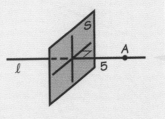

Line $\ell \perp$ plane S. The distance from point A to plane S is 5 units.

A straightedge and a compass are used to make geometric diagrams called *constructions*. You can construct parallel and perpendicular lines, congruent angles, and perpendicular bisectors.

Assessment

VOCABULARY REVIEW

Complete each statement.

1. A quadrilateral with exactly two parallel sides is called a(n) _?_ and the parallel sides are called the _?_ .

2. A(n) _?_ is a line that is coplanar with two other lines and intersects those lines.

3. A(n) _?_ is a type of proof that uses arrows to show the logical connections of the statements.

Use the diagram. Give an example of each of the following.

4. two corresponding angles

5. two alternate interior angles

6. two same-side interior angles

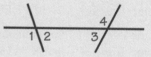

7. **Writing** Compare the definitions for the distance from a point to a line and the distance from a point to a plane. How are they alike? How are they different?

SECTIONS 5.1 *and* 5.2

ALGEBRA Find the value of each variable.

8.   9.   10.

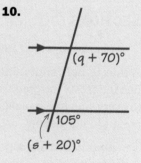

11. Sketch trapezoid *WXYZ* with bases $\overline{WX}$ and $\overline{YZ}$.

   a. What must be true about $\angle X$ and $\angle Y$? Justify your conclusion.

   b. Can you make any conclusions about $\angle Y$ and $\angle Z$? Explain why or why not.

SECTIONS 5.3, 5.4, *and* 5.5

Use the diagram and this theorem: If two parallel lines are cut by a transversal, then the bisectors of two corresponding angles are parallel.

12. Write the Given and Prove statements.

13. Write a flow proof of the theorem.

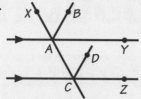

Use the diagram and the given information. State the postulate or theorem that would allow you to conclude that line ℓ and line m are parallel.

14. $\angle 1 \cong \angle 2$

15. $m \angle 3 = 90°$ and $m \angle 4 = 90°$

16. $\angle 2 \cong \angle 6$

17. $m \angle 3 + m \angle 5 = 180°$

18. $\ell \parallel n$ and $m \parallel n$

SECTIONS 5.6 *and* 5.7

19. In the diagram at the right, planes P and Q are parallel and lines j and k are parallel. Explain why lines k and ℓ must also be parallel.

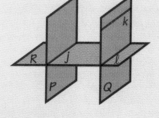

20. **Writing** Suppose plane X is parallel to plane Y, and plane X is parallel to plane Z. Are planes Y and Z *always*, *sometimes*, or *never* intersecting planes? Explain.

21. Draw any line m and a point C not on m. Construct a line that is parallel to m and passes through point C.

PERFORMANCE TASK

22. Write a report that describes how each group of figures could intersect. Include sketches to illustrate each possibility.

- a line and a plane
- a line and two parallel planes
- two parallel lines and a plane
- a line and two intersecting planes

6 Conjectures About Triangles

Shedding *LIGHT* on the Matter

INTERVIEW **Madeleine Fleming**

Madeleine Fleming uses geometry every day in her job as an optical physics specialist at the 3M Optics Technology Center in Minnesota. Fleming designs surfaces covered with microscopic structures—tiny bumps, ridges, indentations, and furrows—that bend and reflect light. "In designing these *microstructures*, I need to understand geometric relationships," Fleming, who has a Ph.D. in optics, says. "In particular, I need to find out what happens to a ray of light when it bumps into or passes through one of these microstructures."

"Math can help us understand the world around us, but we can also use it to build the world around us."

Illuminating Math

Fleming's job is not just about determining the precise shape, angle, and position of microstructures. It also involves people. As part of the Visiting Wizards and TECH programs, Fleming and other 3M scientists travel to elementary, junior high, and high schools, showing students how math and science are applied in the real world. Her advice for getting others interested in math is simple: "Look to the applications!"

> **"Math is interesting in its own right, but what is even more interesting is what you can do with it."**

Some people may not appreciate math for the sake of math, Fleming says, but they can appreciate what it can do and how it can be applied. "The reason I use math every day is because it's practical."

For example, geometry was used to design the microstructures on these light pipes. The pipe is smooth on the inside and grooved on the outside. This design causes light to travel long distances down the pipe, while allowing some light to leak out along the way. A similar surface is used in laptop computer screens to direct the screen's light toward the user.

Knowing All the Angles

Street and highway signs, coated with a thin plastic film made by 3M, can be found all over the world. The back of the clear film contains about 7000 tiny pyramids per square inch. The base of each pyramid is an isosceles triangle. All these triangles are *congruent*, meaning that they are the same size and shape.

The light from the headlights of a car passes through the film and reflects inside the pyramids on the back. The pyramids are designed so that the light reflects back to the driver, no matter where the car is. "In determining the angles, we work out the geometry, figuring out where the sign is, where the headlights are, and where the driver is," Fleming explains.

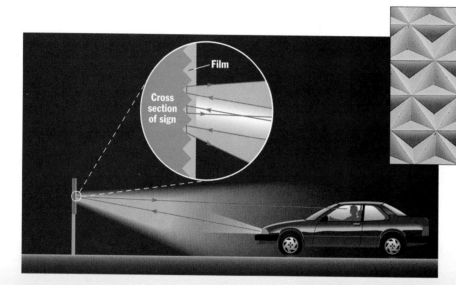

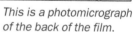

This is a photomicrograph of the back of the film.

Explore and Connect

Madeleine Fleming explains how microstructures are used to make thin lenses.

1. Writing Why do you think it is important for street signs to reflect light from car headlights back to the driver?

2. Research Using a physics textbook or an encyclopedia, look up the meanings of *angle of reflection* and *angle of incidence*. What do these terms mean? How are the angles related?

3. Project Make a collage showing many different triangles, including some congruent triangles. You could cut out illustrations from newspapers and magazines, or draw triangles you see around you. Indicate congruent triangles and any isosceles triangles.

Mathematics & Madeleine Fleming

In this chapter, you will learn more about how mathematics is related to optical microstructures.

Related Exercises

Section 6.2
• Exercises 18–21

Section 6.3
• Exercise 3

Section 6.6
• Exercises 16–18

6.1 Triangle Inequalities

Learn how to...

- apply the Triangle Inequality Theorems

So you can...

- determine if a triangle can be formed from three given lengths

- use the lengths of two sides of a triangle to describe the length of the third side

Can a triangle have sides of lengths 3 in., 4 in., and 5 in.? How about 3 in., 4 in., and 10 in.? If you know the lengths of two sides of a triangle, are there any limits on how long the third side can be? In the Exploration you will investigate the lengths of sides in triangles.

EXPLORATION
COOPERATIVE LEARNING

Comparing Sides of Triangles

Work with another student.
You will need:
- geometry software

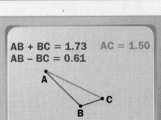

1 Construct two circles with the same center. Label the center *B*. Construct and label point *A* on the larger circle and point *C* on the smaller circle.

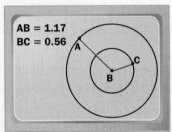

2 Construct and measure $\overline{AB}$ and $\overline{BC}$. Hide the circles so only ∠*ABC* is showing.

3 Calculate $AB + BC$ and $AB - BC$.

4 Using a different type or color of line, construct $\overline{AC}$. Measure *AC*.

5 Compare *AC* with $AB + BC$ as you move points *A* and *C* to different positions on the circles. Compare *AC* with $AB - BC$. What do you notice?

6 When *AC* is at its largest value, do $\overline{AB}$, $\overline{BC}$, and $\overline{AC}$ form a triangle? Explain why or why not. Do the segments form a triangle when *AC* is at its smallest value?

7 If segments $\overline{AB}$, $\overline{BC}$, and $\overline{AC}$ form a triangle, write two inequalities to describe *AC* in terms of *AB* and *BC*.

In the Exploration, you probably noticed that AC is less than $AB + BC$ regardless of the shape of $\triangle ABC$. You may also have noticed how the length of $\overline{AC}$ varies with the measure of $\angle B$. For example, if $\angle B$, which is opposite $\overline{AC}$, is the largest angle of the triangle, then $\overline{AC}$ is the longest side.

Triangle Inequality Theorems

The sum of the lengths of any two sides of a triangle is greater than the length of the third side.

In $\triangle ABC$,
$AB + BC > AC$,
$AB + AC > BC$, and
$AC + BC > AB$.

One side of a triangle is longer than a second side if and only if the angle opposite the first side is larger than the angle opposite the second side.

In $\triangle ABC$, $AB > BC$ if and only if $m \angle C > m \angle A$.

THINK AND COMMUNICATE

1. In the Exploration, you probably discovered that $AC > AB - BC$. How is this inequality related to the first Triangle Inequality Theorem?

2. In $\triangle MHP$, $MH = 4.7$, $HP = 6.3$, $m \angle P < m \angle H$, and $m \angle H < m \angle M$. What can you conclude about the length MP?

EXAMPLE 1   Application: Art

A *stretcher* is the wooden frame that holds the canvas for a painting. The pieces of wood that form the stretcher are called *strips*. Can an artist make a stretcher from strips of the given lengths?

a. 9 in., 18 in., 7 in.   **b.** 10 in., 5 in., 15 in.   **c.** 14 in., 11 in., 8 in.

SOLUTION

Use the first Triangle Inequality Theorem.

a. $9 + 18 > 7$ ✔
$18 + 7 > 9$ ✔
$7 + 9 \not> 18$
These strips cannot form a triangle.

b. $10 + 5 \not> 15$
$5 + 15 > 10$ ✔
$15 + 10 > 5$ ✔
These strips cannot form a triangle.

c. $8 + 11 > 14$ ✔
$11 + 14 > 8$ ✔
$14 + 8 > 11$ ✔
These strips form a triangle.

9 in.   7 in.

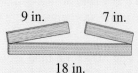

18 in.

10 in.   5 in.

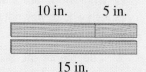

15 in.

11 in.   8 in.

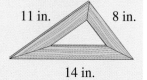

14 in.

EXAMPLE 2

In △*PQR*, *PR* = 15 ft and *RQ* = 12 ft.

a. What do you know about the length *PQ*?

b. Sketch three possible shapes of △*PQR*.

SOLUTION

a. Use the first Triangle Inequality Theorem.

$$PQ + PR > RQ \qquad PQ + RQ > PR \qquad PR + RQ > PQ$$

$$PQ + 15 > 12 \qquad PQ + 12 > 15 \qquad 15 + 12 > PQ$$

$$PQ > -3 \qquad \qquad PQ > 3 \qquad \qquad PQ < 27$$

$\overline{PQ}$ is longer than 3 ft and shorter than 27 ft.

b. Three possible shapes of the triangle are shown.

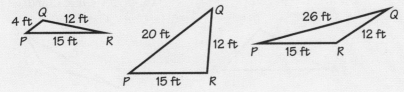

THINK AND COMMUNICATE

3. a. In Example 2, why can't $\overline{PQ}$ be one foot long?

   b. Maureen thinks that the solution to Example 2 should be "$\overline{PQ}$ is longer than −3 ft and shorter than 27 ft." What is wrong with Maureen's statement?

4. a. In Example 2, why can't $\overline{PQ}$ be three feet long?

   b. Why can't $\overline{PQ}$ be 27 ft long?

5. Write an inequality to describe the possible lengths of $\overline{PQ}$ in Example 2.

☑ CHECKING KEY CONCEPTS

Can a triangle be formed from sides of the given lengths?

1. 5 ft, 5 ft, 5 ft   **2.** 15 in., 10 in., 12 in.

3. 10 m, 10 m, 19.9 m   **4.** 5 cm, 13 cm, 6 cm

The lengths of two sides of a triangle are given. What can you conclude about the length of the third side?

5. 13 ft, 19 ft   **6.** 9.9 cm, 10.1 cm   **7.** 8 m, 8 m

8. In △*ABC*, $m \angle C > m \angle B$ and $m \angle B > m \angle A$.
What do you know about the
length of $\overline{AB}$?

6.1 Exercises and Applications

Extra Practice
exercises on
page 670

Tell whether or not a triangle can be formed from sides of the given lengths.

1. 2 in., 7 in., 8 in.

2. 33 cm, 12 cm, 21 cm

3. 13 ft, 14 ft, 15 ft

4. 17 m, 23 m, 7 m

5. 55 ft, 34 ft, 21 ft

6. x, x, $3x$

The lengths of two sides of a triangle are given. What can you conclude about the length of the third side?

7. 4 ft, 3 ft

8. 7 in., 1 in.

9. 9 cm, 12 cm

10. 10.7 m, 21.8 m

11. $7\frac{1}{4}$ in., $6\frac{3}{4}$ in.

12. x, $x + 4$

What can you conclude about the length of $\overline{AB}$?

13.

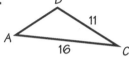

14. $m \angle C > m \angle A$

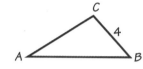

15. $m \angle A < m \angle C$ and $m \angle C < m \angle B$

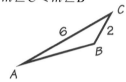

16. $m \angle C < m \angle B$ and $m \angle B < m \angle A$

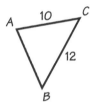

17. $m \angle B < m \angle A$ and $m \angle A < m \angle C$

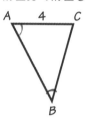

18. $m \angle B > m \angle A$

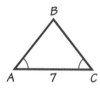

19. SAT/ACT Preview In $\triangle MNP$, if $m \angle 1 < m \angle 2$, then:

A. $MN > NP$ **B.** $MN < NP$ **C.** $MN = NP$

D. Relationship cannot be determined.

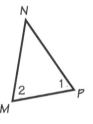

20. Logical Reasoning Is this statement *True* or *False*? If it is false, give a counterexample.

The sum of the measures of two angles of a triangle is always greater than the measure of the third angle.

Can the three points be the vertices of a triangle? If not, explain why not.

21. $D(1, 6)$, $E(1, 9)$, $F(14, 8)$

22. $G(7, -3)$, $H(0, -3)$, $I(-3, -3)$

23. $J(1, 1)$, $K(5, 5)$, $L(-14, -14)$

24. $M(0, 0)$, $N(3, 2)$, $P(-231, 782)$

25. The university is 1 mi from the plaza, and the plaza is $\frac{1}{4}$ mi from the market place. What do you know about the distance from the university to the market place? Explain.

26. POTTERY In her studio, Tammy wants the throwing wheel to be 3 ft from the sink, and the table to be both 5 ft from the sink and 4 ft from the wheel. Is this arrangement possible? Explain why or why not.

In Exercises 27–29, you will estimate the experimental probability of forming a triangle from three segments of random lengths.

> **Toolbox p. 710**
> *Probability*

Cooperative Learning **Work in a group of four people. You will need a ruler, 16 whole pieces of uncooked spaghetti, and a small bag.**

Step 1 Write the lengths "1 cm," "2 cm," . . . , "24 cm" on twenty-four small pieces of paper and put them in the bag.

Step 2 Shake the bag to mix the pieces of paper. Decide where to break a piece of spaghetti by drawing two numbers from the bag. Measure both distances from the same end of the spaghetti and break the spaghetti at each location. Put the numbers back in the bag.

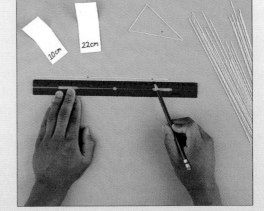

27. Each person should do Step 2 three times. Record the lengths of the spaghetti pieces and whether or not they form a triangle.

28. Share your results from Exercise 27. Use the formula

$$\text{Probability} = \frac{\text{Total number of successes}}{\text{Total number of tries}}$$ to estimate the probability of

forming a triangle from a piece of broken spaghetti.

29. Writing Explain why you should use the pieces of paper to decide where to break the spaghetti instead of breaking the spaghetti wherever you like.

Cooperative Learning **For Exercises 30–32, work in a group of three people. For Exercise 30, one of you should do part (a), another should do part (b), and the third should do part (c). Work together on Exercises 31 and 32. You will need a ruler.**

30. Try to draw a quadrilateral with sides of the given lengths. Is it possible? Explain why or why not.

    a. 2 in., 3 in., 4 in., and 9 in.   **b.** 2 in., 3 in., 4 in., and 12 in.   **c.** 2 in., 3 in., 4 in., and 8 in.

31. Writing Compare your answers to parts (a), (b), and (c) in Exercise 30. Make a conjecture about the relationship between the lengths of the sides of a quadrilateral.

32. Three sides of a quadrilateral are 3 cm, 3 cm, and 3 cm long. Describe all the possible lengths of the fourth side.

33. Challenge Is it possible for the perimeter of $\triangle WBF$ to be 24? Explain why or why not.

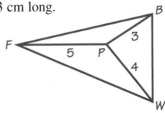

When designing a kitchen, architects look at the *work triangle* to see if the kitchen will be easy to use. The sides of the work triangle are the paths between the sink, the stove, and the refrigerator. The perimeter of a work triangle should be less than 22 ft and more than 12 ft.

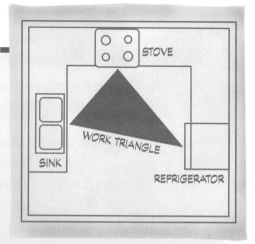

34. What is the range of possible lengths for any side of the work triangle? How did you get your answer?

35. In addition to the guideline above, each side of the work triangle should be between 3.5 ft and 7 ft long. Suppose two sides of the triangle are each 3.5 ft long. What is the range of possible lengths for the third side?

36. Ideally, a work triangle should be equilateral. Give the side lengths of four work triangles that meet this condition as well as the conditions given above.

37. Research Measure the work triangle in a home kitchen. How well does it meet these guidelines?

38. Open-ended Problem Design a kitchen that meets the guidelines described above. Sketch the kitchen, including doorways and dimensions.

ONGOING ASSESSMENT

39. a. Open-ended Problem Give an example of three measurements that *cannot* be the lengths of the sides of a triangle.

b. Give an example of three measurements that can be the lengths of the sides of a scalene triangle, $\triangle ABC$.

c. In your triangle from part (b), which angle has the largest measure? the smallest measure?

SPIRAL REVIEW

Sketch each situation. *(Section 1.4)*

40. Points G, S, and N are collinear.

41. Line ℓ intersects plane R in point T.

Find the slope and *y*-intercept of each line. *(Section 4.2)*

42. $y = -3x + 7$

43. $y = 4$

44. $y = -2x$

6.2 Exploring Congruence

The parts of a car come from many different sources. For all of the pieces to fit together, they must be exactly the right size and shape. For example, the window must match the window frame exactly.

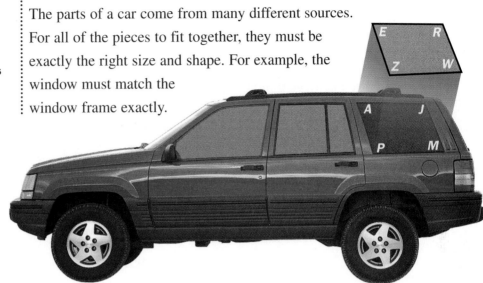

Corner *E* on the window *corresponds* to corner *A* in the car frame. Polygons *AJMP* and *ERWZ* are **congruent polygons** if each part (angle or side) of *AJMP* is congruent to the **corresponding part** of *ERWZ*. Congruent polygons have the same size and shape.

EXAMPLE 1

Are quadrilaterals *AJMP* and *ERWZ* congruent?

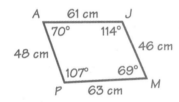
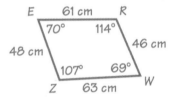

SOLUTION

Use the diagram to tell which pairs of corresponding parts are congruent.

$\angle A \cong \angle E$	$\angle M \cong \angle W$	$\overline{AJ} \cong \overline{ER}$	$\overline{MP} \cong \overline{WZ}$
$\angle J \cong \angle R$	$\angle P \cong \angle Z$	$\overline{JM} \cong \overline{RW}$	$\overline{PA} \cong \overline{ZE}$

All eight parts (four angles and four sides) of quadrilateral *AJMP* are congruent to the corresponding parts of quadrilateral *ERWZ*, so quadrilateral *AJMP* ≅ quadrilateral *ERWZ*.

WATCH OUT! ▶

Corresponding angles in congruent polygons are not the same as *corresponding angles* of parallel lines.

When you identify congruent polygons, list the vertices of one polygon in the same order as the corresponding vertices of the other polygon. For example, quadrilateral *MPAJ* ≅ quadrilateral *WZER* indicates that *M* corresponds to *W*, *P* corresponds to *Z*, *A* corresponds to *E*, and *J* corresponds to *R*.

THINK AND COMMUNICATE

1. List all of the corresponding parts of quadrilaterals *JKLM* and *PRQS* if *JKLM* ≅ *QSPR*.

2. How else can you write *JKLM* ≅ *QSPR* so that it remains true?

3. If △*XYZ* ≅ △*UVW* is $\overline{ZX} \cong \overline{WU}$? Explain how you know.

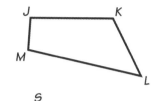

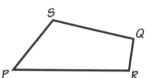

EXAMPLE 2   Connection: Algebra

polygon *MNOPQ* ≅ polygon *VUTSR*

a. Find the value of *x*.

b. Find the value of *d*.

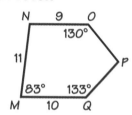

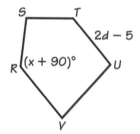

SOLUTION

Since *MNOPQ* ≅ *VUTSR*, the corresponding parts of *MNOPQ* and *VUTSR* are congruent.

a. $\angle R \cong \angle Q$ ∠*R* corresponds to ∠*Q*.

$m\angle R = m\angle Q$

$(x + 90)° = 133°$

$x = 133 - 90$

$x = 43$

b. $\overline{TU} \cong \overline{ON}$ $\overline{TU}$ corresponds to $\overline{ON}$.

$TU = ON$

$2d - 5 = 9$

$2d = 14$

$d = 7$

EXPLORATION
COOPERATIVE LEARNING

Investigating Overlapping Triangles

Work with another student.
You will need:

- scissors

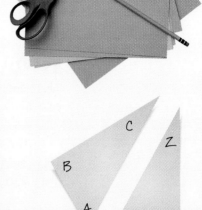

1 Fold in one corner of a piece of paper to form a triangle. Trace the triangle on the paper. What type of triangle is it?

2 Cut along the traced edges and along the fold. Explain why the two triangles are congruent.

3 Label the vertices of each triangle as shown. Label each vertex on both sides of the paper. Complete: △BCA ≅ _?_ .

4 Use your two triangles to form each shape below. After you form the shape, sketch it, label the vertices, and name the congruent triangles.

Example:

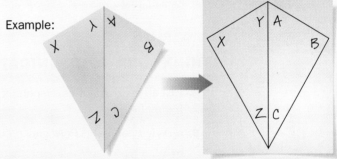

△BCA ≅ △XZY

5 What other shapes can you form with △ABC and △YXZ? Sketch and label at least five new shapes and name the congruent triangles in each one.

EXAMPLE 3

Name all of the triangles that appear to be congruent.

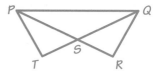

SOLUTION

There are two triangles that appear to be congruent and do not overlap:

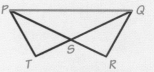

$$\triangle PST \cong \triangle QSR$$

There are two triangles that appear to be congruent and do overlap, just like the triangles in the Exploration. Imagine sliding the figure apart:

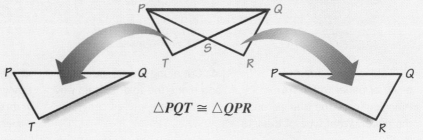

$$\triangle PQT \cong \triangle QPR$$

It appears that $\triangle PST \cong \triangle QSR$ and $\triangle PQT \cong \triangle QPR$.

THINK AND COMMUNICATE

4. List all six pairs of corresponding parts for $\triangle PQT$ and $\triangle QPR$ in Example 3.

5. a. What part do $\triangle PQT$ and $\triangle QPR$ share?

   b. Which property tells you that this part is congruent to itself?

☑ CHECKING KEY CONCEPTS

1. If $\triangle GHJ \cong \triangle BCA$, then $\triangle HGJ \cong \underline{\ ?\ }$.

2. Name all of the triangles that appear to be congruent.

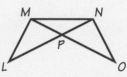

3. $\triangle UQR \cong \triangle VSW$

   a. Find the value of y.

   b. Find the value of t.

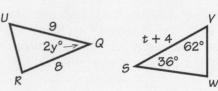

6.2 Exercises and Applications

Extra Practice exercises on page 671

Are any polygons in the figure congruent? If so, name the congruent polygons.

1.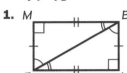

2.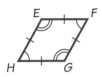

3.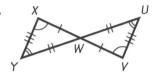

4.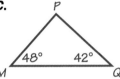

For Exercises 5–7, △MPQ ≅ △WBC. Complete each statement.

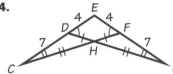

5. $\overline{PQ} \cong$?

6. $m \angle W =$?

7. $MQ =$?

ALGEBRA For Exercises 8–10, △WUN ≅ △TRE.

8. Find the value of x.

9. Find the value of y.

10. **Challenge** Find the value of z.

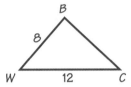

11. **Open-ended Problem** Draw two congruent rectangles on a coordinate grid. Explain how you know that they are congruent.

For Exercises 12–14, tell which triangle appears to be congruent to the given triangle.

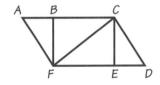

12. △ABF ≅ ?   **13.** △FBC ≅ ?   **14.** △AFC ≅ ?

15. Sketch a scalene obtuse triangle, △KXM. Sketch △NBE so that △KXM ≅ △NBE.

    a. Is △XMK ≅ △BNE?

    b. Is △KMX ≅ △NEB?

16. **Writing** How many different ways can you write △JKL ≅ △MNP? Explain how you found your answer.

17. **BIOLOGY** The Clara Satin moth, which lives in Australia and Tasmania, belongs to the family *geometridae*. The wings of a Clara Satin moth are covered by a geometric pattern. Sketch the moth using only straight lines. Label each vertex and name at least six pairs of polygons that appear to be congruent.

Look back at the article on pages 276–278.

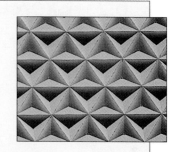

Madeleine Fleming designs surfaces made up of millions of tiny congruent figures. For example, the reflective film at the right contains about 7000 pyramids in each square inch. The film is used to make street signs.

Open-ended Problems For Exercises 18–21, complete each statement.

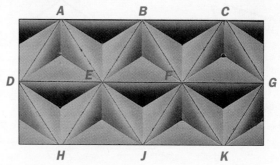

18. It appears that △*AEB* ≅ _?_ .

19. It appears that quadrilateral *ABFJ* ≅ _?_ .

20. It appears that pentagon *AJKGC* ≅ _?_ .

21. Name three other pairs of polygons that appear to be congruent.

22. Copy and complete the proof.

Given: $\overline{AB} \cong \overline{DC}$
$\overline{AD}$ and $\overline{BC}$ bisect each other.
$\overleftrightarrow{AB} \parallel \overleftrightarrow{CD}$

Prove: △*ABE* ≅ △ _?_

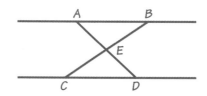

Statements	Reasons
1. $\overline{AB} \cong \overline{DC}$	1. Given
2. $\overline{AD}$ and $\overline{BC}$ bisect each other.	2. _?_
3. $\overline{AE} \cong$ _?_	3. Definition of segment bisector
4. $\overline{BE} \cong \overline{CE}$	4. _?_
5. $\overleftrightarrow{AB} \parallel \overleftrightarrow{CD}$	5. _?_
6. ∠*BAE* ≅ ∠*CDE*	6. _?_
7. ∠_?_ ≅ ∠*DCE*	7. If two ∥ lines are intersected by a transversal, then alternate interior angles are ≅.
8. ∠_?_ ≅ ∠_?_	8. Vertical angles are congruent.
9. _?_ ≅ △*DCE*	9. Definition of congruent polygons

23. Cooperative Learning Work in a group of three people. You will need graph paper and scissors. Two ways to cut a 3 in. by 3 in. square into two congruent polygons are shown. Find at least five more ways. Sketch, label, and name each differently shaped pair of congruent polygons that you cut.

$\triangle ABC \cong \triangle FED$

You may use concave polygons.

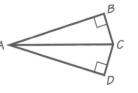

SPIRAL REVIEW

Give the postulate, definition, or property that makes each statement about the diagram true. *(Section 3.2)*

24. $\angle B \cong \angle D$

25. $\overline{AC} \cong \overline{AC}$

ASSESS YOUR PROGRESS

VOCABULARY

congruent polygons (p. 285)   **corresponding part** (p. 285)

Tell whether or not a triangle can be formed from the given lengths. *(Section 6.1)*

1. 1 in., 4 in., 10 in.   **2.** 2 ft, 4 ft, 2 ft   **3.** 5.4 m, 0.5 m, 5.0 m

The lengths of two sides of a triangle are given. What can you conclude about the length of the third side? *(Section 6.1)*

4. 1 in., 4 in.   **5.** 2 ft, 4 ft   **6.** 18 cm, 12 cm

7. Name all of the triangles in the diagram that appear to be congruent. *(Section 6.2)*

8. $\triangle DEF \cong \triangle HKL$ *(Section 6.2)*
   a. Find the value of r.
   b. Find the value of p.

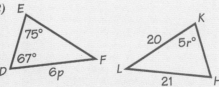

9. Journal Sue says that to tell if three numbers can be the lengths of the sides of a triangle, you only need to check if the sum of the two smaller numbers is greater than the third number. Do you agree? Explain.

6.3 Congruent Triangles: SSS and SAS

Learn how to...

- **use the SSS and SAS Postulates**

So you can...

- **prove that triangles are congruent without proving that all six corresponding parts are congruent**

You probably know some good shortcuts for getting from one place to another. Sometimes shortcuts are better than longer routes. In this section, you will discover two good shortcuts for showing that triangles are congruent.

CONSTRUCTION

CONSTRUCTING A TRIANGLE FROM THREE SIDES

Given △***ABC***, construct a triangle congruent to it by copying only the sides.

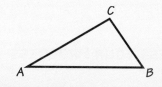

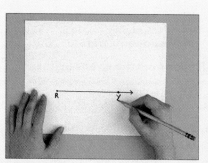

1. Draw a ray. Label it $\overrightarrow{RY}$.

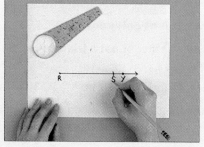

2. Copy segment $\overline{AB}$ onto $\overrightarrow{RY}$ and label the segment $\overline{RS}$.

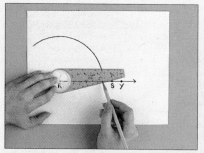

3. Open your compass to radius *AC*. Use this radius to draw a long arc with center *R*.

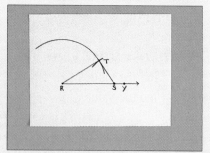

4. Using radius *BC* and center *S*, draw an arc that intersects the arc from Step 3. Label the intersection *T*. Draw △*RST*.

THINK AND COMMUNICATE

Use the construction on page 292.

1. Explain why corresponding sides of △*ABC* and △*RST* are congruent.

2. Measure the corresponding angles. Are the triangles congruent?

3. Use the construction to copy △*ABC* in as many ways as possible. For example, draw the arcs below $\overrightarrow{RY}$, or draw $\overrightarrow{RY}$ diagonally.

4. Is there any way to construct △*RST* so that $\overline{RS} \cong \overline{AB}$, $\overline{RT} \cong \overline{AC}$, and $\overline{ST} \cong \overline{BC}$, but △*RST* and △*ABC* are not congruent? Explain.

As you saw in the construction, triangles with congruent corresponding sides are congruent. This idea is helpful when you want to prove that two triangles are congruent.

Side-Side-Side Postulate (SSS Postulate)

If three sides of a triangle are congruent to three sides of another triangle, then the triangles are congruent.

If $\overline{WN} \cong \overline{FR}$, $\overline{NA} \cong \overline{RO}$, and $\overline{AW} \cong \overline{OF}$, then △*AWN* ≅ △*OFR*.

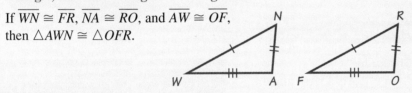

EXAMPLE 1

Given: $\overline{AB} \cong \overline{CB}$; $\overline{AD} \cong \overline{CD}$

Prove: △*ABD* ≅ △*CBD*

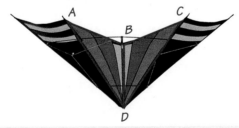

SOLUTION

Plan Ahead: Copy the diagram and mark the parts that you know are congruent. Use the fact that $\overline{BD} \cong \overline{BD}$ and the SSS Postulate.

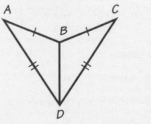

Statements	Reasons
1. $\overline{AB} \cong \overline{CB}$; $\overline{AD} \cong \overline{CD}$	1. Given
2. $\overline{BD} \cong \overline{BD}$	2. Reflexive Property
3. △*ABD* ≅ △*CBD*	3. SSS Postulate

The Side-Angle-Side Postulate

In $\triangle STU$, $\angle S$ is *included* between sides $\overline{ST}$ and $\overline{SU}$. In the Exercises, you will construct congruent triangles by copying two sides and the included angle. No matter how you do the construction, the resulting triangle is always the same size and shape. This leads to another postulate.

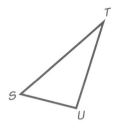

Side-Angle-Side Postulate (SAS Postulate)

If two sides and the included angle of one triangle are congruent to two sides and the included angle of another triangle, then the triangles are congruent.

If $\overline{XY} \cong \overline{UV}$, $\angle X \cong \angle U$, and $\overline{XZ} \cong \overline{UW}$, then $\triangle XYZ \cong \triangle UVW$.

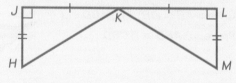

EXAMPLE 2

Given: $\overline{HJ} \cong \overline{ML}$
$\angle J$ is a right angle.
$\angle L$ is a right angle.
K is the midpoint of $\overline{JL}$.

Prove: $\triangle HJK \cong \triangle MLK$

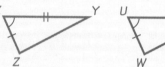
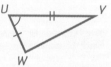

SOLUTION

Plan Ahead: Use the definition of midpoint and what you know about right angles.

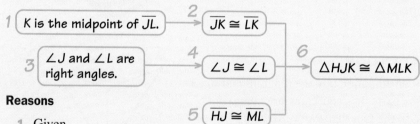

Reasons

1. Given
2. Definition of midpoint
3. Given
4. All right angles are congruent.
5. Given
6. SAS Postulate

☑ CHECKING KEY CONCEPTS

**Decide whether or not you can prove that the triangles are congruent.
If you can, tell which postulate you would use.**

1.

2.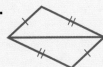

3.

For Questions 4 and 5, use the information in the diagram.

4. What other information do you need to know
in order to use the SSS Postulate to prove
that $\triangle PQR \cong \triangle TQR$?

5. What other information do you need to know
in order to use the SAS Postulate to prove
that $\triangle PQR \cong \triangle TSR$?

6. How are the SSS Postulate and the SAS Postulate
alike? How are they different?

7. Given: $\overline{XY} \cong \overline{ZW}$; $\overline{XY} \parallel \overline{ZW}$
Prove: $\triangle XYZ \cong \triangle ZWX$

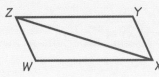

6.3 Exercises and Applications

*Extra Practice
exercises on
page 671*

1. Which angle is included between $\overline{AC}$ and $\overline{BC}$?

2. Open-ended Problem Name two sides and
their included angle in $\triangle ACD$.

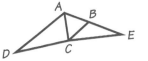

3. OPTICS Madeleine Fleming designs optical surfaces
such as the surface of this light pipe. A mold is used to
make tiny parallel ridges on the outside of the pipe. The
ridges and mold fit together as shown. Which postulate
can you use to explain why $\triangle ABC \cong \triangle DEF$?

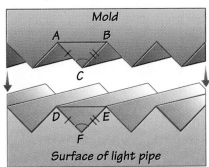

Tell whether you can prove that the triangles are congruent. If you can, name the congruent triangles and tell which postulate you can use.

4.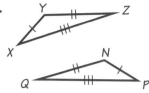

5.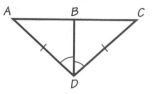

6.

7. $\overline{FH} \cong \overline{GK}$
and $\overline{KL} \cong \overline{HE}$
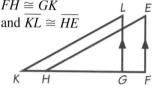

8.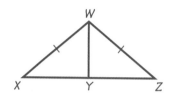

9.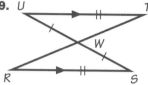

10. Copy and complete the flow proof.

Given: Y is the midpoint of $\overline{XZ}$.
$\overline{XW} \cong \overline{ZW}$

Prove: $\triangle XWY \cong \triangle ZWY$

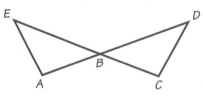

Reasons

1. Given
2. ?
3. ?
4. ?
5. ?

11. Write a two-column proof.
Given: $\overline{AB} \cong \overline{CB}$; $\overline{BE} \cong \overline{BD}$
Prove: $\triangle ABE \cong \triangle CBD$

12. Cooperative Learning Work in a group of four people. You will need a ruler, a compass, and a protractor.

a. Each member in your group should draw quadrilateral $GHJK$ with $GH = 4$ cm, $HJ = 6$ cm, $JK = 8$ cm, and $KG = 10$ cm.

b. Compare your quadrilateral with other quadrilaterals drawn in your group. Are all of the quadrilaterals congruent?

c. Do you think there is an SSSS postulate for proving quadrilaterals congruent? Explain why or why not.

When rescuers enter a partially collapsed building, they must reinforce damaged doorways for safety. If the doorway will not be used during the rescue, diagonal braces can be used.

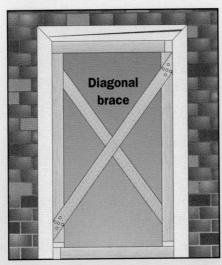

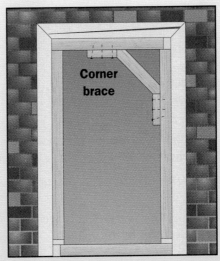

13. a. Writing A quadrilateral might change shape under pressure, but a triangle will not. Use the SSS Postulate to explain why this is true.

b. Explain why diagonal braces make a doorway more stable.

14. Challenge If the doorway must not be blocked, rescuers can strengthen it with a corner brace instead of a diagonal brace. Explain why a corner brace makes the doorway more stable.

Investigation For Exercises 15–19, you will need geometry software, or a compass and straightedge. Draw a triangle and label the vertices *A*, *B*, and *C*. You will construct copies of △*ABC* by copying $\overline{AB}$, ∠*A*, and $\overline{AC}$.

15. Make a copy of △*ABC* by first copying $\overline{AB}$, then ∠*A*, then $\overline{AC}$.

16. Make a copy of △*ABC* by first copying ∠*A*, then $\overline{AB}$, then $\overline{AC}$.

17. Make a copy of △*ABC* by first copying $\overline{AC}$, then ∠*A*, then $\overline{AB}$.

18. Writing How are the triangles you drew in Exercises 15–17 different? How are they alike?

19. Is there a way to construct a copy of △*ABC* so that two of the sides are congruent to $\overline{AB}$ and $\overline{AC}$ and the included angle is congruent to ∠*A*, but the triangle is not congruent to △*ABC*? If so, give an example.

20. SAT/ACT Preview For △*GHJ* and △*PQR*, $\overline{GH} \cong \overline{PQ}$, $\overline{HJ} \cong \overline{QR}$, and ∠*H* ≅ ∠*Q*. △*HGJ* ≅ __?__

A. △*PQR*   **B.** △*QRP*   **C.** △*RQP*   **D.** △*QPR*   **E.** cannot be determined

21. Write a proof using any format.

Given: $\overline{PQ} \cong \overline{RQ}$; $\overline{PS} \cong \overline{RS}$

Prove: $\triangle PQS \cong \triangle RQS$

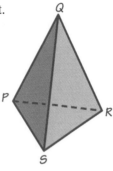

22. The vertices of $\triangle ABC$ are $A(-3, 1)$, $B(-8, 5)$, and $C(-1, 8)$. The vertices of $\triangle DEF$ are $D(0, 1)$, $E(4, 6)$, and $F(7, -1)$.

  a. Graph $\triangle ABC$ and $\triangle DEF$.

  b. Explain how to use the Distance Formula to tell whether the triangles are congruent. Which postulate must you use?

23. TRANSFORMATIONS The vertices of $\triangle ABC$ are $A(1, 7)$, $B(6, 3)$, and $C(3, 1)$.

  a. Graph $\triangle ABC$ and reflect it over the x-axis. What are the coordinates of the vertices of the image?

  b. Reflect $\triangle ABC$ over the y-axis. What are the coordinates of the vertices of the image?

  c. Are the two images of $\triangle ABC$ congruent? Explain why or why not.

ONGOING ASSESSMENT

24. Writing Describe three ways to prove that two triangles are congruent. Choose one of the ways and give an example.

SPIRAL REVIEW

Name all of the triangles that appear to be congruent. *(Section 6.2)*

25.

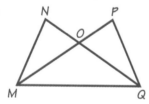

26.

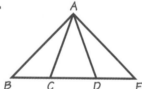

27. The floor plan for a building has a scale of 1 in. = 10 ft. Write this as a ratio in lowest terms. *(Toolbox, page 707)*

Find each unknown angle measure. *(Section 2.2)*

28.

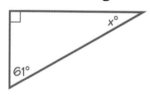

29.

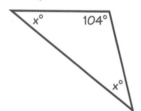

30.

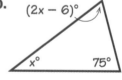

6.4

Congruent Triangles: ASA, AAS, and HL

Learn how to...

- use the ASA Postulate, the AAS Theorem, and the HL Theorem

So you can...

- prove that triangles are congruent

- find the positions of distant objects, such as stars and islands

In 1576, King Frederik II of Denmark gave the astronomer Tycho Brahe an island to encourage him to remain in Denmark. Brahe used *triangulation* to map his island and find its exact location in relation to the rest of Denmark.

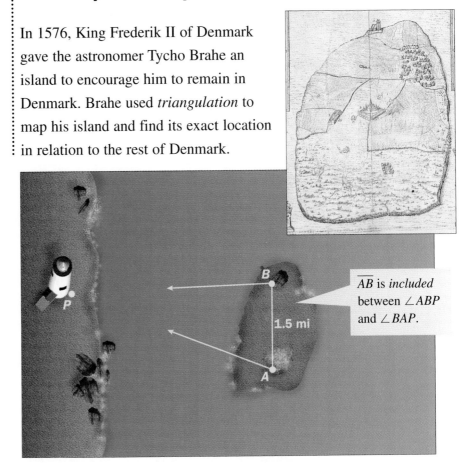

$\overline{AB}$ is *included* between $\angle ABP$ and $\angle BAP$.

Suppose you are on an island and want to find the distance from the island to a known point, *P*, on shore. Choose two points, *A* and *B*, on the island and measure $\overline{AB}$, $\angle PAB$, and $\angle ABP$. You can use these measures to find *BP*.

THINK AND COMMUNICATE

1. Make a scale drawing of △*ABP*: Use a ruler to draw a baseline 1.5 in. long. Then use a protractor to draw the two angles, extending the rays until they intersect at point *P*. Measure the distance from *B* to *P* in your drawing.

2. Compare your triangle with those of your classmates. Is the length *BP* about the same in each triangle? Are all of the triangles congruent?

3. How can you use your scale drawing to estimate the actual distance from point *B* on the island to point *P* on shore?

The Angle-Side-Angle Postulate

When you and your classmates used triangulation to find the distance BP, your triangles should have all been congruent. The method of triangulation leads to another shortcut for proving that triangles are congruent.

Angle-Side-Angle Postulate (ASA Postulate)

If two angles and the included side of one triangle are congruent to two angles and the included side of another triangle, then the triangles are congruent.

If $\angle A \cong \angle L$,
$\overline{AC} \cong \overline{LN}$,
and $\angle C \cong \angle N$,
then $\triangle ABC \cong \triangle LMN$.

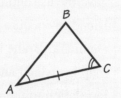 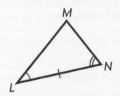

EXAMPLE 1

Given: $\overrightarrow{HK}$ bisects $\angle JHL$.
$\overrightarrow{KH}$ bisects $\angle JKL$.

Prove: $\triangle JHK \cong \triangle LHK$

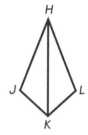

SOLUTION

Plan Ahead: Use the definition of angle bisector to show that $\angle JHK \cong \angle LHK$ and $\angle JKH \cong \angle LKH$. Use the fact that $\overline{HK}$ is a side of both triangles.

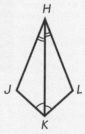

Statements	Reasons
1. $\overrightarrow{HK}$ bisects $\angle JHL$.	1. Given
2. $\angle JHK \cong \angle LHK$	2. Definition of angle bisector
3. $\overline{HK} \cong \overline{HK}$	3. Reflexive Property
4. $\overrightarrow{KH}$ bisects $\angle JKL$.	4. Given
5. $\angle JKH \cong \angle LKH$	5. Definition of angle bisector
6. $\triangle JHK \cong \triangle LHK$	6. ASA Postulate

So far you have learned three shortcuts for proving that two triangles are congruent. SSS, SAS, and ASA are all postulates. You can use these postulates to prove two other shortcuts, which are theorems.

THINK AND COMMUNICATE

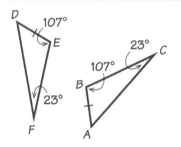

4. Find the measures of $\angle A$ and $\angle D$.

5. If two angles of a triangle are congruent to two angles of another triangle, then what can you tell about the third angles?

6. Explain how you know that $\triangle ABC \cong \triangle DEF$.

Angle-Angle-Side Theorem (AAS Theorem)

If two angles and a non-included side of one triangle are congruent to the corresponding parts of another triangle, then the triangles are congruent.

If $\angle Q \cong \angle X$, $\angle R \cong \angle Y$,
and $\overline{RS} \cong \overline{YZ}$,
then $\triangle QRS \cong \triangle XYZ$.

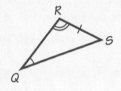

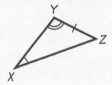

In the triangles at the right, you can use the Pythagorean theorem to see that $MN = 4\sqrt{6}$ and $QP = 4\sqrt{6}$. So $\triangle MNO \cong \triangle QPR$ by the SSS Postulate or the SAS Postulate.

In general, if two pairs of corresponding sides in a right triangle are congruent, you can show that the third pair of corresponding sides is congruent. This leads to a shortcut that works only for right triangles.

Hypotenuse-Leg Theorem (HL Theorem)

If the hypotenuse and a leg of one right triangle are congruent to the corresponding parts of another right triangle, then the triangles are congruent.

If $\triangle ABC$ and $\triangle DEF$ are right
triangles, $\overline{AB} \cong \overline{DE}$, and $\overline{AC} \cong \overline{DF}$,
then $\triangle ABC \cong \triangle DEF$.

EXAMPLE 2

Given: $\overline{EH} \cong \overline{FG}$

$\angle H$ and $\angle G$ are right angles.

Prove: $\triangle EHF \cong \triangle FGE$

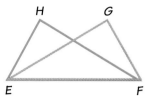

SOLUTION

Plan Ahead: Sketch the triangles separately. Because $\triangle EHF$ and $\triangle FGE$ are right triangles, use the HL Theorem. $\overline{EF}$ is the hypotenuse of both triangles.

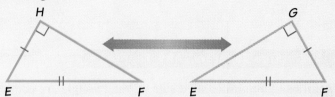

Statements	Reasons
1. $\angle H$ and $\angle G$ are right angles.	**1.** Given
2. $\triangle EHF$ and $\triangle FGE$ are right triangles.	**2.** Definition of right triangle
3. $\overline{EH} \cong \overline{FG}$	**3.** Given
4. $\overline{EF} \cong \overline{FE}$	**4.** Reflexive Property
5. $\triangle EHF \cong \triangle FGE$	**5.** HL Theorem

☑ CHECKING KEY CONCEPTS

1. Draw and label a triangle $\triangle ABC$. Name two angles and their included side. Then name the same two angles and a non-included side.

Tell which method(s) you can use to prove the triangles congruent. If no method can be used, write *none*.

2.

3.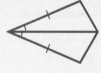

4.

5.

6.

7.

Tell which method(s) you can use to prove that the triangles are congruent. If no method can be used, write *none*.

1.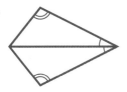

2.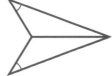

3.

4.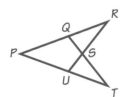

5.

6.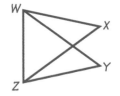

For each exercise, name a pair of overlapping congruent triangles. Tell which method you can use to prove the triangles congruent.

7. Given: $\overline{RU} \cong \overline{TQ}$
$\angle R \cong \angle T$

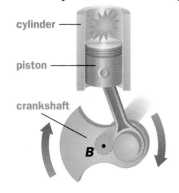

8. Given: $\overline{AD} \cong \overline{BC}$
$\overline{AB} \perp \overline{BC}$
$\overline{AB} \perp \overline{DA}$

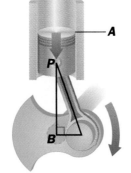

9. Given: $\angle X \cong \angle Y$
$\angle WZX \cong \angle ZWY$

10. Open-ended Problem Choose one of the diagrams in Exercises 7–9. Plan a proof for the exercise that you chose.

11. AUTOMOBILES In a car engine, exploding gas causes pistons to move up and down in a cylinder. The pistons are attached to the crankshaft, which rotates around point *B* as shown below.

a. When the shaft is at the position shown in blue, the piston is at *A*. Where is the piston in relation to *A* when the crankshaft is at the position shown in green? Which postulate or theorem justifies your answer?

b. How close can point *P* get to point *B*? How far can it get from point *B*? Which postulate or theorem justifies your answer?

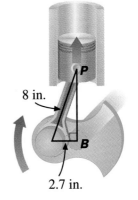

cylinder
piston
crankshaft

8 in.
2.7 in.

12. Given: $\angle A$ is a right angle.
$\angle C$ is a right angle.
$\overrightarrow{BD}$ bisects $\angle ABC$.

Prove: $\triangle ADB \cong \triangle CDB$

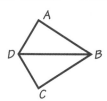

13. Given: $JKLM$ is a rectangle.
$\overline{JP} \cong \overline{KP}$

Prove: $\triangle JMP \cong \triangle KLP$

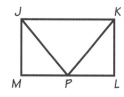

14. Visual Thinking Tell which postulate or theorem about congruent triangles justifies each statement. Explain your choice.

a. If the legs of one right triangle are congruent to the legs of another right triangle, then the triangles are congruent.

b. If a leg and the adjacent acute angle of one right triangle are congruent to the corresponding leg and acute angle of another triangle, then the triangles are congruent.

15. Given: $\overline{PQ} \parallel \overline{VS}$
$\overline{QU} \parallel \overline{ST}$
$\overline{PQ} \cong \overline{VS}$

Prove: $\triangle PQU \cong \triangle VST$

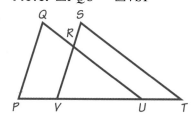

16. Given: $\overline{NH} \perp$ plane EHG
$\overline{GN} \cong \overline{EN}$

Prove: $\triangle GNH \cong \triangle ENH$

17. Investigation You will need geometry software or a compass and straightedge. Draw an acute or obtuse triangle and label the vertices A, B, and C. Follow Steps 1–4.

Step 1 Copy $\angle A$. Label the new angle $\angle XDY$ and construct $\overleftrightarrow{DX}$.

Step 2 Copy $\overline{AB}$ on $\overrightarrow{DY}$. Label the new segment $\overline{DE}$.

Step 3 Construct a **circle** with center E and radius BC.

Step 4 Construct $\overline{EF}$ so that $EF = BC$ and F is on $\overleftrightarrow{DX}$.

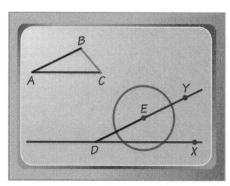

a. How many different ways are there to do Step 4? Do all of the ways result in a triangle that is congruent to $\triangle ABC$? Explain.

b. Draw a right triangle with a right angle at C and hypotenuse $\overline{AB}$. Repeat Steps 1–4 and part (a).

c. Writing Explain why there is an HL Theorem but not an SSA Theorem.

Astronomers can find the distance to a nearby star by noticing how it moves against the background of more distant stars.

18. Challenge Earth is on opposite sides of the sun at different times of the year as shown at the right. Explain how you can tell that astronomers have enough information to find the distance to star *S*. Use at least one postulate or theorem in your answer.

19. 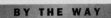 **Technology** You can use the *sine* button on a calculator and the formula

$$SD = \frac{\sin(m\angle J)}{\sin(m\angle S)} \cdot JD$$

to find the distance to star *S*. For *Alpha Centauri*, the nearest star to Earth other than the sun, $m\angle S \approx 0.0004°$.

a. Find $m\angle J$ for *Alpha Centauri*.

b. About how far away is *Alpha Centauri*?

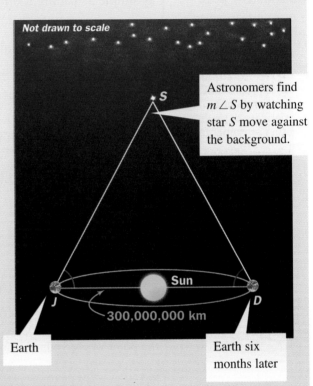

Not drawn to scale

Astronomers find $m\angle S$ by watching star *S* move against the background.

Sun

300,000,000 km

Earth

Earth six months later

BY THE WAY

The Yokut tribe of California used astronomy to time the Mourning Ceremony with the setting of the evening star.

ONGOING ASSESSMENT

20. a. Open-ended Problem List all the methods that you know for proving that two triangles are congruent.

b. Choose two methods from part (a). For each method, sketch and label a pair of triangles and mark the information needed to use that method to prove that the triangles are congruent.

SPIRAL REVIEW

For Exercises 21–23, △ABC ≅ △JLK. Complete each statement. *(Section 6.2)*

21. $\overline{AB} \cong$?

22. $\angle C \cong$?

23. $\overline{KJ} \cong$?

Find the measure of each interior and exterior angle of the polygon. *(Section 2.4)*

24. regular triangle

25. regular 7-gon

26. equiangular 24-gon

6.5

Applying Congruence

Finding the distance between two points can be as simple as pacing it off or using a yardstick or another measuring device. But if the two points are on opposite sides of a busy highway or a stream, you may need to find another method. Congruent triangles can help!

Learn how to...

* **use corresponding parts of congruent triangles**

So you can...

* **write proofs that use congruent triangles to prove other statements**

* **measure distances indirectly**

Here is one way to find the distance across a stream:

Choose a point on the other side of the stream directly opposite the point where you are standing.

Put on a visor or a cap with a visor. Look at the point you chose and adjust the visor until it is in line with your eye and the point.

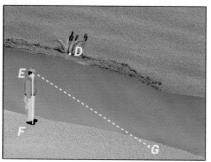

Turn right or left. Note the point on the ground that is now in line with your eye and the visor tip.

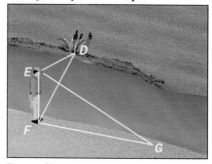

Pace off or measure the distance from your feet to the point.

THINK AND COMMUNICATE

1. Explain why $\triangle EFD$ and $\triangle EFG$ are congruent right triangles.

2. Which segment in $\triangle EFG$ is the same length as the distance across the stream?

The definition of congruent polygons tells you that if two triangles are congruent, then all pairs of corresponding parts are congruent. You can use this idea to prove that the distance across the stream is the same as the distance along the bank.

Given: $\angle EFD$ and $\angle EFG$ are right angles.
$\angle DEF \cong \angle GEF$
Prove: $\overline{DF} \cong \overline{GF}$

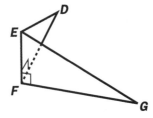

SOLUTION

Plan Ahead: Prove that $\triangle EFD \cong \triangle EFG$. Then $\overline{DF} \cong \overline{GF}$ because they are corresponding parts of congruent triangles.

Statements	Reasons
1. $\angle EFD$ and $\angle EFG$ are right angles.	**1.** Given
2. $\angle EFD \cong \angle EFG$	**2.** All right angles are congruent.
3. $\overline{EF} \cong \overline{EF}$	**3.** Reflexive Property
4. $\angle DEF \cong \angle GEF$	**4.** Given
5. $\triangle EFD \cong \triangle EFG$	**5.** ASA Postulate
6. $\overline{DF} \cong \overline{GF}$	**6.** Definition of congruent triangles

When applying the definition of congruent polygons to triangles in a proof, write *definition of congruent triangles*.

Perpendicular Bisectors

If point A is the same distance from point B as it is from point C, then A is *equidistant* from B and C. You will use this idea to learn more about perpendicular bisectors.

THINK AND COMMUNICATE

$\overleftrightarrow{PR}$ **is the perpendicular bisector of** $\overline{QS}$.

3. Is R equidistant from Q and S? Explain why or why not.

4. Which theorem or postulate tells you that $\triangle PQR \cong \triangle PSR$?

5. Explain why P is equidistant from Q and S.

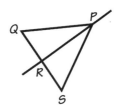

The questions on the previous page lead to the Perpendicular Bisector Theorem. You will prove this theorem in the exercises.

The Perpendicular Bisector Theorem

If a point is on the perpendicular bisector of a segment, then the point is equidistant from the endpoints of the segment.

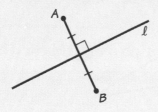

Any point on ℓ is equidistant from A and B.

You can use corresponding parts of congruent triangles to prove statements about midpoints, angle bisectors, segment bisectors, parallel segments, isosceles triangles, and other figures.

EXAMPLE 2

Given: $\overline{JK} \cong \overline{LM}$; $\overline{JM} \cong \overline{LK}$
Prove: $\overline{JM} \parallel \overline{LK}$

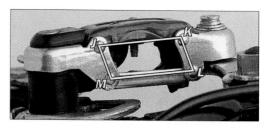

SOLUTION

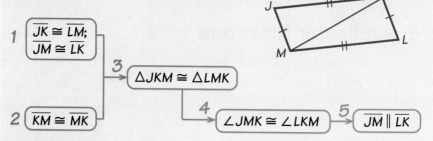

Reasons

1. Given

2. Reflexive Property

3. SSS Postulate

4. Definition of congruent triangles

5. If two lines are intersected by a transversal and alternate interior angles are ≅, then the lines are ∥.

Name a pair of triangles that must be congruent in order for you to use the definition of congruent triangles to prove each statement.

1. $\overline{GH} \cong \overline{GF}$

2. $\overline{HE} \cong \overline{FD}$

3. $\angle EDF \cong \angle DEH$

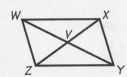

4. $\angle XWY \cong \angle ZYW$

5. $\overline{VX} \cong \overline{VZ}$

6. $\angle WYX \cong \angle YWX$

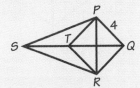

$\overleftrightarrow{QS}$ **is the perpendicular bisector of** $\overline{PR}$.

7. Complete: T is equidistant from points __?__ and __?__.

8. Find the length QR.

9. What type of triangle is $\triangle PSR$?

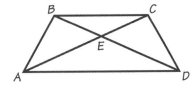

6.5 | **Exercises and Applications**

Extra Practice
exercises on
page 672

Name a pair of triangles that must be congruent in order for you to use the definition of congruent triangles to prove each statement.

1. $\overline{BE} \cong \overline{CE}$

2. $\angle ABE \cong \angle DCE$

3. $\overline{AC} \cong \overline{DB}$

4. $\angle BCA \cong \angle CBD$

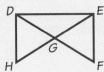

5. Given: $\overline{PS} \parallel \overline{QR}$
$\qquad \angle P \cong \angle R$

Prove: $\overline{PQ} \cong \overline{RS}$

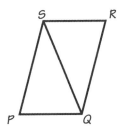

6. Given: $\angle ABC$ and $\angle DEF$
are right angles.
$\overline{AB} \cong \overline{DE}$
$\overline{BC} \cong \overline{EF}$

Prove: $\angle BAC \cong \angle EDF$

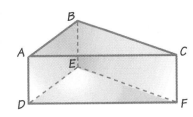

7. Copy and complete the key steps of this proof of the Perpendicular Bisector Theorem.

Given: Line ℓ is the perpendicular bisector of $\overline{AB}$.

Prove: For any point P on ℓ, $PA = PB$.

1. Draw $\overline{PA}$ and $\underline{\ ?\ }$. (Through any two points there is $\underline{\ ?\ }$ line.)

2. $\ell \perp \overline{AB}$ and C is the midpoint of $\overline{AB}$. (Definition of $\underline{\ ?\ }$)

3. $\triangle APC \cong \triangle BPC$ ($\underline{\ ?\ }$)

4. $\underline{\ ?\ }$ ($\underline{\ ?\ }$)

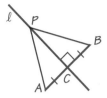

ALGEBRA **In each diagram, ℓ is the perpendicular bisector of $\overline{AB}$.**

8. Find the value of x.

9. Find the value of y.

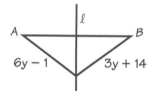

Connection ▶ AERIAL PHOTOGRAPHY

Aerial surveyors use airplanes to survey large areas of land. They fly back and forth over the area, taking pictures with a camera pointed straight down. If the area being photographed is not flat, the surveyor may need to adjust the plane's altitude during the flight.

10. **Writing** Suppose the surveyor takes all of the pictures from the same altitude above sea level. Will photograph 1 show as much of the ground as photograph 2? Explain why or why not.

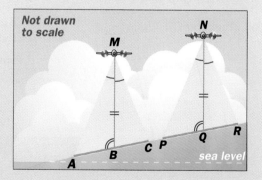

11. To be sure that all of the photographs cover the same amount of ground (and have the same scale), the pilot adjusts the plane's altitude so that the camera is always the same height above the ground.

Given: $\overline{MB} \cong \overline{NQ}$
$\angle AMB \cong \angle BMC \cong \angle PNQ \cong \angle QNR$
$\angle ABM \cong \angle PQN$

Prove: $\overline{AC} \cong \overline{PR}$

(*Hint:* Use congruent triangles to show that $\overline{AB} \cong \overline{PQ}$. Then use congruent triangles to show that $\overline{BC} \cong \overline{QR}$.)

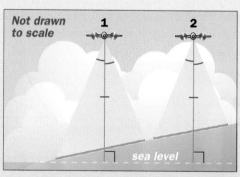

ANGLE BISECTOR

Given an angle, construct its bisector.

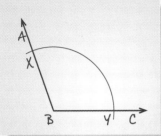
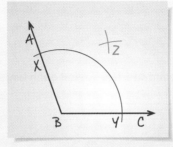
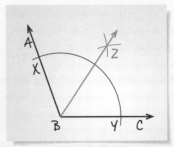

1. Using center B and any radius, draw **an arc** that intersects $\overrightarrow{BA}$ and $\overrightarrow{BC}$. Label the points of intersection X and Y.

2. Draw **an arc** with center X and a convenient radius. Draw **an arc** with center Y and the same radius. Label the intersection of the arcs Z.

3. Draw $\overrightarrow{BZ}$. $\overrightarrow{BZ}$ bisects $\angle ABC$.

12. a. Sketch the figure in Step 3. Include $\overline{ZX}$ and $\overline{ZY}$. How do you know that $\overline{ZX} \cong \overline{ZY}$?

 b. Explain how you know that $\overrightarrow{BZ}$ bisects $\angle ABC$.

13. Open-ended Problem Draw an obtuse angle and construct its bisector.

14. Writing Compare the constructions for an angle bisector and a segment bisector. How are the constructions alike? How are they different?

15. Writing State the converse of the Perpendicular Bisector Theorem. Do you think that it is true? Explain why or why not.

16. Challenge Copy the figure. Use constructions to locate a point that is equidistant from X, Y, and Z.

•Y

X•

•Z

17. Given: $\overline{WQ} \cong \overline{YQ}$; $\overline{XQ} \cong \overline{ZQ}$
 Prove: $\overline{WX} \parallel \overline{YZ}$

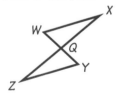

18. Given: $\overleftrightarrow{SQ}$ is the perpendicular bisector of $\overline{PR}$.

 Prove: $\angle SPT \cong \angle SRT$

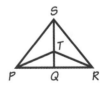

19. SAT/ACT Preview In $\triangle RST$ and $\triangle ABC$, $\overline{RS} \cong \overline{AB}$, $\overline{ST} \cong \overline{BC}$, and $\overline{TR} \cong \overline{CA}$. Which angle is congruent to $\angle T$?

 A. $\angle R$ **B.** $\angle A$ **C.** $\angle B$ **D.** $\angle C$ **E.** cannot be determined

20. Open-ended Problem In $\triangle PQR$, $\overline{PQ} \cong \overline{PR}$ and S is the midpoint of $\overline{QR}$. Sketch the figure and make a list of everything that you can prove about it.

SPIRAL REVIEW

The midpoint of $\overline{RT}$ is S, and the midpoint of $\overline{RU}$ is T. Find each length. *(Section 1.7)*

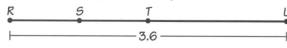

21. RT **22.** ST **23.** SU

$\overrightarrow{NP}$ **bisects** $\angle MNO$**. Find each measure.** *(Section 1.7)*

24. $m \angle MNP$

25. $m \angle MNO$

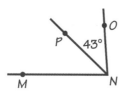

ASSESS YOUR PROGRESS

Tell whether you can prove that the triangles are congruent. If you can, tell which method(s) you can use. *(Sections 6.3 and 6.4)*

1. 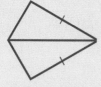   **2.** 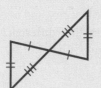   **3.**

4.   **5.** 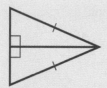   **6.**

Name a pair of triangles that must be congruent in order for you to use the definition of congruent triangles to prove each statement. *(Section 6.5)*

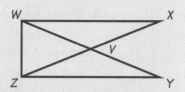

7. $\angle YZV \cong \angle XWV$ **8.** $\overline{ZX} \cong \overline{WY}$

9. Suppose $\overleftrightarrow{AB}$ is the perpendicular bisector of $\overline{CD}$. Sketch the figure using only points A, B, C, and D. Include two pairs of congruent segments and label them. *(Section 6.5)*

10. Journal Describe how to use congruent triangles to find the distance between two points at opposite ends of a lake. Include a sketch in your explanation.

6.6 Properties of Isosceles Triangles

Learn how to...

- apply the Isosceles Triangle Theorem and its converse

So you can...

- find measures in isosceles triangles

- write proofs using isosceles triangles

In one type of rock climbing, called *top roping*, climbers tie themselves to a rope that is supported by *anchors* at the top of the climb. If a climber slips, the anchors catch the fall. For safety, climbers use at least two anchors.

The illustration below shows two anchors in a horizontal crack of a rock face. To be sure that the force on the anchors is equal, the angles formed by the anchors and the crack should be congruent.

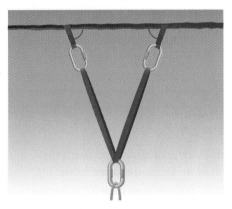

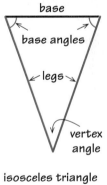

isosceles triangle

The triangle formed by these anchors is an isosceles triangle. The sides opposite the congruent angles are the **legs** and the third side is the **base**. The congruent angles are **base angles** and the third angle is the **vertex angle**.

THINK AND COMMUNICATE

1. a. If the red anchor is longer than the blue anchor, are the base angles congruent? If not, which angle is larger?

b. Use the triangle inequality theorems on page 280. Which theorem supports your answer to part (a)?

2. If a climber adjusts the anchors so they are the same length, do you think that the base angles will be congruent?

Sides and Angles in Isosceles Triangles

The Isosceles Triangle Theorem

If two sides of a triangle are congruent, then the angles opposite the sides are congruent.

In △ABC, if $\overline{AB} \cong \overline{AC}$, then $\angle C \cong \angle B$.

The Converse of the Isosceles Triangle Theorem

If two angles of a triangle are congruent, then the sides opposite the angles are congruent.

In △ABC, if $\angle C \cong \angle B$, then $\overline{AB} \cong \overline{AC}$.

Example 1 shows one way to prove the Isosceles Triangle Theorem. You will prove the Converse of the Isosceles Triangle Theorem in the exercises.

EXAMPLE 1

Prove the Isosceles Triangle Theorem.

SOLUTION

Given: In △XYZ, $\overline{XZ} \cong \overline{YZ}$.

Prove: $\angle X \cong \angle Y$

Plan Ahead: Draw a diagram and mark the given information. Draw the bisector of $\angle Z$ and let P be its intersection with $\overline{XY}$. Use the SAS Postulate to prove that $\triangle XZP \cong \triangle YZP$. Therefore, $\angle X \cong \angle Y$.

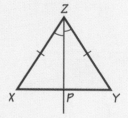

Statements	Reasons
1. $\overline{XZ} \cong \overline{YZ}$	1. Given
2. Draw $\overrightarrow{ZP}$, the angle bisector of $\angle Z$.	2. Angle Bisector Construction (p. 311)
3. $\angle XZP \cong \angle YZP$	3. Definition of angle bisector
4. $\overline{ZP} \cong \overline{ZP}$	4. Reflexive Property
5. $\triangle XZP \cong \triangle YZP$	5. SAS Postulate
6. $\angle X \cong \angle Y$	6. Definition of congruent triangles

EXAMPLE 2   Connection: Algebra

Find $m \angle D.$

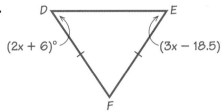

SOLUTION

$\overline{DF} \cong \overline{EF}$ so, by the Isosceles Triangle Theorem, $\angle D \cong \angle E.$

$$m \angle D = m \angle E$$

$$(2x + 6)° = (3x - 18.5)° \quad \text{First find } x.$$

$$6 = 3x - 18.5 - 2x$$

$$6 + 18.5 = x$$

$$24.5 = x$$

$$m \angle D = (2x + 6)° \quad \begin{array}{l}\text{Now use } x \text{ to} \\ \text{find } m \angle D.\end{array}$$

$$= 2(24.5) + 6$$

$$= 49 + 6$$

$$= 55°$$

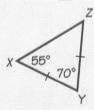

Check

$$m \angle E = m \angle D$$

$$3x - 18.5 = 55$$

$$3(24.5) - 18.5 \overset{?}{=} 55$$

$$73.5 - 18.5 \overset{?}{=} 55$$

$$55 = 55 ✓$$

✓ CHECKING KEY CONCEPTS

Complete.

1. $AB = \underline{\ ?\ }$

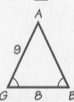

2. $m \angle Z = \underline{\ ?\ }$

ALGEBRA   **Find each length or angle measure.**

3. $m \angle D = \underline{\ ?\ }$

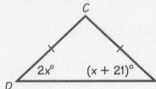

4. $NP = \underline{\ ?\ }$

6.6 Exercises and Applications

Extra Practice
exercises on
page 672

1. **ROCK CLIMBING** A rock climber sets up two anchors in a horizontal crack of a rock face. One anchor is two feet long. How long must the other anchor be for the angles formed by the anchors and the crack to be equal?

For Exercises 2–4, use the diagram.

2. $\angle 1 \cong \angle 2$. Name two congruent segments.

3. $\overline{KP} \cong \overline{PD}$. Name two congruent angles.

4. $\triangle KPD$ and $\triangle SPR$ are both isosceles.

    a. Are $\triangle KPD$ and $\triangle SPR$ congruent? Explain.

    b. Are $\triangle KPS$ and $\triangle DPR$ congruent? Explain.

ARCHITECTURE For Exercises 5 and 6, use the photographs of roofs from around the world.

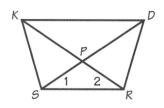
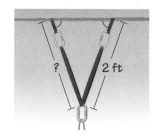

Lithuania   *Japan*   *United States*   *Germany*

5. Which roofs appear to form isosceles triangles?

6. **Open-ended Problem** Give some possible reasons why isosceles triangles are so common in roof building. When might it make more sense to use a scalene triangle instead of an isosceles triangle?

7. Copy and complete this alternate proof of the Isosceles Triangle Theorem.

    Given: In $\triangle ABC$, $\overline{AB} \cong \overline{AC}$.

    Prove: $\angle B \cong \angle C$

Plan Ahead: Imagine picking $\triangle ABC$ up and flipping it over so that $\overline{AB}$ lands on $\overline{AC}$. Use the SAS Postulate to show that $\triangle BAC \cong \triangle CAB$.

Statements	Reasons
1. __?__	1. Given
2. $\angle A \cong \angle A$	2. __?__
3. $\triangle BAC \cong \triangle CAB$	3. __?__
4. __?__	4. __?__

ALGEBRA For Exercises 8–13, find each length or angle measure.

8. AB

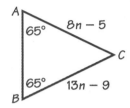

9. PR

10. $m \angle J$

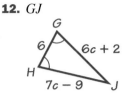

11. $m \angle T$

12. GJ

13. ZW

14. Find the value of n and the length of $\overline{AC}$.

15. Find the value of y and $m \angle D$.

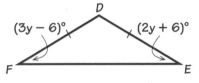

INTERVIEW **Madeleine Fleming**

Look back at the article on pages 276–278.

Madeleine Fleming designs optical surfaces like the reflective film described on page 290. The back of the film has thousands of tiny pyramids on it. One of the pyramids is shown below.

16. Prove that $\triangle ABC \cong \triangle DBC$.

17. **Writing** Explain how you can conclude that $\triangle ABD$ is isosceles.

18. **Writing** Name another isosceles triangle in the pyramid. Explain how you know that it is isosceles.

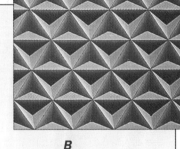

BY THE WAY

In many states and other countries, license plates coated with a retroreflective film have a security mark. The mark can be seen only from a 30° angle above the plate.

19. TRANSFORMATIONS In $\triangle PQR$, $\angle Q \cong \angle R$. Explain why $\triangle PQR$ has reflection symmetry and identify the line of symmetry.

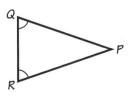

20. Given: $\overline{AB} \cong \overline{BC}$; $\overline{DE} \cong \overline{EC}$

Prove: $\angle A \cong \angle D$

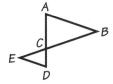

21. Prove the converse of the Isosceles Triangle Theorem.

Given: In $\triangle XYZ$, $\angle X \cong \angle Y$.

Prove: $\overline{XZ} \cong \overline{YZ}$

Plan Ahead: Draw a diagram and mark the given information. Draw the bisector of $\angle Z$ and label its intersection with $\overline{XY}$ point P. Prove that the triangles are congruent. Then use the definition of congruent triangles.

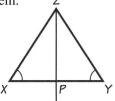

22. Challenge In $\triangle YSW$, $\overline{SY} \cong \overline{SW}$, $\overrightarrow{WR}$ bisects $\angle SWY$, and $\overrightarrow{YT}$ bisects $\angle SYW$. Name three pairs of congruent triangles. Justify your answers.

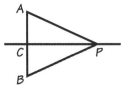

23. Writing Describe how scissors are like an isosceles triangle. Which part of the scissors are the legs of the triangle? the vertex angle? Why wouldn't it make sense for scissors to be like scalene triangles?

$\overleftrightarrow{PC}$ **is the perpendicular bisector of** $\overline{AB}$. **Tell why each statement is true.** *(Section 6.5)*

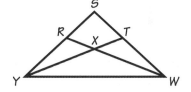

24. $\overline{CA} \cong \overline{CB}$

25. $\overline{AP} \cong \overline{BP}$

For Exercises 26–28, use the diagram at the right.

26. List all of the angle bisectors shown in the diagram. *(Section 1.7)*

27. List all of the bisectors of $\overline{AB}$. *(Section 1.7)*

28. List all of the pairs of perpendicular lines. *(Section 2.1)*

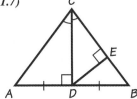

318 Chapter 6 *Conjectures About Triangles*

6.7 Altitudes, Medians, and Bisectors

Learn how to...

- identify and draw medians and altitudes of triangles

So you can...

- explore the properties of these segments in triangles

There are many different kinds of tents, but almost all of them use poles to support the roof. To set up this tent, you put the bottom of the pole in the middle of the bottom edge of the tent. This design ensures that the pole will be perpendicular to the ground.

Both sides of the roof are the same length.

The bottom of the pole is halfway between the corners of the tent.

The tent pole is both a *median* and an *altitude* of the triangular front of the tent. A **median of a triangle** is a segment from a vertex to the midpoint of the opposite side. An **altitude of a triangle** is a perpendicular segment from a vertex to the line that contains the opposite side.

EXAMPLE 1

For each triangle, *M* is the midpoint of $\overline{AB}$. Copy the triangle and draw the median and altitude from vertex *C*.

a.
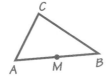

b.
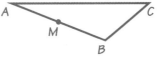

Notice that an altitude can lie outside of the triangle.

SOLUTION

a.
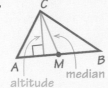

b.
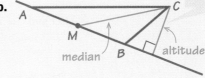

Segments in Isosceles Triangles

The altitude, the median, and the angle bisector from the vertex angle of an isosceles triangle to its base are all the same segment. They are also the same as the perpendicular bisector of the base of the triangle. You can prove these relationships using the definition of congruent triangles.

EXAMPLE 2

In isosceles △*DBC*, $\overline{DB} \cong \overline{DC}$. Prove that the bisector of ∠*D* is also an altitude of △*DBC*.

SOLUTION

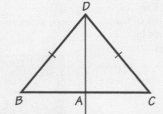

Given: In △*DBC*, $\overline{DB} \cong \overline{DC}$;
$\overrightarrow{DA}$ bisects ∠*BDC*.

Prove: $\overline{DA}$ is an altitude of △*DBC*.

Key steps:

1. △*BDA* ≅ △*CDA* (SAS Postulate, with $\overline{DA} \cong \overline{DA}$ and $\overline{DB} \cong \overline{DC}$)

2. ∠*DAB* ≅ ∠*DAC* (Definition of congruent triangles)

3. $\overline{DA} \perp \overline{BC}$ (∠*DAB* and ∠*DAC* are congruent adjacent angles.)

4. $\overline{DA}$ is an altitude of △*DBC*. (Definition of altitude)

THINK AND COMMUNICATE

1. In Example 2, how would the proof be different if you wanted to prove that $\overline{DA}$ is the perpendicular bisector of $\overline{BC}$?

2. In a scalene triangle, are the angle bisectors also altitudes? Are the angle bisectors also medians? Explain your answers or sketch some examples.

3. Sketch a triangle in which the altitude, the median, and the angle bisector from one vertex are three different segments. Can you sketch a triangle in which they are just two different segments? Explain.

4. In △*ACD*, ∠*ADC* is a right angle and $\overline{AD} \cong \overline{CD}$. $\overline{DB}$ is an altitude of the triangle. How could you prove that △*ABD* and △*CBD* are both isosceles?

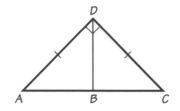

EXAMPLE 3

In this roof truss,
$\overline{WX} \cong \overline{WY}$, $\overrightarrow{WV}$ **bisects**
$\angle XWY$, **and** $\overline{UT}$ **is a**
median of $\triangle UVX$. **Find**
each length or measure.

a. $m\angle WVX$

b. XT

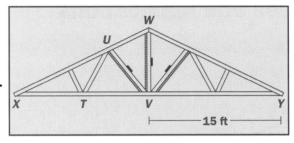

BY THE WAY

Trusses are made up of beams in triangular patterns. They have the strength and rigidity to span large distances with small amounts of material.

SOLUTION

a. Since $\overrightarrow{WV}$ bisects the vertex angle of an isosceles triangle, it is also the perpendicular bisector of the base.

$$WV \perp XY$$

$$m\angle WVX = 90°$$

b. Since $\overrightarrow{WV}$ is the perpendicular bisector of the base,

$$XV = VY$$

$$= 15$$

Because $\overline{UT}$ is a median of $\triangle UVX$, the midpoint of $\overline{XV}$ is T.

$$XT = \frac{1}{2}(XV) \qquad \text{Definition of midpoint}$$

$$= \frac{1}{2}(15)$$

$$= 7.5$$

☑ CHECKING KEY CONCEPTS

In $\triangle DEF$, M is the midpoint of $\overline{DE}$ and $\angle DQE$ is a right angle.

1. Name an altitude of $\triangle DEF$.

2. Name a median of $\triangle DEF$.

3. Could $\overrightarrow{EQ}$ be an angle bisector of $\triangle DEF$? Explain.

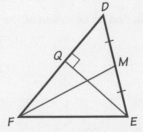

4. Sketch an isosceles triangle with an angle bisector, a median, and an altitude from the same vertex that are not the same segment.

5. In $\triangle PQR$, $\overline{PQ} \cong \overline{QR}$, $\overline{PS}$ is a median, and $\overline{PQ} = 22$ cm. Find the length of $\overline{SR}$.

Exercises and Applications

Extra Practice exercises on page 672

Find each length or measure.

1. In $\triangle JKL$, $\overline{JK} \cong \overline{JL}$, and $\overline{JM}$ is both a median and an angle bisector.

   a. $m\angle KMJ$

   b. KL

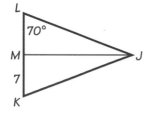

2. ALGEBRA In the figure, $\overline{PQ} \cong \overline{PR}$, and $\overline{PS}$ and $\overline{ST}$ are medians.

   a. QT

   b. QR

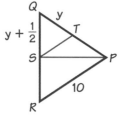

3. a. Write the converse of the theorem in Example 2 on page 320.

   b. Write a plan and prove the statement that you wrote in part (a).

4. a. Copy and complete the proof.

   Given: In $\triangle ABC$, $\overline{AC} \cong \overline{BC}$.
   $\overline{CD}$ is the median to $\overline{AB}$.

   Prove: $\overline{CD}$ is the altitude to $\overline{AB}$.

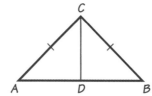

   Plan Ahead: Use the SSS Postulate to show that $\triangle ACD \cong \triangle BCD$. Then $\angle ADC$ and $\angle BDC$ are right angles, so $\overline{CD}$ is the altitude to $\overline{AB}$.

   b. Write the theorem that you proved in part (a) as an if-then statement.

5. Open-ended Problem $\triangle XYZ$ is an isosceles triangle with legs $\overline{XY}$ and $\overline{XZ}$, and W is on $\overline{YZ}$. Choose one of the following statements and write a proof for it.

   If $\overline{XW}$ is the median to $\overline{YZ}$, then it is the perpendicular bisector of $\overline{YZ}$.

   If $\overrightarrow{XW}$ bisects $\angle YXZ$, then it is the perpendicular bisector of $\overline{YZ}$.

   If $\overline{XW}$ is the altitude to $\overline{YZ}$, then it is the median to $\overline{YZ}$.

   If $\overline{XW}$ is the perpendicular bisector of $\overline{YZ}$, then it is an altitude of $\triangle XYZ$.

Sketch each segment or ray. Then find an equation of the line that contains it.

6. the median from P to $\overline{ON}$

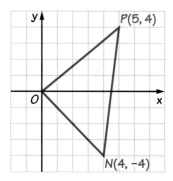

7. the bisector of $\angle P$

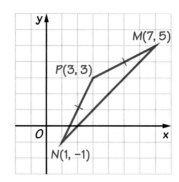

Every object has a *center of gravity*. You can balance an object on a pedestal by putting the pedestal directly under the center of gravity. A freely spinning object spins around its center of gravity.

Investigation **For Exercises 8 and 9, you will need scissors, a compass, a straightedge, and cardboard.**

8. a. Draw a large acute scalene triangle on the cardboard. Cut out the triangle and balance it on the tip of your finger or a pencil. Label the center of gravity.

  b. Construct the perpendicular bisector of each side of the triangle to locate its midpoint. (Tape the triangle to a large sheet of paper if necessary.) Carefully draw each median of the triangle. The intersection of the medians is the triangle's *centroid*. Is the centroid the same as the center of gravity?

9. Draw and cut out several cardboard parallelograms. Find the center of gravity for each parallelogram. What intersecting lines can you draw to find the center of gravity of a parallelogram?

10. Writing Sarah wants to make a wind chime. She will hang three metal rods of equal weight from the vertices of a metal triangle. Describe how Sarah can use geometry to find the point at which she can attach a wire so that the triangle will balance.

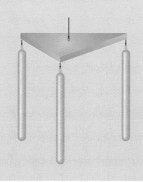

> **BY THE WAY**
>
> Some pagodas in China had more than one hundred wind bells attached to their corners. The sound was believed to drive away evil.

ALGEBRA Find the value of *x*.

11. $\overline{KL}$ is an altitude of $\triangle HJK$.

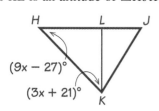

12. $\overline{PO}$ is the perpendicular bisector of $\overline{MN}$.

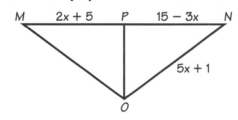

Investigation **For Exercises 13–15, you will need geometry software or a compass and straightedge.**

13. a. Draw a triangle and label it △*XYZ*. Construct the bisectors of the angles and label their intersection *P*. The point *P* is the *incenter* of the triangle. Can the incenter of a triangle ever be outside the triangle? Explain why or why not.

   b. Construct the perpendicular segment from *P* to any side of △*XYZ*. Use the length of the segment as the radius to draw a circle with center *P*. What do you notice? Write a conjecture about the relationship between the incenter and the sides of a triangle.

14. a. Draw a triangle and label it △*ABC*. Construct the three medians of the triangle and label them as shown. The intersection, *G*, of the medians is the *centroid* of the triangle. Can the centroid of a triangle ever be outside the triangle? Explain why or why not.

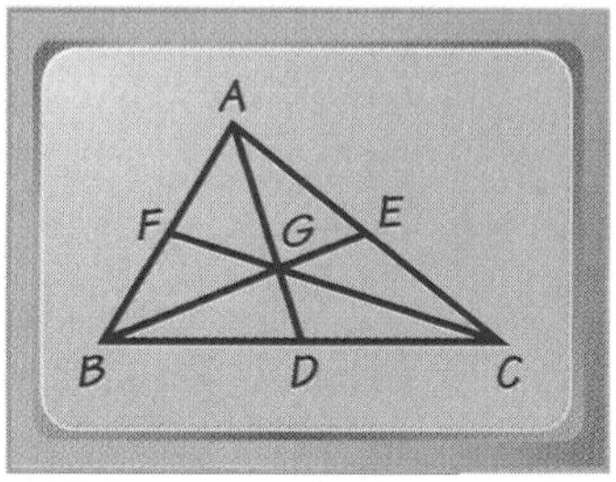

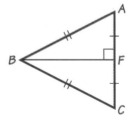

   b. Measure the parts of each median and find $\frac{AG}{GD}$, $\frac{CG}{GF}$, and $\frac{BG}{GE}$. What do you notice?

15. Draw a triangle and label it △*PQR*. Construct the three altitudes of the triangle and label their intersection *C*. *C* is the *orthocenter* of △*PQR*. Can the orthocenter of a triangle ever be outside the triangle? Explain.

16. SAT/ACT Preview For the diagram at the right, which of the following statements must be true?

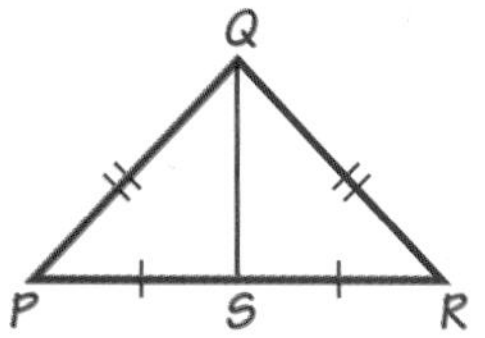

   I. $\overline{QS}$ is a median of △*PQR*.
   II. $\overline{QS}$ is an altitude of △*PQR*.
   III. $\overline{QS}$ is the bisector of ∠*Q*.

   A. I only   **B.** I and II only   **C.** III only   **D.** I, II, and III   **E.** none of the above

17. Challenge Describe the errors in the diagram and plan for this proof. Make an appropriate diagram for the proof and mark what you know.

   Given: △*ABC* is an isosceles triangle with $\overline{AB} \cong \overline{BC}$.
   Prove: The perpendicular bisector of $\overline{AC}$ is the bisector of ∠*B*.

   Plan Ahead: Let $\overline{BF}$ be the perpendicular bisector of $\overline{AC}$. Use the HL Theorem to prove that △*ABF* ≅ △*CBF*. Then ∠*ABF* ≅ ∠*CBF* because of the definition of congruent polygons. So $\overline{BF}$ is the bisector of ∠*ABC*.

ONGOING ASSESSMENT

18. Visual Thinking Imagine a scalene triangle, △*ABC*. What do the segments in each group have in common?

   a. the perpendicular bisector of $\overline{AB}$ and the altitude to $\overline{AB}$
   b. the median to $\overline{AB}$ and the perpendicular bisector of $\overline{AB}$
   c. the median to $\overline{AB}$, the altitude to $\overline{AB}$, and the angle bisector of ∠*C*

Find each length or angle measure. *(Section 2.5)*

19. rhombus *ABCD*
    a. *m ∠ B*
    b. *AD*

20. □*LMNO*
    a. *LO*
    b. *m ∠ M*

21. □*WXYZ*
    a. *QY*
    b. *XQ*

ASSESS YOUR PROGRESS

VOCABULARY

legs of an isosceles triangle (p. 313)

base of an isosceles triangle (p. 313)

base angles of an isosceles triangle (p. 313)

vertex angle of an isosceles triangle (p. 313)

median of a triangle (p. 319)

altitude of a triangle (p. 319)

Find each length or angle measure. *(Section 6.6)*

1. *KM*

2. *m ∠ R*

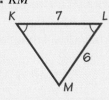

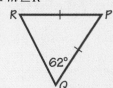

In the diagram, line $n \perp \overleftrightarrow{SH}$, $\overline{JS} \perp \overleftrightarrow{SH}$, **and** $\overline{GT} \cong \overline{TH}$. *(Section 6.7)*

3. Name a perpendicular bisector.

4. Name an altitude of △*GHJ*.

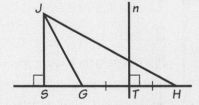

ALGEBRA Find each length or angle measure. *(Sections 6.6 and 6.7)*

5. *m ∠ E*

6. $\overline{BD}$ is a median of △*ABC*. Find *AD*.

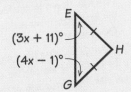

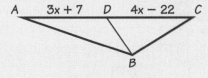

7. **Journal** Write a summary of what you know about angles and segments in isosceles triangles. How are scalene triangles different from isosceles triangles?

Building a Mobile

How can you make a
sculpture that looks like
it's moving? Sculptors
use curves, lines, and texture
to create a feeling of movement.
Some sculptures actually move.
Delicately balanced sculptures that
move in the breeze are called *mobiles*.

Antennae with Red and Blue Dots
by Alexander Calder

PROJECT GOAL Create a mobile that illustrates many of the properties of
triangles that you have learned in this chapter.

Design the Mobile

Work with a partner to design a mobile. Decide how you will meet each
requirement below. Make a sketch of what you would like your mobile
to look like.

1. Create at least one pair of congruent triangles.

2. Include at least one isosceles triangle.

3. Illustrate angle bisectors, altitudes, medians,
and perpendicular bisectors.

4. Balance at least one triangle at
its *centroid*. (See page 323.)

5. Illustrate at least one
of the properties or theorems
you have learned in this
chapter.

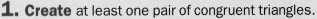

We'll use straws to illustrate
the first triangle inequality.
The green straws don't form
a triangle because the sum of
the lengths of the shorter straws
is less than the length of the
longest straw. The pink straws
don't form a triangle because

Making it Balance

Materials: You will need something to make triangles from, such as cardboard; something to hang the triangles with, such as string; and something sturdy to support the triangles, such as wire or sticks. Tape may also be helpful.

Balance your mobile so it moves in a gentle breeze or when you touch it. It will take some patience to make your mobile balance. Don't give up! If your mobile has more than one level, balance it from the bottom up.

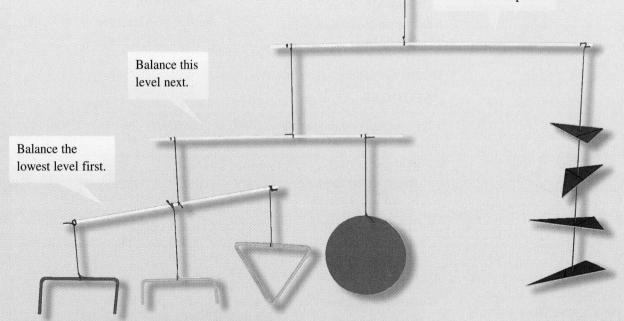

Move the string until you find the point where it balances.

Balance the top last.

Balance this level next.

Balance the lowest level first.

Presenting Your Project

Write a brief description of your mobile. Include an explanation of how your mobile meets each of the requirements on page 326.

Self-Assessment

Compare your completed mobile with your original design and ideas. If you were to make another mobile, what would you do differently? What tips would you give to someone who wants to build a mobile? What do you like best about your mobile?

6 | Review

Work in a group of three students to write a brief summary of Section 6.1. Then each member should write a summary of the major concepts in two more sections of this chapter. Share all of the summaries with your group.

VOCABULARY

congruent polygons (p. 285)

corresponding part (p. 285)

legs of an isosceles triangle (p. 313)

base of an isosceles triangle (p. 313)

base angles of an isosceles triangle (p. 313)

vertex angle of an isosceles triangle (p. 313)

median of a triangle (p. 319)

altitude of a triangle (p. 319)

SECTIONS 6.1 *and* 6.2

You can use two inequality theorems to compare the lengths of the sides and the measures of the angles of a triangle.

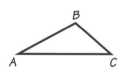

- The sum of the lengths of any two sides of a triangle is greater than the length of the third side.
  In $\triangle ABC$, $AB + BC > AC$, $AB + AC > BC$, and $BC + AC > AB$.

- One side of a triangle is longer than a second side if and only if the angle opposite the first side is larger than the angle opposite the second side.
  In $\triangle ABC$, $AB > BC$ if and only if $m \angle C > m \angle A$.

Congruent polygons have the same size and shape. Two polygons are congruent if and only if their **corresponding angles** and **corresponding sides** are congruent. Be sure to list corresponding vertices in the same order.

polygon *ABCD* ≅ polygon *PQRS*

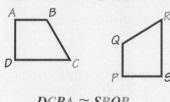

$$\overline{AB} \cong \overline{PQ} \qquad \angle A \cong \angle P$$
$$\overline{BC} \cong \overline{QR} \qquad \angle B \cong \angle Q$$
$$\overline{CD} \cong \overline{RS} \qquad \angle C \cong \angle R$$
$$\overline{DA} \cong \overline{SP} \qquad \angle D \cong \angle S$$

DCBA ≅ *SRQP*

CDAB ≅ *RSPQ*

SECTIONS 6.3 and 6.4

To prove that two triangles are congruent, you can use the definition of congruent triangles or you can use a shortcut.

SSS Postulate	SAS Postulate	ASA Postulate
three sides	2 sides and included angle	2 angles and included side
AAS Theorem	**HL Theorem**	
two angles and non-included side	hypotenuse and leg	There is no SSA shortcut!

For each shortcut, if the indicated corresponding parts are congruent, then the triangles are congruent.

SECTION 6.5

You can use congruent triangles and perpendicular bisectors to prove that parts of triangles are congruent.

If you prove $\triangle RSV \cong \triangle TSV$, then you can prove that $\angle R \cong \angle T$ and $\overline{RS} \cong \overline{TS}$.

If $\overleftrightarrow{AB}$ is the perpendicular bisector of $\overline{CD}$, then $\overline{CB} \cong \overline{DB}$ and $\overline{CA} \cong \overline{DA}$.

SECTIONS 6.6 and 6.7

In an isosceles triangle, the base angles are congruent. Conversely, if two angles of a triangle are congruent, then the sides opposite those angles are congruent.

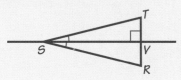 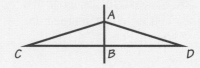

In an isosceles triangle, the bisector of the vertex angle is also a median, an altitude, and the perpendicular bisector of the base.

Assessment

VOCABULARY QUESTIONS

Match each part of the figure with at least one description.

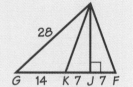

1. $\angle G$

2. $\overline{HK}$

3. $\overline{HJ}$

4. $\overline{GH}$

5. $\angle F$

A. vertex angle of an isosceles triangle

B. base angle of an isosceles triangle

C. median of a triangle

D. altitude of a triangle

E. leg of an isosceles triangle

F. perpendicular bisector

SECTIONS 6.1 *and* 6.2

6. Can you form a triangle using the lengths 4.8 ft, 1.3 ft, and 3.4 ft?

7. The lengths of two sides of a triangle are 6 cm and 8 cm. What can you conclude about the length of the third side of the triangle?

8. In $\triangle ABC$, $m\angle A = 50°$, $m\angle B = 30°$, and $m\angle C = 100°$. Identify the longest and shortest sides of $\triangle ABC$.

9. Name all the triangles that appear to be congruent in the diagram at the right.

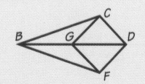

For Exercises 10 and 11, polygon *MRKV* $\cong$ polygon *SBWD*.

10. Complete: polygon *WDSB* $\cong$ _?_

11. ALGEBRA Find the values of *x*, *y*, and *z*.

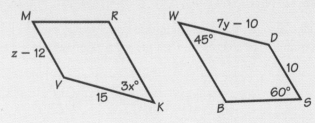

Name two congruent triangles. Tell which method(s) can be used to prove that the triangles are congruent. If no triangles are congruent, write *none*.

12.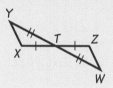

13.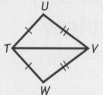

14.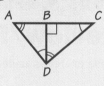

15.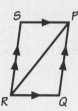

16.

17.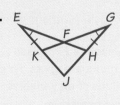

18. Open-ended Problem Sketch and label a pair of overlapping triangles that you could prove congruent using the SAS Postulate. List the given and prove statements for your diagram.

19. Given: Line *m* is the perpendicular bisector of $\overline{CD}$.

Prove: $\angle C \cong \angle D$

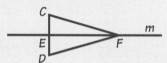

20. In $\triangle QRS$, $\overline{RQ} \cong \overline{RS}$. Find the value of *n*.

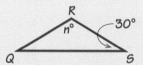

In $\triangle JKL$, $\overline{JM} \perp \overline{KL}$ and $\overline{JN} \cong \overline{NL}$.

21. a. Name a median of $\triangle JKL$.

b. Name an altitude of $\triangle JKL$.

22. If $\overline{KJ} \cong \overline{KL}$, what can you conclude about $\overline{KN}$?

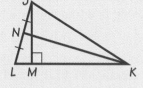

23. Writing Can the lines that contain the altitudes of a triangle intersect at a point on the triangle? Explain why or why not.

PERFORMANCE TASK

24. Open-ended Problem Write a proof: If $\overline{XP}$ is both a median and an altitude of $\triangle XYZ$, then $\triangle XYZ$ is an isosceles triangle.

Algebra
Review/Preview

These exercises review algebra topics you will use in the next chapters.

USING FORMULAS

In each of Exercises 1–6, a formula and the values of some variables in the formula are given. Find the value of the remaining variable. Assume that the values of all variables are positive.

Toolbox p. 694
Evaluating Equations for Given Values

EXAMPLE $V = \pi r^2 h$; $V = 500$ and $h = 7$

Substitute 500 for V and 7 for h in the formula $V = \pi r^2 h$. Then solve for r.

$$500 = \pi r^2 (7)$$

$$\frac{500}{7\pi} = r^2 \qquad \longleftarrow \text{ Divide both sides by } 7\pi.$$

$$\frac{500}{7(3.14)} \approx r^2 \qquad \longleftarrow \text{ Use 3.14 for } \pi.$$

$$22.75 \approx r^2$$

$$\sqrt{22.75} \approx r \qquad \longleftarrow \text{ Take the positive square root.}$$

$$4.77 \approx r$$

1. $V = \pi r^2 h$; $r = 2$ and $h = 3$ **2.** $V = \pi r^2 h$; $V = 72$ and $r = 6$

3. $A = \pi r^2$; $A = 10.6$ **4.** $d = rt$; $d = 150$ and $t = 3.5$

5. $P = 2l + 2w$; $P = 20$ and $l = 4$ **6.** $\frac{1}{f_1} + \frac{1}{f_2} = \frac{1}{f}$; $f_1 = 8$ and $f_2 = 24$

7. PHYSICS After t seconds, the distance d (feet) traveled by an object moving in a straight line and accelerating at a constant rate is $d = vt + \frac{1}{2}at^2$, where v is the initial speed (ft/s) and a is the acceleration (ft/s^2). How far does an object whose initial speed is 20 ft/s and whose acceleration is 4 ft/s^2 travel in 10 s?

Solve each equation for the specified variable.

Toolbox p. 702
Working with Formulas

EXAMPLE $y = mx + b$ for m

$$y - b = mx \qquad \longleftarrow \text{ Subtract } b \text{ from both sides.}$$

$$\frac{y - b}{x} = m \qquad \longleftarrow \text{ Divide both sides by } x.$$

8. $C = 2\pi r$ for r **9.** $A = lw$ for l

10. $A = \frac{1}{2}bh$ for h **11.** $P = 2l + 2w$ for w

12. $d = vt + \frac{1}{2}at^2$ for v **13.** $V = \frac{4}{3}\pi r^3$ for r

RATIOS AND PROPORTIONS

Find each ratio as a fraction in lowest terms.

EXAMPLE	6 days; 3 weeks

$$\frac{6 \text{ days}}{3 \text{ weeks}} = \frac{6 \text{ days}}{21 \text{ days}}$$

Express both quantities in the same units.

$$= \frac{2}{7}$$

Toolbox p. 707
Creating a Ratio

14. 20 lbs; 12 oz   **15.** 30 min; 4 h   **16.** 6 ft; 14 in.

17. 10 km; 800 m   **18.** $4\frac{2}{3}$ kg; 120 g   **19.** 6000 s; $3\frac{1}{2}$ h

Solve each proportion.

EXAMPLE	$\frac{5}{4} = \frac{x}{12}$

$$5 \cdot 12 = 4x$$

If $\frac{a}{b} = \frac{c}{d}$, then $ad = bc$.

$$60 = 4x$$
$$15 = x$$

Toolbox p. 708
Solving Proportions

20. $\frac{9}{y} = \frac{3}{4}$ **21.** $\frac{a}{8} = \frac{5}{12}$ **22.** $\frac{4}{11} = \frac{14}{b}$ **23.** $\frac{15v}{64} = \frac{45}{32}$

SOLVING LINEAR SYSTEMS

Solve each system of equations.

EXAMPLE	$4x - 5y = 23$
	$3x + 10y = 31$

Multiply both sides of the first equation by 2 so the y terms are opposites.

$$2(4x - 5y) = 2(23) \longrightarrow 8x - 10y = 46$$
$$3x + 10y = 31 \longrightarrow + (3x + 10y) = + (31)$$

Add to eliminate y. Then solve for x.

$$11x = 77$$
$$x = 7$$

To find y, substitute 7 for x in either of the original equations.

$$4(7) - 5y = 23$$
$$-5y = -5$$

Subtract 28 from both sides.

$$y = 1$$

The solution (x, y) is (7, 1).

24. $x + 2y = -7$
$\quad\,\,\, 3x - 8y = 7$

25. $2u + 5v = 14$
$\quad\,\,\, 6u + 7v = 10$

26. $18r - 5s = 17$
$\quad\,\,\,\,\,\, 6r + 10s = -6$

27. $2a + 3b = 1$
$\quad\,\,\, 4a + 6b = 5$

28. $7c - 2d = 10$
$\quad\,\,\, -14c + 4d = -20$

29. $6m + 8n = -3$
$\quad\,\,\, 9m + 6n = -7$

Cumulative Assessment

C H A P T E R S $4-6$

CHAPTER 4

For Questions 1–5, use the points $P(3, -5)$ and $Q(-2, 5)$.

1. Find the length of $\overline{PQ}$ and the coordinates of the midpoint of $\overline{PQ}$.

2. ALGEBRA Find the slope of the line that passes through P and Q.

3. Find an equation of the line that contains $(8, 0)$ and is perpendicular to $\overleftrightarrow{PQ}$.

4. Open-ended Problem Find the coordinates of a point R so that $\triangle PQR$ is an isosceles triangle. Use the Distance Formula to show that the triangle is isosceles.

5. Write an equation of a circle whose center is P and contains point Q.

Sketch each circle. Label the coordinates of the center. Draw a radius and label it with its length.

6. $(x - 3)^2 + y^2 = 9$ **7.** $x^2 + y^2 = 25$ **8.** $x^2 + (y + 3)^2 = 2$

Answer Questions 9–11 to complete the proof. Use the coordinates given in the diagram.

Given: $\overleftrightarrow{RS}$ is the perpendicular bisector of $\overline{AB}$.

Prove: $RA = RB$

9. Writing Explain how you know that the origin is the midpoint of $\overline{AB}$.

10. Writing Explain why $\overleftrightarrow{RS}$ must be the y-axis.

11. Find the x-coordinate of point R. Then show that $RA = RB$.

CHAPTER 5

For Questions 12–14, use the diagram at the right.

12. Suppose $m \angle 1 = 70°$ and $m \angle 2 = (9x - 2)°$. Find the value of x.

13. Suppose $m \angle 3 = (6y - 22)°$ and $m \angle 5 = (2y + 2)°$. Find the value of y.

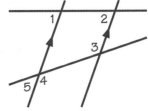

14. Writing What can you conclude about the quadrilateral formed by the four lines in the diagram? Explain how you reached your conclusions.

15. Write a flow proof.

Given: Lines n, p, and t are coplanar.
$n \parallel p$; $t \perp n$

Prove: $t \perp p$

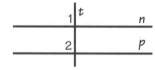

16. Suppose plane P is parallel to plane Q. State a conclusion based on this information and each additional statement.

a. Plane R intersects planes P and Q.

b. Plane R is parallel to plane Q.

17. Using the diagram, write the key steps of a proof.

Given: $\angle 1 \cong \angle 2$
$\angle 3 \cong \angle 4$

Prove: $\overleftrightarrow{AB} \parallel \overleftrightarrow{DE}$

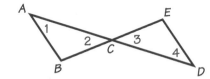

CHAPTER 6

18. Writing Explain why a triangle can be formed with the lengths 5 in., 7 in., and 3 in. Sketch the triangle.

19. $\triangle AMR \cong \triangle LZJ$. Complete each statement.

a. $\overline{ZJ} \cong \underline{\ ?\ }$
b. $\angle M \cong \underline{\ ?\ }$
c. $\triangle ZLJ \cong \underline{\ ?\ }$

Tell which method(s) you can use to prove that the triangles are congruent. If no method can be used, write *none*.

20.

21.

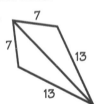

22.

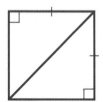

23. Given: $\angle 1 \cong \angle 2$
$\angle 3 \cong \angle 4$

Prove: $\overline{PQ} \cong \overline{RS}$

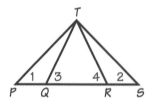

Complete each statement with *always*, *sometimes*, or *never*.

24. An altitude of a triangle is $\underline{\ ?\ }$ a side of the triangle.

25. A median of a scalene triangle is $\underline{\ ?\ }$ an angle bisector of the triangle.

26. If an altitude of a triangle is also a median, then the triangle is $\underline{\ ?\ }$ isosceles.

7 Quadrilaterals, Areas, Volumes

ACCESS for Everyone

Walt Stone

"I never expected to use much math in my job, but now I'm finding that I use it just about every day."

Walt Stone's civic and professional work of the past several years can be summed up in one word: "access." A former alderman in Branson, Missouri, and current chairperson of the town's Advisory Council on Disabilities, Stone is dedicated to a simple proposition. "My main goal is to make Branson accessible to everyone, so that everybody who lives here, or comes here, can enjoy the town," he says.

Branson, a southern Missouri town with some 4500 residents, draws about six million visitors each year because of its picturesque Ozarks setting and its growing reputation as a country music center. Thanks to the efforts of Stone and others, Branson is also becoming an easier place for people with disabilities to get around.

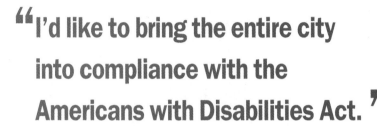

"I'd like to bring the entire city into compliance with the Americans with Disabilities Act."

A Crusade for Ramps

Stone has worked with the local public works department and state highway department to make 400 "curb cuts" (for both wheelchairs and baby strollers) along a four-mile stretch of road that runs through the town's central business district. Of the 34 country music theaters in Branson, about 95 percent are now wheelchair-accessible, according to Stone. "At the depot for the scenic railway, we put in wheelchair ramps from the street so you can roll right onto the train," he says. People in wheelchairs can also freely roam the town's lakefront, as well as a 62-acre park and playground. "I'd like to bring the entire city into compliance with the Americans with Disabilities Act," says Stone.

Personal Motivation

Stone himself has had to rely on a wheelchair since injuring his spine in a motorcycle accident 25 years ago. Although this experience helped motivate his latest campaign, it has not made the task any easier. He's had no formal training for his job and instead has had to figure out on his own how to implement the Americans with Disabilities Act—a highly technical, three-inch thick document. That effort, in turn, has required a lot of measurements and calculations—a surprising development for someone who, until recently, considered himself "math illiterate."

Meeting the Standards

"I never expected to use math much in my job, but now I'm finding that I use it just about every day," Stone says. For instance, Stone finds the number of parking spaces that have to be set aside for drivers or passengers with disabilities. The area of these spaces must meet federal standards. Wheelchair ramps should also meet federal standards. The ramps cannot exceed a specified angle; if they're too steep, most people would not be able to make their way up. "The ramps are made of concrete and the volume of concrete that has to be poured can be determined in advance, depending on the length and height of the ramp," Stone explains.

You can find the volume of a prism-shaped ramp using the following formula:

$$V = Bh$$

where the volume equals the product of the area of the base and the height of the prism.

Base h

Explore and Connect

Mathematics & Walt Stone

In this chapter, you will learn more about how mathematics is related to access for people with disabilities.

Related Exercises

Section 7.4
• **Exercises 31–33**

Section 7.6
• **Exercises 16–18**

Walt Stone plans a curb cut in a Branson sidewalk.

1. Writing   Look at the prism above. What type of prism is it? How would you find the area of the base? Does it matter where you measure the height of the prism? Explain.

2. Research   Contact the person who is responsible for implementing the Americans with Disabilities Act in your city. Find out what the city's implementation goals for the next five years are. How do these goals compare to the goals for Branson, Missouri?

3. Project   Make a net for a ramp. Look back at Chapter 2 where you first learned how to make a net for a prism. Use the methods in Chapter 2 to make your net.

7.1 Classifying Quadrilaterals

Learn how to...

- classify quadrilaterals by using their properties

- recognize the relationships between the diagonals of special quadrilaterals

So you can...

- use properties of quadrilaterals in real-world applications, such as engineering and auto repair

You deal with classifications and properties every day. For instance, since apples are fruits and fruits have seeds, you know that apples have seeds. The same kind of reasoning can be used in geometry.

In Chapter 5 you learned about trapezoids. Another special quadrilateral is the *kite*. A **kite** is a convex quadrilateral that has two pairs of congruent sides, but no pair of opposite sides is congruent.

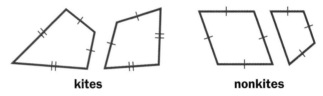

kites   **nonkites**

Many quadrilaterals belong to more than one group. For example, if a given quadrilateral is a rhombus, then it is also a parallelogram. This diagram shows how quadrilaterals can be divided into specific groups.

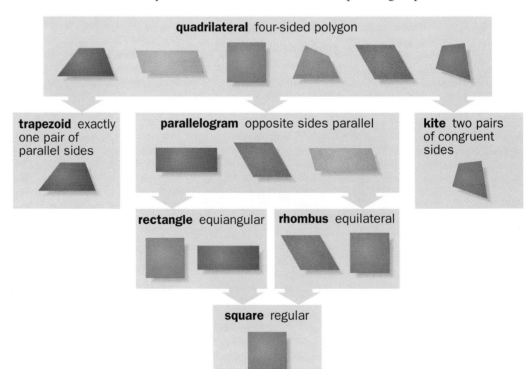

quadrilateral four-sided polygon

trapezoid exactly one pair of parallel sides

parallelogram opposite sides parallel

kite two pairs of congruent sides

rectangle equiangular

rhombus equilateral

square regular

THINK AND COMMUNICATE

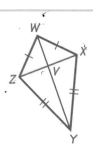

1. How many lines of symmetry does a kite have?

2. Use the fact that $\triangle WZY \cong \triangle WXY$ to show that $\triangle ZWV \cong \triangle XWV$.

3. Explain why $\overline{ZV} \cong \overline{XV}$ and $\angle ZVW \cong \angle XVW$.

4. Tell whether this statement is *True* or *False*: In kite *WXYZ*, $\overline{WY}$ is the perpendicular bisector of $\overline{ZX}$.

When you use the diagram on page 339, it is helpful to remember that a quadrilateral has all the characteristics of all the groups linked above it.

EXPLORATION
COOPERATIVE LEARNING

Investigating Diagonals

**Work with another student.
You will need:**
- four drinking straws
- string
- a protractor

1 Thread the string once through each straw and a second time through the first straw. You should have a movable rhombus.

2 Hold the rhombus steady and use the ends of the string to form its diagonals.

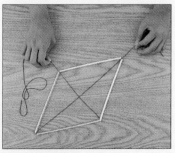

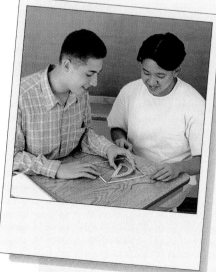

3 Measure the angles that the diagonals form. What do you notice?

4 Measure the adjacent angles at each vertex. How are the measures related?

5 Repeat Steps 2–4 twice more. Adjust the rhombus so that the straws meet at different angles each time.

6 Make two conjectures about the diagonals of a rhombus.

In the Exploration, you discovered the relationship between the diagonals of a rhombus. The diagonals of rectangles and kites also have special relationships. You may recall one of these theorems from Chapter 4.

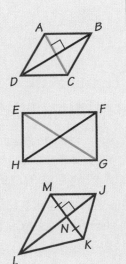

Diagonals of Special Quadrilaterals

The diagonals of a rhombus are perpendicular.

If $ABCD$ is a rhombus, then $\overline{AC} \perp \overline{BD}$.

The diagonals of a rectangle are congruent.

If $EFGH$ is a rectangle, then $\overline{EG} \cong \overline{FH}$.

Exactly one diagonal of a kite is a line of symmetry for the kite and the perpendicular bisector of the other diagonal.

If $JKLM$ is a kite, then $\triangle JKL \cong \triangle JML$, $\overline{MK} \perp \overline{JL}$, and $\overline{MN} \cong \overline{KN}$.

EXAMPLE 1   **Connection: Algebra**

Write a coordinate proof to show that the diagonals of any rectangle are congruent.

SOLUTION

Step 1 To show that two segments in the coordinate plane are congruent, you need to use the Distance Formula. To make the calculations easier, place as many vertices as possible on the axes. One possible choice of vertices is shown.

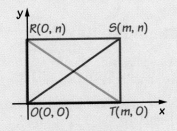

Step 2 Find RT and SO in rectangle $RSTO$.

Remember that $(-m)^2 = m^2$.

$$RT = \sqrt{(0 - m)^2 + (n - 0)^2}$$

$$= \sqrt{m^2 + n^2}$$

$$SO = \sqrt{(m - 0)^2 + (n - 0)^2}$$

$$= \sqrt{m^2 + n^2}$$

Therefore, $RT = SO$. The diagonals of any rectangle are congruent.

EXAMPLE 2

The square in this Samoan cloth is divided by its diagonals. If *MN* = 10 units, find *JK*.

SOLUTION

Since a square is a parallelogram, its diagonals bisect each other. A square is also a rectangle, so its diagonals are congruent.

$$JN = KN = LN = MN = 10 \text{ units}$$

Because a square is a rhombus, the diagonals are perpendicular. Therefore, $\triangle JNK$ is a right triangle with hypotenuse $\overline{JK}$.

Use the Pythagorean theorem to find *JK*.

$$JK^2 = JN^2 + KN^2$$

$$JK^2 = 10^2 + 10^2 \qquad \text{Substitute 10 for } JN \text{ and } KN.$$

$$JK^2 = 200$$

$$JK \approx 14.14$$

The side of the square, $\overline{JK}$, is about 14 units long.

☑ CHECKING KEY CONCEPTS

1. A kite has side lengths of 2 and 8. Sketch what the kite might look like and label all four side lengths.

2. Show that each diagonal of a rhombus is a line of symmetry for the rhombus.

Find each length.

3. rhombus *JKLM*
   a. *MN*
   b. *MK*

4. rectangle *DEFG*
   a. *DF*
   b. *GH*

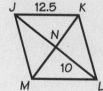

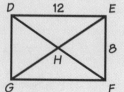

Find the missing side lengths of each kite.

1.

2.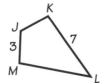

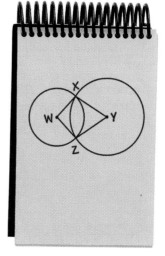

3. a. Show that *WXYZ* at the right is a kite.

 b. CONSTRUCTION Describe a way to construct a kite
 by drawing two circles with different radii. Use your
 answer from part (a) to justify your method.

4. **Writing** The segments that join the midpoints of consecutive
 sides of a rectangle form a rhombus. The segments that join the
 midpoints of consecutive sides of a rhombus form a rectangle.
 What type of quadrilateral is formed by the segments that join
 the midpoints of consecutive sides of a square? Use the diagram
 on page 339 to justify your answer.

5. **Visual Thinking** In the diagram below, $\overline{BH}$ and $\overline{DF}$ divide
 rectangle *ACJG* into four congruent rectangles. Use a theorem
 you learned in this section to show that *DBFH* is equilateral.

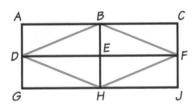

Find each length.

6. rectangle *ABCD*
 a. *BC*
 b. *AB*

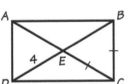

7. rhombus *PQRS*
 a. *SR*
 b. *TQ*

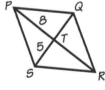

8. square *WXYZ*
 a. *WY*
 b. *XY*

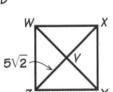

9. kite *FGHJ*
 a. *KF*
 b. *KJ*

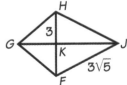

10. **Logical Reasoning** A *right trapezoid* is a trapezoid with
 a right angle. An *isosceles trapezoid* is a trapezoid whose
 legs are congruent. Copy the diagram of types of
 quadrilaterals shown on page 339. Include the right
 trapezoid and the isosceles trapezoid and explain
 how you decided where to place them.

 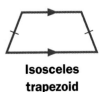

**Right
trapezoid**   **Isosceles
trapezoid**

11. Write the key steps of a proof showing that the diagonals of a rhombus are perpendicular.

12. ALGEBRA Write a coordinate proof to show that the diagonals of an isosceles trapezoid are congruent.

13. Use what you know about the diagonals of a rectangle to prove that for any right triangle, the midpoint of the hypotenuse is equidistant from all three vertices of the triangle.

   Given: $\triangle ABC$ is a right triangle with a right angle at B.
   D is the midpoint of $\overline{AC}$.

   Prove: $AD = BD = CD$

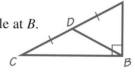

In the diagram at the right, _BDGF_ is a rectangle and _FDHJ_ is a kite. For Exercises 14–19, find each length.

14. FD **15.** FG **16.** BE

17. ED **18.** GH **19.** JH

20. What type of parallelogram is $ADEB$? Explain.

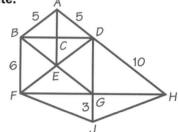

Connection   ENGINEERING

Nearly 2000 years ago, the Persians began building irrigation systems called *qanats*. To measure the slope of a *qanat*, they used a level made of a square wooden frame with a weight at one corner. When the square was hung by the corner opposite the weight, the stick that connected the other two corners was horizontal.

21. Writing Why do you think the weight was necessary for the level to work? Why did the frame have to be hung from the corner opposite the weight?

22. Explain why the stick at diagonal $\overline{DB}$ is always horizontal.

23. Does the frame have to be square? What other shape(s) could be used? Explain your reasoning.

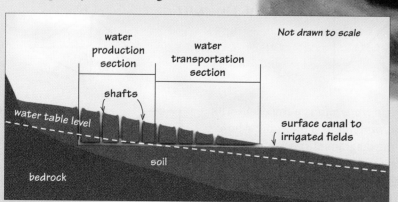

24. **AUTO REPAIR** The jack at the right is used to lift cars off the ground. As the handle is turned, vertices X and Z move together so that $\overline{ZX}$ stays parallel to the ground.

    a. What type of parallelogram is $WXYZ$?

    b. What do you know about the angle $\overline{WY}$ makes with the ground? Explain.

    c. Research Diagonal $\overline{WY}$ represents the direction of *force* that raises the car off the ground. Find out why the jack works best when the direction of the force is perpendicular to the ground.

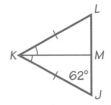

25. **Open-ended Problem** Sketch a rhombus, a rectangle, or a kite. Sketch the diagonals of the figure. Mark all right angles and use the Pythagorean theorem to give possible lengths for each segment.

26. **Open-ended Problem** Sketch a rectangle $ABCD$. Draw $\overline{AC}$ and $\overline{BD}$ and label the point where they intersect E. Measure $\angle BAC$ and use this measure to find the measures of all the angles in your diagram.

27. **Open-ended Problem** Sketch a rhombus $EFGH$. Draw $\overline{EG}$ and $\overline{FH}$. Measure $\angle GEF$ and use this measure to find the measures of all the angles in your diagram.

28. **SAT/ACT Preview** In quadrilateral $WXYZ$, $\overline{WX} \cong \overline{XY}$ and $\overline{WZ} \cong \overline{ZY}$. What type(s) of quadrilateral can $WXYZ$ be?

    I. kite   **II.** rhombus   **III.** square

    A. I only   **B.** I or II   **C.** II or III   **D.** I or III   **E.** I, II, or III

29. **Challenge** In rectangle $JKLM$, $\overline{JL}$ and $\overline{KM}$ intersect at N, and $\overline{JM} \cong \overline{NL}$. Sketch rectangle $JKLM$ and prove that $m \angle LJK + m \angle LMK = m \angle KLN$.

ONGOING ASSESSMENT

30. **Writing** The diagonals of any convex quadrilateral form four non-overlapping triangles. For which quadrilaterals are these triangles right? isosceles? congruent? Explain.

SPIRAL REVIEW

Find each length or angle measure. *(Sections 6.6 and 6.7)*

31. **a.** AC

    b. $m \angle B$

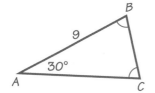

32. **a.** $m \angle R$

    b. $m \angle Q$

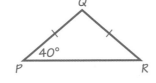

33. **a.** $m \angle KML$

    b. $m \angle LKM$

7.2 | Identifying Parallelograms

Learn how to...

- tell whether a quadrilateral is a parallelogram
- write indirect proofs

So you can...

- reach conclusions about special quadrilaterals in real-world and geometric situations

Suppose you want to make two segments of a structure parallel. One way is to make them the opposite sides of a parallelogram. But how do you make a parallelogram?

EXAMPLE 1   Application: Carpentry

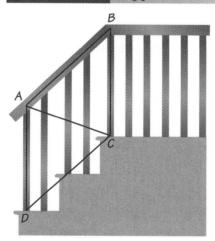

Susan Leonard is building a railing for the front steps of her house. Show that if the end posts are the same length and both perpendicular to the ground, the hand railing will be parallel to the segment that connects the top and bottom steps.

Given: $\overline{AD} \parallel \overline{BC}$; $\overline{AD} \cong \overline{BC}$

Prove: *ABCD* is a parallelogram.

SOLUTION

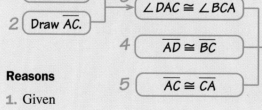

1. $\overline{AD} \parallel \overline{BC}$
2. Draw $\overline{AC}$.
3. $\angle DAC \cong \angle BCA$
4. $\overline{AD} \cong \overline{BC}$
5. $\overline{AC} \cong \overline{CA}$
6. $\triangle DAC \cong \triangle BCA$
7. $\angle DCA \cong \angle BAC$
8. $\overline{AB} \parallel \overline{CD}$
9. *ABCD* is a ▱.

Reasons

1. Given

2. Through any two points, there is exactly one line.

3. If ∥ lines are intersected by a transversal, then alternate interior ∠s are ≅.

4. Given

5. Reflexive Property

6. SAS Postulate

7. Definition of congruent triangles

8. If two lines form ≅ alternate interior ∠s, then the lines are ∥.

9. Definition of parallelogram

In Chapter 2 you learned several properties of parallelograms. If a quadrilateral has one of these properties, you can prove that it is a parallelogram.

Ways to Prove that a Quadrilateral is a Parallelogram

Show that both pairs of opposite sides are parallel. (Definition)

If $\overline{AB} \parallel \overline{CD}$ and $\overline{AD} \parallel \overline{BC}$, then $ABCD$ is a parallelogram.

Show that both pairs of opposite sides are congruent.

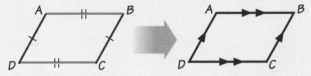

If $\overline{AB} \cong \overline{CD}$ and $\overline{AD} \cong \overline{BC}$, then $ABCD$ is a parallelogram.

Show that both pairs of opposite angles are congruent.

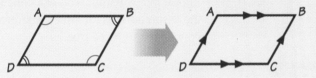

If $\angle A \cong \angle C$ and $\angle B \cong \angle D$, then $ABCD$ is a parallelogram.

Show that one pair of opposite sides is both parallel and congruent.

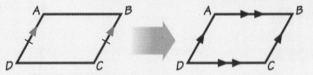

If $\overline{AD} \parallel \overline{BC}$ and $\overline{AD} \cong \overline{BC}$, then $ABCD$ is a parallelogram.

Show that the diagonals bisect each other.

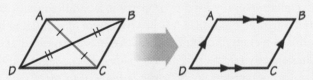

If $\overline{AC}$ bisects $\overline{BD}$ and $\overline{BD}$ bisects $\overline{AC}$, then $ABCD$ is a parallelogram.

THINK AND COMMUNICATE

1. Which method above can you use to show that an equilateral quadrilateral must be a parallelogram?

2. Which method above can you use to show that an equiangular quadrilateral must be a parallelogram?

3. Use one of the methods above to show that any two congruent triangles can be arranged to form a parallelogram.

Indirect Proof

All of the types of proof you have learned in this book are *direct* proofs. These proofs consist of statements that follow directly from each other, leading from the given information to the conclusion. To use *indirect reasoning*, you need to show that all other cases are impossible, so the conclusion must be true.

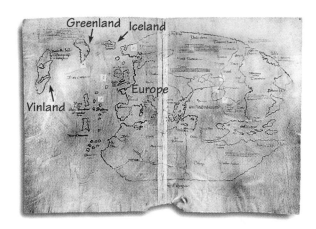

For example, this map found in 1965 implies that the Norse landed in America before Columbus. Some scholars have tried to prove the map invalid. If they can show that the ink is a type used only in the twentieth century, they will know that the map is a forgery.

To write an **indirect proof**, start by assuming that the conclusion is false. When logic leads you to a contradiction or an impossible situation, you can conclude that the assumption is incorrect. Therefore, the conclusion must be true.

EXAMPLE 2

Write an indirect proof to show that the parallel sides of a trapezoid cannot be congruent.

SOLUTION

Given: In trapezoid $JKLM$, $\overline{JK} \parallel \overline{ML}$.
Prove: $\overline{JK} \not\cong \overline{ML}$

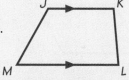

Suppose $\overline{JK} \cong \overline{ML}$. Then quadrilateral $JKLM$ has a pair of opposite sides that are both parallel and congruent. The third theorem on page 347 states that a quadrilateral with this property is a parallelogram. By the definition of a parallelogram, $JKLM$ has two pairs of parallel sides. However, this is a contradiction because $JKLM$ is a trapezoid and therefore has exactly one pair of parallel sides. The assumption that $\overline{JK} \cong \overline{ML}$ must not be true. So $\overline{JK}$ is not congruent to $\overline{ML}$.

THINK AND COMMUNICATE

4. Kiyana has a vase of roses on her desk where she does her homework each night. Give a convincing argument that she is not allergic to roses.

5. Suppose you want to write an indirect proof to show that a given angle is acute. What three cases should you prove are impossible?

6. Sara wrote an indirect proof to show that a given quadrilateral is a trapezoid by proving that it is not a parallelogram or a kite. What is wrong with her reasoning?

☑ **CHECKING KEY CONCEPTS**

Show that each quadrilateral is a parallelogram. Then find each length or measure.

1. a. VX
   b. YZ
   c. WX

2. a. QR
   b. $m \angle SRQ$
   c. $m \angle PSR$

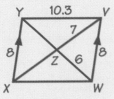
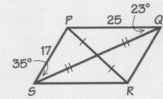

3. Write the key steps of a proof showing that if the diagonals of a quadrilateral bisect each other, then the quadrilateral is a parallelogram.

7.2 **Exercises and Applications**

Extra Practice exercises on page 673

Show that each quadrilateral is a parallelogram. Then find each length or measure.

1. a. $m \angle DAB$
   b. BC
   c. DC

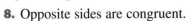

2. a. EJ
   b. HF
   c. HG

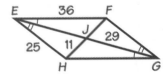

Sketch each quadrilateral. Mark the given information. If the quadrilateral must be a parallelogram, explain why.

3. $ABCD$ with $\overline{AB} \cong \overline{CD}$ and $\overline{BC} \cong \overline{AD}$

4. $WXYZ$ with $\overline{WX} \parallel \overline{YZ}$ and $\overline{XY} \cong \overline{WZ}$

5. $FGHI$ with $\angle F \cong \angle I$ and $\angle G \cong \angle H$

6. $JKLM$ with $\angle J \cong \angle L$ and $\angle K \cong \angle M$

7. Open-ended Problem Draw a parallelogram. Describe the method you used and how you knew the result would be a parallelogram.

Technology Use geometry software to create a quadrilateral with the given property. Then use the slope feature to check that the quadrilateral is a parallelogram.

8. Opposite sides are congruent.

9. Opposite angles are congruent.

10. The diagonals bisect each other.

11. One pair of sides is congruent and parallel.

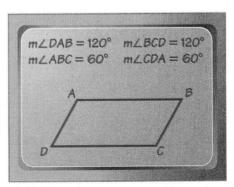

PARALLELOGRAM

Given two segments, construct a parallelogram
with sides congruent to the segments.

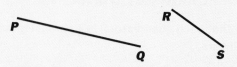

1. Construct $\overline{AB}$ congruent to $\overline{PQ}$.

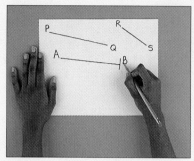

2. Construct $\overline{BC}$ congruent to $\overline{RS}$.

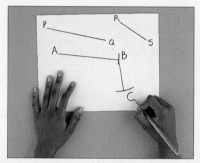

3. Draw an arc with center A and
radius BC.

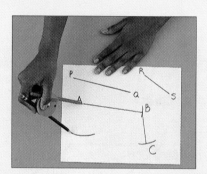

4. Draw an arc with center C and
radius AB. The arc should intersect
the arc you drew in Step 3.

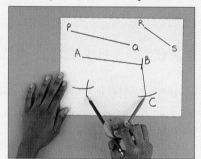

5. Label the point where
the arcs intersect D.
Draw $\overline{AD}$ and $\overline{CD}$.
$ABCD$ is a
parallelogram.

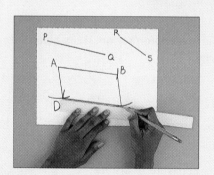

12. Writing Explain why $ABCD$ must be a parallelogram.

13. Open-ended Problem Describe two different ways to check that
$ABCD$ is a parallelogram.

14. Writing How can you use the construction shown above to construct
a rhombus?

15. Challenge How can you use this construction to construct an
isosceles trapezoid, given the length of one base and the length of the
congruent sides? (*Hint:* How would you locate D differently?)

ALGEBRA **Graph the points. Show that they are the vertices of a parallelogram.**

16. $P(-2, 8)$, $Q(2, 8)$, $R(1, 2)$, $S(-3, 2)$

17. $A(4, 6)$, $B(7, 2)$, $C(4, -2)$, $D(1, 2)$

18. $K(4, 7)$, $L(9, 6)$, $M(6, 3)$, $N(1, 4)$

19. $F(0, 4)$, $G(3, 5)$, $H(2, 0)$, $I(-1, -1)$

20. Open-ended Problem Give the coordinates of four points that are the vertices of a nonrectangular parallelogram. Tell which theorem you used to choose the points.

21. Write the key steps of a proof showing that if the opposite angles of a quadrilateral are congruent, then the quadrilateral is a parallelogram.

22. Write an indirect proof to show that $\triangle AEB \not\cong \triangle CEB$ in kite $ABCD$.

Given: kite $ABCD$

Prove: $\triangle AEB \not\cong \triangle CEB$

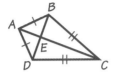

23. Write an indirect proof to show that each side of a kite is congruent to exactly one other side of the kite.

24. Write an indirect proof to show that if a trapezoid has a right angle, then it has exactly two right angles. (*Hint:* You will need to consider more than one case.)

Connection ▶ **M A C H I N E R Y**

The people at the right are photographing an airplane. To get an aerial view, they are standing on a *scissors lift*. Eight congruent metal pieces that bisect each other, $\overline{AE}$, $\overline{BD}$, $\overline{DH}$, $\overline{EG}$, $\overline{GL}$, $\overline{HK}$, $\overline{KP}$, and $\overline{LN}$ move together to raise and lower the platform of the lift.

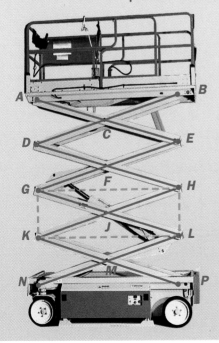

25. a. What type of quadrilaterals are *CEFD*, *FHJG*, and *JLMK*? How do you know?

b. Use what you know about vertical angles to show that these three quadrilaterals remain congruent even while the lift is in motion.

26. You can also think of the metal pieces as the diagonals of quadrilaterals. For instance, $\overline{GL}$ and $\overline{HK}$ are the diagonals of quadrilateral *GKLH*.

a. How do you know that *ADEB*, *DGHE*, *GKLH*, and *KNPL* are parallelograms?

b. When the lift rises, *NL* and *PK* move at the same rate, forcing *KL* to stay horizontal. Show that the platform of the lift always stays horizontal.

27. Use the diagram at the right.

    a. Find the coordinates of point P given that $OP = c$.

    b. Find the y-coordinate of point N so that $\overline{MN} \parallel \overline{OP}$.

    c. Find the x-coordinate of point N so that $\overline{MN} \cong \overline{OP}$.

    d. Which theorem on page 347 ensures that $MNPO$ is a parallelogram? Explain your reasoning.

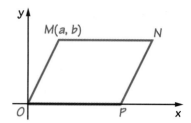

28. DRAFTING Drafting tables are designed so that they can tilt at different angles. You adjust the table shown by joining its legs at different holes. How would you fasten the legs together if you wanted the tabletop to be parallel to the floor? Explain your reasoning.

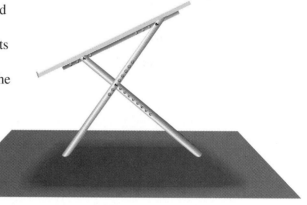

ONGOING ASSESSMENT

29. Open-ended Problem Sketch and label a quadrilateral $WXYZ$. Include both diagonals. Give four different sets of properties that can be used to prove that $WXYZ$ is a parallelogram.

SPIRAL REVIEW

Find each length. *(Section 7.1)*

30. JL in rhombus $JKLM$

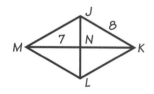

31. RS in kite $PQRS$

32. BD in rectangle $ABCD$

Find the length and midpoint of each segment in the diagram. *(Section 4.1)*

33. $\overline{AB}$ **34.** $\overline{CF}$ **35.** $\overline{DC}$

36. $\overline{HJ}$ **37.** $\overline{EF}$ **38.** $\overline{GB}$

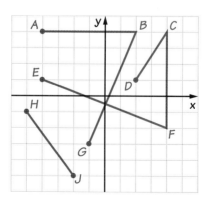

7.3 Conditions for Special Parallelograms

Many chairs and tables have rectangular frames. Since the lengths of the pieces that make up the frames stay the same over time, the frames remain parallelograms. Under the pressure of use, however, the angles may change.

One way to check if a frame is rectangular is to measure its diagonals. If the diagonals are congruent, the frame must be a rectangle.

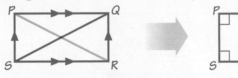

You can keep a frame rectangular by securing one corner at a right angle. When the parallelogram frame has a right angle, it is a rectangle.

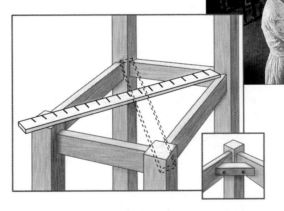

Ways to Prove that a Parallelogram is a Rectangle

If the diagonals of a parallelogram are congruent, then it is a rectangle.

If $\overline{PR} \cong \overline{QS}$ in $\square PQRS$, then $PQRS$ is a rectangle.

If one angle of a parallelogram is a right angle, then it is a rectangle.

If one angle of $\square PQRS$ is a right angle, then $PQRS$ is a rectangle.

THINK AND COMMUNICATE

1. In quadrilateral $JKLM$, $\overline{JL} \cong \overline{KM}$. Explain why the first theorem on page 353 cannot be used to show that $JKLM$ is a rectangle.

2. Use a property of parallelograms to explain why a parallelogram with one right angle must be a rectangle.

There are also ways to tell whether a parallelogram is a rhombus. The example below shows how to do this by looking at the diagonals of the parallelogram.

EXAMPLE 1   Connection: Algebra

Write a coordinate proof to show that if the diagonals of any parallelogram are perpendicular, then it is a rhombus.

SOLUTION

Step 1 Place the diagonals along the x- and y-axes, since they are perpendicular. To make sure the quadrilateral is a parallelogram, choose the vertices so that the origin is the midpoint of both diagonals. For instance, $K(a, 0)$ and $M(-a, 0)$ are both a units from $(0, 0)$.

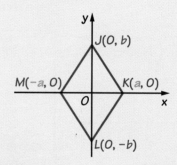

Step 2 Use the Distance Formula to show that $\square JKLM$ is equilateral.

$$JK = \sqrt{(0 - a)^2 + (b - 0)^2}$$

$$= \sqrt{a^2 + b^2}$$

$$KL = \sqrt{(a - 0)^2 + (0 - (-b))^2}$$

$$= \sqrt{a^2 + b^2}$$

Remember that $0 - (-b) = 0 + b$.

$$ML = \sqrt{(-a - 0)^2 + (0 - (-b))^2}$$

$$= \sqrt{a^2 + b^2}$$

$$JM = \sqrt{(0 - (-a))^2 + (b - 0)^2}$$

$$= \sqrt{a^2 + b^2}$$

By substitution, $JK = KL = ML = JM$. Therefore, if the diagonals of a parallelogram are perpendicular, then it is a rhombus.

Ways to Prove that a Parallelogram is a Rhombus

If the diagonals of a parallelogram are perpendicular, then it is a rhombus.

 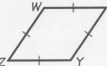

If $\overline{WY} \perp \overline{XZ}$ in $\square WXYZ$, then $WXYZ$ is a rhombus.

If two consecutive sides of a parallelogram are congruent, then it is a rhombus.

 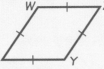

If $\overline{WX} \cong \overline{XY}$ in $\square WXYZ$, then $WXYZ$ is a rhombus.

EXAMPLE 2

**Classify quadrilateral *ABCD*.
Be as specific as possible.**

SOLUTION

Since $\overline{AC}$ and $\overline{BD}$ bisect each other, $ABCD$ is a parallelogram.

The diagram also shows that $\overline{AC} \cong \overline{BD}$. So $\square ABCD$ is a rectangle because its diagonals are congruent.

Since $\overline{BC}$ and $\overline{CD}$ are congruent consecutive sides, $\square ABCD$ is a rhombus.

Since $\square ABCD$ is both a rectangle and a rhombus, it is a square.

✓ CHECKING KEY CONCEPTS

Classify each quadrilateral. Be as specific as possible.

1. 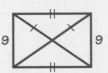   **2.** 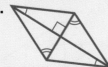   **3.**

4. Write a flow proof to show that if two consecutive sides of a parallelogram are congruent, then it is a rhombus.

Extra Practice exercises on page 673

1. **Logical Reasoning** Explain why you *cannot* use the theorems you learned in this section to classify each quadrilateral below as a rectangle or a rhombus.

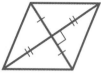
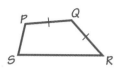
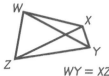

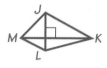

$WY = XZ$

Classify each quadrilateral. Be as specific as possible.

2.

3.

4.

5.
8
8

6.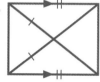

7.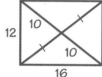
12   10
10
16

8.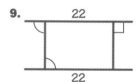

9.
22
22

10. **Open-ended Problem** Sketch an example of a nonsquare quadrilateral whose diagonals are both congruent and perpendicular.

11. **CRAFTS** Two common crafts in Japan are paper-folding (origami) and paper-cutting (kirigami). Since both crafts usually require a square sheet of paper, it is helpful to know how to cut a nonsquare rectangular sheet of paper into a square.

Fold the paper so that a short edge lines up with a long edge.

Then cut along the edge of the two-layer triangle.

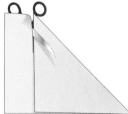

Why is the resulting rectangle a square? Which theorem does this process illustrate?

12. **Visual Thinking** Use the diagram at the right and a theorem you learned in this section to explain why the segments that join the midpoints of consecutive sides of a rhombus form a rectangle.

13. QUILTING Many quilts feature squares that are made up of triangles. Explain why placing four congruent isosceles right triangles as shown at the right always results in a square.

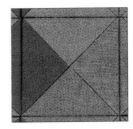

ALGEBRA Give the possible coordinates of the vertices of each quadrilateral so that the diagonals lie on the *x*- and *y*-axes.

14. a kite

15. a nonsquare rhombus

16. a square

17. a rectangle

18. a trapezoid

19. a parallelogram

20. Prove that if two consecutive angles of a parallelogram are congruent, then the parallelogram is a rectangle.

Given: In □*WXYZ*, ∠*W* ≅ ∠*X*.
Prove: □*WXYZ* is a rectangle.

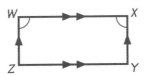

21. ALGEBRA Write a coordinate proof to show that if exactly one diagonal of a quadrilateral is the perpendicular bisector of the other diagonal, then the quadrilateral is a kite.

22. Write the key steps of a proof showing that if the diagonals of a parallelogram are congruent, then the parallelogram is a rectangle.

Connection ▶ SOCIAL STUDIES

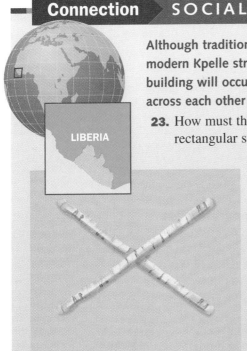

LIBERIA

Although traditional houses of the Kpelle people of Liberia are circular, modern Kpelle structures are rectangular. To determine the space that a building will occupy, the Kpelle sometimes place two long wooden sticks across each other as the diagonals of the rectangle.

23. How must the lengths of the sticks be related in order to produce a rectangular structure?

24. If the sticks are the appropriate lengths, does it matter how they are placed? Why or why not?

25. Writing A square building requires fewer materials than a nonsquare rectangular one that covers the same area. How should the sticks be placed for a square house? Explain your reasoning.

BY THE WAY

Several other groups in Africa whose traditional homes are round are beginning to build rectangular houses instead.

RECTANGLE

1. Construct ⊙E.

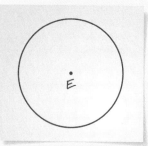

2. Draw two diameters of ⊙E. Label them $\overline{AC}$ and $\overline{BD}$.

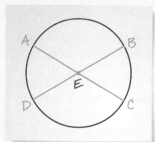

3. Draw $\overline{AB}$, $\overline{BC}$, $\overline{CD}$, and $\overline{AD}$. ABCD is a rectangle.

26. How do you know that quadrilateral ABCD is a parallelogram?

27. What guarantees that ▱ABCD will be a rectangle?

28. Writing The diagonals of two rectangles are congruent, but the rectangles are not congruent. Describe how to construct the rectangles.

29. In Section 5.7, you learned how to construct the perpendicular bisector of a given segment.

    a. Construct two segments, $\overline{WY}$ and $\overline{XZ}$, that are perpendicular bisectors of each other.

    b. Draw $\overline{WX}$, $\overline{XY}$, $\overline{YZ}$, and $\overline{WZ}$. What type of quadrilateral is WXYZ? How do you know?

30. Construct a rectangle JKLM so that $\overline{JL} \cong \overline{RS}$.

31. Challenge Prove that quadrilateral ABCD is a rectangle.

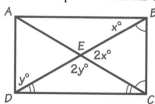

32. SAT/ACT Preview The coordinates of the vertices of a quadrilateral are $(1, 4)$, $(3, 0)$, $(1, -2)$, and $(-3, 0)$. The quadrilateral is:

    A. a rhombus   **B.** a kite   **C.** a square   **D.** a trapezoid   **E.** none of these

ONGOING ASSESSMENT

33. Open-ended Problem Using a ruler and a protractor, use two different methods to draw a square. Describe your methods and prove that they produce squares.

Find each length or angle measure. *(Section 6.6)*

34. *JK*

35. $m \angle Z$

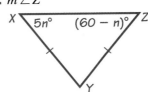

36. $m \angle E$

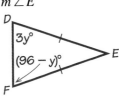

Write an equation for each line described. *(Section 4.2)*

37. contains (6, 2); has slope -3

38. contains (0, -1); has slope 5

ASSESS YOUR PROGRESS

VOCABULARY

kite (p. 339)   **indirect proof** (p. 348)

Find each length. *(Section 7.1)*

1. rectangle *WXYZ*
   a. *VY*
   b. *XZ*
   c. *ZY*

2. rhombus *JKLM*
   a. *JK*
   b. *NK*
   c. *MK*

3. kite *PQRS*
   a. *TQ*
   b. *TR*
   c. *PQ*

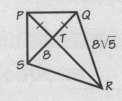

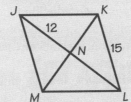

4. Write an indirect proof to show that
   $\angle W \not\cong \angle Y$ in kite *WXYZ*. *(Section 7.2)*
   Given: *WXYZ* is a kite.
   Prove: $\angle W \not\cong \angle Y$

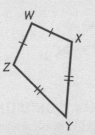

Classify each quadrilateral. Be as specific as possible. Give reasons for your answers. *(Sections 7.2 and 7.3)*

5.

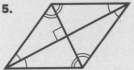

6.

7.

8. **Journal** List all the ways that you have learned to identify each special type of quadrilateral shown on page 339. Which methods do you think are the easiest to remember and to use?

7.4 Areas of Triangles and Quadrilaterals

Learn how to...
- find the areas of triangles and quadrilaterals

So you can...
- use these formulas in real-world applications, such as tiling a floor

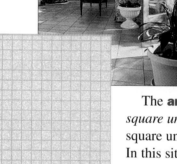

It takes $15\frac{1}{2}$ rows to fill the length of the room.

A row of 12 tiles stretches across the width of the room.

Paul Bornstein's porch is 12 ft wide and $15\frac{1}{2}$ ft long. He wants to cover the floor with 1 ft by 1 ft square tiles. How many tiles does he need? Does it matter what shape the room is? Is there a quick way to figure it out?

Paul needs 12 rows of $15\frac{1}{2}$ tiles. So $12 \times 15\frac{1}{2} = 186$ tiles will cover the full *area* of the floor of his porch.

The **area** of a two-dimensional figure is the number of *square units* enclosed within the boundary of the figure. A square unit is the space enclosed by a 1 unit by 1 unit square. In this situation, the area of each tile is one square foot, written as $1\ \text{ft}^2$.

To find the area of a rectangle, multiply the lengths of two adjacent sides. One length is considered to be the **base** and the other is the **height**.

$$\text{Area of a rectangle} = \text{base} \times \text{height}$$

$$A = bh$$

THINK AND COMMUNICATE

1. Praktisha says that the base of Paul's floor is 12 ft and the height is $15\frac{1}{2}$ ft. Danielle thinks that the base is $15\frac{1}{2}$ ft and the height is 12 ft. Explain why they are both correct.

2. Suppose you want to tile the floor of a 12 ft by 12 ft square room.
   a. What is the height of the square? What is the base of the square?
   b. Find the area of the square.

Toolbox p. 704
Finding Area and Volume

3. Explain how to find the area of the figure at the right. What units will the area be measured in?

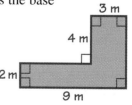

3 m
4 m
2 m
9 m

Discovering Area Formulas

Work in a group of three students.
You will need:
- a rectangular sheet of paper
- scissors

1 Cut the paper into two noncongruent rectangles. Label the sides as shown.

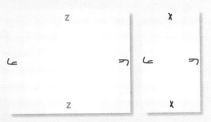

2 Cut along one diagonal of one rectangle to make two congruent right triangles, as shown.

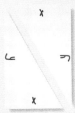

3 How do the base and height (the lengths of the legs) of one triangle compare to the base and height of the rectangle it was cut from? How does the area of one triangle compare with the area of that rectangle? Describe the area of one triangle in terms of its base and height.

4 Use all three pieces to make a rectangle.

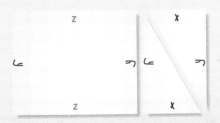

The rectangle's **base** is $x + z$ and its **height** is y. So the area is $A = (x + z)y$.

5 Use all three pieces to make a nonrectangular parallelogram whose *height*, the distance between two of its parallel sides, is y. Write an expression for the base. Does the parallelogram have the same area as the rectangle you made in Step 4? Describe the area of the parallelogram in terms of its base and height.

6 Use all three pieces to make a trapezoid whose height is y. Write an expression for each base and the sum of the bases. Does the trapezoid have the same area as the rectangle you made in Step 4? Describe the area of the trapezoid in terms of its height and the sum of its bases.

In the area formulas given below, the *height* is the distance from the base to the highest point of the figure.

Area Formulas

The area of a rectangle is the product of the base and the height.

$A = bh$

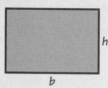

The area of a square is the square of the length of one side.

$A = s^2$

The area of a triangle is half the product of the base and the height.

$A = \frac{1}{2}bh$

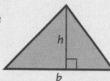

The area of a parallelogram is the product of the base and the height.

$A = bh$

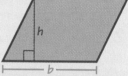

The area of a trapezoid is the product of the height and the mean of the bases.

$A = \frac{1}{2}(b_1 + b_2)h$

EXAMPLE 1

Toolbox p. 703
Finding Perimeter

Find the area of an equilateral triangle with perimeter 18 cm.

SOLUTION

The base of an equilateral triangle with perimeter 18 cm is 6 cm. Sketch the triangle and use the Pythagorean theorem to find its height.

$3^2 + h^2 = 6^2$
$h^2 = 6^2 - 3^2$
$h = 3\sqrt{3}$

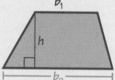

Use the formula given above.

$A = \frac{1}{2}bh$

$= \frac{1}{2}(6)(3\sqrt{3})$ — Substitute 6 for *b* and $3\sqrt{3}$ for *h*.

≈ 15.6

The area of the triangle is about 15.6 cm².

EXAMPLE 2   Connection: Algebra

Find the area of the trapezoid.

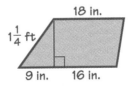

SOLUTION

Use the Pythagorean theorem to find the height h of the trapezoid.

$$9^2 + h^2 = 15^2$$

$$h^2 = 15^2 - 9^2$$

$$h = \sqrt{225 - 81}$$

$$h = 12$$

Make sure the measurements have the same units.

$1\frac{1}{4}$ ft = 15 in.

Use the formula for the area of a trapezoid.

$$A = \frac{1}{2}(b_1 + b_2)h$$

$$= \frac{1}{2}[18 + (9 + 16)](12)$$

$$= \frac{1}{2}(43)(12)$$

$$= 258$$

Use the Segment Addition Postulate to find the length of the lower base.

The area of the trapezoid is 258 in.^{2}.

☑ CHECKING KEY CONCEPTS

Find the area of each polygon.

1.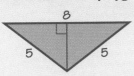

2.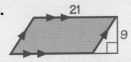

3.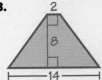

4.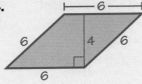

5. a rectangle with side lengths 26 ft and 9 ft

6. a square with side length 19 cm

7. a right triangle with one leg of length 14 in. and hypotenuse 30 in.

7.4 : Exercises and Applications

Extra Practice
exercises on
page 674

Find the area of each polygon.

1.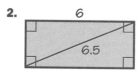
9.5
19

2.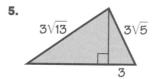
6
6.5

3.
$2\sqrt{2}$

4.
10
8   6

5.
$3\sqrt{13}$ $3\sqrt{5}$
3

6.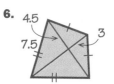
4.5
3
7.5

7. HISTORY Heron, a Greek mathematician who lived over 2000 years ago, developed the following formula to find the area of a triangle with side lengths a, b, and c. In this formula, $s = \frac{1}{2}(a + b + c)$.

$$A = \sqrt{s(s - a)(s - b)(s - c)}$$

Use Heron's Formula to find the area of the triangle in Example 1.

8. CLOTHING A typical Chilean poncho is rectangular, 10 ft long by 6 ft wide. How many square yards of wool cloth are needed for a poncho?

9. Writing The sides of two parallelograms are congruent. Must the areas of the parallelograms be the same? Explain.

10 ft

6 ft

Find the area of each polygon.

10. an isosceles triangle with base 10 cm and perimeter 34 cm

11. a trapezoid with height 2 m and bases that add up to 18 m

12. a square with perimeter 22 ft

13. a rectangle with side length 5 in. and diagonal 13 in.

Investigation The *midline* of a trapezoid is the segment that is parallel to the bases and bisects the legs. Cut a trapezoid out of paper and fold to find its midline.

14. Follow the steps at the right. Show that the pieces form a parallelogram in Step 3.

15. ALGEBRA Find an expression for the length of the midline of the trapezoid in terms of its bases.

16. Write a formula for the area of a trapezoid in terms of the length of its midline, m, and its height, h.

Step 1 Label the midline m.

Step 2 Label the bases b_1 and b_2.

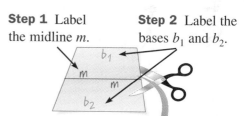

Step 3 Then cut along the midline and arrange the pieces as shown.

ALGEBRA The coordinates of the vertices of a polygon are given. Sketch the polygon and find its area.

17. $S(0, 6)$, $T(0, 9)$, $U(7, 9)$, $V(7, 6)$

18. $D(1, 1)$, $E(1, 5)$, $F(5, 5)$, $G(5, 1)$

19. $A(2, 1)$, $B(4, 8)$, $C(10, 1)$

20. $W(-2, 4)$, $X(0, 4)$, $Y(-3, 0)$, $Z(-5, 0)$

21. $J(0, 3)$, $K(5, 3)$, $L(8, 6)$, $M(3, 6)$

22. $P(-2, 9)$, $Q(-2, -1)$, $R(-5, 5)$, $S(-5, 7)$

23. Sketch a rectangle and a parallelogram that have the same base and the same height. Explain how the formula for the area of a rectangle leads to the formula for the area of a nonrectangular parallelogram.

24. Sketch a parallelogram and one of its diagonals. Explain how the formula for the area of a parallelogram leads to the formula for the area of a triangle.

25. ALGEBRA Write a coordinate proof to show that the area of a rhombus is half the product of the lengths of the diagonals.

26. Open-ended Problem Sketch a trapezoid with area 24 cm². Label all the dimensions of the trapezoid in your sketch.

27. Logical Reasoning The base of $\triangle ABD$ is 24 and the height is 8.

   a. Find the areas of $\triangle ABC$ and $\triangle ACD$.

   b. Writing Describe how to divide a triangle into two or more smaller triangles whose areas are equal. Tell why your method works.

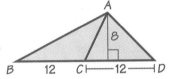

Connection   **GARDENING**

Many communities set aside land for gardens. The ground is usually divided into plots whose areas are equal, but often the plots are different shapes. The plans below show one layout for the Jesse Frey Community Garden in San Jose, California.

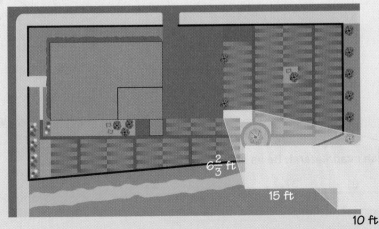

28. A standard plot is a 4 ft by 25 ft rectangle. Find the area of one plot.

29. Writing Describe a way to divide the shaded region into plots with the same area as a 4 ft by 25 ft plot.

30. Open-ended Problem Do you think the shape of a garden plot is important? Why or why not?

Look back at the article on pages 336–338.

One of Walt Stone's many accomplishments was to increase the number of parking places for people with disabilities in Branson. To plan this, he had to calculate the area that the spaces would need.

31. By regulation, each rectangular space for people with disabilities must be 8 ft wide and 20 ft long. What is the area of one space?

32. There must be 5 ft between spaces. Find the minimum area covered by four spaces that are next to each other.

33. How many parking spaces will fit in an area that is 35 ft wide?

34. HISTORY The following formula for the area of a trapezoid was used in calculations on an ancient Babylonian tablet:

$$A = \frac{b_1 + b_2}{2} \cdot \frac{s_1 + s_2}{2}$$

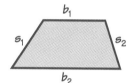

In the formula, s_1 and s_2 are the lengths of the nonparallel sides.

a. How is the formula on the tablet different from the correct formula?

b. For which types of quadrilaterals can this formula be used?

c. Challenge For which trapezoids does this formula give an answer that is approximately correct? Does it *overestimate* or *underestimate* the area? Explain.

ONGOING ASSESSMENT

35. Open-ended Problem Sketch a triangle, a rectangle, a trapezoid, and a nonrectangular parallelogram that all have the same area.

SPIRAL REVIEW

Classify each quadrilateral. Be as specific as possible. *(Sections 7.2 and 7.3)*

36.

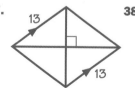

37.

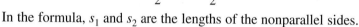

38.

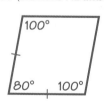

Sketch each figure. Tell how many lines of symmetry it has. *(Section 2.3)*

39. regular pentagon **40.** regular 7-gon **41.** regular 16-gon

7.5 Areas of Regular Polygons and Circles

Learn how to...

- find the areas of regular polygons and circles

So you can...

- use area formulas to find areas of real-world objects, such as buildings or a running track

Some of the 1994 World Cup soccer games were played indoors at the Pontiac Silverdome in Pontiac, Michigan. Since the World Cup officials require a natural grass playing field, local crop and soil researchers grew the grass outdoors and brought it into the Silverdome in regular hexagonal trays.

52 in.

EXAMPLE 1   Application: Sports

Find the area of each hexagonal tray used in the Silverdome.

SOLUTION

A regular hexagon can be divided into six congruent equilateral triangles. Use the Pythagorean theorem to find the height h of one triangle.

$$h^2 + 26^2 = 52^2$$
$$h^2 = 52^2 - 26^2$$
$$h = 26\sqrt{3}$$

26 in.

$26\sqrt{3}$ in.

52 in.

Use this value to find the area of one triangle.

$$\text{Area of hexagon} = 6 \cdot \left(\text{Area of one triangle}\right)$$
$$= 6\left(\frac{1}{2}bh\right)$$
$$= 6 \cdot \frac{1}{2} \cdot 52 \cdot 26\sqrt{3}$$
$$\approx 7025$$

The area of each hexagonal tray is about 7025 in.^{2}.

BY THE WAY

Why were the trays of grass in the Silverdome hexagonal? Why do bees build hexagonal honeycomb cells? The area of a regular hexagon is greater than the area of any other shape with the same perimeter that forms a regular tessellation.

In Example 1 you saw how to divide a hexagon into six congruent triangles. Any regular *n*-gon can be divided into *n* congruent triangles by drawing a segment from the center to each vertex.

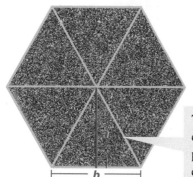

Area of a regular *n*-gon = $n \times$ Area of one triangle

$$= n \times \frac{1}{2}bh$$

$$= \frac{1}{2}h(nb)$$

$$= \frac{1}{2}ap$$

The product of **n** and the side length **b** is the **perimeter, p.**

The distance from the center of a regular polygon to a side is called the **apothem, a.**

Area of a Regular Polygon

The area of any regular polygon is half the product of the apothem and the perimeter.

$$A = \frac{1}{2}ap$$

EXAMPLE 2

The length of a side of a regular pentagon is 12 and the distance from the center to each vertex is 10.2. Find the area of the pentagon.

SOLUTION

Sketch the pentagon. Then use the Pythagorean theorem to find the apothem.

$$6^2 + a^2 = (10.2)^2$$

$$a^2 = (10.2)^2 - 6^2$$

$$a = \sqrt{68.04}$$

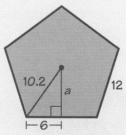

Then use the formula.

$$A = \frac{1}{2}ap$$

$$= \frac{1}{2}\left(\sqrt{68.04}\right)(5 \cdot 12)$$

Multiply the number of sides by the side length to find the perimeter.

$$\approx 247.5$$

The area of the pentagon is about 248.

THINK AND COMMUNICATE

1. You can divide any regular *n*-gon into *n* congruent triangles. Why are the triangles equilateral in a regular hexagon? What type of triangles do you think they are in other regular polygons? Explain your reasoning.

2. In Example 1, you found the apothem of a regular hexagon by using the length of one side. Can you find the apothem of *any* regular polygon using only the side length? Why or why not?

3. The length of each side of a square is *s*. Find expressions for the perimeter and apothem in terms of *s*. Then show that $A = \frac{1}{2}ap = s^2$ for a square.

Area of a Circle

The pattern below shows regular polygons *inscribed* in circles. The radius of a regular polygon is the radius of the circle that the polygon is inscribed in. As the sequence continues and the number of sides of the polygon increases, the length of the apothem approaches the length of the radius. The perimeter and area of the polygon approach the perimeter and area of the circle.

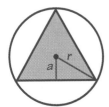 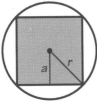  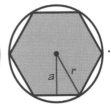 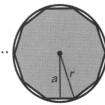

The perimeter of a circle is called the circle's *circumference*. The ratio $\frac{circumference}{diameter}$ is a constant value for all circles. It is an irrational number denoted by the Greek letter π ("pi") and is approximately 3.14, or $\frac{22}{7}$. This leads to the formula $C = 2\pi r$.

You can think of the **radius** of a circle as the **apothem** of a regular polygon, and the **circumference** as the **perimeter**:

$$\text{Area of a regular polygon} = \frac{1}{2}ap$$

$$\text{Area of a circle} = \frac{1}{2}r(2\pi r)$$

$$= \pi r^2$$

Area and Circumference of a Circle

The area of a circle with radius *r* is πr^2.

The circumference of a circle with radius *r* is $2\pi r$.

$A = \pi r^2$
$C = 2\pi r$

EXAMPLE 3   Application: Sports

A high school running track surrounds a field that needs to be watered. Find the area enclosed by the track.

62.84 m
100 m

SOLUTION

The region is a rectangle with a semicircle at each end.

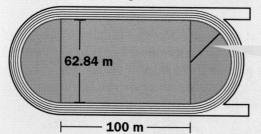
62.84 m
100 m

The diameter of each semicircle is 62.84 m, so the **radius** is **31.42 m**.

$$\begin{pmatrix} \text{Area enclosed} \\ \text{by the track} \end{pmatrix} = \begin{pmatrix} \text{Area of} \\ \text{rectangle} \end{pmatrix} + 2\begin{pmatrix} \text{Area of one} \\ \text{semicircle} \end{pmatrix}$$

$$= (bh) + 2\left(\frac{1}{2}\pi r^2\right)$$

The area of a semicircle is half the area of a circle.

$$= (100 \cdot 62.84) + \pi(31.42)^2$$

$$\approx 9385.43$$

The area enclosed by the track is about 9390 m^2.

☑ CHECKING KEY CONCEPTS

Find the area of each regular polygon or circle.

1.
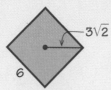
$3\sqrt{2}$
6

2.

12

3.
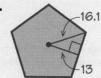
16.1
13

4.

9

5.

13

6.

$C = 8\pi$

7.5 Exercises and Applications

Extra Practice
*exercises on
page 674*

Find the area of each regular polygon.

1.
$P = 80$

 13 13
13
S

2.
38
19
17
36

3.
14

Find the area of each circle.

4.
$\frac{1}{2}$

5.
$C = 22\pi$

6.
10

Find the area of each shaded region.

7.
5

8.
8

9.
1
3

10. Open-ended Problem Sketch two concentric circles. Give an example of what their radii could be if the area enclosed between the circles is 12π.

Connection ▶ BIOLOGY

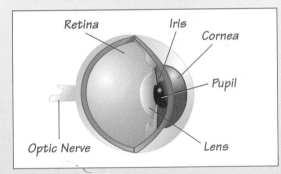

Retina   Iris   Cornea
Pupil
Optic Nerve   Lens

The *pupil* is the opening through which light enters the eye. The *iris*, the colored part of the eye, opens and closes around the pupil to change the amount of light that enters the eye.

Entering a dark room

For Exercises 11 and 12, find the area of the pupil.

11. When you enter a dark room, your iris opens until the pupil is about 8 mm in diameter.

12. In a well-lit room, your iris closes in around the pupil, giving it a diameter of about 4 mm.

13. Writing Look back at Exercises 11 and 12. What is the ratio of the diameters of the pupils? What is the ratio of the areas? Is this what you expected? Explain.

Entering a well-lit room

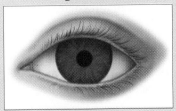

14. 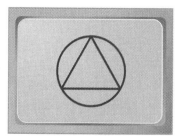 **Technology** Work with two other people. You will need geometry software and a graphing calculator.

   a. Cooperative Learning Each student should construct two different regular polygons, each inscribed in a circle whose radius is 1. Find the perimeter and area of each polygon.

   b. ALGEBRA Make two graphs of your group's data with the number of sides of the polygon on the *x*-axis. For the first graph, put the perimeter of the polygon on the *y*-axis. For the second graph, put the area of the polygon on the *y*-axis. Sketch the two graphs.

   c. Writing Look at your graphs. How do the perimeters and areas change as the number of sides of the polygons increases? Suppose you inscribe a 100-gon in a circle of radius 1. What do you think the perimeter and area would be? Explain.

CONSTRUCTION

REGULAR HEXAGON

1. Construct a circle.

2. Using the same radius, place the point of the compass at any point *A* on the circle and draw an arc. Label the intersection *B*.

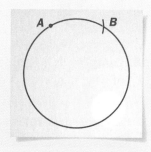

3. Place the point of the compass at *B* and draw an arc that intersects the circle at point *C*. Repeat this process to find *D*, *E*, and *F*.

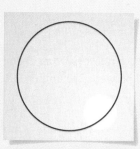

4. Connect the consecutive points around the circle. *ABCDEF* is a regular hexagon.

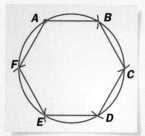

15. How do you know that hexagon *ABCDEF* is equilateral?

16. Explain why hexagon *ABCDEF* is equiangular. (*Hint:* What type of triangles are formed when you connect the opposite vertices of hexagon *ABCDEF*?)

17. How can you change the last step to construct an equilateral triangle instead of a regular hexagon? Explain your reasoning.

18. Construct a regular hexagon.

19. Writing Construct a regular hexagon with side length 3 cm. Write a description of your method. Include enough information so that someone else can follow your steps.

Find the area of each shaded region. Each outer polygon is regular.

20.

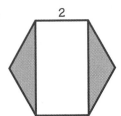

2

21.

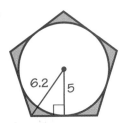

6.2 5

22.

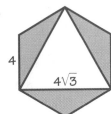

4

$4\sqrt{3}$

23.

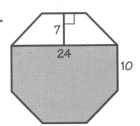

7

24

10

24. Challenge Find the area of the shaded region.

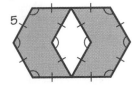

5

ARCHITECTURE The front entrance of the Clore Gallery, part of
the Tate Gallery in London, is shown at the right. The sloped
sides are the same length and make 120° angles with each
other and with the vertical sides. Use this information for
Exercises 25–27.

25. Writing You can think of the shape of the front entrance
as part of a regular hexagon, as shown. Why is this
possible?

26. Use the regular hexagon to find
the area of the front entrance of
the gallery. About what percent
of the area of the hexagon is the
area of the entrance?

27. Describe another method you
can use to find the area of the
front entrance. Which method
do you think is easier? Explain.

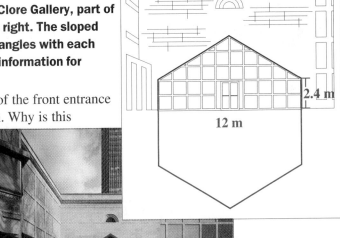

2.4 m

12 m

ONGOING ASSESSMENT

28. Writing Explain how you can find the area of a regular *n*-gon by
finding the area of *n* isosceles triangles.

SPIRAL REVIEW

Find the area of each polygon. *(Section 7.4)*

29. a square with perimeter 28 yd

30. a rectangle with base 5 mm and perimeter 18 mm

Sketch a net for each prism. *(Section 2.6)*

31. an octagonal prism

32. a cube

33. a triangular prism

34. a heptagonal prism

7.6 Prisms and Cylinders

Suppose two tubes of different sizes are made from identical pieces of paper. Do they both hold the same amount? You will answer this question in the Exploration.

Learn how to...

- analyze the parts of prisms and cylinders

So you can...

- find the volumes, lateral areas, and surface areas of prisms and cylinders

EXPLORATION
COOPERATIVE LEARNING

Comparing Volumes

Work with another student. You will need:

- rectangular pieces of paper that are the same size
- tape
- popcorn or dried beans

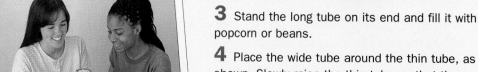

1 Tape the two long sides of one piece of paper together to form a tube.

2 Tape together the two short sides of another piece of paper, identical to the one used in Step 1, to form a shorter, wider tube.

3 Stand the long tube on its end and fill it with popcorn or beans.

4 Place the wide tube around the thin tube, as shown. Slowly raise the thin tube so that the contents fill the wide tube. Which tube holds more? How can you tell?

5 Repeat Steps 1–4 using paper of different dimensions. Do you get the same result?

6 Do you think the *radius* of a tube or the *height* of a tube has more effect on the amount it holds? Explain your reasoning.

The **volume** of a solid is the number of *cubic units* needed to fill the solid. If the edges of a cube are one unit long, then its volume is one cubic unit. To find the volume of a prism, you multiply the number of cubes that "cover" or "make up" the base by the number of layers of cubes that fill the prism.

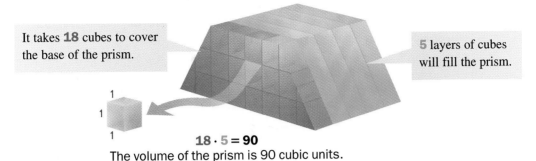

It takes **18** cubes to cover the base of the prism.

5 layers of cubes will fill the prism.

$18 \cdot 5 = 90$
The volume of the prism is 90 cubic units.

The same reasoning can be applied to *cylinders*. A **cylinder** is a space figure whose bases are circles of the same size.

base

height

WATCH OUT!

The height of a figure is not always vertical, and the base is not always the bottom.

Volumes of Prisms and Cylinders

The volume of a prism or a cylinder is the product of the height and the area of the base.

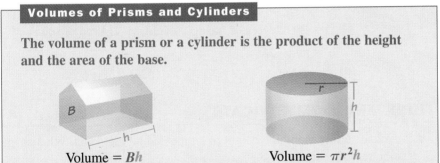

Volume $= Bh$

Volume $= \pi r^2 h$

EXAMPLE 1

Find the volume of this prism.

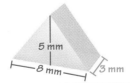

5 mm

8 mm

3 mm

SOLUTION

Multiply the area of the triangular base by the height of the prism.

$$\text{Volume} = Bh$$
$$= \left(\frac{1}{2} \cdot 8 \cdot 5\right) \cdot 3 \qquad \text{Use the formula for the area of a triangle.}$$
$$= 20 \cdot 3$$
$$= 60$$

The volume of the prism is 60 mm^3.

Surface Area

In Chapter 2 you learned how to sketch a net for a prism by drawing the faces connected to one another. The area of a net for a three-dimensional figure is called the **surface area** of the figure, abbreviated *S.A.*

The area of the lateral faces is called the **lateral area**. The lateral faces of a prism can be arranged to form one rectangle.

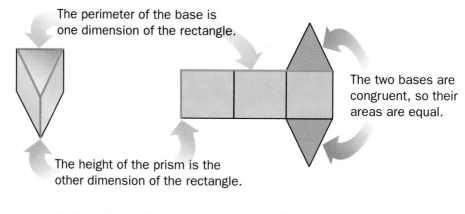

The perimeter of the base is one dimension of the rectangle.

The two bases are congruent, so their areas are equal.

The height of the prism is the other dimension of the rectangle.

Surface Area = Lateral area + Area of bases

= (Perimeter of base × height) + 2(Area of base)

= $ph + 2B$

THINK AND COMMUNICATE

1. Imagine cutting along the height of a tube and laying it flat. What shape is the resulting figure?

2. You learned in Chapter 2 that prisms are named according to their bases. What do you think a *regular prism* is?

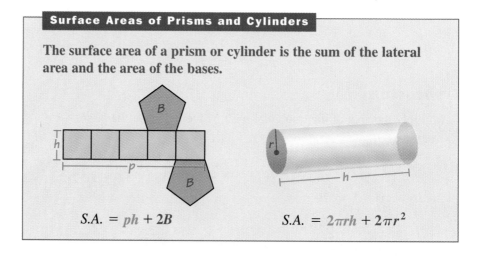

Surface Areas of Prisms and Cylinders

The surface area of a prism or cylinder is the sum of the lateral area and the area of the bases.

$S.A. = ph + 2B$

$S.A. = 2\pi rh + 2\pi r^2$

EXAMPLE 2

Find the volume and surface area of the cylinder that the net folds into.

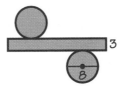

SOLUTION

The diameter of the base is 8, so the radius is 4. Use this value in the formulas for volume and surface area.

$$\text{Volume} = \pi r^2 h \qquad\qquad S.A. = 2\pi rh + 2\pi r^2$$
$$= \pi(4)^2 \cdot 3 \qquad\qquad = (8\pi \cdot 3) + 2\pi(4)^2$$
$$= 48\pi \qquad\qquad\qquad = 56\pi$$
$$\approx 150.8 \qquad\qquad\quad \approx 175.9$$

The volume of the cylinder is approximately 150.8.
The surface area of the cylinder is approximately 175.9.

EXAMPLE 3   Application: Package Design

The gift box at the right is a regular hexagonal prism. How much cardboard was used to make the box?

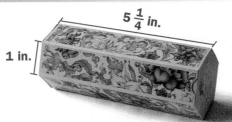

5 $\frac{1}{4}$ in.

1 in.

SOLUTION

Use the Pythagorean theorem to find the apothem of the hexagonal base.

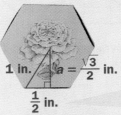

1 in. $a = \frac{\sqrt{3}}{2}$ in.

$\frac{1}{2}$ in.

The length of each side is 1 in., so the length of half of the side is $\frac{1}{2}$ in.

Substitute this value into the formula for surface area.

$$S.A. = ph + 2B$$
$$= \left[(6 \cdot 1) \cdot \frac{21}{4} \right] + 2\left[\frac{1}{2} \cdot \frac{\sqrt{3}}{2} \cdot (6 \cdot 1) \right]$$
$$= \frac{126}{4} + 3\sqrt{3}$$
$$\approx 36.7$$

To find the area of the base, substitute $\frac{\sqrt{3}}{2}$ for a and $6 \cdot 1$ for p in the formula $A = \frac{1}{2}ap$.

About 36.7 in.^{2} of cardboard was used to make the box.

Find the volume and surface area of each regular prism or cylinder.

1.

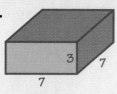

2.

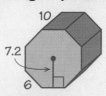

10

7.2

6

3. 6

4√3

4.

2

$C = 20\pi$

5. Sketch the figure that the net folds into. Then find its volume and surface area.

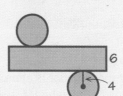

6

4

7.6 **Exercises and Applications**

Extra Practice exercises on page 674

Find the volume and surface area of each prism or cylinder.

1.

9

11

2. rectangular prism

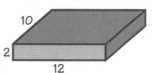

10

2

12

3.

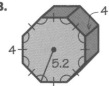

4

4

5.2

Sketch the figure that each net folds into. Then find its volume and surface area.

4.

6

5

5.

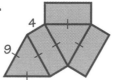

4

9

6.

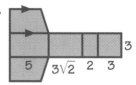

3

5   3√2 2 3

7. NUTRITION According to the package label, each 1 in.³ serving of a certain brand of cheese has 110 calories. Each package contains a block of cheese that is 1 in. by $2\frac{1}{4}$ in. by 6 in.

a. How many servings are in each package?

b. How many calories are in a $\frac{1}{4}$ in. slice off the smallest end of the block?

Find the volume of each space figure described.

8. Each edge of a cube is 3 ft long.

9. The radius of a cylinder is 3 m and the lateral area is 15π m^2.

10. The perimeter of each base of a regular hexagonal prism is 24 cm and the area of one lateral face is 40 cm^2.

11. A triangular prism is 10 in. high. The perimeter of each base is 12 in. and the area of each base is 6 in.^{2}

12. **CARPENTRY** David Taylor plans to reshingle the sides of the house shown at the right. One bundle of shingles covers 25 ft^2.

 a. Sketch the faces of the prism that represent the walls of the house. Are they all lateral faces? Explain why or why not.

 b. Estimate how many bundles of shingles David will need.

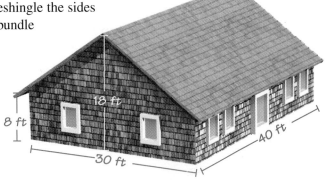

Connection ▶ CONSUMER ECONOMICS

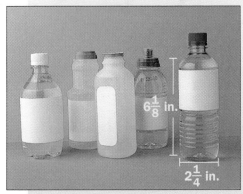

Many beverages are sold in cylindrical bottles. However, some manufacturers use other shapes to make their products more practical or appealing.

13. a. A cylindrical 400 mL bottle has a diameter of $2\frac{1}{4}$ in. and a height of $6\frac{1}{8}$ in. Find the volume of the bottle in cubic inches.

 b. Use the conversion factor $\dfrac{1 \text{ mL}}{0.061 \text{ in.}^3}$ to check your answer to part (a).

14. One brand of spring water comes in a bottle in the shape of a regular triangular prism. The bottle just fits in a cup holder in a car. The cup holder is circular with diameter $3\frac{1}{2}$ in.

 a. The height of an equilateral triangle is one and one-half times the radius of the triangle. Find the length of each side of the base of the triangular bottle.

 b. How tall does the bottle have to be to hold 400 mL of water? How do you know?

15. a. **Open-ended Problem** Estimate the amount of plastic used to make each of the bottles described in Exercises 13 and 14. Explain your reasoning.

 b. **Open-ended Problem** Which bottle do you think costs less to make? Which do you think is easier to store? Overall, which design do you think is more practical and why?

Look back at the
article on pages
336–338.

*Legislators like Walt Stone make sure that wheelchair ramps are
built to meet certain requirements. If they are too steep or too
narrow, they can be dangerous. Some ramps are wooden, but they
can also be made of cement. The shape of a cement ramp is
usually a prism.*

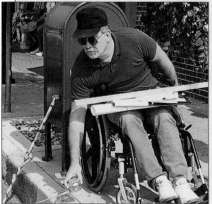

16. A ramp must be at least 3 ft wide and extend at least 12 in.
horizontally for every inch it rises vertically.

  a. How long must a ramp be if it reaches a step that is
  $2\frac{1}{4}$ ft above the ground?

  b. Find the minimum amount of cement needed to make a
  ramp up to a step that is $2\frac{1}{4}$ ft high.

17. Challenge Find the amount of cement needed to make
the ramp shown.

30 in.

36 in.

36 in.

15 in.

60 in. 180 in.

18. Research Measure the dimensions of a cement ramp in your school,
library, or neighborhood and calculate the amount of cement needed
to build it.

19. Writing Look at the diagrams of cylinders in this section. How would
you sketch a cylinder? How is it different from sketching a prism?

20. Open-ended Problem Sketch a cylinder with surface area 72π cm^2.
Give all dimensions of the cylinder and find its volume.

**ALGEBRA The vertices of a prism are given. Sketch the prism and find its
volume.**

21. $T(3, -1, 1)$, $V(5, 2, 1)$, $W(7, -1, 1)$, $X(3, -1, 5)$, $Y(5, 2, 5)$, $Z(7, -1, 5)$

22. $J(0, 0, -2)$, $K(0, -3, -2)$, $L(11, -3, -2)$, $M(11, 0, -2)$, $P(0, 0, 3)$,
$Q(0, -3, 3)$, $R(11, -3, 3)$, $S(11, 0, 3)$

23. $A(-2, 5, 0)$, $B(0, 9, 0)$, $C(6, 9, 0)$, $D(4, 5, 0)$, $E(-2, 5, -8)$, $F(0, 9, -8)$,
$G(6, 9, -8)$, $H(4, 5, -8)$

24. Open-ended Problem Sketch a prism and a cylinder that have approximately the same volume. Give the dimensions and surface area of both objects.

SPIRAL REVIEW

Copy each diagram. Sketch the reflection of the triangle over the red line.
(Section 1.2)

25.

26.

27.

ASSESS YOUR PROGRESS

VOCABULARY

area (p. 360)   **cylinder** (p. 375)
apothem (p. 368)   **surface area** (p. 376)
volume (p. 375)   **lateral area** (p. 376)

Find the area of each figure. *(Sections 7.4 and 7.5)*

1.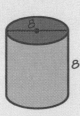

2.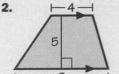

3.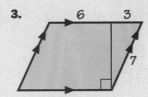

4.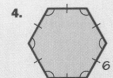

5.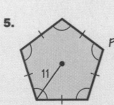
$P = 65$

6.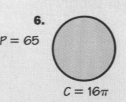
$C = 16\pi$

Find the volume and surface area of each prism or cylinder. *(Section 7.6)*

7.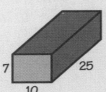

8. rectangular prism

9.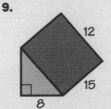

10. Journal List all of the area and volume formulas you learned in this chapter and sketch a diagram to illustrate each one. Show how some formulas can be used to find the others.

Designing a Cottage

An architect who designs a building must work within certain constraints, such as the area the building will cover. Within these restrictions, there is a lot of room for creativity.

PROJECT GOAL Design a floor plan for a one-story cottage, using specified guidelines.

Understanding the Specifications

You and two other students will design a cottage. The floor plan will be a rectangle whose area is between 550 ft² and 650 ft². The cottage should have two entrance doors and at least four rooms, as described in the table. Any hallways you include should be between three and four feet wide.

Minimum Room Areas	
Living Room	300 ft²
Bedroom	90 ft²
Kitchen	65 ft²
Bathroom	50 ft²

Arranging the Floor Plan

1. Determine at least three possibilities for the length and width of the rectangular cottage. For each possibility, make a rough sketch showing the locations of the rooms.

2. Make a detailed cottage floor plan, using one of your sketches. Draw the rectangle on graph paper or dot paper, then determine the exact dimensions and placement of the rooms. Label each room with its name, dimensions, and area.

Indicate the scale of your floor plan.

3. Show the locations of the windows and doors. You may also want to show appliances and furniture.

For nonrectangular rooms, give all the dimensions needed to find the area.

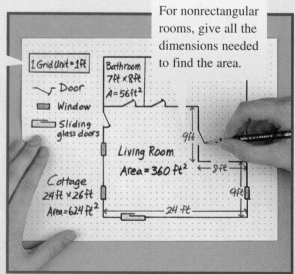

The Third Dimension

Knowing the volume of a building is important for choosing a heating system. The completed cottage will be shaped like a pentagonal prism. Sketch one of the bases, labeling all the lengths you need to find its area. Find the area of the pentagon. Then find the volume of the house.

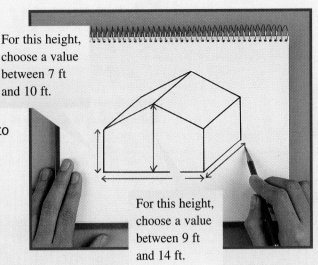

For this height, choose a value between 7 ft and 10 ft.

For this height, choose a value between 9 ft and 14 ft.

Submitting Your Plan

Create a proposal describing the cottage to its future owners. Include the detailed floor plan and explain the features of your design. Also explain how you calculated the volume of the cottage. You can extend the project by exploring one of the following ideas:

- Use a computer with drawing or CAD (Computer-Aided Design) software to help you create your plan.

- Take measurements of rooms where you live or at your school. How do they compare with those in the cottage you designed?

- How would the floor plan be different if it were based on a shape other than a rectangle? Design a floor plan that uses other shapes that you learned about in this chapter.

- Talk to an architect in your area. What other factors affect the design of a house?

Self-Assessment
How did your group decide which of the rough sketches to use for the final floor plan? If you did not have to follow the specifications given in this project, how would you change your design? Explain your reasoning.

7 | Review

STUDY TECHNIQUE

Make a list of the important ideas you learned in Chapter 2 and then revisited and extended in Chapter 7. Also include any new ideas that are developed in Chapter 7.

VOCABULARY

kite (p. 339)
indirect proof (p. 348)
area (p. 360)
apothem (p. 368)

volume (p. 375)
cylinder (p. 375)
surface area (p. 376)
lateral area (p. 376)

SECTIONS | 7.1, 7.2, *and* 7.3

The chart on page 339 describes how quadrilaterals can be classified by their properties. The chart on page 347 shows how to prove that a quadrilateral is a parallelogram.

Diagonals of Quadrilaterals

The diagonals of a rhombus are perpendicular.

The diagonals of a rectangle are congruent.

One diagonal of a kite is a line of symmetry for the kite and the perpendicular bisector of the other diagonal.

The charts on pages 353 and 355 show how to prove that a parallelogram is a rectangle or a rhombus.

An **indirect proof** shows that the conclusion must be true if all other possibilities are false. For example, to prove that two angles of a triangle cannot be obtuse:

Assume that $\angle A$ and $\angle B$ of $\triangle ABC$ are both obtuse. Then $m \angle A > 90°$ and $m \angle B > 90°$, and the sum of the angles of the triangle is greater than $180° + m \angle C$. But the sum of the measures of the angles of a triangle must be $180°$. Therefore two angles of a triangle cannot be obtuse.

SECTIONS 7.4, 7.5, *and* 7.6

The area of a two-dimensional figure is the number of square units enclosed within the boundary of the figure. The chart on page 362 gives the area formulas for several polygons.

The area of a circle with radius r is $A = \pi r^2$ and its circumference is $C = 2\pi r$. If the radius of a circle is 4 cm, then the area is 16π cm^2 and the circumference is 8π cm.

The area A of a regular polygon is given by $A = \frac{1}{2}ap$, where a is the **apothem** of the polygon and p is the perimeter.

For example, each side of a regular heptagon is 5 in. and the distance from the center to each vertex is 5.8 in. To find the area:

Step 1 Sketch the polygon, and use the Pythagorean theorem to find the apothem.

$$2.5^2 + a^2 = 5.8^2$$
$$a^2 = 5.8^2 - 2.5^2$$
$$a \approx 5.2$$

The length of one leg of the triangle is half the length of a side of the heptagon.

Step 2 Use the formula for the area of a regular polygon.

$$A = \frac{1}{2}ap$$
$$\approx \frac{1}{2}(5.2)(7 \cdot 5)$$
$$\approx 91$$

The area of the heptagon is about 91 in.^{2}.

The **volume** of a solid is the number of cubic units needed to fill the solid. The volume of a prism or a **cylinder** is the product of its height and the area of its base, $V = Bh$. The **surface area** of a figure is the area of its net. The **lateral area** of a figure is the sum of the areas of all faces of the figure except its bases.

Volume	Surface Area
$V = Bh$ 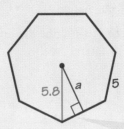 $= (\pi \cdot 49)(3)$ $= 147\pi$ ≈ 461.58 m^3	$S.A. = 2\pi rh + 2\pi r^2$ $= (2\pi \cdot 7)(3) + 2\pi \cdot 7^2$ $= 140\pi$ ≈ 439.60 m^2
$V = Bh$ $= (4 \cdot 8)(2)$ $= 64$ ft^3	$S.A. = ph + 2B$ $= (2 \cdot 4 + 2 \cdot 8)(2) + (2)(32)$ $= 112$ ft^2

7 Assessment

VOCABULARY QUESTIONS

1. **Open-ended Problem** Sketch a kite. Label the lengths of two noncongruent sides.

2. Complete: To write an ? proof, start by assuming that the conclusion is false.

3. Complete: The volume of a cylinder is equal to the area of the ? times the height.

4. **Writing** Describe the difference between the surface area of a prism and its lateral area.

SECTIONS 7.1, 7.2, and 7.3

Find each length or angle measure.

5. kite *ABCD*
   a. *AD*
   b. *m ∠ADC*

6. rectangle *WXYZ*
   a. *WY*
   b. *YZ*

7. square *LMNP*
   a. *LN*
   b. *MN*

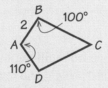

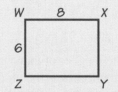

Sketch each quadrilateral. Mark the given information. If the quadrilateral is a parallelogram, explain why.

8. *JKLM* with ∠*J* ≅ ∠*L* and ∠*K* ≅ ∠*M*

9. *ABCD* with ∠*A* ≅ ∠*D* and ∠*B* ≅ ∠*C*

10. *WXYZ* with $\overline{WZ}$ ≅ $\overline{XY}$ and $\overline{WX}$ ≅ $\overline{YZ}$

11. Write an indirect proof to show that a kite cannot be a parallelogram.

Classify each quadrilateral. Be as specific as possible.

12.

13.

14.

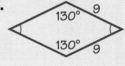

15. Given: $\overline{AB} \parallel \overline{DC}$
$\overline{AB} \cong \overline{DC}$

Prove: $\triangle AED \cong \triangle CEB$

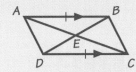

SECTIONS 7.4, 7.5, *and* 7.6

Find the area of each polygon.

16.

7

4

17.

2

4

6

18.

4 6

3

Find the area of each shaded region. Each polygon is a regular polygon.

19.

10

5

4.4

20.

4

3.4

21.

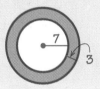

7

3

22. HISTORY In the third century B.C., an ice pit, or refrigerator, was built for the first Ch'in Emperor Shih Huang Ti. It was made of terra cotta rings about 5 ft 8 in. in diameter and 3 ft high, and extended 43 ft below the surface.

a. Find the lateral surface area of one terra cotta ring.

b. Find the volume of a cylindrical ice pit 43 ft deep and with diameter 5 ft 8 in.

23. The area of the base of a cylinder is 7 cm^2 and its height is 3 cm.

a. Find its volume.   **b.** Find its surface area.

24. A hexagonal prism is 8 in. high and the edges of a base of the prism are each 4 in. long.

a. Find its volume.   **b.** Find its lateral area.

PERFORMANCE TASK

25. Design a stained glass window using triangles and quadrilaterals. How do you know which quadrilaterals are rectangles, rhombuses, trapezoids, parallelograms, or squares? How would you find the area of each color of glass in your design? If you know the weight of 1 in.^{3} of glass, how can you find the total weight of the glass in the window?

8 Using Transformations

BLUEPRINTS FOR BALLGAMES

Terri Johnson

Were it not for Terri Johnson's love of baseball, she might never have started designing sports stadiums. Johnson graduated from architecture school in Chicago while construction was starting at the new White Sox stadium, Comiskey Park. "I really wanted to get involved in that project," she says. She contacted HOK Sport, the Kansas City architectural firm handling the job but did not land a position with them for two years. By then, construction on Comiskey Park was completed, though she did get to work on the stadiums where the Cleveland Indians, the Colorado Rockies, and the Carolina Panthers play.

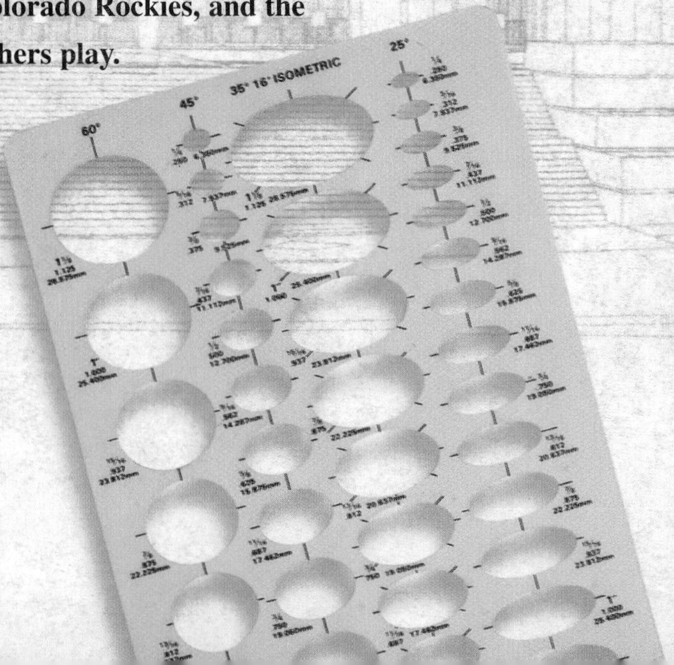

> " GEOMETRY IS A PRACTICAL TOOL FOR AN ARCHITECT. IT'S SOMETHING WE USE ALMOST EVERY DAY. "

A Grid Plan

Every project begins with a "grid geometry plan," an overall scheme that shows how the building will be laid out and supported structurally. Next come more advanced plans that contain the details contractors need to start building a stadium or sports arena. "We get very specific in our construction drawings, down to the team logos in the locker room carpet of Chicago's United Center." Johnson notes.

Building elements such as stairways, doors, windows, seats, and aisles are spelled out precisely in the final drawings submitted by HOK Sport. A computer is a crucial tool throughout this process, used in all designs, presentations, and supporting documents. "A knowledge of geometry and math is also essential for figuring out how three-dimensional spaces work," Johnson says.

Reflecting on Architecture

Symmetry is a basic element used in stadium designs. The entrance to Ericsson Stadium, for example, is symmetrical: If you draw a vertical line through the middle of the building's front facade, the left half and right half are mirror images of each other.

"We use mirror images to design concession stands as well," Johnson explains. "One or two are designed and then reflected over the stadium's lines of symmetry. This cuts down on design time and makes the stadium easier to build."

Clever Repetition

The Ericsson Stadium is divided into four quadrants. After designing Quadrant A, the architects can generate Quadrants B and D by reflecting Quadrant A over the horizontal and vertical axes shown in the diagram. Quadrant C is created by reflecting Quadrant A over both axes, one after the other. "Once you do that, the entire building is composed," Johnson notes.

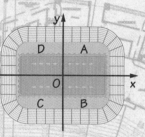

Building elements, such as a section of stairway, can be reproduced on paper via a process called *translation*, which involves reflecting a set of stairs twice over two parallel lines. The result is that a new stairway, identical to the first, appears in another position in the drawing. This process enables architects to use the same set of stairs in their designs and then translate it to various levels and locations in the building.

"WE HAVE TO TURN OUR DESIGNS INTO BUILDINGS. THAT DEPENDS ON GEOMETRY."

Explore and Connect

Terri Johnson uses an architect's scale to measure a drawing.

1. Writing Choose Quadrant B, C, or D in the stadium above. Explain how you can generate the entire stadium plan by reflecting your quadrant.

2. Research Find examples of symmetry in the architecture of other building types, such as schools, theaters, houses, or office buildings. Describe the symmetry you find in each building.

3. Project Build a model of a house or an apartment building that features reflection or rotational symmetry.

Mathematics & Terri Johnson

In this chapter, you will learn more about how mathematics is related to architecture.

Related Exercises

Section 8.1
- Exercises 10–13

Section 8.3
- Exercises 17, 18

Section 8.4
- Exercises 29–32

8.1

Using Reflections

Learn how to...

- **draw the reflection of a polygon**

- **draw a line of reflection for a figure and its image**

So you can...

- **use reflections to solve problems, such as how to project a slide show**

A canoe designer takes great care in shaping the *hull,* the part that touches the water, so that the canoe will be both stable and fast. Because canoes have reflection symmetry, a designer needs to draw only one half of the hull in detail. The other half can be drawn by reflecting the design over the line of symmetry, which is also called the **line of reflection**.

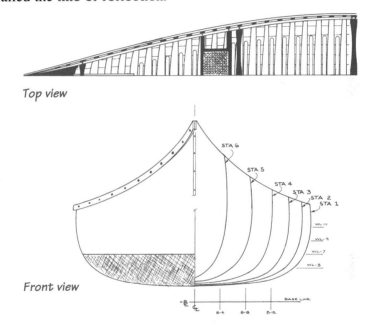

Top view

Front view

Boat builders like Rollin Thurlow, who designed this canoe, call the line of reflection the *centerline* of the boat.

A **reflection** is a transformation in which each point of a figure has an **image** that is the same distance from the line of reflection as the original point. A point and its image lie on opposite sides of the line of reflection unless the point is on the line of reflection itself.

THINK AND COMMUNICATE

1. In the front view of the canoe design above, is the line of reflection horizontal or vertical?

2. Copy the diagram of the front view of the canoe. Mark a point that is *not* on the line of reflection. Describe where the image of that point lies on the diagram.

3. Mark a point on the canoe design that lies on the line of reflection. Describe where the image of that point lies.

Reflecting a Polygon Over a Line

Work in a group of four students.
You will need:
- patty paper, tracing paper, or wax paper
- a ruler and a protractor

1 Fold your paper to make a crease.
Label it line *m*.

2 Draw △*ABC* on one side of line *m*. Reflect
points *A*, *B*, and *C* over line *m* by folding.

3 Mark the image points *A′*, *B′*, and *C′*.
Draw △*A′B′C′*.

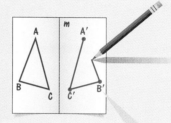

4 Draw $\overline{AA'}$, $\overline{BB'}$, and $\overline{CC'}$. Label the points where these
segments intersect line *m* as *P*, *Q*, and *R*, respectively.

5 Compare the lengths *AP* and *PA′*. Compare *BQ* and *QB′*.
Compare *CR* and *RC′*. What do you notice?

6 Measure ∠*APR*, ∠*BQP*, and ∠*CRQ*. What do you notice?

7 How is line *m* related to $\overline{AA'}$, $\overline{BB'}$, and $\overline{CC'}$?

8 Compare *AB* and *A′B′*. Compare *BC* and *B′C′*. Compare *CA*
and *C′A′*. What do you know about △*ABC* and △*A′B′C′*?

> The image of point *B*
> is often called point
> *B′* (read "*B* prime").

In the Exploration, if you look at the vertices of △*ABC* in alphabetical order,
you see that they are arranged in a *counterclockwise orientation*. The vertices
of △*A′B′C′* are arranged in a *clockwise orientation*.

counterclockwise   **clockwise**
orientation   **orientation**

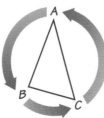

Properties of Reflections

A reflection is a transformation that preserves congruence.

A reflection changes orientation.

The line of reflection is the perpendicular bisector of every segment that connects a point and its image.

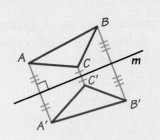

EXAMPLE 1

Copy the diagram and draw the image of △*XYZ* after reflection over line *k*.

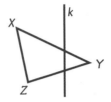

SOLUTION

Reflect each vertex of △*XYZ*. Label the image vertices X', Y', and Z'.

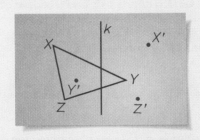

Connect the image vertices to draw the image △$X'Y'Z'$.

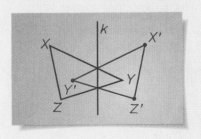

THINK AND COMMUNICATE

4. In Example 1, is the orientation of △*XYZ* counterclockwise or clockwise?

5. What is the orientation of △$X'Y'Z'$?

6. Line *k* is the perpendicular bisector of $\overline{XX'}$. Name two other segments that have line *k* as a perpendicular bisector.

EXAMPLE 2

The diagram shows quadrilateral *JKLM* and
its image after a reflection. Draw the line
of reflection.

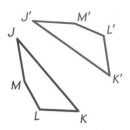

SOLUTION

Draw a segment connecting vertex *K*
and its image *K′*. The line of reflection
is the perpendicular bisector of this
segment, line *n*.

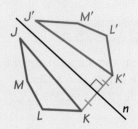

☑ CHECKING KEY CONCEPTS

In the diagram, *P′Q′R′S′* is the image of quadrilateral *PQRS*
after a reflection over line *j*.

1. What is the image of $\overline{PQ}$? of $\overline{RS}$?

2. Name a triangle congruent to △*QRS*.

3. Name a triangle congruent to △*SPQ*.

4. Name two angles in the diagram that
have the same measure.

5. What is the image of *P′Q′R′S′* after a
reflection over line *j*?

6. What are the orientations of *PQRS* and *P′Q′R′S′*?

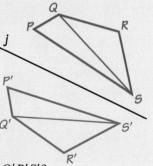

8.1 Exercises and Applications

Extra Practice
exercises on
page 675

Copy each diagram. Sketch the image of the polygon after a reflection over
the given line. Label the image polygon.

1.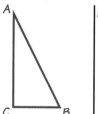

2.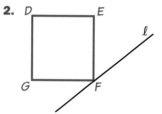

3.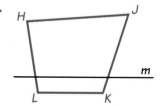

Each diagram shows a polygon and its image after a reflection. Copy each diagram and draw the line of reflection.

4.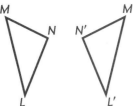

5.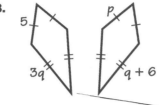

6.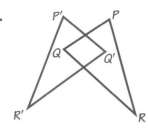

ALGEBRA Each diagram shows a polygon and its image after a reflection. Find the value of each variable.

7.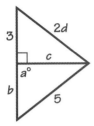

8.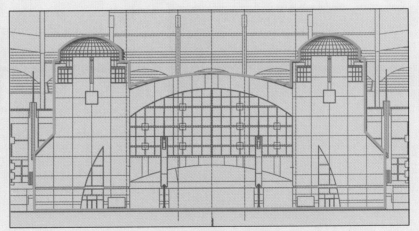

9.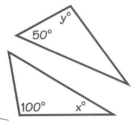

INTERVIEW Terri Johnson

Look back at the article on pages 388–390.

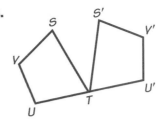

Terri Johnson reviews a presentation of the stadium.

The diagram shows the entrance to Ericsson Stadium, which Terri Johnson helped develop from the design.

10. Trace the items outlined in green on a piece of paper.

11. Fold your paper to find the line of symmetry for the stadium entrance.

12. Open-ended Problem Choose any shape on one side of the entrance. Find its image on the other side.

13. Writing The towers on each side of the entrance are chambers for stairwells. What parts of one tower have reflection symmetry? What parts do not?

For Exercises 14–17, use block letters, as shown.

14. If you write the word OBOE in block letters, as shown, it has a horizontal line of symmetry. What other letters have this kind of symmetry? Use any of these letters to write a word.

15. If you write the word AIM in block letters, as shown, it has a vertical line of symmetry. What other letters have this kind of symmetry? Use any of these letters to write a word.

16. What letters have both horizontal and vertical reflection symmetry?

17. What letters have neither horizontal nor vertical reflection symmetry?

Connection ▶ **AUDIO-VISUAL PRESENTATION**

Rear projection is a way to set up a slide projector so that its picture shines on the back of a screen rather than on the front. Many stages or auditoriums are not large enough to put a projector directly behind the screen, in the position marked *virtual projector* in the diagram. Instead, a mirror is used to reflect the image from the *actual projector* to the screen.

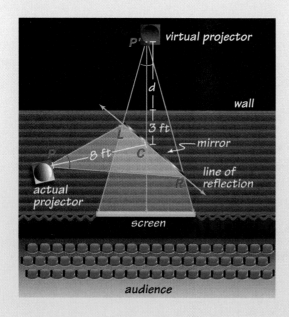

18. What is the image of *P* after the reflection in the mirror?

19. What are the images of *L*, *C*, and *R*, points that lie on the mirror?

20. Name three pairs of congruent triangles shown in the diagram.

21. Explain why $PC = P'C$.

22. Using the distances in the diagram, estimate *d*, the extra stage depth that would be needed if the mirror method were not used.

23. 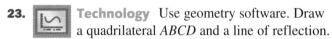 **Technology** Use geometry software. Draw a quadrilateral *ABCD* and a line of reflection.

a. Reflect *ABCD* over the line of reflection. Label its vertices.

b. Measure the corresponding segments and angles of *ABCD* and its image.

c. Explain how you can use the definition of congruent polygons and the results of part (b) to conclude that *ABCD* and its image are congruent.

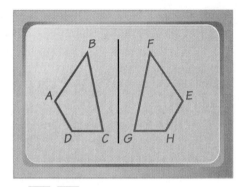

24. Challenge Suppose $\triangle ABC \cong \triangle A'B'C'$ and the midpoints of $\overline{AA'}$, $\overline{BB'}$, and $\overline{CC'}$ lie on a line. Is the line a line of reflection? Make a sketch and explain your answer.

25. Open-ended Problem Make an interesting design that uses a polygon and several reflections. Label the vertices of the polygon and its images and name some congruent polygons.

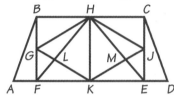

Find the area of the shaded region in each regular polygon or circle.
(Section 7.5)

26.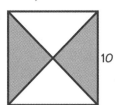

27.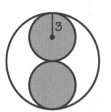

28.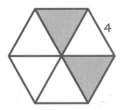

Tell which triangle appears to be congruent to the given triangle. *(Section 6.2)*

29. $\triangle ABF \cong$?

30. $\triangle FGH \cong$?

31. $\triangle JME \cong$?

32. $\triangle CEH \cong$?

Write an equation for each line described. *(Section 4.2)*

33. contains (3, 4); has slope 2

34. contains (−3, 4); has slope −2

35. the horizontal line that contains (8, 7)

36. the vertical line that contains (8, 7)

8.2 Reflections with Coordinates

Learn how to...

- **find the coordinates of a polygon reflected over the x-axis, the y-axis, or the line y = x**

- **describe how coordinates change after reflections**

So you can...

- **explore situations involving reflections that can be modeled on a coordinate plane**

When a polygon is placed on a coordinate plane, you can describe a reflection of the polygon by comparing the coordinates of the polygon and the coordinates of image vertices. Notice what happens to the coordinates of the triangle shown below as it is reflected over the *y*-axis.

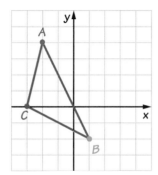 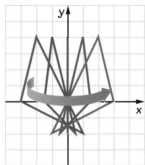 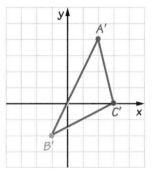

$$A(-2, 4) \rightarrow A'(2, 4)$$
$$B(1, -2) \rightarrow B'(-1, -2)$$
$$C(-3, 0) \rightarrow C'(3, 0)$$

The arrow symbol means "goes to."

THINK AND COMMUNICATE

1. When a point is reflected over the *y*-axis, which of its coordinates changes? How does it change?

2. Choose a point on $\triangle ABC$ that is not a vertex. Estimate its coordinates. Find the coordinates of its image after a reflection over the *y*-axis.

3. Which points on $\triangle ABC$ do not change position when they are reflected over the *y*-axis? Estimate the coordinates of these points.

EXAMPLE 1

Reflect quadrilateral *WXYZ* over the *x*-axis. Give the coordinates of the vertices of the image.

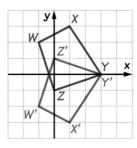

SOLUTION

When a point is reflected over the *x*-axis, only the *y*-coordinate changes.

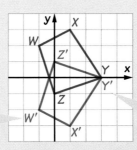

W is 2 units *above* the *x*-axis, so *W'* must be 2 units *below* the *x*-axis.

Points that are on the line of reflection do not change their position.

$$W(-1, 2) \rightarrow W'(-1, -2)$$
$$X(1, 3) \rightarrow X'(1, -3)$$
$$Y(3, 0) \rightarrow Y'(3, 0)$$
$$Z(0, -1) \rightarrow Z'(0, 1)$$

Notice in the Solution to Example 1 that each segment joining a point and its image is vertical. If the reflection were over the *y*-axis, such segments would be horizontal.

Reflecting Over the Coordinate Axes

When a point (a, b) is reflected over a coordinate axis, one coordinate changes to its opposite and the other coordinate does not change.

reflection over *x*-axis   reflection over *y*-axis

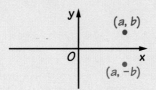

$(a, b) \rightarrow (a, -b)$ $(a, b) \rightarrow (-a, b)$

You can reflect points over lines other than the coordinate axes, but it is not usually easy to write a simple rule showing what happens to the coordinates of a point (a, b). One exception is reflection over the line $y = x$. For reflections over this line, all you have to do is switch the coordinates to find the image of a point.

Reflecting Over the Line $y = x$

When a point (a, b) is reflected over the line $y = x$, the coordinates are reversed.

reflection over the line $y = x$

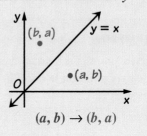

$$(a, b) \rightarrow (b, a)$$

EXAMPLE 2

Reflect △JKL over the line _y_ = _x_. Give the coordinates of the vertices of the image.

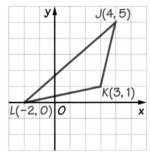

SOLUTION

Draw the line $y = x$. Reverse the coordinates of each point to find its image.

$$J(4, 5) \rightarrow J'(5, 4)$$
$$K(3, 1) \rightarrow K'(1, 3)$$
$$L(-2, 0) \rightarrow L'(0, -2)$$

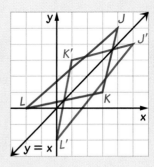

THINK AND COMMUNICATE

4. Use the coordinates of the points in Example 2 to check that the slopes of $\overline{JJ'}$, $\overline{KK'}$, and $\overline{LL'}$ are all -1. Explain why this is so.

5. Find the coordinates of the midpoint of $\overline{KK'}$. Explain how you know that the point is on the line $y = x$.

Copy the diagram and reflect *PQRS* over each line. Give the coordinates of the vertices of the image.

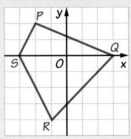

1. the *x*-axis

2. the *y*-axis

3. the line $y = x$

8.2 | **Exercises and Applications**

Extra Practice
exercises on
page 675

Copy each polygon and reflect it over the *y*-axis. Give the coordinates of the vertices of the image.

1.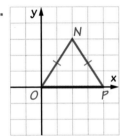

2.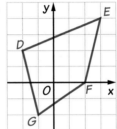

3.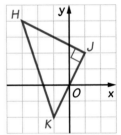

Copy each polygon and reflect it over the *x*-axis. Give the coordinates of the vertices of the image.

4.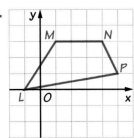

5.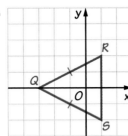

6.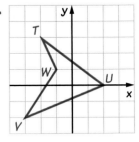

Copy each polygon and reflect it over the line $y = x$. Give the coordinates of the vertices of the image.

7.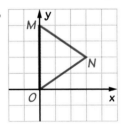

8.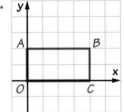

9.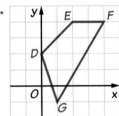

10. ALGEBRA Choose one of the diagrams for Exercises 7–9 on the previous page. Find the midpoints of the segments that join each vertex to its image. Show that each midpoint lies on the line $y = x$.

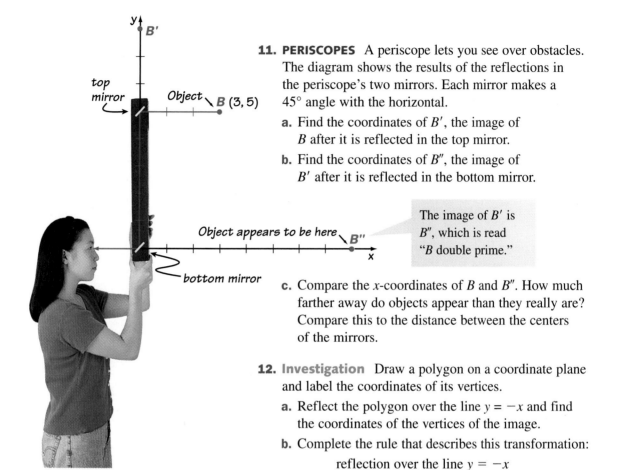

11. PERISCOPES A periscope lets you see over obstacles. The diagram shows the results of the reflections in the periscope's two mirrors. Each mirror makes a 45° angle with the horizontal.

a. Find the coordinates of B', the image of B after it is reflected in the top mirror.

b. Find the coordinates of B'', the image of B' after it is reflected in the bottom mirror.

> The image of B' is B'', which is read "B double prime."

c. Compare the x-coordinates of B and B''. How much farther away do objects appear than they really are? Compare this to the distance between the centers of the mirrors.

12. Investigation Draw a polygon on a coordinate plane and label the coordinates of its vertices.

a. Reflect the polygon over the line $y = -x$ and find the coordinates of the vertices of the image.

b. Complete the rule that describes this transformation:

reflection over the line $y = -x$
$$(a, b) \rightarrow (\underline{\ ?\ }, \underline{\ ?\ })$$

13. OPTICS A person with an eye at E looks at points M and N on a mirror and sees the top and bottom of his face at T' and B', as shown.

a. Use the information in the diagram and what you know about reflections and the midpoints of two sides of a triangle to show that $MN = \frac{1}{2}TB$. (*Hint:* See Exercises 12–15 on page 199.)

b. What does your answer to part (a) suggest is a good size for a hand-held mirror?

c. Open-ended Problem Draw and label a sketch that shows how you can see a reflection of your entire body in a mirror that is half your height.

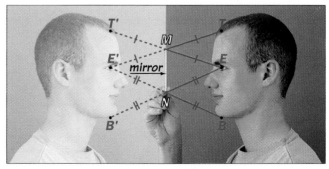

In miniature golf, when a ball with no spin hits a bumpboard, its *angle of reflection* is congruent to its *angle of incidence*. In the diagram below, adapted from a real course,

$$m\angle QPA = m\angle OPH.$$

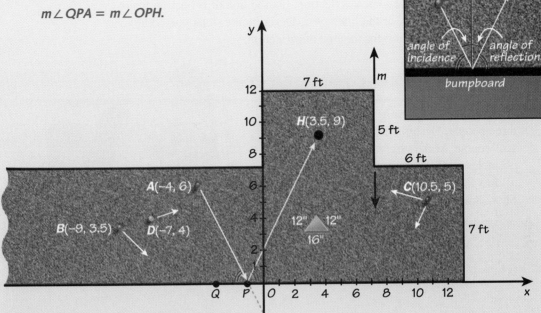

For Exercises 14–19, make a large copy of the diagram on graph paper.

14. Explain why a player at $A(-4, 6)$, $B(-9, 3.5)$, $C(10.5, 5)$, or $D(-7, 4)$ cannot aim directly at the cup at $H(3.5, 9)$.

15. *Writing* Alisa, the player at A, wants to reach H by bouncing a ball off the bumpboard along the x-axis. Explain why aiming at J (the image of H after reflection over the x-axis) is a good idea.

16. Basha, the player at B, aims at J. Draw the path of the ball. What are the coordinates of the point where the ball hits the bumpboard?

17. Carlos, at C, aims at J. Draw the path of the ball. What are the coordinates of the point where the ball hits the bumpboard?

18. Suppose Carlos decides instead to bounce the ball against the bumpboard along the y-axis. Reflect H across the y-axis. Label the image K and give its coordinates. Draw the path of the ball. What are the coordinates of the point where the ball hits the bumpboard?

19. *Challenge* Show how Dianne, the player at D, can bounce a ball off the bumpboard that lies along line m so that it reaches H.

20. 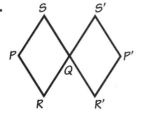 **Technology** Write a program for your graphing calculator that asks for the coordinates of a point and gives the coordinates of the images of that point after reflection in the *x*-axis, the *y*-axis, and the line *y* = *x*.

21. Writing Suppose you know the coordinates of the vertices of a polygon and its image after a reflection, but the line of reflection is not the *x*-axis, the *y*-axis, or the line *y* = *x*. Describe the steps you could take to find an equation for the line of reflection.

ONGOING ASSESSMENT

22. Cooperative Learning Work in a group of three people. Design a hole for a miniature golf course. Make sure your design has obstacles so that a player must bounce a ball against the bumpboards or other surfaces. Show how it is possible to get a hole in one.

SPIRAL REVIEW

Each diagram shows a polygon and its image after a reflection. Copy each diagram and draw the line of reflection. *(Section 8.1)*

23.

24.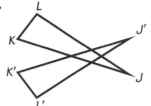

25.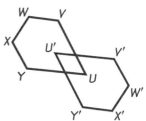

Draw a figure like the one shown. Then complete the construction. *(Section 5.7)*

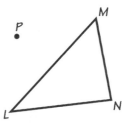

26. Construct the perpendicular bisector of $\overline{MN}$.

27. Construct an angle congruent to $\angle N$.

28. Construct the line through *P* that is parallel to $\overleftrightarrow{LM}$.

Sketch the front, top, and side view of a prism with the given base. Then sketch the prism. *(Section 2.6)*

29. triangular

30. square

31. rectangular

32. hexagonal

8.3

Translations

When you stand between two parallel mirrors, you see many reflections of yourself. Half of them face you and half face the other way. This is a way to see the back of your head! In the Exploration, you will see what happens when you transform a figure by reflecting it twice across parallel lines.

Learn how to...

- **find the coordinates of the image of a figure after a translation**
- **describe a translation based on an original figure and its image**

So you can...

- **describe patterns that involve translation, such as architectural decoration**

EXPLORATION
COOPERATIVE LEARNING

Comparing Translated Polygons

Work in a group of four students.
You will need:

- geometry software or patty paper
- a ruler

1 Draw points *P* and *Q* on a horizontal line and draw vertical lines through them. Label the lines *m* and *n*. Draw △*ABC* to the left of line *m*.

2 Reflect △*ABC* over line *m* and label the image △*A′B′C′*. Reflect △*A′B′C′* over line *n* and label the image △*A″B″C″*.

> The image of **A′** is **A″**, which is read "*A* double prime."

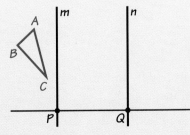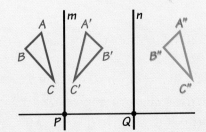

3 What are the orientations of △*ABC* and △*A″B″C″*?

4 Measure *AA″*, *BB″*, *CC″*, and *PQ*. What do you notice?

5 Change the distance between lines *m* and *n* and see what happens to the distances *AA″*, *BB″*, and *CC″*. What do you notice?

If you repeat the Exploration using nonvertical lines, you can compare the coordinates of the vertices of $\triangle ABC$ and $\triangle A''B''C''$ and see that each vertex is shifted the same distance in the same direction.

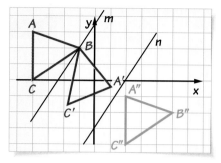

$A(-4, 3) \rightarrow A''(2, -1)$

$B(-1, 2) \rightarrow B''(5, -2)$

$C(-4, 0) \rightarrow C''(2, -4)$

In the diagram above, for example, each vertex is shifted 6 units to the right and 4 units down. This kind of transformation is called a *translation*.

Properties of Translations

A **translation** is a transformation that can be described in coordinate notation this way:

$$(a, b) \rightarrow (a + h, b + k)$$

Every point shifts h units horizontally and k units vertically.

A translation preserves congruence.

A translation preserves orientation.

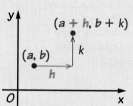

EXAMPLE 1

Sketch the image of $\triangle XYZ$ after the translation $(a, b) \rightarrow (a - 3, b + 4)$.

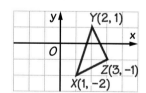

SOLUTION

Substitute and simplify to find the coordinates of the vertices of the image.

$X(1, -2) \rightarrow (1 - 3, -2 + 4) = X'(-2, 2)$

$Y(2, 1) \rightarrow (2 - 3, 1 + 4) = Y'(-1, 5)$

$Z(3, -1) \rightarrow (3 - 3, -1 + 4) = Z'(0, 3)$

Plot the vertices of the image triangle and connect them.

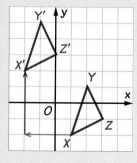

In Chapter 1 you saw that a design has *translational symmetry* if you can translate it so that the design and its image coincide.

EXAMPLE 2

Describe how to translate the pattern so that it matches itself.

SOLUTION

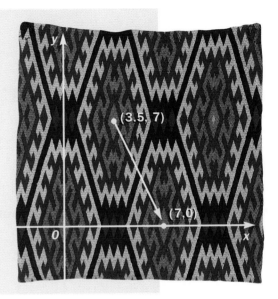

(3.5, 7)

(7, 0)

Step 1 Find the coordinates of any point in the design and a corresponding point in a nearby image.

$$P(3.5, 7) \qquad P'(7, 0)$$

Step 2 Subtract the coordinates to find the horizontal change h and vertical change k.

$$h = 7 - 3.5 \qquad k = 0 - 7$$
$$= 3.5 \qquad\qquad = -7$$

Step 3 Use the values of h and k to describe the translation using coordinate notation:

$$(a, b) \rightarrow (a + 3.5, b - 7)$$

THINK AND COMMUNICATE

1. In Example 2, why is it sufficient to pick a single point and its image?

2. If you translate the design the same distance as in the Solution above but in the opposite direction, does the design exactly match itself? Describe that translation using coordinate notation.

3. Are there other translations that move the design onto itself? Give one or two examples and a general description of these translations.

☑ CHECKING KEY CONCEPTS

Copy the diagram and draw the image of *ABCD* after each translation.

1. $(a, b) \rightarrow (a + 2, b - 5)$

2. $(a, b) \rightarrow (a - 3, b)$

3. $(a, b) \rightarrow (a, b + 4)$

4. $(a, b) \rightarrow (a - 2, b + 5)$

Describe each translation using coordinate notation.

5. Every point moves to the left 3 units and up 4 units.

6. Every point moves to the right 5 units.

8.3 Exercises and Applications

Extra Practice
exercises on
page 676

Copy EFGH and draw its image after each translation.

1. $(a, b) \rightarrow (a - 1, b + 3)$

2. $(a, b) \rightarrow (a, b + 7)$

3. $(a, b) \rightarrow (a + 5, b - 2)$

4. $(a, b) \rightarrow (a - 4, b)$

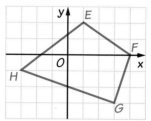

Describe each translation using coordinate notation.

5. Every point moves to the left 7 units and up 7 units.

6. Every point moves to the left 6 units.

7. Every point moves up 4 units.

8. Every point moves to the right 2 units and down 1 unit.

The image of each polygon after a translation is shown in red. Describe each translation using words or coordinate notation.

9.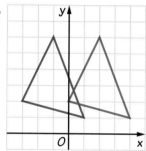

10.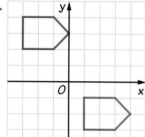

11.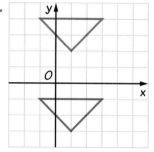

12. **COMPUTERS** On some computers you can design a square 8 pixels wide by 8 pixels high that is translated across the screen to create a background pattern. (A *pixel* is the smallest dot on a computer screen.) Square A is at the upper left corner of the computer screen. The translation

$$(a, b) \rightarrow (a + 8, b)$$

moves square A onto square B.

a. Describe the translations that move square A onto squares C, D, E, and F.

b. **Open-ended Problem** Create your own 8 pixel by 8 pixel pattern on graph paper. Translate it several times to show the background pattern it generates.

13. **SAT/ACT Preview** The distance between a point and its image after the translation $(a, b) \rightarrow (a + 7, b - 5)$ is:

A. 7   B. 5   C. 12   D. $\sqrt{74}$ E. $\sqrt{24}$

Open-ended Problems **Each pattern has translational symmetry. Describe how to translate the pattern so that it matches itself.**

14.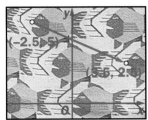

(−2.5, 5)

(5, 2.5)

15.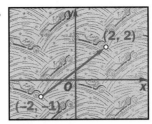

(2, 2)

(−2, −1)

16.  **Technology** Use geometry software to repeat the steps of the Exploration on page 405: Draw parallel lines m and n and $\triangle ABC$. Reflect the triangle over line m and then reflect the result over line n.

  a. Measure AA'', BB'', and CC''.

  b. Change the shape of $\triangle ABC$ by moving its vertices. What happens to AA'', BB'', and CC''?

  c. **Writing** Change the position of $\triangle ABC$ by moving it around on the coordinate plane. Explain what happens to $\triangle A''B''C''$. Does it matter which side of line m or n $\triangle ABC$ is on?

INTERVIEW Terri Johnson

Look back at the article on pages 388–390.

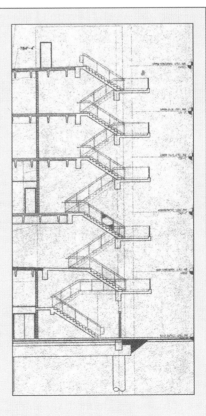

This cross-section view of Ericsson Stadium shows some of the staircases that sports fans climb to reach the tiers of seats.

17. Explain how the top flight of stairs can be translated to become the next lower flight of stairs.

18. Can the bottom flight of stairs be translated to become the next higher flight of stairs? Explain.

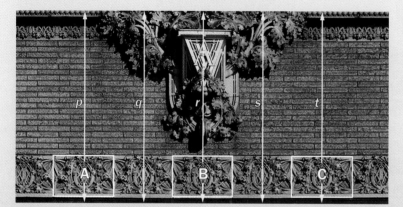

Louis H. Sullivan (1856–1924) was an influential architect who designed some of the earliest skyscrapers in Chicago, Illinois. Sullivan's buildings often had features inspired by plant forms, which he called "organic ornamentation."

The photograph shows terra cotta ornamentation on the Van Allen & Son store in Clinton, Iowa. Reflection lines have been added.

BY THE WAY

Louis Sullivan was the first to use the phrase "form follows function," which modern architects use to describe designs based primarily on usefulness rather than appearance.

19. Name two reflection lines that can be used to translate Block A onto Block C.

20. Name two reflection lines that can be used to translate Block A onto Block B.

21. Open-ended Problem Show that there is more than one correct answer to Exercises 19 and 20.

22. Open-ended Problem Look through some clothing catalogs or newspapers to find three examples of patterns that have translational symmetry. Cut them out and attach them to a piece of paper to show in class. Do any of the patterns also have reflection symmetry?

23. Research A *round*, such as "Row, Row, Row Your Boat," is a song that includes a melody that repeats later in the piece. In classical music, this idea is also used in *canons* and *fugues*. Find some sheet music that displays these forms and explain how the concept of "translation" can be used to describe them.

24. Challenge In the Exploration, you saw that a double reflection over two *parallel* lines is a translation. What kind of transformation happens if you use a double reflection over two *intersecting* lines? Make a sketch that illustrates your answer.

25. a. On graph paper, draw a triangle and label it $\triangle ABC$. Translate $\triangle ABC$ using Translation 1. Find the coordinates of the vertices of $\triangle A'B'C'$.

Translation 1: $(a, b) \rightarrow (a + 3, b + 5)$

b. Translate $\triangle A'B'C'$ using Translation 2. Find the coordinates of the vertices of $\triangle A''B''C''$.

Translation 2: $(a, b) \rightarrow (a - 4, b + 2)$

c. Use coordinate notation to describe a single translation that moves $\triangle ABC$ directly to $\triangle A''B''C''$.

d. Writing Repeat parts (a)–(c) using Translation 2 first, followed by Translation 1. Explain why the order in which you combine two translations does not affect the results.

Tell whether the triangle with sides of the given lengths is *right, obtuse,* or *acute.* *(Section 3.7)*

26. 3, 5, 6   **27.** 2, 7, 7   **28.** 12, 16, 20

29. Draw a pentagon and use a protractor to measure each angle, to the nearest degree. *(Section 1.6)*

ASSESS YOUR PROGRESS

VOCABULARY

line of reflection (p. 391)   **image** (p. 391)
reflection (p. 391)   **translation** (p. 406)

Use the diagram for Exercises 1–4. *K'L'M'N'* is the image of *KLMN.* *(Section 8.1)*

1. What is the image of $\overline{LM}$? of $\overline{KL}$?

2. Name a triangle congruent to $\triangle KMN$.

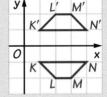

Copy the polygon *KLMN.* Reflect it over each line. Give the coordinates of the vertices of the image. *(Section 8.2)*

3. the line $y = x$ **4.** the y-axis

Describe each translation using coordinate notation. *(Section 8.3)*

5. Every point moves to the right 5 units and down 7 units.

6. Every point moves to the left 3 units and up 4 units.

7. Journal Describe how the number of reflections affects the orientation of a figure that is reflected more than once. Make a sketch that illustrates your conclusions.

8.4 | Applying Rotations

When sails are attached to a windmill, the tilted frames catch the wind and turn around a fixed point.

Learn how to...

- **rotate a figure around a center of rotation**
- **find the coordinates of the vertices of a polygon that has been rotated 90°, 180°, or 270° around the origin**

So you can...

- **describe rotations, such as those found in patterns on quilts**

A transformation in which every point moves along a circular path around a fixed point is called a **rotation**. The fixed point is called the **center of rotation**. Segments drawn from a point and its image to the center of rotation always form the same angle, the **angle of rotation**.

In this book, rotations are measured counterclockwise.

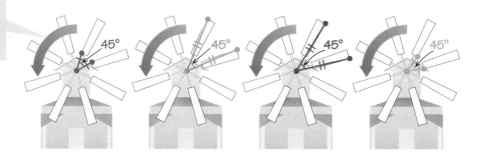

BY THE WAY

In the Netherlands, windmill owners positioned their sails to convey information. Stopping them just before the uppermost sail was vertical meant a celebration of good news.

Properties of Rotations

A rotation is a transformation that preserves congruence.

A rotation preserves orientation.

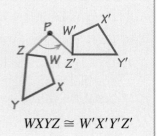

$$WXYZ \cong W'X'Y'Z'$$

You can use a protractor and either a ruler or a compass to help you find the images of points and figures after a rotation.

EXAMPLE 1

Draw the image of the given polygon after a 110° rotation with center P.

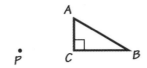

SOLUTION

Step 1 Draw a segment from vertex *A* to the center of rotation point *P*.

Step 2 Measure a 110° angle counterclockwise. Draw a ray.

Step 3 Use a ruler or compass to locate *A'* along the ray so that *PA'* = *PA*.

Step 4 Repeat Steps 1–3 for each vertex. Connect the vertices to form the image polygon.

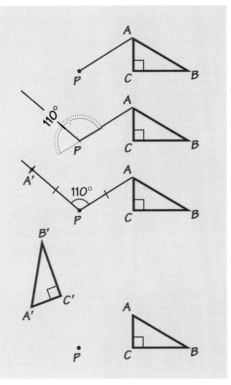

If you rotate a point 90°, 180°, and 270° around the origin on a coordinate plane, you will notice some patterns in the coordinates of the images.

In the diagram at the right, the point (2, 3) has been rotated three times.

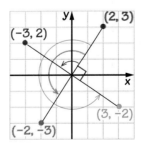

Rotating a Point 90°, 180°, or 270° Around the Origin

When a point (a, b) is rotated counterclockwise around the origin:

90° rotation: $(a, b) \rightarrow (-b, a)$

180° rotation: $(a, b) \rightarrow (-a, -b)$

270° rotation: $(a, b) \rightarrow (b, -a)$

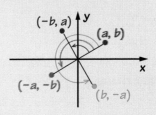

EXAMPLE 2

Sketch the image of the given polygon after each rotation around the origin.

a. **90°**   b. **180°**

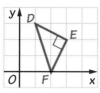

SOLUTION

Substitute to find the coordinates of the vertices of each image.

a. In a 90° rotation around the origin, $(a, b) \rightarrow (-b, a)$.

$D(1, 3) \rightarrow D'(-3, 1)$
$E(3, 2) \rightarrow E'(-2, 3)$
$F(2, 0) \rightarrow F'(0, 2)$

b. In a 180° rotation around the origin, $(a, b) \rightarrow (-a, -b)$.

$D(1, 3) \rightarrow D'(-1, -3)$
$E(3, 2) \rightarrow E'(-3, -2)$
$F(2, 0) \rightarrow F'(-2, 0)$

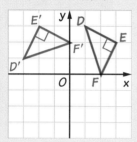   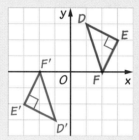

THINK AND COMMUNICATE

1. Sketch the image of $\triangle DEF$ in Example 2 after a 270° rotation.

2. A rotation of 180° is sometimes called a **half-turn**. What could a 90° rotation be called?

3. The center of a 180° rotation is the midpoint of every segment that connects a point and its image. Explain why this is true.

☑ CHECKING KEY CONCEPTS

Name the image of each triangle after the rotation around the origin.

1. Rotate $\triangle ABC$ 90°.

2. Rotate $\triangle DEF$ 180°.

3. Rotate $\triangle JKL$ 270°.

4. Rotate $\triangle GHI$ 180°.

Give the coordinates of the vertices of each triangle.

5. $\triangle DEF$ 6. $\triangle GHI$ 7. $\triangle JKL$

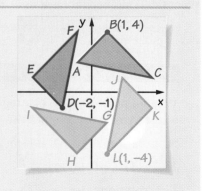

8.4 Exercises and Applications

Extra Practice
exercises on
page 676

Estimate the angle of rotation involved in each situation.

1. turning a doorknob

2. bending your arm at the elbow

3. looking over your shoulder

4. opening a jar of peanut butter

Copy each diagram. Draw the image of the polygon after a rotation with the given measure and center *P*.

5. rotate 45°

P•

6. rotate 120°

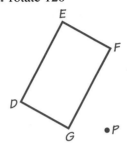

•P

7. rotate 70°

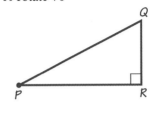

Copy each diagram and sketch the image of each polygon after a 90° rotation around the origin.

8.

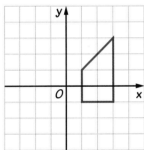

9.

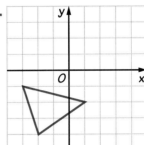

10.

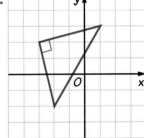

A figure has *rotational symmetry* if it looks the same after a rotation that is less than 360°. Does each object have rotational symmetry? If so, name each angle of rotation that results in an image that matches the original figure.

11.

12.

13.

14. Open-ended Problem Find an object in your home or at school that has rotational symmetry. Make a sketch and describe the symmetry.

15. FURNITURE DESIGN Furniture designer Marc Desplaines of San Francisco built this end table with drawers that are rotated in opposite directions. Two legs support each drawer.

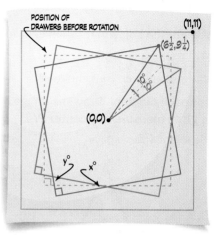

a. The lower square drawer, shown in red, is rotated 10° away from the gray position. One corner is at $\left(6\frac{1}{2}, 9\frac{1}{4}\right)$. Find the coordinates of the other three corners.

In the top view above, measurements are given to the nearest quarter inch.

b. The upper square drawer, shown in green, is rotated −10° away from the gray position. Use what you know about reflections to find the coordinates of its corners.

c. Because the drawers are rotated, you can see a triangular corner of the lower drawer. Find the values of x and y in the triangle.

Copy each polygon on polar graph paper. Draw its image after the given rotation.

16. 50° counterclockwise

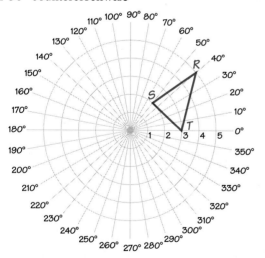

17. 120° counterclockwise

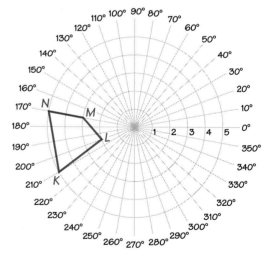

18. Writing The *polar coordinates* of △RST in Exercise 16 are R(5, 40°), S(2, 50°), and T(3, 0°).

a. Write the polar coordinates of △R′S′T′, the image triangle you drew in Exercise 16.

b. Describe how the polar coordinates of △RST and △R′S′T′ are related to each other.

Elizabeth Akana, who lives on the island of Oahu, is an expert in the art of Hawaiian quilting. These quilts are made by folding a square piece of cloth in half three times to form a triangle, and then cutting along the fold lines. When the design is unfolded, it has four lines of symmetry, as shown. The design is then sewn onto other cloth using a pattern of stitches that ripples outward.

The quilt design is based on eight congruent triangles. Describe the rotation that moves each triangle as indicated, or write *no rotation*.

19. Triangle 1 onto Triangle 3   **20.** Triangle 2 onto Triangle 6

21. Triangle 7 onto Triangle 8   **22.** Triangle 4 onto Triangle 2

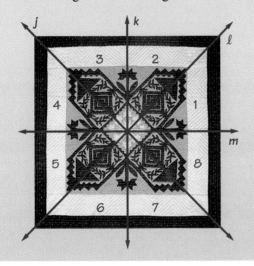

23. Investigation Choose two of the lines of symmetry on the quilt diagram that intersect at an angle of 45°.

  a. What happens to Triangle 1 if it is reflected over one line and the image is reflected over the other line?

  b. Describe the single rotation that has the same effect as the reflections in part (a).

  c. Repeat parts (a) and (b) using two other lines that intersect at an angle of 45°. What do you notice?

24. **Technology** Use geometry software. Draw two intersecting lines and $\triangle ABC$. Reflect $\triangle ABC$ over one line, then reflect the image, $\triangle A'B'C'$, over the other line.

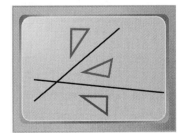

  a. How is the image $\triangle A''B''C''$ related to the original $\triangle ABC$?

  b. Change the angle at which the lines intersect. How is the measure of the angle related to the angle of the rotation you see? Write a conjecture.

Sketch the image of $\overline{AB}$ after a 180° rotation around the origin.

25. 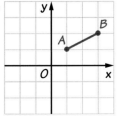   **26.** 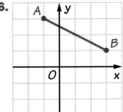   **27.**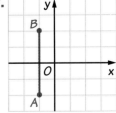

28. Choose one of the sketches you made in Exercises 25–27. Use slopes to show that quadrilateral $ABA'B'$ is a parallelogram.

Look back at the article on pages 388–390.

The diagram shows a simplified drawing of the Ericsson Stadium seating sections. After designing one quadrant of seating, the architects reflect the design over horizontal and vertical axes.

29. What rotation moves quadrant A to quadrant C?

30. What rotation moves quadrant B to quadrant D?

31. Does a 90° rotation move quadrant D to quadrant C? Explain.

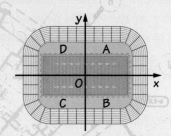

32. Explain how your answers to Exercises 29 and 30 are related to the results you found in Exercise 24.

33. Challenge Suppose you are given only a figure and its image after a rotation. How can you use perpendicular bisectors to find the center of the rotation? Make a sketch and justify your method.

ONGOING ASSESSMENT

34. a. Writing If you reflect a polygon over a line, can you rotate the image back onto the original polygon? Explain why or why not.

   b. Writing If you translate a polygon, can you rotate the image back onto the original polygon? Explain why or why not.

SPIRAL REVIEW

Describe each translation using coordinate notation. *(Section 8.3)*

35. Every point moves to the left 3 units and up 6 units.

36. Every point moves to the right 2 units and down 6 units.

Name a pair of triangles that must be congruent in order to prove each statement. *(Section 6.5)*

37. $\overline{FD} \cong \overline{FE}$ **38.** $\angle BDF \cong \angle BEF$

39. $\overline{AH} \cong \overline{CG}$ **40.** $\angle ABG \cong \angle CBH$

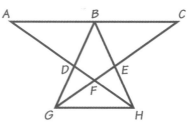

Find the coordinates of the midpoint of the segment with each given pair of endpoints. *(Section 4.1)*

41. $A(5, 2)$, $B(5, -2)$ **42.** $C(-1, 3)$, $D(5, -3)$ **43.** $E(4, 0)$, $F(-4, 3)$

8.5

Glide Reflections

Patterns like the ones below are used to decorate pottery, fabrics, and walls in many cultures. They are called *border* or *frize* patterns. You can see that reflections, translations, and rotations have been used to create these repeating patterns.

Pattern A: Nazca

Pattern B: Oklahoma

Pattern C: Yurok

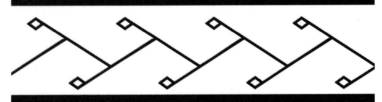

Pattern D: Navaho

THINK AND COMMUNICATE

1. Which of the patterns above use reflections?

2. Which of the patterns use translations?

3. Which of the patterns use rotations?

4. Describe how you could make Pattern D by using a translation followed by a reflection.

A translation followed by a reflection over a line parallel to the translation is called a **glide reflection**.

Properties of Glide Reflections

The line of reflection in a glide reflection is *parallel* to the direction of the translation.

A glide reflection preserves congruence.

A glide reflection reverses orientation.

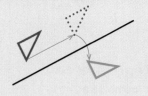

THINK AND COMMUNICATE

5. What does the word "glide" refer to in the phrase *glide reflection*?

6. Which pattern(s) on page 419 can be described using glide reflections?

EXAMPLE 1

Sketch the image of the triangle with the given vertices after a glide reflection using the given translation and reflection.

$A(-5, 1)$, $B(-4, 4)$, $C(1, 2)$

translation: $(a, b) \rightarrow (a + 6, b)$

reflection: over the x-axis

SOLUTION

Plot $\triangle ABC$. Then translate it and reflect it.

> The **translation** moves each point to the right 6 units.

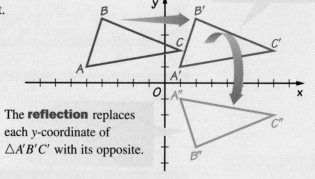

The **reflection** replaces each y-coordinate of $\triangle A'B'C'$ with its opposite.

The coordinates change as shown.

$$
\begin{array}{ccccc}
 & \text{translation} & & \text{reflection} & \\
A(-5, 1) & \rightarrow & A'(1, 1) & \rightarrow & A''(1, -1) \\
B(-4, 4) & \rightarrow & B'(2, 4) & \rightarrow & B''(2, -4) \\
C(1, 2) & \rightarrow & C'(7, 2) & \rightarrow & C''(7, -2)
\end{array}
$$

add 6 to the x-coordinate   change sign of y-coordinate

EXAMPLE 2   Application: Animal Tracks

**The tracks made by the feet of a walking black bear form a glide
reflection. Copy the tracks and describe a translation and reflection
that combine to create the glide reflection.**

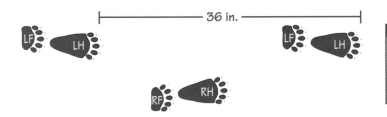

LEGEND
LF → left foreleg
LH → left hind leg
RF → right foreleg
RH → right hind leg

SOLUTION

The distance between footprints made by the same foot, called the *stride*,
is about 36 in. The translation needed for the glide reflection is half this
amount, about 18 in.

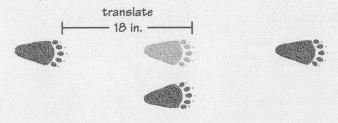

The line of reflection runs along the middle of the tracks, as shown in the
diagram below. The line of reflection is parallel to the direction of the
translation.

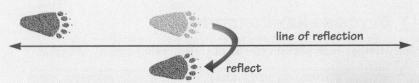

THINK AND COMMUNICATE

7. Is the stride of the hind leg the same as the stride
of the foreleg for a black bear?

8. Explain why the translation needed for the glide
reflection is half the bear's stride.

Transformations that Preserve Congruence

This chart summarizes the four transformations described so far in this chapter. All these transformations preserve congruence.

These transformations reverse orientation.

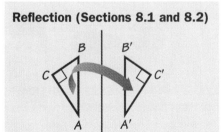

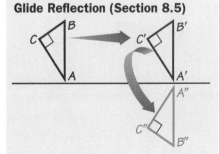

These transformations preserve orientation.

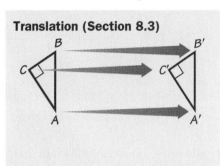

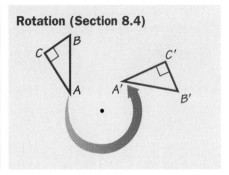

☑ CHECKING KEY CONCEPTS

Tell whether each pattern can be described using a glide reflection.

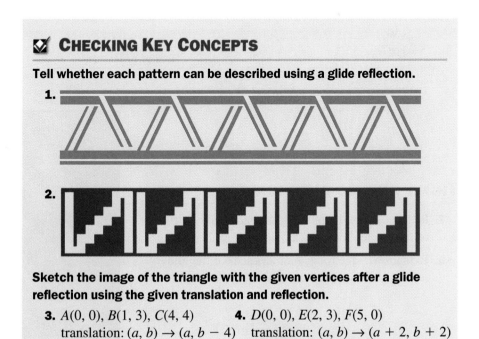

1.

2.

Sketch the image of the triangle with the given vertices after a glide reflection using the given translation and reflection.

3. $A(0, 0)$, $B(1, 3)$, $C(4, 4)$
translation: $(a, b) \rightarrow (a, b - 4)$
reflection: over the y-axis

4. $D(0, 0)$, $E(2, 3)$, $F(5, 0)$
translation: $(a, b) \rightarrow (a + 2, b + 2)$
reflection: over the line $y = x$

8.5 | Exercises and Applications

Extra Practice
exercises on
page 676

Sketch the image of the triangle with the given vertices after a glide reflection using the given translation and reflection.

1.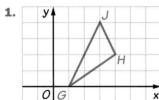

translation: $(a, b) \rightarrow (a - 3, b)$
reflection: over the x-axis

2.

translation: $(a, b) \rightarrow (a - 5, b - 5)$
reflection: over the line $y = x$

3. $P(-3, -2)$, $Q(0, 0)$, $R(2, -4)$
translation: $(a, b) \rightarrow (a, b + 1)$
reflection: over the y-axis

4. $X(-2, 1)$, $Y(0, 4)$, $Z(3, 5)$
translation: $(a, b) \rightarrow (a + 4, b - 4)$
reflection: over the line $y = -x$

5. **Challenge** Choose one of the glide reflections from Exercises 1–4. Describe a rule for the transformation using coordinate notation:

$$(a, b) \rightarrow (\underline{\ ?\ }, \underline{\ ?\ })$$

Draw a sketch of each situation, showing how a glide reflection is involved.

6. rungs on a telephone pole

7. snow-shoe or ski tracks

8. wallpaper or stencil pattern

9. closed zipper

FABRIC DESIGN These patterns are taken from men's ties. Tell whether each pattern uses a glide reflection. If it does, make a sketch, showing the direction and amount of translation, and draw the line of reflection.

10. 11. 12.

13. **Research** Find other examples of border patterns like the ones on page 419. Describe the transformations that are used to create them.

14. **Open-ended Problem** Create a pattern that uses a glide reflection.

15. **SAT/ACT Preview** Which of the following are preserved by a glide reflection?

   I. length   II. orientation   III. angle measure

   A. I only   **B.** I and II   **C.** II and III

   D. I and III   **E.** I, II, and III

A computer game called "Life," developed by John Conway, reveals how complex behavior can arise from simple rules. The game uses a grid of square cells, some filled and others empty.

The number of filled neighbors of each cell determines whether the cell will be filled or empty in the next "generation," according to these rules:

- An empty cell becomes filled if it has exactly three filled neighbors.
- A filled cell stays filled if it has two or three filled neighbors.

Every cell has 8 neighbors. The blue cell has three filled neighbors.

Five generations of an initial population called a "glide-reflection spaceship" are shown.

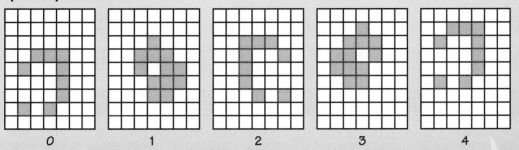

0   1   2   3   4

16. **Cooperative Learning** Work in a group of four people. Each person should choose one of the generations 1–4. Verify that the generation you choose is correctly "born" from the previous one.

17. Which generation is a glide reflection of generation 0? of generation 1?

18. What transformation moves generation 0 to generation 4?

19. **Writing** What happens to this population if the game continues? Explain why you think this population is called a "spaceship."

When counting cell neighbors, imagine the grid extends forever in all directions.

Use graph paper to plot several generations of each initial population using the rules of the game of Life. Describe the glide reflections that appear.

20.   21.   22.

23. **Visual Thinking** In Section 8.3 you saw that a translation can be created by combining two reflections in parallel lines. Show how to create a glide reflection by combining three reflections. How are the lines related?

24. **Challenge** On a coordinate plane, draw a point and its image under a glide reflection that uses the x-axis as the line of reflection.

   a. What do you notice about the midpoint of the segment that joins the point and its image?

   b. Make a conjecture and prove it using coordinates.

   c. Can you extend your conjecture to other kinds of glide reflections?

ANIMAL TRACKS Each set of animal tracks can be described using a glide reflection. Copy the tracks, find the length of the translation, and draw the line of reflection needed.

25. *ant (6 legs)*

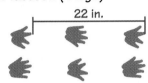

26. *herring gull (2 legs)*

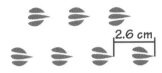

27. *raccoon (4 legs)*

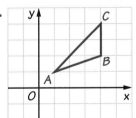

28. *scorpion (8 legs)*

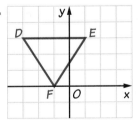

29. **Investigation** Measure your own stride and draw a scale diagram of your footprints, showing how a glide reflection can describe them.

ONGOING ASSESSMENT

30. **Open-ended Problem** Experiment to find out if the order in which you perform the translation and reflection affects the final image of a glide reflection. Are the results different if you reflect first, then translate?

SPIRAL REVIEW

Copy each diagram and sketch the image of each polygon after a 180° rotation around the origin. *(Section 8.4)*

31.

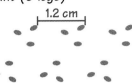

32.

Open-ended Problems Describe at least two objects where you live or at school that have each type of symmetry. *(Section 1.2)*

33. reflection symmetry

34. rotational symmetry

The midpoints of two sides of a triangle are endpoints of a segment. Show that the segment is parallel to the third side and half as long. *(Section 4.5)*

35.

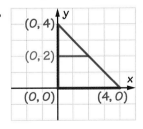

36.

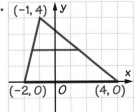

8.6 Dilations

Many photocopiers make reduced or enlarged copies as well as same-size copies. A youth organization used this built-in feature to make a small version of their logo for a notebook cover and a larger version for their T-shirts.

Learn how to...

- **compare lengths in a figure and its image after a dilation**
- **find the scale factor and center of a dilation**

So you can...

- **make reductions and enlargements of figures**
- **describe the relationship between a figure and its image after a dilation**

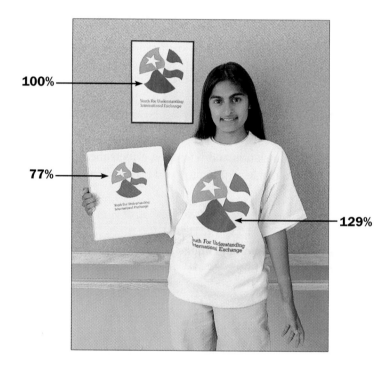

100%

77%

129%

A 77% reduction was used to make the smaller logo. Every segment in this logo is 0.77 times as long as the corresponding segment in the original logo. The number 0.77 is called the *scale factor* for this reduction. A scale factor can be a percent, a decimal, or a fraction.

BY THE WAY

The 64% and 77% settings on a copier allow you to photocopy tabloid- or legal-sized documents onto letter-sized paper. The 129% setting allows you to copy letter paper onto tabloid paper.

THINK AND COMMUNICATE

1. A 129% enlargement was used to make the larger logo. Explain how the lengths of two corresponding segments in the enlarged figure and the original logo are related.

2. What scale factor would you use to produce a same-size copy?

3. The photocopier has 77% and 64% settings. How could you combine them to make an image that is a 50% reduction (half size)?

Making an Enlargement

Work with another student.
You will need:

- a ruler

- a protractor

1 Draw any △*ABC*. Choose any point *O* outside the triangle. Draw rays from *O* through each vertex of the triangle.

2 Use your ruler to find the length *OA*. Double this number and use it to place *A'* on $\overrightarrow{OA}$ so that *OA'* = 2 • *OA*. Use the same process to find *B'* and *C'*. Draw △*A'B'C'*.

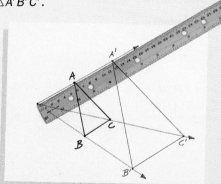

3 Use a protractor to compare the corresponding angles in your two triangles. What do you notice?

4 Use a ruler to compare the corresponding lengths in your two triangles. What do you notice?

5 Choose another point *P*, inside △*ABC*. Repeat Steps 1–4 using *P* instead of *O*. How is your image triangle different? How is it the same?

The enlargements in the Exploration, like the photocopy of the youth organization logo, do *not* preserve congruence, because corresponding lengths are not the same. The transformations in the Exploration are examples of *dilations*.

Properties of Dilations

A **dilation** with **center** O and positive **scale factor** k is a transformation in which every point P has an image P' placed on $\overrightarrow{OP}$ so that

$$\frac{OP'}{OP} = k \quad \text{and} \quad OP' = k \cdot OP.$$

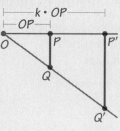

Dilations do not usually preserve congruence:
   If $k > 1$, the dilation is an *enlargement*.
   If $k = 1$, the original figure and its image are
      congruent.
   If $0 < k < 1$, the dilation is a *reduction*.

Dilations preserve angle measures.

The image of *any* segment after a dilation is always k times as long as the original segment. In the diagram above, for example, $P'Q' = k \cdot PQ$.

EXAMPLE 1

Draw the image of *JKLM* after a dilation with center *O* and scale factor $\frac{1}{2}$.

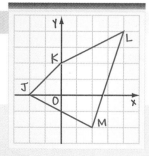

SOLUTION

Copy *JKLM* on graph paper.
Draw $\overrightarrow{OJ}$, $\overrightarrow{OK}$, $\overrightarrow{OL}$, and $\overrightarrow{OM}$.

The scale factor is $\frac{1}{2}$.

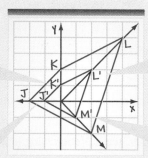

Place K' on $\overrightarrow{OK}$ so
that $OK' = \frac{1}{2}OK$.

Place L' on $\overrightarrow{OL}$ so
that $OL' = \frac{1}{2}OL$.

Place J' on $\overrightarrow{OJ}$ so
that $OJ' = \frac{1}{2}OJ$.

Place M' on $\overrightarrow{OM}$ so
that $OM' = \frac{1}{2}OM$.

THINK AND COMMUNICATE

4. Is the image in Example 1 an *enlargement* or a *reduction*?

5. Use the Distance Formula to find the lengths of two corresponding segments of *JKLM* and *J'K'L'M'*. What is their ratio?

EXAMPLE 2

△*P'Q'R'* is the image of △*PQR* after a dilation. Find the center and the scale factor of the dilation.

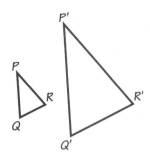

SOLUTION

Draw lines through each pair of corresponding vertices.

Measure $\overline{OP'}$ and $\overline{OP}$.

$OP' = 30$ mm

The scale factor is the ratio $\dfrac{OP'}{OP}$.

$OP = 12$ mm

$\dfrac{OP'}{OP} = \dfrac{30}{12} = 2.5$

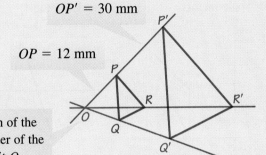

The intersection of the lines is the center of the dilation. Label it *O*.

The dilation has center *O* and scale factor 2.5.

WATCH OUT!

The ratio $\dfrac{PP'}{OP}$ is *not* the scale factor of the dilation.

THINK AND COMMUNICATE

6. In Example 2, what is the ratio $\dfrac{OQ'}{OQ}$? What is the ratio $\dfrac{OR'}{OR}$?

7. Make a conjecture about the perimeters of △*PQR* and △*P'Q'R'*.

☑ CHECKING KEY CONCEPTS

Draw the image of △*ABC* after each dilation.

1. center *C* and scale factor 2

2. center *A* and scale factor $\dfrac{1}{3}$

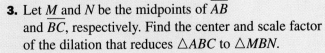

3. Let *M* and *N* be the midpoints of $\overline{AB}$ and $\overline{BC}$, respectively. Find the center and scale factor of the dilation that reduces △*ABC* to △*MBN*.

4. Explain how a dilation is different from the other transformations that you have learned about in this chapter.

8.6 Exercises and Applications

Extra Practice exercises on page 677

Draw the image of each figure after a dilation with center *O* and the given scale factor.

1. scale factor 3

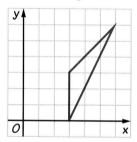

2. scale factor 1.5

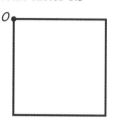

3. scale factor 0.75

4. scale factor 3

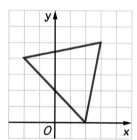

5. scale factor $\frac{1}{3}$

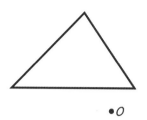

6. scale factor 4

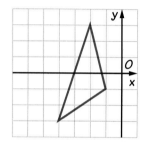

7. a. Use the coordinates of the vertices of the figure in Exercise 4. Copy and complete the table.

b. Complete this conjecture based on your table: The image of (a, b) after a dilation with center *O* and scale factor *k* has coordinates (_?_ , _?_).

c. Test your conjecture by comparing the coordinates of points and their images in Exercises 5 and 6.

Figure	Scale	Image
(_?_ , _?_)	3	(_?_ , _?_)
(_?_ , _?_)	3	(_?_ , _?_)
(_?_ , _?_)	3	(_?_ , _?_)

8. MANUFACTURING European paper has dimensions in the ratio $1 : \sqrt{2}$. For example, letter-size paper, called A4, is 210 mm × 297 mm. Each size can be enlarged to the next size by using a scale factor of $\sqrt{2}$.

a. The next paper size larger than A4 is called A3. What are the dimensions of A3 paper?

b. Challenge What reduction percent would you use to photocopy one size of European paper onto the next *smaller* size?

9. Writing Elena wants to compile some of her family's favorite recipes in a notebook. The recipes are on 3 in. × 5 in. cards. She wants to enlarge them in a photocopier to fit onto 8.5 × 11 in. letter paper. Choose an appropriate enlargement she can use. Explain your choice.

Before the invention of photocopiers, people could use *pantographs* to make enlargements. The title character in Mark Twain's *Pudd'nhead Wilson* uses one to make large drawings of fingerprints to display as evidence at a courtroom trial.

*H*e made fine and accurate reproductions of a number of his [fingerprint] "records," and then enlarged them on a scale of ten to one with his pantograph. [He] made each individual line of the bewildering maze of whorls or curves or loops . . . stand out bold and black by reinforcing it with ink [so] they resembled the markings of a block of wood that has been sawed across the grain, and the dullest eye could detect at a glance, and at a distance of many feet, that no two of the patterns were alike.

In the pantograph shown, *ABCD* is a parallelogram with joints that move so that *O*, *D*, and *E* always lie on a line. Point *O* is fixed to the drawing board.

To enlarge a figure, you put a stylus (a pointed object) at *D* and a pencil at *E*. If you guide the stylus at *D* so it traces your drawing, the pencil at *E* will draw an enlargement.

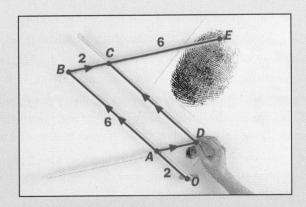

10. Find the value of each ratio.

a. $\dfrac{OB}{OA}$ b. $\dfrac{BE}{AD}$ c. $\dfrac{OE}{OD}$

11. $\triangle OBE$ is the image of $\triangle OAD$ after a dilation. What is the center of the dilation? What is the scale factor?

12. What kind of dilation results if you put the stylus at *E* and a pencil at *D* and trace a drawing with the stylus?

13. Sketch the dimensions of a pantograph that Pudd'nhead Wilson could have used to enlarge his fingerprint collection by a scale factor of 10.

Copy each figure. The red figure is the image of the blue figure after a dilation. Find the center and the scale factor of the dilation.

14.

15.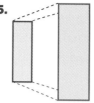

16.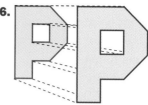

17. Technology Use geometry software to draw a polygon and two points.

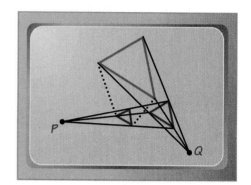

a. Dilate the polygon using one point as the center of dilation. Then dilate the image using the other point as a new center and a different scale factor.

b. Your final polygon in part (a) and the original polygon are related by a single dilation. Find its scale factor.

c. What do you notice about the location of the center of the single dilation from part (b) in relation to the other two centers of dilation?

18. Open-ended Problem Draw any △ABC on graph paper. For each center of dilation, draw the image of △ABC after a dilation of scale factor 2.

a. Use a point outside △ABC as the center of dilation.

b. Use a vertex of △ABC as the center of dilation.

c. Use a point inside △ABC as the center of dilation.

d. Compare the images you drew in parts (a)–(c). What do you notice?

19. Challenge A polygon is dilated with center O and positive scale factor k. Its image is then dilated with center O and positive scale factor j. The image is the original polygon. What must be true of k and j?

20. a. Use a compass to draw two concentric circles. Plot several points on the smaller circle. Connect your points to form a polygon.

b. Draw a ray from the center of the circle through each vertex.

c. Connect the points where the larger circle intersects your rays. What do you notice about this polygon?

d. Writing Explain why your diagram is an example of a dilation.

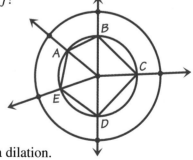

ONGOING ASSESSMENT

21. Cooperative Learning Work with another person. Use a ruler. Each person should draw a figure on plain paper and its image after a dilation. Trace the figure and its image without tracing the center or your construction lines. Exchange papers and find the center and scale factor your partner used.

Complete each construction. *(Section 5.7)*

22. Draw an angle ∠ *KLM*. Then construct an angle congruent to it.

23. Draw any $\overleftrightarrow{AB}$ and a point *R* not on $\overleftrightarrow{AB}$. Construct the line through *R* that is perpendicular to $\overleftrightarrow{AB}$.

Find the lengths of the sides of each polygon whose vertices are given.
(Section 4.1)

24. *P*(0, 0), *Q*(5, 0), *R*(2, 3)   **25.** *X*(1, 1), *Y*(3, 3), *Z*(4, 0)

ASSESS YOUR PROGRESS

VOCABULARY

rotation (p. 412)   **glide reflection** (p. 420)
center of rotation (p. 412)   **dilation** (p. 428)
angle of rotation (p. 412)   **center of a dilation** (p. 428)
half-turn (p. 414)   **scale factor** (p. 428)

Sketch the images of each point after 90° and 180° rotations around the origin. *(Section 8.4)*

1. *X*(−2, 3)   **2.** *Y*(5, 0)

3. *Z*(3, −6)   **4.** *W*(0, −2)

Sketch the image of the triangle with the given vertices after a glide reflection using the given translation and reflection. *(Section 8.5)*

5. *A*(0, 0), *B*(5, 1), *C*(2, 6)   **6.** *D*(−3, 2), *E*(0, −1), *F*(3, 4)
translation: $(a, b) \rightarrow (a - 3, b)$ translation: $(a, b) \rightarrow (a, b - 2)$
reflection: over the *x*-axis   reflection: over the *y*-axis

Draw the image of each figure after a dilation with center *O* and the given scale factor. *(Section 8.6)*

7. scale factor 2   **8.** scale factor $\frac{1}{4}$

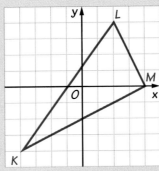

9. Journal You know how to describe *every* translation using coordinate notation. Write a description of the types of reflections, rotations, and dilations that you can describe using coordinate notation.

Tessellating the Plane

A tessellation is a pattern formed by repeating a shape to cover a plane without any gaps or overlapping. A brick walkway is an example of a tessellation that uses rectangles to cover the plane. A tiled wall is another example. So is a honeycomb—not all tessellations are made with quadrilaterals.

PROJECT GOAL Use transformations with different polygons to make tessellations.

Making Quadrilateral Tilings

1. Work with a partner. Cut several congruent rectangles from cardboard to represent bricks, and use transformations to make three different tessellations with them. Discuss with your partner the transformations you used.

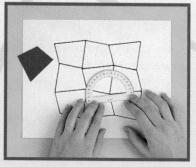

2. Cut a quadrilateral that is *not* a rectangle from a piece of cardboard. Trace the quadrilateral several times to make a tessellation. Discuss the transformations that you used to make your pattern.

3. Choose any vertex on your sketch and measure the angles at that vertex. What do you notice about the sum of the measures of these angles? How might you have predicted this result without measuring?

Exploring with Other Polygons

Cut two identical rectangles from cardboard.

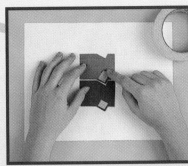

6. Trace the new shape many times to make a tessellation.

4. Place the two rectangles together and cut a piece from the bottom edges of both.

5. Tape a piece to each shape so the piece fills the missing space in the other shape.

Use this method or geometry software to produce three other shapes that will form tessellations when transformations are used. You can start with triangles or other polygons that tessellate. Experiment with reflections, translations, rotations, or combinations of these transformations.

Writing a Report

Write a report that describes your tessellations with quadrilaterals and other polygons.

- Describe the kinds of polygons that can be used to make tessellations and the transformations used with each kind.

- Include sketches or printouts of four of the tessellations that you describe in your report.

You may want to extend your report by including some tessellations that use combinations of two or more polygons to cover the plane.

Self-Assessment

Describe how you and your partner worked together in this project. How did your partner help you understand something about transformations and tessellations? How did you help your partner?

8 Review

Write at least six questions about the chapter. Focus on the concepts that you had the most difficulty learning. Three should be short answer questions about specific situations or details. Three should be more involved questions, justifications, or proofs. Then answer the questions.

VOCABULARY

line of reflection (p. 391)

reflection (p. 391)

image (p. 391)

translation (p. 406)

rotation (p. 412)

center of rotation (p. 412)

angle of rotation (p. 412)

half-turn (p. 414)

glide reflection (p. 420)

dilation (p. 428)

center of a dilation (p. 428)

scale factor (p. 428)

SECTIONS 8.1 *and* 8.2

A *reflection* is a transformation that preserves congruence but reverses orientation. After a **reflection**, each point of a figure and its **image** lie equidistant from the **line of reflection**, but on opposite sides of it.

The line of reflection is the perpendicular bisector of every segment that connects a point and its image.

When reflected over **the y-axis**, the *x*-coordinate becomes its opposite.
$(a, b) \rightarrow (-a, b)$

When reflected over **the line $y = x$**, the coordinates switch places.
$(a, b) \rightarrow (b, a)$

When reflected over **the x-axis**, the *y*-coordinate becomes its opposite.
$(a, b) \rightarrow (a, -b)$

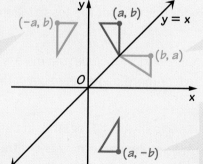

Points that lie on the line of reflection do not change their position.

SECTIONS 8.3 *and* 8.4

A **translation** is a shift that can be described as $(a, b) \rightarrow (a + h, b + k)$. Every point moves h units horizontally and k units vertically. A translation preserves congruence and orientation.

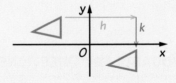

In a **rotation**, every point moves along a circular path around a fixed point called the **center of rotation**. The angle formed by any point, the center of rotation, and the image of the point is called the **angle of rotation**.

A rotation preserves congruence and orientation. When a point is rotated counterclockwise around the origin:

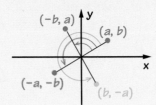

90°: $(a, b) \rightarrow (-b, a)$

180°: $(a, b) \rightarrow (-a, -b)$

270°: $(a, b) \rightarrow (b, -a)$

SECTIONS 8.5 *and* 8.6

A **glide reflection** is the result of a translation followed by a reflection over a line parallel to the translation. A glide reflection preserves congruence but reverses orientation.

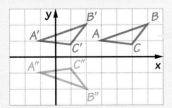

Translated to the left 4 units and then reflected over the x-axis

A **dilation** is a transformation that may enlarge or reduce a figure. In a dilation with **center** O, for every point A and its image A', there is a positive **scale factor** k such that $\dfrac{OA'}{OA} = k$, and $OA' = k \cdot OA$.

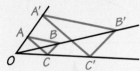

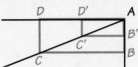

The center of a dilation is at the intersection of lines that connect points and their images.

If $k > 1$, the dilation is an *enlargement*.

If $k < 1$, the dilation is a *reduction*.

The image of any segment after a dilation is k times as long as the original segment. Dilations do not preserve congruence unless $k = 1$.

Assessment

VOCABULARY QUESTIONS

Copy the diagram. In each space, write the name of the transformation indicated by the arrow of the same color.

1. 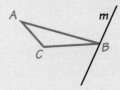 ?

2. ?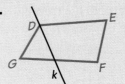

3. ?

4. ?

SECTIONS 8.1 *and* 8.2

Copy each diagram. Sketch the image of each polygon after a reflection over the given line. Label the image polygon.

5. *A* *m* *C* *B*

6. *D* *E* *G* *k* *F*

7. The diagram shows a triangle and its image after a reflection. Draw the line of reflection.

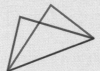

8. Reflect *MNPQ* over the line $y = x$. Give the coordinates of the vertices of the image.

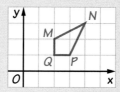

9. On a golf course, a player at *A*(2, 5) wants to reach the cup at *C*(10, 3) by bouncing off the bumpboard at a point *B* along the *x*-axis.
   a. Make a sketch to represent the shot. Draw the path of the ball.
   b. What are the coordinates of point *B* on the bumpboard?

The image of each polygon after a translation is shown in red. Describe each translation using coordinate notation.

10.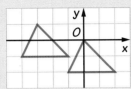

11.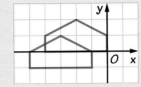

12. $\triangle ABC$ has vertices $A(-2, 1)$, $B(1, 2)$, $C(2, 4)$. Sketch its image after a 180° rotation around the origin.

13. **Open-ended Problem** Draw a triangle and a point P outside it. Rotate the triangle 90° counterclockwise around P.

SECTION 8.5 *and* 8.6

14. For the following glide reflection, find the distance of the translation and draw the line of reflection.

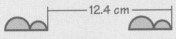

15. Sketch the image of the triangle with vertices $D(2, 2)$, $E(-1, 1)$, $F(3, -2)$ after a glide reflection using the translation: $(a, b) \rightarrow (a + 2, b - 1)$ and reflection over the line $y = -x$. Find the vertices of the image.

Draw the image of each triangle after a dilation with center *P* and the given scale factor.

16. scale factor $\frac{1}{2}$

17. scale factor $\frac{3}{2}$

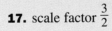
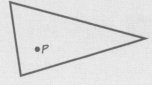

PERFORMANCE TASK

18. On graph paper, prepare two designs for a wallpaper company. Use a simple design element, such as a curved leaf. Make one design that uses rotations and reflections. Make another that uses glide reflections. Describe the transformations using coordinate notation if possible.

Algebra
Review/Preview

These exercises review algebra topics you will use in the next chapters.

Toolbox p. 690

Simplifying and Evaluating Expressions

SIMPLIFYING VARIABLE EXPRESSIONS

Simplify each expression. If it is not possible, write *cannot be simplified*.

EXAMPLES

$$x + \frac{1}{2}x^2 - \frac{5}{6}x^2 + 3x$$

$$= \left(\frac{1}{2}x^2 - \frac{5}{6}x^2\right) + (1x + 3x) \quad \longleftarrow \text{Group the like terms.}$$

$$= \left(\frac{1}{2} - \frac{5}{6}\right)x^2 + (1 + 3)x \quad \longleftarrow \begin{array}{l}\text{Use the distributive} \\ \text{property.}\end{array}$$

$$= \left(\frac{3}{6} - \frac{5}{6}\right)x^2 + 4x \quad \longleftarrow \begin{array}{l}\text{Use the least common} \\ \text{denominator (LCD).}\end{array}$$

$$= -\frac{2}{6}x^2 + 4x$$

$$= -\frac{1}{3}x^2 + 4x \quad \longleftarrow \text{Simplify the answer.}$$

$4y^2 + 5y$ ***cannot be simplified*** because $4y^2$ and $5y$ are not like terms.

1. $2l + 2w$
2. $12y^2 - y^2$
3. $7a + 4b - ab$
4. $\pi r^2 + 4\pi r^2$
5. $n^2 - 3n + 6n^2$
6. $a^2 - a - (a - 4a^2)$
7. $\frac{1}{3}x^2 + \frac{3}{4}x^2$
8. $\frac{1}{5}k^2 + \frac{1}{2}k + \frac{3}{10}k^2$
9. $x^2 + 7.4x^2 - 9.8x^2$

SIMPLIFYING RADICAL EXPRESSIONS

Simplify.

Toolbox p. 700

Simplifying Radicals

EXAMPLES

$$\sqrt{96} = \sqrt{16 \cdot 6} \quad \longleftarrow \begin{array}{l}\text{16 is the largest perfect} \\ \text{square factor of 96.}\end{array}$$

$$= \sqrt{16} \cdot \sqrt{6} \quad \longleftarrow \sqrt{ab} = \sqrt{a} \cdot \sqrt{b}$$

$$= 4\sqrt{6} \quad \longleftarrow \sqrt{16} = \sqrt{4^2} = 4$$

$$(2\sqrt{5})^2 = 2\sqrt{5} \cdot 2\sqrt{5}$$

$$= 2 \cdot 2 \cdot \sqrt{5} \cdot \sqrt{5}$$

$$= 4 \cdot 5 \quad \longleftarrow \sqrt{5} \cdot \sqrt{5} = \sqrt{25} = 5$$

$$= 20$$

$$\sqrt{\frac{8}{81}} = \frac{\sqrt{8}}{\sqrt{81}} = \frac{\sqrt{4 \cdot 2}}{\sqrt{9 \cdot 9}} = \frac{2\sqrt{2}}{9}$$

10. $\sqrt{300}$ **11.** $\sqrt{294}$ **12.** $\sqrt{180}$ **13.** $\sqrt{37^2}$

14. $\sqrt{\dfrac{15}{4}}$ **15.** $\sqrt{\dfrac{45}{16}}$ **16.** $\sqrt{\dfrac{1}{324}}$ **17.** $\left(\sqrt{11}\right)^2$

18. $\left(8\sqrt{3}\right)^2$ **19.** $7\sqrt{225}$ **20.** $5\sqrt{10}\cdot\sqrt{15}$ **21.** $2\sqrt{3}\cdot6\sqrt{6}$

SOLVING QUADRATIC EQUATIONS

Solve each quadratic equation.

EXAMPLES

$$x^2 = 4^2 + \left(4\sqrt{5}\right)^2 \qquad\qquad 5^2 + u^2 = 13^2$$
$$x^2 = 16 + 80 \qquad\qquad\qquad 25 + u^2 = 169$$
$$x^2 = 96 \qquad\qquad\qquad\qquad u^2 = 144$$
$$x = \pm\sqrt{96} \qquad\qquad\qquad u = \pm\sqrt{144}$$
$$x = \pm 4\sqrt{6} \qquad\qquad\qquad u = 12 \text{ or } u = -12$$
$$x = 4\sqrt{6} \text{ or } x = -4\sqrt{6}$$

> **Toolbox p. 701**
> *Solving Simple Quadratic Equations*

22. $x^2 = 9^2 + 12^2$ **23.** $y^2 = 5^2 + \left(5\sqrt{3}\right)^2$ **24.** $z^2 + 4^2 = 8^2$

25. $a\cdot 4a = 12\cdot 18$ **26.** $s^2 = \left(\sqrt{2}\right)^2 + \sqrt{4}$ **27.** $k^2 = 3^2 + 2^2$

28. $x^2 + 9^2 = 13^2$ **29.** $b\cdot 5b = 15\cdot 7$ **30.** $2\sqrt{6^2} + 6\sqrt{2^2} = c^2$

31. $x^2 + x^2 = 6^2$ **32.** $x^2 + \left(9\sqrt{3}\right)^2 = 4x^2$ **33.** $x^2 + \left(x\sqrt{3}\right)^2 = 8^2$

34. Refer to Exercise 32. Suppose x, $9\sqrt{3}$, and $2x$ represent the lengths of the sides of a right triangle. Find the value of x and the length of the hypotenuse of the triangle.

SOLVING MORE PROPORTIONS

Solve each proportion.

EXAMPLES

$$\frac{a}{a+6} = \frac{3}{5} \qquad\qquad 12:x = x:2 \quad\longleftarrow\quad \frac{12}{x} = \frac{x}{2}$$
$$5a = 3(a+6) \qquad\qquad x\cdot x = 12\cdot 2$$
$$5a = 3a + 18 \qquad\qquad\quad x^2 = 24$$
$$2a = 18 \qquad\qquad\qquad\quad x = \pm\sqrt{24}$$
$$a = 9 \qquad\qquad\qquad\qquad = \pm 2\sqrt{6}$$

> **Toolbox p. 707**
> *Ratio and Proportion*

35. $\dfrac{x}{21 - x} = \dfrac{16}{12}$ **36.** $5x:2x = 2x:16$ **37.** $\dfrac{1.8}{0.4} = \dfrac{y + 2}{5}$

38. $\dfrac{12}{r} = \dfrac{9}{15}$ **39.** $\dfrac{t + 6}{t} = \dfrac{7t}{5t}$ **40.** $3^2:5^2 = 225:A$

41. $3:z = z:5$ **42.** $\dfrac{A}{4\pi} = \dfrac{3^2}{(1.2)^2}$ **43.** $\dfrac{63}{x} = \dfrac{x}{7}$

44. $n:8 = 10:n$ **45.** $\dfrac{4}{x^2} = \dfrac{1}{9}$ **46.** $\dfrac{25}{4x} = \dfrac{x}{36}$

9 Similar Polygons

Setting the Scene

Loy Arcenas

Loy Arcenas planned to become a doctor before his career took a dramatic turn. After completing premedical studies at the University of the Philippines, he went to London for training at the Drama Studio and a design course at the English National Opera. He tried both acting and set design. Now, Arcenas is one of the most sought-after set designers in the United States.

When designing a set, Arcenas begins by reading the play. While reading, he has "small ideas that may seem to have nothing to do with one another." Then he finds what these ideas have in common and uses them to sketch the set for the play.

"Much of theater design is based on illusion; there are many tricks involved."

Sculptures that Move

Arcenas relies on careful calculations to guarantee that the pieces in his set will fit together when assembled in the shop. But his job involves a lot more than details. It's an artistic endeavor: the sets are movable sculptures that set the mood for the play. In Shakespeare's comedy *All's Well That Ends Well*, Arcenas used soft blue tones to create a dream-like quality. His set for Shakespeare's history *Henry IV* was dominated by large, sculpted metallic pieces that suggested wealth and power.

The biggest challenge is establishing a mood. The props, backdrops, and lights he chooses can affect the actors and audience emotionally. "It's really about creating an illusion," he says. When he does the job right, the set makes the action on stage seem real to the actors and audience.

Scenes from *All's Well That Ends Well*

"Set design is really about the relation of the actors to the space around them."

Next, he builds a *scale model* out of cardboard and other inexpensive materials. This model helps him understand and improve his plan for the stage. The models always include miniature people so that Arcenas can check the proportions. "Otherwise, we might end up with a well-proportioned room with a door that's just five feet tall," he says. Using the model, he also checks that the audience will be able to see the whole stage.

Scene from *Henry IV*

On the Scale of Things

Arcenas uses his scale models to check that all the pieces of the set will fit together. The pieces in the model are all *similar* to the pieces of the stage set. The lengths all change by the same factor, and the angles do not change at all. The *scale* of the set tells how the lengths will change between the model and the set. He often uses a 1:48 scale, so $\frac{1}{4}$ in. on his model corresponds to 1 ft in the actual set. Arcenas is helped in this process by his facility with math and his knack for visualizing objects in three dimensions. "Initially, all this has to be orchestrated in your mind," he says. "Then you can use the model to help you think about the real thing."

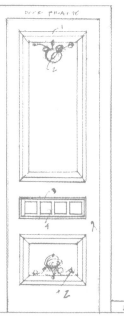

Scale Dimensions for Door *All's Well That Ends Well*			
	Scale	Width (in.)	Height (in.)
Model	1:48	$\frac{13}{16}$	$2\frac{1}{2}$
Blueprint	1:24	$1\frac{5}{8}$	5
Stage Set	1:1	39	120

Explore and Connect

Arcenas paints a piece for a scale model set of *Henry IV*.

1. Research Interview someone who participates in theater in your community. Ask how the available space at the theater affects the number and types of sets that the theater can use. Report your findings.

2. Project Borrow a video of a stage play. Watch how the set is used to provide a setting for each scene. Are some pieces of the set used for several different locations? Analyze the use of the set in a poster or report.

3. Writing Imagine that you are in the studio audience for a taping of a television show. Describe some of the activities that you would be able to see that the television audience does not see.

Mathematics & Loy Arcenas

In this chapter, you will learn more about how mathematics is related to set design and scale models.

Related Exercises

Section 9.3
• Exercises 10–13

Section 9.4
• page 468
• Exercise 17

9.1

Properties of Similar Figures

Learn how to...

- use ratios and proportions to find measures of similar figures
- identify similar figures

So you can...

- use scale models to find the dimensions of an object being modeled, for example

When working on a new sign, a designer will usually use *scale drawings* instead of full-size drawings. The person who builds the sign will use these drawings to make sure that the sign looks exactly as planned.

Matilda Johnson
State Park Signage
Scale 1:4

Read 1:4 as "1 to 4."

Lengths in the scale drawing are different from corresponding lengths on the actual sign, but all angles in the drawing are congruent to corresponding angles on the sign. You can use the *scale* of the drawing to find the size of a shape in the sign. The scale of the drawing is 1:4. Therefore, an object that is 1 in. tall on the drawing will be 4 in. tall on the sign.

THINK AND COMMUNICATE

1. If the designer wants the letters to be 3 in. tall on the sign, how tall should they be in the drawing?

2. Blueprints are scale drawings. A blueprint is labeled "1 cm to 5 m." Is the scale 1 to 5? Explain why or why not.

Toolbox p. 708
Proportions

A **proportion** is an equation that shows that two ratios are equal. The proportions $\frac{a}{b} = \frac{c}{d}$ and $a:b = c:d$ can be read as "*a* is to *b* as *c* is to *d*."

Similar Figures

Two figures are **similar** if these two conditions are true:

1. Corresponding angles are congruent.

2. The lengths of corresponding sides are in proportion.

The symbol $\sim$ means "is similar to."

$PQRS \sim TUVW$

When you name similar figures, be sure to name corresponding vertices in the same order. For example, if $PQRS \sim TUVW$, then:

1. $\angle P \cong \angle T$, $\angle Q \cong \angle U$, $\angle R \cong \angle V$, and $\angle S \cong \angle W$

WATCH OUT! ▶

When you write a proportion, the ratios must compare *corresponding* lengths.

2. $\dfrac{PQ}{TU} = \dfrac{QR}{UV} = \dfrac{RS}{VW} = \dfrac{SP}{WT}$

EXAMPLE 1

Tell whether the shapes in each pair are similar.

a.

b.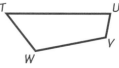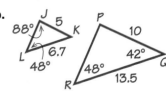

SOLUTION

a. Step 1 The angles are all right angles, so corresponding angles are congruent.

Step 2 Check the ratios of corresponding sides.

$$\frac{AB}{EF} \overset{?}{=} \frac{BD}{FG}$$

You need to check only one proportion, because opposite sides of a rectangle are congruent.

$$\frac{6}{4.5} \overset{?}{=} \frac{12}{9}$$

$$6 \cdot 9 \overset{?}{=} 4.5 \cdot 12$$

Find the cross products.

$$54 = 54 \checkmark$$

Because corresponding angles are congruent and corresponding sides are in proportion, $ABDC \sim EFGH$.

b. Step 1 Are the angles congruent?

Find $m \angle K$ and $m \angle P$.

$$m \angle K = 180° - (88° + 48°) = 44°$$

$$m \angle P = 180° - (42° + 48°) = 90°$$

Corresponding angles are not congruent. The triangles are not similar.

EXAMPLE 2   **Application: Graphic Design**

The hiking symbol in the scale drawing on page 445 is 3 in. wide and 2.5 in. tall. The symbol will be 10 in. tall on the sign. Find the width of the symbol on the sign.

SOLUTION

Set up a proportion. Let w = width of symbol in inches.

$$\frac{\text{height of symbol in drawing}}{\text{height of symbol on sign}} = \frac{\text{width of symbol in drawing}}{\text{width of symbol on sign}}$$

$$\frac{2.5}{10} = \frac{3}{w}$$

$$2.5w = 10 \cdot 3$$

$$w = \frac{30}{2.5}$$

$$w = 12$$

Check:
Use the scale of the drawing. One unit in the drawing is equal to 4 units in the sign, so $3 \cdot 4 = 12$. ✔

The hiking symbol will be 12 in. wide on the sign.

THINK AND COMMUNICATE

3. The scale of the drawing to the park sign is 1:4. What is the scale of the park sign to the drawing of the sign?

4. $\triangle ABC \sim \triangle PRQ$. Is $\frac{AB}{PR} = \frac{RQ}{BC}$?
Explain why or why not. Give a proportion that is true.

5. If two polygons are congruent, must they be similar?

6. If two figures are similar but not congruent, how are they alike? How are they different?

7. If the lengths of the hypotenuses of two right triangles are in the ratio of 1:1, must the triangles be similar? Explain your answer using a sketch.

EXAMPLE 3   Connection: Algebra

Tell whether the two triangles are similar.

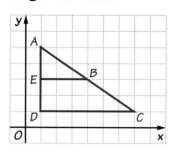

SOLUTION

Step 1   Are the angles congruent?

$\angle A \cong \angle A$, by the Reflexive Property.
Because $\angle ADC$ and $\angle AEB$ are right angles, $\angle ADC \cong \angle AEB$.
$\overline{EB} \parallel \overline{DC}$, so $\angle ABE \cong \angle ACD$.
Corresponding angles of the triangle are congruent.

Step 2   Are the sides in proportion?

Use the Distance Formula or count to find AE, AD, EB, and DC.

$AE = 2$, $AD = 4$, $EB = 3$, and $DC = 6$.

Use the Pythagorean theorem to find AC and AB.

$$AC^2 = AD^2 + DC^2 \qquad\qquad AB^2 = AE^2 + EB^2$$
$$AC = \sqrt{AD^2 + DC^2} \qquad\quad AB = \sqrt{AE^2 + EB^2}$$
$$= \sqrt{4^2 + 6^2} \qquad\qquad\quad = \sqrt{2^2 + 3^2}$$
$$= \sqrt{52} \qquad\qquad\qquad\quad = \sqrt{13}$$
$$= \sqrt{4 \cdot 13} = 2\sqrt{13}$$
$$AC = 2\sqrt{13} \text{ and } AB = \sqrt{13}$$

Check the proportion:

$$\frac{\text{short leg of } \triangle ABE}{\text{short leg of } \triangle ACD} = \frac{\text{long leg of } \triangle ABE}{\text{long leg of } \triangle ACD} = \frac{\text{hypotenuse of } \triangle ABE}{\text{hypotenuse of } \triangle ACD}$$

$$\frac{AE}{AD} \stackrel{?}{=} \frac{EB}{DC} \stackrel{?}{=} \frac{AB}{AC}$$

$$\frac{2}{4} \stackrel{?}{=} \frac{3}{6} \stackrel{?}{=} \frac{\sqrt{13}}{2\sqrt{13}}$$

$$\frac{1}{2} = \frac{1}{2} = \frac{1}{2} \checkmark$$

All angles are congruent and all sides are in proportion. By the definition of similar figures, $\triangle ABE \sim \triangle ACD$.

Tell whether the polygons in each pair are similar. Explain your reasoning.

1.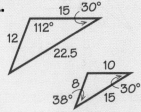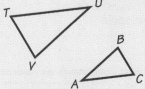

2.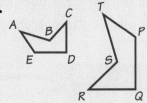

Name each pair of similar figures. Give two proportions that are true.

3.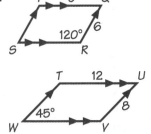

4.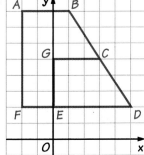

5. If the ratio of one pair of corresponding sides of similar triangles is $\frac{1}{4}$, what are the ratios of other pairs of corresponding sides?

6. Open-ended Problem   Sketch two quadrilaterals that are not similar but whose corresponding sides are in proportion.

9.1 | **Exercises and Applications**

*Extra Practice
exercises on
page 677*

ALGEBRA Solve each proportion.

1. $\frac{2}{4} = \frac{x}{16}$

2. $\frac{2}{5} = \frac{9}{y}$

3. $\frac{4}{z} = \frac{6}{z + 1}$

4. $2:16 = y:20$

5. $2:x = x:8$

6. $z:z + 2 = 30:42$

Tell whether the polygons in each pair are similar. Explain your reasoning.

7.

8.

9.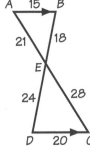

The polygons in each pair are similar. Find the value of each variable.

10.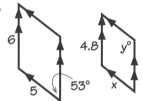

11.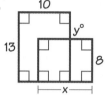

12.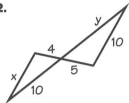

Logical Reasoning Tell whether the polygons in each pair are *always*, *sometimes*, or *never* similar. If a pair is always or sometimes similar, sketch an example. If a pair is never or sometimes similar, sketch a counterexample.

13. two squares

14. two rhombuses

15. two rectangles

16. two equilateral triangles

17. two isosceles trapezoids

18. an obtuse triangle and an acute triangle

Connection ▶ ASTRONOMY

Toolbox p. 709
Scientific Notation

The Lakeview Museum Community Solar System in Peoria, Illinois, is the largest scale model of the solar system in the world. The sizes of and distances between the sun and planets are represented on the same scale. The sun for the model is the Lakeview Museum's planetarium, which is 11 m in diameter.

19. The diameter of the sun is about 1,400,000 km. Find the scale of the model. Remember that the units of the scale must be the same.

 Spreadsheets For Exercises 20 and 21, use a spreadsheet and the scale that you found in Exercise 19.

20. Use a formula to find the mean distance from the planetarium to each planet in the scale model.

21. Find the diameter of each planet in the scale model.

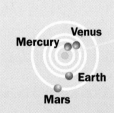

Lakeview Museum (sun)

Mercury   Venus

Earth

Mars

 Peoria Public Library

Jupiter

Regional Airport

Saturn

	A	B	C
1	Planet	Mean distance from sun (km)	Diameter (km)
2	Mercury	5.79 x 10^7	4870
3	Venus	1.08 x 10^8	12100
4	Earth	1.50 x 10^8	12800
5	Mars	2.29 x 10^8	6770
6	Jupiter	7.79 x 10^8	140000
7	Saturn	1.43 x 10^9	117000
8	Uranus	2.87 x 10^9	47000
9	Neptune	4.50 x 10^9	45500
10	Pluto	5.91 x 10^9	11400

Saturn

22. Open-ended Problem Give an example of a scale drawing or a scale model you have seen that is an enlargement of a real-life object or group of objects.

23. a. TECHNICAL DRAWING If you drew the structures in the table below so that 1 in. represents 100 ft, what would the scale be? How tall would each structure be in your drawing?

CN Tower	TMG Offices	Washington Monument	Empire State Building
Toronto, Canada	Tokyo, Japan	Washington, DC	New York City, NY
1815 ft	793 ft	555 ft	1250 ft

b. Open-ended Problem On a sheet of $8\frac{1}{2}$ in. by 11 in. paper, sketch and label a scale drawing of the four structures. Explain how you chose the scale for your drawing.

24. TRANSFORMATIONS Explain why dilations are sometimes called *similarity transformations*.

25. Challenge What do you think must be true for two prisms to be similar? Write a definition for similar prisms.

26. MOVIES For most of the movie *King Kong*, the scale models of King Kong were built to appear about 18 ft tall. When King Kong climbs the Empire State Building, the models make him appear to be 24 ft tall.

a. Make a scale drawing that shows King Kong at both 18 ft and 24 ft tall, a 6 ft tall person, and the Empire State Building (1250 ft).

b. Writing Why do you think the movie makers chose to make King Kong appear taller when he climbs the Empire State Building?

27. SAT/ACT Preview Select the pair of words that *best* expresses a relationship similar to that expressed in the pair *gallon : liquid*.

A. ruler : paper   **B.** week : month   **C.** length : width

D. degree : temperature   **E.** foot : meter

28. GEOGRAPHY The Nile, in Africa, is the world's longest river. It would be about the length of a 24 in. shoelace (61 cm) if drawn so that 1 cm represents 110 km. What is the scale? About how long is the Nile?

29. FRACTALS The pattern at the right is made by adding two squares or an isosceles right triangle at each step. When two squares connect to form a rectangle, you stop adding shapes to that part of the pattern.

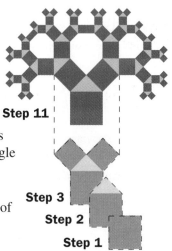

Step 11

Step 3

Step 2

Step 1

a. **Writing** In the diagram at the right, all of the triangles are similar and all of the squares are similar. Are all of the groups of one triangle with three surrounding squares similar? Explain why or why not.

b. Trace or sketch the pattern. Find two groups of more than four shapes in the pattern that are similar but not congruent.

30. Open-ended Problem Use the steps in Exercise 29 to make a design. Instead of using an isosceles right triangle in Step 2, use a different kind of triangle. (For example, you could use a right triangle with legs in a ratio of 1:2, or an isosceles obtuse triangle with a 100° vertex angle.) Draw at least the first five steps of the pattern.

31. SCULPTURE The Crazy Horse Monument, based on a marble sculpture by Korczak Ziolkowski, will be 563 ft high. Lengths for the sculpture in the mountain are taken from the 16 ft tall scale model of the sculpture. What is the scale of the model? Explain how workers can use the scale model to determine lengths on the mountain.

ONGOING ASSESSMENT

32. Open-ended Problem Make a scale drawing of a diagram in this book. On your drawing, give the scale that you used. Show that two of the shapes in your drawing are similar to the shapes in the original drawing. Give the page number of the original diagram.

SPIRAL REVIEW

Tell which method(s) you can use to prove that the triangles are congruent. If no method can be used, write *none*. *(Section 6.4)*

33.

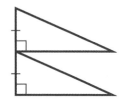

34.

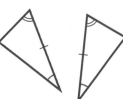

35.

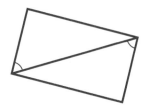

Open-ended Problems Draw a quadrilateral *ABCD* on the coordinate plane. Draw the image of *ABCD* after each dilation. *(Section 8.6)*

36. center *A* and scale factor 4

37. center *C* and scale factor $\frac{1}{2}$

9.2 Similar Triangles

Learn how to...

- **use special rules to identify similar triangles**

So you can...

- **write proofs involving similar triangles**

You know several shortcuts for proving that two triangles are congruent without checking each angle and side. In this lesson you will learn some shortcuts for proving that triangles are similar.

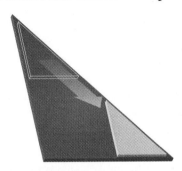

When two triangles have two pairs of congruent angles, the third pair must be congruent, too. If you experiment with two triangles in which all three pairs of angles are congruent, you will find that the lengths of their sides are in proportion.

Angle-Angle (AA) Similarity Postulate

If two angles of a triangle are congruent to two angles of another triangle, then the triangles are similar.

EXAMPLE 1

Given: $\overline{QT} \parallel \overline{RS}$
Prove: $\triangle PRS \sim \triangle PQT$

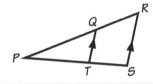

SOLUTION

Given: $\overline{QT} \parallel \overline{RS}$
Prove: $\triangle PRS \sim \triangle PQT$

1 $\overline{QT} \parallel \overline{RS}$ → 2 $\angle PQT \cong \angle PRS$
3 $\angle P \cong \angle P$
2 and 3 → 4 $\triangle PRS \sim \triangle PQT$

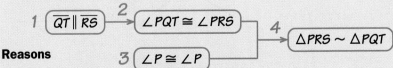

Reasons

1. Given
2. If two $\parallel$ lines are intersected by a transversal, then corresponding angles are $\cong$.
3. Reflexive Property
4. AA Similarity Postulate

THINK AND COMMUNICATE

1. Describe another way to prove that $\triangle PRS \sim \triangle PQT$ in Example 1.

2. Sketch two isosceles right triangles. Are they similar? Why?

As with congruent triangles, there are several ways to prove that two triangles are similar. The proofs of these theorems have been omitted.

Side-Angle-Side (SAS) Similarity Theorem

If an angle of one triangle is congruent to an angle of another triangle, and the sides including these angles are in proportion, then the triangles are similar.

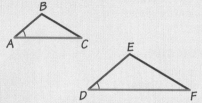

If $\angle A \cong \angle D$ and $\dfrac{AB}{DE} = \dfrac{AC}{DF}$, then $\triangle ABC \sim \triangle DEF$.

Side-Side-Side (SSS) Similarity Theorem

If all corresponding sides of two triangles are in proportion, then the triangles are similar.

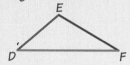

If $\dfrac{AB}{DE} = \dfrac{BC}{EF} = \dfrac{AC}{DF}$, then $\triangle ABC \sim \triangle DEF$.

EXAMPLE 2

Is it possible to prove that the triangles in each pair are similar? Explain why or why not.

a.

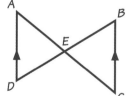

b.

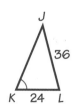

SOLUTION

a. Yes. $\angle AED \cong \angle CEB$ because they are vertical angles. $\angle A \cong \angle C$ because they are alternate interior angles formed by two parallel lines and a transversal. So $\triangle AED \sim \triangle CEB$ by the AA Similarity Postulate.

b. No. The congruent angles are not included between the given sides.

EXAMPLE 3

Tell whether the triangles in each pair are similar. Explain your reasoning.

a.

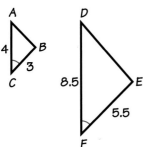

b.

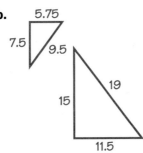

Fenda Gandega paints a wall of her family's home in Djajibinni, Mauritania.

SOLUTION

a. You know that $\angle C \cong \angle F$. Check whether the sides that include the angles are in proportion.

$$\frac{DF}{AC} \stackrel{?}{=} \frac{EF}{BC}$$

$$\frac{8.5}{4} \stackrel{?}{=} \frac{5.5}{3} \qquad \text{Check whether the fractions are equal.}$$

$$2.125 \neq 1.8\overline{3}$$

The sides are not in proportion, so the triangles are not similar.

b. Are the corresponding sides in proportion?

$$\frac{5.75}{11.5} \stackrel{?}{=} \frac{7.5}{15} \stackrel{?}{=} \frac{9.5}{19}$$

$$\frac{1}{2} = \frac{1}{2} = \frac{1}{2} ✔$$

By the SSS Similarity Theorem, the triangles are similar.

THINK AND COMMUNICATE

3. Explain why $\triangle ABC \sim \triangle LJK$. Find the length JK.

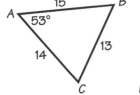

4. How are the SSS and SAS Similarity Theorems like the SSS and SAS Postulates for congruent triangles? How are they different?

5. Two isosceles triangles each have a vertex angle of 50°. Must the triangles be similar? Explain your reasoning.

Explain why the triangles in each pair are similar.

1.

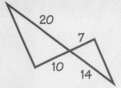

2.

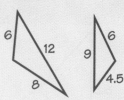

Is it possible to prove that the triangles in each pair are similar? Explain why or why not.

3.

4.

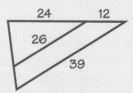

5. If two angles of a triangle are congruent to two angles of another triangle, what can you conclude about the triangles?

6. Complete: You can use the SAS Similarity Theorem to show that two triangles are similar if you know that: __?__ .

7. The length of each side of △ABC is one third the length of the corresponding side in △QRS. What can you conclude about the triangles? Why?

9.2 | **Exercises and Applications**

Extra Practice exercises on page 678

Is it possible to prove that the triangles in each pair are similar? Explain why or why not.

1.

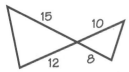

2.

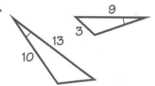

3.

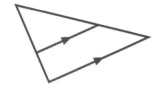

Choose two triangles in each diagram. Explain why they are similar.

4.

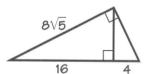

5.

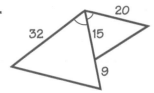

6.

7. Given: $\overline{BC} \parallel \overline{ED}$; $\overline{AB} \parallel \overline{DC}$
   Prove: $\triangle ABC \sim \triangle CDE$

8. Given: $\overline{PT} \parallel \overline{SR}$
   Prove: $\triangle QTP \sim \triangle QRS$

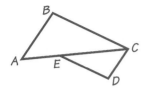

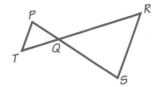

Connection ▶ SURVEYING

Surveyors use modern instruments to measure distances and heights accurately. You can estimate heights and distances using only a tape measure and your knowledge of similar triangles.

9. Writing Carol D'Anjow measured the length of her shadow and the length of the shadow of a tree.

  a. Use the measurements in the diagram to find the height of the tree.

  b. Explain why you must measure both shadows at about the same time for this method to work. Explain why this method may not work for a tree in the forest or a building in the city.

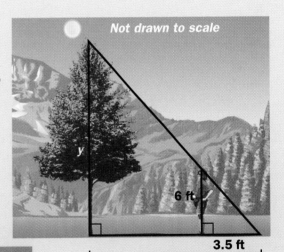

Not drawn to scale

6 ft

3.5 ft

24 ft

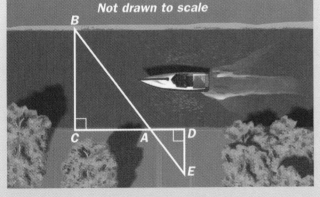

Not drawn to scale

10. a. Explain why $\triangle ABC \sim \triangle AED$ in the diagram at the left.

  b. Clarice Yee wants to estimate the width of a river. She measures three distances, and records that $AC = 5$ m, $AD = 2$ m, and $DE = 8.3$ m. Draw a diagram showing these measurements. How wide is the river?

11. Cooperative Learning Work with another person. Choose one of the methods used in Exercises 9 and 10. Use this method to estimate a height or distance in your neighborhood or at school. Tell which method you chose and explain how you estimated the height or distance.

12. Open-ended Problem In the diagram, $ABCD \sim EFGH$. Name three other pairs of similar figures in the diagram.

13. Challenge Use the diagram for Exercise 12. Name all the similar figures shown in the diagram.

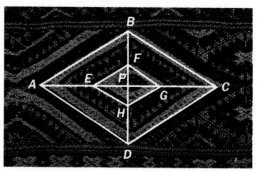

Indian tapestry from Peru

14. Copy and complete the proof.

Given: $\angle A \cong \angle P$
$\overline{AB} \cong \overline{AC}$
$\overline{PR} \cong \overline{PQ}$

Prove: $\triangle ABC \sim \triangle PQR$

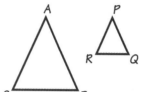

Statements	Reasons
1. $\underline{\ ?\ }$	**1.** Given
2. $\overline{AB} \cong \overline{AC}$; $\overline{PQ} \cong \overline{PR}$ ($AB = AC$; $PQ = PR$)	**2.** $\underline{\ ?\ }$
3. $\dfrac{AB}{PQ} = \dfrac{AC}{PR}$	**3.** Algebra
4. $\triangle ABC \sim \triangle PQR$	**4.** $\underline{\ ?\ }$

15. a. Logical Reasoning Describe how to find the value of c without using the Pythagorean theorem.

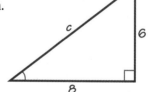

b. Find each missing length without using the Pythagorean theorem.

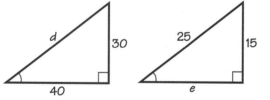

16. Given: $\dfrac{XZ}{PR} = \dfrac{YZ}{QR}$

$\angle Z$ and $\angle R$ are right angles.

Prove: $\triangle XYZ \sim \triangle PQR$

17. Given: $\overline{AB} \perp \overline{AE}$
$\overline{ED} \perp \overline{AE}$

Prove: $\triangle ABC \sim \triangle EDC$

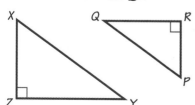

The total *resistance* for a circuit depends on the resistance of each *resistor* in the circuit. Let R = the total resistance for the circuit, R_1 = the resistance of one part of the circuit, and R_2 = the resistance of a second part of the circuit. You can use the formula $\frac{1}{R} = \frac{1}{R_1} + \frac{1}{R_2}$ to find the total resistance for the circuit.

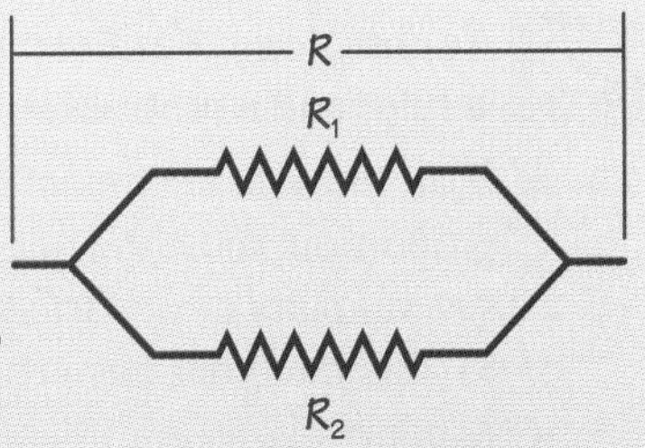

Follow these steps to find the value of R for a circuit with the given values of R_1 and R_2.

Step 1 Plot R_1 and R_2.

Step 2 Draw $\overline{R_1R_2}$.

Step 3 The intersection of $\overline{R_1R_2}$ and the R-axis shows the total resistance.

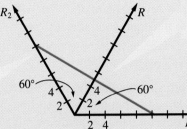

18. $R_1 = 10; R_2 = 10$ **19.** $R_1 = 15; R_2 = 10$ **20.** $R_1 = 24; R_2 = 12$

21. Challenge You can use the diagram below to explain why the graphs you made in Exercises 18–20 give the correct resistance for the circuits. Follow these steps to prove that in the diagram $\frac{1}{R} = \frac{1}{R_1} + \frac{1}{R_2}$. Let $x = R_1$, $y = R_2$, and $z = R$.

a. Extend $\overrightarrow{BA}$ through E so that $\overline{EC} \parallel \overline{AD}$. Explain why $\triangle AEC$ is equilateral. Find the length EC.

b. Explain why $\triangle ECB \sim \triangle ADB$.

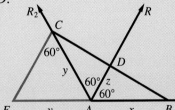

c. Use the proportion $\frac{EB}{AB} = \frac{EC}{AD}$ to show that $xy = xz + yz$.

d. Divide both sides of the equation in part (c) by xyz to show that $\frac{1}{z} = \frac{1}{y} + \frac{1}{x}$. Explain why the proof is now complete.

BY THE WAY

Edith Clarke (1883–1959) developed a method for finding solutions to problems of power loss in electric circuits that is still used by engineers.

22. a. Each side of regular pentagon *ABCDE* is x units long, and each side of regular pentagon *PQRST* is y units long. Explain why $ABCDE \sim PQRST$.

   b. Writing Are all regular polygons with the same number of sides similar? Explain.

23. Writing Which similarity relationship is most closely related to the ASA Theorem for congruent triangles? Explain your choice.

 Technology For Exercises 24–26, use geometry software or a ruler and a protractor.

24. a. If all the side lengths of two quadrilaterals are in proportion, must the quadrilaterals be similar? Draw several pairs of quadrilaterals with sides in proportion to illustrate your answer.

   b. Is your conclusion in part (a) true for polygons with more than four sides? Give an example or a counterexample.

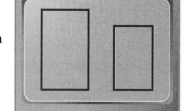

25. a. If all of the corresponding angles of two quadrilaterals are congruent, must the quadrilaterals be similar? Draw several pairs of quadrilaterals with congruent angles to illustrate your answer.

   b. Is your conclusion in part (a) true for polygons with more than four sides? Give an example or a counterexample.

26. If two quadrilaterals have two pairs of corresponding angles that are congruent and these angles are included by sides that are in proportion, must the quadrilaterals be similar? Draw several pairs of quadrilaterals to illustrate your answer.

ONGOING ASSESSMENT

27. a. Open-ended Problem A student made this conjecture based on the proof in Exercise 14. Show that this conjecture is false by finding a counterexample.

   b. Writing Rewrite the conjecture to make it a true statement. Explain why your statement is always true.

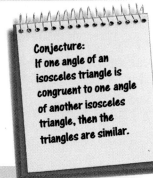

Conjecture:
If one angle of an isosceles triangle is congruent to one angle of another isosceles triangle, then the triangles are similar.

SPIRAL REVIEW

28. Write a two-column proof. *(Section 6.3)*

   Given: $\overline{AD} \cong \overline{DC}$; $\angle ADB \cong \angle BDC$

   Prove: $\triangle ABD \cong \triangle CBD$

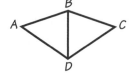

Tell whether the polygons in each pair are similar. Explain your reasoning.
(Section 9.1)

29.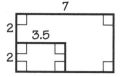

30.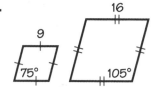

31.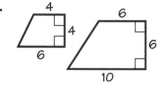

9.3 Proportions and Similarity

Learn how to...

Learn how to...
- **write many different proportions using the sides of similar figures**

So you can...
- **find lengths and measures in similar polygons**

You know that in similar figures the lengths of the sides are in proportion. You can rewrite these proportions in several ways to help you learn more about similar figures. In the Exploration, you will investigate one way that parallel lines and similar figures are related.

EXPLORATION
COOPERATIVE LEARNING

Finding Proportions in Triangles

Work with another student.
You will need:
- geometry software, or patty paper and a ruler

1 Draw a scalene triangle, $\triangle PQR$. Construct $\overline{ST}$ so that $\overline{ST} \parallel \overline{PQ}$ and $\overline{ST}$, $\overline{RT}$, and $\overline{RS}$ form a smaller triangle, $\triangle RST$. Find the lengths of the sides of the triangles.

2 Find $\dfrac{PS}{SR}$ and $\dfrac{QT}{TR}$. Record your results.

3 Explain why $\triangle RPQ \sim \triangle RST$.

4 Each of you should repeat Steps 1 and 2 with three different triangles. Make a conjecture about these triangles based on your results.

5 Compare your conjectures with the conjectures of other groups.

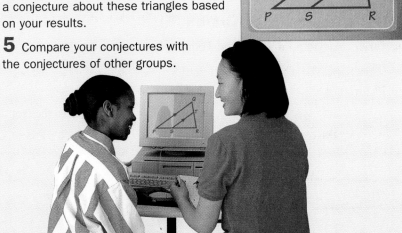

In the Exploration, you saw that a line that intersects two sides of a triangle and is parallel to the third side forms a triangle that is similar to the original one. The proof of this theorem is on page 463.

Triangle Proportionality Theorem

If a segment is parallel to one side of a triangle and intersects the other two sides, then it divides those sides proportionally.

If $\overline{ST} \parallel \overline{QR}$, then $\dfrac{PS}{SQ} = \dfrac{PT}{TR}$.

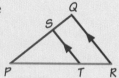

The *midsegment* joins the midpoints of two segments. The Midsegment Theorem is a special case of the Triangle Proportionality Theorem.

Midsegment Theorem

If the midpoints of two sides of a triangle are joined by a segment, then the segment is parallel to the third side of the triangle and its length is half the length of the third side of the triangle.

If D is the midpoint of $\overline{AB}$ and E is the midpoint of $\overline{BC}$, then $\overline{DE} \parallel \overline{AC}$ and $DE = \frac{1}{2}AC$.

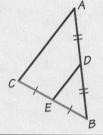

EXAMPLE 1   Connection: Algebra

Find the length JN.

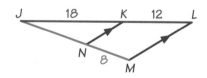

SOLUTION

Use the Triangle Proportionality Theorem.

$$\frac{JN}{NM} = \frac{JK}{KL}$$

$$\frac{JN}{8} = \frac{18}{12} \qquad \text{Find the cross products.}$$

$$12 \cdot JN = 8 \cdot 18$$

$$JN = \frac{144}{12}$$

$$JN = 12$$

Check

$$\frac{12}{8} \overset{?}{=} \frac{18}{12}$$

$$\frac{3}{2} = \frac{3}{2} ✔$$

Proportions

Two equations are *equivalent* if one equation can be changed into the other using algebra. For example, $\frac{a}{b} = \frac{c}{d}$ is equivalent to $ad = bc$ because you can get the second equation by multiplying both sides of the first equation by bd.

Properties of Proportions

All of the proportions below are equivalent to each other.

$$\frac{a}{b} = \frac{c}{d} \qquad \frac{b}{a} = \frac{d}{c} \qquad \frac{a}{c} = \frac{b}{d} \qquad \frac{c}{a} = \frac{d}{b}$$

These proportions are also equivalent to the ones above.

$$\frac{a}{a + b} = \frac{c}{c + d} \qquad \frac{b}{a + b} = \frac{d}{c + d} \qquad \frac{a + b}{a} = \frac{c + d}{c}$$

THINK AND COMMUNICATE

In the diagram, $\overline{EF} \parallel \overline{JG}$. Complete each proportion.

1. $\dfrac{y}{x} = \underline{\ ?\ }$

2. $\dfrac{x}{x + y} = \underline{\ ?\ }$

3. $\dfrac{5}{x} = \underline{\ ?\ }$

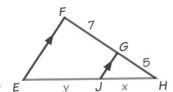

When you use equivalent proportions in proofs, give *A property of proportions* as the reason. Equivalent proportions can be used to prove the Triangle Proportionality Theorem.

Given: $\overline{ST} \parallel \overline{QR}$

Prove: $\dfrac{PS}{SQ} = \dfrac{PT}{TR}$

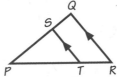

Statements	Reasons
1. $\overline{ST} \parallel \overline{QR}$	**1.** Given
2. $\angle PST \cong \angle PQR$; $\angle PTS \cong \angle PRQ$	**2.** If two $\parallel$ lines are intersected by a transversal, then corresponding angles are $\cong$.
3. $\triangle PST \sim \triangle PQR$	**3.** AA Similarity Postulate
4. $\dfrac{PS}{PQ} = \dfrac{PT}{PR}$	**4.** Definition of similar figures
5. $PQ = PS + SQ$; $PR = PT + TR$	**5.** Segment Addition Postulate
6. $\dfrac{PS}{PS + SQ} = \dfrac{PT}{PT + TR}$	**6.** Substitution Property (Steps 4 and 5)
7. $\dfrac{PS}{SQ} = \dfrac{PT}{TR}$	**7.** A property of proportions

EXAMPLE 2

Find each value.

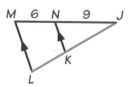

a. $\dfrac{LK}{KJ}$ b. $\dfrac{KN}{LM}$

SOLUTION

Use the Triangle Proportionality Theorem.

a. $\dfrac{LK}{KJ} = \dfrac{MN}{NJ} = \dfrac{6}{9} = \dfrac{2}{3}$ b. $\dfrac{KN}{LM} = \dfrac{JN}{JM} = \dfrac{9}{15} = \dfrac{3}{5}$

☑ CHECKING KEY CONCEPTS

Tell whether each proportion is *True* or *False*.

1. $\dfrac{CD}{DF} = \dfrac{CE}{EG}$ 2. $\dfrac{DE}{FG} = \dfrac{CG}{CE}$

3. $\dfrac{CE + EG}{CG} = \dfrac{CD + DF}{CF}$ 4. $\dfrac{CE}{DF} = \dfrac{CD}{EG}$

Find the value of each ratio or length.

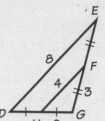

5. a. $\dfrac{GH}{HD}$ 6. a. $\dfrac{PT}{TS}$

   b. EF b. $\dfrac{TQ}{SR}$

   c. DG c. $\dfrac{PS}{PT}$

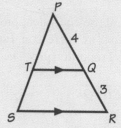

9.3 **Exercises and Applications**

Extra Practice
exercises on
page 678

Find the value of each variable.

1.

2.
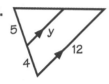

3.
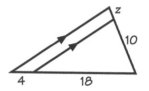

ALGEBRA Complete each equation so it is equivalent to the proportion $\dfrac{x}{y} = \dfrac{3}{4}$.

4. $\dfrac{y}{x} = \underline{\ ?\ }$ 5. $\dfrac{4}{y} = \underline{\ ?\ }$ 6. $\dfrac{4}{4 + y} = \underline{\ ?\ }$

For Exercises 7 and 8, use the diagram at the right. *AB* = 3 cm, *CA* = 4 cm, and *CB* = 4.5 cm.

7. *D* is the midpoint of $\overline{AC}$ and *E* is the midpoint of $\overline{CB}$. Find the lengths *CD*, *CE*, and *DE*.

8. *F* is one fourth the distance from *C* to *A*, and *G* is one fourth the distance from *C* to *B*. Find the lengths *CF*, *CG*, and *FG*.

9. a. Sketch △*KLM* so that $\overline{EF}$ connects the midpoints of $\overline{KM}$ and $\overline{KL}$, $\overline{CD}$ connects the midpoints of $\overline{KE}$ and $\overline{KF}$, and $\overline{AB}$ connects the midpoints of $\overline{KC}$ and $\overline{KD}$.

 b. *AB* = 4 cm. Find the lengths *CD*, *EF*, and *ML*.

 c. *KA* = 3 cm and *KB* = 6 cm. Find the lengths *EM*, *FL*, and *KF*.

Loy Arcenas

Look back at the article on pages 442–444.

Loy Arcenas gives both a scale model and blueprints to the builders of a stage set he has designed. The blueprints and model shown here are for a production of the musical **Once On This Island,** *by Lynn Ahrens and Stephen Flaherty.*

10. **Writing** Loy Arcenas always includes a model of a person in his scale models. Explain how this can help him understand whether he has made the pieces of the set a reasonable size.

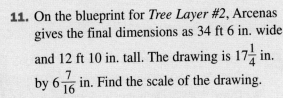

11. On the blueprint for *Tree Layer #2*, Arcenas gives the final dimensions as 34 ft 6 in. wide and 12 ft 10 in. tall. The drawing is $17\frac{1}{4}$ in. by $6\frac{7}{16}$ in. Find the scale of the drawing.

12. The scale of the model stage is 1:48. If Arcenas made *Tree Layer #2* for his scale model, what would its dimensions be? The final dimensions are given in Exercise 11.

13. **Open-ended Problem** In the drawing for a *Typical Flower*, Arcenas does not give a scale. He tells the builder to choose a size for the 25 flowers based on the tree layers. What is a reasonable size for a *Typical Flower*? Explain your reasoning.

This theorem is related to the Triangle Proportionality Theorem: Parallel lines divide transversals into segments that are in proportion. Use this theorem to complete each proportion.

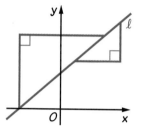

14. $\dfrac{FE}{ED} = \underline{\ ?\ }$ **15.** $\dfrac{CB}{DE} = \underline{\ ?\ }$ **16.** $\dfrac{AC}{AB} = \underline{\ ?\ }$

17. a. Investigation On a coordinate plane, draw a line, ℓ, that is not parallel to an axis. Draw three different right triangles, each with hypotenuse on ℓ and both legs parallel to the axes. Explain why your three triangles are similar.

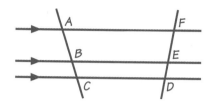

  b. Draw a line j that is parallel to ℓ. Draw a right triangle with hypotenuse on j and legs parallel to the axes. Is this triangle similar to the ones you drew in part (a)? Explain your reasoning.

ALGEBRA Use algebra to show that each statement is true.

18. The proportion $\dfrac{a}{b} = \dfrac{c}{d}$ is equivalent to $\dfrac{a+b}{b} = \dfrac{c+d}{d}$.

19. The proportion $\dfrac{a}{b} = \dfrac{c}{d}$ is equivalent to $\dfrac{b}{a} = \dfrac{d}{c}$.

20. The proportion $\dfrac{a}{b} = \dfrac{c}{d}$ is equivalent to $\dfrac{a}{a-b} = \dfrac{c}{c-d}$.

21. SAT/ACT Preview A 5 ft tall sign has a 4 ft shadow. At the same time, a nearby tree has a 15 ft shadow. How tall is the tree?

  A. $20\dfrac{3}{4}$ ft   **B.** $16\dfrac{1}{2}$ ft   **C.** $18\dfrac{3}{4}$ ft   **D.** 12 ft   **E.** $20\dfrac{1}{2}$ ft

22. Challenge Write a proof in any format.

  Given: $\overline{JS} \parallel \overline{KM};\ \overline{SC} \parallel \overline{EK}$

  Prove: $\dfrac{EJ}{SC} = \dfrac{JS}{CM}$

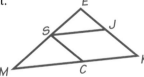

ONGOING ASSESSMENT

23. Cooperative Learning Work with another person. Use geometry drawing software or a compass and ruler.

  a. Draw $\triangle RST$. Draw the bisector of $\angle T$. Label the point where the bisector intersects $\overline{RS}$ as point Y. Find the lengths of the segments and the ratios $\dfrac{ST}{RT}$ and $\dfrac{SY}{RY}$.

  b. Repeat part (a) with at least three other triangles.

  c. Writing Write a summary of your results.

24. An isosceles right triangle with sides of lengths 1, 1, and $\sqrt{2}$ is dilated using a scale factor of 3. What is the perimeter of the image? What is the area of the image? *(Section 8.6)*

Find the area of each polygon. *(Section 7.4)*

25. a rectangle whose sides are 13 ft and 8 ft long

26. a triangle whose base is 5 m and whose height is 6 m

27. a rectangle with a side of length 12 in. and a diagonal 13 in.

Express each probability as a decimal between 0 and 1, inclusive.
(Toolbox, page 710)

28. a 100% chance   **29.** 1 chance in 1000   **30.** a 3 in 8 chance   **31.** a 4% chance

ASSESS YOUR PROGRESS

VOCABULARY

proportion (p. 446)   **similar** (p. 446)

Solve each proportion. *(Section 9.1)*

1. $\dfrac{4}{10} = \dfrac{x}{25}$ **2.** $3:y = 15:21$ **3.** $z:6 = 36:27$

Tell whether the polygons in each pair are similar. *(Sections 9.1 and 9.2)*

4. 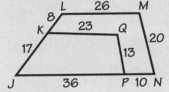   **5.**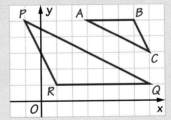

The triangles in each pair are similar. Find the value of each variable.
(Section 9.3)

6. 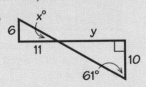   **7.**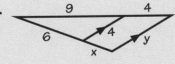

Complete each equation so it is equivalent to the proportion $\dfrac{x}{y} = \dfrac{2}{5}$.
(Section 9.3)

8. $\dfrac{y}{x} = \underline{\ ?\ }$ **9.** $\dfrac{2}{x} = \underline{\ ?\ }$ **10.** $\dfrac{5+y}{y} = \underline{\ ?\ }$

11. Journal The symbol for congruence, $\cong$, is a combination of the symbol for equality, $=$, and the symbol for similarity, $\sim$. Do you think this makes sense? Explain why or why not.

9.4 | Areas and Volumes of Similar Figures

Learn how to...
- compare measures of similar figures

So you can...
- find perimeters, areas, and volumes of similar figures

When Loy Arcenas designs a stage set, he has to consider the amount of space available for the set. The set builder uses Arcenas' model and scale drawings to estimate the amount of material needed to build the set. Relationships between similar figures can help the builder make decisions about area and materials.

The rectangle in the scale drawing is similar to the rectangular platform in the stage set.

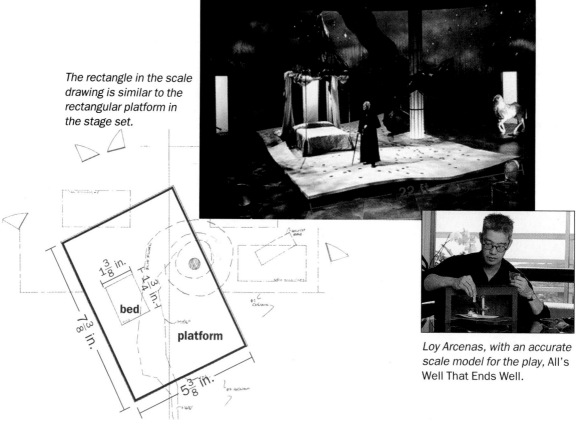

Loy Arcenas, with an accurate scale model for the play, All's Well That Ends Well.

THINK AND COMMUNICATE

1. What is the ratio of a side length in the scale drawing to the corresponding side length in the stage set?

2. Find the perimeters of the two rectangles. What is the ratio of the perimeters?

3. What is the ratio of the areas of the two rectangles?

Examples of linear measures of a figure are lengths of its sides, medians, diameters, altitudes, perimeter, and circumference. In similar figures, all ratios of a linear measure of one figure to the corresponding measure in the other are equal. As you saw on page 468, the areas of similar figures are related by a different ratio.

Perimeters and Areas of Similar Figures

If two similar two-dimensional figures have corresponding lengths whose ratio is $a:b$, then:

- The ratio of all corresponding linear measures is $a:b$.

$$\frac{JK}{PQ} = \frac{KL}{QR} = \frac{JL}{PR} = \frac{2}{5}$$

- The ratio of their perimeters is $a:b$.

$$\frac{\text{Perimeter of } \triangle JKL}{\text{Perimeter of } \triangle PQR} = \frac{2}{5}$$

- The ratio of their areas is $a^2:b^2$.

$$\frac{\text{Area of } \triangle JKL}{\text{Area of } \triangle PQR} = \frac{2^2}{5^2} = \frac{4}{25}$$

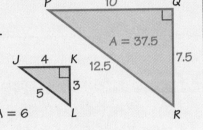

EXAMPLE 1

a. Explain why $\triangle LMN \sim \triangle TUV$.

b. Find the ratio of the lengths of the corresponding sides.

c. Find the ratio of the perimeters.

d. Find the ratio of the areas.

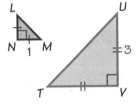

SOLUTION

a. $\frac{LN}{TV} = \frac{MN}{UV}$ and $\angle N \cong \angle V$, so $\triangle LMN \sim \triangle TUV$ by the SAS Similarity Theorem.

b. The triangles are similar and $MN:UV = 1:3$, so $LN:TV = 1:3$ and $LM:TU = 1:3$.

c. The ratio of the perimeters is also $1:3$.

d. The ratio of the areas of $\triangle LMN$ and $\triangle TUV$ is $1^2:3^2 = 1:9$.

THINK AND COMMUNICATE

4. All squares are similar. Do you think that all cubes are similar? Explain your reasoning.

5. Are all circles similar? Explain your reasoning.

6. Two right rectangular prisms have similar bases. If the prisms are similar, what other facts about the prisms must be true?

Areas and Volumes of Similar Figures

If two similar three-dimensional figures have corresponding lengths whose ratio is $a:b$, then:

- The ratio of all corresponding linear measures is $a:b$.
$$\frac{AB}{PQ} = \frac{DE}{ST} = \frac{3}{4}$$

- The ratio of their surface areas (S.A.) is $a^2:b^2$.
$$\frac{S.A. \text{ of I}}{S.A. \text{ of II}} = \frac{3^2}{4^2} = \frac{9}{16}$$

- The ratio of their volumes (V) is $a^3:b^3$.
$$\frac{V \text{ of I}}{V \text{ of II}} = \frac{3^3}{4^3} = \frac{27}{64}$$

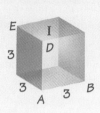
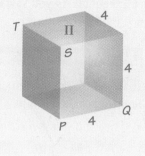

EXAMPLE 2   Connection: Algebra

radius = 2 cm
surface area = 32π cm^2
volume = 24π cm^3

radius = 3 cm

These cylinders are similar. Find the surface area and volume of the larger cylinder.

SOLUTION

Find the ratio of a linear measure: $\frac{a}{b} = \frac{\text{radius of smaller cylinder}}{\text{radius of larger cylinder}} = \frac{2}{3}$.

Let S = the surface area of the larger cylinder.

$$\frac{S.A. \text{ of smaller cylinder}}{S.A. \text{ of larger cylinder}} = \frac{2^2}{3^2} = \frac{32\pi}{S}$$

The ratio of the surface areas is $\frac{a^2}{b^2}$.

$$4S = 9 \cdot 32\pi$$

$$S = \frac{288\pi}{4}$$

$$S = 72\pi$$

The surface area of the larger cylinder is 72π cm^2.
Let V = the volume of the larger cylinder.

$$\frac{\text{Volume of smaller cylinder}}{\text{Volume of larger cylinder}} = \frac{2^3}{3^3} = \frac{24\pi}{V}$$

The ratio of the volumes is $\frac{a^3}{b^3}$.

$$8V = 27 \cdot 24\pi$$

$$V = \frac{648\pi}{8}$$

$$V = 81\pi$$

The volume of the larger cylinder is 81π cm^3.

For Questions 1–3, △*PQR* ∼ △*JKL*.

1. Find the ratio $\frac{QR}{KL}$.

2. Find the ratio of the areas.

3. Find the ratio of the perimeters.

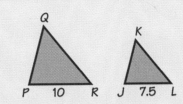

For Questions 4 and 5, use the similar prisms at the right.

4. Find the surface area of II.

5. Find the volume of I.

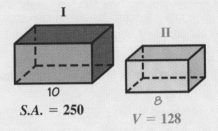

I

II

10

S.A. = 250

8

V = 128

9.4 | **Exercises and Applications**

Extra Practice exercises on page 679

The figures in each pair are similar. Find each missing value.

1. $P = 25.2$ $P = \underline{?}$
   $A = \underline{?}$ $A = \underline{?}$

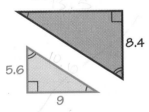

8.4

5.6

9

2. $P = 28$ $P = \underline{?}$
   $A = \underline{?}$ $A = 6$

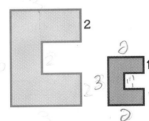

2

1

3. $P = 24$ $P = 36$
   $A = 28$ $A = \underline{?}$
   $EF = \underline{?}$

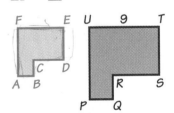

F E U   9   T

C D

A B   R   S

P   Q

4. $C = \underline{?}$ $C = 10\pi$
   $A = 9\pi$ $r = \underline{?}$
            $A = 25\pi$

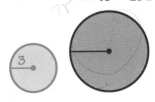

3

5. Area of △*ABE* = 24
   Area of △*ACD* = 54
   $DC = \underline{?}$

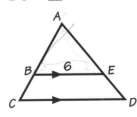

A

B   6   E

C   D

6. $P = 32$ $P = \underline{?}$
   $A = \underline{?}$ $A = 108$

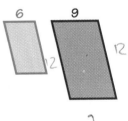

6   9

12   12

7. △*ABC* ∼ △*PQR* and *AB*:*PQ* = 3:5.

   a. What is the ratio of the areas of the triangles?

   b. The lengths of the sides of △*ABC* are 6, 12, and 12. Find the length of the altitude to the shortest side of △*ABC*. What is the corresponding length in △*PQR*?

Tell whether the solids in each pair are *similar*, *not similar*, or if there is *not enough information to determine*.

8.

9.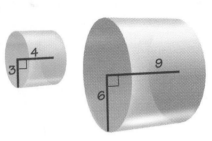

10. Visual Thinking Sketch and label the dimensions of two similar three-dimensional figures that are not congruent. How do you know that they are similar?

For Exercises 11–14, sketch the figures. Mark the given information on your sketch.

11. Two equilateral triangles have altitudes that are 2.4 in. and 5.2 in. long. What is the ratio of corresponding side lengths? of the perimeters? of the areas?

12. Two regular octagons have sides that are 8 in. and 9 in. long, respectively. What is the ratio of two corresponding diagonals?

13. The volumes of two cubes are in the ratio 8 : 125. What is the ratio of the lengths of their sides? of their surface areas?

14. Challenge The sides of a triangle are 12 cm, 16 cm, and 20 cm long. How long are the sides of a similar triangle whose area is 150 cm^2?

15. If the lengths of the sides of a triangle are doubled, how does this change the length of the medians? the perimeter? the area of the triangle?

16. **Technology** Use graphing software or a graphing calculator.

    a. Sketch six cubes with different edge lengths. Record the edge length and find the surface area and volume of each cube. Use a spreadsheet or record your data in a table.

    b. Graph the surface area as a function of edge length. Describe the shape of the graph.

    c. Graph the volume as a function of edge length. Describe the shape of the graph.

Cubes		
A	**B**	**C**
Edge length (cm)	Surface area (cm^2)	Volume (cm^3)
2	24	8
3		
0.5		

    d. Writing What do the graphs in parts (b) and (c) tell you about how surface area and volume change as the edge length changes?

17. SET DESIGN Look at the scale drawing of the stage set for *All's Well That Ends Well* on page 468. What are the dimensions of the bed in the stage set? Explain your reasoning.

18. a. The sides of quadrilateral *ABCD* are dilated by a scale factor of 13. If the area of *ABCD* is *x*, find the area of the image in terms of *x*.

b. If the area of the image is *w*, find the area of *ABCD* in terms of *w*.

19. Challenge The volumes of two similar cylinders are 135π cm^3 and 320π cm^3. Find the ratio of their surface areas.

Connection ▶ CONSUMER ECONOMICS

Manufacturers often sell the same item in different size packages so that you can purchase a *single serving* or an *economy size* package of the same product. The packages often appear to be similar so that consumers will recognize them as the same product.

8 oz   No. 2

20. a. An *8 oz* can has diameter 3 in. and height 4 in. A *No. 2* can is similar to an *8 oz* can and has a diameter of $3\frac{7}{16}$ in. Find the height of a *No. 2* can.

b. How many ounces does a *No. 2* can hold? Round your answer to the nearest ounce. (*Hint:* Remember that an ounce is a measure of volume.)

7.2 cm
10.3 cm
4.2 cm

21. a. Is the large box of cereal similar to the medium size box? Is the large box similar to the single serving box? Explain.

b. The rectangular front faces of all three boxes are similar. Explain why the manufacturer might want the faces to be similar even if the boxes are not similar.

15.8 cm
22.5 cm
4.7 cm

22. a. Find the volume of each of the three boxes of cereal.

b. If the large box costs $3.29, give an appropriate price for each of the other two boxes.

c. Open-ended Problem Often a product in a larger size package costs less per unit than a smaller size package. Give a reason why this is the case and write a convincing argument for your idea.

23. Research Find another product that comes in two different size containers. Are the containers similar? Sketch the containers and give the dimensions of each container. Why might you decide to buy the smaller container? the larger container?

19.4 cm
27.5 cm
5.5 cm

24. TRANSFORMATIONS If the scale factor of a dilation is 0.25, what is the ratio of the perimeter of the original polygon to the perimeter of its image? What is the ratio of the areas?

25. ALGEBRA Let s = the length of a side of a square. Prove using algebra that if the square is dilated by a scale factor of k, the area is changed by a scale factor of k^2.

26. a. Investigation Use two $8\frac{1}{2}$ in. by 11 in. sheets of paper and roll each to form an open-ended cylinder. One cylinder has height $8\frac{1}{2}$ in., and the other has height 11 in. Are the two cylinders similar?

b. Explain why the open-ended cylinders have the same lateral surface area. Do they have the same volume? Explain your reasoning.

27. Writing Many folktales and children's stories describe what it would be like to be very small. Write a paragraph about a one-inch tall person. Describe three ordinary objects that this person could use and what each could be used for.

ONGOING ASSESSMENT

28. Visual Thinking On graph paper, draw a triangle so that its vertices are on grid points.

a. Find the midpoint of each side of the triangle. Connect the midpoints to create four new triangles.

b. Explain why the four new triangles are all congruent to each other and why each one is similar to the original triangle.

c. Give two different reasons why the area of each of the four new triangles is one fourth the area of the original triangle.

SPIRAL REVIEW

29. $ABCD$ is a trapezoid with $\overline{AB} \parallel \overline{CD}$ and diagonals intersecting at E. Prove: $\triangle ABE \sim \triangle CDE$. *(Section 9.2)*

30. In $\triangle JKL$, $\overline{ST} \parallel \overline{LK}$, $JS = 6$, $SL = 3$, and $ST = 5$. Find the length LK. *(Section 9.3)*

31. Miguel studied a population of 43 sea lions. The following year, 32 of these sea lions returned to the area. What is the probability that any given sea lion returned? *(Toolbox, page 710)*

Write each fraction as a percent.

32. $\frac{1}{2}$ **33.** $\frac{8}{7}$ **34.** $\frac{3}{5}$ **35.** $\frac{15}{4}$

Write each percent as a fraction in lowest terms.

36. 18%   **37.** 0.5%   **38.** 210%   **39.** 95%

9.5 Geometric Probability

Scientists may use geometric probability to estimate the likelihood that animals will find their hidden food by searching randomly. This helps scientists decide if the animals are using other techniques, such as memory or smell. In the Exploration, you will use geometric probability to find the likelihood that a randomly thrown object will land on a target.

EXPLORATION
COOPERATIVE LEARNING

Investigating Probability Based on Area

Work in a group of three students.
You will need:
- paper
- dried beans

1 Make a target by drawing several shapes that you can find the area of on a rectangular sheet of paper. Shade in your shapes.

2 Place your target on the floor near a wall. Toss a bean against the wall so that it will land on your target. Record whether the bean lands on a shaded part of the paper. If the bean misses the paper, or lands on a border, do not count that toss.

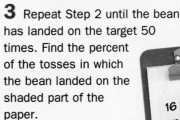

3 Repeat Step 2 until the bean has landed on the target 50 times. Find the percent of the tosses in which the bean landed on the shaded part of the paper.

4 Find the percent of the area of the paper that is shaded.

5 Compare your results in Steps 3 and 4. What do you think is the probability that if a bean lands on the target, it lands in the shaded area? Compare your results and predictions with those of other groups.

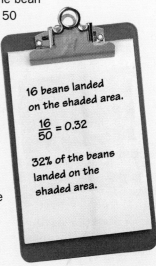

16 beans landed on the shaded area.

$$\frac{16}{50} = 0.32$$

32% of the beans landed on the shaded area.

Toolbox p. 710
Probability

Probability is a ratio that tells how likely an event is to happen. For example, if a teacher randomly chooses 5 students from a class of 20 students:

$$\text{Probability of being chosen} = \frac{\text{Number of students chosen}}{\text{Number of students in class}} = \frac{5}{20}$$

The probability can also be written as $\frac{1}{4}$, or 0.25, or 25%.

In **geometric probability**, the probability of an event is given by ratios that compare lengths, areas, perimeters, or other measures. In the Exploration, you investigated the experimental probability of hitting a target with a randomly tossed object. If you cannot count the number of possible outcomes, you can use areas to find the theoretical probability.

EXAMPLE 1

For each rectangular target, find the probability that a bean tossed at random will land on the shaded area.

a.

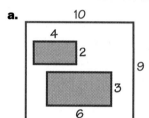

b.

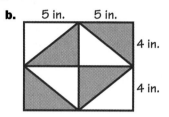

SOLUTION

a. Find the shaded area and the total area.

$$\text{Shaded area} = 4 \cdot 2 + 6 \cdot 3 = \textbf{26 in.}^2$$
$$\text{Total area} = 10 \cdot 9 = \textbf{90 in.}^2$$

$$\begin{aligned} \text{Probability of hitting} \\ \text{a shaded area} \end{aligned} = \frac{\text{Shaded area}}{\text{Total area}}$$

$$= \frac{26}{90}$$

$$= \frac{13}{45}$$

The probability of hitting the shaded area is $\frac{13}{45}$, or about 0.29.

b. The rectangle is divided into 8 congruent triangles. Four of the triangles are shaded.

$$\begin{aligned} \text{Probability of hitting} \\ \text{a shaded area} \end{aligned} = \frac{\text{Number of shaded triangles}}{\text{Total number of triangles}}$$

$$= \frac{4}{8}$$

$$= \frac{1}{2}$$

The probability of hitting the shaded area is $\frac{1}{2}$.

THINK AND COMMUNICATE

1. Give each probability in Example 1 as a percent.

2. In part (a) of Example 1, what is the probability that the bean will land on the unshaded part of the target?

The probabilities in the Exploration and Example 1 use areas. You can also find the probability of an event using lengths. For example, if you chose a point P on $\overline{AB}$ at random, the probability that P is on $\overline{AC} = \dfrac{\text{length of } \overline{AC}}{\text{length of } \overline{AB}}$.

EXAMPLE 2 **Application: Consumer Economics**

A store uses cash register tape that has a red star printed on it every 200 cm. If the register tape for your receipt has a red star on it, you will receive 10% off your next purchase at the store. Joel received 25 cm of cash register tape with his purchase. What is the probability that he received a red star?

SOLUTION

To find the probability, set up a ratio.

$$\begin{aligned} \text{Probability of receiving} \atop \text{a red star} &= \frac{\text{Length of receipt}}{\text{Distance between red stars}} \\ &= \frac{25}{200} \\ &= 0.125 \end{aligned}$$

The probability that Joel received a red star is 12.5%.

THINK AND COMMUNICATE

3. In Example 2, at least how long would Joel's receipt need to be to guarantee that he receives a red star?

4. If the red stars were 150 cm apart, would Joel be *more* or *less* likely to receive a red star than he is in Example 2?

5. If an event has a probability of $\frac{1}{2}$, how often would you expect it to happen? Will it always occur this often?

6. If an event has a probability of 1, how often would you expect it to happen? If an event has a probability of 0, how often would you expect it to happen?

For Questions 1 and 2, find the probability of hitting the shaded area of the target with a bean tossed at random that hits the target.

1.

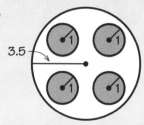

3.5

2.

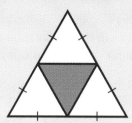

3. If the area of a target is z square units and the area of the shaded area on the target is w square units, what is the probability of a bean tossed at random landing on the shaded area of the target?

4. A store uses register tape that has a green star printed every 75 cm. If the register tape that your receipt is printed on has a green star, you will receive a prize. If your receipt is 9 cm long, what is the probability that you will receive a prize?

5. What is the probability that a point selected at random on $\overline{AB}$ is closer to B than to A?

9.5 | **Exercises and Applications**

Extra Practice exercises on page 679

For Exercises 1–4, find the probability of hitting the shaded area of the target with a bean tossed at random that hits the target.

1.

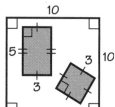

10

5

3

3

10

2.

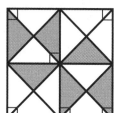

3.

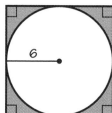

6

4.

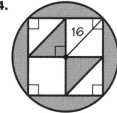

16

5. The midpoint of $\overline{AB}$ is M. What is the probability that a point selected at random on the segment is closer to M than it is to A or to B?

6. a. Visual Thinking In $\triangle ABC$, $\overline{DE}$ connects the midpoints of $\overline{AC}$ and $\overline{BC}$. What is the ratio of the area of $\triangle CDE$ to the area of $\triangle ABC$?

b. If $\triangle ABC$ is a dart board, what is the probability of a dart thrown at random that hits the target landing on $\triangle CDE$? on $ABED$?

7. Open-ended Problem Design a target for which the probability of a bean tossed at random landing on the shaded area is $\frac{1}{3}$.

If you choose a random point on a side of the polygon, what is the probability that it is on the given segment?

8. $\overline{AB}$

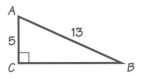

9. $\overline{QR}$

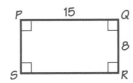

10. a. In $\triangle ABC$, $\overline{XY} \parallel \overline{AB}$, $CA = 3 \cdot CX$, and $CB = 3 \cdot CY$. What is the ratio of the area of $\triangle XYC$ to the area of $\triangle ABC$?

　b. If $\triangle ABC$ is a dart board, what is the probability of a dart thrown at random hitting $ABYX$?

11. a. A square dart board is 3 ft wide. It has three concentric circles on it. The radii of the circles are 15 in., 10 in., and 5 in., respectively. Darts are thrown and hit the target at random. What is the probability of hitting the area inside the smallest circle?

　b. What is the probability of hitting the area inside the largest circle but not inside the 10 in. circle?

Connection ▶ ZOOLOGY

In the wild, many animals collect and hide food when it is plentiful and retrieve it when food sources are scarce. Research suggests that the animals may use landmarks to remember the location of hidden food.

　In one study, a researcher used a rectangular site 116 cm by 238 cm. Landmarks such as rocks and bare twigs were added and chipmunks were allowed to bury seeds in the site.

12. Suppose a chipmunk buried 30 caches of seeds in this rectangular site. The area in which a chipmunk must dig to locate a cache is about 4.5 cm². What is the probability the chipmunk would discover one of its caches by choosing four places to dig at random?

13. Writing In one trial, a chipmunk located three caches in four digs, for a success rate of 75%. Do you think it is likely that the chipmunk searched at random? Explain your reasoning.

14. Open-ended Problem The research suggests that chipmunks use landmarks to find hidden food. What other methods might they use?

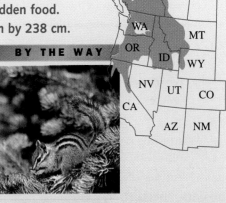

BY THE WAY

Yellow pine chipmunks are common in the western United States. They collect and hide seeds in late summer and early fall, and they return later to eat the seeds or move them to underground nests for use during the winter.

15. BALLOONING In one event at the Albuquerque International Balloon Fiesta, balloon pilots try to hit a target on the ground with a small bag of bird seed or other small object dropped from the balloon. The area of the target is 196 ft^2, and the object must land within 200 ft of the center of the target to be scored in the competition.

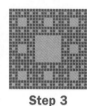

a. A bag of bird seed dropped at random lands within 200 ft of the center of the target. What is the probability that it hits the target?

b. Marilyn Rubin's bag of bird seed lands 5 ft from the center of the target. What is the probability that another pilot's randomly dropped bag will land closer to the center of the target than her bag, given that the pilot's drop scored? Do you think this is related to the actual probability that Marilyn will win the competition? Explain.

FRACTALS The steps below show how the fractal called *Sierpinski's carpet* is created. For Exercises 16–18, start with a square.

Step 1 Divide the square into 9 equal squares, as shown. Remove the middle square.

Step 2 Repeat Step 1 in each of the remaining 8 squares.

Step 3 Repeat Step 1 in each of the remaining 64 squares.

Step n Repeat Step 1 in each of the remaining $8^{(n-1)}$ squares.

16. The area of the original square is 1 square unit. What is the total area remaining after Step 2?

17. After Step 2, what is the probability that a point chosen at random in the original square will be on one of the remaining squares?

18. a. Challenge After Step 3, how many squares will remain? Explain how you know.

| Step 1 | Step 2 | Step 3 |

b. What is the probability that a point chosen at random in the original square will be on one of the remaining squares?

19. SAT/ACT Preview A box contains 10 blue marbles and 5 red marbles. What is the probability that a randomly chosen marble will be red?

A. $\frac{1}{4}$ B. $\frac{1}{3}$ C. $\frac{2}{3}$ D. $\frac{1}{2}$ E. $\frac{3}{4}$

20. ADVERTISING During a one-hour television show, there are eight minutes of commercials. If you turn on the television at a random time during a one-hour show, what is the probability that you will turn on the television during a commercial?

ONGOING ASSESSMENT

21. Open-ended Problem A bean is tossed at random and hits a target. Design a target that meets each condition. Explain your reasoning.

a. The probability that the bean lands on the shaded part is $\frac{1}{2}$.

b. The probability that the bean lands on the shaded part is 0.6.

Complete each equation so that it is equivalent to the proportion $\frac{x}{y} = \frac{5}{8}$. *(Section 9.3)*

22. $\frac{y}{x} = \underline{\ ?\ }$ **23.** $\frac{8 + 5}{8} = \underline{\ ?\ }$ **24.** $\frac{8}{y} = \underline{\ ?\ }$

Tell whether a triangle with sides of the given lengths is *right, obtuse,* or *acute.* *(Section 3.7)*

25. 9, 12, 15   **26.** 5, 16, 20   **27.** 3, 5, 7

ASSESS YOUR PROGRESS

VOCABULARY

geometric probability (p. 476)

The figures in each pair are similar. Find each missing value. *(Section 9.4)*

1. $A = \underline{\ ?\ }$ $A = \underline{\ ?\ }$ **2.** $P = \underline{\ ?\ }$ $P = 540$
    $P = \underline{\ ?\ }$ $P = 21$ $A = 4422.6$ $A = \underline{\ ?\ }$

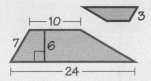

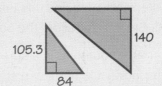

3. The ratio of the volumes of two similar rectangular prisms is $\frac{27}{125}$. What is the ratio of the lengths of any two corresponding sides? *(Section 9.4)*

Find the probability of hitting the shaded area of the target with a dart thrown at random that hits the target. *(Section 9.5)*

4.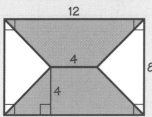

5.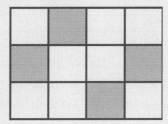

6. On $\overline{AB}$, what is the probability of a point selected at random being within 2 cm of point G? *(Section 9.5)*

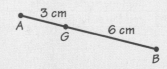

7. Journal Give an example of a situation in which you could use geometric probability to find the probability of an event. Explain why you could not count all of the possible outcomes in this situation.

Scaling the Planets

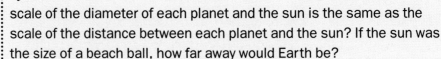

Could you make a scale model of the solar system the size of a football field so that the scale of the diameter of each planet and the sun is the same as the scale of the distance between each planet and the sun? If the sun was the size of a beach ball, how far away would Earth be?

PROJECT GOAL   Create a model of our solar system.

Organizing the Data

1. **Find** the diameter of the sun and each of the nine planets in our solar system. You can find this information in an encyclopedia or almanac. Be sure to record the source of your data.

2. **Organize** your data in a spreadsheet or table. Label everything clearly. Include units of measure. Include columns for the dimensions of the scale model and the actual dimensions of the solar system.

	A	B	C	D	E
1		Diameter at Equator (km)	Mean distance from sun (million km)	Scale diameter (cm)	Scale distance (cm)
2	Sun	1400000	0		
3	Mercury	4870	57.9		
4	Venus				
5	Earth				

3. **Choose** a convenient size for Earth's scale diameter. Use this size to calculate the scale for the model. Your scale must use the same units for Earth's actual diameter and its scale diameter.

21	If the diameter of Earth =	3 cm
22	then the scale factor is	2.35E−09
23	and the scale is 1:	425000000

This is another expression for
2.35 · 10^{-9}.

Making the Model

1. Use the scale you found to calculate the size of the sun and each planet in a model. If the sizes of the models will be unreasonable, choose a new scale for the diameters.

2. Choose a scale to show the distances of the planets from the sun. Can you use the same scale that you chose for the diameters?

3. Create scale models of the sun and each planet in the solar system. Your models may be two- or three-dimensional. Place these models at their scale distance from the sun.

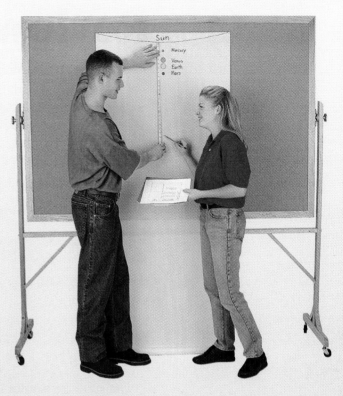

Presenting Your Results

Present your model and data to the class. In your report, you should include:

- the source of your data

- your spreadsheet with an explanation of each formula you used

- a discussion of why you chose the scale(s) that you used for your model

You may want to extend your report by investigating some of these topics:

- Find information about the planets and photos of them to include in your report. Use the Internet, magazines, or other sources. Be sure to include the source(s) you use.

- Find the scale distance between our sun and another star. Include it in your model, if possible.

- Discuss the size of the scale model of the solar system if the diameter of each planet and the distance between each planet and the sun are at the same scale. See the Connection to Astronomy on page 450 for one possibility.

Self-Assessment
Explain how you chose the scale you used for your model. How did making this model change your understanding of the size of the solar system?

9 | Review

Describe any exercises or ideas in this chapter that you found difficult but now understand. How did you resolve your difficulties? Now describe some things you still don't understand. Can you use some of the methods you used before to help you resolve these difficulties?

VOCABULARY

proportion (p. 446)   **geometric probability** (p. 476)
similar (p. 446)

SECTIONS 9.1, 9.2, *and* 9.3

> This means the pentagons are similar.

Two figures are **similar** if corresponding angles are congruent and corresponding sides are in proportion.

Pentagon *ABCDE* ~ pentagon *PQRST*.

Corresponding sides are in the ratio **3 : 1**.

$$\frac{AB}{PQ} = \frac{6}{2} = 3$$

To find the lengths of other sides, set up and solve proportions:

$$\frac{3}{1} = \frac{4.5}{ST} \qquad \frac{3}{1} = \frac{BC}{1.2}$$

$$3ST = 4.5 \qquad (3)(1.2) = BC$$

$$ST = 1.5 \qquad BC = 3.6$$

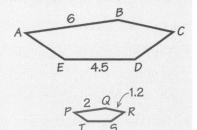

To show that two triangles are similar, you can use:

- AA Similarity Postulate
- SAS Similarity Theorem
- SSS Similarity Theorem

The Triangle Proportionality Theorem and the Midsegment Theorem use similar triangles to give shortcuts to find lengths.

You can use properties of proportions to write proportions that are equivalent to a given proportion.

SECTION 9.4

If two figures are similar and the ratio of the lengths of two corresponding sides is $a:b$, then:

the ratio of the perimeters or any pair of corresponding lengths is $a:b$.

Ratio of perimeters: $\dfrac{5}{9}$

the ratio of the areas is $a^2:b^2$.

Ratio of areas: $\dfrac{25}{81}$

the ratio of the volumes is $a^3:b^3$.

Ratio of volumes: $\dfrac{125}{729}$

The rectangular prisms below are similar, so **rectangle** $ABCD \sim$ **rectangle** $STUW$.

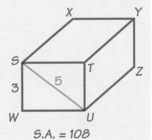

$\dfrac{AD}{SW} = \dfrac{2}{3}$, so $\dfrac{AC}{SU} = \dfrac{2}{3}$ and $AC = 3\dfrac{1}{3}$

The surface area of the larger prism is 108. To find the surface area, A, of the smaller prism:

$$\frac{A}{108} = \frac{2^2}{3^2}$$

$$9A = 4 \cdot 108$$

$$A = 48$$

The surface area of the smaller prism is 48.

The volume of the smaller prism is $21\frac{1}{3}$. To find the volume, V, of the larger prism:

$$\frac{21\frac{1}{3}}{V} = \frac{2^3}{3^3}$$

$$21\frac{1}{3} \cdot 27 = 8V$$

$$V = 72$$

The volume of the larger prism is 72.

SECTION 9.5

In geometric probability, you use ratios of areas, lengths, and other measures to find the probability of an event.

The probability of hitting the shaded area of the target with a bean tossed at random that hits the target is $\dfrac{\text{shaded area}}{\text{total area}} = \dfrac{10}{25} = \dfrac{2}{5}$.

Assessment

VOCABULARY QUESTIONS

1. **Writing** Explain the difference between *congruent* and *similar*. Can two similar figures be congruent? Explain.

2. **Open-ended Problem** Write two equivalent ratios. Then write the ratios as a proportion.

SECTIONS 9.1, 9.2, *and* 9.3

Is it possible to prove that the triangles in each pair are similar? If so, tell which postulate or theorem you would use.

3.

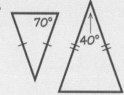

4.

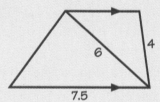

5.

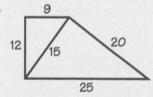

6.

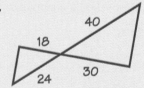

Find the value of each variable.

7.

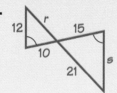

8.

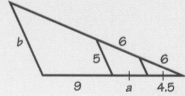

9.

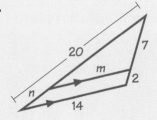

10.

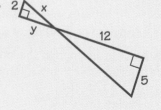

The figures in each pair are similar. Find each missing value.

11. $P = 50$ $P = 20$

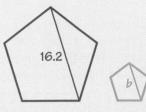

12. $A = 7.2 \text{ m}^2$ $A = \underline{?}$

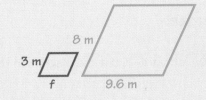

13. $V = \underline{?}$ $V = 682.5 \text{ in.}^3$

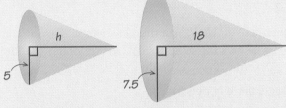

14. The heights of two similar cylinders are in the ratio $2:3$.
   a. The area of the base of the larger cylinder is 54 cm^2. What is the area of the base of the smaller cylinder?
   b. The volume of the smaller cylinder is 72 cm^3. What is the volume of the larger cylinder? Explain your reasoning.

15. If you spin the spinner, what is the probability that it will land on 1?

16. If you toss a bean at random on the target, what is the probability it will land on the shaded area?

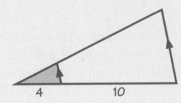

PERFORMANCE TASK

17. Open-ended Problem   Research the prices of pizzas, cakes, ice cream, or other food that is available in similar shapes but different sizes. Calculate and compare the prices for the area or volume. What did you discover? Why do you think there are these differences?

Cumulative Assessment
CHAPTERS 7–9

CHAPTER 7

For Questions 1–3, use the diagram at the right.

1. Show that quadrilateral *WXYZ* is a parallelogram.

2. Find the perimeter and the area of *WXYZ*.

3. If $m \angle WZY = 60°$, find the measures of $\angle WXY$ and $\angle XYZ$.

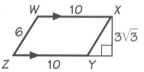

4. **Writing** Quadrilateral *ABCD* is a rhombus whose diagonals are 10 cm and 24 cm long. Sketch the rhombus. Write everything you know about the rhombus. Describe how you could find its area.

Find the area of each circle or polygon.

5.

6.

7.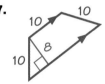

8. a regular hexagon whose perimeter is 60 in.

Find the volume and surface area of each cylinder or prism.

9. a cylinder with radius 12 and height 20

10. the triangular prism shown at the right

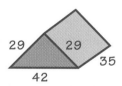

11. **Open-ended Problem** Draw a quadrilateral that is not a parallelogram, but has a diagonal that divides the quadrilateral into two congruent triangles. What kind of quadrilateral must your sketch be? Explain.

CHAPTER 8

The vertices of △*DEF* are *D*(−4, 2), *E*(3, −1), and *F*(0, −2). For Questions 12–15, find the coordinates of the vertices of the image of △*DEF* after each transformation described.

12. after a reflection over the given line:

　　a. the *y*-axis 　　　　**b.** the line $y = x$

13. after a rotation around the origin the given amount:

    a. 270°   **b.** 180°

14. after a dilation with scale factor $\frac{1}{2}$ with the given center:

    a. $(0, 0)$ **b.** $D(-4, 2)$

15. after a glide reflection with translation $(a, b) \rightarrow (a - 2, b)$ and reflection over the *x*-axis

16. Which of the transformations described in Questions 12–15 preserve congruence? Which preserve orientation?

17. **Writing** Suppose you reflect a figure over two parallel lines. How is this like reflecting the figure over two intersecting lines? How is it different?

18. **Open-ended Problem** Draw any polygon. Then draw its image after an enlargement with center at one vertex. Find the area of the original figure and of the image. Make a conjecture about the relationship between the scale factor of the dilation and the ratio of the areas.

CHAPTER 9

19. **Writing** Use the diagram. Write four true statements about the diagram. Explain your reasoning.

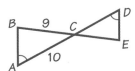

Is it possible to prove that the triangles are similar? Explain why or why not.

20. 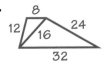   **21.** 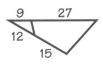   **22.**

23. The ratio of the areas of two regular pentagons is $9:25$. Find the ratio of their perimeters.

In the figure at the right, *M*, *N*, and *P* are the midpoints of the sides of △*ABC*.

24. Find the perimeter of △*MNP*.

25. The area of △*ABC* is 150 square units. Find the area of △*MNP*.

26. The height of a rectangular prism is 12 in. and its surface area is 552 in.^{2}. The height of a similar prism is 18 in. Find the surface area of the larger prism and find the ratio of their volumes.

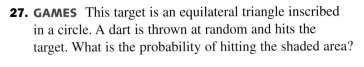

27. **GAMES** This target is an equilateral triangle inscribed in a circle. A dart is thrown at random and hits the target. What is the probability of hitting the shaded area?

10 | Applying Right Triangles

Technology and Tradition

Debby Tewa

You might say Debby Tewa lives in two very different worlds. Like many of her customers, she lives in a traditional sandstone house in Hotevilla, a Hopi village in Arizona. As an electrician, she has mastered the latest in high technology.

Tewa works every day to bring these two worlds together, providing electricity for Hopi, Navajo, and other native American people who would otherwise have none. "I'm fortunate to be in the position to explain the components of these systems to older people in the Hopi community," she says.

Since 1988, when it was founded by the Hopi Foundation, the Solar Electric Enterprise has equipped about 300 homes with photovoltaic panels–rooftop solar panels that convert sunlight into electricity. The electricity, in turn, can be used to run lights, power tools, and kitchen appliances.

"Solar energy teaches people to conserve energy and to become more self-sufficient."

"I never planned this career. It just happened."

Clean Energy

Photovoltaic (or solar) panels produce "clean energy," Tewa says. By using them, the damage to the environment associated with conventional energy sources and the need for power lines is reduced. Solar energy, she adds, is "appropriate not just for Hopi villages, but for everybody. It teaches people to conserve energy and to become more self-sufficient."

Tewa has toured throughout the United States, lecturing about the advantages of solar energy. She has also traveled to Ecuador, where she helped people with their solar installations. It's an exciting, and somewhat surprising, turn of events for her. Tewa never would have become an electrician, or a solar energy expert, if she had not been recruited for trade school shortly after leaving college. "I never planned this career," she explains. "It just happened." And she and her customers are glad that it did.

The Geometry of Solar Energy

Solar panels should be placed at a certain angle to maximize exposure to sunlight throughout the year. The angle depends on location. For example, in the southwestern United States where Debby Tewa lives, the solar panels are typically mounted at a 55° angle. Most Hopi roofs are essentially flat. The panels, roof, and support structures form a triangle. "If you know the length of the panel you need, you can use trigonometry to determine the size of the mounting materials," Tewa says.

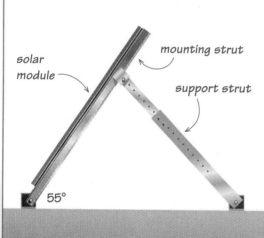

solar module

mounting strut

support strut

55°

"...you can use trigonometry to determine the size of the materials."

Explore and Connect

Debby Tewa demonstrates how she adjusts the solar module to the correct angle.

1. Writing   What do you know about any two right triangles with 55° angles? How could Debby Tewa use this relationship in her calculations? Explain your reasoning.

2. Research   Find out more about solar energy. List some of the environmental and economic benefits of using solar energy.

3. Project   Visit a building that has solar panels and talk with the people who take care of them. At what angle are the panels positioned? Do the solar panels operate differently depending on the weather or the season? How does using solar energy affect the electric bills for the building? Report your findings to your class.

Mathematics
& Debby Tewa

In this chapter, you will learn more about how mathematics is related to solar energy.

Related Exercises

Section 10.2
• Exercises 9–13

Section 10.4
• Exercises 20–22

10.1 Similar Right Triangles

When you draw the altitude to the hypotenuse of a right triangle, you create two smaller right triangles. How are the smaller triangles related to the original triangle? You will find out in the Exploration.

Learn how to...

- recognize relationships among the triangles formed by the altitude to the hypotenuse of a right triangle

- find the geometric mean of two numbers

So you can...

- estimate lengths, such as the height of a house

EXPLORATION
COOPERATIVE LEARNING

Comparing Right Triangles

Work with another student.
You will need:
- a rectangular sheet of paper
- a ruler   • scissors

1 Draw a diagonal of the paper to create two congruent right triangles. Fold the paper to find the altitude to the hypotenuse of one of the triangles.

2 Along the fold, draw the segment that corresponds to $\overline{DE}$ below. Mark the three right angles as shown.

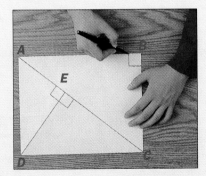

3 Explain why $\angle CAB \cong \angle DCE$ and $\angle ACB \cong \angle DAE$. Mark the triangles to indicate that these pairs of angles are congruent. What can you conclude about $\triangle CAB$ and $\triangle DCE$? about $\triangle CAB$ and $\triangle ADE$? How do you know?

4 Explain why $\angle CDE \cong \angle DAE$. Mark the triangles to indicate that these angles are congruent. What can you conclude about $\triangle DCE$ and $\triangle ADE$?

5 Cut along $\overline{AC}$ and $\overline{DE}$. Arrange the triangles so that three congruent acute angles match up, as shown at the left. How does this arrangement support the angle relationships in Steps 3 and 4?

In the Exploration, you discovered the following result.

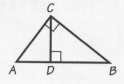

Similar Right Triangles Theorem

If the altitude is drawn to the hypotenuse of a right triangle, then the two triangles formed are similar to the original triangle and to each other.

$\triangle ABC \sim \triangle ACD$
$\triangle ABC \sim \triangle CBD$
$\triangle ACD \sim \triangle CBD$

THINK AND COMMUNICATE

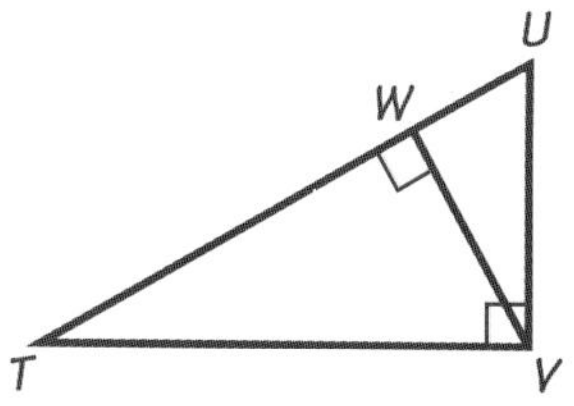

1. a. Which angle in $\triangle TUV$ corresponds to $\angle TVW$ in $\triangle TVW$?

   b. Identify the three pairs of similar triangles in the diagram.

   c. Which side in $\triangle TUV$ corresponds to $\overline{UW}$ in $\triangle VUW$?

2. Use corresponding parts of similar triangles to complete each statement.

   a. $\dfrac{TU}{VU} = \dfrac{VU}{?}$ **b.** $\dfrac{TU}{TV} = \dfrac{?}{TW}$ **c.** $\dfrac{UW}{?} = \dfrac{VW}{TW}$

The Geometric Mean

In the proportion $\dfrac{p}{q} = \dfrac{r}{s}$, the values q and r are called the *means*. If a, b, and x are positive numbers and $\dfrac{a}{x} = \dfrac{x}{b}$, then x is called the **geometric mean** of a and b.

EXAMPLE 1 Connection: Algebra

Find the geometric mean of 2 and 18.

SOLUTION

Let x = the geometric mean of 2 and 18. Write and solve a proportion.

$$\frac{2}{x} = \frac{x}{18}$$

$$36 = x^2$$

$$\sqrt{36} = x \qquad \text{The geometric mean is always positive.}$$

$$6 = x$$

The geometric mean of 2 and 18 is 6.

When you draw the altitude to the hypotenuse of a right triangle, you can use the geometric mean to describe the relationships among the lengths of the segments formed. The diagrams below illustrate these relationships.

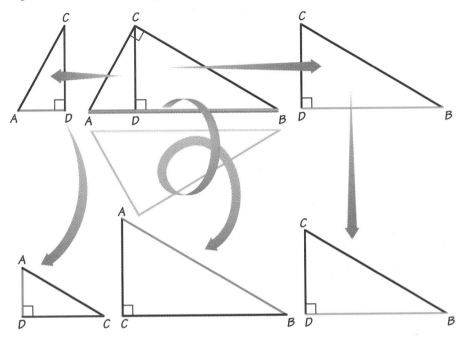

Since $\triangle ACD \sim \triangle CBD$, you can write $\dfrac{AD}{CD} = \dfrac{CD}{BD}$. Therefore, CD is the geometric mean of AD and BD.

Since $\triangle ABC \sim \triangle ACD$, you can write $\dfrac{AB}{AC} = \dfrac{AC}{AD}$. Therefore, AC is the geometric mean of AB and AD.

Since $\triangle ABC \sim \triangle CBD$, you can write $\dfrac{AB}{CB} = \dfrac{CB}{DB}$. Therefore, CB is the geometric mean of AB and DB.

Notice that each value in the proportions above is a length of a segment in $\triangle ABC$: either the altitude, a leg, the hypotenuse, or a segment of the hypotenuse.

Geometric Mean Theorems

If the altitude is drawn to the hypotenuse of a right triangle, then the length of the altitude is the geometric mean of the lengths of the segments of the hypotenuse.

If the altitude is drawn to the hypotenuse of a right triangle, then the length of each leg is the geometric mean of the lengths of the hypotenuse and the segment of the hypotenuse adjacent to that leg.

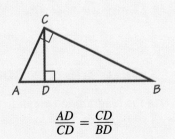

$$\frac{AD}{CD} = \frac{CD}{BD}$$

$$\frac{AB}{AC} = \frac{AC}{AD} \quad \text{and} \quad \frac{AB}{CB} = \frac{CB}{DB}$$

EXAMPLE 2   Application: Home Repair

h − 5 ft

$7\frac{1}{2}$ ft

h

5 ft

Aida Ramos wants to buy a ladder that reaches the roof of her house. To find how tall the ladder should be, she estimates the height of the edge of the roof above the ground as follows:

Step 1   Aida holds a notebook near her eye. She backs away from the house until the edge of the roof and the base of the house are in line with the notebook's edges, as shown in the diagram.

Step 2   Aida measures her distance from the house.

Use the diagram to find h, the height of the edge of the roof.

SOLUTION

Use the Geometric Mean Theorem that includes the length of the altitude.

$$\frac{AD}{CD} = \frac{CD}{BD}$$ *BD* is the distance from Aida's eyes to the ground, about 5 ft.

$$\frac{h - 5}{7.5} = \frac{7.5}{5}$$

$$5(h - 5) = (7.5)^2$$ Use the Distributive Property.

$$5h - 25 = 56.25$$

$$5h = 81.25$$

$$h = 16.25$$

The height of the edge of the roof is about 16 ft.

☑ CHECKING KEY CONCEPTS

Use the diagram to complete each statement.

1. $\angle GEF \cong \angle \underline{\ ?\ }$

2. $\triangle EFG \sim \triangle \underline{\ ?\ } \sim \triangle \underline{\ ?\ }$

3. $\frac{FE}{FG} = \frac{FG}{?}$

4. *GH* is the geometric mean of $\underline{\ ?\ }$ and $\underline{\ ?\ }$.

5. If $EH = 9$ and $EF = 25$, then $EG = \underline{\ ?\ }$.

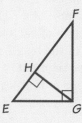

10.1 | Exercises and Applications

Extra Practice
exercises on page 680

Identify the similar triangles in each diagram.

1.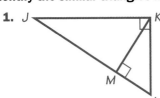

2.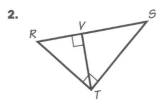

ALGEBRA **Find the geometric mean of the given numbers.**

3. 45 and 5   **4.** 8 and 6   **5.** 12 and 15

6. 4 and $\frac{1}{9}$ **7.** $\frac{3}{7}$ and $\frac{7}{3}$ **8.** 0.2 and 80

Toolbox p. 700
Simplifying Radicals

9. a. **Open-ended Problem** Find two numbers whose geometric mean is 10.

   b. **Writing** Are the numbers you found in part (a) the *only* two numbers whose geometric mean is 10? Explain.

Find the value of each variable.

10.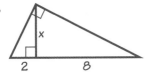

11.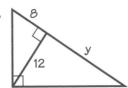

12.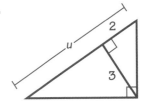

13.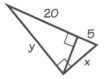

14.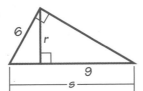

15.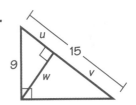

16. Draw a 3-4-5 right triangle and the altitude to the hypotenuse. Use the Geometric Mean Theorems to find the length of the altitude and the lengths of the segments of the hypotenuse.

For Exercises 17 and 18, prove the Geometric Mean Theorems on page 495. Use the diagram at the right for both exercises.

17. Given: In $\triangle ABC$, $\angle BCA$ is a right angle.
$\overline{CD}$ is the altitude to the hypotenuse.

  Prove: $\dfrac{e}{d} = \dfrac{d}{f}$

18. Given: In $\triangle ABC$, $\angle BCA$ is a right angle.
$\overline{CD}$ is the altitude to the hypotenuse.

  Prove: $\dfrac{c}{a} = \dfrac{a}{e}$ and $\dfrac{c}{b} = \dfrac{b}{f}$

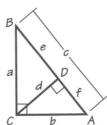

19. One proof of the Pythagorean theorem uses the Geometric Mean Theorem that includes the legs of the triangle. Copy and complete the proof.

Given: In $\triangle ABC$, $\angle ACB$ is a right angle.
$\overline{CD}$ is the altitude to the hypotenuse.

Prove: $a^2 + b^2 = c^2$

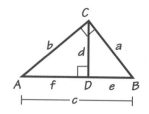

Statements	Reasons
1. In $\triangle ABC$, $\angle ACB$ is a right angle. $\overline{CD}$ is the altitude to the hypotenuse.	**1.** Given
2. $\frac{c}{a} = \frac{a}{e}; \frac{c}{b} = \frac{b}{f}$	**2.** _?_
3. _?_ $= ce$; _?_ $= cf$	**3.** _?_
4. _?_ $+$ _?_ $= ce + cf$	**4.** _?_
5. _?_ $+$ _?_ $= c(e + f)$	**5.** _?_
6. $e + f =$ _?_	**6.** _?_
7. $a^2 + b^2 = c^2$	**7.** _?_

Connection ▶ **LITERATURE**

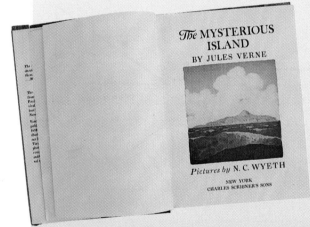

The MYSTERIOUS ISLAND
BY JULES VERNE

Pictures by N. C. WYETH

NEW YORK
CHARLES SCRIBNER'S SONS

In Jules Verne's *The Mysterious Island*, Cyrus Harding, one of several travelers stranded on an island, uses similar right triangles to estimate the height of a cliff.

THE MYSTERIOUS ISLAND

Cyrus Harding had provided himself with a straight stick, twelve feet long. . . . Having reached a spot . . . nearly five hundred feet from the cliff, which rose perpendicularly, Harding thrust the pole two feet into the sand . . . perpendicularly. . . .

That done, he retired the necessary distance, when, lying in the sand, his eye glanced at the same time at the top of the pole and the crest of the cliff. He carefully marked the place with a little stick. . . .

The first distance was fifteen feet between the stick and the place where the pole was thrust into the sand.

The second distance between the stick and the bottom of the cliff was five hundred feet.

20. Visual Thinking Draw and label a diagram that models the situation described in the passage.

21. Identify the similar triangles in the diagram you drew in Exercise 20.

22. Find the height of the cliff.

23. Open-ended Problem What other distances in your diagram can you find? What theorem(s) or postulate(s) would you need to use to calculate these distances?

24. ALGEBRA Let a and b be positive numbers. The *arithmetic mean* of a and b is $\dfrac{a+b}{2}$.

  a. Show that the geometric mean of a and b is $\sqrt{ab}$.

  b. Challenge Show that the arithmetic mean of a and b is greater than or equal to the geometric mean.

  c. When are the arithmetic and geometric means of a and b equal?

25. MUSIC In a *panpipe* based on a *chromatic scale*, the length of the nth pipe is the geometric mean of the lengths of the $(n-1)$th pipe and the $(n+1)$th pipe, where $n = 2, 3, 4, \ldots, 12$. For example, the length of the second pipe is the geometric mean of the lengths of the first and third pipes.

  a. Copy and complete the table. Round your answers to the nearest tenth of a centimeter.

Number of pipe	1	2	3	4	5	6	7	8	9	10	11	12	13
Length (cm)	16.4	15.5	?	?	?	?	?	?	?	?	?	?	?

BY THE WAY

A panpipe consists of several pipes of different lengths attached together. It is played by blowing across the tops of the pipes. Longer pipes produce lower-pitched notes than shorter pipes.

  b. Cooperative Learning Work with a partner to make a panpipe using 13 plastic straws, a ruler, scissors, and masking tape. Cut the straws so that they have the lengths you found in part (a). Place the straws on a flat surface from longest to shortest, and align the top edges. Use the masking tape to attach the straws together.

ONGOING ASSESSMENT

26. Open-ended Problem Choose values for RU and SU in the diagram. Use the values you chose and the Geometric Mean Theorems to find RT, TU, and ST. Use the Pythagorean theorem to check your answers.

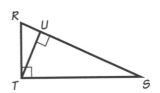

SPIRAL REVIEW

Find the coordinates of the midpoint of the segment with each given pair of endpoints. *(Section 4.1)*

27. $A(0, 0)$, $B(6, -2)$

28. $C(-7, 1)$, $D(11, -5)$

29. $E(3, -9)$, $F(4, 8)$

The legs of a right triangle are a and b units long. The hypotenuse is c units long. Find the unknown length for each right triangle to the nearest hundredth. *(Section 3.6)*

30. $a = 6$, $b = 8$

31. $a = 5$, $c = 11$

32. $b = 48$, $c = 50$

10.2 Special Right Triangles

Learn how to...

- **find relationships among side lengths in 45-45-90 triangles and in 30-60-90 triangles**

So you can...

- **easily find lengths and distances in real-world problems involving special triangles, squares, and equilateral triangles**

If you've ever been to a baseball game, you know that a runner at first base will often try to "steal" second. To prevent the runner from doing so, sometimes the catcher throws the ball from home plate to second base. How far does the catcher need to throw the ball?

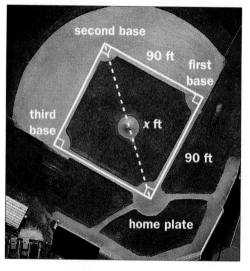

An official baseball diamond is a 90 ft by 90 ft square. The three bases and home plate mark each corner of the square.

Home plate, first base, and second base form an isosceles right triangle. Isosceles right triangles are also called **45-45-90 triangles**. You can use the Pythagorean theorem to find the relationships among the lengths of the sides of a 45-45-90 triangle.

EXAMPLE 1　　Application: Sports

Find the distance from home plate to second base in a baseball diamond.

SOLUTION

Use the Pythagorean theorem.

$$x^2 = 90^2 + 90^2$$
$$x^2 = 2 \cdot 90^2$$
$$x = \sqrt{2 \cdot 90^2}$$
$$x = 90\sqrt{2}$$

The square root of the product of two numbers is the product of the square roots of the numbers.

The distance from home plate to second base is $90\sqrt{2}$ ft, or about 127.28 ft.

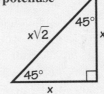
An isosceles right triangle often appears in real-world problems because it is half of a square, as in Example 1. Likewise, it is helpful to know the dimensions of a **30-60-90 triangle**, which is half of an equilateral triangle.

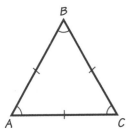

An equilateral triangle is also equiangular.
$m\angle A = m\angle B = m\angle C = 60°$

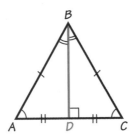

$m\angle ABD = m\angle DBC = 30°$
$AD = CD$

$\triangle DBC$ is a 30-60-90 triangle.

THINK AND COMMUNICATE

1. Explain why $m\angle DBC = 30°$.

2. Explain why all right triangles formed by drawing one altitude of an equilateral triangle are similar.

3. Describe the relationship between the lengths DC and BC. Explain your reasoning.

4. Let $DC = 4$.

   a. Use your answer to Question 3 to find BC.

   b. Use the Pythagorean theorem to find BD.

5. Suppose $DC = k$. Find BC and BD in terms of k.

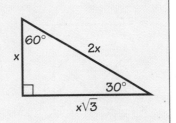
> **WATCH OUT!**
>
> The shorter leg of a 30-60-90 triangle is opposite the 30° angle, and the longer leg is opposite the 60° angle.

EXAMPLE 2   Application: History

The Conimbriga archeological site is located south of Coimbra, Portugal. One structure in the ancient Roman village is this private bath in the shape of a regular hexagon. The apothem of the hexagon is about 2 m. Find the length of each side of the bath.

SOLUTION

Since a regular hexagon can be divided into six congruent equilateral triangles, it can be divided into twelve congruent 30-60-90 triangles.

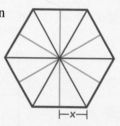

⊢—x—⊣

Let x = the length of the short leg in each 30-60-90 triangle. Then the side opposite the 60° angle is $x\sqrt{3}$, which equals 2 m. Write an equation and solve for x.

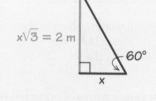

$x\sqrt{3} = 2$ m

60°

x

Divide both sides of the equation by $\sqrt{3}$.

$$x\sqrt{3} = 2$$

$$x = \frac{2}{\sqrt{3}}$$

The length of each side of the hexagon is twice the value of x. So the side length of the hexagon is $\frac{4}{\sqrt{3}}$ m, or about 2.31 m.

☑ CHECKING KEY CONCEPTS

Find the missing side lengths of each triangle.

1.

3

2.

60°
6

3.

10
150°

4.

7√2

5.

8

6.

120°
16

10.2 Exercises and Applications

Extra Practice
*exercises on
page 680*

Find the exact value of each variable.

1.
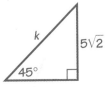
k
$5\sqrt{2}$
$45°$

2.
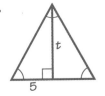
t
5

3.
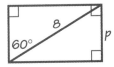
8
$60°$
p

4.

30
d

5.
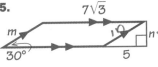
$7\sqrt{3}$
m
$n\sqrt{3}$
$30°$
5

6.
$x°$ $2x°$
y 13
$x°$

7. Open-ended Problem Choose one of the figures in Exercises 1–6 and find its area. Explain your method.

8. Logical Reasoning Suppose a 30-60-90 triangle with leg lengths a and b has the same area as a 45-45-90 triangle with leg length x. Prove that x is the geometric mean of a and b.

INTERVIEW Debby Tewa

Look back at the article on pages 490–492.

One of the solar modules that Debby Tewa uses contains solar cells that are approximately equiangular octagons with the dimensions shown. You can think of the octagon as a square with its corners removed.

65 mm
42 mm

9. Writing What type of triangles must the corners be if the octagon is equiangular? Explain your reasoning.

10. Use your answer to Exercise 9 to find the side length of the square.

11. Find the area of the square.

12. Find the area of each triangular corner.

13. Use your answers to Exercises 11 and 12 to estimate the area of each solar cell.

Throughout history, various cultures have used arches in their buildings and structures. The *Gothic arch* is formed by two arcs that intersect at a vertex. Some Gothic arches are based on the shape of an equilateral triangle.

14. `CONSTRUCTION` Follow these steps to draw the shape of a Gothic arch. Tell which three points of your arch determine the equilateral triangle, and how you know it is equilateral.

1. Draw $\overline{AB}$.

2. Draw an arc with center A and radius AB.

3. Draw an arc with center B and radius AB.

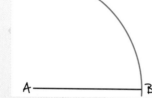

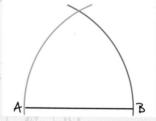

15. Suppose the width of an arch where it starts to curve is 12 in. What is the height h of the curved part of the arch? How do you know?

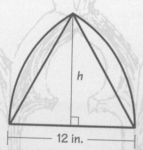

12 in.

Find the total height of each arch.

16.

35 in.

30 in.

Regency Display Cabinet

17.

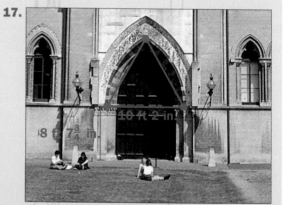

10 ft 2 in.

8 ft 7¾ in.

Oxford College, England

18. Research Find a Gothic arch in your neighborhood or a picture of one that is based on an equilateral triangle. Use the steps in Exercise 14 to draw the arch. Find out the width of the arch and use what you know about special triangles to calculate the height of the arch.

ALGEBRA Find the value of each variable.

19.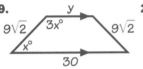

20.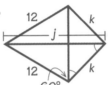

21.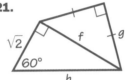

22.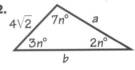

23. SAT/ACT Preview The distance from the center of a square to one of its vertices is 4 cm. What is the area of the square?

A. 16 cm^2 **B.** $4\sqrt{2} \text{ cm}^2$ **C.** 32 cm^2 **D.** $16\sqrt{2} \text{ cm}^2$ **E.** 8 cm^2

24. `CONSTRUCTION` Use the diagram to describe how to construct a 45-45-90 triangle.

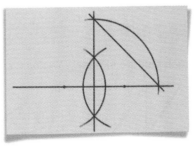

25. Investigation Although it is not possible to trisect every angle with a compass and a straightedge, you can trisect any given angle by folding paper. To trisect a 90° angle, use a rectangular sheet of paper.

Crease the paper by folding it in half and opening it again.

Folding at *A*, bring *C* up to meet the first crease at *D*.

Unfold the paper. $m\angle EAC = 30°$

 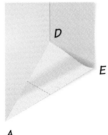 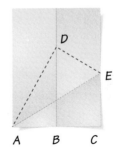

a. Let $AC = x$. Find AB and AD in terms of x. Explain your reasoning.

b. What type of triangle is $\triangle ADB$? How do you know?

c. What relationship does $\overline{AE}$ have to $\angle DAC$? Explain.

d. Use your answers to parts (a)–(c) to write a paragraph proof showing that $m\angle EAC = 30°$.

26. FORESTRY Ed Cushman wants to cut a log into usable boards. To use the log most efficiently, he cuts the cylindrical log into a prism with square bases. Then he cuts the prism into boards. If the diameter of the log is 18 in., how wide can the boards be? Explain.

27. Look back at page 501. Use Questions 1–5 to write a formal proof of the 30-60-90 triangle theorem.

28. Challenge Write an equation for a line that makes a 60° angle with the x-axis. How did you get your answer?

ONGOING ASSESSMENT

29. Writing How can you find the area of a square if you know the length of its diagonal? How can you find the area of an equilateral triangle if you know the length of one side?

SPIRAL REVIEW

Explain why the triangles in each pair are similar. *(Section 9.2)*

30.   **31.**   **32.**

Find the slope of the line that contains each pair of points given. *(Section 4.2)*

33. $(0, 0)$ and $(4, 7)$ **34.** $(-1, 0)$ and $(0, 1)$ **35.** $(2, -3)$ and $(-4, 5)$

ASSESS YOUR PROGRESS

VOCABULARY

geometric mean (p. 494)   **30-60-90 triangle** (p. 501)
45-45-90 triangle (p. 500)

Find the geometric mean of the given numbers. *(Section 10.1)*

1. 3 and 27   **2.** 5 and 8   **3.** 21 and 35

Find the value of each variable. *(Section 10.1)*

4. 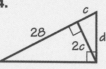   **5.** 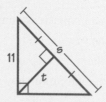   **6.**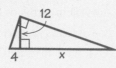

Find the value of each variable. *(Section 10.2)*

7. 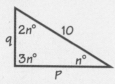   **8.** 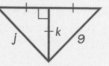   **9.**

10. Journal Draw and label several triangles to illustrate the main ideas you learned in Sections 10.1 and 10.2.

10.3 The Tangent Ratio

If you travel through farming areas, you may see cone-shaped piles of grain. The angle that the grain makes with the ground is called the *angle of repose*. The measure of this angle depends on the type of grain. For example, the angle of repose for wheat measures 27°.

The angle of repose determines the *coefficient of friction of grain on grain*. This coefficient, which measures how easily the kernels of grain slide against each other, is equal to the *tangent* of the angle of repose.

In a right triangle, the **tangent** of an acute angle is the ratio of the length of the leg *opposite the angle* to the length of the leg *adjacent to the angle*.

> Write the tangent of angle *A* as "tan *A*."

$$\tan A = \frac{\text{opposite}}{\text{adjacent}} = \frac{BC}{AC}$$

EXAMPLE 1

Use the diagram to find tan *P* and tan *Q*.

SOLUTION

Use the definition of tangent.

$$\tan P = \frac{\text{length of leg opposite } \angle P}{\text{length of leg adjacent to } \angle P} = \frac{2}{\sqrt{5}}$$

$$\tan Q = \frac{\text{length of leg opposite } \angle Q}{\text{length of leg adjacent to } \angle Q} = \frac{\sqrt{5}}{2}$$

THINK AND COMMUNICATE

1. Look back at the solution to Example 1. How are $m \angle P$ and $m \angle Q$ related? How are tan P and tan Q related? Explain.

2. a. What can you conclude about $\triangle XWV$ and $\triangle XYZ$?

b. Use your answer from part (a) to explain why the tangent of a given angle measure is constant.

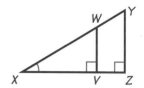

3. a. For each triangle shown, measure $\angle A$ and find tan A.

 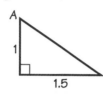 

b. What happens to tan A as $m \angle A$ approaches 90°?

c. Is there a limit to how high the tangent values can be? Explain.

The tangent of an angle in a right triangle depends only on the measure of the angle and not on the size of the triangle. In other words, an expression like tan 27° has a fixed value that is independent of any particular right triangle.

You can approximate the tangent of an angle by measuring. However, many calculators have a tangent key that gives more precise values.

Table p. 733
Trigonometric Ratios

EXAMPLE 2   Application: Agriculture

Suppose the radius of the base of a pile of wheat is 30 ft. Find the height h of the pile.

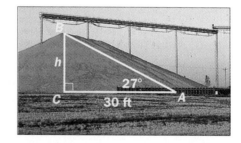

SOLUTION

$$\tan A = \frac{\text{opposite}}{\text{adjacent}}$$

$$\tan 27° = \frac{h}{30}$$

$$30 \tan 27 = h$$

$$30(0.5095) \approx h$$

$$15.3 \approx h$$

```
tan 27
           .5095254495
Ans*30
           15.28576348
```

The height of the pile of wheat is about 15.3 ft.

BY THE WAY

China produces more wheat than any other country. In 1995, about 110 million tons of wheat were produced in China—almost twice the amount produced in the United States.

As well as being able to give tangents of angles, most scientific calculators can determine the measure of an angle given its tangent. If $\angle A$ is an acute angle and $\tan A = x$, then $\tan^{-1} x = m \angle A$. The expression $\tan^{-1} x$ is read as "the inverse tangent of x."

Find $m \angle U$ in $\triangle TUV$.

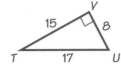

SOLUTION

Use a scientific calculator to find the measure of an angle whose tangent is $\frac{15}{8}$.

$$\tan U = \frac{15}{8}$$

$$m \angle U = \tan^{-1} 1.875$$

$$m \angle U \approx 61.9°$$

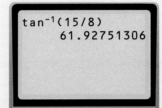

```
tan⁻¹(15/8)
          61.92751306
```

The measure of $\angle U$ is about $61.9°$.

Angles of Elevation and Depression

Suppose a person in a boat sees the top of a lighthouse in the distance. The angle that the person's line of sight makes with the horizontal is called the **angle of elevation**.

Suppose the lighthouse operator spots the boat from the top of the lighthouse at the same time. The angle that the operator's line of sight makes with the horizontal is called the **angle of depression**. You can use parallel lines and alternate interior angles to show that these two angles are congruent.

EXAMPLE 4 Application: Navigation

Suppose the angle of elevation from a ship to the top of a lighthouse is 6°. The lighthouse is 60 ft tall and the cliff is 250 ft high. Find the distance from the ship to the base of the cliff.

SOLUTION

Let *h* = the distance from the base of the cliff to the top of the lighthouse.
Let *d* = the distance from the ship to the base of the cliff.
Draw and label a diagram that models the situation.

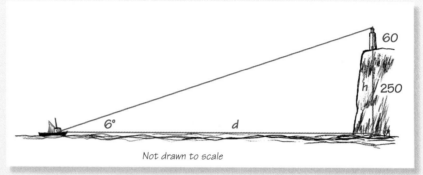

Not drawn to scale

Write an equation involving *h* and *d*:

$$\tan 6° = \frac{h}{d}$$ Substitute **250 + 60 = 310** for *h*.

$$\tan 6 = \frac{310}{d}$$ Multiply both sides by *d*.

$$d \tan 6 = 310$$ Divide both sides by tan 6.

$$d = \frac{310}{\tan 6}$$

$$d \approx 2949$$

The distance from the ship to the base of the cliff is about 2950 ft.

BY THE WAY

During the third century B.C., the Egyptians constructed the Pharos of Alexandria, the tallest lighthouse ever built. This lighthouse was over 400 ft high and guided ships for about 1500 years.

☑ CHECKING KEY CONCEPTS

For Questions 1 and 2, use △JKL.

1. Find tan *J* and tan *K*.

2. Find *m* ∠ *J* and *m* ∠ *K*.

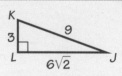

3. **AVIATION** The angle of depression from an airplane flying at an altitude of 5000 ft to the closer end of a runway is 9°.

   a. Find the horizontal distance from the airplane to the runway.

   b. Suppose the runway is 7000 ft long. Find the angle of elevation from the farther end of the runway to the airplane.

Extra Practice
exercises on
page 681

For Exercises 1–3, find tan A and tan B.

1.

2.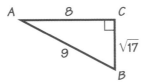

3.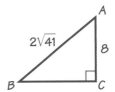

Find the value of each ratio. Round your answers to four decimal places.

4. tan 40°   **5.** tan 68°   **6.** tan 5.7°   **7.** tan 89.9°

8. a. Open-ended Problem Sketch a 45-45-90 triangle and a 30-60-90 triangle. Label each side length.

   b. Use your diagrams from part (a) to find the exact values of tan 30°, tan 45°, and tan 60°.

Find the measure of the acute angle that satisfies the given equation. Round your answers to the nearest tenth of a degree.

9. $\tan A = \frac{2}{7}$ **10.** $\tan B = \frac{19}{5}$ **11.** $\tan R = 1.268$ **12.** $\tan Y = 0.4779$

Find the value of each variable. Round your answers to the nearest tenth.

13. 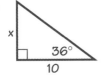   **14.** 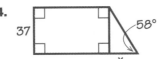   **15.**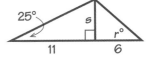

16. Open-ended Problem Draw a right triangle. Measure the legs and use $\tan^{-1}$ to find the measure of each acute angle. Use a protractor to check your answers.

17. PALEONTOLOGY The diagram at the right illustrates several characteristics of a two-legged dinosaur's *gait* (way of walking). In the diagram, *AC* is called the *pace length*, *BD* is called the *height at the hip*, and $\angle ABC$ is called the *angle of gait*. One study suggests that the normal pace length for a two-legged dinosaur was about 0.65 times the dinosaur's height at the hip.

   a. ALGEBRA Let $BD = x$. Find *AC*, *AD*, and *DC* in terms of *x*.

   b. Use your answers to part (a) to find $m \angle ABD$ and $m \angle CBD$.

   c. Writing What was the angle of gait for two-legged dinosaurs? Explain why it did not depend on the size of the dinosaur.

18. MEASUREMENT Two buildings are 75 ft apart. Carolyn, who is standing on the roof of the shorter building, measures the angle of elevation from her eyes to the top of the taller building to be 28°. She measures the angle of depression from her eyes to the bottom of the taller building to be 53°. Find the height of each building, given that Carolyn's eyes are 5 ft above the roof of the shorter building.

Find the measure of each acute angle. Round your answers to the nearest tenth of a degree.

19.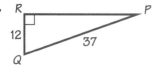

20.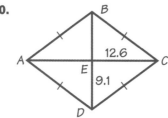

21.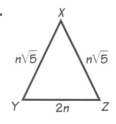

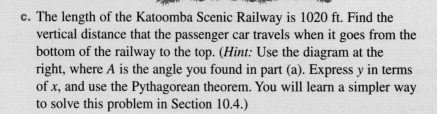

Connection   TRANSPORTATION

A railway's steepness is measured by its *gradient*. The gradient is the ratio of the railway's rise to its run and is usually expressed as a percent. That is, a railway with a gradient of *g*% rises *g* ft vertically for every 100 ft traveled horizontally.

22. **Writing** What is the relationship between a railway's gradient expressed as a percent and the angle that the railway makes with the horizontal? Explain.

23. The steepest railway in the world is the Katoomba Scenic Railway in the Blue Mountains of New South Wales, Australia. A single passenger car is pulled by two steel cables up the length of the railway. The gradient of the railway is 122%.

a. What angle measure does the Katoomba Scenic Railway make with the horizontal?

b. The passenger car traveling on the Katoomba Scenic Railway rises a vertical distance of 550 ft. What horizontal distance does the car travel?

c. The length of the Katoomba Scenic Railway is 1020 ft. Find the vertical distance that the passenger car travels when it goes from the bottom of the railway to the top. (*Hint:* Use the diagram at the right, where *A* is the angle you found in part (a). Express *y* in terms of *x*, and use the Pythagorean theorem. You will learn a simpler way to solve this problem in Section 10.4.)

24. **AGRICULTURE** The table shows the angle of repose for wheat, corn, and oat. Recall that the tangent of the angle of repose is the coefficient of friction of grain on grain.

a. Copy and complete the table.

b. **Writing** What happens to the coefficient of friction of grain on grain as the angle of repose increases? Which grain in the table do you think slides most easily? Explain your reasoning.

Angles of grain piles			
	wheat	corn	oat
$m \angle A$	27.0	27.5	28.0
tan A	?	?	?

25. A tree casts a shadow 25 ft long. The angle of elevation from the tip of the shadow to the top of the tree is 64°. Find the height of the tree.

26. Prove that if $\angle A$ and $\angle B$ are complements, then $(\tan A)(\tan B) = 1$.

27. Prove that if $0 < m \angle A < 45°$, then $\tan A < 1$.

28. **Challenge** A glider is approaching a large river. The angle of depression from the glider to the near side of the river is 14°. The angle of depression from the glider to the far side of the river is 10°. The altitude of the glider is 3000 ft. Find the width w of the river.

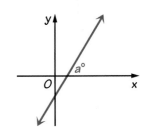

Not drawn to scale

3000 ft

10° 14°

x

w

river

ONGOING ASSESSMENT

29. **Writing** For any line whose slope is positive, let $a°$ be the measure of the angle that the line makes with the x-axis. (Measure this angle counterclockwise from the x-axis to the line, so that the angle is always acute.) Describe how a is related to the slope of the line.

SPIRAL REVIEW

30. The hypotenuse of a 30-60-90 triangle is 8 cm long. What are the lengths of the triangle's legs? *(Section 10.2)*

Can a triangle be formed from sides of the given lengths? *(Section 6.1)*

31. 4, 10, 12

32. 6, 6, 8

33. 1, 3, 5

10.4 Sine and Cosine Ratios

The tangent ratio you learned about in the last section is part of an important branch of mathematics called *trigonometry*. The word "trigonometry" comes from Greek words meaning "triangle measurement," although trigonometry probably originated in ancient Egypt and Mesopotamia. In this section, you will learn about two other trigonometric ratios.

Learn how to...
- **find the sine and cosine of an acute angle**
- **find the measure of an acute angle whose sine or cosine is given**

So you can...
- **find the lengths of the sides of right triangles**
- **solve problems in engineering, for example**

EXPLORATION
COOPERATIVE LEARNING

Analyzing Ratios in Triangles

Work with another student.
You will need:
- geometry software or a ruler and protractor

1 For each angle measure in the table below, complete these steps:

- One student should draw right △ABC so that ∠A has the given measure.

hypotenuse

leg opposite ∠A

leg adjacent to ∠A

$m\angle A$	$\dfrac{\text{opposite}}{\text{hypotenuse}}$	$\dfrac{\text{adjacent}}{\text{hypotenuse}}$
20°	?	?
40°	?	?
60°	?	?
80°	?	?

- The other student should measure the side lengths of △ABC, calculate the ratios $\dfrac{\text{opposite}}{\text{hypotenuse}}$ and $\dfrac{\text{adjacent}}{\text{hypotenuse}}$, and record these values in the table. Round each value to the nearest hundredth.

2 As $m\angle A$ increases, what happens to each ratio?

3 What value do you think each ratio approaches as $m\angle A$ approaches 0°? as $m\angle A$ approaches 90°? Explain your answers.

4 For what value of $m\angle A$ do you think the two ratios will be equal? Draw a right triangle with this angle measure and find out if you are correct.

In a right triangle, the **sine** of an acute angle is the ratio of the length of the leg opposite the angle to the length of the hypotenuse. The **cosine** of an acute angle is the ratio of the length of the leg adjacent to the angle to the length of the hypotenuse. Sines and cosines of acute angles are always between 0 and 1.

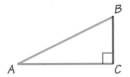

$$\sin A = \frac{\text{opposite}}{\text{hypotenuse}} = \frac{BC}{AB}$$ Write the sine of $\angle A$ as "sin A."

$$\cos A = \frac{\text{adjacent}}{\text{hypotenuse}} = \frac{AC}{AB}$$ Write the cosine of $\angle A$ as "cos A."

EXAMPLE 1

In △JKL, find sin J and cos J.

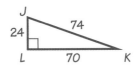

SOLUTION

$$\sin J = \frac{\text{opposite}}{\text{hypotenuse}} = \frac{70}{74} \approx 0.9459$$

$$\cos J = \frac{\text{adjacent}}{\text{hypotenuse}} = \frac{24}{74} \approx 0.3243$$

In Section 10.3, you used a scientific calculator to approximate tangents of angles. Most calculators also have keys for approximating sines and cosines.

EXAMPLE 2

Find the values of x and y in the triangle shown.

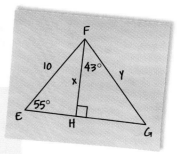

SOLUTION

Step 1 Use a sine ratio in △EFH to find the value of x.

$$\sin 55° = \frac{x}{10}$$

$$10 \sin 55 = x$$

$$10(0.8192) \approx x$$

$$8.19 \approx x$$

Use a scientific calculator to find the value of sin 55°.

Step 2 Use a cosine ratio in △FGH to find the value of y.

$$\cos 43° = \frac{x}{y}$$

$$\cos 43 \approx \frac{8.19}{y}$$ Multiply both sides by y.

$$y \cos 43 \approx 8.19$$

$$y \approx \frac{8.19}{\cos 43}$$ Use a scientific calculator to find the value of cos 43°.

$$y \approx \frac{8.19}{0.7314}$$

$$y \approx 11.20$$

1. a. In Example 2, what is $m \angle EFH$? What is $m \angle FGH$?

   b. Find the value of x in Example 2 using the cosine of $\angle EFH$. Then find the value of y using the sine of $\angle FGH$. Check that your values of x and y match those in Example 2.

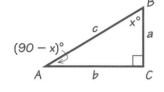

2. Use the diagram at the left to find $\sin x°$, $\cos (90 - x)°$, $\cos x°$, and $\sin (90 - x)°$ in terms of a, b, and c. How are the sine and cosine of an angle related to the sine and cosine of its complement? Explain.

In Section 10.3, you found the measure of an angle given its tangent. You can also find the measure of an angle given its sine or cosine. If $\angle A$ is an acute angle and $\sin A = x$, then $\sin^{-1} x = m \angle A$. Also, if $\angle B$ is an acute angle and $\cos B = y$, then $\cos^{-1} y = m \angle B$. These expressions are read as "inverse sine of x" and "inverse cosine of y."

EXAMPLE 3 | **Application: Theater**

A *raked stage* is a stage that is slanted toward the audience, like a ramp. In general, a raked stage that slants much more than 5° is too steep for the actors to move around on easily. Suppose a theater company builds a raked stage on a stage 29 ft deep. If the raked stage is supported at the back by a post that is 4 ft long and perpendicular to it, at what angle does the stage slant? Is the stage acceptable?

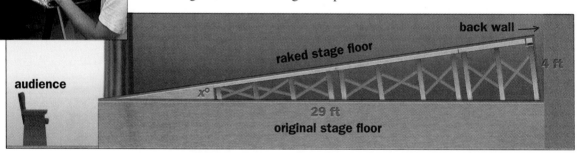

SOLUTION

Use a sine ratio to find the measure of the angle that the stage makes with the horizontal.

$$\sin x° = \frac{4}{29}$$

$$x = \sin^{-1}\left(\frac{4}{29}\right)$$

$$\approx \sin^{-1} 0.1379 \qquad \text{Use a scientific calculator to find } \sin^{-1} 0.1379.$$

$$\approx 7.93°$$

Since the angle measure is greater than 5°, the stage is probably too steep for the actors.

Here is a summary of the three trigonometric ratios that you can use to solve problems.

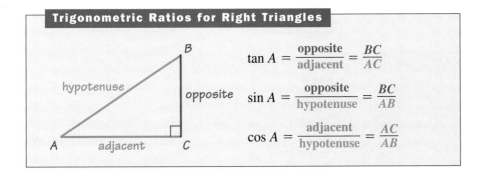

Trigonometric Ratios for Right Triangles

$\tan A = \dfrac{\text{opposite}}{\text{adjacent}} = \dfrac{BC}{AC}$

$\sin A = \dfrac{\text{opposite}}{\text{hypotenuse}} = \dfrac{BC}{AB}$

$\cos A = \dfrac{\text{adjacent}}{\text{hypotenuse}} = \dfrac{AC}{AB}$

☑ CHECKING KEY CONCEPTS

For Questions 1 and 2, use △LMN.

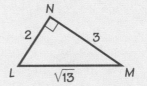

1. Find sin L and cos L. Use sin L to find $m \angle L$ to the nearest degree.

2. Find sin M and cos M. Use cos M to find $m \angle M$ to the nearest degree.

3. Latricia is flying a kite on level ground. Her hands are 4 ft above the ground. The angle between the string and the horizontal measures 50°, and the string is 300 ft long. How high above the ground is the kite? Round your answer to the nearest foot.

4. Find the values of x and y in △PQR.

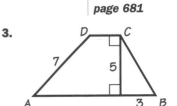

10.4 **Exercises and Applications**

Extra Practice exercises on page 681

For Exercises 1–3, find sin A, cos A, sin B, and cos B.

1.

2.

3.

Find the value of each ratio. Round your answers to four decimal places.

4. sin 59°   5. sin 3.6°   6. cos 14°   7. cos 70.2°

Find the measure of an acute angle that satisfies the given equation. Round your answers to the nearest tenth of a degree.

8. $\sin A = \dfrac{3}{8}$ 9. $\sin M = 0.7874$ 10. $\cos B = 0.1096$ 11. $\cos Y = \dfrac{24}{25}$

Find the value of each variable.

12.

13.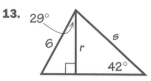

14.

Find the measure of each acute angle. Round your answers to the nearest tenth of a degree.

15.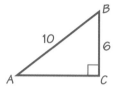

16.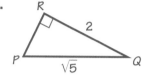

17.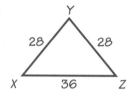

Connection ▶ **ASTRONOMY**

A star appears dimmer when it is near the horizon than when it is directly overhead. This is because light from a star near the horizon must pass through more atmosphere before reaching your eyes, which reduces the light's intensity.

18. Earth's atmosphere is about 1000 km thick. Let $x°$ = the measure of the angle of elevation from an observer to a star, and let d = the distance through the atmosphere that light from the star must travel to reach the observer.

   a. What is the value of d when a star is directly overhead (that is, how many kilometers of atmosphere does the light pass through)?

   b. Write an equation for d in terms of x when a star is not directly overhead.

   c. How far does light from a star travel through the atmosphere when $x = 62$?

19. Research The angle of elevation to an object in space also affects what color it appears to be. Find out how and why an object's apparent color changes as it moves across the sky.

atmosphere

1000 km

Earth

Debby Tewa

Look back at the
article on pages
490–492.

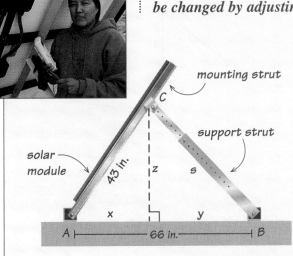

*Debby Tewa uses a special structure like the one below for
supporting a solar module. In the diagram, AC and AB are
permanently set to be 43 in. and 66 in. The measure of ∠A can
be changed by adjusting the length s of the support struts.*

mounting strut

C

support strut

solar
module

43 in.

z   *s*

x

y

A ├────── 66 in. ──────┤ *B*

20. For a home with a flat roof in the
southwestern United States, the exposure to
sunlight is maximized when $m \angle A = 55°$.

   a. What should the value of z be for $\angle A$ to
   have this measure?

   b. Use your answer to part (a) and the
   Pythagorean theorem to find the values
   of x and y.

   c. How long should the support strut be in
   order for $\angle A$ to have the desired measure?

21. Repeat parts (a)–(c) of Exercise 20 for
$m \angle A = 35°$.

22. For this structure, the value of z ranges from 13 in. to 35 in.

   a. Write an inequality that describes the possible values of $m \angle A$ to
   the nearest 5°.

   b. **Open-ended Problem** Describe at least two ways to modify the
   structure so that the range of possible values for $m \angle A$ is expanded.

**Sketch a 45-45-90 triangle and a 30-60-90 triangle. Label each side length.
Then use the triangles to find the exact value of each ratio.**

23. $\sin 45°$ **24.** $\cos 45°$ **25.** $\cos 60°$

26. $\sin 60°$ **27.** $\sin 30°$ **28.** $\cos 30°$

29. AVIATION A jet takes off at a 15° angle. The jet's air speed is 300 ft/s.

   a. Write an equation that gives
   the distance d (in feet) that
   the jet has traveled through
   the air in terms of t, the
   number of seconds it has
   been in the air.

   b. Write equations for the
   horizontal distance h and
   the vertical distance v that
   the jet travels in t seconds.

   c. After 10 s, what horizontal
   distance has the jet traveled?
   What is the jet's altitude?

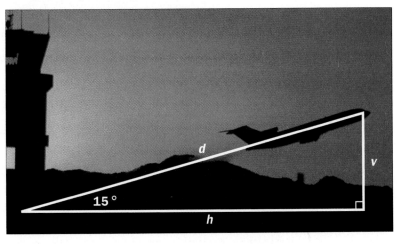

d

v

15°

h

30. Writing Explain how to decide which trigonometric ratio (sine, cosine, or tangent) is best for solving a particular problem.

31. SAT/ACT Preview Suppose that $0° < m \angle X < 45°$. If A = sin X and B = cos X, then:

A. A > B   **B.** B > A   **C.** A = B

D. relationship cannot be determined

For Exercises 32 and 33, use △ABC at the right.

32. Prove that each equation is true.

   a. sin A = cos B **b.** tan $A = \dfrac{\sin A}{\cos A}$

33. ALGEBRA Prove that $(\sin A)^2 + (\cos A)^2 = 1$.
(*Hint:* Use the Pythagorean theorem.)

34. Cooperative Learning Write a problem that you can solve by using a sine or cosine ratio. Have a classmate solve the problem you wrote.

35. Challenge A weight is suspended from a string attached to two vertical poles, as shown. The heights of the poles are h_1 ft and h_2 ft, the distance between the poles is d ft, and the length of the string is ℓ ft. According to the laws of physics, the weight will come to rest at a position such that $\angle ABC$ and $\angle DBE$ have equal measures. Let $n° =$ the measure of each of these angles. Show that $\sin n° = \dfrac{d}{\ell}$. (This means that $m \angle ABE$ does not depend on the heights of the poles.)

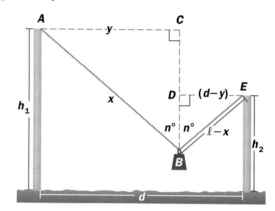

ONGOING ASSESSMENT

36. Open-ended Problem Draw a right triangle. Use a ruler and a protractor to find the length of the hypotenuse and the measure of an acute angle. Use sine and cosine ratios to find the lengths of the triangle's legs. Check your answers by measuring the legs directly.

SPIRAL REVIEW

Find the measure of an acute angle that satisfies the given equation. Round your answers to the nearest tenth of a degree. (*Section 10.3*)

37. tan A = 3   **38.** tan $R = \dfrac{5}{11}$ **39.** tan X = 1.437

40. Find the volume of a cylinder that has diameter 6 in. and height 8 in.
(*Section 7.6*)

Find the coordinates of the midpoint of the segment with each given pair of endpoints. (*Section 4.1*)

41. (0, 0) and (3, 4)   **42.** (−1, 2) and (5, −6)   **43.** (7, 1) and (−3, −5)

10.5 Using Vectors

Orienteering events take place on an established course in a wilderness area. Participants use a map and a magnetic compass to travel between checkpoints.

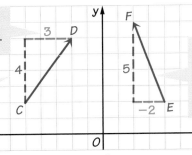

Learn how to...
- **identify vector quantities and express them in component form**

So you can...
- **use vectors and vector sums to analyze real-world situations**

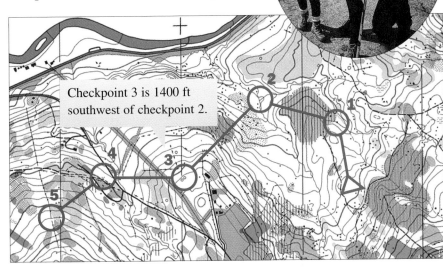

Checkpoint 3 is 1400 ft southwest of checkpoint 2.

In order to plan a course from checkpoint 2 to checkpoint 3, an orienteer needs to know both the direction (southwest) and the distance (1400 ft). This quantity, 1400 ft southwest, is an example of a *vector*.

A **vector** is a quantity that has both *magnitude* (size) and *direction*. A vector is represented by an arrow drawn between two points. For example, the *initial point* of the vector at the right is A, and the *terminal point* is B. The name of the vector is $\overrightarrow{AB}$, read as "vector AB."

By drawing a vector on the coordinate plane, you can easily find its horizontal and vertical components.

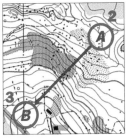

WATCH OUT!

The symbol for a vector looks like the symbol for a ray, but they are *not* the same thing.

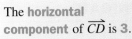

The **horizontal component** of $\overrightarrow{CD}$ is **3**.

The **vertical component** of $\overrightarrow{CD}$ is **4**.

The **magnitude** of $\overrightarrow{EF}$, written as $|\overrightarrow{EF}|$, is the length of this arrow.

In component form, $\overrightarrow{CD} = (3, 4)$ and $\overrightarrow{EF} = (-2, 5)$.

If you know the components of a vector, you can use the Pythagorean theorem to find its magnitude. You can use a trigonometric ratio to find the angle that describes its direction.

EXAMPLE 1

Express $\overrightarrow{PQ}$ in component form.
Find $|\overrightarrow{PQ}|$ and the value of z.

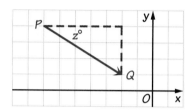

SOLUTION

In component form, $\overrightarrow{PQ} = (5, -3)$.

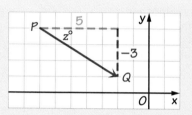

To find $|\overrightarrow{PQ}|$, use the Pythagorean theorem:

$|\overrightarrow{PQ}| = \sqrt{5^2 + 3^2}$ ◀—— Use 3, *not* −3, ——▶
$= \sqrt{34}$
≈ 5.8

because the length of a segment is always positive.

Use the tangent ratio to find the value of z:

$\tan z^\circ = \dfrac{3}{5}$
$z = \tan^{-1}\left(\dfrac{3}{5}\right)$
$z \approx 31.0$

The magnitude and direction of a vector are important, but its location is not. Vectors $\overrightarrow{GH}$ and $\overrightarrow{JK}$ are **equal vectors** because they have the same magnitude and direction.

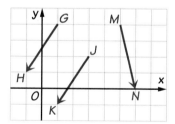

THINK AND COMMUNICATE

Use the graph above.

1. Express each of the three vectors in component form. What do you notice about the components of equal vectors?

2. Give the components of a vector $\overrightarrow{OP}$ that has the same magnitude as $\overrightarrow{MN}$ but a different direction.

3. Give the components of a vector $\overrightarrow{QR}$ that has the same direction as $\overrightarrow{MN}$ but a different magnitude.

Multiplying a Vector by a Number

Suppose the vector $\overrightarrow{AB}$ represents the path taken by a plane traveling for 1 hour at a constant velocity. The path of a plane traveling for 2 hours with the same speed and direction as the first plane is represented by $2\overrightarrow{AB}$. The process of multiplying a vector by a real number is called **scalar multiplication**.

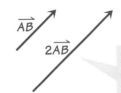

$2\overrightarrow{AB}$ has the same direction as $\overrightarrow{AB}$ and twice the magnitude.

Scalar Multiplication

When a vector is multiplied by a real number k, the length of the vector is multiplied by $|k|$.

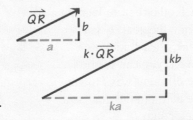

If $\overrightarrow{QR} = (a, b)$, then $k \cdot \overrightarrow{QR} = (ka, kb)$.

EXAMPLE 2   Application: Navigation

A ship travels in a straight path to a location 20 mi west and 15 mi north of its initial point.

a. Graph a vector $\overrightarrow{ST}$ that represents the ship's path. How far did the ship travel?

b. Express $3\overrightarrow{ST}$ in component form. What is the magnitude of $3\overrightarrow{ST}$?

SOLUTION

a.

The distance traveled is the magnitude of the vector.

$$\overrightarrow{ST} = (-20, 15)$$

$$|\overrightarrow{ST}| = \sqrt{20^2 + 15^2}$$

$$= 25$$

The ship traveled 25 mi.

b. $3\overrightarrow{ST} = (3(-20), 3(15))$

$\qquad = (-60, 45)$

$|3\overrightarrow{ST}| = |3| \cdot |\overrightarrow{ST}|$

$\qquad = 3 \cdot 25$

$\qquad = 75$

In component form, $3\overrightarrow{ST} = (-60, 45)$. The magnitude of $3\overrightarrow{ST}$ is 75 mi.

Use the vector $\overrightarrow{OP} = (3, -2)$.

4. Express $-1\overrightarrow{OP}$ in component form.

5. Compare the magnitudes of $\overrightarrow{OP}$ and $-1\overrightarrow{OP}$.

6. Compare the directions of $\overrightarrow{OP}$ and $-1\overrightarrow{OP}$.

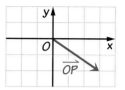

Adding Vectors

Soccer players sometimes have to pass the ball to each other before kicking it into the goal. If you think of each kick as a vector, you can express the different ways to get the ball into the goal as a vector sum.

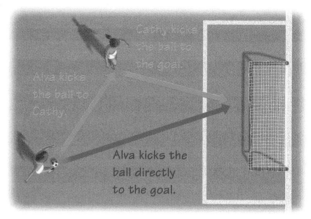

Alva kicks the ball directly to the goal.

$$\overrightarrow{AC} + \overrightarrow{CG} = \overrightarrow{AG}$$

Vector Addition

If $\overrightarrow{GH} = (a, b)$ and $\overrightarrow{HK} = (c, d)$, then $\overrightarrow{GH} + \overrightarrow{HK} = (a + c, b + d)$.

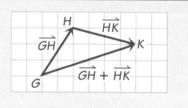

EXAMPLE 3

$\overrightarrow{KL} = (-9, -2)$ and $\overrightarrow{LM} = (3, -3)$. Use a graph to find $\overrightarrow{KL} + \overrightarrow{LM}$. Write the resulting vector in component form.

SOLUTION

Draw $\overrightarrow{KL}$. Then draw $\overrightarrow{LM}$ beginning at the terminal point of $\overrightarrow{KL}$.

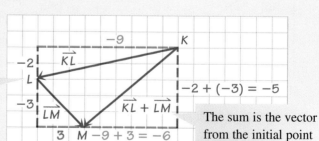

$-2 + (-3) = -5$

$M -9 + 3 = -6$

The sum is the vector from the initial point of $\overrightarrow{KL}$ to the terminal point of $\overrightarrow{LM}$.

$$\overrightarrow{KL} + \overrightarrow{LM} = (-6, -5)$$

☑ CHECKING KEY CONCEPTS

Graph each vector and find its magnitude.

1. $\overrightarrow{AB} = (-5, 0)$ **2.** $\overrightarrow{CD} = (3, 3)$ **3.** $\overrightarrow{EF} = (-3, 6)$

Find the value of each variable.

4.

5.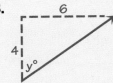

6. In component form, $\overrightarrow{JK} = (3, 7)$. Graph $2\overrightarrow{JK}$ and express $2\overrightarrow{JK}$ in component form.

7. $\overrightarrow{WX} = (-12, 7)$ and $\overrightarrow{YZ} = (5, -21)$. Use these components to find $\overrightarrow{WX} + 4\overrightarrow{YZ}$.

8. $\overrightarrow{NP} = (1, 3)$ and $\overrightarrow{PQ} = (-5, 1)$. Use a graph to find $\overrightarrow{NP} + \overrightarrow{PQ}$ and write the resulting vector in component form.

10.5 **Exercises and Applications**

Extra Practice
exercises on
page 682

Express each vector in component form and find the value of each variable.

1. 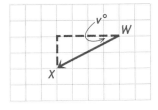   **2.** 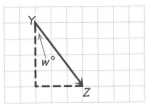   **3.**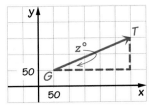

Graph each vector and find its magnitude to the nearest tenth.

4. $\overrightarrow{HJ} = (10, 4)$ **5.** $\overrightarrow{KL} = (-3, 8)$ **6.** $\overrightarrow{MN} = (-4, -4)$ **7.** $\overrightarrow{PQ} = (7, -5)$

8. Graph each scalar multiple of $\overrightarrow{OQ}$ and find its magnitude to the nearest tenth.

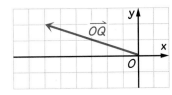

   a. $2\overrightarrow{OQ}$

   b. $\frac{1}{2}\overrightarrow{OQ}$

   c. $-3\overrightarrow{OQ}$

In component form, $\overrightarrow{AB} = (-3, 3)$, $\overrightarrow{CD} = (5, 4)$, and $\overrightarrow{EF} = (-2, 6)$. For Exercises 9–14, use the components to find each vector sum.

9. $\overrightarrow{AB} + \overrightarrow{CD}$ **10.** $\overrightarrow{AB} + \overrightarrow{EF}$ **11.** $\overrightarrow{CD} + \overrightarrow{EF}$

12. $2\overrightarrow{AB} + \overrightarrow{CD}$ **13.** $\overrightarrow{AB} + 2\overrightarrow{CD}$ **14.** $-2\overrightarrow{CD} + \overrightarrow{EF}$

ORIENTEERING Luis must pass through either checkpoint *Q* or checkpoint *R* on his way from checkpoint *P* to checkpoint *S*. Use the map below for Exercises 15 and 16.

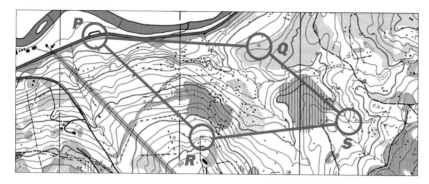

15. The vectors on the map have these components (in feet):

$\overrightarrow{PQ} = (2200, -100)$ $\overrightarrow{QS} = (1400, -1000)$ $\overrightarrow{PR} = (1500, -1300)$ $\overrightarrow{RS} = (2100, 200)$

  a. Express $\overrightarrow{PQ} + \overrightarrow{QS}$ in component form.

  b. Express $\overrightarrow{PR} + \overrightarrow{RS}$ in component form.

  c. Compare your answers to parts (a) and (b). Explain how these answers can be the result of two different vector sums.

16. a. Find $|\overrightarrow{PQ}|$ and $|\overrightarrow{QS}|$. How far will Luis travel if he passes through checkpoint *Q*?

  b. Find $|\overrightarrow{PR}|$ and $|\overrightarrow{RS}|$. How far will Luis travel if he passes through checkpoint *R*?

  c. Compare your answers to parts (a) and (b). Which route is shorter?

  d. **Open-ended Problem** What other factors, besides distance, do you think an orienteer might consider when choosing a path?

Connection ▸ GEOLOGY

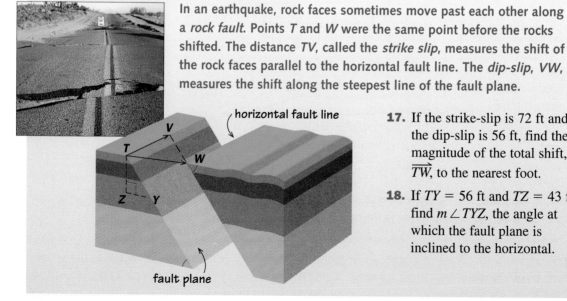

In an earthquake, rock faces sometimes move past each other along a *rock fault*. Points *T* and *W* were the same point before the rocks shifted. The distance *TV*, called the *strike slip*, measures the shift of the rock faces parallel to the horizontal fault line. The *dip-slip*, *VW*, measures the shift along the steepest line of the fault plane.

horizontal fault line

fault plane

17. If the strike-slip is 72 ft and the dip-slip is 56 ft, find the magnitude of the total shift, $\overrightarrow{TW}$, to the nearest foot.

18. If $TY = 56$ ft and $TZ = 43$ ft, find $m\angle TYZ$, the angle at which the fault plane is inclined to the horizontal.

19. You can use the *parallelogram method* to add two vectors. Draw both vectors with the same initial point, as shown. Then draw two segments to complete a parallelogram. The sum of the vectors lies along a diagonal of the parallelogram and has the same initial point as the original vectors.

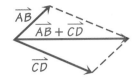

a. In component form, $\overrightarrow{AB} = (2, 2)$ and $\overrightarrow{CD} = (7, -2)$. Use the parallelogram method to find $\overrightarrow{AB} + \overrightarrow{CD}$.

b. Find $\overrightarrow{AB} + \overrightarrow{CD}$ using the method used in Example 3 on page 524.

c. Prove that the methods you used in parts (a) and (b) both yield the same result.

20. Visual Thinking Suppose you want to add three vectors. Sketch what this might look like.

21. Open-ended Problem Give an example of two different vector sums that yield the same result.

22. Suppose A and B are points on the coordinate plane. Prove that $\overrightarrow{BA} = -1\overrightarrow{AB}$.

The magnitude and direction of a vector are labeled on each diagram. Find each unknown length. Then express the vector in component form.

23.

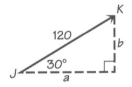

24.

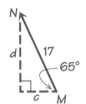

25.

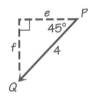

26. Writing Given two points W and Z, explain how the ray $\overrightarrow{WZ}$ and the vector $\overrightarrow{WZ}$ are different.

27. AVIATION In the diagram at the right, $\overrightarrow{VL}$ is an airplane's *velocity vector*. The horizontal component, g, is called the *ground speed*, and the vertical component, r, is the *rate of climb*.

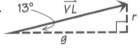

a. Suppose the airplane takes off at an angle of $13°$ to the horizontal, with an air speed of 225 mi/h. Find the airplane's ground speed in miles per hour.

b. Find the rate of climb in miles per hour for the airplane in part (a).

NAVIGATION The *bearing* of a ship or airplane gives its direction of travel. The bearing is the angle, measured clockwise from north, that the velocity vector makes with a vector that points due north.

28. A ship's velocity vector is $\overrightarrow{TU} = (10, 12)$. Draw $\overrightarrow{TU}$ and find its bearing.

29. Challenge In miles per hour, an airplane's velocity vector is $(-180, -130)$. Find the plane's bearing and speed.

30. Cooperative Learning On paper, sketch vectors showing several paths in your classroom. For example, you might show the path from your desk to another desk, or from your teacher's desk to the door. Have another person use the vectors to move around the classroom.

SPIRAL REVIEW

Find the measure of the acute angle that satisfies each equation. Round your answers to the nearest tenth of a degree. *(Sections 10.3 and 10.4)*

31. $\sin D = \dfrac{5}{9}$ **32.** $\tan Z = 3.461$ **33.** $\cos G = 0.1096$

34. $\sin R = \dfrac{\sqrt{3}}{2}$ **35.** $\tan J = 1$ **36.** $\cos B = \dfrac{3}{5}$

37. What is the formula for the volume of a prism? *(Section 7.6)*

ASSESS YOUR PROGRESS

VOCABULARY

tangent of an angle (p. 507)   **vector** (p. 521)
angle of elevation (p. 509)   **magnitude of a vector** (p. 521)
angle of depression (p. 509)   **equal vectors** (p. 522)
sine (p. 515)   **scalar multiplication** (p. 523)
cosine (p. 515)

Find the measure of the acute angle that satisfies each equation. Round your answers to the nearest tenth of a degree. *(Sections 10.3 and 10.4)*

1. $\tan X = 13.73$ **2.** $\sin A = 0.2$ **3.** $\cos K = 0.7501$

Find the value of each variable. Round your answers to the nearest tenth. *(Sections 10.3 and 10.4)*

4.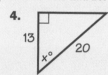

5.

6.

Graph each vector and find its magnitude. *(Section 10.5)*

7. $\overrightarrow{PQ} = (-7, 3)$ **8.** $\overrightarrow{RS} = (25, 40)$

In component form, $\overrightarrow{AB} = (-4, -5)$, $\overrightarrow{CD} = (7, -2)$, and $\overrightarrow{EF} = (3, 1)$. Graph each vector and express it in component form. *(Section 10.5)*

9. $2\overrightarrow{AB}$ **10.** $\overrightarrow{AB} + \overrightarrow{CD}$ **11.** $-3\overrightarrow{EF} + \overrightarrow{CD}$

12. Journal Sketch a right triangle. Label the length of the hypotenuse and the length of one leg. Explain how to find the measures of both acute angles of the triangle.

10.6 Areas and Trigonometry

When you think of a barn, you probably think of a rectangular building. In the early part of the twentieth century, however, barns of many different shapes were built in parts of the United States. A barn in the shape of a regular 12-gon was built in northern Indiana in 1912. The radius of the barn is 28 ft.

EXAMPLE 1   Application: History

Find the area of the floor of the barn described above.

SOLUTION

The 12-gon can be divided into 12 congruent isosceles triangles. The vertex angle of each triangle measures 30°, so each base angle measures 75°.

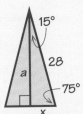

Use trigonometry to find the height and base of each triangle.

$$\sin 75° = \frac{a}{28}$$

$$28 \sin 75 = a$$

$$27.0 \approx a$$

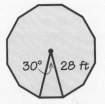

$$\cos 75° = \frac{x}{28}$$

$$28 \cos 75 = x$$

$$7.25 \approx x$$

The apothem of the 12-gon is about 27 ft and each side length is about $14\frac{1}{2}$ ft. Use these values in the formula for the area of a regular polygon.

$$A = \frac{1}{2}ap$$

$$\approx \frac{1}{2}(27)(174)$$

$$\approx 2349$$

The perimeter of the barn is about $12 \cdot 14\frac{1}{2} = 174$.

The area of the floor of the barn is about 2350 ft^2.

THINK AND COMMUNICATE

Look back at Example 1.

1. Explain how dividing the floor into triangles is an important step in the solution.

2. Describe two different ways to find the angle measures of each isosceles triangle in the barn.

As Example 1 shows, sometimes you can find the area of a figure, even if the measures that you need are not given. You can often use trigonometry to find these missing dimensions.

EXAMPLE 2

Find the area of this rhombus.

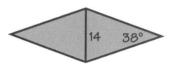

SOLUTION

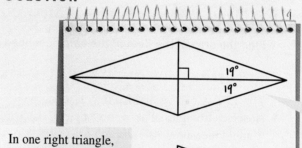

Create four congruent right triangles by drawing the other diagonal.

In one right triangle, write all the measures and lengths that you know.

Use trigonometry to find the value of h.

$$\tan 71° = \frac{h}{7}$$

$$7 \tan 71 = h$$

$$20.33 \approx h$$

The area of the rhombus is four times the area of one right triangle.

$$\text{Area of rhombus} = 4 \cdot \frac{1}{2}bh$$

$$\approx 4 \cdot \frac{1}{2}(7)(20.33)$$

Substitute **7** for **b** and **20.33** for **h**.

$$\approx 284.6$$

The area of the rhombus is approximately 285.

EXAMPLE 3

Each base of this prism is an isosceles trapezoid. Find the volume of the prism.

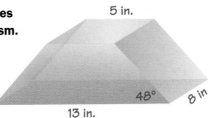

5 in.

48°

8 in.

13 in.

SOLUTION

Step 1 Find the height of the trapezoid.

$$\tan 48° = \frac{h}{4}$$

$$4 \tan 48 = h$$

$$4.44 \approx h$$

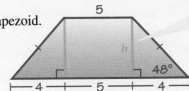

These segments create two congruent right triangles.

Then use the formula for the area of a trapezoid.

Area of base $= \frac{1}{2}(b_1 + b_2)h$

$$\approx \frac{1}{2}(5 + 13)(4.44)$$

$$\approx 40$$

Step 2 Find the volume of the prism.

Volume = (**Area of base**) × (**Height of prism**)

$$\approx 40 \cdot 8$$

$$\approx 320$$

The volume of the prism is about 320 in.^{3}.

☑ **CHECKING KEY CONCEPTS**

Find the area of each polygon.

1.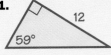
12

59°

2.
9

110°

3.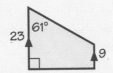
23
61°
9

4. a regular decagon with side length 10 cm

5. a regular pentagon with apothem 8 in.

6. an isosceles triangle with a vertex angle of 40° and leg length 3 ft

10.6 Exercises and Applications

Extra Practice
exercises on
page 682

Find the area of each polygon.

1.
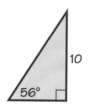

2.
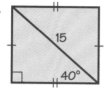
15
40°

3.
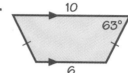
10
63°
6

4.

72°
20

5.
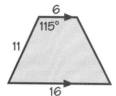
6
115°
11
16

6.
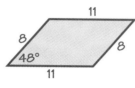
11
8
48°
11
8

For Exercises 7–10, sketch each figure described. Then find its area.

7. The apothem of a regular pentagon is 12 cm.

8. A regular 9-gon has radius 5 mm.

9. A parallelogram has 10 m and 50 m side lengths and a 25° angle.

10. The angle measures of a trapezoid are 90°, 90°, 105°, and 75°. The bases are 8 yd and 9 yd long.

11. In the diagram at the right, $m \angle ABC = 115°$.

  a. Find BD and AD.

  b. Use BD and a trigonometric ratio to find DC.

  c. Use your answers to parts (a) and (b) to find the area of $\triangle ABC$.

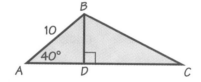

12. Open-ended Problem Draw any $\triangle ABC$. Measure $\angle A$, $\angle B$, and $\overline{AB}$. Then repeat parts (a)–(c) of Exercise 11 to find the area of $\triangle ABC$.

13. Investigation You will need a ruler and a protractor. Copy this graph.

  a. Choose any point on the line and draw a perpendicular segment from the point to the x-axis. Measure the length of the segment.

  b. Use trigonometry and the length you found in part (a) to find the area of the triangle whose sides are the line, the x-axis, and the perpendicular segment.

  c. Graph another line that has a positive slope. Measure the angle that the line makes with the x-axis and repeat parts (a) and (c).

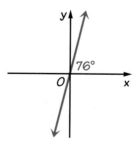

14. Writing Describe two different ways to find the area of this trapezoid. Include diagrams for each method.

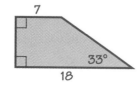
7
33°
18

532 Chapter 10 *Applying Right Triangles*

Find the volume of each prism.

15.
5 m

9 m

39°

16. 68°

15 ft

12 ft

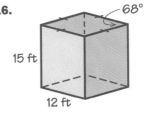

17.
11 cm

8 cm

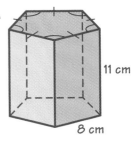

Connection ▸ **HISTORY**

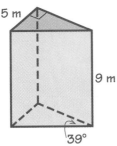

In 1792, George Washington designed a special barn for his new wheat treading process. Horses walked on the wheat on the second level of the barn. Their hooves separated the straw from the grain, which fell through grooves in the floor to the first level. In 1996, this barn, a regular 16-gon, was reconstructed in Mount Vernon, Virginia.

George Washington's original plans

The reconstructed barn

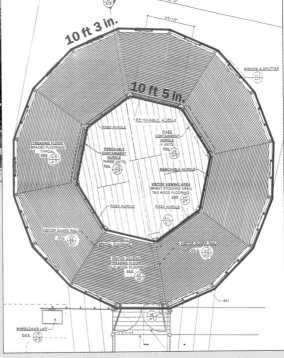

10 ft 3 in.

10 ft 5 in.

The modern plans for the reconstructed barn

18. Each side of the barn is 10 ft 3 in. Find the area of the barn floor.

19. At the center of each level of the barn is a grain storage room in the shape of a regular octagon. The wheat from the field was stored in the upper level storage room until the horses were brought in to walk on it. Each side of the room is about 10 ft 5 in. Find the area of the storage room floor.

20. The horses walked around the outside of the storage room. Use your answers to Exercises 18 and 19 to find the area of the floor where they walked on the wheat.

21. Writing Compare the area of the storage room floor to the area of the floor where the horses walked. Why might Washington have been interested in comparing these areas?

22. ALGEBRA In this section, you learned how to use trigonometry to find the areas of right triangles. You can also use trigonometry to find the area of a non-right triangle. Use $\triangle ABC$ at the right.

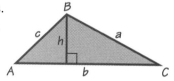

a. Complete: $\sin A = \dfrac{?}{?}$.

b. Solve the equation that you wrote in part (a) for h. Use the result to write an expression for the area of $\triangle ABC$.

c. The formula that you found in part (b) can be used to find the area of a triangle when two side lengths and the measure of the included angle are known. Use this formula to find the area of the triangle below.

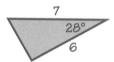

23. Prove that the area of $\triangle XYZ$ with a right angle at Z is $A = \frac{1}{2}XZ^2 \tan X$.

24. Prove that the area of an isosceles triangle with a base angle that measures $x°$ and a base of length b is $A = \frac{1}{4}b^2 \tan x°$.

25. SAT/ACT Preview $ABCD$ is a square, X is the midpoint of $\overline{AD}$, and Y is the midpoint of $\overline{CD}$. Which regions have equal areas?

A. $\triangle ABD$ and $\triangle DYB$

B. $\triangle BDC$ and $\triangle ABX$

C. $\triangle BDY$ and $\triangle BCY$

D. Trapezoid $ABYD$ and quadrilateral $BXDY$

E. None of these

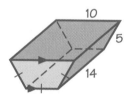

26. Challenge Find the surface area of this prism.

ONGOING ASSESSMENT

27. Writing Write a summary of how to use trigonometry to find areas and volumes. Include sketches and examples if necessary.

SPIRAL REVIEW

Find the magnitude of each vector. *(Section 10.5)*

28. $\overrightarrow{AB} = (3, -4)$

29. $\overrightarrow{HJ} = (1, 1)$

30. $\overrightarrow{MP} = (-7, -11)$

31. The radius of a cylinder is 12 cm, and the height is 3 cm. Find the volume of the cylinder. *(Section 7.6)*

10.7 Pyramids and Cones

Learn how to...

• apply formulas for surface areas and volumes of pyramids and cones

So you can...

• analyze real-world pyramids and cones, such as paper cups

Many major sports and musical events in Memphis, Tennessee, take place at the Pyramid Arena, shown below. Built from 1989 to 1991, the Arena was modeled after the Great Pyramids in Egypt.

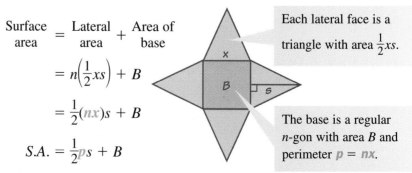

In geometry, a **pyramid** is a three-dimensional figure with one **base** that is a polygon. The other faces of a pyramid, called the **lateral faces**, are triangles that connect the base to the **vertex**.

The Pyramid Arena is a *regular pyramid*. In a **regular pyramid**, the base is a regular polygon and the lateral faces are congruent isosceles triangles.

The **height**, h, of a regular pyramid is the distance from the vertex to the center of the base.

The **slant height**, s, of a regular pyramid is the altitude of a lateral face.

You can use a net to find the formula for the surface area of a pyramid.

$$\text{Surface area} = \text{Lateral area} + \text{Area of base}$$

$$= n\left(\frac{1}{2}xs\right) + B$$

$$= \frac{1}{2}(nx)s + B$$

$$S.A. = \frac{1}{2}ps + B$$

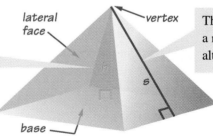

Each lateral face is a triangle with area $\frac{1}{2}xs$.

The base is a regular n-gon with area B and perimeter $p = nx$.

This formula can be used to find the surface area of any regular pyramid.

Volume and Surface Area of Pyramids

The volume of a pyramid is one third the product of the height and the area of the base.

$$V = \frac{1}{3}Bh$$

The surface area of a regular pyramid is the sum of the lateral area and the area of the base.

$$S.A. = \frac{1}{2}ps + B$$

EXAMPLE 1   Application: Archeology

When the pyramid of Khafre was built in Giza, Egypt, it was about 471 ft high. Its base is a square, $707\frac{3}{4}$ ft on a side.

Find the original volume and surface area of the pyramid.

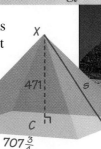

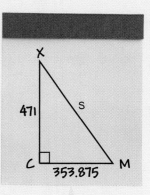

SOLUTION

Find the volume:

$$V = \frac{1}{3}Bh$$

> The area of the square base is $B = (707.75)^2$.

$$= \frac{1}{3}(707.75)^2(471)$$

$$\approx 78{,}643{,}000$$

To find the surface area, first find s, the slant height of the pyramid.

$$s = \sqrt{XC^2 + CM^2}$$

$$= \sqrt{471^2 + (353.875)^2}$$

$$\approx 589$$

Substitute **589** for s in the formula for surface area.

$$S.A. = \frac{1}{2}ps + B$$

> $CM = \frac{1}{2}(707.75)$
> $= 353.875$

$$\approx \frac{1}{2}(4 \cdot 707.75)(589) + (707.75)^2$$

$$\approx 1{,}335{,}000$$

The original volume of the pyramid was about 78,643,000 ft^3. The original surface area was about 1,335,000 ft^2.

Like prisms, pyramids are classified by the shapes of their bases. The Pyramid Arena and the Giza Pyramids are *square pyramids*. You can sketch a square pyramid by following the directions below.

Step 1 Draw the base and locate its center.

Step 2 Draw a vertical segment from the center of the base.

Step 3 Join the vertices of the base to the endpoint of the segment. Make the hidden edges into dashed lines.

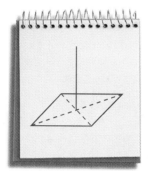
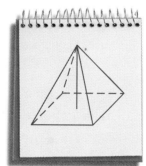

Cones

A **cone** is a three-dimensional figure with one circular base and a vertex. In this book, you will learn about *right cones*. The vertex of a right cone is directly above the center of the base.

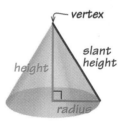

THINK AND COMMUNICATE

1. Describe how a right cone is like a regular pyramid.

2. How would you find the perimeter and area of the base of a cone?

You can use the formulas for volume and surface area of pyramids to find the formulas for volume and surface area of cones.

Volume and Surface Area of Cones

The volume of a cone is one third the product of the height and the area of the base.

$$V = \frac{1}{3}Bh$$

$$= \frac{1}{3}\pi r^2 h$$

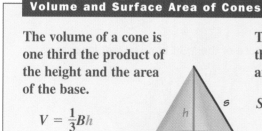

The surface area of a cone is the sum of the lateral area and the area of the base.

$$S.A. = \frac{1}{2}ps + B$$

$$= \frac{1}{2}(2\pi r)s + \pi r^2$$

$$= \pi rs + \pi r^2$$

EXAMPLE 2   Application: Packaging

One type of paper cup is cone-shaped, with the dimensions shown. How much water, in cubic inches, can the cup hold? How much paper is required to make the cup?

SOLUTION

First use trigonometry to find the height of the cup and the radius of its base.

$\sin 70° = \dfrac{h}{3.5}$ $\cos 70° = \dfrac{r}{3.5}$

$(3.5)\sin 70 = h$ $(3.5)\cos 70 = r$

$3.3 \approx h$ $1.2 \approx r$

Substitute these values for h and r into the formulas for volume and lateral surface area.

$$V = \dfrac{1}{3}\pi r^2 h \qquad\qquad\qquad L.A. = \pi r s$$

$$\approx \dfrac{1}{3}\pi(1.2)^2(3.3) \qquad\qquad \approx \pi(1.2)(3.5)$$

$$\approx 5.0 \qquad\qquad\qquad\qquad \approx 13.2$$

The cup can hold about 5.0 in.³ of water. At least 13.2 in.² of paper is required to make the cup.

☑ CHECKING KEY CONCEPTS

1. The volume of this cone is 117π. Find the radius of the base and the slant height.

Find the volume and surface area of the right cone or regular pyramid.

2.

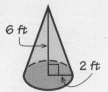

3.

10.7 Exercises and Applications

Extra Practice
exercises on
page 682

Find the volume and surface area of each right cone or regular pyramid.

1.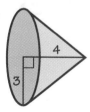

2.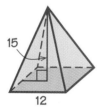
15
12

3.
13
5

4.
31°
10

5.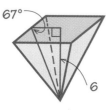
9
12

6. 67°
6

Use the steps shown on page 537 to sketch each pyramid.

7. a square pyramid
8. a triangular pyramid
9. a pentagonal pyramid

10. Writing Write directions that a friend could follow to sketch a cone.

Connection ▶ ARCHITECTURE

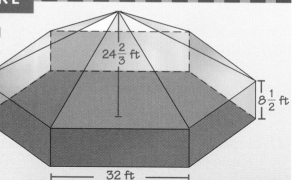

Burt Rutan, an airplane designer, has an unusual home in the Mojave Desert. The top part of his house is a regular hexagonal pyramid. The bottom part is a regular hexagonal prism.

$24\frac{2}{3}$ ft

$8\frac{1}{2}$ ft

32 ft

11. a. Find the lateral area of the pyramid.

  b. How much paint is needed to paint the pyramid's outer surface? (One gallon covers about 400 ft².)

12. a. Find the area of the hexagonal base.

  b. Find the volume of the pyramid.

  c. Find the volume of the prism. Then find the volume of the entire house.

13. Open-ended Problem
Design a house using pyramids and cones. Sketch the house and label its dimensions. Find the volume and the external surface area of the house.

BY THE WAY

Even the furniture in Burt Rutan's house is unusual. The pool table is a parallelogram with two 60° angles and two 120° angles.

house
garage

14. ARCHITECTURE The Misumi Ferry Terminal in Japan is a 16-sided regular pyramid, 25 m high. The radius of the base of the building is 17 m.

a. Find the volume of the terminal.

b. Find the volume of a cone with radius 17 m and height 25 m.

c. **Logical Reasoning** Compare your answers to parts (a) and (b). When does it make sense to estimate the volume of a pyramid by finding the volume of a cone? Explain your reasoning.

15. The volume of a right cone is 144π and the area of its base is 36π. Find the radius, height, and slant height of the cone.

16. A pyramid has a height of 15 ft and a volume of 450 ft^3. What is the area of the base of the pyramid?

Cooperative Learning **Work with another person. You will need heavy paper and dried beans or popcorn kernels. You will compare the volumes of a cone and a cylinder that have the same height and congruent bases.**

Step 1 Cut a rectangle and a semicircle out of paper using the dimensions shown. Tape the edges to make a cylinder and a cone.

Step 2 Fill the cone with beans. Pour the beans from the cone into the cylinder. Repeat until the cylinder is full.

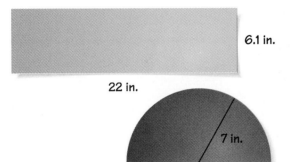

17. Challenge Show that the cylinder and cone have the same height and approximately congruent bases.

18. What is the ratio of the volume of the cone to the volume of the cylinder? Explain your reasoning.

19. Writing Compare the formula for the volume of a cone with the formula for the volume of a cylinder. Is the relationship between the formulas supported by your results of this activity? Explain.

20. Suppose the base of a cone has radius 10 cm and the base angle of the cone is 65°. If the vertex is removed by cutting through the cone along a plane parallel to the base, the resulting figure is called a *truncated cone*.

a. Find the volume of the original cone.

b. The removed vertex is also a cone. What is the base angle of this cone? Find the volume of the vertex if its base has a radius of 5 cm.

c. Find the volume of the truncated cone.

21. The volume formulas you learned can also be applied to *oblique* pyramids and cones. Find the volume of each figure.

a.

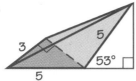

b.

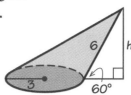

ONGOING ASSESSMENT

22. Cooperative Learning Work with another person. Make a chart showing all of the surface area and volume formulas you have learned so far. Sketch a figure for each, labeling all of its dimensions.

SPIRAL REVIEW

Graph each vector and find its magnitude. *(Section 10.5)*

23. $\overrightarrow{AB} = (0, -7)$ **24.** $\overrightarrow{CD} = (-5, 12)$ **25.** $\overrightarrow{EF} = (550, -400)$

Find the geometric mean of the given numbers. *(Section 10.1)*

26. 4 and 16   **27.** 10 and 1000   **28.** 30 and 60

ASSESS YOUR PROGRESS

VOCABULARY

pyramid (p. 535)   **regular pyramid** (p. 535)
base of a pyramid (p. 535)   **height of a pyramid** (p. 535)
lateral face of a pyramid (p. 535)   **slant height** (p. 535)
vertex of a pyramid (p. 535)   **cone** (p. 537)

Find the area of each polygon. *(Section 10.6)*

1.

2.

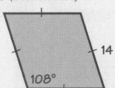

3.

Find the volume and surface area of each right cone or regular pyramid.
(Section 10.7)

4.

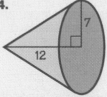

5.

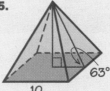

6.

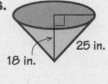

7. Journal Compare the formulas for volume and surface area of pyramids to the formulas for volume and surface area of prisms.

Applying Solar Geometry

The sun can help to heat your home during the winter. The roof of a *passive solar* home allows sunlight to shine in during the winter, when the sun is low in the sky. These roofs also help keep homes cool during the summer when the sun is higher.

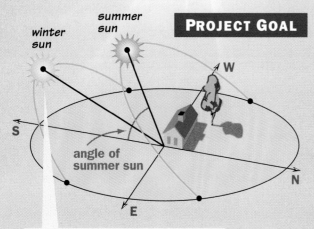

winter sun
summer sun
W
S
angle of summer sun
N
E

At noon on December 21, the sun's angle is 90° − **latitude** − 23.5°.

PROJECT GOAL Use the angle of the sun to plan the roof overhang for a passive solar home.

Analyzing Angles

The house in the diagrams shown is in Chicago, Illinois, at a **latitude of 42°** north . To find the angle of the sun at noon on the longest day of the year, June 21, use this expression:

$$90° − 42° + 23.5°$$

After June 21, the days shorten until the sun reaches its lowest point in the sky on December 21. You can use the angle of the sun to find the best angle for a roof.

The angle of the roof blocks the sun's rays during the summer.

The angle of the roof allows the sun's rays in during the winter.

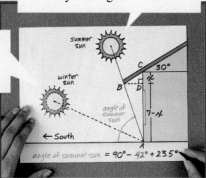

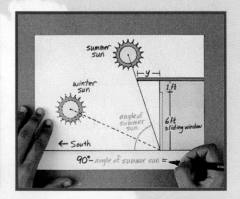

1. Find the amount of overhang, *BC*, of a roof of slope 30° in the Chicago home shown. Explain your method. (*Hint:* Find *BC* in terms of *x* and solve for *x*.)

2. Examine the diagram of a solar home with a flat roof, as shown above. Find the amount of overhang, *y*, that blocks the summer sun.

You can get ideas for your design by looking at homes in your neighborhood.

Designing Your Roof

3. Look up the latitude of your town. Calculate the angle of the sun at noon on the longest and shortest days of the year.

4. Use the angle that you found in Step 3 to design your sloping solar roof. Use trigonometry to find the amount of overhang needed in the summer. Include the dimensions of windows or doors affected by the overhang.

Presenting Your Results

Use diagrams, photographs, or models to illustrate your results. Describe your design and be sure to include these points:

- Describe how a passive solar roof performs different functions at different times of the year.

- Explain the differences between overhangs of flat and sloping roofs. Which type do you prefer, and why?

- How do weather conditions, location, and the direction the house faces affect its solar efficiency?

Extending Your Project

Here are some ideas for extending your project:

- Is there only one correct angle for the roof of a passive solar house? Explain why or why not.

- Why is 23.5° used to calculate the sun's angle? (*Hint:* Find the definition of *ecliptic.*)

- Are there other ways to heat a house using solar energy?

- If possible, talk to an architect to see if your plan is realistic.

Self-Assessment

Did your design for a solar home turn out the way you expected? How did you use trigonometry in your design? Would you consider using some type of passive solar heating if you were building a house? Why or why not?

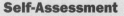

Find pictures or make sketches of different roof designs.

10 Review

STUDY TECHNIQUE

List three techniques you have used in the past to study for a test. Exchange lists with another student and use at least two of the suggestions to study this chapter in preparation for the test.

VOCABULARY

geometric mean (p. 494)
45-45-90 triangle (p. 500)
30-60-90 triangle (p. 501)
tangent of an angle (p. 507)
angle of elevation (p. 509)
angle of depression (p. 509)
sine (p. 515)
cosine (p. 515)
vector (p. 521)
magnitude of a vector (p. 521)

equal vectors (p. 522)
scalar multiplication (p. 523)
pyramid (p. 535)
base of a pyramid (p. 535)
lateral face of a pyramid (p. 535)
vertex of a pyramid (p. 535)
regular pyramid (p. 535)
height of a pyramid (p. 535)
slant height (p. 535)
cone (p. 537)

SECTIONS 10.1 *and* 10.2

You can find the positive number x that is the **geometric mean** of two positive numbers a and b by solving the proportion $\frac{a}{x} = \frac{x}{b}$.

When the altitude is drawn to the hypotenuse of a right triangle:

- the two triangles formed are similar to the original triangle and to each other.

- the length of the altitude is the geometric mean of the lengths of the segments of the hypotenuse.

- the length of each leg is the geometric mean of the lengths of the hypotenuse and the segment of the hypotenuse adjacent to that leg.

These diagrams show the relationships between the side lengths for a **45-45-90 triangle** and for a **30-60-90 triangle**.

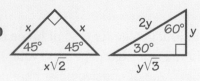

SECTIONS 10.3, 10.4, *and* 10.5

You can find side lengths and angle measures of a right triangle by using these *trigonometric ratios*:

$$\sin A = \frac{a}{c} \qquad \cos A = \frac{b}{c} \qquad \tan A = \frac{a}{b}$$

A vector is a quantity with both magnitude and direction. The **magnitude** of $\vec{RS}$ is the length RS. In the diagram below, $\vec{RS}$ has *horizontal component* -4 and *vertical component* 3.

You can use the Pythagorean theorem to find the magnitude of a vector and trigonometry to find the angle that a vector makes with the horizontal. Vectors with the same magnitude and direction are **equal vectors**.

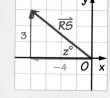

To multiply a vector by a real number k, multiply each component by k. This is called **scalar multiplication**. Multiplying a vector by k multiplies the length of the vector by $|k|$. To add two vectors, add their corresponding components.

SECTIONS 10.6 *and* 10.7

Trigonometry can be used to find the area, surface area, or volume of some figures. For example:

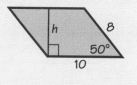

$$\sin 50° = \frac{h}{8} \qquad A = bh$$
$$h \approx 6.1 \qquad \approx 10(6.1)$$
$$= 61$$

A **pyramid** is a figure with one **base** that is a polygon. In a **regular pyramid**, the base is a regular polygon and the **lateral faces** are congruent isosceles triangles.

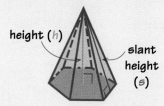

height (h)
slant height (s)

$$V = \frac{1}{3}Bh$$
$$S.A. = \frac{1}{2}ps + B$$

The figure at the right is a **cone**. In a *right cone*, the vertex lies directly above the center of the base.

$$V = \frac{1}{3}\pi r^2 h$$
$$S.A. = \pi rs + \pi r^2$$

10 Assessment

VOCABULARY QUESTIONS

1. Sketch a regular pyramid and a right cone. On each figure, label the base, vertex, height, and slant height.

2. Refer to the right triangle shown. Find the sine, cosine, and tangent of $\angle R$ and $\angle S$ in terms of RS, ST, and RT.
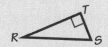

3. Explain how to find the geometric mean of two positive numbers.

4. A train is traveling at a speed of 50 mi/h. Can you represent this situation using a vector? Explain why or why not.

5. Explain the difference between *angle of elevation* and *angle of depression*.

SECTIONS 10.1 *and* 10.2

For Questions 6–11, find the value of each variable.

6.
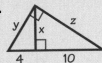

7.

8.
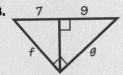

9.
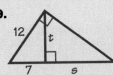

10.
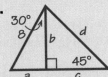

11.
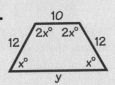

12. Find the geometric mean of the given numbers.

   a. $\dfrac{5}{6}$ and $\dfrac{1}{2}$ **b.** 1 and 1000   **c.** $\sqrt{2}$ and $32\sqrt{2}$

Find the area of each polygon.

13.

14.

15.
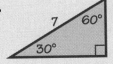

Sections 10.3, 10.4, *and* 10.5

Find the value of each ratio. Round your answers to four decimal places.

16. sin 37.6°   **17.** cos 37.6°   **18.** tan 37.6°

19. A wheelchair ramp that extends 20 ft forms a 4° angle with the level ground. About how high above the ground is the top of the ramp?

20. **MEASUREMENT** When Susan Karp stands 10 ft from a building, the angle of elevation to the top of the building is about 70°. If Susan's eyes are 5 ft above the ground, about how tall is the building?

21. In $\triangle ABC$, $AB = BC = 12$ and $AC = 18$. Find $m \angle A$, $m \angle B$, and $m \angle C$.

In Questions 22 and 23, $\overrightarrow{AB} = (-15, -8)$ and $\overrightarrow{CD} = (9, -3)$.

22. Graph $\overrightarrow{AB}$ and $\frac{1}{3}\overrightarrow{CD}$ and find the magnitude of each vector.

23. Use the components to find the sum $\overrightarrow{AB} + \overrightarrow{CD}$.

24. **Writing** Suppose $\overrightarrow{PQ} = (a, b)$. Explain how to find the magnitude of $\overrightarrow{PQ}$ and the angle $\overrightarrow{PQ}$ makes with the horizontal.

Sections 10.6 *and* 10.7

25. The perimeter of a regular 7-gon is 63. Find the area of the 7-gon.

26. The perimeter of a rhombus is 80 cm, and one of its angles measures 146°. Find the area of the rhombus.

Find the volume and the surface area of the prism, the cone, and the regular pyramid.

27.

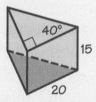

28.

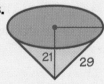

29.

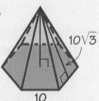

30. **Open-ended Problem** Sketch a square pyramid and choose dimensions for the base and for the slant height. Find the surface area and the volume of the pyramid.

Performance Task

31. Suppose you know the value of one trigonometric ratio of an acute angle. Describe how to find the values of the other two trigonometric ratios of that angle.

Algebra
Review/Preview

These exercises review algebra topics you will use in the next chapters.

SIMPLIFYING MORE EXPRESSIONS

Simplify each expression. If it is not possible, write *cannot be simplified*.

EXAMPLES $2x^3 + 4x^2 - 3x^3$

$$= 2x^3 - 3x^3 + 4x^2 \quad \text{Group the like terms.}$$

$$= -x^3 + 4x^2$$

$2\pi r^2 + 2\pi rh$ *cannot be simplified* because the terms have different variable parts.

1. $y + y^3 - y + y^3$ **2.** $2x^2 + 4xh$ **3.** $9\pi r^2 - 4\pi r^2$

4. $\pi r^2 h + \frac{1}{3}\pi r^2 h$ **5.** $\frac{1}{2}s^2 + 2s^2$ **6.** $\frac{4}{3}\pi r^3 + \frac{1}{2}\pi r^3$

7. $5n + 15n^2 + n^3$ **8.** $(2z)^3 + (3z)^2 + z^3$ **9.** $(x^2 - x) - (x + x^3)$

EXPLORING MATRICES

EXAMPLES The matrix $\begin{bmatrix} -1 & 6 & 5 \\ -3 & -3 & 0 \end{bmatrix}$ has two rows and three columns. Its dimensions are 2 × 3.

10. a. How many rows and how many columns does a 5 × 2 matrix have?

    b. How many elements does a 5 × 2 matrix have?

For Exercises 11–13, use the table and matrix of statistics for hockey teams.

Team	W	L	T
Nordiques	30	13	5
Penguins	29	16	3
Bruins	27	18	3
Sabres	22	19	7
Whalers	19	24	5
Canadiens	18	23	7
Senators	9	34	5

$$M = \begin{bmatrix} 30 & 13 & 5 \\ 29 & 16 & 3 \\ 27 & 18 & 3 \\ 22 & 19 & 7 \\ 19 & 24 & 5 \\ 18 & 23 & 7 \\ 9 & 34 & 5 \end{bmatrix}$$

11. Give the dimensions of matrix M.

12. How many wins did the Bruins have?

13. How many losses did the Canadiens have?

Simplify each expression if possible.

EXAMPLES
$$\begin{bmatrix} 4 & 1 \\ -3 & 2 \end{bmatrix} + \begin{bmatrix} 1 & 0 \\ 3 & -5 \end{bmatrix} = \begin{bmatrix} 4+1 & 1+0 \\ -3+3 & 2+(-5) \end{bmatrix} = \begin{bmatrix} 5 & 1 \\ 0 & -3 \end{bmatrix}$$

$$\begin{bmatrix} 4 & 1 \\ -3 & 2 \end{bmatrix} + \begin{bmatrix} -1 & 6 & 5 \\ -3 & -3 & 0 \end{bmatrix}$$ cannot be evaluated because the matrices have different dimensions.

Use matrices A, B, C, D, and E for Questions 14–17.

$$A = \begin{bmatrix} 1 & -1 \end{bmatrix} \quad B = \begin{bmatrix} 7 \\ -6 \\ 0 \end{bmatrix} \quad C = \begin{bmatrix} 6 & -9 \\ 0 & -2 \\ 1 & 4 \end{bmatrix} \quad D = \begin{bmatrix} -2 \\ -3 \\ 8 \end{bmatrix} \quad E = \begin{bmatrix} 0 & 7 \\ 3 & -2 \\ -1 & 0 \end{bmatrix}$$

14. Give the dimensions of each matrix.

15. Add each pair of matrices, if possible. If the matrices cannot be added, write *not possible*.

a. $A + C$ **b.** $B + D$ **c.** $B + E$ **d.** $C + E$

16. Subtract each pair of matrices, if possible. If the matrices cannot be subtracted, write *not possible*.

a. $B - D$ **b.** $A - E$ **c.** $C - E$ **d.** $E - C$

17. Writing Compare your answers to parts (c) and (d) of Question 16. Does the order in which you subtract matrices affect the result? Do you think that the order in which you add two matrices affects the result? Explain why or why not.

USING FORMULAS

Toolbox sidebar

> **Toolbox p. 702**
> *Working with Formulas*

Solve each formula or equation for the indicated variable.

EXAMPLES

Solve $x^2 + y^2 = r^2$ **for y.**

$$y^2 = r^2 - x^2$$
$$y = \pm\sqrt{r^2 - x^2}$$

Solve $\frac{4}{3}\pi r^3 = 36\pi$ **for r.**

$$\pi r^3 = \left(\frac{3}{4}\right)36\pi$$
$$r^3 = 27 \quad \text{3 \cdot 3 \cdot 3 = 27}$$
$$r = 3$$

Multiply both sides by $\frac{3}{4}$, the inverse of $\frac{4}{3}$.

18. Solve $a^2 + b^2 = (2a)^2$ for b.

19. Solve $x^2 - y^2 = r^2$ for y.

20. Solve $\frac{8}{3}\pi r^2 = 384\pi$ for r.

21. Solve $\frac{24}{21} = \frac{x^3}{7}$ for x.

22. Solve $\frac{x}{360} \cdot 2\pi \cdot 15 = \frac{9\pi}{4}$ for x.

23. Solve $x^2 = 4(4 + 5)$ for x.

24. Solve $S = 4\pi r^2$ for r.

25. Solve $s^3 = 125$ for s.

26. Solve $A = \frac{1}{2}ap$ for p.

27. Solve $V = \frac{1}{3}\pi r^2 h$ for h.

11 | Circles and Spheres

Treating SP●RTS *Injuries*

INTERVIEW **Ron Courson**

As director of sports medicine at the University of Georgia and as the chief athletic trainer for track and field at the 1996 Summer Olympics, Ron Courson has seen a lot of injuries. "There are different challenges every day," Courson says. "Every athlete is different. Their injuries are different and their personalities are different, so the rehabilitation programs we design have to be different, too."

"I use more math on a daily basis than I ever anticipated."

Yet all injuries have some things in common. "Almost everything we do in rehabilitation relates to the *range of motion* of joints like the shoulder, hip, or knee," Courson notes. A joint's range of motion is the arc through which it can move. "One of the first things we do in our evaluation is to measure these arcs and find out whether the athlete has a normal range of motion," Courson explains. He then recommends exercises, such as leg extensions or leg curls, which strengthen the injured part while keeping it within an appropriate arc of motion.

In addition to limiting range of motion, injuries can affect an athlete's sense of balance. To improve balance, Courson sometimes uses a technique developed in Switzerland. The athlete lies on top of a ball and tries to balance as shown. The size of the ball affects how difficult it is to balance. This technique also increases muscle strength.

Educating Athletes

Education, Courson stresses, is an important part of the rehabilitation process. "Even if we see an athlete every day for two hours, the athlete is on his or her own 22 hours a day," he says. "So athletes really need to understand the nature of their injuries and why, for instance, the knee hurts so much when it's extended through the last 45°. We show them a model so they can see all the stresses on the joint and understand the physics and geometry of the situation."

Restricted Zones

For an injury of the *rotator cuff*, which is common to swimmers and baseball pitchers, there may be a painful arc between 130° and 150°. Courson does tests to determine precisely where that painful zone lies. He then recommends exercises that break the shoulder motion down into smaller arcs. The exercises avoid the restricted zone, preventing further injury to the athlete.

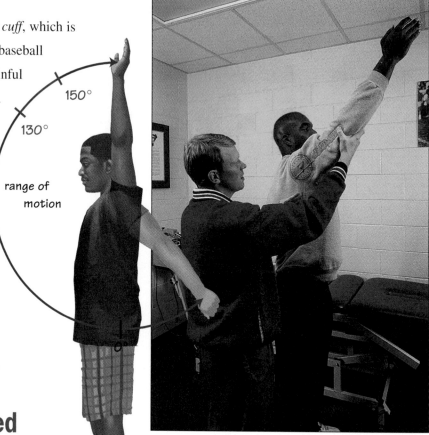

150°

130°

range of motion

"Athletes need to know geometry, too."

Mathematics & Ron Courson

In this chapter, you will learn more about how mathematics is related to athletic training and physical therapy.

Related Exercises

Section 11.1
• Exercises 11 and 12

Section 11.3
• Exercise 16

Section 11.6
• Exercises 10–12

EXPLORE AND CONNECT

Ron Courson

1. Project   Choose one joint of the body and carefully measure the range of motion for at least ten people. Be sure no one moves the joint in a way that hurts. Present your results on a poster. Include a sketch of the joint and an explanation of how you measured the range of motion.

2. Writing   Describe an injury you have had and how your range of motion was affected.

3. Research   One popular exercise is a modified form of *tai chi*, a traditional Chinese series of movements. Learn what some of the movements in *tai chi* are and why practicing them can be beneficial.

11.1 Angles and Circles

The angles formed by lines that intersect circles have special properties. In the diagram, $\angle ACB$ is a **central angle** because the vertex C is the center of a circle. $\angle ADB$ is an **inscribed angle** because the vertex D is on the circle and the sides contain *chords* of the circle. A **chord** of a circle is a segment whose endpoints lie on the circle.

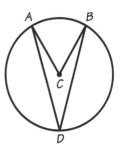

EXPLORATION
COOPERATIVE LEARNING

Comparing Central and Inscribed Angles

Work in a group of three students.
You will need:
- geometry software or a compass and a protractor

m∠ADB	m∠ACB
60°	120°

1 Choose any point on a circle C. From this point, draw two chords to form an acute inscribed angle. Label the angle $\angle ADB$. Find $m\angle ADB$ and record it in a table.

2 Draw $\angle ACB$. Find $m\angle ACB$ and record it in the table.

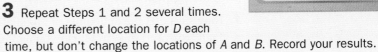

3 Repeat Steps 1 and 2 several times. Choose a different location for D each time, but don't change the locations of A and B. Record your results.

4 Make a conjecture based on your results in Step 3.

5 Repeat Steps 1 and 2 several times, moving only A to form each new acute $\angle ADB$. Record your results and make a conjecture.

6 A *tangent line* intersects a circle in only one point. Draw diameter $\overline{DB}$ and tangent $\overleftrightarrow{AD}$. Use $m\angle ADB$ and your results from Step 5 to make a conjecture about the measure of any angle formed by a tangent and a diameter. Explain your reasoning.

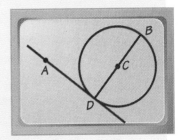

In Step 6 of the Exploration, $\overleftrightarrow{AD}$ is **tangent** to the circle because it is in the
same plane as the circle but intersects the circle in only one point. The
intersection is called the *point of tangency*. $\overrightarrow{AD}$ and $\overline{AD}$ are also tangents.
You may have discovered the following result about tangents in the
Exploration.

> **Tangent Theorem**
>
> **A line, a ray, or a segment is tangent to a circle
> if and only if it is in the same plane as the circle
> and is perpendicular to a radius of the circle at
> the point of intersection.**
>
> $\overleftrightarrow{AC}$ is tangent to $\odot O$ if and only if $\overleftrightarrow{AC} \perp \overline{OB}$ at B.

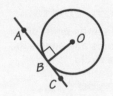

An **arc** is an unbroken part of a circle. In the circle below, points A and B
divide the circle into the *minor arc $\overparen{AB}$* and the *major arc $\overparen{ADB}$*. The arc $\overparen{BCD}$
is an example of a **semicircle**. Notice that three points are used to name a
major arc or a semicircle.

The **minor arc $\overparen{AB}$**
consists of point **A**,
point **B**, and all the
points on the circle in
the interior of $\angle AEB$.

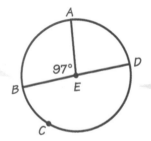

The **major arc $\overparen{ADB}$**
consists of point A, point
B, and all the points of
the circle not on $\overparen{AB}$.

THINK AND COMMUNICATE

1. In the diagram above, $\overparen{BCD}$ is a semicircle. Name another semicircle.

2. Tony says that the endpoints of any diameter of a circle are also the
endpoints of a semicircle. Do you agree? Explain.

The **measure** of a minor arc is equal to the measure of the central angle
that intercepts it. The measure of a major arc is $360°$ minus the measure of
its corresponding minor arc. For example, in $\odot E$ above, $m\overparen{AB} = 97°$ and
$m\overparen{ADB} = 360° - 97° = 263°$. The measure of a semicircle is $180°$. A
complete circle has measure $360°$.

> **Arc Addition Postulate**
>
> **In a circle, the measure of the arc formed by
> two arcs that have exactly one point in common
> is the sum of the measures of the two arcs.**
>
> $m\overparen{AC} = m\overparen{AB} + m\overparen{BC}$

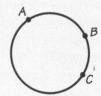

EXAMPLE 1

Find the measure of each arc.

a. $\overgroup{PQR}$

b. $\overgroup{PR}$

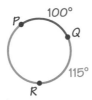

SOLUTION

a. $\overgroup{PQR}$ is formed by $\overgroup{PQ}$ and $\overgroup{QR}$. Use the Arc Addition Postulate.

$$m\,\overgroup{PQR} = m\,\overgroup{PQ} + m\,\overgroup{QR}$$
$$= 100° + 115°$$
$$= 215°$$

So $m\,\overgroup{PQR} = 215°$.

b. $\overgroup{PR}$ is the minor arc with the same endpoints as the major arc $\overgroup{PQR}$.

$$m\,\overgroup{PR} = 360° - m\,\overgroup{PQR}$$
$$= 360° - 215°$$
$$= 145°$$

So $m\,\overgroup{PR} = 145°$.

Two circles are *congruent* if their radii are equal. In the same circle or in congruent circles, arcs whose measures are equal are called **congruent arcs**. In the first circle below, $\angle C$ *intercepts* $\overgroup{AB}$ because A and B lie on the sides of $\angle C$ and the rest of $\overgroup{AB}$ is in the interior of $\angle C$.

The theorems below, which you may have discovered in the Exploration, describe some relationships between congruent arcs and the angles that intercept them. You will prove these theorems in the exercises.

Inscribed Angle Theorems

The measure of an inscribed angle is equal to half the measure of the intercepted arc.

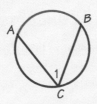

$$m\angle 1 = \frac{1}{2}m\,\overgroup{AB}$$

In a circle or in congruent circles, inscribed angles that intercept the same arc or congruent arcs are congruent.

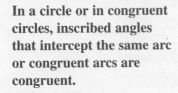

$$\angle 2 \cong \angle 3$$

The measure of an angle formed by a tangent and a chord is equal to half the measure of the intercepted arc.

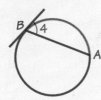

$$m\angle 4 = \frac{1}{2}m\,\overgroup{AB}$$

EXAMPLE 2

Find the measure of ∠*QPT*.

SOLUTION

∠*QSP* is the central angle that intercepts $\overset{\frown}{QP}$, so $m\,\overset{\frown}{QP} = m\angle QSP$. Therefore, $m\,\overset{\frown}{QP} = 136°$.

$$m\angle QPT = \frac{1}{2}m\,\overset{\frown}{QP}$$
$$= \frac{1}{2} \cdot 136°$$
$$= 68°$$

So $m\angle QPT = 68°$.

BY THE WAY

On the surface of Earth, a rainbow appears as a partial or complete semicircle. A rainbow is actually a circle, but we see only part of it.

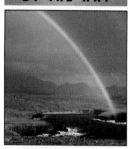

THINK AND COMMUNICATE

Use the diagram at the right.

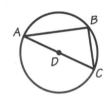

3. a. What is the measure of ∠*ABC*?

b. What is the measure of any angle that intercepts a semicircle? Explain how you know.

4. Sketch an inscribed angle that intercepts a major arc. What type of angle is it? Explain how you know.

CHECKING KEY CONCEPTS

Give an example of each figure in ⊙*B*.

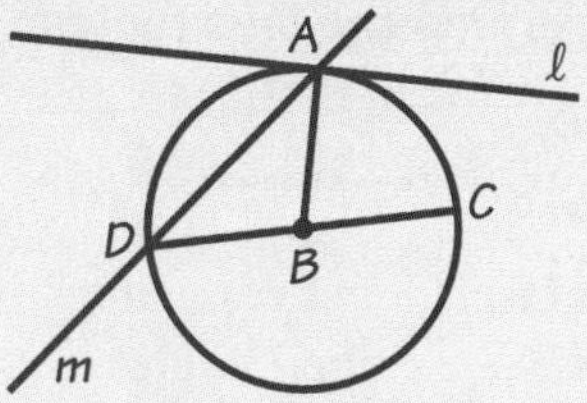

1. a chord

2. a central angle

3. an inscribed angle

4. a tangent to ⊙*B*

5. a minor arc and a major arc that have the same endpoints

Find each unknown measure.

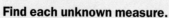

6. $m\,\overset{\frown}{TUS}$ = __?__ **7.** $m\angle QPR$ = __?__ **8.** $m\,\overset{\frown}{DE}$ = __?__

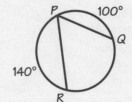

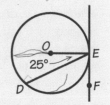

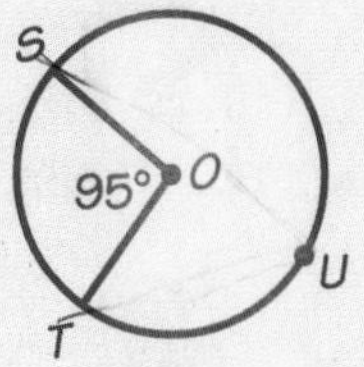

11.1 Exercises and Applications

Extra Practice
exercises on
page 683

Find each measure.

1. $m\,\overparen{QR}$
2. $m\,\overparen{PQR}$
3. $m\angle QPR$
4. $m\,\overparen{PQ}$
5. $m\,\overparen{QPR}$
6. $m\angle PQS$

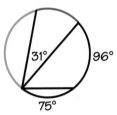

Copy each diagram. Find the measure of each red arc and each blue arc.

7.
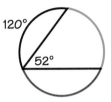
120°
52°

8.
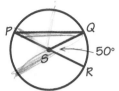
55°

9.
31°
96°
75°

10. **Open-ended Problem** Draw a circle. Locate and label three points on the circle so that two of the points form a 180° arc. Name at least five arcs and give their measures.

INTERVIEW Ron Courson

Look back at the article on pages 550–552.

A physical therapist may measure range of motion to make a diagnosis or treatment plan. Range of motion is the arc through which a part of the body can move. The diagram shows maximum range of motion for certain movements of the shoulder.

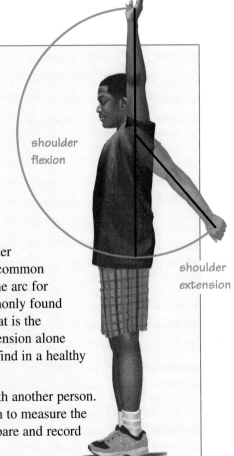
shoulder flexion

shoulder extension

elbow flexion

11. A total arc of about 225° for shoulder flexion and extension combined is common in healthy adults. The measure of the arc for shoulder flexion alone that is commonly found in healthy adults is about 180°. What is the measure of the arc for shoulder extension alone that Ron Courson would expect to find in a healthy adult?

12. **Cooperative Learning** Work with another person. Each of you should use the diagram to measure the other person's elbow flexion. Compare and record your results.

ALGEBRA **Find the value of each variable and the measure of each labeled arc or angle.**

13.
$(4x + 20)°$
$80°$
$(10z + 120)°$

14.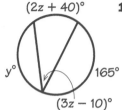
$(2z + 40)°$
$y°$
$165°$
$(3z - 10)°$

15.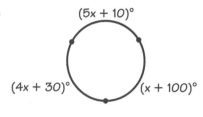
$(5x + 10)°$
$(4x + 30)°$
$(x + 100)°$

16. SAT/ACT Preview Points P, Q, and R lie on the same circle. If $A = m\ \overset{\frown}{PR} + m\ \overset{\frown}{PQR}$, and $B = 360°$, then:

A. $A > B$ **B.** $B > A$ **C.** $A = B$ **D.** relationship cannot be determined

ALGEBRA **Find the value of each variable.**

17.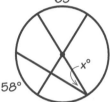
$65°$
$x°$
$58°$

18.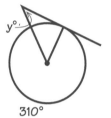
$y°$
$310°$

19.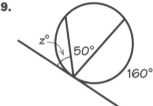
$z°$
$50°$
$160°$

20. Copy and complete the proof of the first Inscribed Angle Theorem on page 555 for the case in which the center of the circle is *on* a side of the inscribed angle.

Given: $\angle ABD$ is inscribed in $\odot C$.

Prove: $m\angle ABD = \frac{1}{2}m\overset{\frown}{AD}$

Statements	Reasons
1. Draw radius $\overline{AC}$.	**1.** ?
2. ?	**2.** All radii of a circle are congruent.
3. $\triangle ABC$ is isosceles.	**3.** ?
4. $m\angle ABD = m\angle BAC$	**4.** ?
5. $m\angle ACD = m\angle ABD + m\angle BAC$	**5.** ?
6. $m\angle ACD = 2(m\angle ABD)$	**6.** ?
7. $m\overset{\frown}{AD} = m\angle ACD$	**7.** ?
8. $m\overset{\frown}{AD} = 2(m\angle ABD)$	**8.** ?
9. $\frac{1}{2}m\overset{\frown}{AD} = m\angle ABD$	**9.** Algebra

21. Challenge The center of the circle may be *outside* or *in the interior* of an inscribed angle. Write a paragraph proof of the first Inscribed Angle Theorem in one of these cases. (*Hint:* Draw a diameter from the vertex of the inscribed angle. Use your results from Exercise 20 for the two inscribed angles formed.)

22. If the vertices of a polygon lie on a circle, then the polygon is *inscribed* in the circle. Prove that the opposite angles of an inscribed quadrilateral are supplementary.

Given: Quadrilateral *PQRS* is inscribed in a circle.

Prove: $m \angle P + m \angle R = 180°$

23. Prove the second Inscribed Angle Theorem on page 555.

NAVIGATION Suppose you are on a boat in the ocean and want to find your location, *T*, on the map at the right. From the boat you can see the two lighthouses, *M* and *N*, and the buoy *B*. Using a navigator's compass, you find that $m \angle MTB = 36°$ and $m \angle BTN = 56°$. Trace the shoreline, including *M*, *B*, and *N*. Use your diagram for Exercises 24–26. You will need a protractor, a compass, and a straightedge.

24. Draw $\triangle BMC$ so that $m \angle B = m \angle M = 90° - m \angle MTB$ and *C* is in the water. Construct $\odot C$ with radius *CM*. Explain why $m \angle C = 2m \angle MTB$.

25. a. Writing Circle *C* in Exercise 24 is called a *circle of position*. Use the first and second Inscribed Angle Theorems and $m \angle C$ from Exercise 24 to explain why your boat may be anywhere on $\odot C$.

b. Could the boat not be on $\odot C$? Explain your reasoning.

26. Use points *B* and *N* and $m \angle BTN$ to draw another circle of position for your boat. Label the center of the circle *D*.

a. How can you use $\odot D$ and $\odot C$ from Exercise 24 to find your exact position?

b. Label your position *T* and measure $\angle MTB$ and $\angle BTN$ to check your results.

27. Open-ended Problem Copy the diagram. Without measuring, label the measures of as many of the arcs and angles in the diagram as you can.

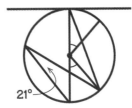

21°

Find the volume and surface area of each right cone or regular pyramid. *(Section 10.7)*

28.

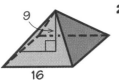

9
16

29.

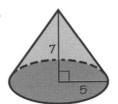

7
5

30.

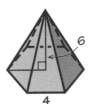

6
4

31. If you choose a point at random on $\overline{AB}$, what is the probability that the point is on $\overline{AD}$? *(Section 9.5)*

144
A   92   D   B

SECTION

11.2 Tangents, Secants, and Chords

Learn how to...

- **find arc and angle measures when segments intersect circles**

So you can...

- **solve problems such as how far a migrating bird can see**

Hikers can use the Global Positioning System (GPS) to keep from getting lost. A hand-held GPS receiver can calculate its exact location by comparing broadcasts from at least four satellites. Scientists chose an altitude for the satellites that minimizes the number of satellites needed to cover the globe completely.

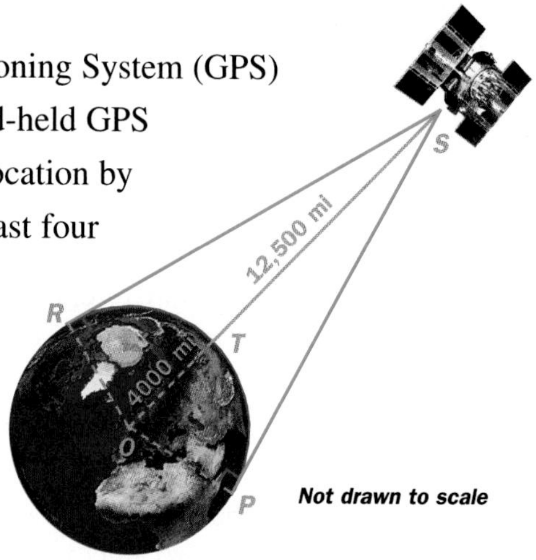

S
12,500 mi
R
T
4000 mi
O
P
Not drawn to scale

EXAMPLE 1 Application: Global Positioning

Find $m\widehat{RTP}$ and use it to describe how far around Earth the satellite's signal can reach.

SOLUTION

$$m\widehat{RTP} = m\angle ROP$$
$$= m\angle ROS + m\angle POS$$
$$= 2\, m\angle ROS$$

$\triangle RSO \cong \triangle PSO$ by the HL Theorem so $\angle ROS \cong \angle POS$.

Use trigonometry to find $m\angle ROS$.

$$\cos \angle ROS = \frac{RO}{OS}$$
$$= \frac{4000}{4000 + 12{,}500}$$
$$\approx 0.2424$$
$$m\angle ROS \approx 76^\circ$$

$$m\widehat{RTP} \approx 2 \cdot 76$$
$$\approx 152^\circ$$

The measure of a full circle is 360°, so the satellite's signal reaches about $\frac{152^\circ}{360^\circ}$, or about 42% of the way around Earth along $\odot O$.

In Example 1, trigonometry and congruent triangles were used to find the measure of an arc. You can use the rules below to find the measures of arcs formed by tangents or *secants*. A **secant** is a line, a ray, or a segment that contains a chord.

Angles Formed by Tangents and Secants

The measure of an angle formed by the intersection of two tangents, two secants, or a secant and a tangent, at a point outside a circle, is half the difference of the measures of the intercepted arcs.

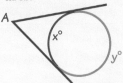

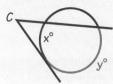

$m\angle A = \frac{y° - x°}{2}$ $m\angle B = \frac{y° - x°}{2}$ $m\angle C = \frac{y° - x°}{2}$

THINK AND COMMUNICATE

1. Use the solution to Example 1 and the theorem above to find $m\angle S$ in the diagram on page 560.

2. If the GPS satellites were closer to Earth, more satellites would be needed for the system. Explain why.

BY THE WAY

GPS is used by many people, including pilots, weather forecasters, and farmers. From any point on Earth, at least five GPS satellites are always in view.

EXAMPLE 2

Find $m\angle GJM$ and $m\overset{\frown}{GM}$.

SOLUTION

Find $m\overset{\frown}{GM}$ first.

$$m\overset{\frown}{GM} + 75° = 180°$$

$$m\overset{\frown}{GM} = 105°$$

$\overline{HM}$ is a diameter of the circle, so $m\overset{\frown}{HGM} = 180°$.

Find $m\angle GJM$. Use the theorem above.

$$m\angle GJM = \frac{m\overset{\frown}{GM} - 75°}{2}$$

$$= \frac{105 - 75}{2}$$

$$= 15°$$

Therefore, $m\angle GJM = 15°$ and $m\overset{\frown}{GM} = 105°$.

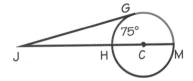

You know how to find angle and arc measures if two lines intersect either *on* a circle or *outside* it. You can also find angle and arc measures if two chords intersect *inside* a circle.

Arcs Formed by Intersecting Chords

The measure of an angle formed by two chords is equal to half the sum of the measures of the intercepted arcs.

$$m \angle 1 = m \angle 2 = \frac{x^\circ + y^\circ}{2}$$

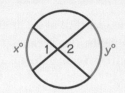

EXAMPLE 3   Connection: Algebra

Find the value of *z* and the measure of each labeled arc.

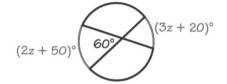

SOLUTION

Use the theorem above.

$$\frac{(3z + 20)^\circ + (2z + 50)^\circ}{2} = 60^\circ$$

$$\frac{5z + 70}{2} = 60 \qquad \text{Multiply both sides of the equation by 2.}$$

$$5z + 70 = 120$$

$$5z = 50$$

$$z = 10$$

So $(3z + 20)^\circ = 50^\circ$ and $(2z + 50)^\circ = 70^\circ$.

☑ CHECKING KEY CONCEPTS

ALGEBRA Find the values of *x* and *y*.

1.
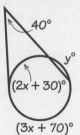

2.
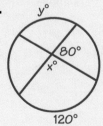

3.
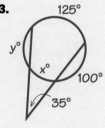

11.2 Exercises and Applications

Extra Practice exercises on page 683

Give an example of each figure in $\odot O$.

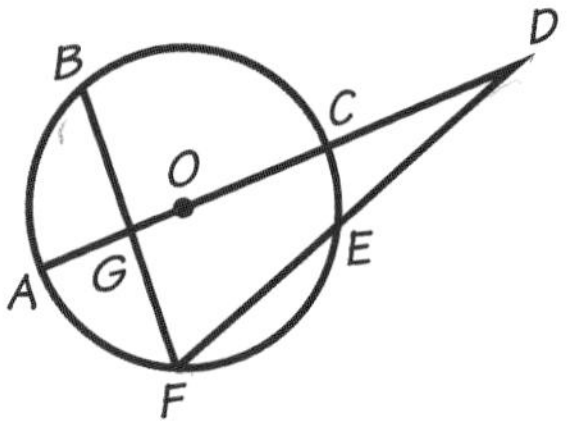

1. a secant
2. an angle formed by two secants
3. an inscribed angle
4. an angle formed by two chords

ALGEBRA Find the value of each variable.

5.

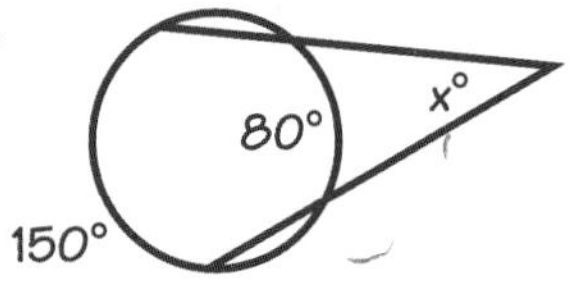

6.

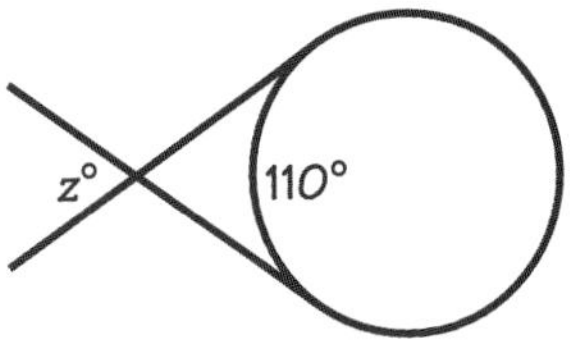

7.

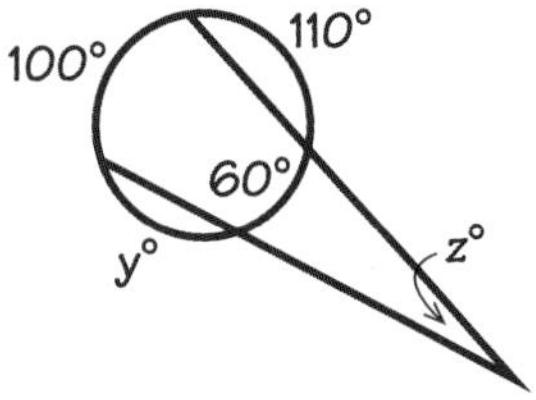

8.

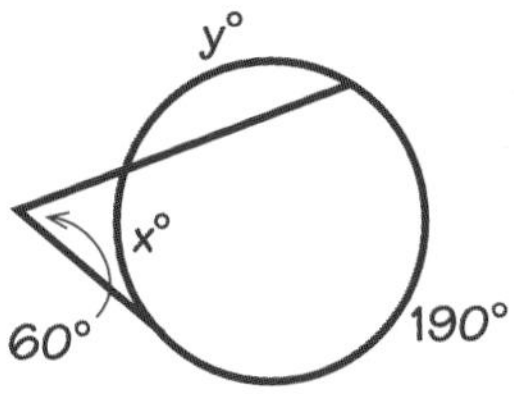

9.

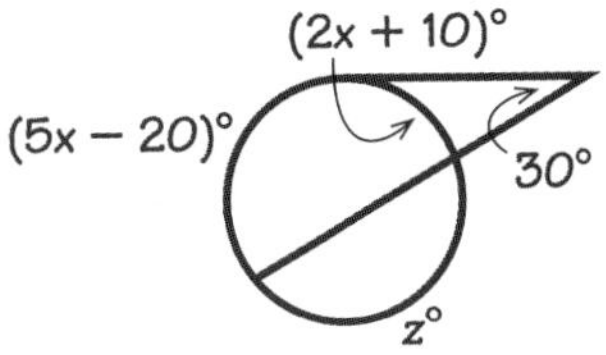

10.

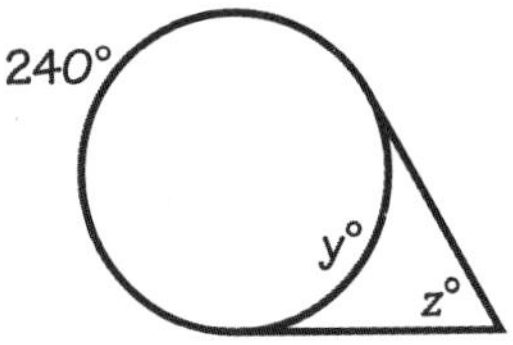

11. Logical Reasoning Arrange the statements below and give reasons to write a proof of the second case of the theorem on page 561.

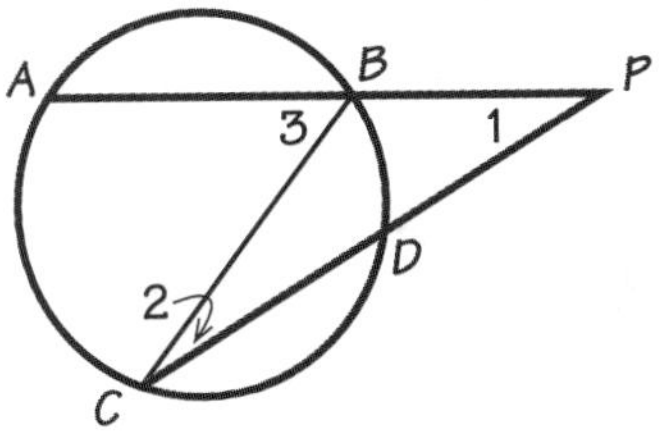

Given: Secants $\overline{AP}$ and $\overline{CP}$ intersect as shown.

Prove: $m\angle 1 = \dfrac{m\widehat{AC} - m\widehat{BD}}{2}$

$m\angle 1 = \frac{1}{2}m\widehat{AC} - \frac{1}{2}m\widehat{BD}$

$m\angle 1 = \dfrac{m\widehat{AC} - m\widehat{BD}}{2}$

Secants $\overline{AP}$ and $\overline{CP}$ intersect as shown.

$m\angle 3 = \frac{1}{2}m\widehat{AC}$; $m\angle 2 = \frac{1}{2}m\widehat{BD}$

$m\angle 1 = m\angle 3 - m\angle 2$

Draw chord $\overline{BC}$.

$m\angle 3 = m\angle 2 + m\angle 1$

$m\angle 1 = \frac{1}{2}(m\widehat{AC} - m\widehat{BD})$

ALGEBRA Copy each diagram. Find the measure of each labeled angle or arc.

12.

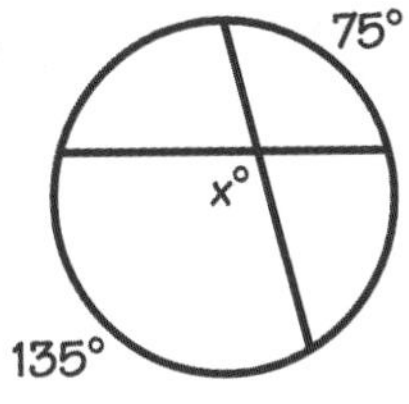

13.

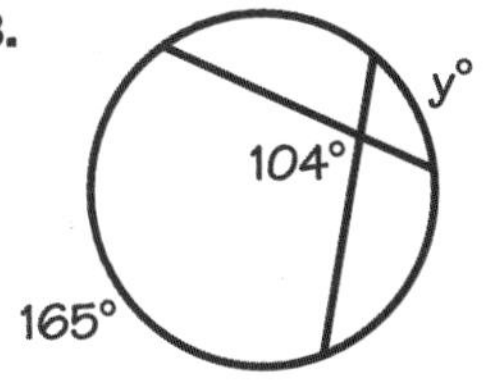

14.

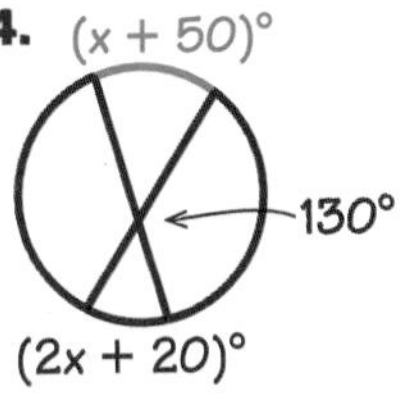

In Exercises 15 and 16, you will prove two cases of the theorem on page 561.

15. Given: Secant $\overline{CP}$ and tangent $\overline{AP}$

Prove: $m\angle P = \dfrac{m\widehat{AC} - m\widehat{AB}}{2}$

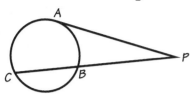

16. Given: Tangents $\overline{AP}$ and $\overline{BP}$

Prove: $m\angle P = \dfrac{m\widehat{ACB} - m\widehat{AB}}{2}$

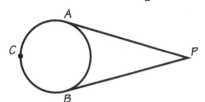

17. Open-ended Problem Create a problem like the ones in Exercises 5–10. The value of one of the variables should be 35.

Connection ▸ BIOLOGY

Many factors affect the altitude at which birds fly while migrating between their summer and winter habitats. The amount of land that a bird can see while flying depends on its altitude.

Use the fact that 1 mi = 5280 ft.

18. During migration, a bald eagle flies at an altitude between 100 ft and 1000 ft.

  a. What is the measure of the arc of Earth that an eagle can see when flying at 100 ft? at 1000 ft?

  b. What percent of the circumference of Earth can an eagle see at each height?

  c. The circumference of Earth is about 25,000 mi. Use your answer to part (b) to determine the length of the arc the eagle can see at each height. Explain what these numbers represent.

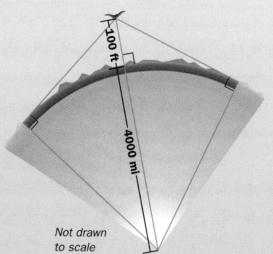

Not drawn to scale

19. Whooping cranes usually fly at an altitude between 900 ft and 4500 ft during migration. How much of the circumference of Earth can the crane see at each of these altitudes? Give your answer as a percent and as a length. (*Hint:* Look at Exercise 18 for the steps to answer this question.)

20. **Research** Find out why whooping cranes migrate at a higher altitude than bald eagles do.

SPORTS In the game of soccer, the goalkeeper tries to prevent an opponent from kicking the ball into the net and scoring a goal.

21. a. Writing The goalkeeper may try to stay on the line that bisects the angle formed by the opponent and the goal posts as shown below. Explain why this will help the goalkeeper prevent a goal.

b. The goalkeeper will also approach an opponent who is trying to score. Explain how this helps the goalkeeper prevent a goal.

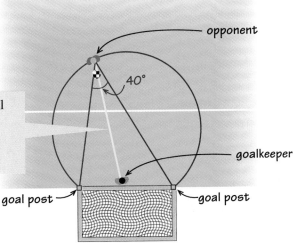

The goalkeeper will move toward the opponent along the angle bisector.

22. a. Explain why the angle formed by the goal posts and the opponent is 40° for every point on the circle in the diagram.

b. Open-ended Problem An approaching opponent wants to make the angle 40° or larger. Sketch some examples and explain why a wider angle provides a better opportunity for making a goal. What is the best position for the opponent to attempt a goal? Explain your reasoning.

23. a. Challenge Find the value of *x*.

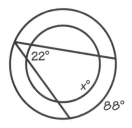

b. Write an equation for *c* in terms of *a* and *b*.

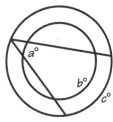

---

24. Writing Write a note to a student who missed class for this lesson. Explain what information is needed to find each arc or angle measure: $m\,\widehat{AB}$, $m\angle C$, $m\angle CPD$.

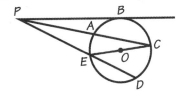

25. Describe this translation using coordinate notation: Every point moves to the left 5 units and up 2 units. *(Section 8.3)*

Sketch each figure. *(Section 11.1)*

26. a circle and a line that is tangent to it

27. a circle and two congruent inscribed angles

ASSESS YOUR PROGRESS

VOCABULARY

central angle (p. 553)

inscribed angle (p. 553)

chord (p. 553)

tangent to a circle (p. 554)

arc (p. 554)

semicircle (p. 554)

minor arc (p. 554)

major arc (p. 554)

measure of an arc (p. 554)

congruent arcs (p. 555)

secant (p. 561)

Find each unknown arc or angle measure. *(Section 11.1)*

1. $m\widehat{ABC} = $?

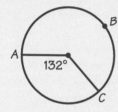

2. $m\angle PQR = $?

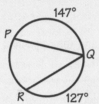

3. $m\angle SMT = $?

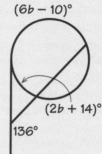

4. Each vertex of equilateral $\triangle ABC$ lies on $\odot M$. Explain why $m\widehat{AB} = m\widehat{BC} = m\widehat{AC}$.

ALGEBRA Find the measure of each labeled arc. *(Section 11.2)*

5. $(2x + 5)°$

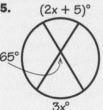

6.

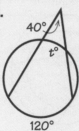

7. $(6b - 10)°$

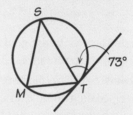

8. **Journal** Compare the theorem that describes how to find the measure of an inscribed angle and the theorem that describes how to find the measure of an angle formed by two secants. Are these theorems related? Explain your reasoning.

11.3 Applying Properties of Chords

Learn how to...

- **identify relationships among arcs and chords of circles**

So you can...

- **solve problems involving arcs and chords of circles**

What is the "center" of a triangle? One way you could define the "center" is the point that is an equal distance from each vertex. This point is called the *circumcenter*. It is the center of the circle that passes through the vertices of the triangle. In the Exploration, you will *circumscribe a circle about a triangle.*

EXPLORATION
COOPERATIVE LEARNING

Circumscribing a Triangle

Work with another student.
You will need:

- patty paper
- compass
- straightedge

1 Draw any △*DEF* on a piece of patty paper.

2 Carefully fold the paper to form the perpendicular bisector of each side of the triangle.

3 The three perpendicular bisectors should meet at a point. Label this point *C*.

4 Draw a circle with *C* as the center and *CD* as the radius. Describe your results. How do you know that all three vertices are the same distance from *C*?

5 Compare your results from Step 4 with the results from other groups. Make a conjecture.

6 Check whether your conjecture in Step 5 is true for other triangles. Be sure to try right, obtuse, and acute triangles.

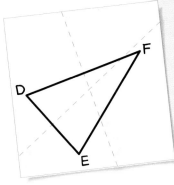

You saw in the Exploration that each side of a triangle is a chord of the circumscribed circle, and the perpendicular bisector of each chord contains the center of the circle. The arc *corresponding to* the chord is also bisected.

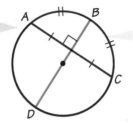

$\overset{\frown}{AC}$ **corresponds to** $\overline{AC}$.

$\overline{AB} \cong \overline{BC}$, so $\overline{BD}$ bisects $\overset{\frown}{AC}$.

Chord Bisector Theorems

The perpendicular bisector of a chord passes through the center of the circle.

If $\overline{AC} \perp \overline{BD}$ and $\overline{AE} \cong \overline{CE}$, then $\overline{BD}$ contains the center of $\odot P$.

A diameter that is perpendicular to a chord bisects the chord and its corresponding arc.

If $\overline{AC} \perp \overline{BD}$ and $\overline{BD}$ contains the center of $\odot P$, then $\overline{AE} \cong \overline{CE}$ and $\overset{\frown}{AB} \cong \overset{\frown}{CB}$.

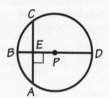

EXAMPLE 1   Connection: Algebra

Find $m\overset{\frown}{AB}$ and $m\overset{\frown}{BC}$.

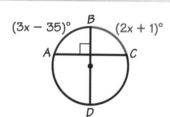

SOLUTION

Because $\overline{BD}$ is a diameter that is perpendicular to $\overline{AC}$, it bisects $\overset{\frown}{ABC}$.

$$m\overset{\frown}{AB} = m\overset{\frown}{BC}$$

$$(3x - 35)^\circ = (2x + 1)^\circ$$

$$x = 36$$

Substitute 36 for x to find $m\overset{\frown}{AB}$.

$$m\overset{\frown}{AB} = (3x - 35)^\circ$$

$$= 3 \cdot 36 - 35$$

$$= 73$$

So $m\overset{\frown}{AB} = 73°$. Because $m\overset{\frown}{BC} = m\overset{\frown}{AB}$, $m\overset{\frown}{BC} = 73°$ also.

Check:
$m\overset{\frown}{AB} = m\overset{\frown}{BC}$
$73° \overset{?}{=} (2x + 1)°$
$\overset{?}{=} 2 \cdot 36 + 1$
$= 73$ ✔

You can use the second Chord Bisector Theorem to prove that two congruent chords of a circle are an equal distance from the center of the circle.

EXAMPLE 2

Write the key steps of a proof of the theorem:

In a circle, two congruent chords are equidistant from the center of the circle.

SOLUTION

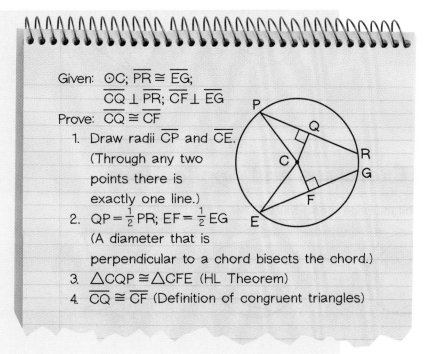

Given: $\odot C$; $\overline{PR} \cong \overline{EG}$;
$\overline{CQ} \perp \overline{PR}$; $\overline{CF} \perp \overline{EG}$
Prove: $\overline{CQ} \cong \overline{CF}$

1. Draw radii $\overline{CP}$ and $\overline{CE}$. (Through any two points there is exactly one line.)
2. $QP = \frac{1}{2} PR$; $EF = \frac{1}{2} EG$ (A diameter that is perpendicular to a chord bisects the chord.)
3. $\triangle CQP \cong \triangle CFE$ (HL Theorem)
4. $\overline{CQ} \cong \overline{CF}$ (Definition of congruent triangles)

The converse of the theorem in Example 2 is also true. A proof of the converse is in the exercises.

THINK AND COMMUNICATE

1. Suppose you want to prove that two congruent chords of congruent circles are equidistant from the center of each circle. Explain how the proof would be different from the proof in Example 2.

2. State the converse of the theorem in Example 2.

3. **a.** In the diagram, $\overarc{QR} \cong \overarc{RS}$. Explain how you know that $\overline{QR} \cong \overline{RS}$.

   b. If two chords of the same circle are congruent, are their corresponding arcs also congruent? Explain why or why not.

As you may have discovered in Question 3, in the same circle or congruent circles, congruent chords have congruent corresponding arcs. The converse is also true. You will prove the converse in the exercises.

☑ CHECKING KEY CONCEPTS

Find each length or measure.

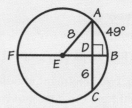

1. $m\widehat{BC}$ **2.** AD

3. $m\widehat{CFA}$ **4.** BF

Find the value of each variable.

5.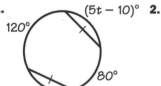

6.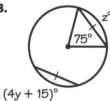

7.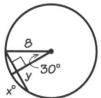

11.3   Exercises and Applications

Extra Practice exercises on page 684

ALGEBRA Find the value of each variable.

1.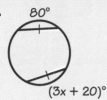

2.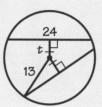

3.

4.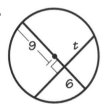

5.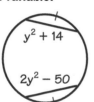

6.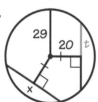

7. Open-ended Problem   Draw a large right triangle, a large acute triangle, and a large obtuse triangle. Circumscribe a circle about each of the three triangles you drew.

8. Write a complete proof of the theorem in Example 2.

9. Write a proof of the theorem: In a circle, congruent arcs have congruent chords.

    Given: $\odot P, \widehat{AB} \cong \widehat{CD}$

    Prove: $\overline{AB} \cong \overline{CD}$

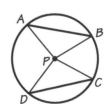

570   Chapter 11 *Circles and Spheres*

10. 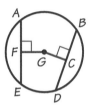 **Technology** Use geometry software or patty paper and a straightedge. Draw any △*XYZ*. Find the circumcenter, the centroid, the incenter, and the orthocenter of the triangle. (See Exercises 13–15 on page 324 for definitions of *centroid, incenter,* and *orthocenter.*) Are the four points collinear? Do you think this will be true for any triangle? Explain.

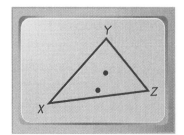

11. a. Challenge A circle is circumscribed about a square. The radius of the circle is *r*. Write an expression for the perimeter of the square in terms of *r*.

  b. Write an expression for the area of the square in part (a).

12. Write a proof of the theorem: In a circle, chords equidistant from the center are congruent.

> **Given:** $\overline{GC} \perp \overline{BD}$
> $\overline{GF} \perp \overline{AE}$
> $\overline{GF} \cong \overline{GC}$
> **Prove:** $\overline{AE} \cong \overline{BD}$

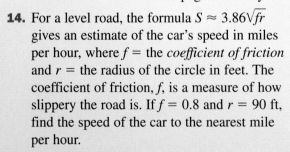

Connection ▸ HIGHWAY SAFETY

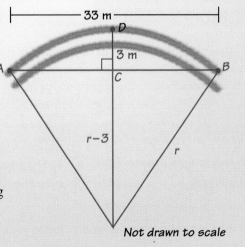

An accident investigator may need to determine how fast a car was moving before it was involved in an accident. If the tracks left by a car are an arc of a circle, the radius of the circle can be used to estimate the speed of the car.

13. Sheila Martinez is investigating an accident. She chooses two points *A* and *B* on the tire marks and measures *AB*. Then she finds the midpoint, *C*, of $\overline{AB}$ and measures *CD*, as shown in the diagram. Use her measurements and the Pythagorean theorem to find the radius, *r*, of the circle. Which of the theorems on page 568 did you use?

14. For a level road, the formula $S \approx 3.86\sqrt{fr}$ gives an estimate of the car's speed in miles per hour, where *f* = the *coefficient of friction* and *r* = the radius of the circle in feet. The coefficient of friction, *f*, is a measure of how slippery the road is. If *f* = 0.8 and *r* = 90 ft, find the speed of the car to the nearest mile per hour.

15. Writing Use the formula in Exercise 14. Will the marks from a fast-moving car have a *larger* or *smaller* radius than those from a slow-moving car? Explain your reasoning.

16. PHYSICAL THERAPY A physical therapist may have a client use a tilt board as a way to improve balance and flexibility. A cross section of a tilt board shows that the bottom is an arc of a circle.

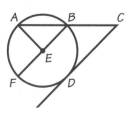

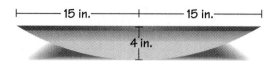

a. Imagine that the cross section is part of a circle. Use the Pythagorean theorem to find the radius of the circle to the nearest inch.

b. Use trigonometry to find the measure of the central angle that intercepts the arc of the tilt board.

c. What is the maximum angle that the flat surface of the tilt board can make with the ground while a person is standing on it? Explain your reasoning.

17. Copy the flow proof of the theorem: A diameter that is perpendicular to a chord bisects the chord. Give the reason for each step.

Given: $\odot C$; $\overline{PR} \perp \overline{QS}$

Prove: $\overline{PT} \cong \overline{RT}$

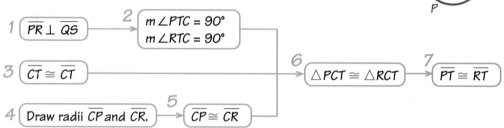

18. Write a proof of the theorem: A diameter that is perpendicular to a chord bisects the arc corresponding to the chord. (*Hint:* The proof is similar to the one in Exercise 17.)

ONGOING ASSESSMENT

19. Open-ended Problem Explain how you could use geometry software or patty paper to demonstrate that a diameter perpendicular to a chord bisects the corresponding arc.

SPIRAL REVIEW

Give an example of each figure in $\odot E$. (*Sections 11.1 and 11.2*)

20. a secant **21.** an inscribed angle

22. a central angle **23.** a tangent

Find the value of each expression. Round your answers to four decimal places. (*Sections 10.3 and 10.4*)

24. sin 36° **25.** cos 67° **26.** tan 15°

11.4 Segment Lengths in Circles

Learn how to...
- find lengths of segments formed by chords, tangents, and secants

So you can...
- solve problems such as explaining why Earth appears to be flat

In earlier sections, you learned about the measures of angles and arcs formed by chords, tangents, and secants of circles. The lengths of the segments formed by chords, tangents, and secants of a circle are also related.

THINK AND COMMUNICATE

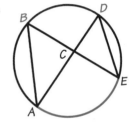

1. Explain why $\angle B \cong \angle D$ and $\angle A \cong \angle E$.
2. Explain why $\triangle ABC \sim \triangle EDC$.
3. Complete this proportion:

$$\frac{AC}{EC} = \frac{?}{DC}$$

4. **a.** Complete this equation:
   $AC \cdot DC = \underline{\ ?\ } \cdot EC$.

   b. Will the equation in part (a) be true for any pair of chords that intersect inside a circle? Explain.

The conclusion that you reached by answering the questions above is one of several theorems about the relationships between the lengths of segments that intersect circles.

Lengths of Chords and Secants

If $\overline{AD}$ and $\overline{BE}$ are chords of a circle and $\overline{AD}$ intersects $\overline{BE}$ at C, then $AC \cdot DC = BC \cdot EC$.

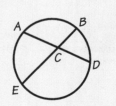

If points A, B, D, and E lie on a circle and $\overrightarrow{DA}$ and $\overrightarrow{EB}$ intersect outside the circle at C, then $CA \cdot CD = CB \cdot CE$.

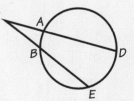

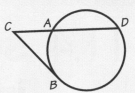

Lengths of Tangents and Secants

If points A, B, and D lie on a circle and $\overrightarrow{DA}$ and tangent $\overline{BC}$ intersect at C, then $AC \cdot CD = BC^2$.

If $\overline{CA}$ and $\overline{CB}$ are tangent to a circle at A and B respectively, then $AC = BC$.

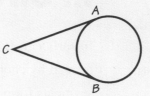

You will prove some of these theorems in the exercises. You can also use them to solve algebra problems like the one in Example 1.

| EXAMPLE 1 | **Connection: Algebra** |

Find RT and ST.

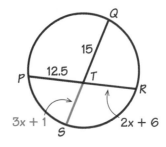

SOLUTION

$\overline{QS}$ and $\overline{PR}$ are chords that intersect inside the circle at T. Use the first theorem on page 573.

$$QT \cdot ST = PT \cdot RT$$

$$15(3x + 1) = 12.5(2x + 6)$$ Use the distributive property. Then solve for x.

$$45x + 15 = 25x + 75$$

$$20x = 60$$

$$x = 3$$

Toolbox p. 690
Operations with Variable Expressions

To find **RT** and **ST**, substitute 3 for x.

$$\begin{aligned} RT &= 2x + 6 & ST &= 3x + 1 \\ &= 2 \cdot 3 + 6 & &= 3 \cdot 3 + 1 \\ &= 12 & &= 10 \end{aligned}$$

$RT = 12$ and $ST = 10$.

| EXAMPLE 2 | Application: Global Positioning |

When the GPS system was designed, engineers wanted to minimize both the number of satellites and the distance that the transmitters needed to send signals. If the satellites are at an altitude of 12,500 mi, how far do the transmitters need to be able to send signals?

SOLUTION

Points R and P are the farthest points on Earth that the satellite's signal can reach without being blocked by Earth, so
$PS = RS =$ the farthest that the transmitters need to send signals.

The diameter of Earth is about 8000 mi, so $SQ \approx 8000 + 12{,}500 = 20{,}500$. Use the formula for segment lengths formed by a tangent and a secant.

$$RS^2 = TS \cdot SQ$$
$$RS^2 = 12{,}500 \cdot 20{,}500$$
$$RS^2 = 256{,}250{,}000$$
$$RS = \sqrt{256{,}250{,}000}$$
$$RS \approx 16{,}000$$

The transmitters on the GPS satellites need to be able to transmit signals about 16,000 mi.

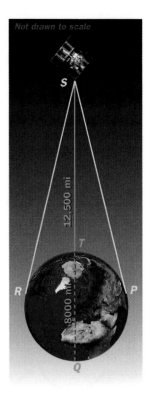

Not drawn to scale

☑ CHECKING KEY CONCEPTS

1. Write three true statements about the lengths of the segments in the diagram.

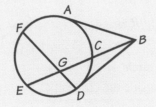

ALGEBRA Find each unknown length.

2.

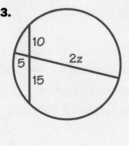

18 cm

x

3.

10

5

2z

15

4.

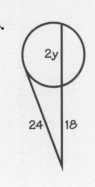

2y

24   18

11.4 | Exercises and Applications

Extra Practice exercises on page 684

For Exercises 1–10, use the diagram at the right. For Exercises 1–6, tell whether each statement is *True* or *False*.

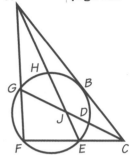

1. $AB^2 = AG \cdot AF$

2. $AF^2 = AB^2$

3. $CD \cdot DG = CE \cdot EF$

4. $JH \cdot JE = JG \cdot JD$

5. $AB^2 = CB^2$

6. $AH \cdot HJ = AG \cdot AF$

ALGEBRA Explain why each statement is true.

7. $\dfrac{CD}{CE} = \dfrac{CF}{CG}$

8. $CD^2 + CD \cdot DG = CE^2 + CE \cdot EF$

9. $CB = \dfrac{CE \cdot CF}{CB}$

10. $\dfrac{JG}{JH} = \dfrac{EJ}{JD}$

Copy the diagram and find each unknown segment length.

11.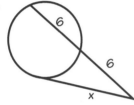

12.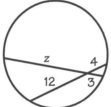

13.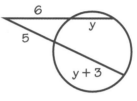

14.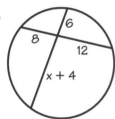

15.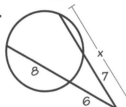

16.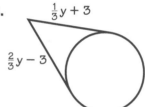

17. Copy and complete the proof of the second theorem on page 573.

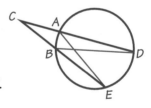

Given: A, B, D, and E lie on a circle.
$\overrightarrow{DA}$ and $\overrightarrow{EB}$ intersect outside the circle at C.

Prove: $CA \cdot CD = CB \cdot CE$

Statements	Reasons
1. Draw segments $\overline{AE}$ and $\overline{BD}$.	1. _?_
2. $\angle ADB \cong$ _?_	2. In a circle or in $\cong$ circles, inscribed angles that intercept the same arc are $\cong$.
3. $\angle C \cong$ _?_	3. Reflexive Property
4. $\triangle DBC \sim \triangle EAC$	4. AA Similarity Postulate
5. $\dfrac{DC}{EC} = \dfrac{BC}{AC}$	5. _?_
6. _?_ $\cdot$ _?_ $= CB \cdot CE$	6. _?_

18. SAT/ACT Preview In the diagram, $\overline{AB}$ is tangent to the circle. $BC = \underline{\ ?\ }$

    A. 20   **B.** 22   **C.** -22 **D.** 16   **E.** 18

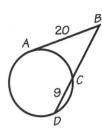

19. GLOBAL POSITIONING Use the diagram in Example 2 on page 575. One way to reduce the number of satellites needed is to place the satellites in a high orbit. A typical high orbit for a satellite is 22,400 mi. If a GPS satellite were placed in orbit at this altitude, how far would the transmitter need to send signals?

In Exercises 20 and 21, you will prove two of the theorems in this section.

20. Given: $\overline{AD}$ and $\overline{BE}$ are chords of a circle. $\overline{AD}$ intersects $\overline{BE}$ at C.

    Prove: $AC \cdot DC = BC \cdot EC$

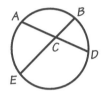

21. Given: Points A, B, and D lie on a circle. $\overrightarrow{DA}$ and tangent $\overline{BC}$ intersect at C.

    Prove: $AC \cdot CD = BC^2$

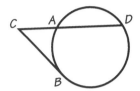

22. Writing Look at the theorems on page 574. Explain why the second theorem is a special case of the first theorem.

Connection   EARTH SCIENCE

For many centuries, people believed that the world is flat, but it is really round. Why does the world look flat? If you stand at C and look a mile away to A, the ground between you and A is higher than $\overline{AC}$, as shown.

23. a. Use the first theorem on page 573 to find BE in inches. (1 mi = 63,360 in.)

    b. Writing If you are standing at point C looking toward point A, do you think that you will notice the difference in height at point B caused by the curvature of Earth? Explain your reasoning.

24. a. Challenge Use trigonometry to find $m\,\widehat{AB}$ and $m\,\widehat{AC}$.

    b. What percent of the entire measure of a circle is $m\,\widehat{AC}$?

    c. The circumference of Earth is about 25,000 mi. Use your answer to part (b) to find the length of $\widehat{AC}$.

    d. Writing Explain the meaning of the length you found in part (c).

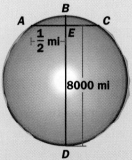

Not drawn to scale

25. Writing Is ⊙P intersected by *two secants* or by *a secant and a tangent*? Explain how you know.

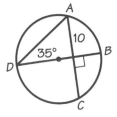

26. a. Open-ended Problem Make up a problem like Exercises 11–16. Be sure that your sketch is accurate and give the answer. (*Hint:* To draw the circle, you may want to use one of the Chord Bisector Theorems on page 568.)

b. Describe how you drew the sketch to make it accurate.

ONGOING ASSESSMENT

27. Visual Thinking Mary used geometry software to draw a circle with two lines that contained chords of the circle. She moved the endpoints of the chords around the circle to create the pictures below. Use her diagrams to explain how the four theorems in this lesson are related.

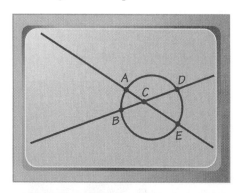

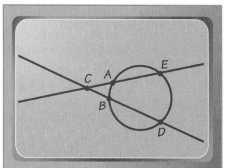

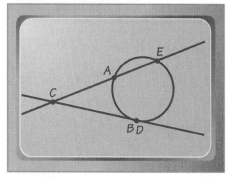

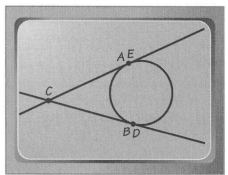

SPIRAL REVIEW

Find the area of the circle with each measure. (*Section 7.5*)

28. radius = 7 in.   **29.** circumference = 18 cm   **30.** diameter = 12 ft

Find each length or measure. (*Section 11.3*)

31. $m\overset{\frown}{AB}$

32. AC

33. $m\overset{\frown}{ADC}$

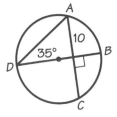

11.5 Sectors and Arc Lengths

Suppose you stop at a pizza shop for lunch. The shop sells pizza by the slice. You can afford two slices from a small pizza or one slice from a large pizza. Which will satisfy your hunger better?

Diameter = 10 in.

Diameter = 14 in.

EXAMPLE 1

What is the area of one slice of the large pizza?

SOLUTION

The large pizza is cut into eight pieces, so each piece is $\frac{1}{8}$ of the whole pizza. The area of each slice is $\frac{1}{8}$ of the area of the whole pizza.

$$\text{area of slice} = \frac{1}{8} \cdot \text{area of large pizza}$$

$$= \frac{1}{8} \cdot \pi (7)^2 \qquad \text{The area of a circle is } \pi r^2. \text{ The diameter of the large pizza is 14 in., so its radius } r \text{ is 7 in.}$$

$$= \frac{49}{8} \pi$$

The area of a slice of the large pizza is $\frac{49}{8} \pi$ in.^{2}, or about 19.2 in.^{2}

THINK AND COMMUNICATE

1. What is the total area of two slices from the small pizza?

2. Which lunch has the greater area (which will satisfy your hunger better), *one slice of the large pizza* or *two slices of the small pizza*?

Each pizza slice represents a geometric figure called a *sector*. A **sector** is a region of a circle that is bounded by two radii and an arc of the circle. The area of the sector depends on the measure of the arc and the radius of the circle.

Area of a Sector

If x represents the measure of the arc of a sector and r is the radius of the circle, then

Area of the sector $= \dfrac{x}{360} \cdot \pi r^2.$

EXAMPLE 2

The area of the shaded sector is 8π cm^2 and $BC = 6$ cm. Find $m\angle ABC$.

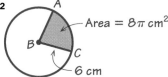

Area $= 8\pi$ cm^2

6 cm

SOLUTION

Use the formula for the area of a sector.

Area of sector $= \dfrac{x}{360} \cdot \pi r^2$

$8\pi = \dfrac{m\widehat{AC}}{360} \cdot \pi(6)^2$ Solve for $m\widehat{AC}$.

$8 = \dfrac{m\widehat{AC}}{360} \cdot 36$

$80 = m\widehat{AC}$

Because $m\angle ABC = m\widehat{AC}$, $m\angle ABC = 80°$.

The **length of an arc** is the distance from one endpoint to the other along the circle. Just as the area of a sector is a fraction of the area of the entire circle, the length of an arc is a fraction of the circle's circumference. The fraction depends on the arc's measure.

Length of an Arc

If x represents the measure of an arc and r is the radius of the circle, then

Length of the arc $= \dfrac{x}{360} \cdot 2\pi r.$

EXAMPLE 3 Application: Irrigation

A pivot irrigation system waters a sector of a field by spraying water from a moving arm. If the moving arm is 1800 ft long and traces an arc with measure 240°, how far does the end of the arm travel?

SOLUTION

$$\text{Length of the arc} = \frac{x}{360} \cdot 2\pi r$$

$$= \frac{240}{360} \cdot (2 \cdot \pi \cdot 1800)$$

$$= \frac{240}{360} \cdot 3600\pi$$

$$= 2400\pi$$

$$\approx 2400(3.14) \qquad \text{Use } \pi \approx 3.14.$$

$$\approx 7536 \text{ ft}$$

The end of the arm travels about 7500 ft.

BY THE WAY

In circular irrigation systems, a long arm of sprinklers pivots around a central point. An arm that is $\frac{1}{4}$ mi long can complete a full revolution in half a day.

THINK AND COMMUNICATE

3. Describe two methods you could use to find the length of the minor arc in the situation in Example 3. What is that length?

4. Which is longer, a 50° arc in a circle of radius 7 cm, or a 50° arc in a circle of radius 12 cm? Explain.

☑ CHECKING KEY CONCEPTS

Find the area of each shaded sector and the length of each red arc. Give your answers to the nearest whole number.

1.

10 in.
85°

2.
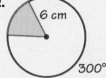
6 cm

3.

15 m
120°

300°

4. Find the measure of an arc with length 2π ft if the radius of the circle is 4 ft.

5. The measure of an arc is 45° and the arc's length is 8 in. What is the circumference of the circle?

Extra Practice exercises on page 684

Find the area of each shaded sector to the nearest whole number.

1.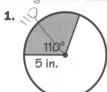
110°
5 in.

2.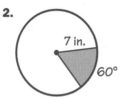
7 in.
60°

3.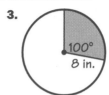
100°
8 in.

4.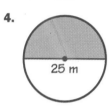
25 m

5.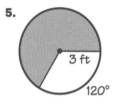
3 ft
120°

6.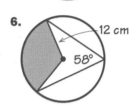
12 cm
58°

7. CATERING Wedding cakes are often constructed in *tiers*. For example, the bottom tier may have a 12 in. diameter, the middle tier a 9 in. diameter, and the top tier a 6 in. diameter. Suppose one third of the 12 in. tier is left at the end of the party.

a. If the remaining cake is a sector, what is the central angle?

b. How many wedges with arc length 2.5 in. can be cut from this remaining cake?

Find the length of the red arc in each circle. Give your answers to the nearest whole number.

8.
90°
7 m

9.
10 cm
45°

10.
75°
15 yd

11.
9 in.
240°

12.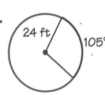
24 ft
105°

13.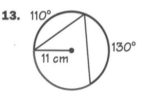
110°
11 cm
130°

14. In ⊙*O*, the area of the sector with central angle ∠*AOB* is 30π cm^2 and radius *OA* is 6 cm. Find *m* $\widehat{AB}$ and the length of $\widehat{AB}$.

15. In ⊙*O*, *m* $\widehat{RS}$ = 45°. If the area of the sector with central angle ∠*ROS* is 20 in.^{2}, find the length of $\widehat{RS}$.

16. The radius of a semicircle is 5 ft. What is the length of the semicircle?

17. Open-ended Problem What are some possible values for the radius and central angle of a sector that has an area of 36π cm^2?

18. FASHION Fans have been used in many cultures for many reasons. Folding fans can be used either fully or partly opened. For each fan, estimate the length of the arc at the edge of the fan.

a.

├── 9 in. ──┤

b.

├─ 21 cm ─┤

c.
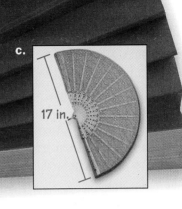
17 in.

19. AUTOMOBILES The diameter of the steering wheel of a car is 15.5 in.

a. If you turn the steering wheel through an arc of 90° without taking your hands from the wheel, how far do your hands move?

b. If you move your hands 10 in., what angle does the steering wheel pass through?

20. Visual Thinking Suppose you split open and flatten an ice cream cone with radius 2 in. and height 5 in.

a. Explain how to use the radius and height of the cone to find its slant height, ℓ. Find the circumference of the base of the cone.

b. What geometric figure does the flattened cone represent? Make a sketch of the figure. Label any dimensions you know.

c. How can you use the slant height and circumference of the base of the cone to find the area of the figure that you sketched in part (b)?

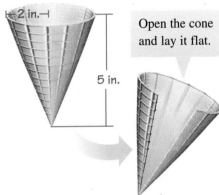
├─2 in.─┤
5 in.

Open the cone and lay it flat.

21. TRACK AND FIELD On a track like the one on page 370, the racers' starting positions are staggered so the racer in the inside lane travels the same distance over the course of the race as the racer in the outside lane.

a. For the outside lane, suppose the radius of the curve at each end of the track is 36.5 m. How far does the racer in the outside lane travel around the 180° curve?

b. What is the radius of the curve for the inside lane if the track has 8 lanes, each 1.22 m wide? How far does the racer in the inside lane travel around the curve?

c. **Writing** In some races the participants all race in the inside lane and must pass on the outside. Why does it make more sense to pass on a straight section of the track than on a curve?

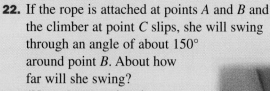

Safety is important to experienced rock climbers. In the illustrations at the right, the lead climber has attached the rope to the rock face at convenient intervals. The rope attachments, along with the second climber, provide protection if the lead climber slips from the rock.

22. If the rope is attached at points *A* and *B* and the climber at point *C* slips, she will swing through an angle of about 150° around point *B*. About how far will she swing? (*Note:* Assume that the rope does not stretch.)

23. If the rope is attached at point *A* only and the climber at *C* slips, she will swing about 150° around point *A*. About how far will she swing?

24. Writing Why is it important for rock climbers to attach rope to the rock face often?

25. Challenge In the figure, $\widehat{AD}$ and $\widehat{BC}$ are arcs of concentric circles. $\overline{AB}$ and $\overline{CD}$ lie on radii of the larger circle. What is the area of the figure?

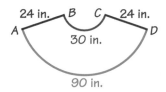

A *segment* of a circle is the portion of a sector outside of the triangle formed by the sector's two radii and a chord.

26. Find the area of each segment.

a.

b.

c.

27. Writing Describe a general method for finding the area of a segment of a circle.

28. In part (b) of Exercise 26, what is the perimeter of the sector bounded by the major arc?

29. **Cooperative Learning** Work in a group of at least six people. One person should stand in the middle of a circle of people standing shoulder to shoulder.

 a. The person in the center should approximate a central angle by holding out his or her arms. How many people in the circle make up the arc intercepted by this central angle?

 b. Reduce the radius of the circle by moving closer to the student in the center. You may need to change the number of people in the circle. How many students are in the intercepted arc now?

 c. **Writing** Make some sketches of the results of this activity. Describe how the radius of a circle affects the lengths of arcs on the circle.

SPIRAL REVIEW

Find the surface area and volume of each figure. *(Sections 7.6 and 10.4)*

30.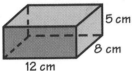
 12 cm, 8 cm, 5 cm

31.
 15 ft, 10 ft, 10 ft

32.
 20 in., 9 in.

ASSESS YOUR PROGRESS

VOCABULARY

sector (p. 580)   **length of an arc** (p. 580)

ALGEBRA Find the value of each variable. *(Sections 11.3 and 11.4)*

1.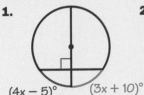
 $(4x - 5)°$ $(3x + 10)°$

2.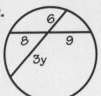
 6, 8, 9, 3y

3.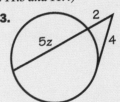
 5z, 2, 4

4. Find the area of the shaded sector and the length of the red arc. Give your answers to the nearest whole number. *(Section 11.5)*

 12 m, 80°

5. **Journal** Suppose you know the area of a sector of a circle. Describe what else you need to know to find the length of the arc of the sector.

SECTION

11.6 Surface Areas and Volumes of Spheres

Learn how to...

- calculate the surface area and volume of a sphere

So you can...

- compare spheres of different sizes, such as sports balls or planets

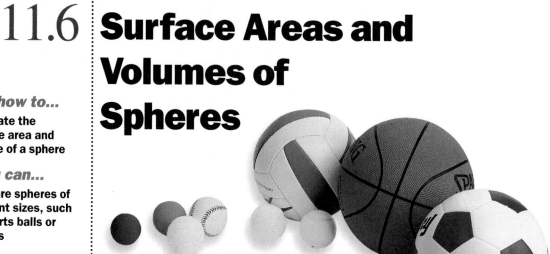

How many of the sports that you play or watch use a round ball? Balls are examples of *spheres*. A **sphere** is the set of all points in space equidistant from a given point.

Many of the terms used with spheres are the same as those used with circles.

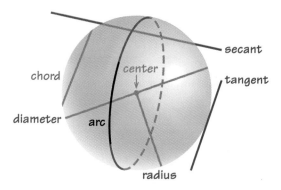

In the third century B.C., Archimedes discovered formulas for the surface area and volume of a sphere by imagining cutting a sphere, cone, and cylinder into very thin slices and comparing the slices.

Surface Area and Volume of a Sphere

The surface area of a sphere with radius r is $4\pi r^2$.

The volume of a sphere with radius r is $V = \frac{4}{3}\pi r^3$.

EXAMPLE 1   Application: Sports

The radius of a *boule* is 5.5 cm. Find the surface area and volume of a boule.

← boule

SOLUTION

Use **5.5 cm** for r in the formulas for surface area and volume of a sphere.

$$\text{Surface Area} = 4\pi r^2 \qquad\qquad \text{Volume} = \frac{4}{3}\pi r^3$$

$$= 4\pi(5.5)^2 \qquad\qquad\qquad = \frac{4}{3}\pi(5.5)^3$$

$$\approx 380 \text{ cm}^2 \qquad\qquad\qquad \approx 697 \text{ cm}^3$$

The surface area of a boule is about 380 cm^2 and its volume is about 700 cm^3.

THINK AND COMMUNICATE

The length of a side of a cube is d and the radius of a sphere is $\frac{d}{2}$.

1. Which do you think is greater, the volume of the cube or the volume of the sphere? Why? Check your answer by writing an expression for the volume of each and comparing the expressions.

2. Which do you think is greater, the surface area of the cube or the surface area of the sphere? Why? Check your answer.

EXAMPLE 2   Connection: Algebra

The surface area of a sphere is 24 cm^2. Find the radius of the sphere.

SOLUTION

$$A = 4\pi r^2 \qquad \text{Use the formula for the surface area of a sphere.}$$

$$24 = 4\pi r^2$$

$$\frac{6}{\pi} = r^2 \qquad \text{To solve for } r \text{, find the square root of } \frac{6}{\pi}.$$

$$\sqrt{\frac{6}{\pi}} = r$$

$$1.38 \approx r$$

The radius of the sphere is about 1.4 cm.

☑ CHECKING KEY CONCEPTS

For Questions 1–6, name a point, a segment, or a line that fits each description.

1. a diameter

2. a chord

3. a secant

4. a tangent

5. a radius

6. the center of the sphere

7. What is the volume of the sphere?

8. What is the surface area of the sphere?

9. *Jai alai* is a fast game that originated in northern Spain. The surface area of the ball (*pelota*) is about 13 in.² What is the radius?

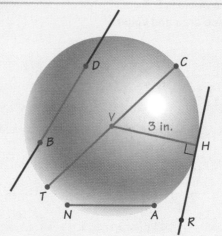

11.6 **Exercises and Applications**

Extra Practice exercises on page 685

For Exercises 1–5, match each segment or line with the best description.

1. $\overline{VC}$
2. $\overline{HG}$
3. $\overline{DC}$
4. $\overleftrightarrow{EF}$
5. $\overleftrightarrow{AB}$

A. a secant

B. a radius

C. a diameter

D. a tangent

E. a chord

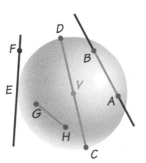

Give the surface area and the volume of each sphere to the nearest whole number.

6.

7.

10 in.

8.

golf ball

1.7 in.

9.

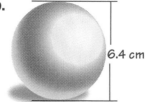

6.4 cm

lacrosse ball

BY THE WAY

Lacrosse players use a stick with a net pocket on the end to throw or scoop the ball into the opposing team's goal. Lacrosse was invented by several North American peoples.

Look back at the
article on pages
550–552.

*Different types of balls are used by athletic trainers and physical
therapists like Ron Courson to improve patients' strength, balance,
and proprioception.*

26.5 cm

10. Patients squeeze small putty balls to strengthen their
hands and forearms. The diameter of a ball is about
2.5 in. About how much putty is in the ball?

11. When an athlete is injured, his or her sense of exactly
where the injured part is or what it is doing may be
thrown off. This sense is called *proprioception*. To
improve proprioception for an injured arm, the
physical therapist and injured athlete may play catch
with a ball like
the one at the
left. What is the
surface area of
the ball?

12. Injuries may also impair an athlete's sense of balance.
To improve balance, the physical therapist may ask
the patient to balance on a large ball. The diameter of
the ball is 80 cm. What is the volume of the ball?

13. The surface area of a sphere is 35 ft². What is the radius?

14. The surface area of a sphere is 16 m². What is the diameter?

15. The volume of a sphere is 36π ft³. What is the radius?

16. ARCHITECTURE A basic igloo is a shell of snow in the shape of half of
a sphere. The size of an igloo depends on how it will be used.

a. The diameter of a small igloo is
6 ft. Find the volume of the igloo,
disregarding the entrance.

b. If the diameter of the igloo is
doubled, how is the volume
affected? Disregard the entrance.

├─── 6 ft ───┤

ASTRONOMY For Exercises 17–19, assume that the planets are spheres.

17. a. The diameter of Earth is about 8000 mi. Find the surface area
of Earth.

b. About 29% of Earth's surface is land. Find the area of the land.

18. The diameter of Mars is about 4200 mi. What is the surface area
of Mars?

19. Research In the solar system, which planet's surface area is the
greatest? What is the surface area?

An official basketball is an inflated sphere with a leather, rubber, or synthetic casing. The ball weighs 20 oz to 22 oz and the circumference is about 30 in. Basketball players throw the ball through a cast-iron rim. The diameter of the rim is 18 in.

20. Find the radius of a basketball.

21. If a player dunks the basketball right through the center of the rim, how much clearance is there between the rim and the basketball?

22. The casing of a basketball is attached with an adhesive, so no extra material is needed at the seams. How much leather is needed to cover a basketball?

23. What is the volume of a basketball?

A volleyball is an inflated sphere with a leather, rubber, or synthetic casing. It is smaller than a basketball and weighs 9 oz to 10 oz. The circumference is between 25 in. and 27 in.

24. What is the radius of a volleyball if the circumference is 25 in.? 27 in.? Use your answers to describe the range of radii for volleyballs.

25. Describe the range of surface areas for volleyballs.

26. Describe the range of volumes for volleyballs.

27. a. 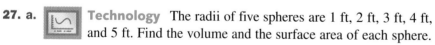 **Technology** The radii of five spheres are 1 ft, 2 ft, 3 ft, 4 ft, and 5 ft. Find the volume and the surface area of each sphere.

b. Find the ratio of volume to surface area $\left(\dfrac{V}{S.A.}\right)$ for each sphere in part (a).

c. Use a graphing calculator to graph $\dfrac{V}{S.A.}$ as a function of the radius. What do you notice?

28. Open-ended Problem Find an object shaped like a sphere at school or where you live. Estimate the surface area and volume of the object. Explain how you made your estimate.

29. PHYSICS A soap bubble 4 cm in diameter is blown out until its diameter is 8 cm.

   a. What is the ratio of the new volume to the old volume?

   b. What is the ratio of the new surface area to the old surface area?

   c. If the amount of soap film doesn't change, the product of the film's thickness and its surface area is always constant. How does the thickness of the soap film change as the diameter increases from 4 cm to 8 cm?

30. a. Visual Thinking Describe the intersection of a sphere and a plane through the center of the sphere.

   b. Can the intersection of a sphere and a plane be a point? Explain.

31. Challenge The volume of a cylinder is the same as the volume of a sphere. The radius of the sphere is 1 in.

   a. Give three possibilities for the dimensions of the cylinder.

   b. Writing Is the surface area of the cylinder *always*, *sometimes*, or *never* greater than the surface area of the sphere? Explain.

ONGOING ASSESSMENT

32. Visual Thinking A *hemisphere* is half of a sphere.

   a. Which do you think is greater, the volume of the hemisphere or the volume of the cone? Why? Check your answer by finding the volume of each.

   b. Which do you think is greater, the surface area of the hemisphere or the surface area of the cone? Why? Check your answer.

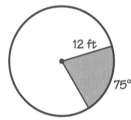

hemisphere

SPIRAL REVIEW

Find the area of each shaded sector. Give your answer to the nearest whole number. *(Section 11.5)*

33.
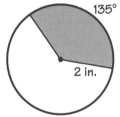
135°
2 in.

34.
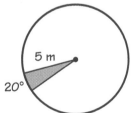
5 m
20°

35.

12 ft
75°

For Exercises 36–38, △ABC ~ △DEC. *(Section 9.4)*

36. Find the ratio $\dfrac{BC}{EC}$.

37. Find the ratio of the perimeters of △ABC and △DEC.

38. Find the ratio of the areas of △ABC and △DEC.

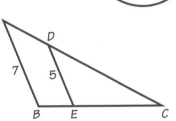

11.7 Volumes of Similar Solids

Learn how to...
- compare volumes and surface areas of similar solids

So you can...
- estimate measures of objects such as weather balloons, scale models, food, and salamanders

Weather balloons are often used to send instruments high into the atmosphere. Data from the instruments help meteorologists predict the weather. There are several different sizes of weather balloons. Larger balloons can lift more equipment than smaller balloons.

EXAMPLE 1   Connection: Data Analysis

The table shows how much weight can be lifted by two balloons of different sizes.

Diameter of balloon	Weight balloon can lift
8 ft	17 lb
16 ft	137 lb

a. Assume both balloons are spheres. Find the volume of each balloon.

b. Compare the ratio of the diameters to the ratio of the volumes.

SOLUTION

a. Use the formula $V = \frac{4}{3}\pi r^3$ to find the volume of each balloon.

Smaller balloon: $V = \frac{4}{3}\pi(4)^3 \approx 268 \text{ ft}^3$ The diameter of the smaller balloon is **8**, so $r = 4$.

Larger balloon: $V = \frac{4}{3}\pi(8)^3 \approx 2145 \text{ ft}^3$

b. Find the ratio of the diameters.   Find the ratio of the volumes.

$$\frac{8}{16} = \frac{1}{2}$$

$$\frac{\frac{4}{3}\pi(4)^3}{\frac{4}{3}\pi(8)^3} = \frac{(4)^3}{(8)^3} = \left(\frac{1}{2}\right)^3$$

The ratio of the volumes is the cube of the ratio of the diameters.

THINK AND COMMUNICATE

1. For each balloon in Example 1, find the ratio of the volume to the weight that the balloon can lift. What do you notice?

2. How can you estimate the weight that a 12 ft diameter balloon can lift?

Recall from Section 9.4 that if the ratio of corresponding lengths of two similar three-dimensional figures is $a:b$, then the ratio of their surface areas is $a^2:b^2$ and the ratio of their volumes is $a^3:b^3$. You can use these ratios to find unknown measures of similar three-dimensional figures.

EXAMPLE 2   Connection: Estimation

A souvenir model of the Statue of Liberty has a volume of **26 mL**, excluding the pedestal. The scale of the model is about 1 : 200. Estimate the volume of the actual Statue of Liberty, excluding the pedestal. Assume that the model and the real statue are similar.

SOLUTION

Since the scale of the model is about 1 : 200, the ratio of corresponding lengths of the statue and the model is about $\frac{200}{1}$. So the ratio of their volumes is about $\left(\frac{200}{1}\right)^3$.

$$\frac{\text{Volume of statue}}{\text{Volume of model}} \approx \left(\frac{200}{1}\right)^3$$

$$\text{Volume of statue} \approx \text{Volume of model} \cdot \left(\frac{200}{1}\right)^3$$

$$\approx 26 \text{ mL} \cdot \left(\frac{200}{1}\right)^3$$

$$\approx 208{,}000{,}000 \text{ mL}$$

$$\approx 208{,}000 \text{ L} \qquad 1 \text{ L} = 1000 \text{ mL}$$

The volume of the Statue of Liberty is about 208,000 L.

BY THE WAY

You can find the volume of an object by putting it under water and measuring how much water it displaces.

THINK AND COMMUNICATE

3. The right arm of the Statue of Liberty is 42 ft long. A large model of the statue has a right arm that is $\frac{1}{2}$ ft long.

   a. What is the scale of the model?

   b. What is the ratio of the surface areas of the model and the statue?

4. The ratio of the areas of the bases of two similar cones is 4 : 9. Find the ratio of each measure of the cones.

   a. the radii   **b.** the heights   **c.** the surface areas   **d.** the volumes

ARCHITECTURE Architects use scale models to design structures. A model of a shopping mall has the scale 1:100.

1. The model has 30 ft^2 of floor space. How many square feet of floor space will the shopping mall have?

2. One of the stores in the mall will have a back room with 1000 ft^3 of space for storage. How much space for storage is there in the back room in the model?

The height of a cylinder is 4 in. and the height of a similar cylinder is 9 in. Find the ratio of each measure of the cylinders.

3. the radii of the bases   **4.** the areas of the bases

5. the surface areas   **6.** the volumes

11.7 | **Exercises and Applications**

Extra Practice exercises on page 685

1. ASTRONOMY You can think of Earth and the moon as spheres. The diameter of Earth is about 12,800 km and the diameter of the moon is about 3500 km. Find the ratio of each measure.

   a. the lengths of their equators   **b.** their surface areas   **c.** their volumes

2. The cones below are similar. Find each measurement.

   a. surface area ≈ 280 cm^2
   $V \approx \underline{\ ?\ }$

   b. surface area ≈ $\underline{\ ?\ }$
   $V \approx 1$ L

10 cm

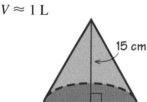
15 cm

3. The heights of two similar mugs are 3.5 in. and 4 in. If the larger mug holds 12 oz, find the capacity of the smaller mug.

4. $\triangle ABC \sim \triangle EDF$.

   a. Find the lengths *AB*, *BC*, and *CA*.

   b. What is the area of $\triangle ABC$?

   c. Find *FE*. What is the ratio of the lengths of corresponding sides of the two triangles?

   d. What is the area of $\triangle EDF$? Explain how you found your answer.

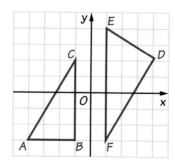

5. **SPORTS** A sports trophy is a bronze cup 10 in. tall. The winners of the event for which the trophy is awarded are given replicas of it that are 6 in. tall. If the replica holds 8 oz, find the capacity of the trophy.

6. **Open-ended Problem** Describe two objects in your school that are roughly similar. Estimate the ratio of corresponding lengths, the ratio of the areas, and the ratio of the volumes of the two objects.

7. a. **ALGEBRA** What is the ratio of any two corresponding lengths of the two cones? What should the ratio of the volumes be?

   b. Write an expression for the volume of each solid.

   c. Write and simplify an expression for the ratio of the volumes. Does your answer agree with your answer to part (a)?

8. **SAT/ACT Preview** The volumes of two similar cones are 8π and 27π. What is the ratio of the lateral areas of the cones?

   A. $\dfrac{2}{3}$ **B.** $\dfrac{\sqrt{8}}{\sqrt{27}}$ **C.** $\dfrac{4}{9}$ **D.** $\dfrac{8}{27}$ **E.** $\dfrac{8^2}{27^2}$

Connection ▶ BIOLOGY

Salamanders that don't have lungs must get all of the oxygen they need by breathing through the skin. In 1961, Juliusz Czopek published a study of the respiratory surfaces of various species of lungless salamanders. Salamanders from the same species, but of different sizes, are roughly similar.

9. The ratio of the volumes of two salamanders is the same as the ratio of their masses. For parts (a)–(d), use the data for the largest salamander and the smallest salamander in the table.

   a. What is the ratio of the volumes of the two salamanders?

   b. Use your calculator and the formula below to find the ratio of the lengths of the two salamanders.

   ratio of lengths = (ratio of volumes)$^{1/3}$

   c. What is the ratio of the surface areas of the two salamanders?

   d. Estimate the surface area of the smallest salamander.

10. Estimate the surface area of each salamander. Copy and complete the table.

11. How does the ratio of surface area to volume change as size varies?

12. **Writing** Explain why a lungless salamander the size of a horse would not be able to get enough oxygen for all of its cells.

Measurements of Slimy Salamanders (*Plethodion glutinesus*)

Mass (g)	Surface Area (mm²)
3.40	?
3.85	?
3.96	?
5.50	?
6.80	2980

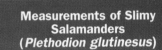

STUFFED EGGPLANT
Wendy Heagney

1 large eggplant (about 10 in. long)
stuffing (see below)
½ cup Parmesan cheese

Cut the eggplant in half lengthwise and scoop out the middle.
Prepare the stuffing. Fill the shell with stuffing and bake,
covered, at 375° for 25 minutes. Sprinkle with cheese and bake,
uncovered, for another 5–10 minutes or until cheese melts.

Stuffing:
2 cups cooked rice   1 tbsp fresh basil
1 cup onion, finely chopped 1 tbsp garlic
2 cups carrots, chopped   ½ tsp oregano (dried)
3 cups zucchini,
4 cups eggpla

Stuffing direc

Cooks often modify recipes to fit their needs.
The recipe describes how to make a stuffed
eggplant that is 10 in. long. Suppose you want
to make six individual portions by stuffing six
small eggplants that are each 5 in. long.

13. Assume that the stuffing from the
recipe would exactly fill a 10 in. long
eggplant. If you don't modify the
stuffing recipe, will there be *too much*
or *too little* stuffing to fill six small
eggplants?

14. What fraction of the original stuffing
recipe should you make to fill the
six small eggplants nearly exactly?

15. Challenge A recipe for
apple pie calls for 8 large apples.
According to the cookbook, the
diameter of a large apple is about
four inches. If the only apples you
have are about three inches in diameter,
how many should you use so the pie
will contain the same volume of apples?
Explain how you found your answer.

16. Writing List the surface area and volume formulas for cubes,
cylinders, cones, and spheres. What patterns do you notice? How are
the surface area formulas alike? How are the volume formulas alike?

17. While writing a proof, Farzana
introduced an auxiliary line.
Which postulate or theorem can
she use to justify drawing the
line? *(Section 5.3)*

18. The surface area of a soccer ball
is about 240 in.^{2} *(Section 11.6)*

  a. What is the radius?

  b. What is the volume?

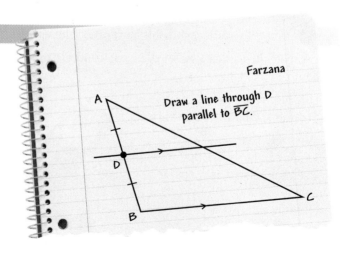

Farzana

Draw a line through D
parallel to $\overline{BC}$.

11.8 Spherical Geometry

Learn how to...
- represent lines and shapes on a sphere

So you can...
- explore geometry on a sphere such as Earth

Because Earth is like a sphere, the shortest way to fly from Seattle, Washington, to Zurich, Switzerland, is to fly over Iceland! Geometry on a sphere is often surprising. You can use a basketball or a globe to explore the nature of *spherical geometry*.

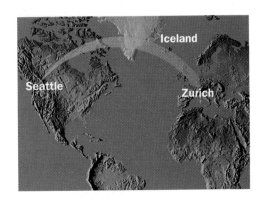

EXPLORATION
COOPERATIVE LEARNING

Triangles on a Sphere

Work in a group of two or three students.
You will need:
- a basketball, globe, or other sphere-shaped object
- five long strips of paper
- tape
- a protractor

1 Tape one end of a strip of paper to the sphere. Run your finger along the strip to fit it to the sphere. Tape the strip at the other end.

2 Repeat Step 1 for two more strips to form a triangle on the sphere. Label the vertices of the triangle A, B, and C.

3 To measure ∠B, hold one of the remaining strips of paper along $\overline{AB}$ and the other along $\overline{BC}$. Tape the two strips together at B. Lay these strips on a flat surface and measure the angle. Repeat to measure ∠A and ∠C.

4 What is the sum of the measures of the angles of △ABC?

5 Form at least three other triangles of different sizes on the sphere. Find the sum of the measures of the interior angles for each triangle. What do you notice?

The geometry you have learned so far in this book is called *Euclidean geometry*. As you saw in the Exploration, geometry on a ball is different from Euclidean geometry. For example, on a ball, the sum of the measures of the interior angles of a triangle is always greater than 180°. Geometry on a ball is an example of *spherical geometry*.

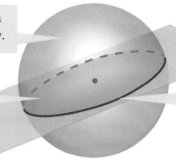

The surface of a sphere is a plane in spherical geometry.

Great circles are the lines in spherical geometry.

A **great circle** is the intersection of a sphere with a plane containing the center of the sphere.

EXAMPLE 1

The illustration shows a line ℓ in spherical geometry. Sketch another line that is perpendicular to ℓ.

SOLUTION

Sketch a great circle that is perpendicular to ℓ at the point of intersection.

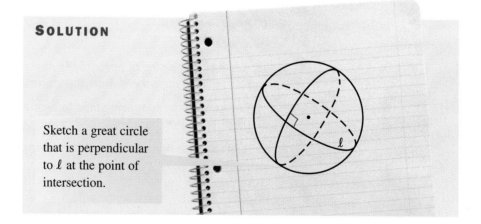

THINK AND COMMUNICATE

The illustration shows a rubber band around a ball.

1. Is the rubber band a great circle of the sphere?

2. Imagine running your finger along the rubber band. Does the band follow the surface of the sphere in a straight line? Does it ever curve to the right or the left?

EXAMPLE 2

Show that this theorem from Euclidean geometry is *not* true for spherical geometry.

In a plane, if two lines are both perpendicular to a third line, then the two lines are parallel.

SOLUTION

The theorem is not true for spherical geometry because there is a counterexample. Imagine a globe and two different great circles that pass through both poles. Both lines are perpendicular to the equator, but they are not parallel to each other.

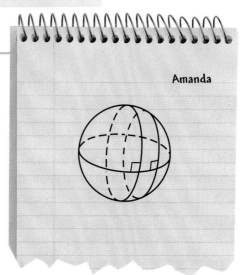

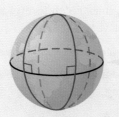

The surface of the globe is a plane in spherical geometry.

THINK AND COMMUNICATE

3. Explain how you know that the two lines in Example 2 that are perpendicular to the equator are not parallel to each other.

4. Amanda tried to justify the theorem in Example 2 by drawing this sketch. What mistake did Amanda make?

5. In Chapter 5, you learned Euclid's Parallel Postulate:

   Through a point not on a given line, there is exactly one line parallel to the given line.

   Is this postulate true in spherical geometry? Explain.

Amanda

✓ CHECKING KEY CONCEPTS

For Questions 1 and 2, draw a sphere and sketch each situation in spherical geometry. If it is not possible, explain why.

1. △*ABC* is equiangular.

2. The measure of each angle of a triangle is 60°.

Describe each situation in spherical geometry.

3.

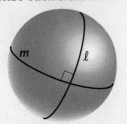

4.

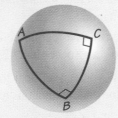

5. In Euclidean geometry, from a point not on a line there is exactly one line that is perpendicular to the given line. Do you think this is true on a sphere? Explain.

11.8 : Exercises and Applications

1. **Writing** When the concept of a line is extended to a sphere, the result is a great circle of the sphere. Describe how you would extend the concept of *segment* to a sphere.

2. In Euclidean geometry, lines go on forever and have no ends. Do lines in spherical geometry have ends? Explain your reasoning.

3. Can you draw a ray on a sphere? Explain your reasoning.

Describe each figure. Is it possible to sketch the figure on a Euclidean plane? Explain your reasoning.

4.
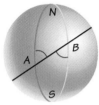

5.

6.

7. In a plane, the measure of each angle of an equiangular triangle is 60°. Show that this is not true for spherical triangles by sketching a triangle that includes three 90° angles.

In Exercises 8 and 9, show that the given statement from Euclidean geometry is *not* true for spherical geometry.

8. A triangle always has at least two acute angles.

9. If two lines are cut by a transversal and the corresponding angles formed are congruent, then the lines are parallel.

10. **Investigation** You will need a spherical object, three long strips of paper, tape, and a ruler. Use the strips of paper to make a triangle on the sphere. Mark the vertices of the triangle on the strips of paper, and untape the strips to measure the sides. Repeat several times. Are the Triangle Inequality Theorems on page 280 true on a sphere?

SPORTS The soccer ball at the right is a sphere covered by regular pentagons and hexagons. Two hexagons and one pentagon fit around each vertex without overlapping or leaving any gaps. On the soccer ball, as on a plane, the sum of the angles at each vertex is 360°.

11. **a.** On a plane, what is the measure of each interior angle of a regular hexagon? of a regular pentagon?

   b. On a plane, can two regular hexagons and a regular pentagon fit perfectly around a vertex as they do on a soccer ball? Explain.

12. **Open-ended Problem** Explain how you can tell that the measures of the interior angles of regular hexagons and pentagons are not the same on a sphere as on a plane.

13. The measure of each interior angle of a hexagon on a soccer ball is about 124.3°. What is the measure of each interior angle of a pentagon? Explain your reasoning.

Extra Practice exercises on page 685

Any point on Earth can be located by its latitude and longitude measured in degrees (°) and minutes ('). One degree equals sixty minutes.

Chicago is **north** of the equator and $m\angle CTE = $ **41°50'**. The *latitude* of Chicago is **41°50' north**.

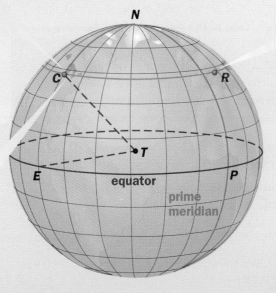

The latitude and longitude of Rome are 41°50' north, 12°40' east.

Chicago is **west** of the prime meridian, and $m\angle CNP = $ **87°40'**. The *longitude* of Chicago is **87°40' west**. *Note:* $\angle CNP$ is on the surface of Earth.

14. a. The orange line on the globe above is called a *line of latitude* because all of the points on it have the same latitude. Are lines of latitude the same as lines in spherical geometry? Explain.

   b. The prime meridian is a *line of longitude*. Does the prime meridian line lie on a great circle of Earth? Explain your reasoning.

The shortest distance between two points on the globe is the length of the great-circle arc between them. This distance can be found using a *terrestrial triangle* whose third vertex is *N*, the North Pole.

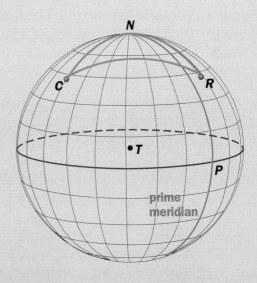

15. Find the measure of $\angle CNR$ in spherical triangle *NCR*. (*Hint:* Let *P* represent a point on the prime meridian. $m\angle CNR = m\angle CNP + m\angle PNR$.)

16. $\triangle NCR$ is an isosceles triangle with $m\angle NCR = m\angle NRC \approx 51°20'$. Use your result from Exercise 15 to show that the sum of the measures of the angles of $\triangle NCR$ is greater than 180°.

17. $m\angle CTR = m\overset{\frown}{CR} \approx 70°00'$

   a. Convert the arc measure to minutes.

   b. A *nautical mile* is the length of a one-minute arc of a great circle on Earth. Find the distance from Chicago to Rome in nautical miles.

   c. A nautical mile is about 1.15 miles. Find the distance from Chicago to Rome in miles.

Working in Alexandria, Egypt, in the third century B.C., Euclid tried to prove geometric concepts using a small number of postulates. Using only five postulates, he was able to prove most of the theorems that you have learned in this book. One of the five postulates he used is the *Parallel Postulate:*

Euclid: Through a point not on a given line, there is **exactly one** line parallel to the given line.

In the eighteenth century, mathematicians began to explore two different parallel postulates:

Spherical geometry: Through a point not on a given line, there is **no** line parallel to the given line.

Hyperbolic geometry: Through a point not on a given line, there is **more than one** line parallel to the given line.

Felix Klein, a German mathematician, developed a model to make it easier to explore hyperbolic geometry. In his model, the *hyperbolic plane* consists of only the interior points of a Euclidean circle. A *hyperbolic line* is the part of a Euclidean line contained in the circle.

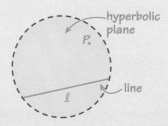

For Exercises 18–20, use Klein's model of the hyperbolic plane.

18. How do you know that two points determine a single line?

19. The figure above shows line ℓ and a point P not on line ℓ. Copy the figure and sketch several different lines through P that are parallel to (do not intersect) line ℓ.

20. In this model, angle measures and distances are not, in general, measured in the standard Euclidean manner. However, if line m is a diameter of the Euclidean circle, then another line k is perpendicular to m if k and m are perpendicular in the Euclidean sense. This figure shows two perpendicular lines. Copy the figure and sketch a third line parallel to both m and k. Which postulate of Euclidean geometry does your sketch contradict?

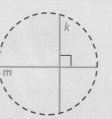

21. Research   Write a report about one of the following mathematicians: G. F. Bernhard Riemann (1826–1866), Janos Bolyai (1802–1860), Nikolai Ivanovich Lobachevsky (1792–1856), or Carl Friedrich Gauss (1777–1855).

BY THE WAY

The Dutch artist M. C. Escher (1898–1972) used hyperbolic geometry to produce figures that get smaller as they approach the edge of the art piece.

22. Challenge Are the SSS and SAS Postulates true on a sphere? What if the two triangles are on spheres with different diameters? Explain your reasoning.

23. Open-ended Problem Choose a postulate or theorem from Euclidean geometry. Is it true on a sphere? Explain. Include a sketch with your explanation.

The heights of two similar pyramids are 10 cm and 7 cm. Find the ratio of each measure of the pyramids. *(Section 11.7)*

24. the areas of the bases   **25.** the lateral areas   **26.** the volumes

27. The vertices of $\triangle ABC$ are $A(-2, -1)$, $B(0, 3)$, and $C(5, -3)$. $\triangle ABC$ is reflected over the *x*-axis. Give the coordinates of the vertices of the image. *(Section 8.2)*

28. If $\overline{PQ}$ and $\overline{RQ}$ are tangent to $\odot C$ at P and R respectively, then: *(Section 11.4)*

 A. $PQ > RQ$ **B.** $RQ > PQ$ **C.** $PQ = RQ$

 D. relationship cannot be determined

ASSESS YOUR PROGRESS

VOCABULARY

sphere (p. 586)   **great circle** (p. 598)

1. Sketch a sphere. Draw and label a diameter, a secant, and a tangent of the sphere. *(Section 11.6)*

Find the surface area and volume of each sphere. *(Section 11.6)*

2.

3 in.

3.

5.8 cm

4. The slant heights of two similar cones are 5 ft and 12 ft. Find the ratio of each measure of the cones. *(Section 11.7)*

 a. the lateral areas   **b.** the volumes

5. In spherical geometry, is the sum of the measures of the interior angles of a quadrilateral 360°? Explain your reasoning. *(Section 11.8)*

6. Journal Some people think that spherical geometry is exciting and others find it unbelievable. Describe your reaction to learning that the rules of geometry change if you change the original postulates.

BY THE WAY

Another field of geometry is *projective geometry*, the study of the properties involved in transferring a three-dimensional object to a two-dimensional plane. Mathematicians in this field, such as Ruth Moufang (1905–1977), combine both algebraic and geometric concepts in their work.

Modeling Cavalieri's Principle

Archimedes discovered the formula for the volume of a sphere by imagining cutting a sphere, cone, and cylinder into very thin slices and comparing the slices. This method was developed further by Cavalieri, and became part of the foundation for calculus. Imagine slicing each object at the same height. How do you think the areas of the cross sections are related?

$2\frac{1}{8}$ in.   3 in.   3 in.   3 in.   3 in.

3 in.   3 in.

PROJECT GOAL   Compare the volumes of a sphere, cylinder, and two cones by comparing cross sections of the objects.

Making the Shapes

Work with a partner to make the sphere, cylinder, and two cones with the dimensions shown above. Use modeling clay.

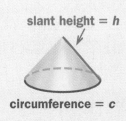

slant height = *h*

circumference = *c*

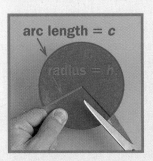

arc length = *c*

radius = *h*

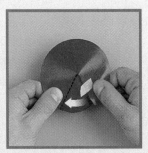

CONE To make a mold for the cone above, cut a circle out of strong plastic and draw a sector as shown. Tape the sector's edges together.

SPHERE You can check the diameter of your sphere by using a circular cutout with a diameter that is slightly larger than 3 in.

CYLINDER You might want to make your cylinder too large and then trim it carefully by using a compass and plastic knife.

Comparing Cross Sections

1. Cut each object carefully into six slices of equal thickness. Use fishing line or dental floss.

2. Combine the clay from the bottom slice of the cones and the bottom slice of the sphere to make a new disk of equal thickness.

3. Compare the new disk with the bottom slice of the cylinder. How were the cross sectional areas of the bottom slices of the objects related?

4. Repeat Steps 2 and 3 five times, comparing the cross sections of the slices at each height.

Presenting Your Project

Write a report describing your experiment. Include answers to the following questions.

- How are the areas of the cross sections of the objects related? How are the volumes of the objects related? How do you know?

- How can you use the volume formulas for a cylinder and a cone to find the volume formula for a sphere?

You may want to extend your project to include some of the ideas below:

- Look up *Cavalieri's Principle* in an encyclopedia or a mathematics dictionary. Summarize it and explain how you used it in your project.

- If x is the height at which you slice each object, write an expression for the area of each cross section. Show that the relationship you discovered is true for any length x that is less than 3 in.

Self-Assessment
What grade would you give yourself for this project? Why? If you did the experiment again, what would you do differently? Why?

11 Review

STUDY TECHNIQUE

Make a study sheet. Include definitions, equations, diagrams, and hints that you want to remember. Review your sheet twice a day for a week.

VOCABULARY

central angle (p. 553)
inscribed angle (p. 553)
chord (p. 553)
tangent to a circle (p. 554)
arc (p. 554)
minor arc (p. 554)
major arc (p. 554)
semicircle (p. 554)

measure of an arc (p. 554)
congruent arcs (p. 555)
secant (p. 561)
sector (p. 580)
length of an arc (p. 580)
sphere (p. 586)
great circle (p. 598)

SECTIONS 11.1 and 11.2

When both rays of an angle intersect a circle, the formula for finding the measure of the angle depends on the location of the angle's vertex.

Vertex at the center of the circle   **Vertex on the circle**

The angle measure equals the measure of the intercepted arc.

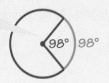

The angle measure equals half the measure of the intercepted arc.

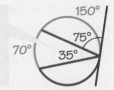

Vertex outside the circle   **Vertex inside the circle**

The angle measure equals half the *difference* of the measures of the intercepted arcs.

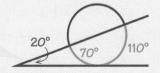

The angle measure equals half the *sum* of the measures of the intercepted arcs.

Inscribed angles that intercept the same arc or congruent arcs are congruent. A **tangent** to a circle is perpendicular to a radius at the point of tangency.

SECTIONS 11.3 and 11.4

The perpendicular bisector of a chord passes through the center of the circle. A diameter that is perpendicular to a chord bisects the chord and its corresponding arc.

You can find the lengths of some segments that intersect a circle:

Two chords
$AC \cdot CD = BC \cdot CE$

Two secants
$AC \cdot CD = BC \cdot CE$

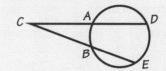

A tangent and a secant
$AC \cdot CD = BC^2$

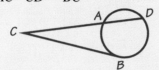

Two tangents
$AC = BC$

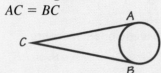

In the same circle or congruent circles, congruent chords are an equal distance from the center of the circle. Their corresponding arcs are congruent.

SECTIONS 11.5, 11.6, and 11.7

In a circle:

> Area of a sector = **fraction of circle · area of circle**
> $$= \frac{x}{360} \cdot \pi r^2$$
> Length of an arc = **fraction of circle · circumference of circle**
> $$= \frac{x}{360} \cdot 2\pi r$$

For a sphere with radius r:

> Surface area $= 4\pi r^2$ Volume $= \frac{4}{3}\pi r^3$

For two similar solids, if the ratio of two corresponding lengths is $a:b$, then:
ratio of corresponding areas $= a^2:b^2$ ratio of the volumes $= a^3:b^3$

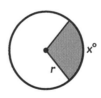

SECTION 11.8

In spherical geometry, a line is represented by a **great circle** of a sphere. The sum of the measures of the interior angles of a triangle is more than 180°, and there are no parallel lines. As a result, many of the theorems of Euclidean geometry are not true for spherical geometry.

11 | Assessment

VOCABULARY REVIEW

Define each term and sketch an example.

1. congruent arcs   **2.** major arc   **3.** inscribed angle

4. secant   **5.** tangent of a circle   **6.** great circle

SECTIONS 11.1 *and* 11.2

Find each angle or arc measure.

7. $m \angle A$

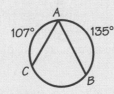

8. $m \widehat{EF}$

9. $m \widehat{MP}$

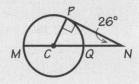

10. Given: $\widehat{AB} \cong \widehat{CD}$
Prove: $\overline{AD} \parallel \overline{BC}$

SECTIONS 11.3 *and* 11.4

Find the length of each labeled segment.

11.

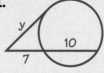

12.

13.

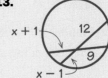

14. Given: $\widehat{MN} \cong \widehat{RS}$
Prove: $\triangle MNP \cong \triangle RSP$

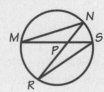

SECTIONS 11.5, 11.6, *and* 11.7

Find the area of each shaded sector and the length of each red arc.

15.

16.

17.

Find the surface area and the volume of each sphere.

18.

19.

6.5 cm

20. The radii of a sphere and a cone are both 7 cm. If their volumes are equal, what is the height of the cone?

21. The heights of two similar cylinders are 6 in. and 10 in. Find the ratio of each measure of the cylinders.

   a. the radii of the bases

   b. the areas of the bases

   c. the volumes

SECTION 11.8

22. How is the sum of the measures of the interior angles of a triangle different in spherical geometry than in Euclidean geometry?

23. **Open-ended Problem** Give an example of a statement from Euclidean geometry that is not true for spherical geometry. Explain why it is not true in spherical geometry.

24. **Writing** Is it possible to sketch a parallelogram on a sphere? Explain why or why not.

PERFORMANCE TASK

25. Make a list summarizing the theorems you know about angles and segments in circles. Choose two of the theorems. Do you think that they are also true for spheres? Explain your reasoning.

CHAPTER

12 | Coordinates for Transformations

Plotting EARTH *and* SKY

INTERVIEW **Adriana Ocampo**

In 1969, at age fourteen, Adriana Ocampo saw how dreams can come true when she saw humans walk on the moon for the first time. "I immediately realized that I wanted to be part of that," she recalls. "It's always been my dream to get involved in space research and space missions." Now a planetary geologist at the Jet Propulsion Laboratory (JPL) in Pasadena, California, Ocampo is lucky enough to be fulfilling her dreams.

Adriana Ocampo stands among scale models of spacecraft.

"Space exploration is part of human destiny, and science makes it possible."

610

"Studying the processes that occur on other planets helps us better understand Earth."

An Extended Project

Born in Colombia and raised in Argentina, Ocampo came to the United States in 1970 and went to high school in South Pasadena. In a Caltech/JPL program for high school students, she learned about climatology and environmental monitoring while using sophisticated equipment. She continued working at the laboratory as a college undergraduate and as a graduate student, and has been there ever since. "This high school project extended into a career," she says.

From Earth to the Far Planets

Geology is the study of the processes that change the surface of Earth and its interior. Planetary geology, Ocampo's speciality, involves looking at the same processes on other planets and making comparisons. "By understanding how these processes occur on other planets, we gain a better understanding of our own planet," she explains.

Ocampo has worked on the Viking mission to Mars; the Voyager mission to Jupiter, Saturn, Neptune, and Uranus; and the Galileo mission to Jupiter. Her research also includes the exploration of a giant crater in the Yucatán peninsula in Mexico.

In space flight missions such as Galileo, Ocampo works chiefly as a planetary geologist. Her work relies heavily on tools developed using math and geometry.

Back to School

Ocampo participates in mentoring programs to encourage high school girls to get the training they need to pursue careers in science. "I think it's everyone's responsibility to give back a little of what they received," she says. "I understand how important it is to follow your dreams."

The Galileo mission examined Jupiter's atmosphere and some of its moons.

Space Travel with Matrices

"We use matrices to define the coordinate system of the stars that we navigate the spacecraft by," Ocampo explains. Other matrices and coordinate systems define the position of the spacecraft relative to Earth. There's even a matrix to describe the orientation of the spacecraft. That's important to know so that cameras and sensors are pointed in the right direction. Navigation specialists at NASA/JPL use matrices to describe the translations and rotations that link these coordinate systems together. These calculations help keep the spacecraft on the correct trajectory.

The Galileo spacecraft was launched from the space shuttle Atlantis. Black and gold fabric protects it from the heat of the sun and the cold of interplanetary space.

Galileo used the gravity fields of Venus and Earth to boost itself toward Jupiter. When it was 60,000 miles above Venus, its Near Infrared Mapping Spectrometer (NIMS) gathered data about radiant heat, shown in this false-color image.

Explore and Connect

Remote-sensing techniques for obtaining data, developed to study other planets, are now used to study our own.

1. Project   Find out about the trajectory of and the planets visited by one of these spacecraft: Viking, Voyager, or Galileo. Present a report about some of the results of the mission, including some images.

2. Writing   Imagine a spacecraft traveling through space. What transformations might it undergo and why? Explain.

3. Research   Find out how far the large planets in the solar system are from Earth and how long it takes for spacecraft to reach them. Explain how the orbits of the planets influence the timing of the launches of spacecraft.

Mathematics
& Adriana Ocampo

In this chapter, you will learn more about how mathematics relates to space science.

Related Exercises

Section 12.2
• Exercises 20 and 21

Section 12.5
• Exercises 8 and 9

12.1 Dilations with Matrices

In computer games, such as driving simulators, the illusion of depth is created by enlarging objects on the screen as a player nears them. The screens shown are from a bike riding simulator in which users feel they are moving down a road because objects along the roadside get bigger.

Programmers can use *matrices* to store the coordinates of objects they want to display. Then they transform the coordinates using dilations to draw larger objects.

A **matrix** is a rectangular arrangement of numbers, called **elements**, in **rows** and **columns**. *Matrices* is the plural of *matrix*. Matrices are enclosed in square brackets.

The number of rows and number of columns in a matrix are its **dimensions**. A matrix that has m rows and n columns is called an $m \times n$ (read "m by n") matrix.

You can represent a point in the coordinate plane with a **point matrix**. List the coordinates vertically. The x-coordinate is at the top. The y-coordinate is at the bottom.

$$\text{2 rows} \rightarrow \begin{bmatrix} 3 \\ 2 \end{bmatrix} \begin{matrix} \leftarrow x\text{-coordinate} \\ \leftarrow y\text{-coordinate} \end{matrix}$$

1 column

This is a 2×1 matrix.

You can represent a triangle with a **polygon matrix**. Each column represents a vertex. List the vertices in consecutive order, just as you would for a polygon.

$$\text{2 rows} \rightarrow \begin{bmatrix} 1 & 9 & 8 \\ 2 & 0 & 6 \end{bmatrix} \begin{matrix} \leftarrow x\text{-coordinates} \\ \leftarrow y\text{-coordinates} \end{matrix}$$

3 columns

This is a 2×3 matrix.

EXAMPLE 1

Write a polygon matrix that represents pentagon *VWXYZ*. What are the dimensions of the matrix?

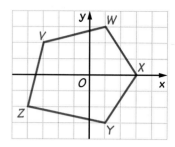

SOLUTION

Write the coordinates of the vertices in consecutive order.

$$\begin{array}{ccccc} V & W & X & Y & Z \\ \begin{bmatrix} -3 & 1 & 3 & 1 & -4 \\ 2 & 3 & 0 & -3 & -2 \end{bmatrix} \end{array}$$

The columns are written from left to right in the order used to name the polygon.

The dimensions of the matrix are 2×5.

THINK AND COMMUNICATE

1. Write a matrix for pentagon *ZYXWV* in Example 1. How is it different from the matrix in the solution above?

2. What are the dimensions of a matrix that represents a quadrilateral in the coordinate plane?

To dilate a polygon using (0, 0) as the center of the dilation and a scale factor of 1.5, multiply each coordinate of the polygon by 1.5. Multiplying a matrix by a real number is called **scalar multiplication**.

In this chapter, light blue is used to represent transformations.

$$1.5 \cdot \begin{array}{ccc} A & B & C \\ \begin{bmatrix} -54 & -54 & -26 \\ 40 & 0 & 16 \end{bmatrix} \end{array} = \begin{array}{ccc} A' & B' & C' \\ \begin{bmatrix} -81 & -81 & -39 \\ 60 & 0 & 24 \end{bmatrix} \end{array}$$

scale factor · polygon matrix = image matrix

The **image matrix** represents the image polygon.

Each element of the image matrix is 1.5 times the value of the corresponding element in the original polygon matrix. The original triangle and its image are shown on the screens below.

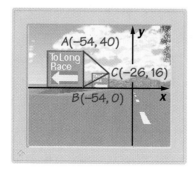

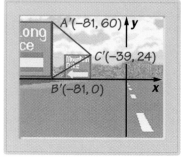

To multiply a matrix by a number k, multiply each element of the matrix by k:

$$k \begin{bmatrix} a & b & c \\ d & e & f \end{bmatrix} = \begin{bmatrix} ka & kb & kc \\ kd & ke & kf \end{bmatrix}$$

EXAMPLE 2

The matrix $\begin{bmatrix} -1 & 2 & 2 & 0 \\ 2 & 2 & 0 & -1 \end{bmatrix}$ represents quadrilateral *PQRS*.

a. Find the product: $3 \begin{bmatrix} -1 & 2 & 2 & 0 \\ 2 & 2 & 0 & -1 \end{bmatrix}$.

b. Graph the polygons represented by $\begin{bmatrix} -1 & 2 & 2 & 0 \\ 2 & 2 & 0 & -1 \end{bmatrix}$ and $3 \begin{bmatrix} -1 & 2 & 2 & 0 \\ 2 & 2 & 0 & -1 \end{bmatrix}$.
Label them *PQRS* and *P'Q'R'S'*.

c. Describe the transformation that changes *PQRS* to *P'Q'R'S'*.

SOLUTION

a.
$$3 \begin{array}{c} P \quad Q \quad R \quad S \\ \begin{bmatrix} -1 & 2 & 2 & 0 \\ 2 & 2 & 0 & -1 \end{bmatrix} \end{array} = \begin{bmatrix} 3(-1) & 3(2) & 3(2) & 3(0) \\ 3(2) & 3(2) & 3(0) & 3(-1) \end{bmatrix}$$

$$= \begin{array}{c} P' \quad Q' \quad R' \quad S' \\ \begin{bmatrix} -3 & 6 & 6 & 0 \\ 6 & 6 & 0 & -3 \end{bmatrix} \end{array}$$

b. Graph the original polygon and its image using the same pair of axes.

c. The transformation is a dilation with center at $(0, 0)$ and scale factor 3.

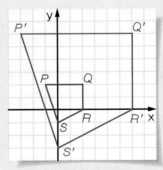

THINK AND COMMUNICATE

3. Find $\frac{1}{3} \begin{bmatrix} -3 & 6 & 6 & 0 \\ 6 & 6 & 0 & -3 \end{bmatrix}$. Explain what this product represents.

4. After a matrix is multiplied by a number, are the dimensions of the image matrix different from the dimensions of the original polygon matrix? Explain why or why not.

5. Suppose you multiply a polygon matrix by 1. Describe the image matrix and the image polygon.

6. Suppose you multiply a polygon matrix by 0. Describe the image matrix. What do you think the graph of the image matrix looks like?

1. How many elements are in a 2 × 6 matrix? What kind of polygon could such a matrix represent?

2. Write a polygon matrix that represents parallelogram *DEFG*.

3. Multiply the matrix for *DEFG* by a scale factor of 1.5. Graph the image and label it *D'E'F'G'*.

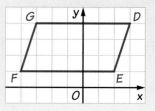

12.1 Exercises and Applications

Extra Practice exercises on page 685

Use the graph. Write a matrix that represents each figure.

1. *M*

2. *P*

3. $\overline{KN}$

4. △*MNK*

5. △*NPK*

6. *MNPK*

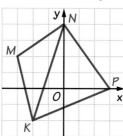

Find each product.

7. $5\begin{bmatrix} 3 \\ -1 \end{bmatrix}$

8. $4\begin{bmatrix} 1 & -3 \\ -2 & 5 \end{bmatrix}$

9. $3\begin{bmatrix} 0 & 3 & -2 \\ 0 & -1 & -4 \end{bmatrix}$

10. $2.5\begin{bmatrix} 1 & 0 & -2 & 3 \\ 1 & 5 & 0 & -4 \end{bmatrix}$

11. $1\begin{bmatrix} 10 & 5 & 15 \\ 5 & 5 & 10 \end{bmatrix}$

12. $\frac{2}{3}\begin{bmatrix} 9 & -12 \\ 0 & 3 \end{bmatrix}$

13. Describe the transformation represented by each product in Exercises 7–12.

Copy and complete each scalar multiplication problem.

14. $\underline{?}\begin{bmatrix} 9 & -3 \\ 21 & 0 \end{bmatrix} = \begin{bmatrix} 3 & -1 \\ 7 & 0 \end{bmatrix}$

15. $4\begin{bmatrix} 2 & \underline{?} & 0 \\ \underline{?} & 4 & \underline{?} \end{bmatrix} = \begin{bmatrix} 8 & 12 & \underline{?} \\ -4 & \underline{?} & 1 \end{bmatrix}$

The red polygon in each diagram is the image of the blue polygon after a dilation. Write a product to represent each dilation.

16.

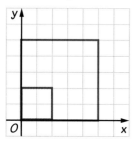

17.

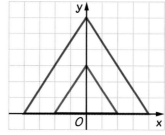

For each scale factor and polygon given, write a product to represent the dilation. Then graph each polygon and its image.

18. scale factor 3

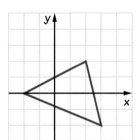

19. scale factor 2

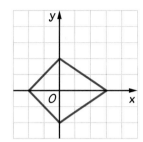

20. scale factor $\frac{3}{4}$

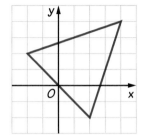

Connection ▶ DIGITAL MAPS

Digital maps available on the Internet can show a location you are interested in and then let you zoom in or out to see more or less detail. These maps show the area around Chinatown in San Francisco. From Map B, the user can zoom out to see Map A, or zoom in to see Map C. The origin is labeled X on each map.

Map A

For Exercises 21–23, the coordinates of each landmark on Map B are given. Write a point matrix for each landmark.

21. Cable Car Museum $(-58, 30)$

22. Chinatown Gate $(20, -36)$

23. Transamerica Pyramid $(54, 34)$

24. Write a triangle matrix that represents the three landmarks on Map B.

25. A dilation with scale factor $\frac{1}{2}$ reduces Map B onto Map A. Use scalar multiplication to find a triangle matrix that represents the coordinates of the landmarks on Map A.

Map B

26. A dilation with scale factor 2 enlarges Map B onto Map C. Use scalar multiplication to find a triangle matrix that represents the coordinates of the landmarks on Map C.

27. Writing Describe the dilation that enlarges Map A directly onto Map C. Explain your reasoning.

28. What point does not change its location during all of the dilations described above?

Map C

29. Suppose a matrix that represents an octagon in the coordinate plane is multiplied by a number. Give the dimensions of the image matrix.

30. 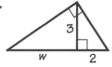 **Technology** Use a calculator or software with matrix capabilities. Matrices are named with capital letters. (See the *Technology Handbook*, pages 723–725, for more information about matrix calculations.)

> ```
> [A]
>    [[2 7 4 -1]
>     [1 4 9  6 ]]
> 5*[A]
> ```

  a. Store $\begin{bmatrix} 2 & 7 & 4 & -1 \\ 1 & 4 & 9 & 6 \end{bmatrix}$ as matrix A.

  b. Predict the elements in the matrices for $5A$, $3A$, and $0A$.

  c. Check your predictions using the calculator.

31. **Challenge** A triangle in a three-dimensional coordinate system is represented by a 3×3 matrix. The matrices below represent $\triangle ABC$.

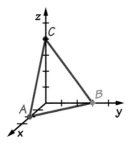

  a. Find the product: $2\begin{array}{c} A\ B\ C \\ \begin{bmatrix} 2 & 0 & 0 \\ 0 & 3 & 0 \\ 0 & 0 & 4 \end{bmatrix} \end{array}$.

  b. Find the product: $\dfrac{1}{2}\begin{array}{c} A\ B\ C \\ \begin{bmatrix} 2 & 0 & 0 \\ 0 & 3 & 0 \\ 0 & 0 & 4 \end{bmatrix} \end{array}$.

  c. Graph your results from parts (a) and (b). Describe the transformations.

ONGOING ASSESSMENT

32. **Open-ended Problem** Multiply any triangle matrix by several negative numbers. Graph each image triangle. Explain what a dilation with a *negative scale factor* looks like.

SPIRAL REVIEW

33. The volumes of two similar bottles are 27 oz and 8 oz. Find the ratio of their heights. *(Section 11.7)*

Find the value of each variable. *(Section 10.1)*

34.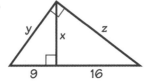

35.

Copy square *JKLM* and draw its image after each translation. *(Section 8.3)*

36. $(a, b) \rightarrow (a + 2, b)$

37. $(a, b) \rightarrow (a, b + 2)$

38. $(a, b) \rightarrow (a - 3, b + 1)$

39. $(a, b) \rightarrow (a + 1, b - 3)$

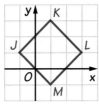

12.2 Translations with Matrices

Dilations are not the only transformations you can represent using matrices. You can also express translations with matrices. In Chapter 8, you described a translation by describing the change in the coordinates of a point. A translation to the right 4 units and up 7 units, for example, shifts $(3, -2)$ to $(7, 5)$.

Add **4** to the
x-coordinate.

$(3, -2) \rightarrow (7, 5)$

Add **7** to the
y-coordinate.

To use matrices to describe this translation, you must use **matrix addition**, the operation that adds corresponding elements of two matrices.

To add two matrices, they must have the same dimensions.

Add corresponding elements of the matrices.

In this chapter, transformations are written on the left.

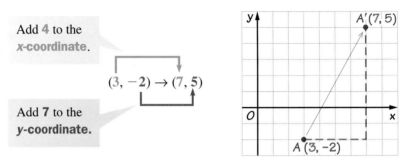

$$\begin{bmatrix} 4 \\ 7 \end{bmatrix} + \begin{bmatrix} 3 \\ -2 \end{bmatrix} = \begin{bmatrix} 4 + 3 \\ 7 + (-2) \end{bmatrix} = \begin{bmatrix} 7 \\ 5 \end{bmatrix}$$

translation   point   image
matrix   matrix   matrix

THINK AND COMMUNICATE

1. Write a translation matrix that shifts a point to the left 5 units and up 3 units. Find the image of point $A(3, -2)$ after this translation.

2. Write a translation matrix that shifts a point to the right 2 units. Find the image of point $A(3, -2)$ after this translation.

3. Write a translation matrix that shifts a point to the right h units and up k units. Find the image of point $A(3, -2)$ after this translation.

When you use matrices to translate a polygon, the dimensions of the translation matrix must match the dimensions of the polygon matrix. Each column of the translation matrix is the same as the column you would use to translate a point.

EXAMPLE 1

The matrix $\begin{bmatrix} -1 & 2 & 4 \\ 2 & 3 & 0 \end{bmatrix}$ represents $\triangle ABC$.

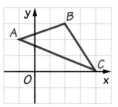

a. Write a translation matrix that shifts $\triangle ABC$ to the left 3 units and down 5 units.

b. Find the image matrix representing $\triangle A'B'C'$.

c. Graph $\triangle ABC$ and its image.

SOLUTION

a. Because the polygon matrix has 3 columns, the translation matrix must also have 3 columns. Because the shift is to the left 3 units and down 5 units, the translation matrix is:

The negative signs indicate movement *to the left* and *down*.

$$\begin{bmatrix} -3 & -3 & -3 \\ -5 & -5 & -5 \end{bmatrix}$$

$$\begin{array}{ccc} & A\ B\ C & A'\ B'\ C' \end{array}$$

b. $\begin{bmatrix} -3 & -3 & -3 \\ -5 & -5 & -5 \end{bmatrix} + \begin{bmatrix} -1 & 2 & 4 \\ 2 & 3 & 0 \end{bmatrix} = \begin{bmatrix} -4 & -1 & 1 \\ -3 & -2 & -5 \end{bmatrix}$

translation   polygon   image
matrix   matrix   matrix

c.

Each vertex of $\triangle ABC$ moves **to the left 3 units** and **down 5 units**.

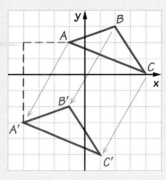

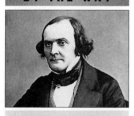
THINK AND COMMUNICATE

4. To translate $\triangle A'B'C'$ to $\triangle ABC$, what translation matrix would you use?

5. To shift a pentagon using the same translation as in Example 1, what translation matrix would you use?

You can use **matrix subtraction** to find an original polygon matrix if you are given an image matrix and a translation matrix. The matrices must have the same dimensions. Subtract the corresponding elements of the matrices.

EXAMPLE 2

The matrix $\begin{bmatrix} 3 & 4 & 6 \\ -2 & 0 & -4 \end{bmatrix}$ represents $\triangle D'E'F'$, the image of $\triangle DEF$ after a translation to the right 5 units and down 3 units. Find the polygon matrix for $\triangle DEF$.

SOLUTION

Write the translation matrix and the image matrix.

The translation is to the right 5 units and down 3 units.

$$\begin{bmatrix} 5 & 5 & 5 \\ -3 & -3 & -3 \end{bmatrix} + \overset{\substack{D \quad E \quad F}}{\begin{bmatrix} ? & ? & ? \\ ? & ? & ? \end{bmatrix}} = \overset{\substack{D' \quad E' \quad F'}}{\begin{bmatrix} 3 & 4 & 6 \\ -2 & 0 & -4 \end{bmatrix}}$$

translation   polygon   image
matrix   matrix   matrix

$$\begin{bmatrix} ? & ? & ? \\ ? & ? & ? \end{bmatrix} = \begin{bmatrix} 3 & 4 & 6 \\ -2 & 0 & -4 \end{bmatrix} - \begin{bmatrix} 5 & 5 & 5 \\ -3 & -3 & -3 \end{bmatrix}$$

polygon   image   translation
matrix   matrix   matrix

Subtract the translation matrix from both sides of the equation.

$$= \begin{bmatrix} 3-5 & 4-5 & 6-5 \\ -2-(-3) & 0-(-3) & -4-(-3) \end{bmatrix}$$

$$= \begin{bmatrix} -2 & -1 & 1 \\ 1 & 3 & -1 \end{bmatrix}$$

$\triangle DEF$ has vertices $D(-2, 1)$, $E(-1, 3)$, and $F(1, -1)$.

☑ **CHECKING KEY CONCEPTS**

Add the matrices.

1. $\begin{bmatrix} 4 \\ 1 \end{bmatrix} + \begin{bmatrix} 0 \\ 3 \end{bmatrix}$
       2. $\begin{bmatrix} -1 & -1 & -1 \\ 5 & 5 & 5 \end{bmatrix} + \begin{bmatrix} 0 & 3 & 7 \\ 3 & -4 & -2 \end{bmatrix}$

3. $\begin{bmatrix} 3 & 3 \\ -2 & -2 \end{bmatrix} + \begin{bmatrix} 1 & 4 \\ 5 & 2 \end{bmatrix}$
       4. $\begin{bmatrix} 0 & 0 & 0 & 0 \\ 4 & 4 & 4 & 4 \end{bmatrix} + \begin{bmatrix} -2 & 3 & 2 & -5 \\ 0 & -1 & 6 & 5 \end{bmatrix}$

5. Suppose each matrix sum in Questions 1–4 represents a translation of a figure. Describe the figures and the translations.

6. Describe how to subtract matrices.

12.2 Exercises and Applications

Extra Practice
exercises on
page 686

Add or subract the matrices.

1. $\begin{bmatrix} -3 \\ 5 \end{bmatrix} + \begin{bmatrix} 5 \\ -1 \end{bmatrix}$

2. $\begin{bmatrix} 1 & 2 & -1 \\ -3 & 4 & 6 \end{bmatrix} - \begin{bmatrix} 3 & 3 & 3 \\ 2 & 2 & 2 \end{bmatrix}$

3. $\begin{bmatrix} 1 & 1 \\ 0 & 0 \end{bmatrix} + \begin{bmatrix} -1 & 3 \\ 3 & -1 \end{bmatrix}$

4. $\begin{bmatrix} 2 & 2 & 2 & 2 \\ -4 & -4 & -4 & -4 \end{bmatrix} + \begin{bmatrix} 0 & 3 & 0 & -5 \\ -2 & 2 & 3 & 1 \end{bmatrix}$

5. Suppose each matrix sum or difference in Exercises 1–4 represents a translation of a figure. Describe the figures and the translations.

Write the matrix you would use to translate each figure as indicated.

6. a point; to the right 3 units and up 4 units

7. a segment; to the left 2 units

8. a triangle; down 5 units

9. a quadrilateral; to the right 6 units and down 1 unit

For Exercises 10–13, use the method shown in Example 1.

10. The matrix $\begin{bmatrix} 3 & 5 \\ 1 & 1 \end{bmatrix}$ represents $\overline{XY}$.

   a. Write a matrix that translates $\overline{XY}$ to the right 3 units and up 1 unit.

   b. Find the image matrix that represents $\overline{X'Y'}$.

   c. Graph $\overline{XY}$ and its image.

11. The matrix $\begin{bmatrix} -4 & -8 & 0 \\ -3 & 2 & 3 \end{bmatrix}$ represents $\triangle ABC$.

   a. Write a matrix that translates $\triangle ABC$ to the left 4 units and up 4 units.

   b. Find the image matrix that represents $\triangle A'B'C'$.

   c. Graph $\triangle ABC$ and its image.

12.

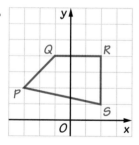

   a. Write a matrix that represents quadrilateral *PQRS*.

   b. Write a matrix that translates *PQRS* to the right 5 units and down 3 units.

   c. Find the image matrix that represents *P'Q'R'S'*. Graph the image.

13.

   a. Write a matrix that represents pentagon *JKLMN*.

   b. Write a matrix that translates *JKLMN* to the left 4 units and down 8 units.

   c. Find the image matrix that represents *J'K'L'M'N'*. Graph the image.

14. Writing Does it matter whether you write a translation matrix on the left side or the right side of a polygon matrix before you add the matrices to find an image matrix? Explain why or why not.

The red polygon in each diagram is the image of the blue polygon after a translation. Write each translation as a matrix sum.

15.
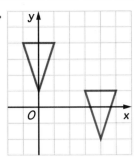

16.
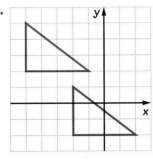

17.
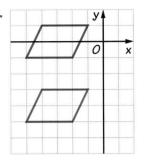

For Exercises 18 and 19, use the method shown in Example 2.

18. The matrix $\begin{bmatrix} 3 & 6 \\ 4 & 3 \end{bmatrix}$ represents $\overline{J'K'}$, the image of $\overline{JK}$ after a translation to the right 2 units and up 4 units. Find the matrix for $\overline{JK}$.

19. The matrix $\begin{bmatrix} -2 & 0 & 6 & 4 \\ 0 & -3 & 0 & 3 \end{bmatrix}$ represents $W'X'Y'Z'$, the image of quadrilateral $WXYZ$ after a translation down 6 units. Find the matrix for $WXYZ$.

INTERVIEW # Adriana Ocampo

Look back at the article on pages 610–612.

Ganymede, Europa, and Callisto are three of Jupiter's moons. To help the Galileo spacecraft take pictures of and get data from the moons, scientists such as Adriana Ocampo must know the moons' locations in relation to the spacecraft.

20. In a three-dimensional coordinate system whose origin is at the center of Jupiter, the positions of the three moons at a given time are in the matrix on the right below. The translation matrix converts the positions into a coordinate system whose origin is at the spacecraft.

10×10^5 represents 1,000,000 km.

x-coordinate →
y-coordinate →
z-coordinate →
$$\begin{bmatrix} -7.4 \times 10^5 & -7.4 \times 10^5 & -7.4 \times 10^5 \\ -2.7 \times 10^5 & -2.7 \times 10^5 & -2.7 \times 10^5 \\ -1.4 \times 10^5 & -1.4 \times 10^5 & -1.4 \times 10^5 \end{bmatrix} + \begin{matrix} G & E & C \\ \begin{bmatrix} 10 \times 10^5 & 6.4 \times 10^5 & 18 \times 10^5 \\ 3.1 \times 10^5 & 1.6 \times 10^5 & 2.0 \times 10^5 \\ 2.1 \times 10^5 & 1.4 \times 10^5 & 4.9 \times 10^5 \end{bmatrix} \end{matrix}$$

translation matrix

a. Add the matrices to find a matrix for $G'E'C'$, the moons' positions in a coordinate system whose origin is at the spacecraft.

b. Use the distance formula in three dimensions to find EC and $E'C'$.

c. Writing What does each of the distances you found in part (b) represent? Compare the two values and explain your results.

21. Challenge Write a matrix for the positions of Callisto, Jupiter, and the spacecraft in a coordinate system whose origin is at Europa.

Because a color photocopier scans a photograph and stores it in digital form, the copier can shift the image to a different location on the page that it prints out. Some photocopiers use this feature to center items on the output page.

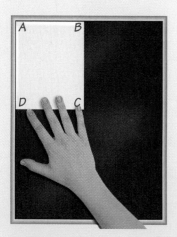

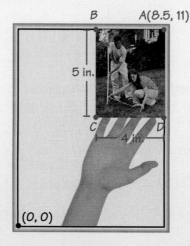

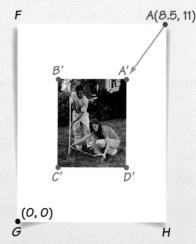

what you see   *what the scanner sees*   *output page*

22. Suppose you want to photocopy a 4 in. by 5 in. color photograph onto 8.5 in. by 11 in. paper. You place the photograph against the upper left corner of the glass plate, as shown above. Write a matrix that represents photograph *ABCD* as viewed by the scanner.

23. Explain why (4.25, 5.5) is the center of paper *AFGH*.

24. Explain why (6.5, 8.5) is the center of photograph *ABCD*.

25. To center the image, the photocopier must shift *ABCD* to *A'B'C'D'*. What translation matrix could the photocopier use?

26. Write a matrix that represents *A'B'C'D'*.

27. **Open-ended Problem** Choose dimensions different from 4 in. by 5 in. for a photograph. Describe the translation matrix the photocopier could use to center the photograph on a piece of 8.5 in. by 11 in. paper.

28. **Investigation** Use square *OPQR* shown at the right.
    a. Write a matrix that represents *OPQR*.
    b. Translate *OPQR* using a translation matrix of your choice. Then dilate the result using center (0, 0) and scale factor 2. Graph the image of *OPQR* after these two transformations.
    c. Dilate *OPQR* using center (0, 0) and scale factor 2. Then translate the result using the matrix you used in part (b). Graph the image of *OPQR* after these two transformations.
    d. Compare your answers to parts (b) and (c). Does the order in which the two transformations are applied affect the final outcome?

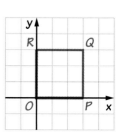

29. Cooperative Learning Work in a group of four people.
Each person should use a different 2 × 4 translation matrix.

The fourth person translates quadrilateral 3 to quadrilateral 4, then returns the paper to its owner.

  a. Each person should draw a quadrilateral on a different
  piece of graph paper. Write a matrix for the
  quadrilateral. Apply your translation matrix to the
  quadrilateral matrix and graph the result. Label
  the image "1."

  b. Pass your papers in a circle. Each time you receive
  a paper, apply your translation matrix to the latest
  quadrilateral on it. Number the new quadrilateral in
  sequence. Continue until you get your original paper.

  c. Find a translation matrix that will take quadrilateral
  4 to the original quadrilateral on your paper. Compare
  your answer with others in your group. What do you
  notice? Can you explain the results?

30. Find the area of a rectangle with a side of length 12 cm and a diagonal
of length 20 cm. *(Section 7.4)*

31. Find the area of a sector of a circle with a central angle of 84° and
radius 2. *(Section 11.5)*

ASSESS YOUR PROGRESS

VOCABULARY

matrix (p. 613)	**point matrix** (p. 613)
elements (p. 613)	**polygon matrix** (p. 613)
rows (p. 613)	**scalar multiplication** (p. 614)
columns (p. 613)	**matrix addition** (p. 619)
dimensions of a matrix (p. 613)	**matrix subtraction** (p. 621)

Find each product. *(Section 12.1)*

1. $5\begin{bmatrix} 0 & 4 \\ 5 & 3 \end{bmatrix}$

2. $1.5\begin{bmatrix} 1 & 3 & 4 \\ 0 & 2 & 3 \end{bmatrix}$

3. $0.5\begin{bmatrix} 2 & 4 & 3 & 1 \\ 0 & 4 & 6 & -2 \end{bmatrix}$

4. $10\begin{bmatrix} 0 & 5 & 1 \\ 0 & -2 & -3 \end{bmatrix}$

Write the matrix you would use to translate each figure. *(Section 12.2)*

5. a pentagon; to the left 1 unit and up 6 units

6. a parallelogram; to the right 5 units

7. Journal Suppose you dilate a quadrilateral using (0, 0) as the center.
If the quadrilateral has one vertex in each quadrant, will the image
also have one vertex in each quadrant? Explain using matrices.

12.3 Multiplying Matrices

Learn how to...

- multiply matrices

So you can...

- transform points and polygons using matrix multiplication

- analyze special effects such as morphing

Many movies and music videos use a special effect called *morphing*, in which one picture is gradually transformed into another. Some morphing techniques use *matrix multiplication* to transform images.

You can use **matrix multiplication** to multiply a point matrix or a polygon matrix by a transformation matrix. To transform the point $P(3, 5)$ using the transformation matrix $\begin{bmatrix} 2 & 0 \\ -7 & 4 \end{bmatrix}$, multiply the transformation matrix and the point matrix. The product is the image matrix.

Multiply elements of row 1 and row 2 of the transformation matrix by elements of the **point matrix**.

Add the products in each row.

$$\underset{\substack{\text{trans-}\\\text{formation}\\\text{matrix}}}{\begin{bmatrix} 2 & 0 \\ -7 & 4 \end{bmatrix}} \cdot \overset{P}{\underset{\substack{\text{point}\\\text{matrix}}}{\begin{bmatrix} 3 \\ 5 \end{bmatrix}}} = \underset{\text{image matrix}}{\begin{bmatrix} 2 \cdot 3 + 0 \cdot 5 \\ -7 \cdot 3 + 4 \cdot 5 \end{bmatrix}}$$

$$= \overset{P'}{\begin{bmatrix} 6 \\ -1 \end{bmatrix}}$$

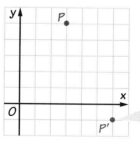

$P'(6, -1)$ is the image of the point $P(3, 5)$.

EXAMPLE 1

Use the transformation matrix
$\begin{bmatrix} -3 & 4 \\ -1 & 2 \end{bmatrix}$ **to transform** $\triangle ABC$.
Graph $\triangle ABC$ **and its image.**

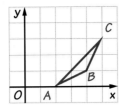

SOLUTION

Set up a matrix product with the transformation matrix on the left
and the polygon matrix on the right.

Use a matrix to
represent $\triangle ABC$.

To find the value in **row 1,
column 3**, use **row 1** of the
left matrix and **column 3** of
the right matrix.

$$\begin{array}{c} \\ \begin{bmatrix} -3 & 4 \\ -1 & 2 \end{bmatrix} \end{array} \cdot \begin{array}{c} A\ B\ C \\ \begin{bmatrix} 2 & 4 & 5 \\ 0 & 1 & 3 \end{bmatrix} \end{array} = \begin{bmatrix} -3(2) + 4(0) & -3(4) + 4(1) & -3(5) + 4(3) \\ -1(2) + 2(0) & -1(4) + 2(1) & -1(5) + 2(3) \end{bmatrix}$$

trans-   polygon
formation matrix
matrix

image
matrix

$$= \begin{array}{c} A'\ \ B'\ \ C' \\ \begin{bmatrix} -6 & -8 & -3 \\ -2 & -2 & 1 \end{bmatrix} \end{array}$$

$C'(-3, 1)$ is the
image of $C(5, 3)$.

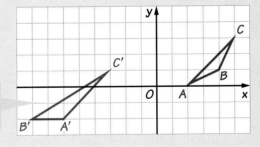

$\triangle A'B'C'$ is the
image of $\triangle ABC$.

**2 elements
in each row**

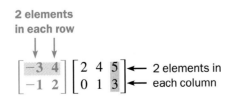

$\begin{bmatrix} -3 & 4 \\ -1 & 2 \end{bmatrix} \begin{bmatrix} 2 & 4 & 5 \\ 0 & 1 & 3 \end{bmatrix}$ ← 2 elements in
← each column

Notice that in the matrix product shown above, the number of elements in
one **row** of the left matrix is equal to the number of elements in one **column**
of the right matrix. This is true for all matrix multiplication.

THINK AND COMMUNICATE

Use this matrix equation.

$$\underbrace{\begin{bmatrix} 1 & 3 \\ 6 & -4 \end{bmatrix}}_{A} \cdot \underbrace{\begin{bmatrix} -4 & 2 & 5 & 0 & -3 \\ 6 & 4 & -1 & -2 & 0 \end{bmatrix}}_{B} = \underbrace{\begin{bmatrix} 14 & 14 & 2 & ? & -3 \\ -48 & -4 & 34 & 8 & -18 \end{bmatrix}}_{C}$$

1. Tell which matrix represents the:

   a. original polygon   **b.** image polygon   **c.** transformation

2. How many vertices does the original polygon have? Explain how you know.

3. Give the dimensions of matrices A, B, and C.

4. Explain which elements from matrix A and matrix B were used to find the value 34 in matrix C.

5. What value should go in the blank in row 1, column 4 of matrix C? Describe how to find this value.

EXAMPLE 2

The matrix $\begin{bmatrix} 0 & 4 & 4 & 0 \\ 0 & 0 & 6 & 6 \end{bmatrix}$ represents rectangle *ORST*. Use the

transformation matrix $\begin{bmatrix} 2 & 3 \\ 0 & 1 \end{bmatrix}$ to transform *ORST*. Graph *ORST*

and its image.

SOLUTION

Method 1

Use paper and pencil.

$$\begin{matrix} & O\ R\ S\ T \\ \begin{bmatrix} 2 & 3 \\ 0 & 1 \end{bmatrix} & \begin{bmatrix} 0 & 4 & 4 & 0 \\ 0 & 0 & 6 & 6 \end{bmatrix} \end{matrix}$$

Put the **transformation matrix** on the left and the polygon matrix on the right.

$$= \begin{bmatrix} 2(0) + 3(0) & 2(4) + 3(0) & 2(4) + 3(6) & 2(0) + 3(6) \\ 0(0) + 1(0) & 0(4) + 1(0) & 0(4) + 1(6) & 0(0) + 1(6) \end{bmatrix}$$

$$= \begin{matrix} O\ R'\ S'\ T' \\ \begin{bmatrix} 0 & 8 & 26 & 18 \\ 0 & 0 & 6 & 6 \end{bmatrix} \end{matrix}$$

Choose an appropriate scale for the graph.

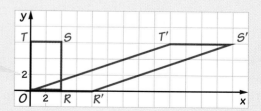

SOLUTION

Method 2

Use a graphing calculator or software with matrix capabilities.

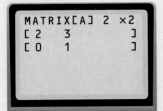

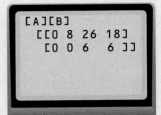

Step 1 Enter the transformation matrix as matrix A.

Step 2 Enter the polygon matrix as matrix B.

Step 3 Multiply A and B to find the image matrix.

The image matrix is $\begin{array}{cccc} O & R' & S' & T' \\ \begin{bmatrix} 0 & 8 & 26 & 18 \\ 0 & 0 & 6 & 6 \end{bmatrix} \end{array}$.

Graph $ORST$ and $OR'S'T'$ as shown in Method 1.

THINK AND COMMUNICATE

6. Can either the transformation in Example 1 or the transformation in Example 2 be described as a dilation, translation, reflection, or rotation? Explain why or why not.

7. In Example 2, which point in the polygon is its own image?

For more information about matrix calculations, see the *Technology Handbook*, pp. 723–725.

☑ CHECKING KEY CONCEPTS

1. Complete this matrix product.

$$\begin{bmatrix} 6 & -1 \\ 4 & 7 \end{bmatrix} \begin{bmatrix} 3 & -4 & 10 \\ 2 & 0 & -5 \end{bmatrix} = \begin{bmatrix} 16 & \underline{?} & 65 \\ \underline{?} & -16 & \underline{?} \end{bmatrix}$$

2. a. Write the polygon matrix for quadrilateral $ABCD$.

b. Use the transformation matrix $\begin{bmatrix} 3 & 1 \\ 0 & -1 \end{bmatrix}$ to transform $ABCD$.

c. Graph $ABCD$ and $A'B'C'D'$ on the same pair of axes.

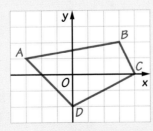

3. In order to solve Example 2, Cameron set up this matrix product. Explain what is wrong with his approach.

$$\begin{array}{cccc} O & R & S & T \\ \begin{bmatrix} 0 & 4 & 4 & 0 \\ 0 & 0 & 6 & 6 \end{bmatrix} \end{array} \begin{bmatrix} 2 & 3 \\ 0 & 1 \end{bmatrix}$$

Exercises and Applications

Extra Practice
exercises on
page 686

For Exercises 1–4:

a. Complete each matrix product.

b. Graph the original figure and its image.

1. $$\begin{bmatrix} 5 & 2 \\ 1 & 0 \end{bmatrix} \begin{matrix} Q \\ \begin{bmatrix} 1 \\ 3 \end{bmatrix} \end{matrix}$$

2. $$\begin{bmatrix} 1 & -4 \\ 2 & -3 \end{bmatrix} \begin{matrix} A\ B\ C \\ \begin{bmatrix} 6 & 4 & 5 \\ 1 & 3 & 0 \end{bmatrix} \end{matrix}$$

3. $$\begin{bmatrix} -1 & 0 \\ 1 & -2 \end{bmatrix} \begin{matrix} D\ \ E\ \ F\ G\ \ H \\ \begin{bmatrix} -1 & -3 & 1 & 5 & 4 \\ -4 & -1 & 2 & 1 & -3 \end{bmatrix} \end{matrix}$$

4. $$\begin{bmatrix} 3 & 0.5 \\ 2.4 & 1 \end{bmatrix} \begin{matrix} K\ \ L \\ \begin{bmatrix} -1 & 3 \\ 2 & 8 \end{bmatrix} \end{matrix}$$

For each transformation matrix and polygon given, use the matrix to transform the polygon. Then graph the polygon and its image.

5. $$\begin{bmatrix} 2 & -1 \\ 1 & -2 \end{bmatrix}$$

6. $$\begin{bmatrix} -3 & 1 \\ 0 & -4 \end{bmatrix}$$

7. $$\begin{bmatrix} 1 & 0 \\ 2 & -3.5 \end{bmatrix}$$

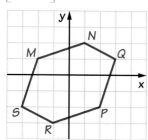
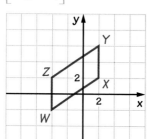
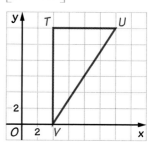

8. **a. Open-ended Problem** Write a matrix for any polygon, and use the matrix $\begin{bmatrix} 1 & 0 \\ 0 & 1 \end{bmatrix}$ to transform your polygon.

   b. Why do you think $\begin{bmatrix} 1 & 0 \\ 0 & 1 \end{bmatrix}$ is called the *identity matrix*?

9. **Cooperative Learning** Work in a group of three people. Each group member should choose a different one of these transformation matrices:

   $$\begin{bmatrix} 2 & 0 \\ 0 & 2 \end{bmatrix} \qquad \begin{bmatrix} 3 & 0 \\ 0 & 3 \end{bmatrix} \qquad \begin{bmatrix} 4 & 0 \\ 0 & 4 \end{bmatrix}$$

   a. The matrix $\begin{bmatrix} 2 & 4 & -4 \\ 3 & -3 & 0 \end{bmatrix}$ represents $\triangle ABC$. Use your transformation matrix to transform $\triangle ABC$. Graph $\triangle ABC$ and its image, $\triangle A'B'C'$.

   b. Writing As a group, compare your graphs from part (a). What is the geometric effect of the transformation $\begin{bmatrix} n & 0 \\ 0 & n \end{bmatrix}$? Explain.

10. **ALGEBRA** Find the values of m and n.

    $$\begin{bmatrix} 3 & 0 \\ 0 & 5 \end{bmatrix} \begin{bmatrix} m \\ n \end{bmatrix} = \begin{bmatrix} 12 \\ -35 \end{bmatrix}$$

11. **Challenge** Find the values of p and q.

    $$\begin{bmatrix} 1 & -4 \\ 2 & 7 \end{bmatrix} \begin{bmatrix} p \\ q \end{bmatrix} = \begin{bmatrix} 10 \\ 5 \end{bmatrix}$$

Morphing is the gradual transformation of one picture into another. In order to make a morph, a special effects artist selects pairs of *key points*, one in the starting picture and one in the final picture, which correspond to each other.

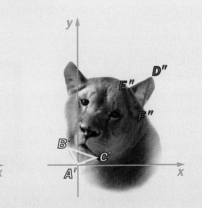

 Technology **In the morph above, the transformation matrix**
$\begin{bmatrix} 1.435 & 0.473 \\ -0.647 & 0.888 \end{bmatrix}$ **is used to transform the key points A(0, 0),**

B(−39, 38), and C(41, 49) into their images A′, B′, and C′.

12. Use matrix multiplication to find the coordinates of A', B', and C'. Round coordinate values to the nearest integer.

The key points D(94, 456), E(47, 390), and F(77, 345) are transformed into corresponding key points D″, E″, and F″ by a combination of matrix multiplication and matrix addition.

13. a. Multiply the polygon matrix for $\triangle DEF$ by the transformation

matrix $\begin{bmatrix} 1.55 & 0.078 \\ -0.844 & 1.237 \end{bmatrix}$ to find the matrix that represents $\triangle D'E'F'$.

 b. Add the matrix $\begin{bmatrix} 109 & 109 & 109 \\ -157 & -157 & -157 \end{bmatrix}$ to the matrix that represents

$\triangle D'E'F'$ to find a matrix for the final image, $\triangle D''E''F''$.

 c. Describe the transformation that is caused by the matrix addition you performed in part (b).

14. Open-ended Problem Sketch any quadrilateral *KLMN* on a coordinate grid.

  a. Write a polygon matrix for *KLMN*.

  b. Using the transformation matrix $\begin{bmatrix} 3 & -2 \\ -14 & 10 \end{bmatrix}$, transform *KLMN* to find a matrix that represents *K′L′M′N′*.

  c. Multiply the matrix that represents *K′L′M′N′* by the transformation matrix $\begin{bmatrix} 5 & 1 \\ 7 & 1.5 \end{bmatrix}$ to find a matrix for the image, *K″L″M″N″*.

  d. Writing Compare the coordinates of *K″L″M″N″* to those of *KLMN*. What can you conclude about the effects of the two transformation matrices in parts (b) and (c)?

15. 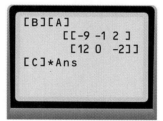 **Technology** Enter the polygon matrix *A* and the transformation matrices *B* and *C*, shown below, into your graphing calculator.

$$A = \begin{matrix} & Q & R & S \\ & \begin{bmatrix} 0 & -2 & 1 \\ 3 & 1 & -1 \end{bmatrix} \end{matrix} \qquad B = \begin{bmatrix} -1 & -3 \\ 2 & 4 \end{bmatrix} \qquad C = \begin{bmatrix} 0 & 2 \\ 1 & -4 \end{bmatrix}$$

  a. Transform △*QRS* using matrix *B*. Then transform the result using matrix *C*. What is the image of △*QRS* after these two transformations?

  b. Transform △*QRS* using matrix *C* first. Then transform the result using matrix *B*. What is the image of △*QRS* after these two transformations?

  c. Compare your answers to parts (a) and (b). Does the order in which the transformations are applied affect the final image? Explain your answer.

16. In the partially completed matrix product shown below, a triangle is being transformed in a three-dimensional coordinate system.

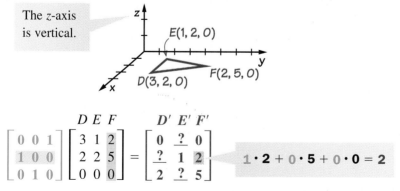

The *z*-axis is vertical.

E(1, 2, 0)
D(3, 2, 0)
F(2, 5, 0)

$$\begin{bmatrix} 0 & 0 & 1 \\ 1 & 0 & 0 \\ 0 & 1 & 0 \end{bmatrix} \begin{matrix} \begin{matrix} D & E & F \end{matrix} \\ \begin{bmatrix} 3 & 1 & 2 \\ 2 & 2 & 5 \\ 0 & 0 & 0 \end{bmatrix} \end{matrix} = \begin{matrix} \begin{matrix} D' & E' & F' \end{matrix} \\ \begin{bmatrix} 0 & ? & 0 \\ ? & 1 & 2 \\ 2 & ? & 5 \end{bmatrix} \end{matrix}$$

$1 \cdot 2 + 0 \cdot 5 + 0 \cdot 0 = 2$

  a. Write the three-dimensional coordinates of point *F* and its image, *F′*.

  b. Complete the matrix product.

  c. Graph △*DEF* and its image, △*D′E′F′*.

17. SAT/ACT Preview Use matrices A and B. Which one of these products and sums can be done?

$$A = \begin{bmatrix} 2 & -1 & 3 \\ 0 & 4 & 9 \end{bmatrix} \quad B = \begin{bmatrix} 5 & 2 \\ -1 & 4 \end{bmatrix}$$

A. AB **B.** BA **C.** $3 + A$ **D.** $A + B$

18. Open-ended Problem Complete this matrix with any two numbers to create a transformation matrix: $\begin{bmatrix} 0 & 0 \\ ? & ? \end{bmatrix}$

a. Use matrix multiplication and your matrix to transform point $R(3, 2)$. Graph R and its image.

b. Use your matrix to transform $\triangle JKL$, which is represented by the matrix $\begin{bmatrix} 4 & 2 & 2 \\ 0 & 3 & -3 \end{bmatrix}$. Graph $\triangle JKL$ and its image.

c. Describe the geometric effect of your transformation matrix.

ONGOING ASSESSMENT

19. Cooperative Learning Work in a group of four people. Each group member should choose a different one of these transformation matrices:

$$\begin{bmatrix} 1 & 1 \\ 0 & 1 \end{bmatrix} \quad \begin{bmatrix} 1 & 2 \\ 0 & 1 \end{bmatrix} \quad \begin{bmatrix} 1 & 3 \\ 0 & 1 \end{bmatrix} \quad \begin{bmatrix} 1 & 4 \\ 0 & 1 \end{bmatrix}$$

a. The matrix $\begin{bmatrix} 0 & 0 & 3 & 3 \\ 0 & 3 & 3 & 0 \end{bmatrix}$ represents square $ABCD$. Use your transformation matrix to find the image of $ABCD$.

b. On one pair of axes, graph $ABCD$ and the image that you found, $A'B'C'D'$.

c. Find the area of $ABCD$ and the area of $A'B'C'D'$.

d. Compare your answers to parts (b) and (c) with others in your group. Describe how the four transformation matrices affect square $ABCD$ and its area.

SPIRAL REVIEW

Add the matrices. *(Section 12.2)*

20. $\begin{bmatrix} 1 & 1 & 1 \\ 3 & 3 & 3 \end{bmatrix} + \begin{bmatrix} 5 & 3 & -3 \\ 0 & -1 & 1 \end{bmatrix}$

21. $\begin{bmatrix} 1 & 3 & 4 \\ 0 & 2 & 3 \end{bmatrix} + \begin{bmatrix} 1 & 3 & 4 \\ 0 & 2 & 3 \end{bmatrix}$

22. List four different ways to prove that a quadrilateral is a parallelogram. *(Section 7.2)*

23. Rectangle $ABCD$ has vertices $A(-2, 1)$, $B(4, 1)$, $C(4, 3)$, and $D(-2, 3)$. Reflect the rectangle over each line and find the coordinates of the vertices of the image. *(Section 8.2)*

a. the x-axis **b.** the y-axis **c.** the line $y = x$

12.4 Reflections with Matrices

Learn how to...

- **represent reflections using matrices**

So you can...

- **reflect points and polygons using matrices**

- **analyze images such as those produced with computers**

Computer software programs for manipulating photographs allow you to reflect parts of a photograph across a line. In the images below, each half of the photograph on the left is flipped across a vertical line of reflection to create a new portrait. Because human faces are not perfectly symmetrical, the portraits look quite different.

Left   *Right*   *Left*   *Left*   *Right*   *Right*

EXPLORATION

COOPERATIVE LEARNING

Reflections Using Matrix Multiplication

Work in a group of four students.
You will need:
- graph paper

1 Each member of the group should draw a different △*ABC* and label the coordinates of its vertices. Each of you should write a polygon matrix that represents your triangle.

2 Multiply your triangle matrix by each transformation matrix below, putting the transformation matrix on the left. Graph the image of △*ABC* after each multiplication.

$$\begin{bmatrix} 1 & 0 \\ 0 & -1 \end{bmatrix} \quad \begin{bmatrix} -1 & 0 \\ 0 & 1 \end{bmatrix} \quad \begin{bmatrix} 0 & 1 \\ 1 & 0 \end{bmatrix} \quad \begin{bmatrix} 0 & -1 \\ -1 & 0 \end{bmatrix}$$

3 Compare your results. Describe the effect each transformation matrix has on a triangle.

The table below summarizes the reflection matrices and the effect each one has on a polygon or a point.

Summary of Matrices for Reflections

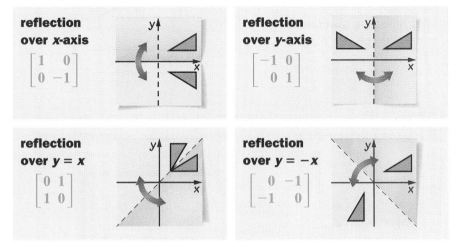

reflection over x-axis
$$\begin{bmatrix} 1 & 0 \\ 0 & -1 \end{bmatrix}$$

reflection over y-axis
$$\begin{bmatrix} -1 & 0 \\ 0 & 1 \end{bmatrix}$$

reflection over y = x
$$\begin{bmatrix} 0 & 1 \\ 1 & 0 \end{bmatrix}$$

reflection over y = -x
$$\begin{bmatrix} 0 & -1 \\ -1 & 0 \end{bmatrix}$$

THINK AND COMMUNICATE

1. How is the matrix for reflection over the x-axis like the matrix for reflection over the y-axis? How is it different?

2. How is the matrix for reflection over the line $y = x$ like the matrix for reflection over the line $y = -x$? How is it different?

EXAMPLE 1

Use matrix multiplication to find the image of △**ABC** after reflection over the y-axis. Graph the image.

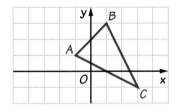

SOLUTION

Refer to the table of matrices above to find the matrix that represents reflection over the y-axis.

$$
\underset{\substack{\text{reflection} \\ \text{over y-axis}}}{\begin{bmatrix} -1 & 0 \\ 0 & 1 \end{bmatrix}}
\underset{\substack{\text{polygon} \\ \text{matrix}}}{\overset{A \quad B \quad C}{\begin{bmatrix} -1 & 1 & 3 \\ 1 & 3 & -1 \end{bmatrix}}}
=
\begin{bmatrix} \overset{A'}{-1(-1) + 0(1)} & \overset{B'}{-1(1) + 0(3)} & \overset{C'}{-1(3) + 0(-1)} \\ 0(-1) + 1(1) & 0(1) + 1(3) & 0(3) + 1(-1) \end{bmatrix}
$$

$$
= \overset{A' \quad B' \quad C'}{\begin{bmatrix} 1 & -1 & -3 \\ 1 & 3 & -1 \end{bmatrix}}
$$
image matrix

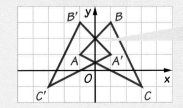

Graph the image, △**A'B'C'**.

EXAMPLE 2

Use matrix multiplication to find the image of quadrilateral *DEFG* after reflection over the line $y = -x$. Graph the image.

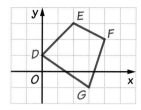

SOLUTION

If you enter the reflection matrix as Matrix *A* and the polygon matrix as Matrix *B* in a graphing calculator, it can do the multiplication for you.

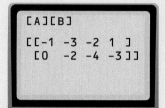

$$\underset{\substack{\text{reflection}\\\text{over } y = -x}}{\begin{bmatrix} 0 & -1 \\ -1 & 0 \end{bmatrix}} \underset{\substack{\text{polygon}\\\text{matrix}}}{\begin{bmatrix} 0 & 2 & 4 & 3 \\ 1 & 3 & 2 & -1 \end{bmatrix}} =$$

[A][B]
[[-1 -3 -2 1]
 [0 -2 -4 -3]]

Graph the image, $D'E'F'G'$.

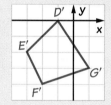

✓ CHECKING KEY CONCEPTS

1. $\begin{bmatrix} 3 \\ 5 \end{bmatrix}$ is the result after the point matrix $\begin{bmatrix} 3 \\ -5 \end{bmatrix}$ is multiplied by a reflection matrix. What reflection matrix was used?

2. $\begin{bmatrix} 1 \\ 2 \end{bmatrix}$ is the result after the point matrix $\begin{bmatrix} 2 \\ 1 \end{bmatrix}$ is multiplied by a reflection matrix. What reflection matrix was used?

Use matrix multiplication to find the image of trapezoid *JKLM* after reflection over each line. Then graph the image. Use the table on page 635 for the transformation matrices.

3. the *x*-axis

4. the *y*-axis

5. the line $y = x$

6. the line $y = -x$

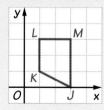

12.4 Exercises and Applications

Extra Practice exercises on page 686

Complete each matrix product to find the image of (7, −2) after a reflection, and tell which line is the line of reflection.

1. $\begin{bmatrix} 1 & 0 \\ 0 & -1 \end{bmatrix} \begin{bmatrix} 7 \\ -2 \end{bmatrix}$
2. $\begin{bmatrix} -1 & 0 \\ 0 & 1 \end{bmatrix} \begin{bmatrix} 7 \\ -2 \end{bmatrix}$
3. $\begin{bmatrix} 0 & 1 \\ 1 & 0 \end{bmatrix} \begin{bmatrix} 7 \\ -2 \end{bmatrix}$
4. $\begin{bmatrix} 0 & -1 \\ -1 & 0 \end{bmatrix} \begin{bmatrix} 7 \\ -2 \end{bmatrix}$

5. $\begin{bmatrix} -4 \\ 2 \end{bmatrix}$ is the result after $\begin{bmatrix} 4 \\ 2 \end{bmatrix}$ is multiplied by a reflection matrix.
What reflection matrix was used?

6. $\begin{bmatrix} -7 \\ 3 \end{bmatrix}$ is the result after $\begin{bmatrix} -3 \\ 7 \end{bmatrix}$ is multiplied by a reflection matrix.
What reflection matrix was used?

Use matrix multiplication to find the image of trapezoid *AEUO* after reflection over each line. Then graph the image. Use the table on page 635 for the transformation matrices.

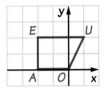

7. the *x*-axis
8. the *y*-axis
9. the line $y = x$
10. the line $y = -x$

Connection ▶ ART PHOTOGRAPHY

Marika Barnett takes photographs of buildings and then reflects them over horizontal and vertical axes using a computer. In her "Done with Mirrors" series, each composition presents the same image four ways, as if reflected in two perpendicular mirrors.

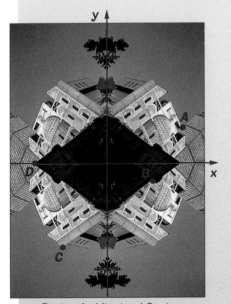

Boston Architectural Center

BY THE WAY

Hungarian-born Marika Barnett is a software engineer who enjoys creating kaleidoscopic collages using her own photographs.

Use matrix multiplication to find the image of each point after reflection over the *x*-axis and over the *y*-axis.

11. $A(24, 10)$
12. $B(15, -5)$
13. $C(-15, -27)$
14. $D(-23, 0)$

15. Open-ended Problem Choose one of the points in Exercises 11–14. Reflect this point over the *x*-axis and then reflect the result over the *y*-axis. Is the image the same as if you reflect the point over the *y*-axis first and then reflect the result over the *x*-axis? Explain.

**Write a polygon matrix for each polygon. Then use matrix mutiplication to
find the image after reflection over the given line. Use the table on page 635
for the transformation matrices.**

16. the *x*-axis

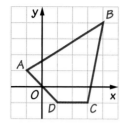

17. the *y*-axis

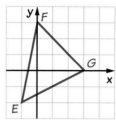

18. the line *y* = *x*

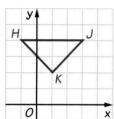

19. the line *y* = −*x*

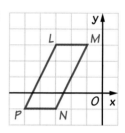

For Exercises 20 and 21, use trapezoid *PQRS*, at the right.

20. a. Write a polygon matrix for *PQRS*.

    b. Use matrix multiplication to reflect *PQRS* over the line *y* = *x*.
Label the image *P′Q′R′S′*.

    c. Use matrix multiplication to reflect *P′Q′R′S′* over the *y*-axis.
Label the image *P″Q″R″S″*.

    d. Describe how *P″Q″R″S″* is related to *PQRS*.

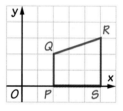

21. a. Repeat Exercise 20 but reflect *PQRS* over the *y*-axis first, then reflect
the result over the line *y* = *x*. How is the image related to *PQRS*?

    b. Writing Compare your results in part (a) with those in Exercise 20.
Does the order in which the reflections are applied affect the image?

**Challenge Each matrix product reflects rectangle *ABCD* in a three-
dimensional coordinate system. Complete each product, graph the image,
and describe the reflection.**

22.
$$\begin{bmatrix} 1 & 0 & 0 \\ 0 & 1 & 0 \\ 0 & 0 & -1 \end{bmatrix} \overset{A\ B\ C\ D}{\begin{bmatrix} 3 & 1 & 1 & 3 \\ 0 & 0 & 3 & 3 \\ 2 & 2 & 0 & 0 \end{bmatrix}} = \overset{A'\ B'\ C'\ D'}{\begin{bmatrix} 3 & 1 & 1 & \underline{?} \\ 0 & 0 & \underline{?} & \underline{?} \\ -2 & \underline{?} & \underline{?} & \underline{?} \end{bmatrix}}$$

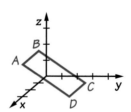

23.
$$\begin{bmatrix} 1 & 0 & 0 \\ 0 & -1 & 0 \\ 0 & 0 & 1 \end{bmatrix} \overset{A\ B\ C\ D}{\begin{bmatrix} 3 & 1 & 1 & 3 \\ 0 & 0 & 3 & 3 \\ 2 & 2 & 0 & 0 \end{bmatrix}} = \overset{A'\ B'\ C'\ D'}{\begin{bmatrix} 3 & \underline{?} & \underline{?} & 3 \\ 0 & \underline{?} & \underline{?} & -3 \\ 2 & \underline{?} & \underline{?} & 0 \end{bmatrix}}$$

24.
$$\begin{bmatrix} -1 & 0 & 0 \\ 0 & 1 & 0 \\ 0 & 0 & 1 \end{bmatrix} \overset{A\ B\ C\ D}{\begin{bmatrix} 3 & 1 & 1 & 3 \\ 0 & 0 & 3 & 3 \\ 2 & 2 & 0 & 0 \end{bmatrix}} = \overset{A'\ B'\ C'\ D'}{\begin{bmatrix} \underline{?} & \underline{?} & \underline{?} & -3 \\ \underline{?} & \underline{?} & 3 & 3 \\ \underline{?} & 2 & 0 & 0 \end{bmatrix}}$$

Jack is a computer-generated three-dimensional human figure based on measurements of actual people. Jack is used as a model to predict how humans might behave in a given environment, such as in a space shuttle, a submarine, or a tractor.

Unlike real people, Jack has perfect symmetry. One half of his body is reflected to create the other half. For each point on Jack's face, use matrix multiplication to find the corresponding point on the other side.

25. $A(237, 572)$

26. $B(167, 224)$

27. $C(-365, 374)$

28. $D(-433, 1004)$

The top half of Jack's eyeball can be reflected to create the lower half.

29. Write a matrix for hexagon *TUVWXY*.

30. Use matrix multiplication to find a hexagon for the eyeball's lower half.

31. Writing Do you think Jack's eyeball has any other symmetry? Explain.

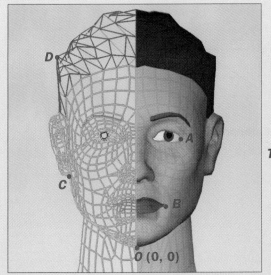

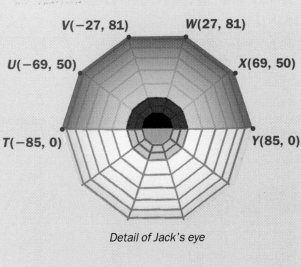

$V(-27, 81)$ $W(27, 81)$

$U(-69, 50)$ $X(69, 50)$

$T(-85, 0)$ $Y(85, 0)$

Detail of Jack's eye

Describe each reflection using a matrix product. Include the transformation matrix, polygon matrix, and image matrix. Use the table on page 635 for the transformation matrices.

32.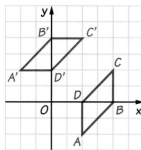

33.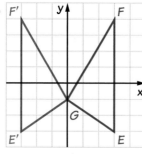

34.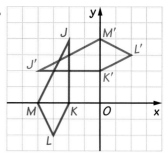

ONGOING ASSESSMENT

35. Writing Graph any segment. Use each of the matrices in the table on page 635 to reflect your segment. Compare the slope of each image segment with the slope of the original segment. Summarize your results. Write a conjecture about how different reflections affect slope.

SPIRAL REVIEW

For each transformation matrix and polygon given, use the matrix to transform the polygon. Then graph the polygon and its image. *(Section 12.3)*

36. $\begin{bmatrix} 2 & 1 \\ 1 & 2 \end{bmatrix}$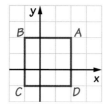

37. $\begin{bmatrix} -2 & 0 \\ 0 & -2 \end{bmatrix}$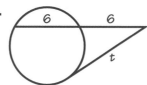

ALGEBRA Find the value of each variable. *(Section 11.4)*

38.

39.

Sketch the image of each polygon after a 90° rotation around the origin.
(Section 8.4)

40.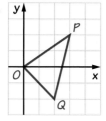

41.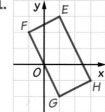

12.5 | Rotations with Matrices

If a matrix has the same number of rows and columns, it has two *diagonals*: one from upper left to lower right, and one from lower left to upper right.

There are two elements on each diagonal of a 2 × 2 matrix.

There are three elements on each diagonal of a 3 × 3 matrix.

EXPLORATION

COOPERATIVE LEARNING

Matrices with Zeros on a Diagonal

Work in a group of four students. You will need:

- graph paper

1 As a group, make a list of every possible 2 × 2 matrix for which both elements on one diagonal are 0 and each element on the other diagonal is either 1 or −1.

$$\begin{bmatrix} ? & 0 \\ 0 & ? \end{bmatrix} \quad \text{or} \quad \begin{bmatrix} 0 & ? \\ ? & 0 \end{bmatrix}$$

Each of the missing elements is either 1 or −1.

2 Divide the list of matrices among the members of your group.

Multiply each matrix that you are assigned by $\begin{bmatrix} 3 & 4 & 7 \\ 1 & 3 & 1 \end{bmatrix}$, which represents △ABC.

3 For each matrix product in Step 2, graph △ABC and its image. Describe each transformation.

4 Make a chart that shows your group's results.

Each of the transformation matrices that you used in the Exploration is either a reflection matrix or a rotation matrix. The reflection matrices were shown on page 635; the rotation matrices are shown in the table below.

WATCH OUT!

In this section, all rotations are around the origin and are measured in degrees counterclockwise.

Summary of Matrices for Rotations

90° rotation

$$\begin{bmatrix} 0 & -1 \\ 1 & 0 \end{bmatrix}$$

180° rotation

$$\begin{bmatrix} -1 & 0 \\ 0 & -1 \end{bmatrix}$$

270° rotation

$$\begin{bmatrix} 0 & 1 \\ -1 & 0 \end{bmatrix}$$

360° rotation

$$\begin{bmatrix} 1 & 0 \\ 0 & 1 \end{bmatrix}$$

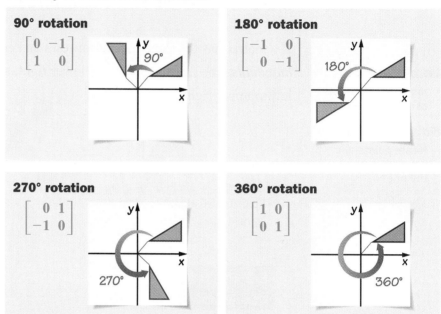

EXAMPLE 1

The matrix $\begin{bmatrix} -2 & -5 & -5 & -2 \\ 1 & 1 & 3 & 3 \end{bmatrix}$ represents rectangle *EFGH*. Use matrix multiplication to rotate it 270°. Graph *EFGH* and its image.

SOLUTION

Find the matrix for a 270° rotation in the table above, and write it to the left of the polygon matrix. Use paper and pencil or a graphing calculator to find the matrix product.

$$\begin{bmatrix} 0 & 1 \\ -1 & 0 \end{bmatrix} \overset{\displaystyle E \quad F \quad G \quad H}{\begin{bmatrix} -2 & -5 & -5 & -2 \\ 1 & 1 & 3 & 3 \end{bmatrix}} = \overset{\displaystyle E' \ F' \ G' \ H'}{\begin{bmatrix} 1 & 1 & 3 & 3 \\ 2 & 5 & 5 & 2 \end{bmatrix}}$$

matrix for 270° rotation

E'F'G'H' is the image of *EFGH* after a 270° rotation.

Combining Rotations and Reflections

In the Exploration, you found some transformation matrices for reflections and some for rotations. As shown below, you can apply one type of transformation to a polygon, then apply a different type of transformation to the result.

EXAMPLE 2

Use the transformation matrix $\begin{bmatrix} -1 & 0 \\ 0 & -1 \end{bmatrix}$ to transform $\triangle QRS$. Then transform the result using $\begin{bmatrix} 1 & 0 \\ 0 & -1 \end{bmatrix}$. What is the combined effect of these two transformations?

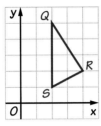

SOLUTION

Use the first matrix to transform $\triangle QRS$.

$$\begin{bmatrix} -1 & 0 \\ 0 & -1 \end{bmatrix} \overset{Q\ R\ S}{\begin{bmatrix} 2 & 4 & 2 \\ 5 & 2 & 1 \end{bmatrix}} = \overset{Q'\ R'\ S'}{\begin{bmatrix} -2 & -4 & -2 \\ -5 & -2 & -1 \end{bmatrix}}$$

Then use the second matrix to transform $\triangle Q'R'S'$.

$$\begin{bmatrix} 1 & 0 \\ 0 & -1 \end{bmatrix} \overset{Q'\ R'\ S'}{\begin{bmatrix} -2 & -4 & -2 \\ -5 & -2 & -1 \end{bmatrix}} = \overset{Q''\ R''\ S''}{\begin{bmatrix} -2 & -4 & -2 \\ 5 & 2 & 1 \end{bmatrix}}$$

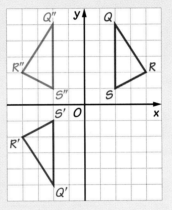

The combined effect of the two transformations is equivalent to a reflection over the y-axis.

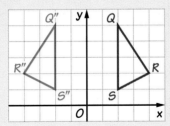

THINK AND COMMUNICATE

Use the solution to Example 2.

1. What transformation is represented by the matrix $\begin{bmatrix} -1 & 0 \\ 0 & -1 \end{bmatrix}$?

2. What transformation is represented by the matrix $\begin{bmatrix} 1 & 0 \\ 0 & -1 \end{bmatrix}$?

1. The matrix $\begin{bmatrix} 5 & 5 & 3 & 3 \\ 1 & -2 & -2 & 3 \end{bmatrix}$ represents quadrilateral *ABDE*.

   a. Use the matrix $\begin{bmatrix} 0 & -1 \\ 1 & 0 \end{bmatrix}$ to rotate *ABDE*.

   b. Graph *ABDE* and its image. Describe the rotation.

2. The matrix $\begin{bmatrix} -3 & -5 & -2 \\ 1 & 0 & -3 \end{bmatrix}$ represents $\triangle CGS$.

   a. Use the matrix $\begin{bmatrix} 0 & -1 \\ 1 & 0 \end{bmatrix}$ to transform $\triangle CGS$. Then transform

   the result using the matrix $\begin{bmatrix} -1 & 0 \\ 0 & -1 \end{bmatrix}$.

   b. Graph $\triangle CGS$, $\triangle C'G'S'$, and $\triangle C''G''S''$. What is the combined effect of the two transformations?

12.5 Exercises and Applications

Extra Practice
exercises on page 686

Use matrix multiplication to find the image of each polygon after the given rotation. Then graph the polygon and its image. Use the table on page 642 for the transformation matrices.

1. 90° rotation

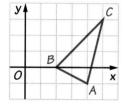

2. 270° rotation

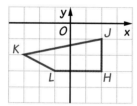

3. 360° rotation

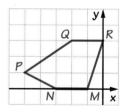

4. 270° rotation

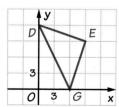

5. 180° rotation

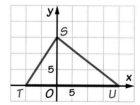

6. 90° rotation

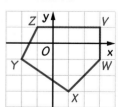

7. Open-ended Problem Choose any two rotation matrices.

   a. Use each of your rotation matrices to transform $F(2, 7)$ and $H(3, 6)$.

   b. For each of your rotation matrices, complete the following:
   $(a, b) \rightarrow (\underline{?}, \underline{?})$.

INTERVIEW # Adriana Ocampo

Look back at the
article on pages
610–612.

*While traveling through space, the
Galileo spacecraft spins around a
central axis. In the spacecraft's frame
of reference, it appears as if the
spacecraft is stationary at the origin
and the universe is rotating around it.*

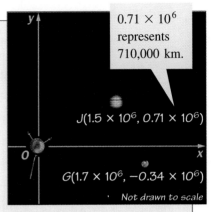

0.71 × 10⁶ represents 710,000 km.

$J(1.5 \times 10^6, 0.71 \times 10^6)$

$G(1.7 \times 10^6, -0.34 \times 10^6)$

Not drawn to scale

8. At time 12:00:00, Ganymede
(one of Jupiter's moons) and Jupiter
are located at points G and J, respectively.

  a. Write point matrices for G and J. Multiply the transformation
  matrix $\begin{bmatrix} 0 & -1 \\ 1 & 0 \end{bmatrix}$ by these point matrices to find G' and J',
  Ganymede's and Jupiter's positions five seconds later, at 12:00:05.

  b. Multiply the transformation matrix $\begin{bmatrix} 0 & -1 \\ 1 & 0 \end{bmatrix}$ by your results
  from part (a) to find G'' and J'', the locations at 12:00:10.

  c. On one pair of axes, graph Ganymede's and Jupiter's locations at
  12:00:00, 12:00:05, and 12:00:10.

  d. Based on your answers to parts (a)–(c), about how much time does
  it take the spacecraft Galileo to make one full (360°) turn?

9. Open-ended Problem Why might a Galileo mission scientist
such as Adriana Ocampo need to know the position of Ganymede
in relation to the spacecraft at a particular time?

10. The matrix $\begin{bmatrix} 5 & 5 & 3 \\ 0 & -4 & -4 \end{bmatrix}$ represents $\triangle PQR$.

  a. Use the matrix $\begin{bmatrix} -1 & 0 \\ 0 & 1 \end{bmatrix}$ to transform $\triangle PQR$.

  b. Use the matrix $\begin{bmatrix} -1 & 0 \\ 0 & -1 \end{bmatrix}$ to transform the result from part (a).

  c. Graph $\triangle PQR$, $\triangle P'Q'R'$, and $\triangle P''Q''R''$. What is the combined
  effect of the two transformations in parts (a) and (b)?

11. Challenge The product of two transformation matrices is also a
transformation matrix.

  a. Complete the matrix product $\begin{bmatrix} -1 & 0 \\ 0 & -1 \end{bmatrix} \begin{bmatrix} 0 & -1 \\ 1 & 0 \end{bmatrix} = \begin{bmatrix} ? & ? \\ ? & ? \end{bmatrix}$.

  180°   90°

  Use the table on page 642 to find the rotation caused by the product.

  b. Complete the product $\begin{bmatrix} 0 & 1 \\ -1 & 0 \end{bmatrix} \begin{bmatrix} ? & ? \\ ? & ? \end{bmatrix} = \begin{bmatrix} -1 & 0 \\ 0 & -1 \end{bmatrix}$.

BY THE WAY

Ganymede was
discovered by Italian
scientist Galileo Galilei
(1564–1642), after
whom the spacecraft
Galileo is named.

In *virtual reality simulators*, such as some computer games, players can feel as if they are traveling through a three-dimensional world and interacting with virtual people. When a player turns, the world appears to rotate, as shown. The new coordinates of objects are found using rotation matrices.

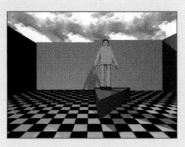
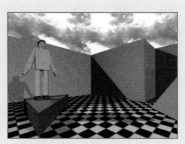
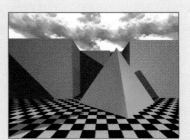

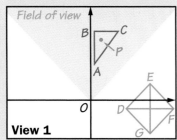

View 1

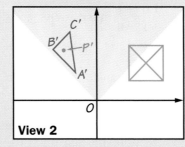

View 2

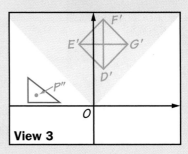

View 3

In each view, the player is at (0, 0). In View 1, the coordinates of platform *ABC*, virtual person *P*, and base *DEFG* of the pyramid are given by these matrices:

$$\begin{array}{ccc} A & B & C \end{array}$$
$$\begin{bmatrix} 2 & 2 & 19 \\ 24 & 46 & 46 \end{bmatrix}$$

$$\begin{array}{c} P \end{array}$$
$$\begin{bmatrix} 7 \\ 40 \end{bmatrix}$$

$$\begin{array}{cccc} D & E & F & G \end{array}$$
$$\begin{bmatrix} 24 & 41 & 58 & 41 \\ -6 & 11 & -6 & -23 \end{bmatrix}$$

12. a. Multiply the matrix representing *DEFG* by the rotation matrix $\begin{bmatrix} 0 & -1 \\ 1 & 0 \end{bmatrix}$ to find a matrix for *D'E'F'G'*, the pyramid in View 3.

　b. Describe the rotation caused by the transformation matrix in part (a).

13. 📈 **Technology** Use a graphing calculator or computer software with matrix capabilities.

　a. Multiply the matrix that represents *ABC* by the rotation matrix
$\begin{bmatrix} 0.707 & -0.707 \\ 0.707 & 0.707 \end{bmatrix}$ to find a matrix for *A'B'C'*, the platform in View 2.

　b. Multiply the point matrix for *P* by the rotation matrix $\begin{bmatrix} 0.707 & -0.707 \\ 0.707 & 0.707 \end{bmatrix}$

　to find the coordinates of *P'*. Then multiply the result by the same rotation matrix to find the coordinates of *P''*, the person in View 3.

　c. **Writing** What is the combined effect of the two rotations in part (b)? Find the degree measure of each rotation and explain your reasoning.

14. Research Find out more about how virtual reality is used. Present your results in a report or visual display.

15. Open-ended Problem Write a matrix for any quadrilateral *TUVW*.

  a. Use one of the rotation matrices from page 642 to transform *TUVW*. Then transform the result using a different rotation matrix.

  b. Transform *TUVW* using the same two rotation matrices from part (a), but reverse the order in which you apply them.

  c. Do you think that the order in which the two rotations are applied has any impact on the final image? Explain your answer.

Find the value of each ratio. Round your answers to four decimal places.
(Sections 10.3 and 10.4)

  16. $\tan 35°$ **17.** $\sin 75°$ **18.** $\cos 30°$

Solve each proportion to find a geometric mean. *(Sections 9.1 and 10.1)*

  19. $\dfrac{7}{x} = \dfrac{x}{14}$ **20.** $\dfrac{1}{x} = \dfrac{x}{25}$ **21.** $\dfrac{10}{x} = \dfrac{x}{40}$

ASSESS YOUR PROGRESS

VOCABULARY

matrix multiplication (p. 626)

Find the values of the variables in each matrix product. *(Section 12.3)*

  1. $\begin{bmatrix} 4 & 1 \\ 3 & 2 \end{bmatrix} \begin{bmatrix} x \\ 3 \end{bmatrix} = \begin{bmatrix} 11 \\ y \end{bmatrix}$ **2.** $\begin{bmatrix} 1 & -1 \\ 3 & 5 \end{bmatrix} \begin{bmatrix} 2 \\ g \end{bmatrix} = \begin{bmatrix} 4 \\ h \end{bmatrix}$

Use matrix multiplication to find the image of each polygon after reflection over the given line. Then graph the image. Use the table on page 635 for the transformation matrices. *(Section 12.4)*

  3. the line $y = x$ **4.** the line $y = -x$

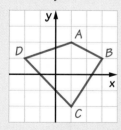   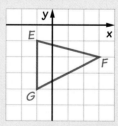

Use matrix multiplication to find the image of △*EFG* from Question 4 after the given rotation. Then graph △*EFG* and its image. Use the table on page 642 for the transformation matrices. *(Section 12.5)*

  5. 90° rotation **6.** 270° rotation

  7. Journal Write a letter to a student who will take this geometry course next year. Describe some of the projects or problems you found interesting. Include advice about how to do well in the course.

Creating Fractals

Like many fractals, this *Sierpinski Triangle* is *self-similar*. It is made up of smaller pieces which are geometrically similar to the whole fractal. Each of these pieces is made up of even smaller pieces that are also similar to the whole.

PROJECT GOAL **Work with another student to analyze fractals and generate them on your graphing calculator.**

Analyzing the Sierpinski Triangle

The Sierpinski Triangle contains three half-size copies of itself. Each copy is the image of the whole triangle after a transformation.

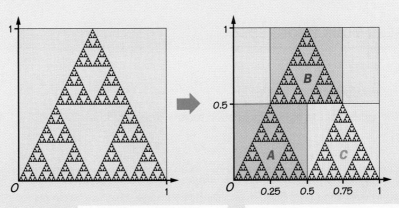

Transformation *A* is a dilation with center (0, 0) and scale factor 0.5.

Transformation *B* is a dilation followed by a translation to the right 0.25 units and up 0.5 units.

This *similarity diagram* shows how the three transformations that describe the Sierpinski Triangle affect the unit square.

Transformation *C* is a dilation followed by a translation to the right 0.5 units.

Finding Values for the Transformations

Each transformation used to create the Sierpinski Triangle can be represented by two matrix operations. For point $P(a, b)$:

$$\underset{P}{\begin{bmatrix} p & q \\ s & t \end{bmatrix} \begin{bmatrix} a \\ b \end{bmatrix}} + \begin{bmatrix} r \\ u \end{bmatrix} = \underset{P'}{\begin{bmatrix} ? \\ ? \end{bmatrix}}$$

For each of the three transformations, you can use algebra to find the values of p, q, r, s, t, and u.

To find the values for transformation B, use the similarity diagram to see how transformation B affects the points $(0, 0)$, $(0, 1)$, and $(1, 0)$.

The image of $(0, 0)$ is **(0.25, 0.5)**.

The image of $(0, 1)$ is **(0.25, 1)**.

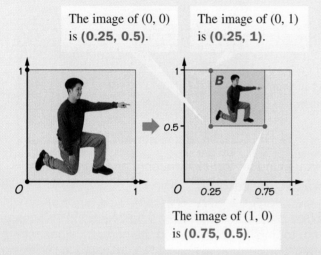

The image of $(1, 0)$ is **(0.75, 0.5)**.

Step 1 Use the image of $(0, 0)$ to solve for r and u.

$(0, 0) \rightarrow$ **(0.25, 0.5)**

$$\begin{bmatrix} p & q \\ s & t \end{bmatrix} \begin{bmatrix} 0 \\ 0 \end{bmatrix} + \begin{bmatrix} r \\ u \end{bmatrix} = \begin{bmatrix} 0.25 \\ 0.5 \end{bmatrix}$$

$$\begin{bmatrix} p \cdot 0 + q \cdot 0 \\ s \cdot 0 + t \cdot 0 \end{bmatrix} + \begin{bmatrix} r \\ u \end{bmatrix} = \begin{bmatrix} 0.25 \\ 0.5 \end{bmatrix}$$

$$\begin{bmatrix} 0 + r \\ 0 + u \end{bmatrix} = \begin{bmatrix} 0.25 \\ 0.5 \end{bmatrix}$$

So $r = 0.25$ and $u = 0.5$.

Step 2 Use the images of $(0, 1)$ and $(1, 0)$ to solve for p, q, s, and t. Substitute the values of r and u that you found in Step 1.

$(0, 1) \rightarrow$ **(0.25, 1)**

$$\begin{bmatrix} p & q \\ s & t \end{bmatrix} \begin{bmatrix} 0 \\ 1 \end{bmatrix} + \begin{bmatrix} r \\ u \end{bmatrix} = \begin{bmatrix} 0.25 \\ 1 \end{bmatrix}$$

$$\begin{bmatrix} p \cdot 0 + q \cdot 1 \\ s \cdot 0 + t \cdot 1 \end{bmatrix} + \begin{bmatrix} \mathbf{0.25} \\ \mathbf{0.5} \end{bmatrix} = \begin{bmatrix} 0.25 \\ 1 \end{bmatrix}$$

$$\begin{bmatrix} q + 0.25 \\ t + 0.5 \end{bmatrix} = \begin{bmatrix} 0.25 \\ 1 \end{bmatrix}$$

So $q = 0$ and $t = 0.5$.

$(1, 0) \rightarrow$ **(0.75, 0.5)**

$$\begin{bmatrix} p & q \\ s & t \end{bmatrix} \begin{bmatrix} 1 \\ 0 \end{bmatrix} + \begin{bmatrix} r \\ u \end{bmatrix} = \begin{bmatrix} 0.75 \\ 0.5 \end{bmatrix}$$

$$\begin{bmatrix} p \cdot 1 + q \cdot 0 \\ s \cdot 1 + t \cdot 0 \end{bmatrix} + \begin{bmatrix} \mathbf{0.25} \\ \mathbf{0.5} \end{bmatrix} = \begin{bmatrix} 0.75 \\ 0.5 \end{bmatrix}$$

$$\begin{bmatrix} p + 0.25 \\ s + 0.5 \end{bmatrix} = \begin{bmatrix} 0.75 \\ 0.5 \end{bmatrix}$$

So $p = 0.5$ and $s = 0$.

Use the same method to find the values of p, q, r, s, t, and u for transformation A and for transformation C.

```
PROGRAM: FRACTAL
:PlotsOff
:AxesOff
:FnOff
:ClrDraw
:{3,1} →dim [D]
:[[0][0][1]]→[D]
:0→K
:0→Xmin
:1→Xmax
:0→Ymin
:1→Ymax
:ZSquare
:While (K<2000)
:int (rand*3+1)→R
:If R=1
:[A]*[D]→[E]
:If R=2
:[B]*[D]→[E]
:If R=3
:[C]*[D]→[E]
:[E](1,1)→[D](1,1)
:[E](2,1)→[D](2,1)
:Pt-On([E](1,1),[E](2,1))
:K+1→K
:End
```

Displaying the Fractal

Using the program at the left and a TI-82 graphing calculator, you can display the Sierpinski Triangle using the three transformations that you found.

1. Enter the program into your graphing calculator directly, or download it from another calculator or computer. See your graphing calculator manual for more information on entering or transmitting programs.

2. Store the values you found for transformation B into matrix B on your graphing calculator. Enter the six values into a 2×3 matrix in this order:

$$\begin{bmatrix} p & q & r \\ s & t & u \end{bmatrix}$$

For transformation B, store this matrix in your graphing calculator.

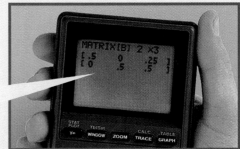

3. Store the values you found for transformation A and transformation C into matrices A and C, respectively.

4. Execute the program to display the fractal. It may take several minutes for the program to finish. If you entered the correct values into the matrices, your graphing calculator screen should look like this:

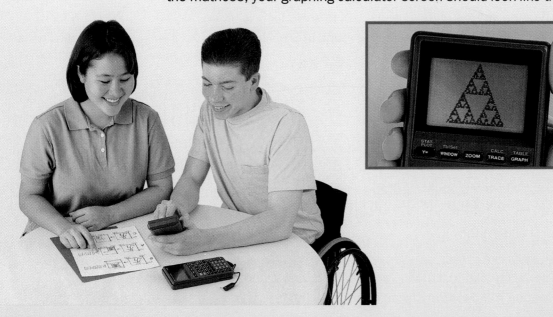

Generating Other Fractals

Choose one of the fractals shown below. Use the corresponding similarity diagram to find the values of p, q, r, s, t, and u for each transformation. Store these values in your graphing calculator as matrices A, B, and C. Then run the program to display the fractal.

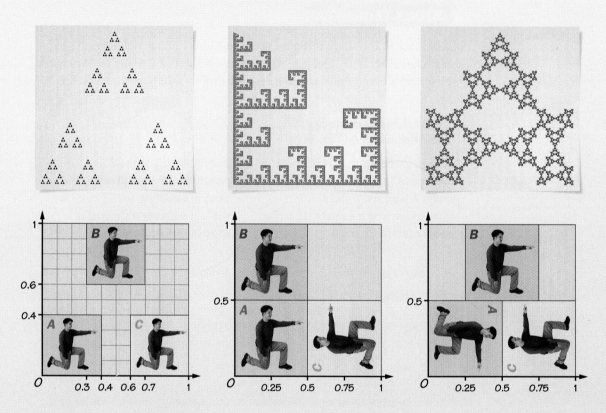

Writing a Report

In your report, give the values that you found for transformations A and C in the Sierpinski Triangle, and show the steps that you used to find them. Give the values for all three transformations in one of the fractals above, and show the steps that you used to find them. Describe how the image on your graphing calculator screen compares to the picture of the fractal above. If possible, include printouts of your calculator screen. To extend the project, draw a new similarity diagram and use your graphing calculator to display the fractal it describes.

Self-Assessment

In this project, you used some techniques that were not emphasized in the chapter. What was the most difficult part of the project? Explain.

12 | Review

Look back at the study techniques you used in previous Review and Assessment sections. Decide which technique helped you the most. Choose the technique you think would be most helpful in studying Chapter 12.

VOCABULARY

matrix (p. 613)
elements (p. 613)
rows, columns (p. 613)
dimensions of a matrix (p. 613)
point matrix (p. 613)

polygon matrix (p. 613)
scalar multiplication (p. 614)
matrix addition (p. 619)
matrix subtraction (p. 621)
matrix multiplication (p. 626)

SECTIONS 12.1 *and* 12.2

A point, segment, or polygon can be represented by a **matrix**.

$$A: \begin{bmatrix} -2 \\ -2 \end{bmatrix} \quad \overline{AB}: \begin{bmatrix} -2 & 0 \\ -2 & 2 \end{bmatrix} \quad ABCDE: \begin{bmatrix} -2 & 0 & 3 & 3 & 1 \\ -2 & 2 & 2 & 0 & -2 \end{bmatrix} \begin{matrix} \leftarrow x\text{-coordinates} \\ \leftarrow y\text{-coordinates} \end{matrix}$$

You can use **scalar multiplication** to dilate a polygon. The center of the dilation is (0, 0).

scale factor

$$\frac{1}{2}\begin{bmatrix} \overset{A}{-2} & \overset{B}{0} & \overset{C}{3} & \overset{D}{3} & \overset{E}{1} \\ -2 & 2 & 2 & 0 & -2 \end{bmatrix} = \begin{bmatrix} \overset{A'}{-1} & \overset{B'}{0} & \overset{C'}{\frac{3}{2}} & \overset{D'}{\frac{3}{2}} & \overset{E'}{\frac{1}{2}} \\ -1 & 1 & 1 & 0 & -1 \end{bmatrix}$$

polygon matrix   **image matrix**

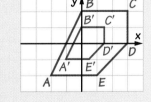

You can translate a polygon using **matrix addition**. The top row of the translation matrix translates horizontally, and the bottom row translates vertically. To add matrices, add corresponding elements.

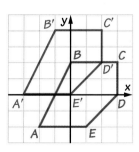

$$\begin{bmatrix} -1 & -1 & -1 & -1 & -1 \\ 2 & 2 & 2 & 2 & 2 \end{bmatrix} + \begin{bmatrix} \overset{A}{-2} & \overset{B}{0} & \overset{C}{3} & \overset{D}{3} & \overset{E}{1} \\ -2 & 2 & 2 & 0 & -2 \end{bmatrix} = \begin{bmatrix} \overset{A'}{-3} & \overset{B'}{-1} & \overset{C'}{2} & \overset{D'}{2} & \overset{E'}{0} \\ 0 & 4 & 4 & 2 & 0 \end{bmatrix}$$

translation matrix   **polygon matrix**   **image matrix**

You can use **matrix multiplication** to transform polygons. For example, the matrix $\begin{bmatrix} 1 & 0 \\ 0.5 & 1 \end{bmatrix}$ can transform a square into a parallelogram.

To multiply matrices, multiply the rows in the left matrix by the **columns** in the right matrix.

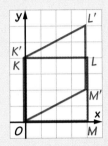

$$\begin{bmatrix} 1 & 0 \\ 0.5 & 1 \end{bmatrix} \quad \overset{O\ K\ L\ M}{\begin{bmatrix} 0 & 0 & 4 & 4 \\ 0 & 4 & 4 & 0 \end{bmatrix}} = \overset{O\ K'L'M'}{\begin{bmatrix} 0 & 0 & 4 & 4 \\ 0 & 4 & 6 & 2 \end{bmatrix}}$$

trans-   **polygon**   image
formation   **matrix**   matrix
matrix

$0.5 \cdot 4 + 1 \cdot 0 = 2$

You can use matrix multiplication to reflect a polygon over certain lines, or to rotate a polygon around the origin. Use the appropriate transformation matrix from the tables below.

Matrix	Reflection
$\begin{bmatrix} 1 & 0 \\ 0 & -1 \end{bmatrix}$	reflection over x-axis
$\begin{bmatrix} -1 & 0 \\ 0 & 1 \end{bmatrix}$	reflection over y-axis
$\begin{bmatrix} 0 & 1 \\ 1 & 0 \end{bmatrix}$	reflection over $y = x$
$\begin{bmatrix} 0 & -1 \\ -1 & 0 \end{bmatrix}$	reflection over $y = -x$

Matrix	Rotation
$\begin{bmatrix} 0 & -1 \\ 1 & 0 \end{bmatrix}$	90° rotation
$\begin{bmatrix} -1 & 0 \\ 0 & -1 \end{bmatrix}$	180° rotation
$\begin{bmatrix} 0 & 1 \\ -1 & 0 \end{bmatrix}$	270° rotation
$\begin{bmatrix} 1 & 0 \\ 0 & 1 \end{bmatrix}$	360° rotation

These rotations are measured in degrees counterclockwise.

For example, to rotate $\triangle XYZ$ 90° around the origin, multiply by the appropriate transformation matrix.

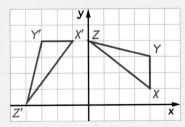

$$\begin{bmatrix} 0 & -1 \\ 1 & 0 \end{bmatrix} \quad \overset{X\ Y\ Z}{\begin{bmatrix} 4 & 4 & 0 \\ 1 & 3 & 4 \end{bmatrix}} = \overset{X'\ Y'\ Z'}{\begin{bmatrix} -1 & -3 & -4 \\ 4 & 4 & 0 \end{bmatrix}}$$

90°   **polygon**   image
rotation   **matrix**   matrix

12 Assessment

VOCABULARY QUESTIONS

1. Give an example of a matrix and state its dimensions.

2. Give an example of scalar multiplication.

3. Give an example of matrix addition.

SECTIONS 12.1 *and* 12.2

Write a matrix that represents each figure.

4. B

5. E

6. $\overline{BC}$

7. $\overline{DE}$

8. $\triangle ABC$

9. pentagon $ABCDE$

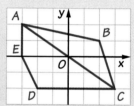

Find each product.

10. $\dfrac{5}{3}\begin{bmatrix} -6 & 3 & 15 \\ 0 & 9 & -3 \end{bmatrix}$

11. $8\begin{bmatrix} 5 & -2 \\ 3 & 9 \end{bmatrix}$

12. $0\begin{bmatrix} 1 & 8 & -5 \\ 1 & 4 & -9 \end{bmatrix}$

Add or subtract the matrices.

13. $\begin{bmatrix} -2 \\ -5 \end{bmatrix} - \begin{bmatrix} 13 \\ -4 \end{bmatrix}$

14. $\begin{bmatrix} -4 & -4 & -4 \\ 0 & 0 & 0 \end{bmatrix} + \begin{bmatrix} 1 & 2 & 3 \\ 1 & -2 & 2 \end{bmatrix}$

15. The matrix $\begin{bmatrix} -2 & 1 & 3 & 1 \\ 0 & 2 & 0 & -2 \end{bmatrix}$ represents quadrilateral $A'B'C'D'$, the image of quadrilateral $ABCD$ after a translation to the right 5 units and down 2 units. Find the matrix for $ABCD$.

The red polygon in each diagram is the image of the blue polygon after a translation. Write each translation as a matrix sum.

16.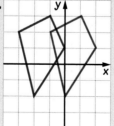

17.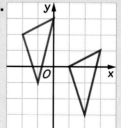

18.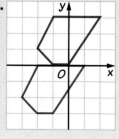

Complete each matrix product.

19. $\begin{bmatrix} 0 & -3 \\ -2 & 5 \end{bmatrix} \begin{bmatrix} 2 & 4 \\ -1 & -6 \end{bmatrix}$

20. $\begin{bmatrix} -1 & 0 \\ 4 & 4 \end{bmatrix} \begin{bmatrix} -5 & 0 & 5 & 1 & -3 \\ 0 & 2 & -2 & -6 & -2 \end{bmatrix}$

21. The matrix $\begin{bmatrix} -6 & -4 & 4 & 2 \\ -2 & 4 & 3 & -4 \end{bmatrix}$ represents quadrilateral *JKLM*. Use the

matrix $\begin{bmatrix} 0.5 & 0 \\ 1 & 1 \end{bmatrix}$ to transform *JKLM*. Graph *JKLM* and its image.

22. Use the diagram at the right.

   a. Write a matrix for $\overline{AB}$.

   b. Multiply the matrix for $\overline{AB}$ by the

   transformation matrix $\begin{bmatrix} 0 & 1 \\ 1 & 0 \end{bmatrix}$ to find

   a matrix for $\overline{A'B'}$.

   c. Graph $\overline{AB}$ and its image. Describe the transformation caused by the matrix in part (b).

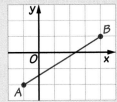

23. Open-ended Problem Sketch a polygon on graph paper.

   a. Create a design by reflecting or rotating the polygon more than once.

   b. Writing Explain how a computer could draw your design using the coordinates of the polygon's vertices and transformations.

Use each matrix given to transform quadrilateral *PRST*. Then graph the image of *PRST* and describe the transformation.

24. $\begin{bmatrix} 1 & 0 \\ 0 & -1 \end{bmatrix}$ **25.** $\begin{bmatrix} 0 & -1 \\ -1 & 0 \end{bmatrix}$

26. $\begin{bmatrix} -1 & 0 \\ 0 & -1 \end{bmatrix}$ **27.** $\begin{bmatrix} 1 & 0 \\ 0 & 1 \end{bmatrix}$

28. $\begin{bmatrix} 0 & -1 \\ 1 & 0 \end{bmatrix}$ **29.** $\begin{bmatrix} 0 & 1 \\ -1 & 0 \end{bmatrix}$

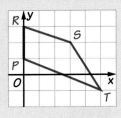

PERFORMANCE TASK

30. Open-ended Problem Write a matrix to represent a polygon. Use matrices to dilate, translate, "morph," reflect, and rotate your polygon. Make a poster displaying your results. Include matrix equations and graphs for all of the transformations.

Cumulative Assessment

CHAPTERS $10-12$

CHAPTER 10

For Questions 1–3, use the diagram. Complete each statement.

1. $\triangle JKL \sim \triangle\underline{?} \sim \triangle\underline{?}$.

2. If $LM = 9$ and $MJ = 24$, then $KL = \underline{?}$.

3. If $JK = 15$ and $KL = 9$, then
   $m\angle J \approx \underline{?}$ and $m\angle L \approx \underline{?}$.

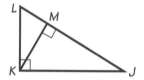

4. **Writing** Explain why any triangle with side lengths in the ratio
   $1:\sqrt{3}:2$ must be a 30-60-90 triangle. (*Hint:* Use the Converse of the
   Pythagorean Theorem and trigonometry.)

5. The perimeter of a rhombus is 80 cm, and one diagonal is 14 cm long.
   a. Find the measures of the four angles of the rhombus. Give your
      answers to the nearest tenth of a degree.
   b. Find the area of the rhombus.

6. If $\overrightarrow{AB} = (-10, -8)$, graph $-\frac{1}{2}\overrightarrow{AB}$ and express it in component form.

7. **Open-ended Problem** Give two vectors whose sum is $(0, 0)$. Make a
   conjecture about the components and the magnitudes of any two vectors
   whose sum is $(0, 0)$.

8. Find the area of a regular decagon whose sides are 8 in. long.

**Find the surface area and volume of the prism, right cone, and regular
pyramid.**

9.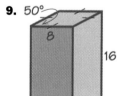

10.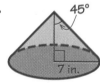

11.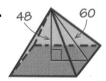

CHAPTER 11

12. **Writing** In $\odot O$, $m\widehat{AB} = 100°$ and $m\widehat{BC} = 100°$. What type of triangle
    is $\triangle ABC$? Find the measure of each angle of the triangle. Explain your
    reasoning.

ALGEBRA Find the value of each variable.

13.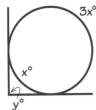

14.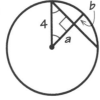

15.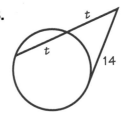

16. Open-ended Problem Two chords of a circle intersect to form a 105° angle. Find a pair of possible measures for the intercepted arcs.

17. In $\odot O$, $m\widehat{RST} = 200°$ and $OR = 18$. Find the lengths of $\widehat{RT}$ and of $\widehat{RST}$. Also find the area of the sector with central angle $\angle ROT$.

18. SPORTS The diameter of a tennis ball is 2.6 in. and the radius of a baseball is 1.4 in. What is the ratio of their volumes? of their surface areas?

19. Writing Explain why the following statement from Euclidean geometry is not true for spherical geometry: If two lines intersect, then they intersect in exactly one point.

CHAPTER 12

20. Open-ended Problem Draw any $\triangle ABC$ and label the coordinates of its vertices. Write a matrix that represents $\triangle ABC$.

   a. Choose any scale factor. Write a matrix to represent the image of $\triangle ABC$ after a dilation with your scale factor and center at (0, 0).

   b. Choose any translation. Write a matrix to represent the image of $\triangle ABC$ after your translation.

The vertices of quadrilateral *PQRS* are *P*(4, 0), *Q*(−1, 3), *R*(−3, −1), and *S*(0, −2).

21. Write the polygon matrix for quadrilateral $PQRS$.

22. Multiply the polygon matrix for $PQRS$ by the transformation matrix $\begin{bmatrix} 1 & 1 \\ 1 & 1 \end{bmatrix}$. Graph $PQRS$ and its image matrix. Describe the effect of the transformation matrix.

23. Use matrix multiplication to find the image of $PQRS$ after a rotation of 90°. Graph $PQRS$ and its image. Use the table on page 642.

24. Use matrix multiplication to find the image of $PQRS$ after reflection over the line $y = -x$. Use the table on page 635.

25. Writing Name at least two types of transformations that *cannot* be represented by multiplying a polygon matrix by a 2 × 2 transformation matrix. Explain your reasoning.

Contents of Student Resources

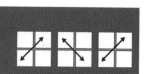

Extra Practice

CHAPTER 1

Use inductive reasoning to find the next two numbers in each pattern. *Section 1.1*

1. $-2, -5, -8, -11, \underline{?}, \underline{?}$

2. $0, 2, 5, 9, \underline{?}, \underline{?}$

3. $0, 1, 4, 9, \underline{?}, \underline{?}$

4. $1, \frac{1}{2}, \frac{2}{3}, \frac{3}{4}, \underline{?}, \underline{?}$

5. $7, -14, 28, -56, \underline{?}, \underline{?}$

6. $2, 7, 12, 17, \underline{?}, \underline{?}$

Use inductive reasoning to sketch the next shape in each pattern. *Section 1.1*

7.

8.

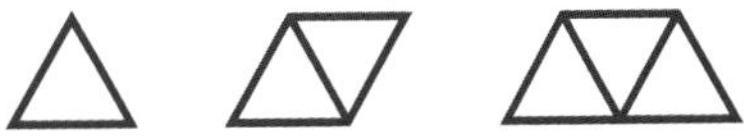

Write a formula for the value of the *n*th term in each pattern. *Section 1.1*

9.

Term	1	2	3	4	5	6	...	n
Value	2	1	0	1	2	3	...	?

10.

Term	1	2	3	4	5	6	...	n
Value	2	5	8	11	14	17	...	?

Perform each transformation. *Section 1.2*

11. Reflect the triangle over the line.

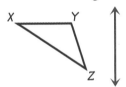

12. Rotate the triangle around point C by a half-turn.

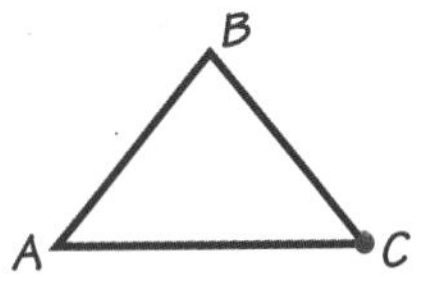

13. Translate the triangle any distance to the right.

14. Reflect the triangle over the line.

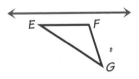

Describe the symmetry of each figure. *Section 1.2*

15.

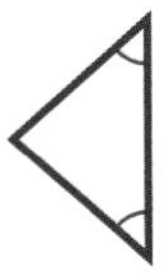

16.

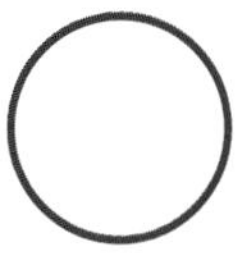

17.

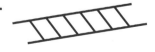

Rewrite each statement using _if_ and _then_. *Section 1.3*

18. Every square is a rectangle.

19. My car runs out of gas when I forget to fill the tank.

20. Two lines are perpendicular if they meet to form right angles.

Identify the hypothesis and the conclusion of each conditional statement. *Section 1.3*

21. $2x - 5 = 15$ if $x = 10$. **22.** If the weather gets cold, then birds fly south.

Tell whether each statement is _True_ or _False_. If it is false, give a counterexample.
Section 1.3

23. If it is 6:00 P.M. in California, then it is 9:00 P.M. in Maryland.

24. If a banana is ripe, then it is yellow.

25. If Carmine is in Arizona, then he is in Tucson.

**Use the diagram at the right, showing plane _ABC_
parallel to plane _FGH_.** *Section 1.4*

26. What is the intersection of $\overleftrightarrow{AB}$ and $\overleftrightarrow{FB}$?

27. What is the intersection of $\overleftrightarrow{BF}$ and $\overleftrightarrow{GF}$?

28. Name three lines parallel to $\overleftrightarrow{DH}$.

29. Name a line through point H that is skew to $\overleftrightarrow{FG}$.

30. What is the intersection of planes ABC and CGF?

31. Name two planes that intersect in $\overleftrightarrow{AE}$.

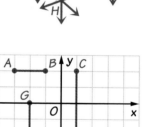

Give the length of each segment. *Section 1.5*

32. $\overline{AB}$ **33.** $\overline{CD}$

34. $\overline{DE}$ **35.** $\overline{FG}$

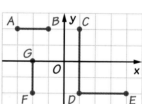

Give the length of each segment. *Section 1.5*

36. $\overline{JP}$ **37.** $\overline{QM}$

38. $\overline{MN}$ **39.** $\overline{JM}$

40. $\overline{MP}$ **41.** $\overline{LJ}$

Find the measure of each angle. *Section 1.6*

42. $\angle LMP$ **43.** $\angle LMQ$

44. $\angle PMQ$ **45.** $\angle QMN$

46. $\angle PMN$ **47.** $\angle PMR$

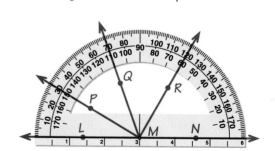

Tell whether each statement is *True* or *False*. *Section 1.7*

48. C is the midpoint of $\overline{AD}$.

49. C is the midpoint of $\overline{BD}$.

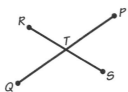

A   B   C   D

50. If $BC = 3$, then $CD = 6$.

51. If $BD = 10$, then $BC = 5$.

52. If $BC = 10$, then $AD = 30$.

In the diagram, $\overline{PQ}$ and $\overline{RS}$ bisect each other. *Section 1.7*

53. If $RT = 3$, find the length of $\overline{RS}$.

54. If $PT = 7$, find the length of $\overline{TQ}$.

55. If $PT = 3x + 1$ and $TQ = 4x - 1$, find the length of $\overline{PQ}$.

56. If $TR = 2x + 2$ and $TS = 3x - 3$, find the length of $\overline{RT}$.

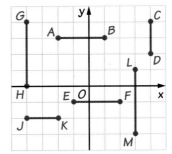

Find the midpoint of each segment. *Section 1.7*

57. $\overline{AB}$

58. $\overline{CD}$

59. $\overline{EF}$

60. $\overline{GH}$

61. $\overline{JK}$

62. $\overline{LM}$

CHAPTER 2

Complete each statement about the diagram at the right. *Section 2.1*

1. $\angle EFG$ and $\underline{\ ?\ }$ are vertical angles.

2. $\angle EGF$ and $\underline{\ ?\ }$ are congruent adjacent angles.

3. $\angle FGJ$ and $\underline{\ ?\ }$ are complementary angles.

4. $\angle KGJ$ and $\underline{\ ?\ }$ are supplementary angles.

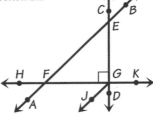

Find the measure of each angle in the diagram at the right. *Section 2.1*

5. $\angle CGH$

6. $\angle HCD$

7. $\angle HBC$

8. $\angle GCB$

9. $\angle GHC$

10. $\angle GCF$

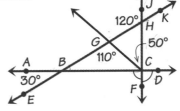

Find each unknown angle measure. *Section 2.2*

11.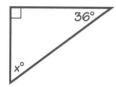

12.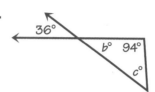

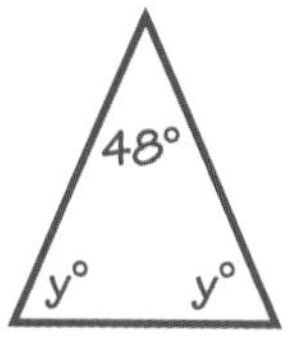

13.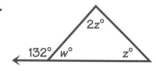

14.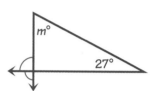

15.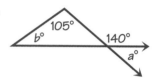

16.

17.

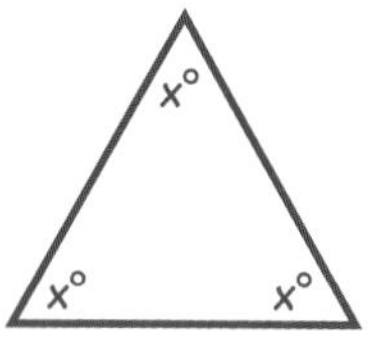

18.

19.

Classify each polygon. Be as specific as possible. *Section 2.3*

20.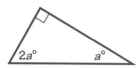

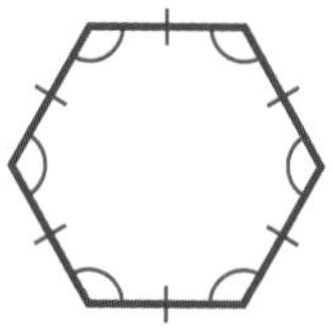

21.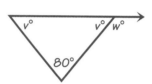

22.

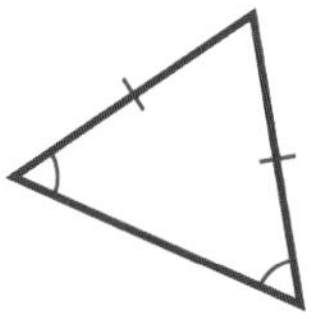

23.

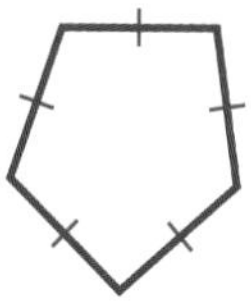

24.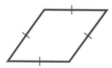

25.

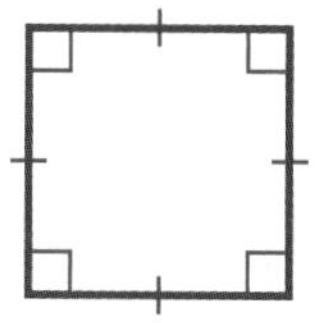

Find the sum of the measures of the interior angles of each polygon. *Section 2.4*

26. hexagon

27. 9-gon

28. 22-gon

Find each unknown angle measure. *Section 2.4*

29.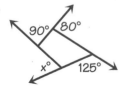

30.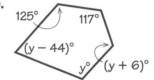

31.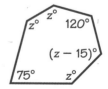

Find the measure of each interior and exterior angle of the polygon. *Section 2.4*

32. regular octagon

33. regular 10-gon

34. regular 28-gon

Find each length or angle measure. *Section 2.5*

35. rhombus *ABCD*
   a. *AB*
   b. *AD*
   c. *m∠C*

36. rectangle *KLMN*
   a. *NL*
   b. *ML*
   c. *m∠KLM*

37. parallelogram *PQRS*
   a. *m∠SQR*
   b. *m∠SPQ*
   c. *m∠SPT*
   d. *m∠QSR*

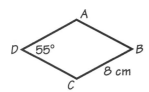

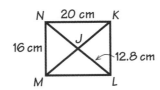

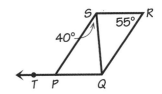

Find each length or angle measure of rectangle *ABCD*. *Section 2.5*

38. *AB*

39. *AE*

40. *m∠ADE*

41. *m∠ADC*

42. *m∠AEB*

43. *m∠EAB*

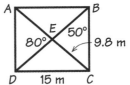

Use the prism at the right. *Section 2.6*

44. Name the two bases of the prism.

45. What type of prism is it?

46. How many lateral faces does the prism have?

47. How many edges meet at vertex *A*?

48. How many edges does the prism have?

49. How many vertices does the prism have?

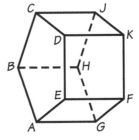

CHAPTER 3

Tell whether each argument uses *inductive* or *deductive* reasoning. *Section 3.1*

  1. Rainbows only appear after a rain. It just stopped raining, so there will be a rainbow.

  2. Vertical angles are congruent. ∠3 and ∠4 are vertical angles, so ∠3 ≅ ∠4.

  3. All squares have four right angles. Polygon *ABCD* is a square, so *ABCD* has four right angles.

  4. African violets need special light conditions. My violets are not growing well, so I don't have the correct light conditions.

Use inductive reasoning to predict the next number in each pattern. Justify your prediction. *Section 3.1*

  5. 5, 8, 11, 14, $\underline{\ ?\ }$

  6. 20, 27, 36, 47, $\underline{\ ?\ }$

  7. 3, 1, $\frac{1}{3}$, $\frac{1}{9}$, $\underline{\ ?\ }$

Identify the property that makes each statement true. *Section 3.2*

8. If $AB = CD$, then $CD = AB$.

9. $m\angle WXY = m\angle WXY$

10. If $XY = CD$ and $CD = 2$, then $XY = 2$.

For Exercises 11–16, give the postulate, definition, property, or previous statement that makes the statement about the diagram true. *Section 3.2*

11. $PQ = ST$ **12.** $2PQ = 2ST$

13. $QR = RS$ **14.** $QR = QR$

15. $PR = RT$ **16.** $3QR = 3RS$

Draw an obtuse triangle, an acute triangle, and a right triangle. Tell whether each statement is *True* or *False*. *Section 3.3*

17. All six exterior angles of an obtuse triangle are obtuse.

18. Both exterior angles of one angle of a triangle may be obtuse.

Find the measure of each indicated angle. *Section 3.4*

19. $m\angle NPQ$ **20.** $m\angle BCD$ **21.** $m\angle WYZ$

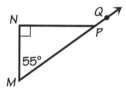 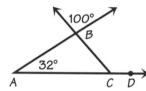 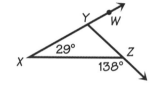

Rewrite each statement so that it is always true. *Section 3.4*

22. Angles that are complementary to the same angle are complementary.

23. Supplementary angles are angles in a linear pair.

24. Complete the proof. *Section 3.4*

    Given: $m\angle 2 = m\angle 3$

    Prove: $m\angle 1 = m\angle 4$

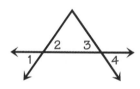

Statements	Reasons
1. $m\angle 2 = m\angle 3$	**1.** ?
2. $m\angle 2 = m\angle 1$; $m\angle 3 = m\angle 4$	**2.** ?
3. ?	**3.** Substitution Property (Step(s) ? and ?)

Write the converse of each statement. Tell whether the converse is *True* or *False*.
Section 3.5

25. If a wildflower is goldenrod, then it is yellow.

26. If a woman lives in Massachusetts, then she lives in Boston.

27. If it rains, I will take my umbrella.

For Exercises 28–30: *Section 3.5*

a. **Rewrite the statement as a conditional.**   b. **Write the converse of the statement.**

28. The opposite sides of a rectangle are congruent.

29. The complements of congruent angles are congruent.

30. The diagonals of a rhombus are perpendicular.

Use a calculator to approximate each square root to the nearest hundredth. Then simplify each expression. *Section 3.6*

31. $\sqrt{20}$

32. $-\sqrt{144}$

33. $\pm\sqrt{32}$

34. $\pm\sqrt{10}$

35. $-\sqrt{150}$

36. $\sqrt{256}$

Tell whether the given lengths can be the sides of a right triangle. *Section 3.6*

37. 1.5 m, 2 m, 2.5 m

38. 6 cm, 8 cm, 12 cm

39. 5 in., 12 in., 14 in.

40. $\sqrt{5}, \sqrt{3}, \sqrt{8}$

41. 5 ft, 3 ft, 8 ft

42. 5.5 cm, 4 cm, 9.5 cm

The legs of a right triangle are *a* and *b* units long. The hypotenuse is *c* units long. Find the unknown length for each right triangle, to the nearest hundredth. *Section 3.6*

43. $a = 12, c = 13$

44. $a = 24, b = 7$

45. $b = 12, c = 20$

46. $a = 8, c = 12$

47. $a = 119, b = 120$

48. $a = 4, b = 5$

Tell whether the triangle with sides of the given lengths is *right*, *obtuse*, or *acute*. *Section 3.7*

49. 5, 9, 12

50. 13, 15, 17

51. 7, 24, 25

52. 7, 24, 26

53. 10, 24, 26

54. 25, 25, 30

Write the inverse and the contrapositive of each statement. *Section 3.7*

55. If Jo studies hard, she will do well on the test.

56. If the piano is moved, then it will need to be tuned.

57. If Bill was born in France, then he can speak French.

CHAPTER 4

Find the length of the segment with the given endpoints. *Section 4.1*

1. $A(2, -2)$ and $B(-3, -2)$

2. $A(5, 8)$ and $C(10, 4)$

3. $D(-6, 2)$ and $B(1, -4)$

4. $F(2, 8)$ and $G(-7, 8)$

5. $M(-3, 4)$ and $N(3, 1)$

6. $R(4, -6)$ and $S(0, 3)$

Find the coordinates of the midpoint of the segment with each given pair of endpoints. *Section 4.1*

7. $(2, 3)$ and $(6, 5)$

8. $(5, -2)$ and $(7, -4)$

9. $(-6, 2)$ and $(-2, 8)$

10. $(-1, 5)$ and $(3, -1)$

11. $(8, 2)$ and $(2, -4)$

12. $(11, -7)$ and $(15, 8)$

Find the lengths of each of the sides of each polygon whose vertices are given.
Give the most specific name for each polygon. *Section 4.1*

13. $A(0, 4), B(3, 0), C(0, -4)$

14. $A(0, 3), B(8, 3), C(5, -1), D(-3, -1)$

15. $A(-1, -4), B(-3, 0), C(0, 4), D(2, 0)$

Find the slope of the line that contains each pair of given points. *Section 4.2*

16. $(2, 3)$ and $(5, 6)$

17. $(5, -2)$ and $(6, -4)$

18. $(-6, 2)$ and $(-2, 3)$

19. $(-1, 5)$ and $(2, -1)$

20. $(8, 2)$ and $(3, -4)$

21. $(10, -7)$ and $(15, 8)$

22. $(3, -3)$ and $(5, 7)$

23. $(-4, -4)$ and $(3, 3)$

24. $(-7, -9)$ and $(1, 1)$

Write an equation for each line. *Section 4.2*

25. contains $(1, 3)$; has slope 2

26. contains $(4, -5)$; has slope -3

27. contains $(3, -2)$; has slope $\dfrac{1}{4}$

28. contains $(9, 4)$; has slope $-\dfrac{2}{3}$

29. the vertical line that contains $(5, -2)$

30. the horizontal line that contains $(-4, -2)$

31. contains $(6, 1)$; has slope $\dfrac{7}{2}$

32. contains $(8, -12)$; has slope 6

Tell whether each pair of lines is *parallel, perpendicular,* or *neither.* *Section 4.3*

33. $x = 2$
   $y = 3$

34. $y = \dfrac{1}{3}x + 2$
   $y = -3x - 2$

35. $y = 2$
   $y = -\dfrac{1}{2}$

Find the slope of each line. *Section 4.3*

36. a line parallel to the line $y = 2x - 1$

37. a line parallel to the line $y = -2$

38. a line perpendicular to the line $y = -2x + 4$

39. a line perpendicular to the line $y = -2$

40. a line perpendicular to the line $y = 5x$

41. a line parallel to the line $3x - y = 1$

Use slopes to tell whether each triangle with the given vertices is a right triangle.
Section 4.3

42. $A(0, 4), B(3, 0), C(0, -4)$

43. $C(1, 4), D(3, -1), E(6, 6)$

44. $E(2, 2), F(8, 5), G(6, 9)$

45. $J(-2, -5), K(2, -2), L(-2, 1)$

Write an equation of the circle with each given center and radius. *Section 4.4*

46. center $(0, 0)$, radius 6

47. center $(2, 5)$, radius 4

48. center $(2, -3)$, radius 1

49. center $(-2, -5)$, radius 4.5

**Sketch each circle. Label the coordinates of the center and draw a radius and
label it with its length.** *Section 4.4*

50. $x^2 + y^2 = 144$

51. $(x - 2)^2 + (y - 1)^2 = 16$

52. $(x + 1)^2 + (y + 4)^2 = 12$

53. $(x - 5)^2 + (y + 6)^2 = 169$

54. Tell whether the point $(3, 4)$ is on the circle with equation $x^2 + y^2 = 5$. *Section 4.4*

Find the missing coordinates without using any new variables. *Section 4.5*

55. parallelogram

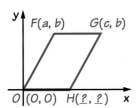

56. rectangle

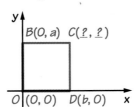

57. isosceles triangle

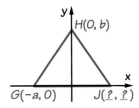

For each $\triangle ABO$, find the coordinates of the midpoint M of the hypotenuse. Then find MA, MB, and MO. *Section 4.5*

58.

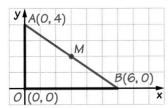

59.

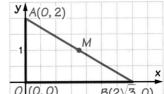

Find the coordinates of the midpoint of the segment with the given endpoints.
Section 4.6

60. $(3, 4, -1)$ and $(5, 2, 1)$ **61.** $(-2, -3, 6)$ and $(-4, 3, 6)$ **62.** $(1, 0.5, 5)$ and $(5, 1.5, -4)$

Find the length of the segment with the given endpoints. *Section 4.6*

63. $(1, 4, -1)$ and $(5, 1, -1)$ **64.** $(3, 4, -1)$ and $(5, 2, 1)$ **65.** $(13, 0, 1)$ and $(5, 2, 11)$

CHAPTER 5

Classify each pair of angles as *corresponding angles*, *alternate interior angles*, or *same-side interior angles*. *Section 5.1*

1. $\angle 3$ and $\angle 6$ **2.** $\angle 3$ and $\angle 7$

3. $\angle 3$ and $\angle 5$ **4.** $\angle 1$ and $\angle 5$

5. $\angle 5$ and $\angle 4$ **6.** $\angle 2$ and $\angle 6$

7. $\angle 4$ and $\angle 11$ **8.** $\angle 6$ and $\angle 14$

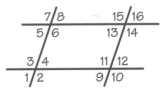

Find the values specified in each diagram. *Section 5.1*

9. Find the values of x, y, and z.

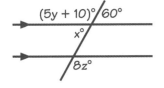

10. Find the value of w and $m\angle 1$.

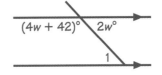

Find the measure of each numbered angle. *Section 5.2*

11.

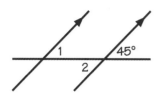

12.

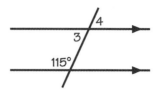

13.

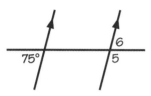

14.

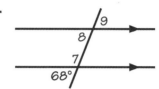

Use trapezoid *KLMN*. *Section 5.2*

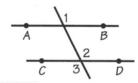

15. Tell which sides are the bases.

16. Which sides are the legs?

17. Find the measure of $\angle L$.

18. Find the measure of $\angle M$.

For Exercises 19–21, copy and complete each proof. *Section 5.3*

19. Given: $\overline{AB} \parallel \overline{CD}$

   Prove: $\angle 1 \cong \angle 3$

1. $\angle 1 \cong \angle 2$ (If two $\parallel$ lines are intersected by a transversal, then $\underline{?}$ are $\cong$.)

2. $\underline{?}$ (Vertical angles are congruent.)

3. $\angle 1 \cong \angle 3$ ($\underline{?}$ Property (Steps 1 and 2))

20. Given: $j \parallel k; m \parallel n$

   Prove: $\angle 1 \cong \angle 4$

Statements	Reasons
1. $m \parallel n$	**1.** Given
2. $\angle 1 \cong \angle 2$	**2.** If two $\parallel$ lines are intersected by a transversal, then $\underline{?}$ angles are congruent.
3. $j \parallel k$	**3.** $\underline{?}$
4. $\angle 2 \cong \angle 4$	**4.** If two $\parallel$ lines are intersected by a transversal, then $\underline{?}$ angles are congruent.
5. $\angle 1 \cong \angle 4$	**5.** $\underline{?}$

21. Given: $\overline{MN} \parallel \overline{BC}$; $\angle B \cong \angle C$

Prove: $\angle MKB \cong \angle NKC$

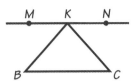

Because $\overline{MN} \parallel \overline{BC}$, $\angle MKB \cong \angle B$ and $\angle NKC \cong \angle C$, since they are
_?_ angles. Because $\angle B \cong \angle C$, it must be true that $\angle MKB \cong \angle NKC$
by the _?_ Property.

**For each diagram, find the value of the variable that will allow you to prove that
two lines are parallel.** *Section 5.4*

22.

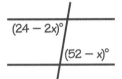

23.

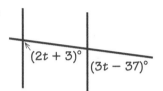

24.

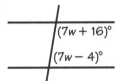

**For each pair of lines shown, tell whether they are *parallel* or *not parallel* and
explain your reasoning.** *Section 5.4*

25. ℓ and m

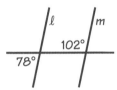

26. c and d

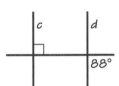

27. p and q

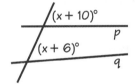

Use the diagram at the right. *Section 5.5*

28. What do you know about $\overline{PM}$ and $\overline{RS}$?

29. Using the Dual Parallels Theorem you can tell
that $\overline{AB} \parallel$ _?_ .

30. Find the value of z.

31. Find the measure of $\angle RTX$.

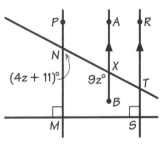

Find each distance. *Section 5.5*

32. Find the distance from X to $\overline{MN}$.

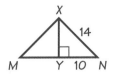

33. Find the distance from A to $\overline{PQ}$.

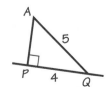

In the diagram, plane X ∥ plane Z. *Section 5.6*

34. How are $\overline{CD}$ and $\overline{FG}$ related?

35. Suppose $m\angle CDE = 76°$. Find $m\angle FED$.

36. Name two angles that are congruent to $\angle CDE$.

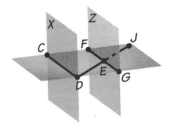

Draw two angles like the ones shown. For Exercises 37–40, construct an angle with the indicated measure. *Section 5.7*

37. $2w°$
38. $(w + v)°$

39. $(180 + v)°$
40. $(2v - w)°$

CHAPTER 6

Tell whether or not a triangle can be formed from sides of the given lengths.
Section 6.1

1. 3 ft, 8 ft, 10 ft
2. 22 cm, 11 cm, 11 cm
3. 8 in., 9 in., 7 in.

4. 25 m, 35 m, 11 m
5. $x, 3x, 4x$
6. 16 ft, 9 ft, 25 ft

7. 5 ft, 6 ft, 4 ft
8. 2 in., 16 in., 12 in.
9. 8 cm, 14 cm, 25 cm

The lengths of two sides of a triangle are given. What can you conclude about the length of the third side? *Section 6.1*

10. 3.8 m, 14.5 m
11. $4\frac{1}{2}$ in., $3\frac{1}{4}$ in.
12. $y + 2, y$

13. 5 ft, 6 ft
14. 2 in., 16 in.
15. 8 cm, 14 cm

16. 3 ft, 8 ft
17. 20 cm, 11 cm
18. 9 in., 7 in.

What can you conclude about the length AB in each triangle? *Section 6.1*

19.

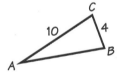

20. $m\angle B > m\angle C$

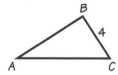

21. $m\angle A < m\angle B$

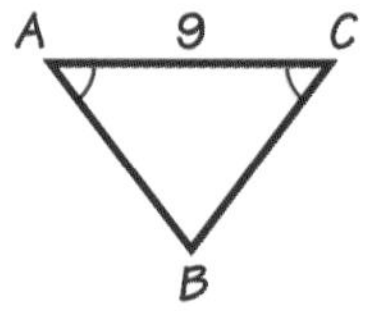

22. $m\angle B < m\angle A$ and $m\angle C < m\angle B$

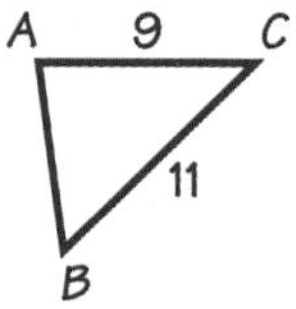

23. $m\angle C < m\angle A$

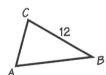

24. $m\angle B < m\angle C$ and $m\angle C < m\angle A$

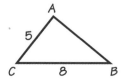

Use the diagram at the right. *Section 6.1*

25. $\triangle DEC \cong$ _?_

26. $\angle CDE \cong$ _?_

27. $\angle DCE \cong$ _?_

28. $\angle DEC \cong$ _?_

29. $\overline{DC} \cong$ _?_

30. $\overline{CF} \cong$ _?_

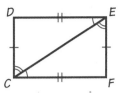

Use the diagram at the right. $\triangle ABC \cong \triangle WXY$. *Section 6.2*

31. $\triangle CBA \cong$ _?_

32. $\angle A \cong$ _?_

33. $\angle C \cong$ _?_

34. $\overline{AC} \cong$ _?_

35. Find the value of x.

36. Find the value of y.

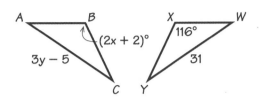

Decide whether or not you can prove that the triangles are congruent. If you can, tell which postulate you would use. *Section 6.3*

37.

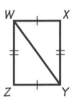

38.

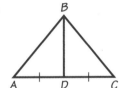

39.

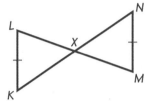

40.

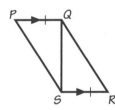

41.

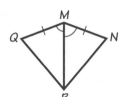

42.

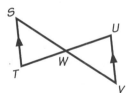

43. Copy and complete the flow proof. *Section 6.3*

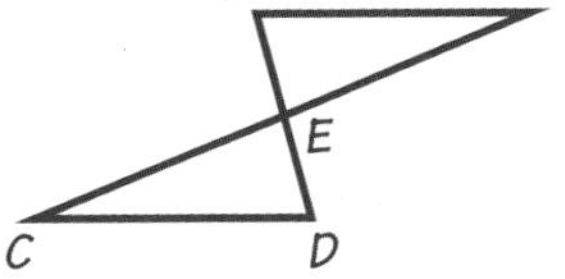

Given: E is the midpoint of $\overline{BC}$.
E is the midpoint of $\overline{AD}$.

Prove: $\triangle ABE \cong \triangle DCE$

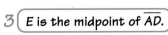

Reasons

1. Given

2. Definition of _?_

3. _?_

4. _?_

5. _?_

6. _?_

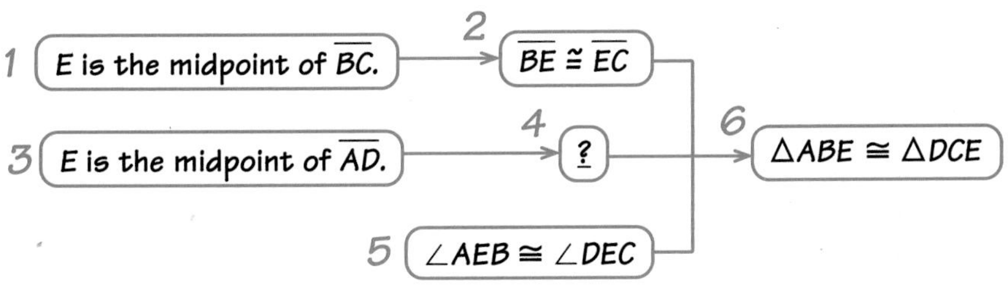

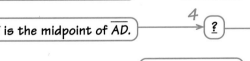

Tell which method you can use to prove that the triangles are congruent. If no method can be used, write *none*. *Section 6.4*

44.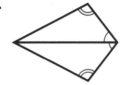

45.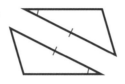

46.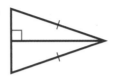

47.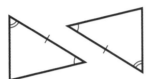

48.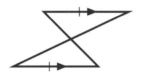

49.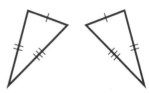

Tell which pair of triangles must be congruent in order to prove each statement. *Section 6.5*

50. $\overline{CE} \cong \overline{DE}$

51. $\overline{AD} \cong \overline{BC}$

52. $\angle ABC \cong \angle BAD$

53. $\angle ACB \cong \angle BDA$

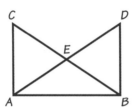

Use the diagram at the right. *Section 6.6*

54. $\angle 2 \cong \angle 3$. Name two congruent segments.

55. $\angle 7 \cong \angle 6$. Name two congruent segments.

56. $\overline{PT} \cong \overline{QT}$. Name two congruent angles.

57. $\overline{ST} \cong \overline{RT}$. Name two congruent angles.

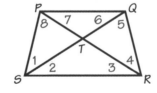

In △ABC, AB = AC, $\overline{AD}$ and $\overline{CE}$ are medians and AC = 16. Find each length or angle measure. *Section 6.7*

58. BA

59. $m\angle BDA$

60. BE

61. EA

CHAPTER 7

Find each length. *Section 7.1*

1. Rectangle *PQRS*
   a. *PT*
   b. *QT*
   c. *SR*

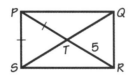

2. Kite *ABCD*
   a. *DE*
   b. *BD*
   c. *BC*

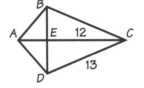

3. Rhombus *EFGH*

a. *FJ*

b. *FH*

c. *FG*

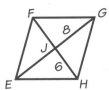

4. Square *KLMN*

a. *KN*

b. *PK*

c. *LN*

5. *ABCD* is a rectangle and *ABGH* is a parallelogram. Find each length.

a. *BC*

b. *AH*

c. *DB*

d. *HC*

e. *AC*

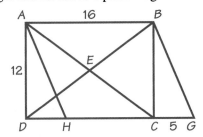

6. Use the diagram for Exercise 5 above.

Given: *ABCD* is a rectangle; *ABGH* is a parallelogram.

Prove: △*ADH* ≅ △*BCG*

Show that each quadrilateral is a parallelogram. Then find each length or measure. *Section 7.2*

7. a. *m∠NQP*

b. *NR*

c. *QN*

8. a. *m∠ABC*

b. *m∠ADC*

c. *AD*

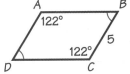

9. a. *m∠GHJ*

b. *m∠FGJ*

c. *JH*

10. a. *SR*

b. *TP*

c. *QS*

Classify each quadrilateral. Be as specific as possible. *Section 7.3*

11.

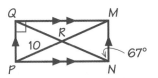

12.

13.

14.

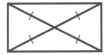

15.

16.

Find the area of each polygon. *Section 7.4*

17.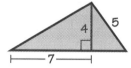

18.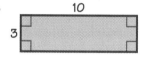

19.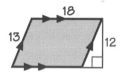

20.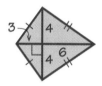

21.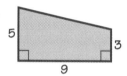

22.

Find the area of each polygon. *Section 7.4*

23. a square with diagonal 22 cm

24. a rectangle with side 12 ft and diagonal 20 ft

25. an equilateral triangle with perimeter of 24 in.

Find the area of each regular polygon. *Section 7.5*

26.

27.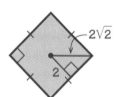

28.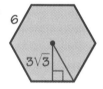

Find the area of each circle. *Section 7.5*

29.

30.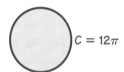

31.

Find the area of each shaded region. *Section 7.5*

32.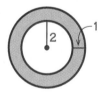

33.

34.

Find the volume and surface area of each prism or cylinder. *Section 7.6*

35.

36.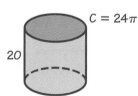

37.

Find the volume of each three-dimensional figure described. *Section 7.6*

38. Each edge of a cube is 12.5 cm long.

39. A pentagonal prism is 8 m long. The area of the base is 12 m^2.

40. The radius of a cylinder is 5 in. The height of the cylinder is 8 in.

CHAPTER 8

Copy each diagram. Sketch the image of the polygon after a reflection over the given line. Label the image polygon. *Section 8.1*

1.

2.

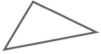

3.

Each diagram shows a polygon and its image after a reflection. Copy each diagram and draw the line of reflection. *Section 8.1*

4.

5.

6.

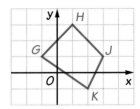

Copy each polygon and reflect it over the *y*-axis. Give the coordinates of the vertices of the image. *Section 8.2*

7.

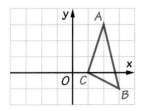

8.

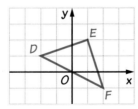

9.

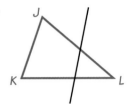

Copy each polygon and reflect it over the *x*-axis. Give the coordinates of the vertices of the image. *Section 8.2*

10.

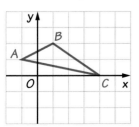

11.

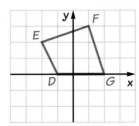

12.

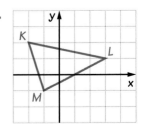

Copy each polygon and reflect it over the line $y = x$. Give the coordinates of the vertices of the image. *Section 8.2*

13.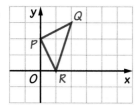

14.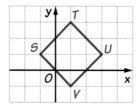

15.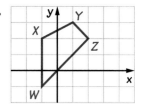

Copy ABCD and draw its image after each translation. *Section 8.3*

16. $(a, b) \rightarrow (a + 1, b - 2)$

17. $(a, b) \rightarrow (a, b + 3)$

18. $(a, b) \rightarrow (a - 2, b)$

19. $(a, b) \rightarrow (a - 1, b + 4)$

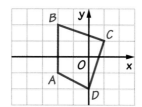

Describe each translation using coordinate notation. *Section 8.3*

20. Every point moves 5 units to the left and 3 units down.

21. Every point moves 2 units to the right and 1 unit up.

22. Every point moves 9 units to the left.

Copy each diagram and sketch the image of each polygon after the given rotation around the origin. *Section 8.4*

23. rotate 90°

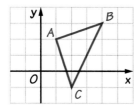

24. rotate 180°

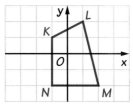

25. rotate 270°

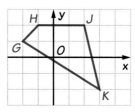

26. rotate 90°

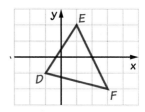

Sketch the image of the triangle with the given vertices after a glide reflection using the given translation and reflection. *Section 8.5*

27. $A(2, 0)$, $B(5, 2)$, $C(4, 4)$; translation: $(a, b) \rightarrow (a - 2, b)$; reflection: over the x-axis

28. $A(2, 6)$, $B(3, 5)$, $C(1, 4)$; translation: $(a, b) \rightarrow (a + 2, b - 1)$; reflection: over the y-axis

29. $A(-1, 0)$, $B(3, 2)$, $C(2, 4)$; translation: $(a, b) \rightarrow (a, b + 3)$; reflection: over the line $y = x$

30. $A(-3, 3)$, $B(1, 3)$, $C(-1, 0)$; translation: $(a, b) \rightarrow (a - 2, b)$; reflection: over the line $y = -x$

Draw the image of each polygon after a dilation with center *O* and the given scale factor. *Section 8.6*

31. scale factor 2

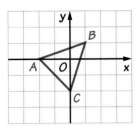

32. scale factor $\frac{1}{2}$

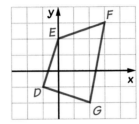

33. scale factor $\frac{2}{3}$

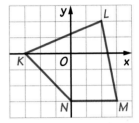

CHAPTER 9

ALGEBRA Solve each proportion. *Section 9.1*

1. $\dfrac{3}{5} = \dfrac{x}{20}$

2. $\dfrac{3}{4} = \dfrac{y}{10}$

3. $\dfrac{5}{z} = \dfrac{z}{9}$

4. $3 : 1 = y : 20$

5. $3 : x = x : 27$

6. $z : 5 = 12 : 30$

Tell whether the polygons in each pair are similar. Explain your reasoning. *Section 9.1*

7.
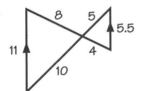

8.
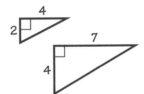

9.

The polygons in each pair are similiar. Find the value of each variable. *Section 9.1*

10.
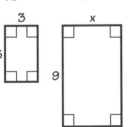

11.
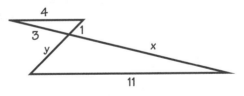

12.
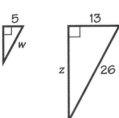

13.
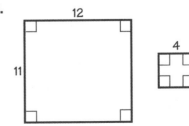

Is it possible to prove that the triangles in each pair are similar? Explain why or why not. *Section 9.2*

14.

15.
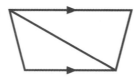

16.
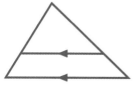

17.
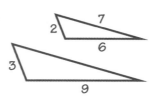

Explain why the triangles in each pair are similar. *Section 9.2*

18.
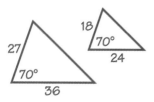

19.
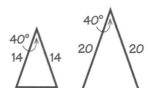

20.
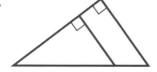

21.
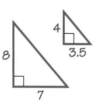

22.
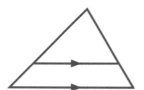

23.
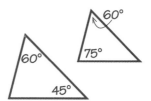

Find the value of each variable. *Section 9.3*

24.
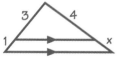

25.
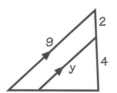

26.

27.
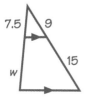

28.
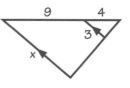

29.
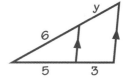

ALGEBRA Complete the equation so that it is equivalent to the proportion $\frac{x}{y} = \frac{4}{5}$. *Section 9.3*

30. $\frac{y}{x} = \underline{\ ?\ }$

31. $\frac{x}{4} = \underline{\ ?\ }$

32. $\underline{\ ?\ } = \frac{y + 5}{5}$

The figures in each pair are similar. Find each missing value. *Section 9.4*

33. $A = 5$ $A = \underline{\ ?\ }$

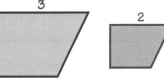

34. $P = 24$ $P = \underline{\ ?\ }$
$A = \underline{\ ?\ }$ $A = 54$

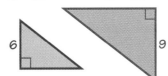

35. $C = 6\pi$ $C = \underline{\ ?\ }$

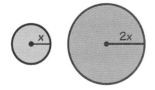

36. Area of $\triangle DEC = 16$
Area of $\triangle ABC = \underline{\ ?\ }$

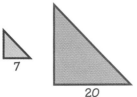

37. $P = \underline{\ ?\ }$ $P = 36$

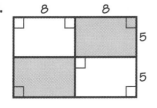

38. $A = \underline{\ ?\ }$ $A = 200$

Find the probability of hitting the shaded area of the target with a bean that is tossed at random that hits the target. *Section 9.5*

39.

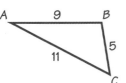

40.

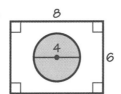

41.

If you choose a random point on a side of the polygon, what is the probability that it is on the given segment? *Section 9.5*

42. $\overline{AB}$

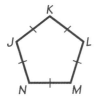

43. $\overline{PQ}$

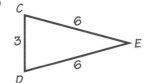

44. $\overline{JK}$

45. $\overline{CD}$

CHAPTER 10

Use the diagram to complete each statement. *Section 10.1*

1. $\angle QPR \cong \angle \underline{\,?\,}$

2. $\triangle PQR \sim \triangle \underline{\,?\,} \sim \triangle \underline{\,?\,}$

3. $\dfrac{PQ}{QR} = \dfrac{PS}{?}$

4. QS is the geometric mean of $\underline{\,?\,}$ and $\underline{\,?\,}$.

ALGEBRA   Find the geometric mean of the given numbers. *Section 10.1*

5. 3 and 27

6. 8 and 3

7. 9 and $\dfrac{1}{4}$

8. 23.04 and 34.81

9. $\dfrac{2}{5}$ and $\dfrac{5}{2}$

10. 0.5 and 50

Find the value of each variable. *Section 10.1*

11.

12.

13.

14.

15.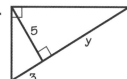

16.

Find the value of each variable. *Section 10.2*

17.

18.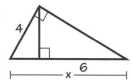

19.

20.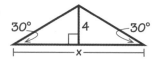

21.

22.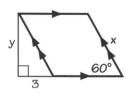

23.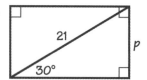

24.

25.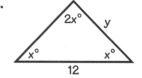

680   Extra Practice

For Exercises 26–28, find tan *A* and tan *B*. *Section 10.3*

26.

27.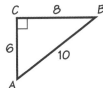

28.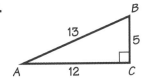

Find the value of each expression. Round your answers to four decimal places.
Section 10.3

29. $\tan 28°$

30. $\tan 12.5°$

31. $\tan 85°$

Find the value of each variable. Round your answers to the nearest tenth.
Section 10.3

32.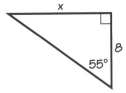

33.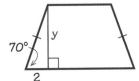

34.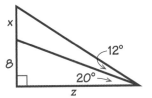

For Exercises 35–37, find sin *A*, cos *A*, sin *B*, and cos *B*. *Section 10.4*

35.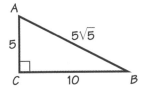

36.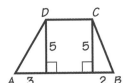

37.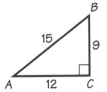

Find the value of each expression. Round your answers to four decimal places.
Section 10.4

38. $\sin 48°$

39. $\sin 8.5°$

40. $\cos 68.3°$

41. $\cos 12°$

Find the measure of an acute angle that satisfies the given equation. Round your answers to the nearest tenth of a degree. *Section 10.4*

42. $\sin A = \dfrac{9}{13}$

43. $\cos B = 0.2554$

44. $\sin M = \dfrac{10}{17}$

45. $\cos W = 0.9797$

46. $\cos D = 0.0175$

47. $\sin V = \dfrac{10}{13}$

48. $\sin C = 0.500$

49. $\cos Z = \dfrac{5}{9}$

Find the value of each variable. *Section 10.4*

50.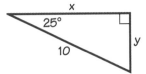

51.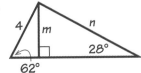

52.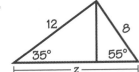

Find the value of each variable. Round your answers to the nearest tenth of a degree. *Section 10.4*

53.

54.

55.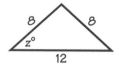

Express each vector in component form and find the value of each variable. *Section 10.5*

56.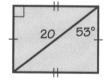

57.

58.

Graph each vector and find its magnitude. *Section 10.5*

59. $\overrightarrow{HJ} = (4, 9)$ **60.** $\overrightarrow{JK} = (-2, 7)$ **61.** $\overrightarrow{MN} = (20, -3)$ **62.** $\overrightarrow{PR} = (-3, -3)$

$\overrightarrow{AB} = (2, -4)$ and $\overrightarrow{CD} = (-3, 7)$. **Use a graph to find each vector sum. Write the resulting vector in component form.** *Section 10.5*

63. $\overrightarrow{AB} + \overrightarrow{CD}$ **64.** $2\overrightarrow{AB} + \overrightarrow{CD}$ **65.** $\overrightarrow{AB} + 2\overrightarrow{CD}$

Find the area of each polygon. *Section 10.6*

66.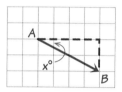

67.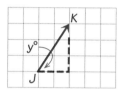

68.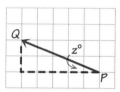

Find the volume of each prism. *Section 10.6*

69.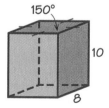

70.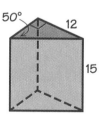

71.

Find the volume and surface area of each right cone or regular pyramid. *Section 10.7*

72.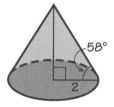

73.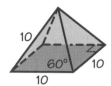

CHAPTER 11

Find each measure. *Section 11.1*

1. $m\,\widehat{ABC} = \underline{\ ?\ }$

2. $m\angle WXY = \underline{\ ?\ }$

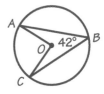

3. $m\,\widehat{AB} = \underline{\ ?\ }$

4. $m\,\widehat{BC} = \underline{\ ?\ }$

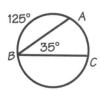

5. $m\,\widehat{ACB} = \underline{\ ?\ }$

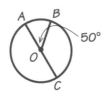

6. $m\,\widehat{AC} = \underline{\ ?\ }$

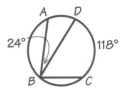

7. $m\,\widehat{BAC} = \underline{\ ?\ }$

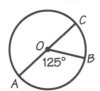

8. $m\angle AOC = \underline{\ ?\ }$

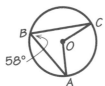

9. $m\,\widehat{ABC} = \underline{\ ?\ }$

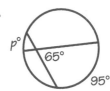

Find the value of each variable. *Section 11.2*

10.

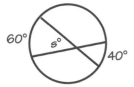

11.

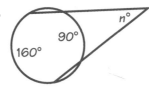

12.

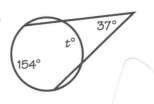

13.

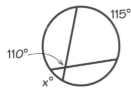

14.

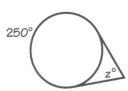

15.

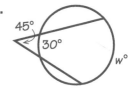

16.

17.

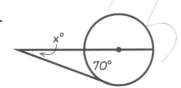

18.

ALGEBRA Find the value of each variable. *Section 11.3*

19.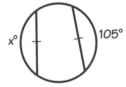

20.

21.

22.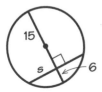

23.

24.

25.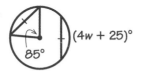

26.

27.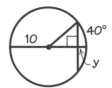

Find each unknown length. *Section 11.4*

28.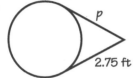

29.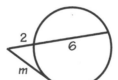

30.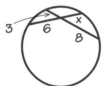

31.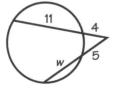

32.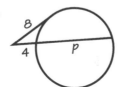

33.

Find the area of each shaded sector to the nearest tenth. *Section 11.5*

34.

35.

36.

37.

38.

39.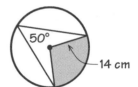

Find the length of the red arc in each circle. Give your answers to the nearest tenth. *Section 11.5*

40.

120°
15 m

41.

7 in.
80°

42.

120°
10 m

Give the surface area and the volume of each sphere to the nearest tenth. *Section 11.6*

43.

12 in.

44.

12 ft

45.

8 m

For the similar figures, find each missing volume. *Section 11.7*

46. $V = 4500$ in.^{3} $V = \underline{\ ?\ }$

5x
2x

47. $V = \underline{\ ?\ }$ $V = \dfrac{16\pi}{3}$ ft^3

2x
x

Tell whether each statement is *true* or *false*. *Section 11.8*

48. On a sphere, parallel lines intersected by a transversal form corresponding angles that are congruent.

49. On a sphere, an equilateral triangle has angles that measure 60°.

50. On a sphere, the sum of the interior angles of a quadrilateral is 360°.

CHAPTER 12

Use the graph. Write a matrix that represents each figure. *Section 12.1*

1. A

2. B

3. AC

4. $\triangle ABC$

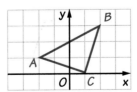

Find each product. *Section 12.1*

5. $2\begin{bmatrix} 4 \\ -3 \end{bmatrix}$

6. $1.5\begin{bmatrix} 4 & -12 \\ 6 & 14 \end{bmatrix}$

7. $\dfrac{1}{2}\begin{bmatrix} 8 & 6 & -6 \\ -4 & 2 & 10 \end{bmatrix}$

8. $7\begin{bmatrix} 2 & -1 \\ -3 & -2 \end{bmatrix}$

9. $2.5\begin{bmatrix} 8 & -16 \\ 6 & 12 \end{bmatrix}$

10. $\dfrac{2}{3}\begin{bmatrix} 9 & 6 & -6 \\ -3 & 3 & 12 \end{bmatrix}$

Add the matrices. *Section 12.2*

11. $\begin{bmatrix} 4 \\ -1 \end{bmatrix} + \begin{bmatrix} -7 \\ 10 \end{bmatrix}$

12. $\begin{bmatrix} -1 & -1 \\ 6 & 6 \end{bmatrix} + \begin{bmatrix} 4 & -2 \\ -2 & -5 \end{bmatrix}$

13. $\begin{bmatrix} 2 & 2 & 2 \\ -3 & -3 & -3 \end{bmatrix} + \begin{bmatrix} 0 & 2 & 3 \\ -1 & 3 & 2 \end{bmatrix}$

Write the matrix you would use to translate each figure as indicated. *Section 12.2*

14. a point; to the right 2 units and down 1 unit

15. a segment; to the left 3 units

16. a triangle; to the right 2 units and up 1 unit

17. a quadrilateral; down 4 units

Complete each matrix product. *Section 12.3*

18. $\begin{bmatrix} 3 & -1 \\ 5 & 0 \end{bmatrix} \begin{bmatrix} 2 \\ -2 \end{bmatrix}$

19. $\begin{bmatrix} 1 & 0 \\ 0 & 1 \end{bmatrix} \begin{bmatrix} -4 & -1 \\ 10 & 3 \end{bmatrix}$

20. $\begin{bmatrix} 2 & -1 \\ 1 & -2 \end{bmatrix} \begin{bmatrix} 1 & 3 & 2 \\ 2 & 1 & -1 \end{bmatrix}$

21. $\begin{bmatrix} 0 & 0.5 \\ 1.2 & 1 \end{bmatrix} \begin{bmatrix} 2 & 4 & 8 \\ 0 & -1 & 3 \end{bmatrix}$

For each transformation matrix and polygon given, use the matrix to transform the polygon. Then graph the polygon and its image. *Section 12.3*

22. $\begin{bmatrix} -1 & 2 \\ 2 & -1 \end{bmatrix}$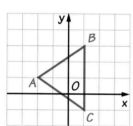

23. $\begin{bmatrix} -3 & 1 \\ 0 & 4 \end{bmatrix}$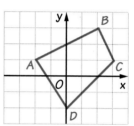

Complete each matrix product to find the image of (2, –1) after a reflection, then state the line over which the point has been reflected. *Section 12.4*

24. $\begin{bmatrix} 1 & 0 \\ 0 & -1 \end{bmatrix} \begin{bmatrix} 2 \\ -1 \end{bmatrix}$

25. $\begin{bmatrix} 0 & 1 \\ 1 & 0 \end{bmatrix} \begin{bmatrix} 2 \\ -1 \end{bmatrix}$

26. $\begin{bmatrix} -1 & 0 \\ 0 & 1 \end{bmatrix} \begin{bmatrix} 2 \\ -1 \end{bmatrix}$

27. $\begin{bmatrix} 0 & -1 \\ -1 & 0 \end{bmatrix} \begin{bmatrix} 2 \\ -1 \end{bmatrix}$

28. The matrix $\begin{bmatrix} 2 & 4 & 5 \\ 0 & -1 & 3 \end{bmatrix}$ represents $\triangle ABC$. *Section 12.5*

   a. Use the matrix $\begin{bmatrix} 0 & -1 \\ 1 & 0 \end{bmatrix}$ to transform $\triangle ABC$.

   b. Use the matrix $\begin{bmatrix} -1 & 0 \\ 0 & -1 \end{bmatrix}$ to transform the result from part (a).

   c. Graph $\triangle ABC$, $\triangle A'B'C'$, and $\triangle A''B''C''$. Describe the result of each transformation. What is the combined effect of the two transformations in parts (a) and (b)?

Toolbox

USING GEOMETRIC TOOLS AND TRANSFORMATIONS

Using a Protractor

A *protractor* is used to measure angles between 0° and 180°. To measure an angle, place the center mark of the protractor over the vertex of the angle and the 0° mark on one side of the angle. The numbers on one curve of the protractor measure angles counterclockwise. The numbers on the other curve of the protractor measure angles clockwise.

EXAMPLE

Use a protractor to measure the angles.

a.

b.
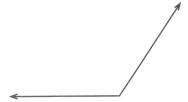

SOLUTION

a.
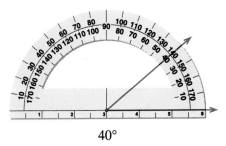

40°

b.
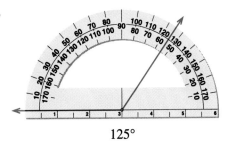

125°

PRACTICE

Use a protractor to measure the angles.

1.

2.

3.

4.

5.

6.
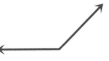

7.

8.
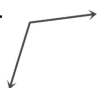

Using a Compass

A *compass* is used to draw a circle of a certain radius. Use a ruler to mark the center of the circle and a point one radius away from the center. Place the point of the compass on the center, and adjust the compass so that the pencil is on the point one radius away. Rotate the pencil about the center to draw the circle.

EXAMPLE

Draw a circle with the given radius.

a. 1.5 in.

b. 3 cm

SOLUTION

a.

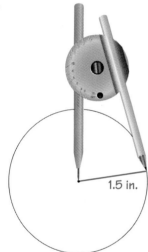

1.5 in.

b.

You can also use the measurements marked on the compass.

3 cm

PRACTICE

Draw a circle with the given radius.

1. 2 in.

2. 0.75 in.

3. $\frac{1}{2}$ in.

4. $1\frac{1}{4}$ in.

5. 6 cm

6. 3 cm

7. 4 cm

8. 2.5 cm

Transformations

A *transformation* of a geometric figure is a change in the position or size of the figure. The result of a transformation is called an *image*.

EXAMPLE

a. Translate line segment $\overline{AB}$ down 2 units and to the left 1 unit.

b. Reflect $\overline{AB}$ over the y-axis.

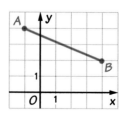

a.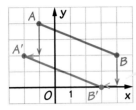

Move each
point **down**
two units and
left one unit.

b.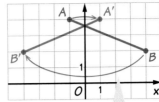

Reflect each point over the *y*-axis.

Copy the diagram of △DEF. Sketch the image of each transformation.

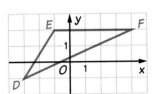

1. Translate 2 units up.

2. Reflect over the *y*-axis.

3. Translate 3 units left.

4. Reflect over the *x*-axis.

5. Translate 1 unit down and 4 units right.

6. Translate 1 unit up, then reflect over the *y*-axis.

Recognizing Symmetry

A geometric figure has symmetry if the figure can be drawn by rotating or
reflecting one part.

A flower has *rotational symmetry*.

A cap has *reflection symmetry*.

The red petal is
rotated and
copied to draw
the other petals.

The red side
of the cap is
reflected to
draw the
other side.

**State whether the figure has symmetry. If it does, tell whether it is *rotational*
or *reflection symmetry*.**

a.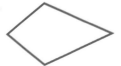
b.
c.
d.

SOLUTION

a. reflection

b. reflection
and rotational

c. no symmetry

d. rotational
and reflection

State whether the figure has symmetry. If it does, tell whether it is *rotational* or *reflection symmetry*.

1.  **2.**   **3.**  **4.**

5. Draw an object with reflection symmetry.

6. Draw an object with rotational symmetry.

7. Draw an object that has no symmetry.

OPERATIONS WITH VARIABLE EXPRESSIONS

Simplifying and Evaluating Expressions

When you simplify a numerical expression, you must use the *order of operations*, a set of rules that guarantees that an expression has just one value.

> ### Order of Operations
>
> **1.** First simplify expressions inside parentheses or other grouping symbols.
> **2.** Then evaluate powers.
> **3.** Next, do multiplications and divisions in order from left to right.
> **4.** Last, do additions and subtractions in order from left to right.

Also, remember the following rules:

$$a - (-b) = a + b \qquad (-1)(-a) = a$$

EXAMPLE 1

Simplify each expression.

a. $-5(-2 - (-3))$

b. $\dfrac{3^3 + 2^2 - (2 - 4)}{12 \div 2 + (1 \cdot 6) + 3}$

SOLUTION

a. $-5(-2 - (-3)) = -5(-2 + 3)$

$$= -5(1)$$

$$= -5$$

b. $\dfrac{3^3 + 2^2 - (2 - 4)}{12 \div 2 + (1 \cdot 6) + 3} = \dfrac{27 + 4 - (-2)}{12 \div 2 + 6 + 3}$

> The fraction bar acts like parentheses. Simplify the numerator and the denominator, then simplify the fraction.

$$= \dfrac{27 + 4 + 2}{6 + 6 + 3}$$

$$= \dfrac{33}{15} = \dfrac{11}{5}$$

A *variable expression* is an expression formed using variables, numbers, and operation symbols. For a variable expression like $x^2 + 2x^2 - 5x + 1$, the parts that are joined by plus signs and minus signs are called *terms*. Terms with the same variable parts, such as x^2 and $2x^2$, are called *like terms*.

To write an expression in simplest form, find an equivalent expression that has no parentheses and has all like terms combined. In simplest form, the expression $x^2 + 2x^2 - 5x + 1$ becomes $3x^2 - 5x + 1$.

EXAMPLE 2

Simplify each expression.

a. $-3(-t + t^2 - 5t)$

b. $\frac{1}{2}(2x + 4) + 7x - 8$

SOLUTION

a. $-3(-t + t^2 - 5t) = -3(t^2 - 6t)$ ← Combine like terms.

$\qquad = -3(t^2) - 3(-6t)$ ← Use the distributive property.

$\qquad = -3t^2 - (-18t)$

$\qquad = -3t^2 + 18t$

b. $\frac{1}{2}(2x + 4) + 7x - 8 = \frac{1}{2}(2x) + \frac{1}{2}(4) + 7x - 8$

$\qquad = x + 2 + 7x - 8$

$\qquad = 8x - 6$

To *evaluate* a variable expression, substitute a value for each variable and simplify the resulting expression using the *order of operations*.

EXAMPLE 3

Evaluate each expression when $a = -4$ and $b = 6$.

a. $(a + b)^3 - a^2 + 2b$

b. $3(2a + b) - \frac{2}{3}b(7 - b)$

SOLUTION

a. $(a + b)^3 - a^2 + 2b = (-4 + 6)^3 - (-4)^2 + 2(6)$ ← Substitute **−4** for *a* and **6** for *b*.

$\qquad = 2^3 - (-4)^2 + 2(6)$ ← Use the order of operations.

$\qquad = 8 - (16) + 2(6)$

$\qquad = 8 - 16 + 12 = 4$

b. $3(2a + b) - \frac{2}{3}b(7 - b) = 3(2(-4) + 6) - \frac{2}{3} \cdot 6(7 - 6)$

$\qquad = 3(-8 + 6) - \frac{2}{3} \cdot 6(1)$

$\qquad = 3(-2) - \frac{2}{3}(6)$

$\qquad = -6 - 4 = -10$

Simplify each expression.

1. $-3(-2 + 5) + 1$　　　**2.** $13 + 11 \cdot 4 - 8$　　　**3.** $(-2 - 5) \cdot (3 \cdot 8)$　　　**4.** $4 \cdot 2 - (3 - 9)$

5. $42 \div (14 \div 2) - 15$　　**6.** $(18 - 4^2) \cdot (-3^2)$　　　**7.** $12 - 18 + 15 \div 3$　　　**8.** $-7(-3) + 8 \cdot 4$

9. $(63 \div 3) \div (15 - 8)$　**10.** $\dfrac{20}{5^3 \div 5 - 5 \cdot 3}$　　**11.** $\dfrac{1 + 3 + 8 - 2}{-15 + 3 - 3 \cdot 2}$　　**12.** $8 \cdot 2 - \dfrac{(2 + 2)3}{3^2 - 5}$

Simplify each expression.

13. $8x + 2 - 5x - 3$　　　　**14.** $4x^2 + x^3 - 2x^3 - 3x^3$　　**15.** $-x^3 + 5x(x^2 - 2)$

16. $17x + 3(7 - 2x)$　　　　**17.** $12(t + 3) + t(t - 3)$　　　**18.** $-3x(-x - 5 - 2x)$

19. $x(x^2 + 1) - x^3$　　　　**20.** $13m - 12m(5 - 2)$　　　　**21.** $5x^3 + 7(x^2 + x + 1)$

22. $\dfrac{3x + 5 - 2}{3}$　　　　**23.** $\dfrac{3}{2}(-4x + 8) - x$　　　　**24.** $\dfrac{14x}{2}(x - 2) + 3(-x)$

Evaluate each expression when $t = -7$.

25. $t + 2(t - 1)$　　　**26.** $3t^2 - 12t + 8$　　　**27.** $8t$　　　　　　**28.** $5(t + 4)^2 + (2t + 1)$

29. $\dfrac{1}{2}(t + 13) - t$　　**30.** $11 - 2t$　　　　　**31.** $3t - t^2 + (t + 2)^2$　　**32.** $5(t + 9)^3 + \dfrac{2t - 1}{t + 4}$

Evaluate each expression when $x = 3$ and $y = -1$.

33. $2(x + 2y)$　　　　**34.** $(x + y)(x - y)$　　　**35.** $12(13y + 4x) - 1$　　**36.** $(2x + 3y)^3 + 2x^2$

37. $x^2 + y^2 + 12 - y$　**38.** $x + y + 3x - y$　　　**39.** $\dfrac{2(8x - y)}{x - 7y}$　　　**40.** $\dfrac{2x - 8y}{x + y} + (x + 1)^2$

Translating Phrases into Variable Expressions

Often a mathematics problem is presented in words. The words must be translated into a variable expression to solve the problem.

EXAMPLE

Rewrite each statement as a variable expression.

a. Oranges cost $1 per dozen.

b. Each player brings three markers.

c. A science class splits into lab groups with five students in each group.

d. The total cost depends on the number of tickets, at $22 each, and the number of cars parked, at $10 each.

SOLUTION

a. You pay 1 dollar for every 12 oranges, so the cost of x oranges is $\dfrac{x}{12}$.

b. The number of markers is three times the number of players, or $3p$.

c. The number of students in the class is five times the number of lab groups, or $5g$.

d. The cost for t tickets and c cars is $22t + 10c$.

Rewrite each statement as a variable expression.

1. A child must pick up three toys for each year of his or her age.

2. Lunch costs $4.25 for each main dish and $.75 for each side dish. Drinks are $1 each.

3. Lionel wants to pour orange juice evenly into 8 glasses.

4. Monica charges $20 an hour for lost data retrieval.

5. The hurricane is moving northeast at 45 miles per hour.

6. The number of tiles used in each mosaic are 44 yellow tiles, 125 white tiles, 90 blue tiles, 78 red tiles, and 13 green tiles.

Algebraic Properties

The following properties are useful when you are calculating with numbers or simplifying algebraic expressions.

Property	Commutative $a + b = b + a$ $ab = ba$	Associative $(a + b) + c = a + (b + c)$ $(ab)c = a(bc)$	Distributive $a(b + c) = ab + ac$
Summary	You can add or multiply numbers in any order without changing the result. $5 + 2 = 2 + 5$ $5 \cdot 2 = 2 \cdot 5$	When you add or multiply three or more numbers, you can regroup the numbers without changing the result. $(-4 + 5) + 3 = -4 + (5 + 3)$ $(-4 \cdot 5) \cdot 3 = -4 \cdot (5 \cdot 3)$	When a sum is multiplied by a number, you can distribute the multiplication to each of the numbers being added. $8(4 + 1) = 8 \cdot 4 + 8 \cdot 1$

EXAMPLE

Tell what property is used in the following sequence of steps:
$-3 + 16 = 16 - 3 = 13.$

SOLUTION

The order of the factors has been changed. The commutative property is used.

PRACTICE

Tell what property is used in each lettered step.

1. a. $-4(x - 1) + 6 = 6 - 4(x - 1)$
   b. $\qquad = 6 - 4x + 4$
   $\qquad\quad = 10x - 4$

2. a. $(3x)(5y) = (3x \cdot 5)y$
   b. $\qquad\quad = (3 \cdot 5x)y$
   c. $\qquad\quad = (3 \cdot 5)xy$
   $\qquad\quad = 15xy$

3. a. $4(2 + a) + 7 = 4(2) + 4(a) + 7$
   b. $\qquad\qquad = 4a + 4(2) + 7$
   $\qquad\qquad = 4a + 15$

LINEAR EQUATIONS

Evaluating Equations for Given Values

To evaluate an equation for a given value, substitute the value for the variable.

EXAMPLE

For $y = 3x + 2$, find y when $x = -5$.

SOLUTION

$$y = 3x + 2$$
$$= 3(-5) + 2$$
$$= -15 + 2$$
$$= -13$$

PRACTICE

Find y when $x = 2$.

1. $y = x - 6$
2. $y = 3x - 5$
3. $y = 8 - 4x$
4. $y = 4(2x - 1)$

5. $y = x^2 + 7$
6. $y = 3x^2 + 8x - 14$
7. $y = \dfrac{-x + 1}{4}$
8. $y = \dfrac{3}{2}(x + 2)^2$

Translating Sentences into Equations

Often a mathematics problem is presented in words. The words must be translated into an equation to solve the problem.

EXAMPLE

Rewrite the following sentences as equations.

a. What is the cost of a bunch of bananas if bananas sell for $.70 per pound?

b. How many books will Maura read if she reads 2 books a week?

c. How far will Zachary travel if he drives at an average speed of 45 miles per hour?

d. How can you measure a distance in yards with a ruler that is only 1 ft long?

SOLUTION

a. The cost c in dollars equals 0.70 times the weight w of the bananas in pounds.
$c = 0.70w$

b. The total number of books b equals 2 times the number of weeks w.
$b = 2w$

c. The distance d in miles equals 45 times the time t in hours.
$d = 45t$

d. The distance y in yards equals 3 times the number of feet f.
$y = 3f$

Rewrite the following sentences as equations.

1. How much will Manuel pay for soda if soda costs $.80 per bottle?

2. A potter makes seventeen pieces a week. How many pieces are produced over a period of time?

3. How fast does Elaine run if she covers a distance in 15 seconds?

4. How much will Laura pay for using a computer program, if the cost is $1 for using the program plus $.05 per minute used?

5. What is the total weight of a container of food from the deli, if the container weighs 0.01 lb?

6. If a chemical reaction requires two hydrogen atoms for every oxygen atom, how much oxygen is needed?

Solving One-Step Equations

Any value of a variable that makes an equation true is called a *solution* of the equation. To find a solution, you need to get the variable alone on one side of the equation. You can do this by adding or subtracting the same value from each side of the equation.

EXAMPLE 1

Solve each equation.

a. $x + 8 = 0$

b. $y - 5 = 11$

SOLUTION

a. $x + 8 = 0$
$x + 8 - 8 = 0 - 8$ — Subtract **8** from each side.
$x = -8$

b. $y - 5 = 11$
$y - 5 + 5 = 11 + 5$ — Add **5** to each side.
$y = 16$

You can also isolate the variable on one side by multiplying or dividing each side of the equation by the same value.

EXAMPLE 2

Solve each equation.

a. $-3t = 21$

b. $\frac{3}{4}w = 12$

SOLUTION

a. $-3t = 21$
$\frac{-3t}{-3} = \frac{21}{-3}$ — Divide each side by **-3**.
$t = -7$

b. $\frac{3}{4}w = 12$
$\frac{4}{3} \cdot \frac{3}{4}w = \frac{4}{3} \cdot 12$ — Multiply each side by $\frac{4}{3}$, the inverse of $\frac{3}{4}$.
$w = 16$

Solve each equation.

1. $y + 2 = 7$ **2.** $x - 5 = 8$ **3.** $y + 4 = 13$

4. $3 + t = -1$ **5.** $3x = 8$ **6.** $\frac{1}{3}z = 15$

7. $t \div 8 = 4$ **8.** $-1 + x = -2$ **9.** $\frac{6}{5}y = 6$

10. $8w = 9$ **11.** $2 = \frac{1}{2}y$ **12.** $2x = -6$

Solving Two-Step Equations

Sometimes it takes more than one step to solve an equation. You may need to add or subtract, and then multiply or divide.

EXAMPLE

Solve each equation.

a. $6x + 4 = 16$ **b.** $-0.2x - 8.1 = 0.3$

SOLUTION

a.
$$6x + 4 = 16$$
$$6x + 4 - 4 = 16 - 4$$
Subtract **4** from each side.
$$6x = 12$$
$$\frac{6x}{6} = \frac{12}{6}$$
Divide each side by **6**.
$$x = 2$$

b.
$$-0.2x - 8.1 = 0.3$$
Add **8.1** to each side.
$$-0.2x - 8.1 + 8.1 = 0.3 + 8.1$$
$$-0.2x = 8.4$$
$$\frac{-0.2x}{-0.2} = \frac{8.4}{-0.2}$$
Divide each side by **-0.2**.
$$x = -42$$

PRACTICE

Solve each equation.

1. $4x + 5 = 7$ **2.** $-x + 3 = 2$ **3.** $8 \div t = 2$ **4.** $12x + 8 = -6$

5. $\frac{1}{2}y + 2 = 3$ **6.** $-\frac{5}{3}x + 3 = -12$ **7.** $\frac{1}{x} = \frac{1}{9}$ **8.** $12 = 4 - 2y$

9. $1.2x + 3 = -4.7$ **10.** $18 = -5 + 3x$ **11.** $9.90 = 3.35 + 1.2t$ **12.** $4 - 5y = 1.8$

GRAPHING ON THE COORDINATE PLANE

Graphing Points

A *coordinate plane* consists of a horizontal *x-axis* and a vertical *y-axis* that intersect at a point called the *origin*, labeled *O*. The axes divide the coordinate plane into four *quadrants* as shown at the top of page 697.

Each point in a coordinate plane is associated with an *ordered pair* (*a*, *b*) of real numbers. The first number, *a*, is the *x-coordinate*. The second number, *b*, is the *y-coordinate*.

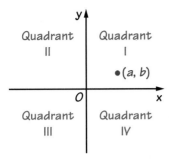

EXAMPLE

Graph the points *A*(−4, −2) and *B*(0, 3) in a coordinate plane. Name the quadrant, if any, in which each point lies.

SOLUTION

To graph *A*(−4, −2), start at the origin and move **left** 4 units and **down** 2 units. The point is in Quadrant III.

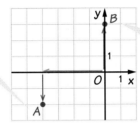

To graph *B*(0, 3), start at the origin and move horizontally 0 units and **up** 3 units. The point is not in any quadrant.

PRACTICE

For Exercises 1–8, graph each point in the same coordinate plane. Name the quadrant, if any, in which each point lies.

1. $A(0, 1)$ **2.** $B(-3, 6)$ **3.** $C(2, -2)$ **4.** $D(7, 0)$

5. $E(0, 0)$ **6.** $F(-2, -2)$ **7.** $G(-3, 0)$ **8.** $H(4, 3)$

Give the coordinates of a point that satisfies each condition.

9. The point is in the fourth quadrant.

10. The point is on the negative *x*-axis.

Graphing Linear Equations

A *linear equation* is an equation whose graph is a line. One way to write a linear equation is $y = mx + b$. The slope is represented by the value *m*, and the *y-intercept* is represented by the value *b*.

The slope of a line is the vertical change divided by the horizontal change. The slope of a line through points (x_1, y_1) and (x_2, y_2) is:

$$m = \frac{y_2 - y_1}{x_2 - x_1}$$

EXAMPLE 1

Graph the equation.

a. $y = 2x - 3$

b. $y = -\frac{2}{3}x + 1$

SOLUTION

Graph the y-intercept. Use the slope to graph one or two more points.

a.

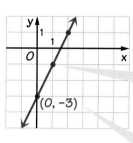

The slope is $\frac{2}{1}$; go up 2 and right 1.

The y-intercept is -3.

b.

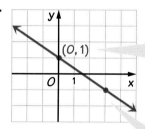

The y-intercept is 1.

The slope is $-\frac{2}{3}$; go down 2 and right 3.

You can find the slope of a line by putting an equation into slope-intercept form and then reading m from the equation. You can determine the slope of a graph by choosing two points and finding the vertical change and the horizontal change between them.

EXAMPLE 2

Find the slope of each line.

a. $y = 4x + 2$

b. $y = 3 - \frac{1}{2}x$

c. $x + y = 1$

SOLUTION

Subtract x from each side.

Rewrite the equations in slope-intercept form.

a. $y = 4x + 2$

slope: 4

b. $y = 3 - \frac{1}{2}x$

$y = -\frac{1}{2}x + 3$

slope: $-\frac{1}{2}$

c. $x + y = 1$

$x + y - x = 1 - x$

$y = -x + 1$

slope: -1

EXAMPLE 3

Find the slope of the line.

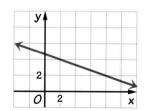

SOLUTION

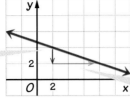

The **vertical change** is **–2**.

The **horizontal change** is **6**.

The slope is $\dfrac{-2}{6}$, or $-\dfrac{1}{3}$.

You can find the y-intercept of a line by putting an equation into slope-intercept form and then reading b from the equation. You can determine the y-intercept of a graph by inspection.

EXAMPLE 4

Find the y-intercept of each line.

a. $y = -x + 2$

b. $y = 4 - 5x$

c. $3x + y = 0$

SOLUTION

Rewrite the equations in slope-intercept form.

a. $y = -x + 2$
  y-intercept: 2

b. $y = 4 - 5x$
  $y = -5x + 4$
  y-intercept: 4

c. $3x + y = 0$
  $y = -3x$
  y-intercept: 0

EXAMPLE 5

Find the y-intercept of the line in Example 3.

SOLUTION

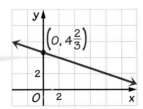

The y-intercept is $4\dfrac{2}{3}$.

$\left(0, 4\dfrac{2}{3}\right)$

PRACTICE

Find the slope and y-intercept of each line.

1. $y = 3x$

2. $y = -2x$

3. $y = -x + 2$

4. $y = -4$

5. $y = -5x - 1$

6. $y = -3 + 8x$

7. $y - x = 2$

8. $y = 122 + 13.8x$

9. $y = x - 3 + 2x$

10. $y = -\dfrac{5}{4}x - \dfrac{3}{2}$

11. $y = \dfrac{1}{2}x + \dfrac{1}{2}$

12. $y + \dfrac{2}{3}x - 3 = 0$

13–24. Graph each of the equations in Exercises 1–12.

For each graph in Exercises 25–33, find the slope and the *y*-intercept.

25.

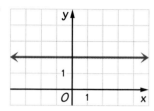

26.

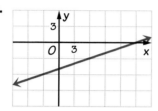

27.

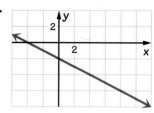

28.

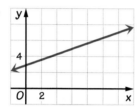

29.

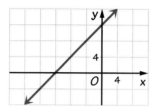

30.

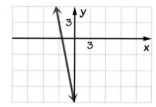

31.

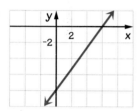

32.

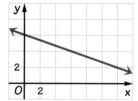

33.

QUADRATIC EQUATIONS

Simplifying Radicals

An expression written using the symbol $\sqrt{}$ is in *radical form*. An expression in radical form is in simplest form when the expression under the radical sign is in simplest form and:

1. there is no integer under the radical sign with a perfect square factor,

2. there are no fractions under the radical sign, and

3. there are no radicals in the denominator.

Properties of Square Roots

For all nonnegative numbers a and b:

$$\sqrt{a^2} = a$$

$$\sqrt{ab} = \sqrt{a} \cdot \sqrt{b}$$

$$\sqrt{\frac{a}{b}} = \frac{\sqrt{a}}{\sqrt{b}}, b \neq 0$$

$$\left(\sqrt{a}\right)^2 = a$$

For example:

$$\sqrt{36} = 6$$

$$\sqrt{9 \cdot 5} = \sqrt{9} \cdot \sqrt{5} = 3\sqrt{5}$$

$$\sqrt{\frac{2}{9}} = \frac{\sqrt{2}}{\sqrt{9}} = \frac{\sqrt{2}}{3}$$

$$\left(\sqrt{16}\right)^2 = (4)^2 = 16$$

EXAMPLE

Simplify each expression.

a. $\sqrt{\dfrac{27}{4}}$

b. $8\sqrt{12} - \sqrt{3}$

SOLUTION

a. $\sqrt{\dfrac{27}{4}} = \dfrac{\sqrt{27}}{\sqrt{4}}$

$= \dfrac{\sqrt{9 \cdot 3}}{\sqrt{4}}$

$= \dfrac{\sqrt{3^2} \cdot \sqrt{3}}{\sqrt{2^2}}$

$= \dfrac{3\sqrt{3}}{2}$

b. $8\sqrt{12} - \sqrt{3} = 8\sqrt{4 \cdot 3} - \sqrt{3}$

$= 8\sqrt{2^2} \cdot \sqrt{3} - \sqrt{3}$

$= 8 \cdot 2 \cdot \sqrt{3} - \sqrt{3}$

$= 16\sqrt{3} - \sqrt{3}$

$= (16 - 1)\sqrt{3}$

$= 15\sqrt{3}$

> Simplify each term in the expression.

> Use the distributive property.

PRACTICE

Simplify each expression.

1. $\sqrt{16}$

2. $\sqrt{75}$

3. $\sqrt{2}$

4. $\sqrt{9 \cdot 4}$

5. $\sqrt{\dfrac{8}{9}}$

6. $\sqrt{\dfrac{1}{64}}$

7. $\sqrt{\dfrac{15}{4}}$

8. $\sqrt{\dfrac{48}{100}}$

9. $\sqrt{8} + \sqrt{18}$

10. $\sqrt{125}$

11. $-\sqrt{45} - \sqrt{20}$

12. $\sqrt{9} - \sqrt{81}$

13. $\left(\sqrt{49}\right)^2$

14. $\sqrt{92 - 11}$

15. $\sqrt{\dfrac{2 \cdot 5}{40}}$

16. $\dfrac{\sqrt{54}}{6}$

Solving Simple Quadratic Equations

You can solve some simple equations using radicals.

EXAMPLE 1

Solve.

a. $x^2 = 16$

b. $x^2 = 15$

c. $2x^2 + 1 = 9$

SOLUTION

Rewrite the equation, if necessary, so that it is in the form $x^2 = a$, where a is positive. Find the square root of both sides.

a. $x^2 = 16$

$x = \pm\sqrt{16}$

$x = 4 \text{ or } x = -4$

b. $x^2 = 15$

$x = \pm\sqrt{15}$

$x = \sqrt{15} \text{ or } x = -\sqrt{15}$

c. $2x^2 + 1 = 9$

$2x^2 = 8$

$x^2 = 4$

$x = \pm 2$

$x = 2 \text{ or } x = -2$

Solve.

1. $x^2 = 25$ 2. $x^2 = 50$ 3. $2x^2 = 12$ 4. $3x^2 = 108$

5. $x^2 + 5 = 6$ 6. $-x^2 + 2 = -14$ 7. $2x^2 - 8 = 12$ 8. $2x^2 - 1 = -1$

9. $2x^2 = x^2 + 4$ 10. $3x^2 = 36 - x^2$ 11. $x^2 = 2^2 + 5^2$ 12. $3x^2 - 2 = 16 + x^2$

FORMULAS FOR GEOMETRIC FIGURES

Working with Formulas

A *formula* is a statement of a relationship between two or more quantities.
Some common formulas include

$$\text{speed} = \frac{\text{distance}}{\text{time}}, \text{ circumference} = 2\pi \cdot \text{radius, and distance} = \frac{1}{2} \cdot \text{acceleration} \cdot (\text{time})^2.$$

When you work with formulas, you may need to rewrite the formula to isolate a
variable on one side of an equation. To use a formula, substitute a number for each
known variable and solve.

EXAMPLE 1

Rewrite the formula $s = \dfrac{d}{t}$ to find an equation for (a) d and (b) t.

SOLUTION

a. $s = \dfrac{d}{t}$ To solve for d, multiply each side by t.

$st = d$

$d = st$ Write the equation with d on the left. This is called an equation for d in terms of s and t.

b. $s = \dfrac{d}{t}$ To get t out of the denominator, multiply by t.

$st = d$

$t = \dfrac{d}{s}$ To isolate t, divide each side by s.

EXAMPLE 2

**Use the formula $s = \dfrac{d}{t}$, where s is speed in mi/h, d is distance in miles, and
t is time in hours, to find each value.**

a. Find s when $d = 30$ and $t = 4$.

b. Find t when $s = 40$ and $d = 20$.

SOLUTION

a. $s = \dfrac{d}{t}$

$s = \dfrac{30}{4}$ Put in the values you know.

$s = \dfrac{15}{2}$, or 7.5 mi/h Include units in your answer.

b. $t = \dfrac{d}{s}$ Use the formula for t from Example 1.

$t = \dfrac{20}{40}$

$t = 0.5$ h

Use the formula $s = \dfrac{d}{t}$, where s is speed in mi/h, d is distance in miles, and t is time in hours, to find each value.

1. Find t when $d = 4$ and $s = 12$. **2.** Find d when $s = 50$ and $t = 0.5$.

3. Find d when $s = 35$ and $t = 1$. **4.** Find s when $d = 100$ and $t = 1.5$.

Use the formula $A = \pi r^2$, where A is the area in cm^2 and r is the radius in cm, to find each value.

5. Find A when $r = 0.4$. **6.** Find r when $A = 9\pi$.

7. Use the formula $a = 4\dfrac{bc}{d}$.

  a. Write an equation for c.

  b. Find a when $b = 2$ in., $c = 12$ in., and $d = 6$ in.

Finding Perimeter

The *perimeter* of a geometric figure is the distance around the figure, or the sum of the length of the edges.

EXAMPLE

Find the perimeter of each figure.

a.

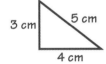

b.

SOLUTION

a. Add together the lengths of the edges: $3 + 4 + 5 = 12$ cm.

b. $3 + 2 + 5 + 1 = 11$

PRACTICE

Find the perimeter of each figure.

1.

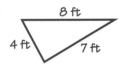

2.

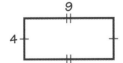

3.

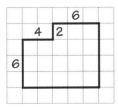

4. A square whose sides are 4 in. long.

5. A rectangle whose sides are 2 ft and 5 ft long.

6. A polygon whose sides are 3, 7, 2, 1, and 8 units long.

Finding Circumference

The perimeter of a circle, called its *circumference*, is the distance around the circle. The formula for circumference is $C = 2\pi r$, where r is the *radius*. Circumference and radius are both in units of length.

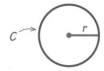

EXAMPLE

Find the circumference of a circle with radius 4 in.

SOLUTION

$$C = 2\pi r = 2\pi(4) = 8\pi \text{ in.}$$

PRACTICE

Find the circumference of a circle with each radius.

1. 8 in.	**2.** 2 cm	**3.** 1 in.	**4.** 12 in.
5. 2	**6.** 14 m	**7.** 30	**8.** 4.5

Find the radius of a circle with each circumference.

9. 18π in.	**10.** 34π	**11.** 1π	**12.** 2 mi

Finding Area and Volume

The *area* of a figure is a measure of the number of square units of space its surface covers; for a rectangle this is length times width, $A = \ell w$. The *volume* of a figure is the number of cubic units of space it occupies. For a rectangular prism this is length times width times height, $V = \ell w h$.

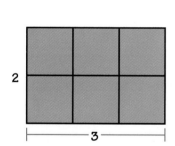

$A = 2 \cdot 3 = 6$ square units

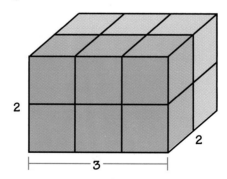

$V = 2 \cdot 3 \cdot 2 = 12$ cubic units

EXAMPLE

Find the area of a rectangle that is 2 in. long and 4 in. wide.

SOLUTION

$$A = \ell w = (2)(4) = 8 \text{ in.}^2$$

Find the area of a rectangle with each length and width.

1. $\ell = 4, w = 3$ **2.** $\ell = 1, w = 2$ **3.** $\ell = 14, w = 12$ **4.** $\ell = 7, w = 0.5$

5. $\ell = 1.5, w = 1.5$ **6.** $\ell = 5, w = 12$ **7.** $\ell = 2, w = 2$ **8.** $\ell = 1, w = 20$

Find the volume of a rectangular prism with the given dimensions.

9. $\ell = 1, w = 2, h = 5$ **10.** $\ell = 3, w = 3, h = 3$ **11.** $\ell = 10, w = 10, h = 14$

12. $\ell = 3, w = 1, h = 2$ **13.** $\ell = 4, w = 4, h = 6$ **14.** $\ell = 0.5, w = 0.5, h = 0.5$

For a rectangle, find the value of the given variable.

15. ℓ, when $A = 12$ and $w = 6$ **16.** w, when $A = 15$ and $\ell = 4$

17. w, when $A = 3$ and $\ell = 9$ **18.** A, when $\ell = 5$ and $w = 10$

INEQUALITIES

Solving Inequalities

You solve a *linear inequality* such as $2n + 1 \le 7$ in much the same way as a linear equation. One important difference is this: When you multiply or divide by a negative number, you must reverse the direction of the inequality sign.

EXAMPLE

Solve each inequality. Graph the solution on a number line.

a. $2n + 1 \le 7$ **b.** $-2x > 10 + 3x$

SOLUTION

a. $2n + 1 \le 7$

$2n + 1 - 1 \le 7 - 1$ — Subtract **1** from each side.

$2n \le 6$

$\dfrac{2n}{2} \le \dfrac{6}{2}$ — Divide each side by **2**.

$n \le 3$

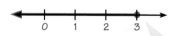

The closed circle shows that 3 *is* a solution.

b. $-2x > 10 + 3x$

$-2x - 3x > 10 + 3x - 3x$ — Subtract **3x** from each side.

$-5x > 10$

$\dfrac{-5x}{-5} < \dfrac{10}{-5}$ — Divide each side by **−5**, reversing the inequality sign.

$x < -2$

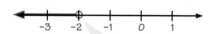

The open circle shows that −2 *is not* a solution.

PRACTICE

Solve each inequality. Graph on a number line.

1. $\dfrac{1}{2}y - 12 < -4$ **2.** $\dfrac{5}{3}n + 2 > -\dfrac{2}{3}n + 5$ **3.** $8n \le 18 - n$ **4.** $1 - n > 1 - 2n$

5. $-3x + 2 \ge -x$ **6.** $12x < 15 - 2x$ **7.** $\dfrac{1}{4}y + 3 < 1$ **8.** $-4n - 3 \ge -2n - 1$

Graphing Systems of Equations and Inequalities

A *system of linear equations* is two or more linear equations that state relationships between the same variables. The solution of a system of equations is the point where the lines intersect.

A *system of linear inequalities* is two or more inequalities that state relationships between the same variables. The solution of a system of inequalities is the area included in all the graphs.

EXAMPLE 1

Graph the system $y = 2x + 1$ and $x + y = -2$. Use the graph to find the solution of the system.

SOLUTION

Put both equations into slope-intercept form.

Graph the equations.

$y = 2x + 1$

$x + y = -2$

$x + y - x = -2 - x$

$y = -x - 2$

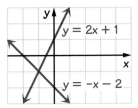

The solution is $(-1, -1)$. You can check this by substituting $(-1, -1)$ into the two equations.

EXAMPLE 2

Graph the system $y \leq \frac{1}{2}x + 1$ and $y + 3 > 0$.

SOLUTION

Put both inequalities into slope-intercept form.

Graph the inequalities.

Graph the area below and including $y = \frac{1}{2}x + 1$.

$y \leq \frac{1}{2}x + 1$

$y + 3 > 0$

$y + 3 - 3 > 0 - 3$

$y > -3$

Graph the area above $y = -3$.

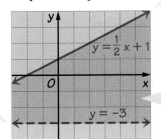

The solution is the area shaded in purple.

The solutions of the system $y \leq \frac{1}{2}x + 1$ and $y + 3 > 0$ are all points that are above the line $y = -3$ and are below or on the line $y = \frac{1}{2}x + 1$.

PRACTICE

Graph each system, and find the solution of the system.

1. $y = \frac{3}{2}x - 6$

$y = -\frac{1}{2}x + 6$

2. $y = -2$

$y = -\frac{1}{2}x$

3. $y - 2 = \frac{2}{5}x$

$x + 3y + 5 = 0$

4. $2x + y = 1$

$x + y = -1$

Graph each system. Label the solution of the system on your graph.

5. $y > x - 2$
$y < \frac{1}{2}x + 2$

6. $y < \frac{2}{3}x + 2$
$y > \frac{2}{3}x$

7. $y \leq \frac{1}{4}x - 2$
$y - \frac{1}{2}x + 2 \leq 0$

8. $-x + 3y < 18$
$x < 3$

RATIO AND PROPORTION

Creating a Ratio

You can use *ratios* to compare two numbers using division. A ratio can be written in three ways: a to b, $a:b$, and $\frac{a}{b}$. Ratios are often used to give *rates*, ratios that compare two different quantities. A *unit rate* is a rate per one unit of a given quantity.

EXAMPLE 1

Write each ratio in lowest terms.

a. 10 to 15

b. $\frac{45 \text{ s}}{1 \text{ min}}$

SOLUTION

a. Write as a fraction in lowest terms:

$\frac{10}{15} = \frac{2}{3}$

b. Convert minutes to seconds:

$\frac{45 \text{ s}}{1 \text{ min}} = \frac{45 \text{ s}}{60 \text{ s}} = \frac{3}{4}$

EXAMPLE 2

What is the unit rate if you read 8 articles in 2 hours?

SOLUTION

Write the comparison as a fraction: $\frac{\text{articles}}{\text{hours}} \rightarrow \frac{8}{2} = 4$ articles per hour

PRACTICE

Write each ratio in lowest terms.

1. $12:3$

2. 18 to 40

3. $\frac{20}{100}$

4. $\frac{3}{7}$

5. 20 min to 1 h

6. 500 lb to 2 tons

7. 3 ft : 15 in.

8. 1 cm : 1 m

9. 6 months to three years

Write each unit rate.

10. 20 mi in 6 h

11. $40 in 1 h

12. $15,000 in 2 years

13. 45 h in 7 days

14. 13 cats in 4 hours

15. 120 mi in 5 days

16. 7200 cycles in 60 s

17. 18 in. in 72 min

18. 240 apples in 8 baskets

Solving Proportions

To solve a proportion, you must isolate the variable on one side.

EXAMPLE

Solve $\frac{3}{4} = \frac{x}{8}$.

SOLUTION

$$\frac{3}{4} = \frac{x}{8}$$

$$8 \cdot \frac{3}{4} = 8 \cdot \frac{x}{8} \quad \text{Multiply each side by 8.}$$

$$6 = x$$

PRACTICE

Solve.

1. $\frac{x}{5} = \frac{4}{10}$

2. $\frac{1}{2} = \frac{x}{14}$

3. $\frac{3}{n} = 9$

4. $\frac{y}{8} = \frac{13}{4}$

5. $\frac{4}{9} = \frac{12}{x}$

6. $\frac{x}{100} = \frac{12}{60}$

7. $\frac{4}{15} = \frac{m}{3}$

8. $\frac{2}{7} = \frac{10}{x}$

DATA ANALYSIS AND PROBABILITY

Finding Averages

There are two types of averages often taken of a data set. The *mean* is the sum of all the data divided by the number of data items. The *median* is the middle number when you put the data in order from smallest to largest. If the number of data items is even, the median is the mean of the two middle numbers.

EXAMPLE

For the data set 2, 4, 8, 7, 1, 6, find each value.

a. the mean

b. the median

SOLUTION

a. Find the value: $\dfrac{\text{sum of the data}}{\text{number of data items}}$

$$\frac{2 + 4 + 8 + 7 + 1 + 6}{6} = \frac{28}{6}$$

$$= \frac{14}{3}, \text{ or } 4\frac{2}{3}$$

The mean of the data set is $4\frac{2}{3}$.

b. Put the data in order: 1, 2, 4, 6, 7, 8

The two middle numbers are 4 and 6.

Find the mean of the two middle numbers:

$$\frac{4 + 6}{2} = \frac{10}{2} = 5$$

The median of the data set is 5.

Find the mean and the median of each data set.

1. 5, 7, 7, 10 **2.** 18, 20, 19, 20, 20 **3.** 1, 1, 3, 5, 7, 8, 8, 8 **4.** 12, 14, 11, 10

5. 14, 20, 23, 18, 7, 3 **6.** 1, 0, 3, 4, 3, 2, 0 **7.** 8, −5, 3, −4, −6, −2 **8.** 25, 22, 18, 17, 13

Scientific Notation

Very large and very small numbers are often expressed using scientific notation. A number is expressed in scientific notation if it is in the form $a \times 10^n$, where $1 \le a < 10$ and n is an integer.

EXAMPLE 1

Write each number as a decimal.

a. 2.8×10^{-4} **b.** 9.21×10^6

SOLUTION

a. $2.8 \times 10^{-4} = 0.00028$ **b.** $9.21 \times 10^6 = 9,210,000$

Move the decimal point 4 places to the left.

Move the decimal point 6 places to the right.

EXAMPLE 2

Write each number in scientific notation.

a. 15,400,000,000 **b.** 0.007

SOLUTION

a. $15,400,000,000 = 1.54 \times 10^{10}$ **b.** $0.007 = 7 \times 10^{-3}$

Factoring out 10^{10} moves the decimal point 10 places to the left.

Factoring out 10^{-3} moves the decimal point 3 places to the right.

Write each number as a decimal.

1. 1.8×10^{-5} **2.** 4.55×10^{-11} **3.** 5.2×10^9 **4.** 3.812×10^2

5. 9.84×10^1 **6.** 5.7×10^{-8} **7.** 3.9×10^{-1} **8.** 4.20×10^4

Write each number in scientific notation.

9. 0.00088 **10.** 0.12 **11.** 1,245,000 **12.** 0.0131

13. 0.00000000006 **14.** 45,000 **15.** 158,000,000 **16.** 230,000

17. 6200 **18.** 0.005 **19.** 0.0000000000011 **20.** 3,486,510

Probability

Probability is a ratio that measures how likely an event is to happen. *Experimental probability* is the ratio of the number of times an event actually occurs to the number of times an experiment is done. *Theoretical probability* is the ratio of the number of favorable outcomes to the total number of possible outcomes.

EXAMPLE 1

Karen makes 18 out of 30 free throws. What is the experimental probability that she will make a free throw?

SOLUTION

$$\frac{\text{number made}}{\text{number attempted}} = \frac{18}{30} = \frac{3}{5}, \text{ or } 0.60$$

A probability is usually written as a number between 0 and 1.

EXAMPLE 2

A 6-sided die is rolled. What is the theoretical probability of rolling a 1 or a 2?

SOLUTION

probability of rolling a 1 or a 2 = probability of a 1 + probability of a 2

$$= \frac{1}{6} + \frac{1}{6} = \frac{1}{3} \approx 0.33$$

PRACTICE

Find each experimental probability.

1. What is the experimental probability that Mark stands on bare floor at a random point in the room, if carpet covers $\frac{1}{4}$ of the floor?

2. José got a hit in 20 out of 58 at-bats. What is the experimental probability of getting a hit?

3. A coin is flipped and lands heads up on 2 out of 5 flips. What is the experimental probability that the coin will land heads up?

4. You have found an empty seat on the train on 8 out of 14 days. What is the experimental probability of finding an empty seat?

5. If it rains on 12 out of 20 days, what is the experimental probability that you need to bring a raincoat to school on any given day?

Find each theoretical probability.

6. Getting a head on the flip of a fair coin.

7. Drawing any 4 red cards from a standard deck of 52 cards.

8. Rolling a number greater than 3 on a 6-sided die.

9. Rolling a 7 on a 6-sided die.

10. Rolling a 10, 11, or 12 on a 20-sided die.

MATRICES

A *matrix* is a group of numbers arranged in rows and columns. If a matrix has
m rows and n columns, we say its *dimensions* are m by n, or $m \times n$. If two
matrices have the same dimensions, you can add or subtract them by adding or
subtracting corresponding elements.

EXAMPLE 1

Evaluate each matrix expression.

a. $\begin{bmatrix} 1 & 2 \\ 3 & 4 \end{bmatrix} + \begin{bmatrix} 5 & -6 \\ 9 & 0 \end{bmatrix}$

b. $\begin{bmatrix} 3 & -2 \\ 0 & 1 \end{bmatrix} - \begin{bmatrix} -4 & 8 \\ -7 & 5 \end{bmatrix}$

SOLUTION

a. $\begin{bmatrix} 1 & 2 \\ 3 & 4 \end{bmatrix} + \begin{bmatrix} 5 & -6 \\ 9 & 0 \end{bmatrix} = \begin{bmatrix} 1+5 & 2-6 \\ 3+9 & 4+0 \end{bmatrix} = \begin{bmatrix} 6 & -4 \\ 12 & 4 \end{bmatrix}$

b. $\begin{bmatrix} 3 & -2 \\ 0 & 1 \end{bmatrix} - \begin{bmatrix} -4 & 8 \\ -7 & 5 \end{bmatrix} = \begin{bmatrix} 3-(-4) & -2-8 \\ 0-(-7) & 1-5 \end{bmatrix} = \begin{bmatrix} 7 & -10 \\ 7 & -4 \end{bmatrix}$

You can multiply a matrix by a number, or *scalar*. Multiply each element of
the matrix by the scalar.

EXAMPLE 2

Evaluate $3\begin{bmatrix} 5 \\ -4 \end{bmatrix}$.

SOLUTION

$$3\begin{bmatrix} 5 \\ -4 \end{bmatrix} = \begin{bmatrix} 3 \cdot 5 \\ 3 \cdot (-4) \end{bmatrix} = \begin{bmatrix} 15 \\ -12 \end{bmatrix}$$

PRACTICE

Evaluate each matrix expression.

1. $\begin{bmatrix} 1 & 0 \\ 0 & 1 \end{bmatrix} + \begin{bmatrix} 5 & 5 \\ 2 & 12 \end{bmatrix}$

2. $\begin{bmatrix} 10 \\ 4 \end{bmatrix} + \begin{bmatrix} -3 \\ 6 \end{bmatrix}$

3. $\begin{bmatrix} 3 & -3 \\ 2 & 1 \end{bmatrix} - \begin{bmatrix} 1 & 1 \\ 0 & 0 \end{bmatrix}$

4. $\begin{bmatrix} -7 \\ -2 \\ -5 \end{bmatrix} + \begin{bmatrix} 8 \\ -3 \\ 0 \end{bmatrix}$

5. $\begin{bmatrix} 8 & 8 \\ 8 & 8 \end{bmatrix} + \begin{bmatrix} 2 & -1 \\ 0 & 8 \end{bmatrix}$

6. $\begin{bmatrix} 5 & 2 & 0 \\ -1 & -2 & 3 \end{bmatrix} - \begin{bmatrix} 4 & -9 & 6 \\ -3 & 0 & 7 \end{bmatrix}$

7. $2\begin{bmatrix} 1 & 0 \\ 0 & 1 \end{bmatrix}$

8. $\dfrac{1}{2}\begin{bmatrix} -8 & 3 \\ 6 & -4 \end{bmatrix}$

9. $8\begin{bmatrix} 9 \\ -5 \end{bmatrix}$

10. $3\begin{bmatrix} 1 & 4 & -2 \\ -4 & -1 & 1 \end{bmatrix}$

11. $-3\begin{bmatrix} 1 & 0 \\ 0 & 1 \end{bmatrix}$

12. $2\begin{bmatrix} -3 & 0 & 5 \\ 8 & -2 & -1 \\ 9 & 7 & 0 \end{bmatrix}$

13. $-\dfrac{5}{3}\begin{bmatrix} 5 & 3 \\ 3 & -3 \\ -1 & 6 \end{bmatrix}$

14. $\dfrac{3}{10}\begin{bmatrix} 15 \\ 10 \\ -40 \end{bmatrix}$

Postulates and Properties

Patterns, Lines, and Planes

▨ For any two points, there is exactly one line through the points. **p. 22**

▨ For any three noncollinear points, there is exactly one plane through the points. **p. 22**

▨ If two planes intersect, their intersection is a line. **p. 23**

▨ If a line and a plane intersect, then their intersection is a point or a line. **p. 23**

▨ If two lines intersect, then they are coplanar. **p. 23**

▨ **Segment Addition Postulate**   Point Y is between points X and Z if and only if $XY + YZ = XZ$. **p. 29**

▨ **Angle Addition Postulate**   If D is in the interior of $\angle ABC$, then $m\angle ABD + m\angle DBC = m\angle ABC$. **p. 36**

Reasoning in Geometry

An example of each property is given.

▨ **Reflexive Property** $\overline{PQ} \cong \overline{PQ}$. **p. 119**

▨ **Symmetric Property**   If $\overline{PQ} \cong \overline{RS}$, then $\overline{RS} \cong \overline{PQ}$. **p. 119**

▨ **Transitive Property**   If $\angle 1 \cong \angle 2$ and $\angle 2 \cong \angle 3$, then $\angle 1 \cong \angle 3$. **p. 119**

▨ **Addition Property**   If $PQ = RS$, then $PQ + QR = QR + RS$. **p. 119**

▨ **Subtraction Property**   If $m\angle JNL = m\angle KNM$, then $m\angle JNL - m\angle KNL = m\angle KNM - m\angle KNL$. **p. 119**

▨ **Substitution Property**   If $m\angle 1 + m\angle 2 = 180°$ and $m\angle 1 = m\angle 3$, then $m\angle 3 + m\angle 2 = 180°$. **p. 119**

Parallel Lines

▨ **Corresponding Angles Postulate**   If two parallel lines are intersected by a transversal, then corresponding angles are congruent. **p. 222**

▨ **Parallel Postulate**   Through a point not on a given line, there is exactly one line parallel to the given line. **p. 235**

▨ **Perpendicular Postulate**   Through a point not on a given line, there is exactly one line perpendicular to the given line. **p. 235**

Converse of the Corresponding Angles Postulate   If two lines are intersected by a transversal and corresponding angles are congruent, then the lines are parallel. **p. 243**

Conjectures About Triangles

SSS Postulate   If three sides of a triangle are congruent to three sides of another triangle, then the triangles are congruent. **p. 293**

SAS Postulate   If two sides and the included angle of one triangle are congruent to two sides and the included angle of another triangle, then the triangles are congruent. **p. 294**

ASA Postulate   If two angles and the included side of one triangle are congruent to two angles and the included side of another triangle, then the triangles are congruent. **p. 300**

Using Transformations

A reflection preserves congruence. **p. 393**

A reflection changes orientation. **p. 393**

The line of reflection is the perpendicular bisector of every segment that connects a point and its image. **p. 393**

A translation preserves congruence and orientation. **p. 406**

A rotation preserves congruence and orientation. **p. 412**

The line of reflection in a glide reflection is parallel to the direction of the translation. **p. 420**

A glide reflection preserves congruence. **p. 420**

A glide reflection changes orientation. **p. 420**

A dilation does not usually preserve congruence. **p. 428**

A dilation preserves angle measures. **p. 428**

Similar Polygons

AA Similarity Postulate   If two angles of a triangle are congruent to two angles of another triangle, then the two triangles are similar. **p. 453**

Circles and Spheres

Arc Addition Postulate   In a circle, the measure of the arc formed by two arcs that have exactly one point in common is the sum of the measures of the two arcs. **p. 554**

Theorems

Triangles and Polygons

- The angles that form a linear pair are supplementary. **p. 59**
- Vertical angles are congruent. **pp. 59, 131**
- Two lines are perpendicular if and only if they form congruent adjacent angles. **p. 60**
- **The Triangle Sum Theorem** The sum of the angle measures of a triangle is 180°. **pp. 66, 236–237**
- In a polygon with n sides, the sum of the angle measures is $(n-2)180°$. **p. 80**
- The sum of the measures of the exterior angles of any polygon is 360°. **p. 81**
- The opposite sides of a parallelogram are congruent. **p. 88**
- The opposite angles of a parallelogram are congruent. **p. 88**
- The diagonals of a parallelogram bisect each other. **p. 89**

Reasoning in Geometry

- The measure of an exterior angle of a triangle is equal to the sum of the measures of the two interior angles not adjacent to it. **p. 126**
- **The Pythagorean Theorem** In a right triangle, the square of the length of the hypotenuse is equal to the sum of the squares of the lengths of the legs. **p. 142**
- **Converse of the Pythagorean Theorem** If a, b, and c are the lengths of the sides of a triangle and $a^2 + b^2 = c^2$, then the triangle is a right triangle. **p. 143**

Coordinates in Geometry

- **The Distance Formula** The distance between the points $A(x_1, y_1)$ and $B(x_2, y_2)$ is $AB = \sqrt{(x_2 - x_1)^2 + (y_2 - y_1)^2}$. **p. 166**
- **The Midpoint Formula** The midpoint of the segment joining the points $A(x_1, y_1)$ and $B(x_2, y_2)$ has the coordinates $\left(\dfrac{x_1 + x_2}{2}, \dfrac{y_1 + y_2}{2} \right)$. **p. 167**
- **The Slope Formula** The slope m of the line containing the points (x_1, y_1) and (x_2, y_2) is $m = \dfrac{y_2 - y_1}{x_2 - x_1}$. **p. 174**

The equation of a line with slope m and y-intercept b is $y = mx + b$. **p. 175**

Two nonvertical lines are parallel if and only if their slopes are equal. **p. 181**

Two nonvertical lines are perpendicular if and only if the product of their slopes is -1. **p. 181**

An equation of the circle with center $(0, 0)$ and radius r is $x^2 + y^2 = r^2$. **p. 188**

An equation of the circle with center (h, k) and radius r is $(x - h)^2 + (y - k)^2 = r^2$. **p. 189**

The midpoint of the segment that joins any two points $A(x_1, y_1, z_1)$ and $B(x_2, y_2, z_2)$ has the coordinates $\left(\dfrac{x_1 + x_2}{2}, \dfrac{y_1 + y_2}{2}, \dfrac{z_1 + z_2}{2} \right)$. **p. 203**

The distance AB between two points in space $A(x_1, y_1, z_1)$ and $B(x_2, y_2, z_2)$ is $\sqrt{(x_2 - x_1)^2 + (y_2 - y_1)^2 + (z_2 - z_1)^2}$. **p. 203**

Parallel Lines

If two parallel lines are intersected by a transversal, then alternate interior angles are congruent. **p. 227**

If two parallel lines are intersected by a transversal, then same-side interior angles are supplementary. **p. 227**

If two lines are intersected by a transversal and alternate interior angles are congruent, then the lines are parallel. **p. 243**

If two lines are intersected by a transversal and same-side interior angles are supplementary, then the lines are parallel. **p. 244**

In a plane, if two lines are both perpendicular to a third line, then the two lines are parallel. **p. 249**

If two lines are both parallel to a third line, then the two lines are parallel. **p. 250**

If two parallel planes are intersected by a third plane, then the lines of intersection are parallel. **p. 257**

If two planes are both parallel to a third plane, then the two planes are parallel. **p. 257**

Conjectures About Triangles

The sum of the lengths of any two sides of a triangle is greater than the length of the third side. **p. 280**

One side of a triangle is longer than a second side if and only if the angle opposite the first side is larger than the angle opposite the second side. **p. 280**

- **AAS Theorem**   If two angles and a non-included side of one triangle are congruent to the corresponding parts of another triangle, then the triangles are congruent. **p. 301**

- **HL Theorem**   If the hypotenuse and a leg of one right triangle are congruent to the corresponding parts of another right triangle, then the triangles are congruent. **p. 301**

- If a point is on the perpendicular bisector of a segment, then the point is equidistant from the endpoints of the segment. **p. 308**

- **The Isosceles Triangle Theorem**   If two sides of a triangle are congruent, then the angles opposite the sides are congruent. **p. 314**

- **The Converse of the Isosceles Triangle Theorem**   If two angles of a triangle are congruent, then the sides opposite the angles are congruent. **p. 314**

Quadrilaterals, Areas, and Volumes

- The diagonals of a rhombus are perpendicular. **p. 341**

- The diagonals of a rectangle are congruent. **p. 341**

- Exactly one diagonal of a kite is a line of symmetry for the kite and the perpendicular bisector of the other diagonal. **p. 341**

- If both pairs of opposite sides of a quadrilateral are congruent, then the quadrilateral is a parallelogram. **p. 347**

- If both pairs of opposite angles of a quadrilateral are congruent, then the quadrilateral is a parallelogram. **p. 347**

- If one pair of opposite sides of a quadrilateral is both parallel and congruent, then the quadrilateral is a parallelogram. **p. 347**

- If the diagonals of a quadrilateral bisect each other, then the quadrilateral is a parallelogram. **p. 347**

- If the diagonals of a parallelogram are congruent, then the parallelogram is a rectangle. **p. 353**

- If one angle of a parallelogram is a right angle, then the parallelogram is a rectangle. **p. 353**

- If the diagonals of a parallelogram are perpendicular, then the parallelogram is a rhombus. **p. 355**

- If two consecutive sides of a parallelogram are congruent, then the parallelogram is a rhombus. **p. 355**

- The area of a rectangle is the product of the base and the height. ($A = bh$) **p. 362**

- The area of a square is the square of the length of one side. ($A = s^2$) **p. 362**

- The area of a triangle is half the product of the base and the height. ($A = \frac{1}{2}bh$) **p. 362**

The area of a parallelogram is the product of the base and the height. ($A = bh$) **p. 362**

The area of a trapezoid is the product of the height and the mean of the bases. ($A = \frac{1}{2}(b_1 + b_2)h$) **p. 362**

The area of any regular polygon is half the product of the apothem and the perimeter. ($A = \frac{1}{2}ap$) **p. 368**

The circumference of a circle with radius r is $2\pi r$. ($C = 2\pi r$) **p. 369**

The area of a circle with radius r is πr^2. ($A = \pi r^2$) **p. 369**

The volume of a prism is the product of the height and the area of the base. ($V = Bh$) **p. 375**

The volume of a cylinder is the product of the height and the area of the base. ($V = \pi r^2 h$) **p. 375**

The surface area of a prism is the sum of the lateral area and the areas of the bases. ($S.A. = ph + 2B$) **p. 376**

The surface area of a cylinder is the sum of the lateral area and the areas of the bases. ($S.A. = 2\pi rh + 2\pi r^2$) **p. 376**

Similar Polygons

SAS Similarity Theorem   If an angle of one triangle is congruent to an angle of another triangle, and the sides including these angles are in proportion, then the triangles are similar. **p. 454**

SSS Similarity Theorem   If all corresponding sides of two triangles are in proportion, then the triangles are similar. **p. 454**

If a segment is parallel to one side of a triangle and intersects the other two sides, then it divides those sides proportionally. **p. 462**

If the midpoints of two sides of a triangle are joined by a segment, then the segment is parallel to the third side of the triangle and its length is half the length of the third side of the triangle. **p. 462**

If two similar two-dimensional figures have corresponding lengths whose ratio is $a:b$, then:

the ratio of all corresponding linear measures is $a:b$ **p. 469**

the ratio of their perimeters is $a:b$ **p. 469**

the ratio of their areas is $a^2:b^2$ **p. 469**

If two similar three-dimensional figures have corresponding lengths whose ratio is $a:b$, then:

the ratio of all corresponding linear measures is $a:b$ **p. 470**

the ratio of their surface areas ($S.A.$) is $a^2:b^2$ **p. 470**

the ratio of their volumes (V) is $a^3:b^3$ **p. 470**

Applying Right Triangles

- If the altitude is drawn to the hypotenuse of a right triangle, then the two triangles formed are similar to the original triangle and to each other. **p. 494**

- If the altitude is drawn to the hypotenuse of a right triangle, then the length of the altitude is the geometric mean of the lengths of the segments of the hypotenuse. **p. 495**

- If the altitude is drawn to the hypotenuse of a right triangle, then the length of each leg is the geometric mean of the lengths of the hypotenuse and the segment of the hypotenuse adjacent to that leg. **p. 495**

- In a 45-45-90 triangle, the hypotenuse is $\sqrt{2}$ times the length of each leg. **p. 501**

- In a 30-60-90 triangle, the length of the hypotenuse is twice the length of the shorter leg and the length of the longer leg is $\sqrt{3}$ times the length of the shorter leg. **p. 501**

- The volume of a pyramid is one third the product of the height and the area of the base. ($V = \frac{1}{3}Bh$) **p. 536**

- The surface area of a regular pyramid is the sum of the lateral area and the area of the base. ($S.A. = \frac{1}{2}ps + B$) **p. 536**

- The volume of a cone is one third the product of the height and the area of the base. ($V = \frac{1}{3}\pi r^2 h$) **p. 537**

- The surface area of a cone is the sum of the lateral area and the area of the base. ($S.A. = \pi rs + \pi r^2$). **p. 537**

Circles and Spheres

- A line, ray, or segment is a tangent of a circle if and only if it is in the same plane as the circle and is perpendicular to a radius of the circle at the point of intersection. **p. 554**

- The measure of an inscribed angle is equal to half the measure of the intercepted arc. **p. 555**

- In a circle or in congruent circles, inscribed angles that intercept the same arc or congruent arcs are congruent. **p. 555**

- The measure of an angle formed by a tangent and a chord is equal to half the measure of the intercepted arc. **p. 555**

- The measure of an angle formed by the intersection of two tangents, two secants, or a secant and a tangent, at a point outside a circle, is half the difference of the measures of the intercepted arcs. **p. 561**

- The measure of an angle formed by two chords is equal to half the sum of the measures of the intercepted arcs. **p. 562**

- The perpendicular bisector of a chord passes through the center of the circle. **p. 568**

- A diameter that is perpendicular to a chord bisects the chord and its corresponding arc. **p. 568**

- If two chords intersect in the interior of a circle, then the product of the lengths of the segments of one chord is equal to the product of the lengths of the segments of the other chord. **p. 573**

- If two secant segments are drawn to a circle from an exterior point, then the product of the lengths of one secant segment and its external segment is equal to the product of the lengths of the other secant segment and its external segment. **p. 573**

- If a tangent segment and a secant segment are drawn to a circle from an exterior point, then the product of the lengths of the secant segment and its external segment is equal to the square of the length of the tangent segment. **p. 574**

- If two tangent segments are drawn to a circle from an exterior point, then the lengths of the tangent segments are equal. **p. 574**

- If x represents the measure of the arc of a sector and r is the radius of the circle, then the area of the sector is $\dfrac{x}{360} \cdot \pi r^2$. **p. 580**

- If x represents the measure of an arc and r is the radius of the circle, then the length of the arc is $\dfrac{x}{360} \cdot 2\pi r$. **p. 580**

- The surface area of a sphere with radius r is $4\pi r^2$. ($S.A. = 4\pi r^2$) **p. 586**

- The volume of a sphere with radius r is $\dfrac{4}{3}\pi r^3$. ($V = \dfrac{4}{3}\pi r^3$) **p. 586**

Constructions

Triangles and Polgons

- Construct an equilateral triangle. **p. 69**

Parallel Lines

- Given an angle, construct an angle congruent to the given angle. **p. 263**
- Given two separate angles, construct an angle whose measure is equal to the sum of the measures of the given angles. **p. 264**
- Given a line, construct a line parallel to the given line. **p. 264**
- Given a segment, construct its perpendicular bisector. **p. 265**
- Given a line and a point on the line, construct a line perpendicular to the given line and through the given point. **p. 267**
- Given a line and a point not on the line, construct a line perpendicular to the given line and through the given point. **p. 267**

Conjectures About Triangles

- Given three sides of a triangle, construct the triangle. **p. 292**
- Given two sides and their included angle, construct a triangle. **p. 297**
- Given an angle, construct its bisector. **p. 311**

Quadrilaterals, Areas, and Volumes

- Construct a kite. **p. 343**
- Given two segments, construct a parallelogram with sides congruent to the segments. **p. 350**
- Construct a rectangle. **p. 358**
- Construct a regular hexagon. **p. 372**

Applying Right Triangles

- Construct a 45-45-90 triangle. **p. 505**

Circles and Spheres

- Given a triangle, circumscribe a circle about the triangle. **p. 567**

Technology Handbook

This handbook introduces features of graphing calculators, spreadsheets, and geometry software that you can use with material in this book.

USING A GRAPHING CALCULATOR

This section discusses features common to most graphing calculators. Check your calculator's instruction manual for any details not provided here.

PERFORMING CALCULATIONS

The Keyboard

Look closely at your calculator's keyboard. Notice that most keys serve more than one purpose. Each key is labeled with its primary purpose, and labels for any secondary purposes appear somewhere near the key. You may need to press **2nd**, **SHIFT**, or **ALPHA** to use a key for a secondary purpose.

On the TI-82, for example, the **x^2** key can be used as follows:

- Press **x^2** to square a number.
- Press **2nd** and then **x^2** to take a square root.
- Press **ALPHA** and then **x^2** to get the letter I.

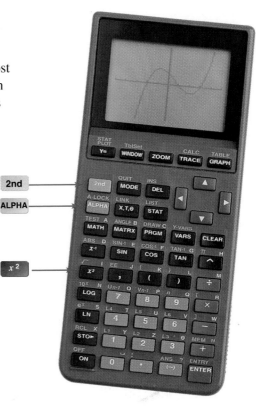

The Home Screen

Your calculator has a *home screen* where you can do calculations. You can usually enter a calculation on a graphing calculator just as you would write it on a piece of paper.

Your calculator may recognize implied multiplication and have built-in constants like pi.

Don't confuse the subtraction key, **–**, and the negation key, **(–)** .

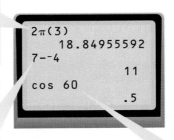

```
2π(3)
        18.84955592
7--4
              11
cos 60
             .5
```

When doing trigonometric calculations, be sure your calculator is in the degree mode.

Use your calculator to find the value of each expression.

1. $-3 - 9$ **2.** $5(6.37)$ **3.** $\sqrt{11.56}$ **4.** $\sqrt{5^2 + 9^2}$

5. 8^4 **6.** $\frac{4}{3}\pi(7^3)$ **7.** $\tan 61°$ **8.** $\sin^{-1}\left(\frac{18}{23}\right)$

DISPLAYING GRAPHS

Making a Scatter Plot

Most graphing calculators will make a scatter plot for a set of data pairs. For example, suppose you want to make a scatter plot of the data in the table.

x = length of an edge of a cube (in.)	1	2	3	4	5	6
y = surface area of the cube (in.²)	6	24	54	96	150	216

On the TI-82, you first need to enter the data into *lists*. Follow these steps:

Enter the lengths in list L1. Enter the surface areas in list L2.

Tell the calculator to display a scatter plot using the data in lists L1 and L2.

Press GRAPH **to display the scatter plot. Adjust the viewing window as needed (see below).**

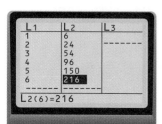      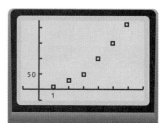

Adjusting the Viewing Window

When making a scatter plot, think of the calculator screen as a *viewing window* that lets you look at part of the coordinate plane. On the TI-82, you can adjust the viewing window by pressing WINDOW and entering appropriate values for the window variables. The values given below define a good viewing window for the scatter plot of the cube data (see above).

The interval $-1 \le x \le 7$ will be shown on the *x*-axis.

Tick marks will be **1** unit apart on the *x*-axis.

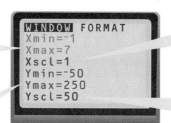

The interval $-50 \le y \le 250$ will be shown on the *y*-axis.

Tick marks will be **50** units apart on the *y*-axis.

9. The table shows the numbers of diagonals that can be drawn for polygons with different numbers of sides. Make a scatter plot of the data in the table. Adjust the viewing window so that you can see all the data points. Describe the shape of the scatter plot.

x = number of sides	3	4	5	6	7	8	9
y = number of diagonals	0	2	5	9	14	20	27

USING MATRICES

Entering a Matrix

Many graphing calculators will let you enter and perform operations on matrices. To enter a matrix on the TI-82, follow these steps:

Press MATRX **, and select EDIT. Choose a name for the matrix, such as [A].**

Set the dimensions of the matrix, such as 4 × 3. A matrix full of zeros appears.

Replace the zeros with the desired matrix elements.

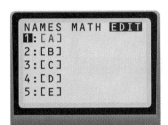

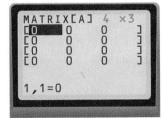

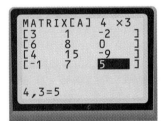

Performing Scalar Multiplication

You can use a graphing calculator to perform scalar multiplication. For example, suppose you want to find the image matrix for a dilation of a triangle, where the coordinates of the triangle's vertices are $(-1, 0)$, $(3, 4)$, and $(6, -5)$, the center of the dilation is $(0, 0)$, and the scale factor is 2. Follow these steps:

Create a 2 × 3 polygon matrix [A]. Enter the coordinates of each vertex of the triangle into a column of the matrix.

Multiply [A] by the scale factor 2 to find the image matrix for the dilation.

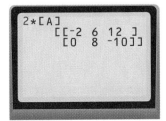

Adding and Subtracting Matrices

You can use a graphing calculator to add matrices that have the same dimensions. For example, suppose you want to find the image matrix for the translation of a triangle 2 units right and 4 units down. The coordinates of the triangle's vertices are $(-3, 1)$, $(0, 5)$, and $(7, 0)$. Follow these steps:

Create a 2 × 3 translation matrix [A] that shifts a triangle 2 units right and 4 units down.

Create a 2 × 3 polygon matrix [B] containing the coordinates of the triangle's vertices.

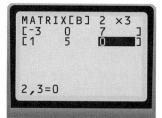

Add [A] and [B] to find the image matrix for the translation.

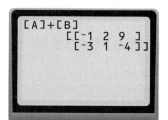

You can also use a graphing calculator to subtract matrices. For example, suppose a triangle has been translated 5 units right and 3 units up, and the coordinates of the image's vertices are $(-3, 4)$, $(2, 1)$, and $(4, 8)$. To find the coordinates of the original triangle's vertices, first create the translation matrix [C] and the image matrix [D], and then find [D] – [C]. The screen at the right shows that the original triangle's vertices are at $(-8, 1)$, $(-3, -2)$, and $(-1, 5)$.

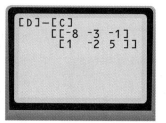

Finding the Product of Two Matrices

You can use a graphing calculator to multiply two matrices when the number of columns in the first matrix is the same as the number of rows in the second matrix. For example, suppose you want to find the image matrix for the reflection of a triangle over the y-axis. The coordinates of the triangle's vertices are $(-5, 4)$, $(-3, -3)$, and $(-2, -6)$. Follow these steps:

Create a 2 × 2 matrix [A] that reflects polygons over the y-axis. (See page 635 for a list of reflection matrices.)

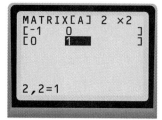

Create a 2 × 3 polygon matrix [B] containing the coordinates of the triangle's vertices.

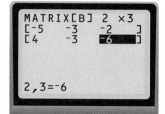

Multiply [A] and [B] to find the image matrix for the reflection.

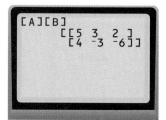

The coordinates of the vertices of a triangle are (1, 3), (4, 7), and (9, 2).
Find the image matrix for each transformation of the triangle.

10. a translation 4 units left and 1 unit up

11. a dilation with center (0, 0) and scale factor 5

12. a reflection over the line $y = -x$

13. a 90° rotation counterclockwise around (0, 0)

USING A SPREADSHEET

You can use a computer spreadsheet to perform repetitive calculations quickly and easily. For example, the spreadsheet below has been set up to calculate the surface areas of balls used in different sports. (In the spreadsheet, r is the radius of a ball and A is the ball's surface area.)

A spreadsheet is made up of cells named by a column letter and a row number, such as A3 or B4. You can enter a label, a number, or a formula into a cell.

For example, row 2 of the spreadsheet at the left contains the label "**Golf ball**" in cell A2, the number **2.1** in cell B2, and the formula "**=4*PI()*B2^2**" in cell C2. This formula tells the computer to multiply the square of the number in cell B2 by 4π and store the result in cell C2.

You don't have to type every formula in column C. Instead, you can type just the first formula and then use the *copy* or *fill down* command to generate the others.

row numbers   column letters

Ball Data

	A	B	C
1	Type of ball	r (cm)	A (cm^2)
2	Golf ball	2.1	=4*PI()*B2^2
3	Tennis ball	3.3	=4*PI()*B3^2
4	Softball	4.8	=4*PI()*B4^2
5	Volleyball	10.5	=4*PI()*B5^2
6	Basketball	11.9	=4*PI()*B6^2

The computer replaces all formulas in the spreadsheet with calculated values, as shown at the right. For example, you can see that the surface area of a basketball is about 1780 cm^2.

Ball Data

	A	B	C
1	Type of ball	r (cm)	A (cm^2)
2	Golf ball	2.1	55.42
3	Tennis ball	3.3	136.85
4	Softball	4.8	289.53
5	Volleyball	10.5	1385.44
6	Basketball	11.9	1779.52

14. Use a spreadsheet to find the volumes of the balls listed in the spreadsheets above.

15. For any integer n greater than 1, the set of numbers $2n$, $n^2 - 1$, and $n^2 + 1$ is a Pythagorean triple (a set of integers that can be the lengths of the sides of a right triangle). Use a spreadsheet to generate Pythagorean triples for $n = 2, 3, \ldots, 10$.

USING GEOMETRY SOFTWARE

This section discusses features common to most geometry software. Check your software's instruction manual for any details not provided here.

CONSTRUCTING GEOMETRIC OBJECTS

Constructing Points

You can construct a point either by moving the cursor to where you want the point to be and clicking the mouse button, or by entering the point's coordinates. Points are given one-letter names, starting with *A*. Several more complicated constructions involving points are described below.

You can define a point to be on an object. (The point can't be moved off the object.)

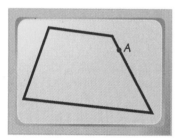

You can construct the point or points where two objects intersect.

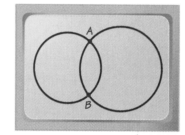

You can construct one or more points that subdivide a segment into equal parts.

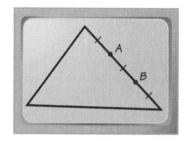

Constructing Lines, Segments, Rays, and Vectors

With most software, you can draw a line, segment, ray, or vector "freehand" by clicking the mouse button at some location on the screen and dragging away from that location to form the object. You can also construct a line by entering its equation, and a vector by specifying its tail and components. Several other constructions are described below.

You can construct a segment joining two named points, such as *B* and *D* below.

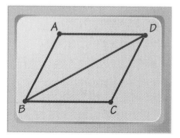

You can construct a ray that bisects an angle, such as ∠ *A* below.

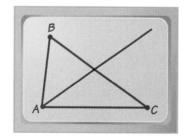

You can construct a line parallel or perpendicular to a given line or segment and through a given point.

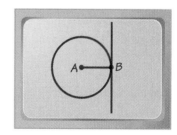

Constructing Polygons

With most software, you can construct and modify a polygon as follows:

Think of the polygon as a group of connected segments, and construct each segment individually.

The last segment should end at the vertex where you started the polygon construction.

You can change the shape of the polygon by dragging its sides and vertices.

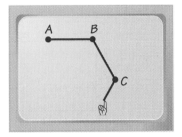

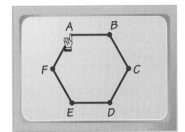

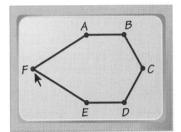

Many software programs also have tools for constructing special kinds of polygons, such as right and isosceles triangles, trapezoids, parallelograms, and regular polygons. Some programs will let you change the shape of a polygon in such a way that its perimeter or area remains constant.

Constructing Circles

Most software lets you construct a circle in one of three ways.

- You can draw the circle "freehand." This is usually done by moving the cursor to where you want the circle's center to be, clicking the mouse button, and dragging away from the center to form the circle.

- You can enter the name of a point to use for the circle's center and a number to use for the radius. For example, you can tell the computer to construct a circle whose center is point *A* and whose radius is 3.

- You can enter an equation for the circle.

Many software programs will also perform more complicated constructions involving circles. Two such constructions are shown below.

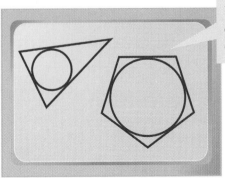

You can inscribe a circle in a triangle or regular polygon.

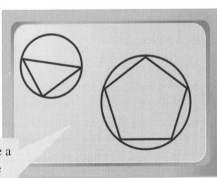

You can circumscribe a circle about a triangle or regular polygon.

16. Construct a line and a point *A* not on the line. Construct a second line parallel to the first line and passing through point *A*.

17. Construct a pentagon. Then construct each diagonal of the pentagon.

18. Construct a point *A* and a circle with center *A* and radius 4. Construct a point *B* on the circle, the radius joining *A* and *B*, and a line tangent to the circle at *B*.

FINDING LENGTHS AND ANGLE MEASURES

You can use most software to find and display lengths of segments and measures of angles. The length and angle-measure displays are automatically updated if you change the corresponding segments and angles. This is useful when you want to construct segments and angles having specified lengths and measures.

For example, suppose you want to construct a triangle such that two of its sides are 10 cm and 5 cm long and have an included angle measuring 40°. Follow these steps:

Construct any triangle *ABC*. (See "Constructing Polygons" on page 727.) Define three measurement displays for ∠*BAC*, $\overline{AB}$, and $\overline{AC}$.

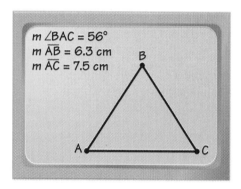

Drag vertex *B* up and to the right until m∠*BAC* = 40° and *AB* = 10 cm. Then drag vertex *C* to the left until *AC* = 5 cm.

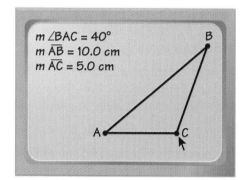

19. Construct a rhombus having a 115° angle and sides that are 8 cm long.

20. **a.** Construct a triangle *ABC* and the bisectors of the angles of the triangle. Construct the point *D* where the bisectors intersect.

b. For each side of the triangle, construct a line perpendicular to the side through point *D*. Construct points *E*, *F*, and *G* where the perpendicular lines intersect the triangle's sides. Measure $\overline{DE}$, $\overline{DF}$, and $\overline{DG}$.

c. Drag the sides and vertices of △*ABC* and observe how *DE*, *DF*, and *DG* change. Based on your observations, make a conjecture about the angle bisectors of a triangle.

PERFORMING TRANSFORMATIONS

Most software will let you perform the transformations described below.
(In the computer screens shown, the original object is blue and the image
after the transformation is red.)

You can reflect an object over a line.

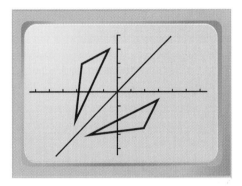

You can rotate an object around a point.

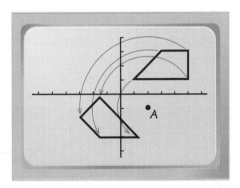

You can translate an object in the direction of an existing vector. The object is translated a distance equal to the vector's magnitude.

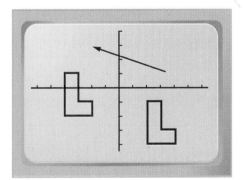

You can dilate an object using a specified center of dilation and scale factor.

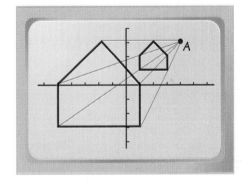

SOFTWARE PRACTICE

**Let *ABCD* be a quadrilateral with vertices at (1, 1), (2, 6), (4, 5), and (5, 2).
Construct the image of the quadrilateral after each transformation.**

21. a reflection over the line $y = -3$

22. a 200° rotation counterclockwise around $(-1, 0)$

23. a translation 5 units left and 6 units down (*Hint:* First construct a vector whose component form is $(-5, -6)$.)

24. a dilation with center $(-2, 4)$ and scale factor 1.5

25. a glide reflection where the translation is given by $(a, b) \rightarrow (a, b - 3)$ and the reflection is over the *y*-axis

Table of Measures

Time

60 seconds (s) = 1 minute (min)	$\left.\begin{array}{l}\text{365 days} \\ \text{52 weeks (approx.)} \\ \text{12 months}\end{array}\right\}$ = 1 year
60 minutes = 1 hour (h)	
24 hours = 1 day	10 years = 1 decade
7 days = 1 week	100 years = 1 century
4 weeks (approx.) = 1 month	

Metric

Length

10 millimeters (mm) = 1 centimeter (cm)

$\left.\begin{array}{l}\text{100 cm} \\ \text{1000 mm}\end{array}\right\}$ = 1 meter (m)

1000 m = 1 kilometer (km)

Area

100 square millimeters = 1 square centimeter
(mm^2)   (cm^2)

10,000 cm^2 = 1 square meter (m^2)

10,000 m^2 = 1 hectare (ha)

Volume

1000 cubic millimeters = 1 cubic centimeter
(mm^3)   (cm^3)

1,000,000 cm^3 = 1 cubic meter (m^3)

Liquid Capacity

1000 milliliters (mL) = 1 liter (L)

1000 L = 1 kiloliter (kL)

Mass

1000 milligrams (mg) = 1 gram (g)

1000 g = 1 kilogram (kg)

1000 kg = 1 metric ton (t)

Temperature — Degrees Celsius (°C)

0°C = freezing point of water

37°C = normal body temperature

100°C = boiling point of water

United States Customary

Length

12 inches (in.) = 1 foot (ft)

$\left.\begin{array}{l}\text{36 in.} \\ \text{3 ft}\end{array}\right\}$ = 1 yard (yd)

$\left.\begin{array}{l}\text{5280 ft} \\ \text{1760 yd}\end{array}\right\}$ = 1 mile (mi)

Area

144 square inches (in.^{2}) = 1 square foot (ft^2)

9 ft^2 = 1 square yard (yd^2)

$\left.\begin{array}{l}\text{43,560 ft}^2 \\ \text{4840 yd}^2\end{array}\right\}$ = 1 acre (A)

Volume

1728 cubic inches (in.^{3}) = 1 cubic foot (ft^3)

27 ft^3 = 1 cubic yard (yd^3)

Liquid Capacity

8 fluid ounces (fl oz) = 1 cup (c)

2 c = 1 pint (pt)

2 pt = 1 quart (qt)

4 qt = 1 gallon (gal)

Weight

16 ounces (oz) = 1 pound (lb)

2000 lb = 1 ton (t)

Temperature — Degrees Fahrenheit (°F)

32°F = freezing point of water

98.6°F = normal body temperature

212°F = boiling point of water

Table of Symbols

Symbol		Page
n^2	n to the 2nd power	4
$-a$	the opposite of a	5
$\dots$	and so on	6
(a, b)	ordered pair	7
$=$	is equal to	13
$\geq$	is greater than or equal to	19
$\leq$	is less than or equal to	19
$\overleftrightarrow{AB}$	line AB	22
$\parallel$	is parallel to	22
$\overrightarrow{AB}$	ray AB	28
$\overline{AB}$	segment AB	28
AB	the length of segment AB	28
$\cong$	is congruent to	28
$\circ$	degree(s)	34
$\angle ABC$	angle ABC	34
$m \angle ABC$	the measure of angle ABC	35
$\perp$	is perpendicular to	58
$\triangle ABC$	triangle ABC	65
n-gon	polygon with n sides	73
$\square ABCD$	parallelogram $ABCD$	88
$\cdot$	multiplication, times ($\times$)	112
$\sqrt{a}$	the nonnegative square root of a	142
$\pm$	plus or minus	142
$\approx$	is approximately equal to	142
$\stackrel{?}{=}$	is this statement true?	143
$\neq$	is not equal to	143
x_1	x sub 1	166

Symbol		Page
m	slope	174
$\odot C$	circle C	187
(x, y, z)	ordered triple	202
$\lvert x \rvert$	the absolute value of x	208
$\angle s$	angles	228
$>$	is greater than	280
$<$	is less than	280
$\not\cong$	is not congruent to	348
π	pi; an irrational number, about 3.14	369
A'	A prime	392
$\rightarrow$	goes to	398
A''	A double prime	405
$a:b$	the ratio of a to b	430
$\sim$	is similar to	446
$\tan A$	the tangent of angle A	507
$\tan^{-1} A$	the inverse tangent of angle A	509
$\sin A$	the sine of angle A	515
$\cos A$	the cosine of angle A	515
$\sin^{-1} A$	the inverse sine of angle A	516
$\cos^{-1} A$	the inverse cosine of angle A	516
$\overrightarrow{AB}$	vector AB	521
$\lvert \overrightarrow{AB} \rvert$	the magnitude of vector AB	523
$\overarc{AB}$	arc AB	554
$m \overarc{AB}$	the measure of arc AB	554
$\begin{bmatrix} 1 & 0 \\ 0 & 1 \end{bmatrix}$	matrix	613
a^{-n}	$\frac{1}{a^n}, a \neq 0$	709

Table of Squares and Square Roots

No.	Square	Sq. Root	No.	Square	Sq. Root	No.	Square	Sq. Root
1	1	1.000	51	2,601	7.141	101	10,201	10.050
2	4	1.414	52	2,704	7.211	102	10,404	10.100
3	9	1.732	53	2,809	7.280	103	10,609	10.149
4	16	2.000	54	2,916	7.348	104	10,816	10.198
5	25	2.236	55	3,025	7.416	105	11,025	10.247
6	36	2.449	56	3,136	7.483	106	11,236	10.296
7	49	2.646	57	3,249	7.550	107	11,449	10.344
8	64	2.828	58	3,364	7.616	108	11,664	10.392
9	81	3.000	59	3,481	7.681	109	11,881	10.440
10	100	3.162	60	3,600	7.746	110	12,100	10.488
11	121	3.317	61	3,721	7.810	111	12,321	10.536
12	144	3.464	62	3,844	7.874	112	12,544	10.583
13	169	3.606	63	3,969	7.937	113	12,769	10.630
14	196	3.742	64	4,096	8.000	114	12,996	10.677
15	225	3.873	65	4,225	8.062	115	13,225	10.724
16	256	4.000	66	4,356	8.124	116	13,456	10.770
17	289	4.123	67	4,489	8.185	117	13,689	10.817
18	324	4.243	68	4,624	8.246	118	13,924	10.863
19	361	4.359	69	4,761	8.307	119	14,161	10.909
20	400	4.472	70	4,900	8.367	120	14,400	10.954
21	441	4.583	71	5,041	8.426	121	14,641	11.000
22	484	4.690	72	5,184	8.485	122	14,884	11.045
23	529	4.796	73	5,329	8.544	123	15,129	11.091
24	576	4.899	74	5,476	8.602	124	15,376	11.136
25	625	5.000	75	5,625	8.660	125	15,625	11.180
26	676	5.099	76	5,776	8.718	126	15,876	11.225
27	729	5.196	77	5,929	8.775	127	16,129	11.269
28	784	5.292	78	6,084	8.832	128	16,384	11.314
29	841	5.385	79	6,241	8.888	129	16,641	11.358
30	900	5.477	80	6,400	8.944	130	16,900	11.402
31	961	5.568	81	6,561	9.000	131	17,161	11.446
32	1,024	5.657	82	6,724	9.055	132	17,424	11.489
33	1,089	5.745	83	6,889	9.110	133	17,689	11.533
34	1,156	5.831	84	7,056	9.165	134	17,956	11.576
35	1,225	5.916	85	7,225	9.220	135	18,225	11.619
36	1,296	6.000	86	7,396	9.274	136	18,496	11.662
37	1,369	6.083	87	7,569	9.327	137	18,769	11.705
38	1,444	6.164	88	7,744	9.381	138	19,044	11.747
39	1,521	6.245	89	7,921	9.434	139	19,321	11.790
40	1,600	6.325	90	8,100	9.487	140	19,600	11.832
41	1,681	6.403	91	8,281	9.539	141	19,881	11.874
42	1,764	6.481	92	8,464	9.592	142	20,164	11.916
43	1,849	6.557	93	8,649	9.644	143	20,449	11.958
44	1,936	6.633	94	8,836	9.695	144	20,736	12.000
45	2,025	6.708	95	9,025	9.747	145	21,025	12.042
46	2,116	6.782	96	9,216	9.798	146	21,316	12.083
47	2,209	6.856	97	9,409	9.849	147	21,609	12.124
48	2,304	6.928	98	9,604	9.899	148	21,904	12.166
49	2,401	7.000	99	9,801	9.950	149	22,201	12.207
50	2,500	7.071	100	10,000	10.000	150	22,500	12.247

Table of Trigonometric Ratios

Angle	Sine	Cosine	Tangent	Angle	Sine	Cosine	Tangent
1°	.0175	.9998	.0175	46°	.7193	.6947	1.0355
2°	.0349	.9994	.0349	47°	.7314	.6820	1.0724
3°	.0523	.9986	.0524	48°	.7431	.6691	1.1106
4°	.0698	.9976	.0699	49°	.7547	.6561	1.1504
5°	.0872	.9962	.0875	50°	.7660	.6428	1.1918
6°	.1045	.9945	.1051	51°	.7771	.6293	1.2349
7°	.1219	.9925	.1228	52°	.7880	.6157	1.2799
8°	.1392	.9903	.1405	53°	.7986	.6018	1.3270
9°	.1564	.9877	.1584	54°	.8090	.5878	1.3764
10°	.1736	.9848	.1763	55°	.8192	.5736	1.4281
11°	.1908	.9816	.1944	56°	.8290	.5592	1.4826
12°	.2079	.9781	.2126	57°	.8387	.5446	1.5399
13°	.2250	.9744	.2309	58°	.8480	.5299	1.6003
14°	.2419	.9703	.2493	59°	.8572	.5150	1.6643
15°	.2588	.9659	.2679	60°	.8660	.5000	1.7321
16°	.2756	.9613	.2867	61°	.8746	.4848	1.8040
17°	.2924	.9563	.3057	62°	.8829	.4695	1.8807
18°	.3090	.9511	.3249	63°	.8910	.4540	1.9626
19°	.3256	.9455	.3443	64°	.8988	.4384	2.0503
20°	.3420	.9397	.3640	65°	.9063	.4226	2.1445
21°	.3584	.9336	.3839	66°	.9135	.4067	2.2460
22°	.3746	.9272	.4040	67°	.9205	.3907	2.3559
23°	.3907	.9205	.4245	68°	.9272	.3746	2.4751
24°	.4067	.9135	.4452	69°	.9336	.3584	2.6051
25°	.4226	.9063	.4663	70°	.9397	.3420	2.7475
26°	.4384	.8988	.4877	71°	.9455	.3256	2.9042
27°	.4540	.8910	.5095	72°	.9511	.3090	3.0777
28°	.4695	.8829	.5317	73°	.9563	.2924	3.2709
29°	.4848	.8746	.5543	74°	.9613	.2756	3.4874
30°	.5000	.8660	.5774	75°	.9659	.2588	3.7321
31°	.5150	.8572	.6009	76°	.9703	.2419	4.0108
32°	.5299	.8480	.6249	77°	.9744	.2250	4.3315
33°	.5446	.8387	.6494	78°	.9781	.2079	4.7046
34°	.5592	.8290	.6745	79°	.9816	.1908	5.1446
35°	.5736	.8192	.7002	80°	.9848	.1736	5.6713
36°	.5878	.8090	.7265	81°	.9877	.1564	6.3138
37°	.6018	.7986	.7536	82°	.9903	.1392	7.1154
38°	.6157	.7880	.7813	83°	.9925	.1219	8.1443
39°	.6293	.7771	.8098	84°	.9945	.1045	9.5144
40°	.6428	.7660	.8391	85°	.9962	.0872	11.4301
41°	.6561	.7547	.8693	86°	.9976	.0698	14.3007
42°	.6691	.7431	.9004	87°	.9986	.0523	19.0811
43°	.6820	.7314	.9325	88°	.9994	.0349	28.6363
44°	.6947	.7193	.9657	89°	.9998	.0175	57.2900
45°	.7071	.7071	1.0000				

1

A Brief History of Geometric Systems

Geometry is primarily the study of the properties of space and figures in space. Like most of mathematics, geometry developed over thousands of years. It is the product of contributions made by many individuals and civilizations.

The Practical Origins of Geometry

The first peoples known to have written about geometry were the ancient Egyptians and Babylonians (about 4000 B.C. to 300 B.C.). They developed accurate methods for finding areas and volumes of many plane figures and solids. The Babylonians knew the Pythagorean theorem more than 1000 years before the Greek mathematician Pythagoras. Scholars believe that geometric accomplishments similar to those of the Egyptians and Babylonians also took place in ancient India and China in the same period.

Both the Egyptians and Babylonians viewed geometry as a tool for solving real-world problems, such as constructing buildings and canals and finding land areas. Their approach to geometry was based on inductive reasoning, generalizing from trial-and-error and everyday experience.

Euclid's Deductive Approach

The Greeks were responsible for two major changes in geometry. The first was from applications to the study of abstract geometric figures. The second, more important change was from inductive methods toward deductive methods. Both changes are in the work of the Greek geometer Euclid.

Euclid wrote what is probably the most important book in the history of deductive geometry, *Elements*, around 300 B.C. Euclid began by defining common geometric objects, such as points and segments, and then stated a small number of seemingly obvious geometric truths called *postulates*. For example, one of the postulates states that it is always possible to draw a segment joining two points. Euclid then used deductive reasoning to prove geometric results called *theorems* from his postulates and definitions.

Mathematical Systems

The postulate-theorem approach of geometry as a *mathematical system* that Euclid used is still vital to mathematics today. Research in the foundations of mathematics began when mathematicians looked at the relationships among postulates and appropriate ways of reasoning from them.

Non-Euclidean Geometry

One of the postulates in Euclid's *Elements*—called the *parallel postulate*—is usually stated: In a plane, given a line ℓ and a point P not on ℓ, there is exactly one line parallel to ℓ through P.

Mathematicians after Euclid were uncomfortable with the parallel postulate because it didn't seem as obvious as the other postulates. The postulate could not be verified by everyday experience because it is not possible to construct two parallel lines that extend infinitely in both directions and confirm that the lines never meet.

Many mathematicians tried to show that the parallel postulate was not needed or to replace it. In the early 1800s, Carl Friedrich Gauss of Germany, Nikolai Lobachevsky of Russia, and Janos Bolyai of Hungary independently discovered *non-Euclidean geometry*.

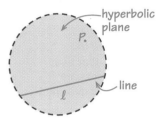

These three mathematicians replaced Euclid's parallel postulate with the following postulate: In a plane, given a line ℓ and a point P not on ℓ, there are *infinitely* many lines parallel to ℓ through P. You can visualize the postulate by defining a plane to be the interior of a circle and a line to be a chord of the circle, excluding the chord's endpoints. Then given a point P in this new plane and a line ℓ not containing P, there are in fact an infinite number of lines through P that are parallel to (do not intersect) line ℓ.

The non-Euclidean geometry formulated by Gauss, Lobachevsky, and Bolyai is called *hyperbolic geometry*. Another non-Euclidean geometry known as *spherical geometry* was derived by the German mathematician G. F. Bernhard Riemann. In spherical geometry, given any line ℓ and point P not on ℓ, there is *no* line parallel to ℓ through P.

Which type of geometry, Euclidean or non-Euclidean, is the best model for the real world? In some fields, such as engineering and surveying, Euclidean geometry still works best as a representation of the world around us. On the other hand, navigation uses spherical geometry, and Albert Einstein used non-Euclidean geometry to describe gravity in his general theory of relativity. Different non-Euclidean models are essential to physics and other sciences.

The following pages discuss particular non-Euclidean geometries:

• Taxicab Geometry (pp. 208–209)   • Spherical Geometry (pp. 597–603)

RESEARCH QUESTIONS

1. **Writing** Some mathematicians who made significant contributions to geometry are Thales, Hypatia, Hipparchus, Maria Gaetana Agnesi, René Descartes, Mohammad Abu'l-Wafa, and Felix Klein. Write a report on the life and mathematical work of one of these people.

2. **Writing** An important topic not discussed in this history is projective geometry. Write a general overview of projective geometry.

2

Truth Tables and Logic

Learn how to...

- **represent statements using logical symbols for** *and, or, not,* **and** *implies*

- **make a truth table for a logical statement or argument**

So you can...

- **tell whether two statements are logically equivalent**

- **recognize a tautology or a contradiction**

- **tell whether a logical argument is valid**

Does the argument below seem logical to you?

> *If ABCD is a rhombus, then it is a parallelogram.*
> *If ABCD is a rectangle, then it is a parallelogram.*
> *Therefore, if ABCD is a rhombus, it is a rectangle.*

If something seems wrong to you about that reasoning, how could you convince someone it is wrong? In this appendix, you will learn one way to show whether an argument is *logically valid*.

To study logical arguments, statements are often represented by letters, such as p, q, and r.

> p: "Gene Kelly is in the movie *Singing in the Rain*."
> q: "It never rains in southern California."
> r: "The rain in Spain stays mainly in the plain."

If a statement is known to be true or false, it can be assigned a **truth value** of T or F. Statement p is true, so its truth value is T. Statement q is false, so its truth value is F. If you lack information about the weather and geography of Spain, then you will not be able to assign statement r a truth value.

To decide the truth value of a compound statement, you examine the truth values of its components.

BY THE WAY

The logical symbol $\vee$ is an abbreviation of the Latin word "vel," which means "or."

A **conjunction** joins two statements with the word "and." The symbol $\wedge$ stands for "and."

$$p \wedge q$$

"Gene Kelly is in the movie *Singing in the Rain* **and** it never rains in southern California."

This conjunction is false, because the second statement is false.

A **disjunction** joins two statements with the word "or." The symbol $\vee$ stands for "or."

$$p \vee q$$

"Gene Kelly is in the movie *Singing in the Rain* **or** it never rains in southern California."

This disjunction is true, because at least one of the statements is true.

THINK AND COMMUNICATE

Use statements *p*, *q*, and *r* above. Express each statement in words.

1. $p \wedge r$ **2.** $q \vee r$

3. Suppose the truth value of statement r is T. What are the truth values of the statements in Questions 1 and 2?

A **truth table** shows how the truth value of an expression depends on the truth values of its components. The truth tables below have four rows to allow for all combinations of truth values for statements p and q.

Truth table for conjunction		
p	q	$p \wedge q$
T	T	**T**
T	F	**F**
F	T	**F**
F	F	**F**

Truth table for disjunction		
p	q	$p \vee q$
T	T	**T**
T	F	**T**
F	T	**T**
F	F	**F**

A conjunction is true only when both of its components are true.

A disjunction is false only when both of its components are false.

This book uses "or" in an *inclusive* sense. The statement $p \vee q$ means "p is true and q is false, q is true and p is false, or both p and q are true."

Two other truth tables that are useful in logic are the table for *negation* and the table for a *conditional* statement. A curly symbol ($\sim$) is used for negation. An implication arrow ($\rightarrow$) is used for a conditional.

Truth table for negation	
p	$\sim p$
T	**F**
F	**T**

Truth table for a conditional		
p	q	$p \rightarrow q$
T	T	**T**
T	F	**F**
F	T	**T**
F	F	**T**

The **negation** of the statement p is the statement $\sim p$, which means "not p" or "It is not the case that statement p is true." A statement and its negation have opposite truth values.

You can express the meaning of the conditional statement $p \rightarrow q$ in many ways: "If p, then q," "p implies q," and "q follows from p." In other words, it never happens that p occurs without q also occurring.

Notice that by convention, a conditional $p \rightarrow q$ is considered true if its hypothesis p is false. For example, the statement "If the moon is made of green cheese, then $2 + 2 = 5$" is *true*.

THINK AND COMMUNICATE

Let p stand for "$\triangle ABC$ is an isosceles triangle," and q stand for "$\triangle ABC$ is an equilateral triangle." Write each sentence using symbolic notation.

4. $\triangle ABC$ is an equilateral triangle or it is not an equilateral triangle.

5. If $\triangle ABC$ is an equilateral triangle, then it is an isosceles triangle.

EXAMPLE 1

Make a truth table for $\sim p \wedge \sim q$.

SOLUTION

Step 1 Make a table with all combinations of T and F for p and q.

Step 2 Add a column for $\sim p$ using truth values for the negation of p. Add a column for $\sim q$.

Step 3 Use the truth table for conjunction to combine the columns for $\sim p$ and $\sim q$.

Truth table for $\sim p \wedge \sim q$				
p	q	$\sim p$	$\sim q$	$\sim p \wedge \sim q$
T	T	F	F	F
T	F	F	T	F
F	T	T	F	F
F	F	T	T	T

THINK AND COMMUNICATE

6. Make a truth table for $q \rightarrow p$. Is it different from the truth table for $p \rightarrow q$? Explain.

7. Make a truth table for $\sim q \wedge \sim p$. Is it different from the truth table for $\sim p \wedge \sim q$? Explain.

If two logical statements have truth tables with identical final columns, the statements are called **logically equivalent**. In the following example, parentheses are used for grouping. As in algebra, you should evaluate statements in parentheses first.

EXAMPLE 2

Tell whether $\sim(p \vee q)$ and $\sim p \wedge \sim q$ are logically equivalent.

SOLUTION

Build a truth table for $\sim(p \vee q)$ and compare it with the truth table for $\sim p \wedge \sim q$.

Truth table for $\sim(p \vee q)$			
p	q	$(p \vee q)$	$\sim(p \wedge q)$
T	T	T	F
T	F	T	F
F	T	T	F
F	F	F	T

Truth table for $\sim p \wedge \sim q$				
p	q	$\sim p$	$\sim q$	$\sim p \wedge \sim q$
T	T	F	F	F
T	F	F	T	F
F	T	T	F	F
F	F	T	T	T

The statements are logically equivalent because the final columns of the truth tables are identical.

THINK AND COMMUNICATE

8. Describe the statements $\sim(p \vee q)$ and $\sim p \wedge \sim q$ in words. Explain why they mean the same thing.

9. Tell whether $\sim(p \vee q)$ and $\sim p \vee q$ are logically equivalent.

If a statement has all F's in the final column of its truth table, it is called a **contradiction**. If a statement has all T's in the final column of its truth table, it is called a **tautology**. Two simple examples are shown.

Truth table for $p \wedge \sim p$		
p	$\sim p$	$p \wedge \sim p$
T	F	**F**
F	T	**F**

a contradiction

Truth table for $p \vee \sim p$		
p	$\sim p$	$p \vee \sim p$
T	F	**T**
F	T	**T**

a tautology

A tautology is always true, regardless of the truth values of its component statements. When a tautology involves an implication, you can say that the argument it represents is **logically valid**.

In geometry and other forms of mathematics, you often use this reasoning:

If p **is true**, and p implies q, then q is true.

$$[p \quad \wedge \quad (p \rightarrow q)] \quad \rightarrow \quad q$$

You can show that this reasoning, called *modus ponens*, is logically valid by making a truth table and showing that it has all T's in its final column.

EXAMPLE 3

Show that $[p \wedge (p \rightarrow q)] \rightarrow q$ is a logically valid argument.

SOLUTION

Make a truth table. You can add intermediate columns to help you. Use columns 1 and 3 to calculate column 4. Repeating column 2 as column 5 may help you calculate column 6.

1	2	3	4	5	6
p	q	$p \rightarrow q$	$p \wedge (p \rightarrow q)$	q	$[p \wedge (p \rightarrow q)] \rightarrow q$
T	T	T	T	T	**T**
T	F	F	F	F	**T**
F	T	T	F	T	**T**
F	F	T	F	F	**T**

Because the last column contains only T's, the argument is logically valid.

Exercises and Applications

Using the statements below, express each symbolic statement in words.

p: "I win." q: "You win." r: "They win."

1. $p \wedge q$

2. $p \vee q$

3. $\sim p \wedge \sim q$

4. $q \rightarrow \sim r$

5. $(q \vee r) \rightarrow \sim p$

6. $\sim(p \wedge q)$

7. $p \vee (q \wedge r)$

8. $\sim(q \vee r)$

Let p stand for "$ABCD$ has two pairs of parallel sides," q stand for "$ABCD$ has four right angles," r stand for "$ABCD$ is a rectangle," and s stand for "$ABCD$ is a square." Express each sentence in symbolic notation, using parentheses if necessary.

9. If $ABCD$ has four right angles, then it is a rectangle.

10. If $ABCD$ is a square, then it has two pairs of parallel sides and four right angles.

11. If $ABCD$ does not have two pairs of parallel sides, then it is not a square or a rectangle.

Make a truth table for each symbolic statement. Identify any tautologies or contradictions.

12. $p \vee \sim p$

13. $p \wedge \sim p$

14. $p \rightarrow p$

15. $\sim(\sim p)$

16. $p \vee \sim q$

17. $\sim p \vee q$

18. $p \wedge \sim q$

19. $\sim p \wedge q$

In Exercises 20 and 21, use *De Morgan's rules*:

The statement $\sim(p \vee q)$ is logically equivalent to $\sim p \wedge \sim q$.

The statement $\sim(p \wedge q)$ is logically equivalent to $\sim p \vee \sim q$.

20. The first rule was proved in Example 2. Use truth tables to prove the second.

21. Writing Express De Morgan's rules in words.

22. Investigation On page 149, these four statements are summarized:

Conditional	Inverse
If p, then q.	If not p, then not q.

Converse	Contrapositive
If q, then p.	If not q, then not p.

a. Write each statement using symbolic notation.

b. Make a truth table for each statement.

c. Tell which statements are logically equivalent.

23. The statement "p if and only if q" is called a *biconditional*. In symbols, it can be expressed as $(p \rightarrow q) \wedge (q \rightarrow p)$. Make a truth table for the biconditional.

Tell whether each statement is a *contradiction*, a *tautology*, or *neither*. Explain your answer.

24. $(p \wedge q) \to p$

25. $(p \wedge q) \wedge (\sim p \vee \sim q)$

26. $(p \vee q) \to \sim q$

27. $[(p \vee q) \wedge \sim p] \to q$

28. Open-ended Problem Write a symbolic statement that represents a tautology or a contradiction. Show its truth table.

29. The following form of reasoning is known as *modus tollens*:

If p implies q, and q is not true, then p is not true.

$$[(p \to q) \wedge \sim q] \to \sim p$$

Make a truth table to show that *modus tollens* is a logically valid form of reasoning.

30. The following form of reasoning is known as *sorites*:

If p implies q, and q implies r, then p implies r.

$$[(p \to q) \wedge (q \to r)] \to (p \to r)$$

Complete the truth table to show that *sorites* is a logically valid form of reasoning.

1	2	3	4	5	6	7	8
p	q	r	$p \to q$	$q \to r$	$(p \to q) \wedge (q \to r)$	$p \to r$	$[(p \to q) \wedge (p \to r)] \to (p \to r)$
T	T	T	T	T	T	?	?
T	T	F	T	F	F	?	?
T	F	T	F	T	?	?	?
T	F	F	F	T	?	?	?
F	T	T	T	?	?	?	?
F	T	F	T	?	?	?	?
F	F	T	T	?	?	?	?
F	F	F	T	?	?	?	?

31. Challenge You can extend the *sorites* form of reasoning. Make a truth table to show that $[(p \to q) \wedge (q \to r) \wedge (r \to s)] \to (p \to s)$. How many rows does your truth table require?

32. The word *sorites* is derived from a Greek root meaning "heap." Explain why this is appropriate.

33. Let r stand for "*ABCD* is a rhombus," p stand for "*ABCD* is a parallelogram," and t stand for "*ABCD* is a rectangle." The argument at the top of page 736 can be phrased this way:

If r implies p, and t implies p, then r implies t.

$$[(r \to p) \wedge (t \to p)] \to (r \to t)$$

Make an eight-row truth table and use it to show that the above argument is *not* logically valid.

34. Research Find out how to use truth tables to model the behavior of electricity in circuits that have switches placed *in parallel* or *in series*.

Glossary

acute angle (p. 57) An angle that measures between 0° and 90°.

acute triangle (p. 65) A triangle with three acute angles.

adjacent angles (p. 58) Two coplanar angles that share a vertex and a side but do not overlap.

alternate interior angles (p. 219) Two angles that lie on opposite sides of a transversal between the two lines that the transversal intersects.

altitude of a triangle (p. 319) A perpendicular segment from a vertex of a triangle to the line that contains the opposite side.

angle (p. 34) A figure formed by two rays (called *sides*) with a common endpoint (called the *vertex*).

angle of depression (p. 509) When a point is viewed from a higher point, the angle that the person's line of sight makes with the horizontal.

angle of elevation (p. 509) When a point is viewed from a lower point, the angle that the person's line of sight makes with the horizontal.

apothem (p. 368) The distance from the center of a regular polygon to a side.

arc (p. 554) An unbroken part of a circle.

auxiliary line (p. 235) A line, ray, or segment added to a diagram to help complete a proof.

base angles of an isosceles triangle (p. 313) The angles opposite the two congruent sides. The third angle is the *vertex angle*.

biconditional (p. 118) A statement that contains the words "if and only if." This single statement is equivalent to writing both "If p, then q" and its converse "If q, then p."

bisector of a segment (p. 41) A line, segment, ray, or plane that intersects the segment at its midpoint.

bisector of an angle (p. 42) A line or ray that divides the angle into two congruent angles.

central angle of a circle (p. 553) An angle whose vertex is the center of a circle.

chord (p. 553) A segment whose endpoints lie on a circle.

circle (p. 187) The set of all points in a plane that are an equal distance (the *radius*) from a given point (the *center*), which is also in the plane.

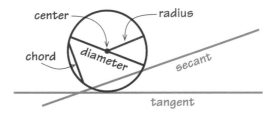

circumference of a circle (p. 369) The perimeter of a circle.

circumscribed circle (p. 369) A circle is *circumscribed* about a polygon when each vertex of the polygon lies on the circle. The polygon is *inscribed* in the circle.

circumscribed circle   inscribed circle
inscribed polygon   circumscribed polygon

circumscribed polygon (p. 324, Ex. 13) A polygon is *circumscribed* about a circle when each of the sides of the polygon is tangent to the circle. The circle is *inscribed* in the polygon.

collinear points (p. 22) Points on the same line.

complementary angles, complements (p. 58) Two angles whose measures add up to 90°.

concentric circles (p. 189) Two or more circles that lie in the same plane and have the same center.

conditional statement (p. 15) A statement that can be written in the form "If p, then q." Statement p is the *hypothesis* and statement q is the *conclusion*.

cone (p. 537) A three-dimensional figure with one circular base and a vertex. In a right cone, the *vertex* is directly above the center of the base. In the right cone shown, *h* is the height, *s* is the slant height, and *r* is the radius.

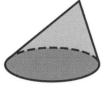

right cone   oblique cone

congruent angles (p. 35) Two angles that have the same measure.

congruent arcs (p. 555) Arcs, in the same circle or in congruent circles, that have equal measures.

congruent polygons (p. 285) Polygons that have congruent corresponding sides and angles.

conjecture (p. 15) Something believed to be true but not yet proved.

consecutive angles (p. 74) In a polygon, two angles that share a side.

consecutive sides (p. 74) In a polygon, two sides that share a vertex.

contrapositive (p. 149) The contrapositive of the conditional statement "If *p*, then *q*" is the statement "If not *q*, then not *p*."

converse (p. 137) The converse of the conditional statement interchanges the hypothesis and conclusion. "If *p*, then *q*" becomes "If *q*, then *p*."

convex polygon (p. 73) A polygon in which no segment that connects two vertices can be drawn outside the polygon.

coordinate plane (pp. 696–697) A grid formed by two axes that intersect at the *origin*. The axes divide the coordinate plane into four *quadrants*.

coplanar (p. 22) Points and other figures that lie in the same plane.

corollary of a theorem (p. 239) A statement that can be easily proved using the theorem.

corresponding angles (p. 219) Two angles that lie on the same side of a transversal, in corresponding positions with respect to the two lines that the transversal intersects.

corresponding parts (pp. 285, 445) A side (or angle) of a polygon that is matched up with a side (or angle) of a congruent or similar polygon.

cosine of an acute angle, cosine ratio (p. 515) In a right triangle, the ratio of the length of the leg adjacent to the angle to the length of the hypotenuse.

counterexample (p. 16) An example for which the hypothesis of a conditional statement is true, but the conclusion is false. A counterexample shows that the conditional statement is not always true.

cube (p. 94) A rectangular prism whose faces are all squares.

cylinder (p. 375) A space figure whose *bases* are circles of the same size. The right cylinder shown has height *h* and radius *r*.

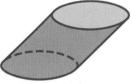

right cylinder   oblique cylinder

deductive reasoning (p. 112) Using facts, definitions, and accepted properties in a logical order to reach a conclusion or to show that a conjecture is always true.

diagonal of a polygon (p. 74) In a polygon, a segment that connects nonconsecutive vertices.

diameter of a circle (p. 187) A segment through the center of the circle with endpoints on the circle; also the length of such a segment. *See* circle.

dilation (p. 428) A transformation with center *O* and positive *scale factor k* in which every point *P* has an image *P'* on $\overrightarrow{OP}$ so that $OP' = k \cdot OP$. If $k > 1$, the dilation is an *enlargement*. If $k < 1$, the dilation is a *reduction*.

distance (pp. 166, 203, 249, 258) The length of the segment that connects two points, or the length of the perpendicular segment from a point to the object.

endpoint of a ray (p. 28) The point where a ray begins. A ray is named by the endpoint and another point on the ray. *See* ray.

endpoints of a segment (p. 28) The points where a segment begins and ends. These points are used to name the segment.

equiangular polygon (pp. 65, 73) A polygon whose angles are all congruent.

equidistant (pp. 209, 307) At the same distance.

equilateral polygon (pp. 65, 73) A polygon whose sides are all congruent.

Euclidean geometry (p. 598) A geometric system based on the postulates of Euclid.

exterior angle of a polygon (pp. 81, 125) An angle formed when one side of the polygon is extended. The angle is adjacent to an *interior angle* of the polygon.

geometric mean (p. 494) If a, b, and x are positive numbers and $\frac{a}{x} = \frac{x}{b}$, then x is the geometric mean of a and b.

geometric probability (p. 476) The probability of an event is given by a ratio that compares lengths, areas, perimeters, or other measures.

glide reflection (p. 420) A translation followed by a reflection over a line parallel to the translation.

great circle (p. 598) The intersection of a sphere with a plane containing the center of the sphere.

half-turn (pp. 9, 414) A rotation of 180°.

height *See* cone, cylinder, prism, *and* pyramid.

hexagon (p. 73) A polygon with six sides.

hyperbolic geometry (p. 602) A non-Euclidean geometric system based on the postulate that through a point not on a line, there is more than one line parallel to the given line.

hypotenuse (p. 141) The side opposite the right angle in a right triangle. The other two sides are the *legs*.

hypothesis in a proof (p. 125) The given information.

image (pp. 8, 391) The figure resulting from applying a transformation to a figure.

image matrix (p. 614) A matrix whose elements represent the coordinates of the image of a transformed polygon.

included angle (p. 294) An angle of a polygon whose vertex is the shared point of two sides of the polygon.

included side (p. 300) A side of a polygon whose endpoints are the vertices of two angles of the polygon.

inductive reasoning (p. 4) A type of reasoning in which a prediction or conclusion is based on an observed pattern.

inscribed angle (p. 553) An angle whose vertex is on a circle and whose sides are chords of the circle.

inscribed circle (p. 324, Ex. 13) *See* circumscribed polygon.

inscribed polygon (p. 369) *See* circumscribed circle.

intercept (p. 175) The distance from the origin to the point where a line crosses one of the axes.

interior angle of a polygon (p. 81) See exterior angle of a polygon.

inverse (p. 149) The inverse of the conditional statement "If p, then q" is the statement "If not p, then not q."

isosceles triangle (p. 65) A triangle with at least two congruent sides. The two congruent sides are the *legs*. The third side is the *base*. The angles opposite the legs are the *base angles*, and the third angle is the *vertex angle*.

kite (p. 339) A quadrilateral that has two pairs of consecutive congruent sides, but whose opposite sides are not congruent.

lateral area (pp. 376, 536, 537) For a prism and a pyramid, the area of the lateral faces. For a cylinder and a cone, the area of the curved lateral surface.

length of an arc (p. 580) Along a circle, the distance between the endpoints of an arc.

line of reflection (p. 391) The line over which a figure is flipped, or reflected.

line of symmetry (p. 10) The line over which a figure is flipped, resulting in a figure that coincides exactly with the original figure.

line perpendicular to a plane (p. 258) A line that intersects a plane and is perpendicular to every line in the plane that passes through the point of intersection.

linear pair (p. 58) Two adjacent angles form a linear pair if their nonshared rays form a straight angle.

magnitude of a vector (p. 521) The size of the vector; also the length of the arrow that represents the vector.

major arc (p. 554) Two points on a circle and all the points of the circle not on the minor arc between the two given points. A major arc measures greater than 180° and is named by three points, its endpoints, and a third point on the major arc.

matrix (matrices) (p. 613) A rectangular arrangement of numbers, called *elements*, in rows and columns. The matrix below has 2 rows and 3 columns, or *dimension* 2 × 3.

$$\begin{bmatrix} 1 & 0 & -2 \\ 5 & -3 & 4 \end{bmatrix}$$

measure of an angle (p. 35) A number between 0° and 180° that is paired with the angle.

measure of an arc (p. 554) The measure of a minor arc is equal to the measure of the central angle that intercepts it. The measure of a major arc is equal to 360° minus the measure of the corresponding minor arc. The measure of a semicircle is 180°. The measure of a complete circle is 360°.

median of a triangle (p. 319) A segment from a vertex to the midpoint of the opposite side.

midline of a trapezoid (p. 364, Ex. 16) The segment that is parallel to the bases of a trapezoid and connects the midpoints of the legs.

midpoint (p. 41) The point on a segment that divides it into two congruent segments.

minor arc (p. 554) Two points on a circle and all the points of the circle in the interior of the central angle between the two endpoints. A minor arc measures less than 180° and is named by its two endpoints.

n-gon (p. 73) A polygon with *n* sides.

negation (p. 149) The negation of a statement *p* is the statement "not *p*," or "*p* is not true."

net (p. 95) A plane figure that can be folded to form a three-dimensional figure, such as a prism.

non-Euclidean geometry (pp. 208, 597, 602) A geometric system not based on the postulates of Euclid.

obtuse angle (p. 57) An angle that measures between 90° and 180°.

obtuse triangle (p. 65) A triangle with an obtuse angle.

octagon (p. 73) A polygon with eight sides.

ordered triple (p. 202) The coordinates (*x*, *y*, *z*) that give the position of a point in three dimensions.

orientation in a transformation (p. 392) The order and direction in which the vertices of a given figure are arranged.

origin (pp. 696–697) On a coordinate plane, the point where the axes intersect.

parallel lines (p. 22) Lines in the same plane that do not intersect.

parallel planes (p. 23) Planes that do not intersect.

parallelogram (p. 88) A quadrilateral with both pairs of opposite sides parallel.

pentagon (p. 73) A polygon with five sides.

perpendicular (p. 58) Two lines, segments, rays, or planes that intersect to form right angles.

perpendicular bisector of a segment (p. 181) A line, ray, or segment that bisects the segment and is perpendicular to it.

point matrix (p. 613) A matrix that represents a point.

point of tangency (p. 554) The point where a tangent intersects a circle.

polar coordinates (p. 416) A system of coordinates based on distance from the origin and an angle of rotation.

polygon (p. 73) A closed plane figure whose sides are segments that intersect only at their endpoints, with each segment intersecting exactly two other segments.

polygon matrix (p. 613) A matrix that represents a polygon. Each column represents the coordinates of a vertex.

postulate (p. 117) A mathematical statement that is accepted without proof.

prism (pp. 93, 375) A three-dimensional figure with two congruent faces, called *bases*, that lie in parallel planes. The other faces, called *lateral faces*, are rectangles that connect corresponding vertices of the bases. A prism's vertices are connected by segments called *edges*. In the prisms below, *h* is the height.

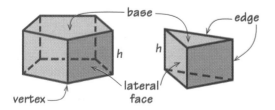

proof (p. 124) A convincing argument that can be used to show that a conjecture is true.

proportion (p. 446) An equation that shows that two ratios are equal.

pyramid (p. 535) A three-dimensional figure with one *base* that is a polygon. The other faces, called *lateral faces*, are triangles that connect the base to the *vertex*. In the right pyramid shown, *h* is the height, *s* is the slant height, *a* is the apothem, and *r* is the radius.

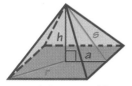

right pyramid   oblique pyramid

quadrilateral (p. 73) A polygon with four sides.

radius of a circle (p. 187) A segment whose endpoints are the center of the circle and a point on the circle; also the length of such a segment. *See* circle.

ray (p. 28) A part of a line that starts at a point, the endpoint, and continues forever in one direction.

rectangle (p. 88) An equiangular parallelogram.

rectangular prism (p. 94) A prism whose bases and faces are rectangles.

reflection (pp. 8, 391) A transformation that flips a figure over a line.

regular polygon (p. 73) A polygon that is both equilateral and equiangular.

regular pyramid (p. 535) A pyramid with a base that is a regular polygon and with lateral faces that are congruent isosceles triangles.

rhombus (p. 88) An equilateral parallelogram.

right angle (p. 57) An angle that measures 90°.

right triangle (p. 65) A triangle with a right angle.

rotation (pp. 8, 412) A transformation in which every point moves along a circular path around a fixed point called the *center of rotation*.

same-side interior angles (p. 219) Two angles that lie on the same side of a transversal between the two lines that the transversal intersects.

scalar multiplication of a matrix (p. 614) The process of multiplying each element of a matrix by a real number.

scalar multiplication of a vector (p. 523) The process of multiplying each component of a vector by a real number.

scalene triangle (p. 65) A triangle with no congruent sides.

secant of a circle (p. 561) A line, ray, or segment that contains a chord. *See* circle.

sector of a circle (p. 580) A region of a circle that is bounded by two radii and an arc of the circle.

segment (p. 28) A part of a line with two endpoints.

semicircle (p. 554) Either of the two arcs of a circle intersected by a diameter of the circle. A semicircle is named by three points, its endpoints and a third point on the semicircle.

similar (p. 446) Two figures are similar if corresponding angles are congruent and the lengths of corresponding sides are in proportion.

sine of an acute angle, sine ratio (p. 515) In a right triangle, the ratio of the length of the leg opposite the angle to the length of the hypotenuse.

skew lines (p. 22) Two lines that do not intersect and are not parallel.

slope (pp. 174, 697) The measure of the steepness of a line, or vertical change divided by horizontal change.

slope-intercept form (p. 175) The equation of a line in the form $y = mx + b$, where m is the slope of the line and b is the y-intercept.

sphere (p. 586) The set of all points in space equidistant from a given point.

spherical geometry (pp. 597–598, 602) A geometric system based on the statement that in a plane, through a point not on a given line, there is no line parallel to a given line. The surface of a sphere is a plane in spherical geometry, and a line is a *great circle*.

square (p. 88) A regular parallelogram.

straight angle (p. 57) An angle that measures 180°.

supplementary angles, supplements (p. 58) Two angles whose measures add up to 180°.

surface area (p. 376) The area of a net for a three-dimensional figure.

symmetry (p. 10) A figure has symmetry if the figure and its image coincide after a transformation.

tangent of an acute angle, tangent ratio (p. 507) In a right triangle, the ratio of the length of the leg opposite the angle to the length of the leg adjacent to the angle.

tangent to a circle (p. 554) A line in the plane of the circle that intersects the circle in only one point. *See* circle.

tangent to a sphere (p. 586) A line or plane that intersects the sphere in only one point.

taxicab geometry (p. 208) A non-Euclidean geometry based on travel on a rectangular grid.

tessellation (p. 83) A pattern of polygons that covers a plane without gaps or overlaps.

theorem (p. 125) A conjecture that can be proved to be true.

three-dimensional coordinate system (p. 201) A system in which the position of each point in space is given using three coordinate axes. The z-axis is perpendicular to the x-axis and the y-axis.

transformation (p. 8) A change made to the size or position of a figure.

translation (pp. 8, 406) A transformation that slides each point of a figure the same distance in the same direction.

transversal (p. 219) A line that intersects two or more other lines in the same plane at different points.

trapezoid (p. 228) A quadrilateral with exactly one pair of parallel sides, called the *bases*. The two non-parallel sides are called the *legs*.

triangle (pp. 64, 73) A polygon with three sides.

vector (p. 521) A quantity that has both magnitude, or size, and direction.

vertex (vertices) of a polygon (pp. 65, 73) An endpoint of a side of a polygon.

vertical angles (p. 58) Non-adjacent, non-overlapping angles formed by two intersecting lines.

y-intercept (p. 175) The y-coordinate of a point where a graph crosses the y-axis.

z-axis (p. 201) *See* three-dimensional coordinate system.

Index

sector of, 579–580
segment of, 584
segment lengths in, 573–575
tangent to, 554–555, 560–561, 574–575
Circumcenter, 311 (Ex. 16), 567, 571 (Ex. 10)
Circumference, 369, 704
Circumscribed circle, 369, 372 (Ex. 14), 567
Circumscribed polygon, 324 (Ex. 13)
Classifying
arcs, 554, 555
information, 154–155
quadrilaterals, 88–89, 138, 339–342
triangles, 64–65, 148, 168, 172 (Ex. 35)
Coefficient of friction, 507, 571
Collaborative learning *See* Cooperative learning, Explorations, *and* Portfolio Projects.
College entrance exams, preparing for *See* SAT/ACT Preview.
Communication
discussion, *in every lesson exposition as* Think and Communicate *questions, for example,* 29, 40, 57, 72, 74, 130, 132, 143, 174, 188, 235, 244, 265, 288, 306, 313, 340, 347, 369, 414, 419, 454, 468, 501, 522, 530, 561, 569, 581, 599, 615, 628
making a poster, display, or video, 2, 49, 53, 101, 105, 110, 146, 155, 209, 271, 278, 327, 387, 390, 444, 483, 492, 543, 552, 651
making a presentation or writing a report, 49, 56, 110, 164, 209, 275, 327, 383, 435, 444, 483, 602, 605, 612, 646, 651
writing and journal, *See* Journal *and* Writing.
See also Performance Task assessment, Portfolio Projects, *and* Research.
Compass, 69, 190, 263, 688
Complementary angles, 58, 63 (Exs. 29, 30), 137
Components of vector, 521–524
Composition of transformations, 405–406, 409, 411, 417 (Exs. 23, 24), 418 (Ex. 32), 419–422, 424 (Ex. 23), 425 (Ex. 30), 432 (Ex. 17), 624 (Ex. 28), 632 (Ex. 15), 637 (Ex. 15), 638 (Exs. 20, 21), 643–647
Computer
computer-aided design, 93, 271, 383
digital maps, 617
games, 424, 613–614, 646
geometry software, 9, 27, 38, 70, 125, 132, 147, 164, 168, 172, 180, 200, 221, 267, 279, 297, 304, 324, 349, 372, 397, 405, 409, 417, 432, 435, 460, 461, 466, 514, 553, 571, 572 (Ex. 19), 578, 726–727

graphing software, 168, 180, 472, 721–723
matrix software, 629, 646, 723–725
patterns and, 408
photographs, 634
simulation, 424
spreadsheets, 16, 146, 169, 450, 472, 482, 725
system designer, 162–164, 170, 198, 206
virtual reality, 639, 646
See also Calculator *and* Technology.
Conclusion
of a conditional statement, 15, 137
of a proof, 125–126, 132
Conditional statement, 15, 17–19, 135, 137, 149, 152 (Ex. 39)
Cone, 537–538
base of, 537
height of, 537
oblique, 541 (Ex. 18)
radius of, 537
right, 537
slant height of, 537
surface area of, 537–538
vertex of, 537
volume of, 537–538
Congruence
of angles, 35, 119
applying, 306–308
of arcs, 554
of circles, 554
determining, 28–30, 35, 60, 69, 292–294, 296 (Ex. 12), 299–302, 307
of polygons, 285–288
properties of, 60, 285–286, 307
of right angles, 60 (Question 4), 114 (Ex. 13), 128 (Ex. 10), 134 (Ex. 15)
of segments, 28, 119
and transformations, 41, 298, 393, 406, 412, 420, 422
of triangles, applying, 306–308
of triangles, methods of proof, 121, 292–295, 299–302, 304 (Ex. 14)
Conjecture, 15
making a, 3–4, 14–20, 25 (Ex. 30), 26 (Ex. 34), 38 (Ex. 24), 63 (Exs. 30, 31), 70 (Exs. 31, 32), 79, 87, 125, 148, 172 (Ex. 33), 180, 186 (Ex. 33), 206 (Ex. 30), 221, 242, 244 (Question 2), 283 (Ex. 31), 324 (Ex. 13), 340, 417 (Ex. 24), 424 (Ex. 24), 429 (Question 7), 430 (Ex. 7), 460 (Exs. 27, 28), 461, 553, 567, 640 (Ex. 36)
Conjunction (and), 736
Connections
among mathematics topics
algebra, 4, 16, 41, 43, 80, 81, 88, 112, 141, 142, 196, 222, 227, 243, 286, 315, 341, 354, 363, 448, 462, 470, 494, 500, 562, 568, 574, 587
data analysis, 592
measurement, 122, 379
probability, 283

to cross-curricular subjects
art, 26, 27, 83, 602
astronomy, 7, 305, 450, 518
biology, 61, 371, 564, 595
earth science, 577
geography, 44, 57, 70, 136, 138
history, 19, 70, 146, 533
language, 139
literature, 127, 431, 498
physics, 323
science, 39 *See also* Science.
social studies, 357
sports, 524, 587, 588, 590
to real-world applications
aerial photography, 310
architecture, 192, 232, 410, 504, 539
art photography, 637
audio-visual presentation, 396
auto maintenance, 151
botany, 224
city planning, 178
communications, 13, 62
computer games, 613
computer simulation, 424
consumer economics, 379, 473
cooking, 596
crafts, 40, 45, 77
dance, 247
design, 254
digital maps, 617
electronics, 259
engineering, 177, 240
gardening, 44, 57, 70, 136, 138, 365
geology, 526
Hawaiian quilts, 417
highway safety, 571
horticulture, 246
hydraulics, 344
interior design, 284
machinery, 351
manufacturing, 31, 285
meteorology, 592
miniature golf, 403
navigation, 121, 597, 601
nutrition, 149
optics, 91
paper folding, 266
photocopiers, 624
puzzles, 171
quilt patterns, 185
rescue safety, 297
rock climbing, 584
special effects, 631
surveying, 457
technical drawing, 261
transportation, 512
video games, 646
virtual reality, 639, 646
zoology, 115, 479
See also Applications, By the Way, Career Interview, Historical notes, History of mathematics, Multicultural connections, Portfolio Projects, Research, *and* Science.
Constructions, 69, 263, 720
angle bisector, 311

centroid of a triangle, 323 (Ex. 8), 324 (Ex. 14)

circle that circumscribes a given triangle, 567

congruent angles, 263

equilateral triangle, 69, 372 (Ex. 17)

45-45-90 triangle, 505 (Ex. 24)

Gothic arch, 504

incenter of a triangle, 324 (Ex. 13)

isosceles trapezoid, 350 (Ex. 15)

kite, 343

orthocenter of a triangle, 324 (Ex. 15)

parallel lines, 264

parallelogram, 350

perpendicular bisector, 265

perpendicular lines, 267

rectangle, 358

regular hexagon, 372

sum of angles, 264

triangle given three sides, 292

triangle given two sides and the included angle, 297

using geometry software, 726–729

Contradiction, 739

Contrapositive, 149, 151

Converse, 137, 149, 243–244

of Pythagorean theorem, 143, 148

Convex polygon, 73

Convincing argument, 112, 117–118, 120, 124, 237

writing a, 114 (Ex. 14), 115 (Exs. 17, 18), 116 (Ex. 25)

Cooperative learning

Exercises, 12, 46, 62, 98, 169, 171, 186, 193, 200, 231, 269, 283, 291, 296, 372, 404, 424, 432, 457, 466, 499, 528, 540, 541, 557, 585, 625, 630, 633

Explorations, 3, 14, 21, 64, 79, 87, 125, 141, 148, 180, 194, 221, 242, 256, 279, 287, 340, 361, 374, 392, 405, 427, 461, 475, 493, 514, 553, 567, 597, 634, 641

Portfolio Projects, 48–49, 100–101, 154–155, 208–209, 270–271, 326–327, 382–383, 434–435, 482–483, 542–543, 604–605, 648–651

Coordinate geometry

distances and midpoints, 165–168

equations of circles, 187–190

equations of lines, 173–175

parallel and perpendicular lines, 180–183

polar, 416

proof, 194–197, 198 (Ex. 11), 199 (Exs. 15, 24, 25), 200 (Exs. 27, 28), 207 (Questions 3–6), 213 (Questions 25–27, 32), 341, 344 (Ex. 12), 354, 357 (Ex. 21), 365 (Ex. 25), 424 (Ex. 24)

in three dimensions, 201–204, 380, 618 (Ex. 31), 632 (Ex. 16), 638 (Exs. 22–24)

See also Matrix *and* Transformations.

Coordinate plane, 197, 202, 696–700

Corollary, 239

Corresponding angles, 219

Corresponding Angles Postulate, 222, 712

Corresponding parts, of congruent figures, 285–288, 306–308

Cosine ratio, 514–517, 529, 538, 560, 733

Counterexample, 16

finding a, 18 (Exs. 11–14), 24 (Exs. 7–10), 27 (Exs. 38, 39), 90 (Questions 2–4), 152 (Ex. 39), 179 (Exs. 45–47), 193 (Exs. 42–45), 225 (Exs. 34, 35), 262 (Ex. 31), 282 (Ex. 20), 460 (Exs. 24, 25, 27)

Critical thinking *See* Reasoning, Study techniques, *and* Visual Thinking.

Cross-curriculum connections *See* Applications *and* Connections.

Cross sections, 206 (Ex. 28), 379 (Ex. 14), 540 (Ex. 18), 572 (Ex. 16), 604–605

Cube, 94, 101, 145 (Ex. 30)

Cubic units, 375, 730

Cultural diversity *See* Career Interview *and* Multicultural connections.

Cumulative Assessment, 160–161, 334–335, 488–489, 654–655

Customary units, 730

Cylinder, 375–377

base of, 375

height of, 375

lateral area of, 376

radius of, 374

surface area of, 376–377

volume of, 375

Data analysis, 3, 4, 6 (Ex. 9), 16, 76 (Ex. 5), 79, 96 (Ex. 11), 111, 124, 155, 169 (Exs. 1–3), 201 (Questions 1, 2), 221, 372 (Ex. 14), 450, 451, 472, 482–483, 508 (Question 3), 514, 590 (Ex. 27), 592, 595, 708–709

See also Matrix, Patterns, Portfolio Projects, *and* Technology Handbook.

Decagon, 73

Deductive reasoning, 112–113, 734–735

See also Reasoning.

Definition, 118

Degree measure, 35–36

Diagonals

of a kite, 341

of a parallelogram, 89, 138, 194–195, 347, 353–355

of a polygon, 74, 76 (Ex. 5)

of a prism, 98 (Exs. 27, 28), 145 (Ex. 30), 206 (Ex. 29)

of a rectangle, 199 (Ex. 16), 341

of a rhombus, 340–341

of special quadrilaterals, 340–342

of a square, 138, 196, 342

Diagrams

drawing, 23, 67, 94, 118, 155, 242, 258, 261, 537, 539 (Exs. 7–9)

interpreting, 22, 28, 30, 35, 58, 220

isometric, 270–271

nets, 95, 101, 147 (Ex. 45), 376–378, 535

oblique, and perspective, 261

scale drawing, 299, 445, 468

technical, 261, 270–271, 451

Diameter

of a circle, 187, 192 (Exs. 23–25)

perpendicular to chord, 568

of a sphere, 586

Dilation, 426–429, 451 (Ex. 24), 473 (Ex. 18); 474 (Ex. 24)

center of, 428

with matrices, 613–615

properties of, 428

scale factor of, 428, 614–615

Dimensions of a matrix, 613

Direction of vector, 521–522

Discrete mathematics

cellular automata (game of Life), 424

classifying and sorting, 64, 128 (Exs. 12–14) 133 (Exs. 5–9), 151 (Exs. 25–28), 154–155, 339–342

Euler's formula, 96

fractals, 3, 452, 480

geometric probability, 475–477

Heron's formula, 364 (Ex. 7)

networks, 14

Platonic solids, 100–101

writing a program, 404

See also Matrix.

Discussion *See* Communication.

Disjunction (or), 736

Distance

between two parallel lines, 249

between two points, 28, 166, 203

from a point to a line, 249

from a point to a plane, 258

taxicab, 208–209

Distance formula

on a coordinate plane, 166, 168, 188–189, 192 (Ex. 22), 196, 208, 714

in three dimensions, 203–204

Distributive property, 496, 574

Dodecahedron, 101

Drawings *See* Diagrams.

Edge

of a network, 14

of a prism, 93

Element of a matrix, 613

Elevation, angle of, 91, 509–510

Endpoint

of an arc, 554

of a ray, 28

of a segment, 28

Enlargement, 427, 428

Enrichment *See* Challenge exercises *and* Portfolio Projects.

Equation(s)

of circles, 188–190

Galilei, Galileo, 645
Gauss, Carl Friedrich, 602, 735
Heron, 364
Hipparchus, 735
Hypatia, 735
Klein, Felix, 602, 735
Lobachevsky, Nikolai Ivanovich, 602, 735
Moufang, Ruth, 603
Pythagoras, 141
Riemann, G. F., 602, 735
Thales, 735
Wiles, Andrew, 127
Hyperbolic geometry, 602
Hypotenuse, 141, 144 (Exs. 13–18)
midpoint of, 198 (Ex. 11), 344 (Ex. 13)
Hypotenuse-Leg (HL) Theorem, 301, 716
Hypothesis
of a conditional, 15
of a proof, 125–126, 132

Icosahedron, 101
If and only if, 118
If-then statement, 15, 17–19, 135, 137, 149, 152 (Ex. 39)
Image, in a transformation, 8, 391
Incenter, 324 (Ex. 13), 571 (Ex. 10)
Included angle, 294
Included side, 300
Indirect measurement, 306, 312 (Question 10), 457, 496, 498
Indirect proof, 348, 351 (Exs. 22–24), 359 (Question 4), 386 (Question 9)
Inductive reasoning, 4, 112–113, 734
See also Conjecture *and* Reasoning.
Inequalities
linear, 705–707
for triangles, 279–281, 600 (Ex. 10)
Inscribed angle, 553–555
Inscribed circle, 324 (Ex. 13)
Inscribed polygon, 369, 559 (Ex. 22), 567
Intercept, 175, 179 (Ex. 34), 697–699
Interior angles, 81
alternate, 219
same-side, 219–221, 227–229, 244
Intersecting lines, 22, 120
Intersection, of geometric figures, 22–23, 256–258, 275 (Ex. 22), 591 (Ex. 30)
Interview *See* Career Interview.
Inverse of a conditional, 149, 151
Investigation exercises, 18, 26, 84, 91, 170, 172, 297, 304, 323, 324, 364, 402, 417, 425, 466, 474, 505, 532, 600, 624, 740
Isometric drawing, 261, 270–271
Isosceles trapezoid, 200 (Ex. 28), 343 (Ex. 10), 344 (Ex. 12)
Isosceles triangle, 65, 70 (Ex. 32)
Isosceles Triangle Theorem, 314, 716
parts of, 313
properties of, 138, 313–315, 320–321, 322 (Exs. 3–5)

Journal, 20, 33, 47, 71, 86, 99, 123, 135, 153, 186, 207, 233, 255, 269, 291, 312, 325, 359, 381, 411, 433, 467, 481, 506, 528, 541, 566, 585, 603, 625, 647
See also Writing.
Justification *See* Convincing argument, Logical reasoning, Proof, *and* Reasoning.

Key steps of a proof, 228, 230 (Ex. 13), 232 (Ex. 30), 236, 240 (Ex. 13), 245 (Ex. 8), 310 (Ex. 7), 320, 344 (Ex. 11), 349 (Question 3), 351 (Ex. 21), 357 (Ex. 22), 569
Kite, 339–341, 357 (Ex. 21)

Lateral area, 376–377, 535–538
Lateral face
of a prism, 93
of a pyramid, 535
Latitude, 601
Law of sines, 305 (Ex. 19)
Legs
of an isosceles triangle, 313
of a right triangle, 141
of a trapezoid, 228
Length(s)
of an arc, 580–581
proportional, 446, 461–464
of a segment, 28
using geometry software, 728
Limit, concept of, 152, 369, 508 (Question 3), 514
Line(s), 21–22, 118
auxiliary, 235, 254 (Ex. 24), 596 (Ex. 17)
concurrent, 46 (Ex. 34), 311 (Ex. 16), 323 (Ex. 8), 324 (Exs. 13–15), 571 (Ex. 10)
contain two points, 21, 22, 120
coplanar, 22
distance to a point, 249
equations of, 173–175
horizontal, 174–175
intersecting, 22, 120
parallel, 22, 135, 137, 181, 219–223, 226–229, 234–237, 242–245, 249–251, 256–258, 263–265, 461–464, 466, 602
constructing, 264
slopes of, 180–182
perpendicular, 58, 60, 181–182, 186 (Ex. 33), 235, 249–250, 267
constructing, 267
slopes of, 180–182
perpendicular, to plane, 258, 260
proving parallel, 242–244, 249–251, 257–258
of reflection, 391–394
in a glide reflection, 420–422
skew, 22, 137

slopes of, 173–174, 180–182, 640 (Ex. 36), 697–700
of symmetry, 10, 74, 318 (Ex. 19), 341, 342 (Question 2), 391, 396
vertical, 174–175
Linear equations *See* Equation(s).
Linear inequalities, 705–707
Linear pair, 58–59, 126, 130
Literature
The Mysterious Island, by Jules Verne, 498
Pudd'nhead Wilson, by Mark Twain, 431
Where Do You Stop?, by Eric Kraft, 127
Locus, concept of, 209
Logic *See* Proof *and* Reasoning.
Logical reasoning exercises, 68, 76, 85, 92, 160, 282, 343, 356, 365, 458, 736–741
See also Reasoning.
Longitude, 121, 601

Magnitude of vector, 521–524
Manipulatives
Exercises, 18, 43, 45, 77, 91, 170, 171, 172, 184, 266, 283, 291, 323, 356, 364, 431, 474 (Ex. 26), 499 (Ex. 25), 505, 540, 583, 600 (Ex. 10)
in concept development, 40, 42, 59, 65, 376, 535, 598
in Explorations and Projects, 21, 64, 79, 100–101, 141, 148, 194, 242, 256, 287, 340, 361, 374, 392, 475, 493, 567, 597
See also Constructions *and* Scale model.
Mathematical systems, 734–735
Mathematicians *See* History of mathematics.
Matrix (matrices), 613, 711
addition, 619, 711
columns of, 613
diagonal of, 641
dilations with, 613–615
dimensions of, 613, 711
element of, 613
identity, 630
image, 614
multiplication, 626–629
point, 613
polygon, 613–615
reflections with, 634–636
rotations with, 641–643
rows of, 613
scalar multiplication of, 614–615, 711
subtraction, 621, 711
translations with, 619–621
Mean
arithmetic, 499 (Ex. 24), 708
geometric, 494–496
Measure(s)
of an angle, 35–36
of angles of polygons, 79–82

of an arc, 554–555
of a segment, 28
table of, 730
See also Distance formula.
Median, 319–321, 323 (Ex. 8),
324 (Ex. 14)
Metric units, 730
Mid-chapter review *See* Assessment,
Assess Your Progress.
Midline of a trapezoid, 364
Midpoint formula
on a coordinate plane, 167, 170
(Ex. 19), 194–195, 198–199,
714
in three dimensions, 203
Midpoint of a segment, 41, 181
Model, scale, 443–444, 450, 468,
482–483, 593
Modeling, mathematical, 3–5, 17
(Ex. 1), 21, 34, 38 (Ex. 20), 39,
46 (Ex. 28), 61, 79, 85, 91, 98,
121, 122, 128, 133, 136, 146,
147, 162–164, 165, 171, 173,
176, 182, 185, 188, 191, 201,
206, 231, 240, 299, 305, 379,
396, 402, 403, 444, 445, 447,
450–452, 457, 459, 465, 468,
473, 475–480, 496, 502, 510,
516, 523, 524, 538, 559, 560,
565, 571, 575, 584, 595,
597–602, 617, 639, 645
Morphing, with matrices, 626–629,
631
Multicultural connections, 13, 17, 26,
27, 40, 45, 57, 70, 72, 77, 83,
84, 100, 121, 122, 136–138,
139, 145, 146, 147, 171, 173,
174, 177, 185, 188, 192, 226,
231, 240, 251, 252, 268, 299,
305, 316, 342, 344, 348, 356,
357, 364, 369, 373, 417, 419,
452, 455, 458, 474, 502, 510,
536, 540, 587, 588, 602
See also Career Interview.
Multiple methods, 4, 82, 95, 130, 198
(Ex. 11), 236–237, 267
(Exs. 13, 15), 293 (Question 3),
297 (Exs. 15–17), 316 (Ex. 7),
344 (Ex. 13), 373 (Ex. 27),
454 (Question 1), 474 (Ex. 28),
527 (Ex. 19), 534 (Ex. 23),
572 (Ex. 19), 628–629
See also Open-ended Problems.
Multiple representations, 6 (Ex. 12),
17 (Ex. 2), 79, 95, 146, 194,
201, 361, 376, 406, 413, 444,
445, 450, 465, 468, 482–483,
495, 521, 602, 614, 632, 635,
642

Negation, 149, 737
Net, 95, 101, 147, 376–378, 535
Network, 14
***n*-gon,** 73, 368
Non-Euclidean geometry
comparison with Euclidean
geometry, 734–735

hyperbolic, 602, 735
spherical, 597–603, 735
taxicab, 208–209, 735
Number line, 705

Oblique drawing, 261
Octagon, 73
Octahedron, 101
Ongoing assessment *See* Assessment.
Open-ended Problems
Exercises, *throughout the book, for
example,* 26, 53, 84, 97, 134,
145, 184, 205, 241, 303, 334,
345, 356, 410, 425, 452, 465,
511, 557, 564, 565, 633, 647
Chapter Assessment, 53, 104, 159,
213, 331, 439, 486, 487, 547,
609, 655
Or (disjunction), inclusive, 736, 737
Ordered pair, 697
Ordered triple, 202
Orientation, in a transformation,
392–393, 406, 412, 420, 422
Origin, 696–697
Origins of geometry, Babylonian,
Egyptian, and Greek, 734–735
Orthocenter, 324 (Ex. 15), 571
(Ex. 10)
Orthographic projection, 270–271

Pantograph, 431
Paper cutting, 77 (Exs. 11–16), 91
(Exs. 11, 12), 141, 148, 194,
242, 256, 287, 356 (Ex. 11),
361, 434–435, 493
Paper folding, 18 (Exs. 15–18),
42, 45 (Exs. 23–27), 74, 77
(Exs. 11–16), 91 (Exs. 11, 12),
100–101, 170 (Exs. 16, 17), 266
(Exs. 8–11), 287, 356 (Ex. 11),
364 (Exs. 14–16), 392, 493,
505 (Ex. 25), 567
Paragraph proof, 124–126,
128 (Ex. 10), 129 (Exs. 20, 21),
135 (Question 1), 229, 232
(Ex. 34), 237, 239 (Exs. 2, 5),
241 (Ex. 16), 246 (Ex. 15), 250,
505 (Ex. 25), 558 (Ex. 21)
Parallelogram, 88–89
angle(s) of, 87–88, 232 (Ex. 30), 353
area of, 361–362, 365 (Ex. 23)
conditions for, 138, 182, 346–348,
353–355
construction of, 194, 350
diagonals of, 89, 135, 194–195,
199 (Ex. 25), 347, 353–355
properties of, 88–89, 194–195,
232 (Ex. 30)
special, 88–90, 339–342
See also Rectangle, Rhombus, *and*
Square.
Parallel Postulate, 235, 599, 602, 712,
735
Patterns
frieze, 419

identifying, 3–7, 13 (Ex. 18), 20,
76 (Ex. 5), 79, 96 (Ex. 11),
101, 112–115, 125, 146, 149,
171, 182, 184, 185, 198, 369,
372 (Ex. 14), 424 (Ex. 19),
430 (Ex. 7), 514, 596 (Ex. 16)
See also Conjecture.
in tessellations, 83, 434–435
and transformations, 419
Pentagon, 73, 368
Performance Task assessment, 53,
105, 159, 213, 275, 331, 387,
439, 487, 547, 609, 655
Perimeter, 368, 703
Perpendicular bisector, 170 (Ex. 17),
181, 184 (Exs. 19–22), 265,
307–308, 311 (Ex. 15)
Perpendicular lines, 58, 60, 180–182,
186 (Ex. 33), 196, 235,
249–250, 258, 260, 267
Perpendicular Postulate, 235, 712
Perspective drawing, 26, 261
Pi (π), 369
Plan for proof *See* Proof.
Plane(s), 21, 23, 118
coordinate, 202
intersecting, 23, 120–121, 256–257
line perpendicular to a, 258
models of, 21–23, 256–257
parallel, 23, 256–257
Platonic solids, 100–101
Point(s), 21–22, 118
collinear, 22
on a coordinate plane, 696–697
coplanar, 21–22
distance between, 28, 166, 203
matrix, 613
in space, 201–204
of tangency, 554
Polar coordinates, 416
Polygon(s), 73–74
angles of, 73–74, 79–82, 132
circumscribed, 324 (Ex. 13)
concave, 73
congruent, 285–288
convex, 73
diagonals of, 74, 76 (Ex. 5)
equiangular, 73
equilateral, 73
inscribed, 369
regular, 73–74, 100, 367–369, 529
sides of, 73–74, 170 (Exs. 4–9)
similar, 446–448
similar, perimeters and areas of,
468–470
sum of measures of angles, 79–80
vertices of, 73–74
Portfolio Projects
applying solar geometry, 542–543
building a mobile, 326–327
building the platonic solids, 100–101
classifying information, 154–155
creating fractals, 648–651
creating technical drawings, 270–271
designing a cottage, 382–383
exploring taxicab geometry, 208–209
investigating symmetry, 48–49
modeling Cavalieri's principle,
604–605

properties of, 393, 406, 412, 420, 422
shear, 628
and symmetry, 8–10, 318, 391, 396,
    407, 409, 410, 415
using geometry software, 729
See also Dilation, Glide reflection,
    Matrix, Morphing, Reflection,
    Rotation, *and* Translation.
Translation, 8–10, 12 (Ex. 14),
    25 (Ex. 30), 41, 405–407
with matrices, 619–621
properties of, 406, 422
and symmetry, 8–10, 48–49, 407,
    409, 410
Transversal, 219
Trapezoid, 228–229, 233 (Ex. 35),
    339, 348
area of, 361–363, 364 (Ex. 16),
    366 (Ex. 34), 531
bases of, 228
isosceles, 200 (Ex. 28), 343 (Ex. 10),
    344 (Ex. 12)
legs of, 228
midline of, 364
right, 343
Triangle(s), 64, 73
acute, 65, 127, 148
altitudes of, 319–320, 324 (Ex. 15),
    571 (Ex. 10)
angle bisectors of, 320–321,
    324 (Ex. 13), 466 (Ex. 23),
    571 (Ex. 10)
angles of, 65–66, 125–126, 235–237,
    240 (Ex. 11), 280
area of, 141, 361–362, 364 (Ex. 7),
    365 (Ex. 24), 534 (Ex. 22)
centroid of, 323 (Ex. 8), 324 (Ex. 14)
circumcenter of, 311 (Ex. 16), 567,
    571 (Ex. 10)
classifying, 64–65, 148, 168,
    172 (Ex. 35)
congruent, 240 (Ex. 11), 292–295,
    299–302, 306–308
constructing, 69, 292,
    297 (Exs. 15–19)
equiangular, 65, 66 (Questions 3–5),
    70 (Ex. 31), 239 (Ex. 6)
equilateral, 65, 69, 70 (Ex. 31),
    129, 152 (Ex. 35), 168, 362,
    372 (Ex. 17)
exploring, 46 (Ex. 34), 64,
    70 (Exs. 31, 32), 141, 148, 152
    (Ex. 35), 198–199 (Exs. 8–15),
    297, 304, 323, 324,
    466 (Ex. 23), 571 (Ex. 10)
exterior angles of, 125–126,
    129 (Ex. 21)
45-45-90, 500–501
incenter of, 324 (Ex. 13),
    571 (Ex. 10)
inequalities for, 279–281,
    600 (Ex. 10)
inscribed, 567
isosceles, 65, 70 (Ex. 32),
    114 (Ex. 11), 121, 129, 138,
    168, 313–315, 320–321,
    322 (Exs. 3–5)
medians of, 319–320, 323 (Ex. 8),
    324 (Ex. 14)

obtuse, 65, 127, 148, 152 (Ex. 35)
orthocenter of, 324 (Ex. 15),
    571 (Ex. 10)
perpendicular bisectors of sides,
    311 (Ex. 16), 320, 567
right, 65, 68 (Ex. 7), 127, 137,
    139 (Ex. 11), 141–143, 148,
    149, 198 (Ex. 11), 301,
    344 (Ex. 13), 493–531
    See also Right triangle(s).
scalene, 65, 138
segment joining midpoints of two
    sides, 199 (Exs. 12–15), 462
sides of, 65, 148, 279–281
similar, 453–455, 461
spherical, 597–601
sum of measures of angles, 65–66,
    236–237
terrestrial, 601
30-60-90, 501–502
vertices of, 65, 183
Triangle Sum Theorem, 66, 236–237,
    714
Triangulation, 299
Trigonometry
applications of, 522, 529–531, 538,
    560, 572 (Ex. 16), 577 (Ex. 24)
and area, 529–531
law of sines, 305
sine and cosine ratios, 514–517, 733
summary of trigonometric ratios, 517
table of trigonometric ratios, 733
tangent ratio, 507–510, 733
Truth table, 737
Truth value, 736
Two-column proof, 130–132, 133
    (Ex. 4), 134 (Exs. 17, 18), 135
    (Question 2, 3), 139 (Ex. 11),
    140 (Ex. 18), 159 (Question 11),
    225 (Ex. 28), 232 (Ex. 33), 236,
    239 (Ex. 4), 240 (Ex. 11), 241
    (Exs. 15, 18), 244, 248 (Ex. 21),
    254 (Ex. 21), 293, 296 (Ex. 11),
    300, 302, 307, 314, 316 (Ex. 7),
    458 (Ex. 14), 463, 498 (Ex. 19),
    558 (Ex. 20), 563 (Ex. 11),
    576 (Ex. 17)

Undefined terms, 118
Units of measurement, 730

Valid argument, 739
Vanishing point, 26
Vector(s), 521–524
addition of, 524, 527 (Exs. 19, 20)
equal, 522
scalar multiplication of, 523
Vertex (vertices)
of an angle, 34
of a cone, 537
consecutive, 74
of a network, 14
of a polygon, 73
of a prism, 93
of a pyramid, 535

of a triangle, 65
Vertex angle, of an isosceles triangle,
    313
Vertical angles, 58–59, 130–131
Visual thinking, 18, 91 (Exs. 11, 12),
    96, 97, 115, 147, 225, 247, 253,
    262, 304, 324, 343, 356, 424,
    472, 474, 478, 498, 505, 527,
    578, 583, 591
    See also Manipulatives.
Vocabulary, 20, 33, 47, 50, 52, 71, 86,
    99, 102, 104, 123, 135, 153,
    156, 158, 186, 207, 210, 212,
    233, 255, 269, 272, 274, 291,
    325, 328, 330, 359, 381, 384,
    386, 411, 433, 436, 438, 467,
    481, 484, 486, 506, 528, 541,
    544, 546, 566, 585, 603, 606,
    608, 625, 647, 652, 654,
    742–747
Volume(s), 375
of a cone, 537–538
of a cylinder, 375
of a prism, 375, 531, 704–705
of a pyramid, 536
of similar solids, 469–470, 592–593
of a sphere, 586–587

Writing
Chapter Assessment, 52, 104, 212,
    213, 274, 275, 331, 334–335,
    386, 486, 547, 609, 655
Examples and Exercises, *throughout
    the book, for example,* 6, 15, 69,
    86, 97, 116, 129, 145, 176, 192,
    239, 261, 267, 289, 304, 311,
    364, 372, 403, 418, 424, 457,
    465, 466, 506, 511, 512, 559,
    565, 571, 640, 656
introductory questions, 2, 56, 110,
    164, 218, 278, 338, 390, 444,
    492, 552, 612
Journal, 20, 33, 47, 71, 86, 99, 123,
    135, 153, 186, 207, 233, 255,
    269, 291, 312, 325, 359, 381,
    411, 433, 467, 481, 506, 528,
    541, 566, 585, 603, 625, 647
Portfolio Projects, 49, 101, 155, 209,
    271, 327, 383, 434, 482, 543,
    651
    See also Research.

x-axis, 696
x-coordinate, 697

y-axis, 696
y-coordinate, 697
y-intercept, 175, 179 (Ex. 34)

z-axis, 201
z-coordinate, 201–202

Credits

Selected Answers

CHAPTER 1

Page 5 Checking Key Concepts

1. The number of triangles is 2 less than the number of sides. **3.** $n - 2, n \geq 3$

Pages 5–7 Exercises and Applications

3. 24, 35 **5.** $\frac{5}{4}, \frac{6}{5}$ **11.** $3n + 1$

19, 21.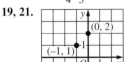

23. 17

25. $\frac{1}{6}$

Page 11 Checking Key Concepts

1.

3.

5. reflection symmetry **7.** translation symmetry

Pages 11–13 Exercises and Applications

5.

7.

9.

11. translational; Each gymnast's position is an image of the first gymnast's position. **13.** rotational; Each point of the gymnast's body rotates around some point by the same amount. **15.** The bicycles have translational symmetry. **17.** The starfish has rotational symmetry (about its center) and reflection symmetry. It has five planes of symmetry, one running down the middle of each leg. **25.** –36 **27.** $1.5n + 9$, where n = the number of bags

Page 16 Checking Key Concepts

1. a figure is a square; it has four lines of symmetry
3. people live in glass houses; they shouldn't throw stones
5. True.

Pages 17–20 Exercises and Applications

3. If you're at the bottom of a human pyramid, then you have to be steady. hypothesis: you're at the bottom of a human pyramid; conclusion: you have to be steady **5.** If a tumbler doesn't run fast enough before hitting the trampoline, he or she won't make it to the other side of the pyramid. hypothesis: a tumbler doesn't run fast enough; conclusion: he or she won't make it to the other side of the pyramid **7.** If the figure has reflection symmetry, then the hidden part of the figure looks like: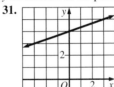

9. hypothesis: you reflect a shape; conclusion: the image is the same size as the original shape **11.** False; for example, the network in Figure A on page 14 is traceable, but two of its five vertices are odd. **13.** True. **23, 25, 27.** Answers may vary. Examples are given. **23.** If today is January 1, then tomorrow is January 2. **25.** If a point is located on the x-axis, then the y-coordinate of the point is 0. **27.** If you translate a point up one unit on a coordinate plane, then the y-coordinate of the point increases by 1.

31.

Page 20 Assess Your Progress

1.

Number of squares	3	4	5	6	7
Number of toothpicks	10	13	16	19	22

Each additional square requires 3 additional toothpicks.
2. 31 toothpicks; Answers may vary. An example is given. I noticed that for n squares, the number of toothpicks is $3n + 1$. **3.** If you want to make a toothpick figure having n squares, you need $3n + 1$ toothpicks. **4.** The first, third, and fifth figures have reflection symmetry. The second and fourth have rotational symmetry.

5. a. 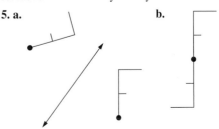 **b.**

c.

6. False; for example, if $a = -2$ and $b = -1$, then $a \div b = 2$ and $a \div b > a$.

29, 31, 33. Answers may vary. Examples are given.

29.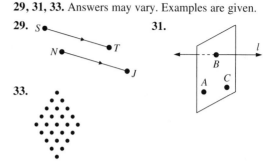

31.

33.

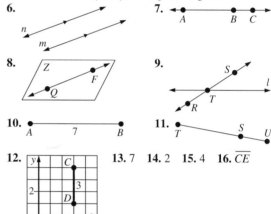

Page 33   Assess Your Progress

1. any two of $\overleftrightarrow{EC}$, $\overleftrightarrow{DC}$, and $\overleftrightarrow{BC}$ **2.** point D **3.** point A
4. plane AED and plane ACD **5.** $\overleftrightarrow{ED}$ or $\overleftrightarrow{EC}$
6–12. Answers may vary. Examples are given.

6.

7.

8.

9.

10.

11.

12.

13. 7   **14.** 2   **15.** 4   **16.** $\overline{CE}$

Page 24   Checking Key Concepts

1. point A **3. a.** coplanar but noncollinear   **b.** collinear
c. coplanar but noncollinear   **d.** coplanar but noncollinear
e. noncoplanar   **f.** noncoplanar   **5.** point B **7.** plane ADC

Pages 24–27   Exercises and Applications

1. point J **3.** $\overleftrightarrow{IK}$ **5.** plane ABG, plane ABI **7.** True.
9. False; in the figure, points A, B, and C are collinear and
lie in two of the planes shown.   **11.** parallel
13. intersecting   **19.** C   **21, 23, 25.** Answers may vary.
Examples are given.

21.

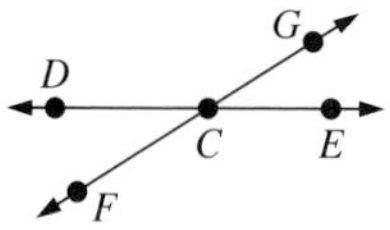

23.

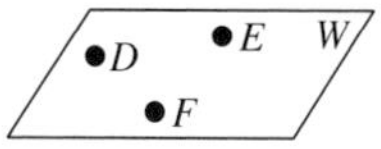

25.

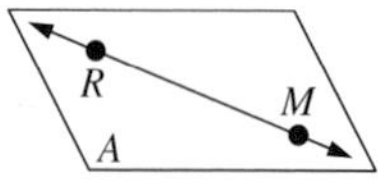

27. $\overleftrightarrow{CF}$ **29.** $\overleftrightarrow{BX}$, $\overleftrightarrow{EX}$, and $\overleftrightarrow{XY}$
39. True.   **41.** 1   **43.** $\frac{1}{2}$; 0
45. -2; -2

Page 30   Checking Key Concepts

1. $\overline{NX}$ **3.** No; $\overrightarrow{PQ}$ and $\overrightarrow{QP}$ have different endpoints and
extend in different directions.   **5.** 5   **7.** $\overline{WF}$ and $\overline{KM}$
9. Answers may vary. An example is given.

Page 30–33   Exercises and Applications

1. 2   **3.** 2   **5.** 2   **7.** $\overline{AB} \cong \overline{EG}$; $\overline{BC} \cong \overline{CE}$; $\overline{AC} \cong \overline{CG}$
9, 11, 13. Answers may vary. Examples are given.

9.

11.

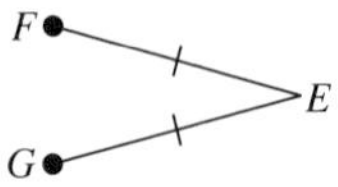

13.

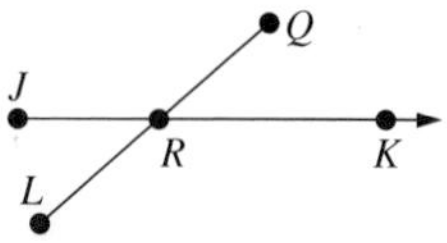

19. A   **21.** B   **23.** F
25. 3; 22

Page 37   Checking Key Concepts

1. $\angle WYZ$ or $\angle ZYW$ **3.** $\angle ZYX$ **5.** 110°   **7.** 70°

Pages 37–39   Exercises and Applications

1. Answers may vary. An example is given.

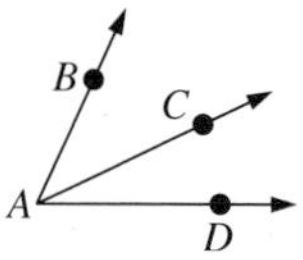

In the example, none of the angles can be
named $\angle A$ because that name could refer
to $\angle BAC$, $\angle BAD$, or $\angle CAD$.

3, 5, 7, 9. Answers may vary slightly.   **3.** 41°   **5.** 65°
7. 115°   **9.** 74°   **11.** 32°   **13.** 45°   **15.** 27°   **17.** $\angle HED$
31. 6   **33.** 12

Page 43   Checking Key Concepts

1. 3.4   **3.** 6.8   **5.** 38°   **7.** Fold the paper so that J falls on
top of L. Unfold the paper. The point where the crease
meets $\overline{JL}$ is the midpoint of $\overline{JL}$.

Pages 44–47   Exercises and Applications

1. 11   **3.** 9   **5.** False.   **7.** False.   **11.** $(-2, 3)$ **13.** $(4, 1)$
15. $(1, 0)$ **17, 19.** Answers may vary. Examples are given.

17.

2   2   4

C   B   A   D

19.

21. $3\frac{1}{2}$ **29.** 8   **31.** 16   **35.** 78°
37. 31°   **39.** 26, 37

Page 47   Assess Your Progress

1. $\angle FGH$; $\angle HGF$; $\angle G$ **2.** $\angle FHE$ **3.** 130°   **4.** 27°
5. 52°   **6.** 13; 26   **7.** 1   **8.** (–1, –2)   **9.** 4; 94°

Pages 52–53   Chapter 1 Assessment

1. inductive   **2.** the hypothesis (the "if") and the conclusion
(the "then")   **3.** image; rotation, reflection, translation
5. 324, 972   **6.**

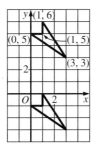

7. a. If a person is tall, then he or she is a good basketball
player. hypothesis: a person is tall; conclusion: he or she is a
good basketball player   **b.** False; there are many counter-
examples, including tall people who do not play basketball
at all.
8.

9.

10.

11. True.   **12.** True.   **13.** False; two lines in the same plane
are either parallel or they intersect.   **14.** False; if two planes
intersect, they intersect in a line.   **15.** plane S, plane ABD,
plane CBD (The order of the letters can also be changed; for
example, the plane could also be called plane ACB.)
16. point E **17.** point D **18.** Answers may vary. An
example is given. $\overleftrightarrow{AB}$ and $\overleftrightarrow{CE}$ **19.** $\overleftrightarrow{AB}$ and $\overleftrightarrow{DC}$; $\overleftrightarrow{AD}$ and $\overleftrightarrow{BC}$
20–22. Answers may vary. Examples are given.
20.

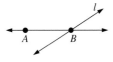

21.

22.

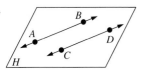

23. $\overline{RV} \cong \overline{RS}$, $\angle RVS \cong \angle RSV$, $\overline{SV} \cong \overline{TU}$, $\overline{ST} \parallel \overline{VU}$
24. Answers may vary. Examples are given. $\overline{ST} \cong \overline{VU}$,
$\overline{SV} \parallel \overline{TU}$; Angles VST, T, U, and UVS are all congruent to one
another.   **25. a.** 5   **b.** 10   **c.** 8   **26.** 1; 5   **27.** 60°; 120°

CHAPTER 2

Page 61   Checking Key Concepts

1. complementary: 50°; supplementary: 140°; vertical: 40°
3. complementary, $x < 90°$: $90° - x°$; supplementary,
$x < 180°$: $180° - x°$; vertical: $x°$ **5.** always

Pages 61–63   Exercises and Applications

1. $\angle FHD$ **3.** $\angle CJE$ **11.** 80°; Linear Pair Property
13. 45°; If two lines form congruent adjacent angles, then
the lines are perpendicular; Linear Pair Property or Vertical
Angles Property; definition of complementary angles.
15. 135°; Linear Pair Property (using Ex. 13)   **17.** True.
19. False.   **27.** 140°   **31. a.** 50°; by the Linear Pair
Property   **b.** 50°   **c.** Two angles that are supplementary to
the same angle are congruent. This is because if $m \angle 1 +
m \angle 2 = 180°$ and $m \angle 1 + m \angle 3 = 180°$, then $m \angle 2 =
180° - m \angle 1$ and $m \angle 3 = 180° - m \angle 1$, so $m \angle 2 = m \angle 3$.
35. a. 70°   **b.** 40°   **c.** 70°

Page 67   Checking Key Concepts

1. scalene, acute triangle   **3.** Answers may vary. An
example is given.   **5.** 33°

Pages 68–71   Exercises and Applications

1. Answers may vary. An example is given.

3. not possible; An equiangular triangle has three angles that
measure 60°, so it has only acute angles.   **7. a.** 90°
b. They are complementary. Because the right angle
measures 90°, the other two angle measures must total
$180° - 90° = 90°$.   **9.** $b° = 30°$; $4b° = 120°$ **11.** $w° = 42°$;
$z° = 25°$ **19.** never   **21.** always   **23.** sometimes   **25.** 55°
27. 67.5°   **29.** 72.5°   **35. a.** 5 segments   **b.** $\overline{FG} \cong \overline{GH} \cong
\overline{HI} \cong \overline{IJ} \cong \overline{JF}$ **c.** $\angle F \cong \angle G \cong \angle H \cong \angle I \cong \angle J$

Page 71   Assess Your Progress

1. 37; Vertical Angles Property   **2.** 143; Linear Pair
Property   **3.** 90; If two lines form congruent adjacent
angles, then the lines are perpendicular; Linear Pair Property
or Vertical Angles Property.   **4.** 53°; The sum of the angle
measures of a triangle is 180°.

5–9. Answers may vary. Examples are given.

5. a. 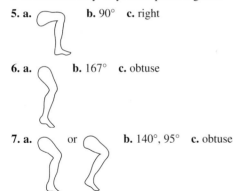 **b.** 90° **c.** right

6. a. **b.** 167° **c.** obtuse

7. a. or **b.** 140°, 95° **c.** obtuse

pedal   pedal
at front at back

8. △ABD: scalene, acute triangle; △BDE: equilateral triangle (thus, equiangular and acute); △BEF: isosceles, obtuse triangle; △BCF: scalene, right triangle

Page 75   Checking Key Concepts

1, 3. Names may vary. Examples are given.   **1.** convex, equiangular pentagon; *JKLMN*   **3.** (convex) regular quadrilateral (a square); *WXYZ*   **5.** Its borders include curves, and not just line segments.   **7.** False; a rectangle is an example of an equiangular polygon that is not regular.   **9.** False; the equilateral polygon shown at the bottom of page 73 is not convex.

Pages 76–78   Exercises and Applications

1. convex quadrilateral   **3.** concave, equilateral 12-gon
5. pentagon: 5 sides, 5 diagonals; hexagon: 6 sides, 9 diagonals; heptagon: 7 sides, 14 diagonals; octagon: 8 sides, 20 diagonals; Prediction—nonagon: 9 sides, 27 diagonals
7. a. 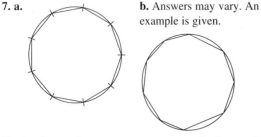 **b.** Answers may vary. An example is given.

To sketch a regular nonagon, you must mark nine evenly spaced points around the circle before connecting consecutive points. For a nonagon that is not regular, you can connect any nine points consecutively around the circle.   **9. a.** 4 vertices
b. 2 vertices   **c.** 8 names   **17.** 6 lines of symmetry
19. 2 lines of symmetry   **21, 23.** Answers may vary. Examples are given.

21. 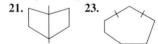   **23.**

25. Sketches may vary. Examples are given.

Polygon	Triangle	Quadrilateral	Pentagon
Concave	not possible		
Convex			
Regular			
Equilateral only	not possible		
Equiangular only	not possible		

27. hypothesis: a polygon is regular; conclusion: it is equiangular   **29.** 60°

Page 82   Checking Key Concepts

1. 720°   **3.** $(2x - 2)180°$, or $(x - 1)360°$ **5.** 29-gon   **7.** 63°

Pages 83–86   Exercises and Applications

1. 360°   **3.** 13,140°   **5.** 135°   **7.** interior: 108°; exterior: 72°   **9.** interior: 168.75°; exterior: 11.25°   **13. a.** 90°
b. 4   **c.** Yes.   **15. a.** 120°   **b.** 3   **c.** Yes.   **19.** 22-gon
21. 14-gon   **25.** $z° = 51°$; $4z° = 204°$ **27.** $w° = 25°$; $4w° = 100°$; $(6w - 15)° = 135°$ **29.** $4z° = 60°$; $9z° = 135°$; $(5z - 10)° = 65°$; $(5z + 25)° = 100°$ **31.** 20
35. a. Answers may vary. An example is given.

 7 sides; 7 triangles
b. 1260°
c. the angles with vertex at the interior point; 360°

d. The sum of the measures of the angles of the *n*-gon is the sum of the measures of the angles of the *n* triangles, $180n$, minus the sum of the measures of those angles of the triangles that are not part of the angles of the polygon, 360. This gives $180n - 360°$ for the sum of the measures of the angles of the polygon. So, for a heptagon, for example, the sum is $180(7) - 360 = 1260 - 360 = 900°$, which agrees with earlier results.   **43.** Parallel lines are coplanar lines that do not intersect, while skew lines are not coplanar and also do not intersect.   **45.** Answers may vary. An example is given.

Page 86   Assess Your Progress

1. concave hexagon   **2.** convex, equilateral pentagon
3. convex, equiangular quadrilateral (a rectangle)
4. convex hexagon

5. a. 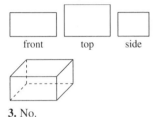 **b.** 6 lines of symmetry
c. $\overline{AB}$ and $\overline{BC}$, $\overline{BC}$ and $\overline{CD}$, $\overline{CD}$ and $\overline{DE}$, $\overline{DE}$ and $\overline{EF}$, $\overline{EF}$ and $\overline{AF}$ or $\overline{AF}$ and $\overline{AB}$

d. $\angle A$ and $\angle B$, $\angle B$ and $\angle C$, $\angle C$ and $\angle D$, $\angle D$ and $\angle E$, $\angle E$ and $\angle F$, or $\angle F$ and $\angle A$ **6.** 140° **7.** interior: 165°; exterior: 15°

Page 90   Checking Key Concepts

1. a. 78° **b.** 102° **c.** 20 mm **3.** False; a rectangle is a square only if it is also equilateral.

Pages 90–92   Exercises and Applications

1. a. 15 m **b.** 90° **c.** 12 m **3. a.** 37° **b.** 85° **c.** 85°
5. 13 **7.** 108° **9.** 36° **17.** parallelogram; rectangle; rhombus; square **19.** parallelogram; rectangle; rhombus; square **21.** rectangle; square **23.** parallelogram; rectangle; rhombus; square **25.** B **27.** C **29.** G

Page 96   Checking Key Concepts

1. Answers may vary. An example is given.

front   top   side

3. No.

Pages 96–99   Exercises and Applications

1. *ABCDEF*, *GHIJKL*; hexagonal prism **3.** 6 lateral faces
5. Answers may vary. Examples are given. $\overline{AF}$ and $\overline{CI}$, $\overline{HG}$ and $\overline{EK}$ **7.** 6 faces; 8 vertices; 12 edges **9.** 8 faces; 12 vertices; 18 edges **11.** Answers may vary. An example is given. The number of faces is two more than the number of sides of a base. The number of vertices is twice the number of vertices of a base. The number of edges is three times the number of sides of a base. For a prism whose base is an *n*-gon, $F = n + 2$, $V = 2n$, and $E = 3n$. **13.** Answers may vary. An example is given. For an octagonal prism, $F = 10$, $V = 16$, and $E = 24$, so $F + V - 2 = 26 - 2 = E$. This will work for any prism, because the values determined in Ex. 11 have the relationship determined in Ex. 12 for any value of *n*.
15. Answers may vary. An example is given.

17. Yes.

19, 21. Answers may vary. Examples are given.
19.

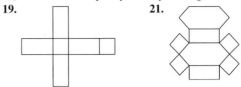

21.

27. a. any two of $\overline{CE}$, $\overline{BH}$, and $\overline{GA}$ **b.** Both points lie on the front face of the prism. **c.** 4 diagonals **31.** An isosceles triangle is a triangle in which at least two of the sides are congruent. An example is given. **33.** 25, 36

Page 99   Assess Your Progress

1. a. 60° **b.** 120° **c.** 27 **2. a.** 90° **b.** 14 **c.** 50 **3. a.** 9
b. 45° **c.** 135° **4, 5.** Answers may vary. Examples are given.

4.

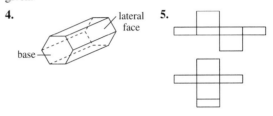

lateral face

base

5.

Pages 104–105   Chapter 2 Assessment

1. obtuse **2.** diagonal **3.** square **4.** exterior **5.** $\angle BEG$ and $\angle DEF$, or $\angle BED$ and $\angle GEF$ **6.** $\angle ACF$ and $\angle ACB$
7. A prism is a three-dimensional figure formed by connecting the corresponding vertices of two congruent faces that lie in parallel planes. A net is a two-dimensional representation of this three-dimensional figure formed by drawing the faces of the prism as if the prism had been cut along some of its edges (without disconnecting any of the faces), unfolded, and then laid out flat. **8.** rhombus (a convex, equilateral parallelogram) **9.** convex, equiangular pentagon
10. isosceles right triangle **12. a.** 90 **b.** 30 **c.** 60
13. complementary: 19°; supplementary: 109°; vertical: 71°
14. complementary: 51°; supplementary: 141°; vertical: 39°
15. complementary: 5°; supplementary: 95°; vertical: 85°
16. $4x° = 40°$; $5x° = 50°$ **17.** $6x° = 54°$; $9x° = 81°$
18. $x° = 34°$; $2x° = 68°$; $y° = 78°$ **20.** regular hexagon (convex, equiangular, and equilateral) **21.** any three of the following: *ABCDEF, BCDEFA, CDEFAB, DEFABC, EFABCD, FABCDE, AFEDCB, FEDCBA, EDCBAF, DCBAFE, CBAFED, BAFEDC* **22.** $\overline{AC}$, $\overline{AD}$, $\overline{AE}$
23. $5x° = 100°$; $6x° = 120°$ **24.** $x° = 60°$; $2x° = 120°$; $(x + 10)° = 70°$; $(2x - 10)° = 110°$ **25.** 30 **26.** 165°
27. a. 16 m **b.** 16 m **c.** 60° **28. a.** 110° **b.** 25° **c.** 25°
29. Answers may vary. An example is given.

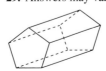

30. a. a hexagonal prism **b.** 6 lateral faces **c.** Answers may vary. Examples are given.

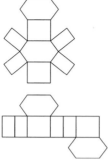

Pages 106–107 Algebra Review/Preview

1. –44 **3.** –10 **5.** –4.8 **7.** 22 **9.** 12 **11.** 26 weeks
13. $3\sqrt{3}$ **15.** $5\sqrt{3}$ **17.** $60\sqrt{2}$ **19.** $4\sqrt{2}$ **21.** ±9
23. ±3 **25.** ±2

27.

29.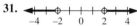

31.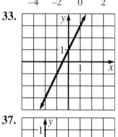

33. 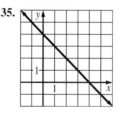 **35.**

37.

CHAPTER 3

Page 113 Checking Key Concepts

1. inductive reasoning **3.** deductive reasoning

Pages 114–116 Exercises and Applications

1. inductive reasoning **3.** inductive reasoning
5, 7, 9. Justifications may vary. Examples are given.
5. 10 and 15; The number of dots in each row after the first is one more than the number of dots in the previous row.
7. 16, 32; Each term after the first is obtained by doubling the previous term. **9.** 720, 5040; For $n > 1$, the nth term is found by multiplying the previous term by n. **11.** is
13. is not **27.** right **29.** obtuse

Page 120 Checking Key Concepts

1. Reflexive Property **3.** Symmetric Property **5.** A

Pages 121–123 Exercises and Applications

1. postulate **3.** definition **5.** Transitive Property
7. Reflexive Property **9.** given, definition of congruent segments **11.** given, definition of congruent angles
13. Ex. 11, Angle Addition Postulate, Subtraction Property
23. given information **25.** Vertical angles are congruent.
27. Ex. 26 and the definition of congruent angles **31.** 27°;
117°; 63° **33.** $(90 - 2y)°$ for $y \le 45$; $(180 - 2y)°$ for $y \le 90$;
$2y°$

Page 123 Assess Your Progress

1. deductive **2.** inductive **3.** is **4.** might have been
5. definition of angle bisector **6.** Vertical angles are congruent. **7.** Ex. 6, definition of congruent angles, given, and the Addition Property **8.** Vertical angles are congruent.

Page 126 Checking Key Concepts

1. a. ☐ **b.** hypothesis: A quadrilateral is a square. conclusion: It has four right angles.

3. a. **b.** hypothesis: A quadrilateral is a rhombus. conclusion: Its diagonals bisect each other.

5. False; the two exterior angles at each vertex are congruent.

Pages 127–129 Exercises and Applications

1. False; False; False. **3.** False; False; True. **5.** 110°
11. a. $\angle AGF$ and $\angle CGD$ **b.** $\angle BGD$ by the Angle Addition Postulate **c.** 180°; because C, G, and F are collinear points **d.** $\angle AGF \cong \angle CGD$ and $\angle EGF \cong \angle BGC$, since vertical angles are congruent. By the definition of congruent angles and the Substitution Property, $\angle BGC \cong \angle CGD$. **15.** Yes; since the triangle is equilateral, two of its sides are congruent. **17.** No; for example, let $m \angle 1 = 60°$. Then $m \angle 3 = 120°$ and $m \angle 2 = 120°$. $\angle 1$ and $\angle 2$ are not congruent angles. **19.** B **23.** Transitive Property **25.** Answers may vary. Examples are given. 2, 4, 8, 16, 32, 64, … (The nth term is $2n$.); 2, 4, 8, 14, 22, 32, … (For $n > 1$, the nth term is found by adding $2n - 2$ to the previous term.)

Page 132 Checking Key Concepts

1. Substitution Property (Step 1); $m \angle 2 = m \angle 3$; $m \angle 1 = m \angle 4$ **3.** 19; If $(n - 2)180 = 3060$, where n is the number of sides, then $n = 19$.

Pages 132–135 Exercises and Applications

1. The statement is incorrect. Angles that are supplementary to the same angle are congruent. **3.** The measure of an exterior angle of a triangle is not necessarily greater than the measure of the adjacent interior angle. Since the measure of an exterior angle of a triangle is equal to the sum of the measures of the two nonadjacent interior angles and each angle measure is positive, the measure of an exterior angle is greater than the measure of each nonadjacent interior angle. **11.** 94° **13.** 159° **15.** Answers may vary. An example is given.

a.

b. Given: $j \perp k$. Prove: $\angle 1 \cong \angle 2$

19. If <u>two angles are congruent</u>, then (their measures are equal.)

21. If <u>a quadrilateral is a parallelogram</u>, then (its diagonals bisect each other.)

23.

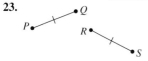

Page 135   Assess Your Progress

1. Given: $\angle 1$ and $\angle 2$ are supplementary.
$\angle 1$ and $\angle 3$ are supplementary.
Prove: $\angle 2 \cong \angle 3$.
By the definition of supplementary angles, $m \angle 1 + m \angle 2 = 180°$ and $m \angle 1 + m \angle 3 = 180°$. Therefore, by the Substitution Property, $m \angle 1 + m \angle 2 = m \angle 1 + m \angle 3$. Subtracting $m \angle 1$ from both sides gives $m \angle 2 = m \angle 3$, so $\angle 2 \cong \angle 3$.

2.

Statements	Reasons
1. $m \angle 1 + m \angle 2 + m \angle 3 = 180°$	1. The sum of the angle measures of a triangle is 180°.
2. $m \angle 4 + m \angle 3 = 180°$	2. Angles in a linear pair are supplementary.
3. $m \angle 1 + m \angle 2 + m \angle 3 = m \angle 4 + m \angle 3$	3. Substitution Property (Steps 1 and 2)
4. $m \angle 1 + m \angle 2 = m \angle 4$	4. Subtraction Property

3. Reasons

1. Given
2. Definition of congruent angles
3. Angles in a linear pair are supplementary.
4. Substitution Property (Step 3)
5. Substitution Property (Steps 2 and 4)
6. Subtraction Property
7. Definition of congruent angles

Page 138   Checking Key Concepts

1. If you are in Guyana, then the official language is English; True; True.   **3.** If the official language is Portuguese, then you are in Brazil; True; True.   **5. a.** If a quadrilateral is a square, then it is a rectangle.   **b.** If a quadrilateral is a rectangle, then it is a square.   **7. a.** If a figure is a triangle, then the sum of the angle measures is 180°.   **b.** If the sum of the angle measures of a figure is 180°, then the figure is a triangle.

Pages 138–140   Exercises and Applications

1. If the American League team is the championship team, then it won the first four games of the World Series; False.
3. If the area of a rectangle is 6 ft², then the rectangle is 2 ft long and 3 ft wide; False.

5. a. If two angles are congruent, then supplements of those angles are congruent.   **b.** If the supplements of two angles are congruent, then the two angles are congruent.   **7. a.** If a triangle is isosceles, then it is not scalene.   **b.** If a triangle is not scalene, then it is isosceles.   **9. a.** If a quadrilateral is a square, then its diagonals are congruent and perpendicular.
b. If the diagonals of a quadrilateral are congruent and perpendicular, then the quadrilateral is a square.   **21.** 90°
23. 10   **25.**

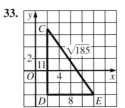

27.

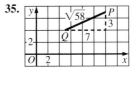

Page 144   Checking Key Concepts

1. 7; 7   **3.** -4.90; $-2\sqrt{6}$ **5.** 8   **7.** Yes.   **9.** No.

Pages 144–147   Exercises and Applications

1. 11; 11   **3.** ±5.20; $\pm3\sqrt{3}$ **5.** -20.12; $-9\sqrt{5}$ **7.** Yes.
9. No.   **11.** Yes.   **13.** 10.91   **15.** 12   **17.** 3.32
19. $\sqrt{3} \approx 1.73$ **21.** $\sqrt{130} \approx 11.40$ **23.** $\dfrac{3\sqrt{2}}{2} \approx 2.12$
25. Yes.   **27.** No.

33.

35.

49. If (two lines are perpendicular,) then <u>they intersect at right angles</u>.

Page 150   Checking Key Concepts

1. acute   **3.** obtuse   **5.** False; a triangle with exactly two congruent sides is not equilateral, yet it is isosceles.

Pages 150–153   Exercises and Applications

1. obtuse   **3.** acute   **5.** acute   **7.** obtuse   **11.** False.
13. False.   **17.** Inverse: If it does not snow, then classes are not canceled. Contrapositive: If classes are not canceled, then it does not snow.   **19.** Inverse: If a quadrilateral is not a parallelogram, then its opposite sides are not congruent. Contrapositive: If the opposite sides of a quadrilateral are not congruent, then the quadrilateral is not a parallelogram.
21. If two lines are not perpendicular, then they do not form congruent adjacent angles; True.   **23.** If two angles are not vertical angles, then they are not congruent; False.
29. Yes; the lengths are the lengths of the sides of a right triangle and the triangle appears to be a right triangle.
31. No; the lengths are not the lengths of the sides of a right triangle.   **33.** No; the lengths are not the lengths of the sides of a right triangle.   **37.** Sketches and labels may vary.
a. If two angles are not congruent, then they are not both right angles.

b.

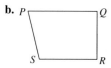

c. Given: $\angle Y$ is not congruent to $\angle X$.
   Prove: $m \angle Y \neq 90°$ or $m \angle X \neq 90°$.
41. 2   **43.** Yes.   **45.** Yes.

Page 153   Assess Your Progress

1. $2\sqrt{6} \approx 4.90$ **2.** $6\sqrt{5} \approx 13.42$ **3.** $\dfrac{5\sqrt{5}}{2} \approx 5.59$

4. right   **5.** obtuse   **6.** obtuse   **7.** acute   **8.** acute
9. obtuse   **10.** Inverse: If a quadrilateral is not a rectangle, then its diagonals are not congruent. Converse: If the diagonals of a quadrilateral are congruent, then the quadrilateral is a rectangle. Contrapositive: If the diagonals of a quadrilateral are not congruent, then the quadrilateral is not a rectangle.
11. Inverse: If the exterior rays of two adjacent angles are not perpendicular, then the angles are not complements. Converse: If two adjacent angles are complements, then the exterior rays of the angles are perpendicular. Contrapositive: If two adjacent angles are not complements, then the exterior rays of the angles are not perpendicular.

12. a. In quadrilateral $PQRS$, if $\angle P$ is not congruent to $\angle R$, then $PQRS$ is not a parallelogram.

b.   **c.** Given: $\angle P$ is not congruent to $\angle R$.
   Prove: $PQRS$ is not a parallelogram.

Pages 158–159   Chapter 3 Assessment

1. Sketches and labels may vary. An example is given. For the triangle in the sketch, the Pythagorean theorem can be stated as $a^2 + b^2 = c^2$.

2. proof; hypothesis; conclusion   **3.** If point Y is not between points X and Z, then $XY + YZ \neq XZ$.   **4.** must
5. may have   **6.** inductive   **7.** Through any two points there is exactly one line.   **8.** definition of midpoint
9. Transitive Property   **10.** The sum of the angle measures of a triangle is 180°.   **11.** Given; Vertical angles are congruent; Substitution Property (Steps 1 and 2); $\angle 1 \cong \angle 4$; Substitution Property (Steps 2 and 3)
12. a. Sketches and labels may vary.

b. Given: $ABCD$ is a rhombus.
   Prove: $\overline{AC} \perp \overline{BD}$

13. a. Converse: If the diagonals of a quadrilateral are perpendicular, then it is a rhombus. Inverse: If a quadrilateral is not a rhombus, then its diagonals are not perpendicular. Contrapositive: If the diagonals of a quadrilateral are not perpendicular, then it is not a rhombus.

b. For any of the statements, the diagram given in the answer to Ex. 12 can be used. Converse—Given: $\overline{AC} \perp \overline{BD}$; Prove: $ABCD$ is a rhombus. Inverse—Given: $ABCD$ is not a rhombus; Prove: $\overline{AC}$ is not perpendicular to $\overline{BD}$. Contrapositive—Given: $\overline{AC}$ is not perpendicular to $\overline{BD}$; Prove: $ABCD$ is not a rhombus.   **15.** $2\sqrt{41} \approx 12.81$
16. $3\sqrt{3} \approx 5.20$ **17.** obtuse   **18.** right   **19.** acute

CHAPTERS 1–3

Pages 160–161   Cumulative Assessment

1. 15 handshakes; 21 handshakes   **7.** $\angle BAE, \angle BAC,$
$\angle EAB, \angle CAB$ **9.** one of the following: $\overrightarrow{BD}, \angle ABD,$ and $\angle CBD,$ or $CA, \angle BCA,$ and $\angle DCA$ **11.** always
13. always   **15.** $x = 6; y = \dfrac{1}{3}$ **17.** $x = 36; y = 30$
21. The measure of an exterior angle of a triangle is equal to the sum of the measures of the two interior angles that are not adjacent to it.   **23.** the Segment Addition Postulate
25. The sum of the angle measures of a triangle is 180°. Since the angles of an equiangular triangle are congruent, their measures are equal. Thus, the measure of each angle is $\dfrac{180°}{3} = 60°$.   **27.** $\sqrt{39} \approx 6.24$ **29.** 1

CHAPTER 4

Page 169   Checking Key Concepts

1. 6   **3.** 5   **5.** $(2, -4)$ **7.** $(2.5, 0)$

Pages 169–172   Exercises and Applications

5. $DE = EF = \sqrt{13}; DF = \sqrt{26};$ isosceles right triangle
7. $JK = LM = 3\sqrt{2}; JM = KL = 2\sqrt{2};$ rectangle
9. $WX = XY = YZ = WZ = \sqrt{26};$ rhombus   **11.** $(-5, 8)$
13. $(3.5, 3.5)$ **15.** $(2, 2.5)$ **21.** $(0, 2), (0, 1), (1, 1), (-2, 0),$ $(0, 0), (1, 0), (2, 0), (0, -1), (1, -1), (0, -2)$ **37.** obtuse
39. Answers may vary.   **41.** -4 **43.** -1
An example is given.

Page 176   Checking Key Concepts

1. 0   **3.** $y = -x + 2$

Pages 176–179   Exercises and Applications

1. 4   **3.** $-\dfrac{5}{4}$ **5.** $\dfrac{1}{2}$ **7. a.** 36 in. or 3 ft   **b.** 3.5 in.
9. $y = 2x + 5$ **11.** $y = -0.75x + 1$ **13.** $x = 13$
15. $y = 0.25x + 1.5$ **23.** 3; 5   **25.** $-5; -25$ **27.** $-\dfrac{2}{3}; 0$
29. 4; -3 **31.** slope of $\overline{AB} =$ slope of $\overline{CD} = 0;$ slope of $\overline{AD} =$ slope of $\overline{BC} = -2$ **33.** slope of $\overline{FO} = \dfrac{1}{4};$ slope of $\overline{OH} = 4;$ slope of $\overline{GH} = -\dfrac{1}{4};$ slope of $\overline{FG} = 2$

37. (6, 3) **39.** (−7, 1.5) **41.** $\overleftrightarrow{CD}$, $\overleftrightarrow{EF}$, and $\overleftrightarrow{GH}$ **43.** $\overleftrightarrow{AD}$, $\overleftrightarrow{EH}$, and $\overleftrightarrow{FG}$ **45.** $\overleftrightarrow{BC}$, $\overleftrightarrow{DC}$, $\overleftrightarrow{GC}$, $\overleftrightarrow{HG}$, $\overleftrightarrow{FG}$ **47.** False; if the rhombus is not a square, the angles are not right angles.

Page 183 Checking Key Concepts
1. neither **3.** parallel **5.** 7 **7.** 0

Pages 183–186 Exercises and Applications
1. perpendicular **3.** parallel **5.** 0 **7.** $\frac{1}{3}$ **9.** Yes. **11.** No.
15. quadrilateral **17.** square **19.** Yes. **21.** No; $\overleftrightarrow{CD}$ does not bisect $\overline{AB}$. **29.** $y = 7$ **31.** $x = 2$ **35.** 1; 0
37. Answers may vary. An example is given. **39.** $b = 15.5$

Page 186 Assess Your Progress
1. a. $12 + 3\sqrt{2}$ **b.** Let M, N, and P be the midpoints of $\overline{AB}$, $\overline{AC}$, and $\overline{BC}$, respectively. $M = (2.5, 0)$, $N = (1, 1.5)$, $P = (4.5, 1.5)$ **c.** $6 + \frac{3}{2}\sqrt{2}$; The perimeter of $\triangle MNP$ is half the perimeter of $\triangle ABC$. **2.** $-\frac{3}{5}$ **3.** 0 **4.** $\frac{5}{2}$ **5.** 4 **6.** $\frac{1}{5}$

Page 190 Checking Key Concepts
1. 5 **3.** $(x - 2)^2 + (y + 3)^2 = 16$

Pages 191–193 Exercises and Applications
1. $x^2 + y^2 = 9$ **3.** $(x + 2)^2 + (y - 3)^2 = 9$
5. $(x - 5)^2 + (y - 2)^2 = 49$ **7.** $(x + 2)^2 + (y - 1)^2 = 1$
9. $(x - 10)^2 + (y + 20)^2 = 400$
11.

13.

15.

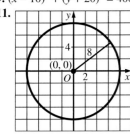

17, 19. Answers may vary. Examples are given.
17.

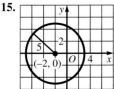

19.

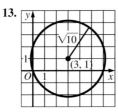

23. (3, 0); 2; $(x - 3)^2 + y^2 = 4$ **25.** (2, 2); $\sqrt{10}$; $(x - 2)^2 + (y - 2)^2 = 10$ **29.** Yes; $(2\sqrt{2})^2 + 2^2 = 12$.
31. Yes; $(7 - 2)^2 + (5\sqrt{3})^2 = 100$. **35.** rhombus
37. square **39.** 6 **41.** 3 **43.** True.

Page 197 Checking Key Concepts
1. The fourth vertex is (b, h).
3. Answers may vary. An example is given. Place a rectangle on a coordinate plane. Use the Distance Formula to show that the diagonals are congruent.

Pages 197–200 Exercises and Applications
1. $(0, b)$, $(0, -b)$, $(a, -b)$ **3.** $(2a, 0)$ **9.** (2.5, 2.5); $MA = MB = MO = \frac{5}{2}\sqrt{2}$
11. Answers may vary. An example is given.

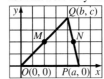

Using the Midpoint Formula, $M = \left(\frac{a}{2}, \frac{b}{2}\right)$. By the Distance Formula, $MA = MB = MO = \frac{1}{2}\sqrt{a^2 + b^2}$.

13. Let M and N be the midpoints of the sides of the triangle. $M = (2, 2)$ and $N = (4.5, 2)$. The slope of $\overline{MN} = 0 =$ slope of the base, so $\overline{MN}$ is parallel to the base. $MN = 2.5 = \frac{1}{2} \times$ length of base
15. Answers may vary. An example is given.

Using the Midpoint Formula, $M = \left(\frac{b}{2}, \frac{c}{2}\right)$ and $N = \left(\frac{a + b}{2}, \frac{c}{2}\right)$. The slope of $\overline{MN} = 0 =$ slope of $\overline{OP}$, so $\overline{MN} \parallel \overline{OP}$. By the Distance Formula, $MN = \frac{a}{2} = \frac{1}{2} \times OP$. This proves that the segment that joins the midpoints of two sides of a triangle is parallel to the third side and half as long as the third side.
17. $E(a, 0)$, $F(a + b, c)$, $G(b + d, c + e)$, $H(d, e)$ **19.** The segments joining the midpoints of the opposite sides of a quadrilateral bisect each other. **29.** $(x - 7)^2 + y^2 = 36$
31. $(x + 1)^2 + (y - 4)^2 = 49$ **33.** 720° **35.** $\sqrt{89} \approx 9.43$
37. 5

Page 204 Checking Key Concepts
1. (4, 0, 2) **3.** (0, 3, −1) **5.** (−1, 4, −4) **7.** $\left(2, -3, -\frac{1}{2}\right)$
9. $\sqrt{14} \approx 3.74$

Pages 205–207 Exercises and Applications
1. (4, 0, 3) **3.** (4, 6, 3) **5.** (0, 0, 0) **7.** the xy-plane
9. the yz-plane **11.** the xz-plane **13.** (5, −2, 4.5)
15. (4, −0.5, −1) **17.** $\sqrt{21} \approx 4.58$ **19.** isosceles
21. scalene **31.** If all the sides of a triangle are congruent, then the triangle is equilateral. **33.** Answers may vary. An example is given.

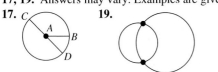

Page 207 Assess Your Progress

1.   2.

3. $D\left(-\dfrac{k}{2}, \dfrac{3k}{2}\right)$; $E\left(\dfrac{k}{2}, \dfrac{3k}{2}\right)$ 4. slope of $\overline{AE} = 1$;

slope of $\overline{CD} = -1$ 5. -1 6. If the product of the slopes of two lines is -1, then the lines are perpendicular.
7. equilateral   8. isosceles

Pages 212–213 Chapter 4 Assessment

1. B   2. E   3. C   4. D   5. A   6. a. $PR = 3\sqrt{4}$; $QR = 3\sqrt{4}$; $PQ = 2\sqrt{2}$ b. isosceles triangle
7. a. $AB = 4\sqrt{2}$; $CD = 4\sqrt{2}$; $BC = 2\sqrt{2}$; $AD = 2\sqrt{2}$
b. rectangle   8. $-\dfrac{1}{2}$ 9. -4 10. undefined   11. 3; -12

12. $\dfrac{3}{4}$; 1   13. -1; 4   14. $y = 2x - 11$ 15. $y = -\dfrac{3}{2}x$

16. $x = 1$ 17. $y = -1$ 18. The slope of $\overline{JK}$ is -3 and the slope of $\overline{KL}$ is $\dfrac{1}{3}$. Since $(-3)\left(\dfrac{1}{3}\right) = -1$, $\overline{JK} \perp \overline{KL}$, so $\triangle JKL$ is a right triangle.   20. $x^2 + y^2 = 9$ 21. $x^2 + y^2 = 5$
22. $(x - 4)^2 + (y - 2)^2 = 81$ 23. $(x - 6)^2 + (y + 2)^2 = 6.25$
24. a.   b. $(-3, 1)$ c. See part (a).

25. $(-2a, 0)$ 26. $S(a, b)$; $T(-a, b)$; $O(0, 0)$ 27. a. $TS = 2a$; $TO = SO = \sqrt{a^2 + b^2}$ b. $TO = SO$; $\triangle STO$ is isosceles by definition.   29. $(-1, 6, -3)$ 30. $MN = 2\sqrt{14}$

Pages 214–215 Algebra Review/Preview

1. 2   3. 0   5. $\dfrac{12}{7}$ 7. -8 9. 1   11. 3   13. $\dfrac{97}{11}$

15. $(10, 20)$ 17. $\left(-1, \dfrac{2}{3}\right)$ 19. no solution

21. 1550 adults and 2700 children

23. $z < 18$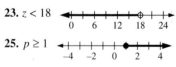

25. $p \geq 1$

27. $k \geq -7$

CHAPTER 5

Page 223 Checking Key Concepts

1. j 3. $\angle 3$ and $\angle 6$, $\angle 4$ and $\angle 5$ 5. a. $60°$ b. $60°$
c. 58

Pages 223–225 Exercises and Applications

1. $\angle 2$ and $\angle 5$, $\angle 3$ and $\angle 8$ 3. $\angle 1$ and $\angle 5$, $\angle 2$ and $\angle 6$, $\angle 3$ and $\angle 7$, $\angle 4$ and $\angle 8$ 5. corresponding angles
7. corresponding angles   9. same-side interior angles
17. $x = 55$; $y = 70$; $z = 27$ 19. same-side interior angles
21. none of these   23. none of these   25. same-side interior angles   29. $\overline{AB}$ and $\overline{DE}$; $\overline{BC}$ and $\overline{EF}$; $\overline{CD}$ and $\overline{FA}$
33. $(2.5, -3.5, 5)$; $3\sqrt{6}$ 35. False; answers may vary. An example is given.

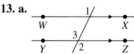

Page 229 Checking Key Concepts

1. $z = 50$; $m\angle 1 = 130°$ 3. Answers may vary. An example is given.

Pages 230–233 Exercises and Applications

1. $m\angle 1 = m\angle 2 = 125°$ 3. $m\angle 5 = m\angle 7 = m\angle 9 = 140°$; $m\angle 6 = m\angle 8 = 40°$ 5. 26   7. always   9. sometimes
11. $115°$
13. a.   b. (1) $\angle 1 \cong \angle 3$
(2) Vertical $\angle s$ are $\cong$.
(3) $\angle 1 \cong \angle 2$
15. Answers may vary. An example is given.

17. not possible; Answers may vary. For example, Example 2 on page 229 showed that the angles of a trapezoid consist of two pairs of supplementary angles. Two acute angles cannot be supplementary, so a trapezoid cannot have three acute angles.   23. $\angle 2$, $\angle 4$, and $\angle 6$ 25. Answers may vary. An example is given. The sum of the measures is the same for both pairs of angles.   31. $m\angle 1 = 140°$;
$m\angle 2 = 40°$ 33. Answers may vary. An example is given.
Given: $k \parallel l$
Prove: $\angle 1 \cong \angle 3$

Statements	Reasons
1. $k \parallel l$	1. Given
2. $\angle 1 \cong \angle 2$	2. If two $\parallel$ lines are intersected by a transversal, then corresponding $\angle s$ are $\cong$.
3. $\angle 2 \cong \angle 3$	3. Vertical $\angle s$ are $\cong$.
4. $\angle 1 \cong \angle 3$	4. Transitive Property (Steps 2 and 3)

37. $w° = 115°$; $z° = 65°$; $y° = 68°$

Page 233 Assess Your Progress

1. same-side interior angles: $\angle 3$ and $\angle 7$, $\angle 2$ and $\angle 6$; alternate interior angles: $\angle 3$ and $\angle 6$, $\angle 2$ and $\angle 7$; corresponding angles: $\angle 1$ and $\angle 7$, $\angle 3$ and $\angle 5$, $\angle 4$ and $\angle 6$, $\angle 2$ and $\angle 8$ 2. 29   3. $y = 135$; $z = 38$

4. Answers may vary. For the example shown, $\overline{JK}$ and $\overline{ML}$ are the bases and $\overline{KL}$ and $\overline{JM}$ are the legs.

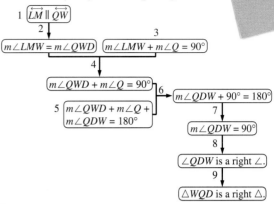

Page 238   Checking Key Concepts

1. $m\angle 1 + m\angle 2 + m\angle 4 = 180°$

Pages 239–241   Exercises and Applications

1. (1) $\angle A$ and $\angle B$ are complements; (3) $m\angle A + m\angle B + m\angle C = 180°$; (4) $180°$; Substitution Property (Steps 2 and 3), (5) $m\angle C = 90°$; Subtraction Property.   **7.** If a triangle contained more than one angle with measure 90° or greater, then the sum of the measures of the three angles would be greater than 180°. Therefore, if one angle is right or obtuse, the other two angles must be acute in order for the angle sum to be exactly 180°.

13. 1. $m\angle 1 + m\angle 2 = 180°$; $m\angle 4 + m\angle 5 = 180°$
(The $\angle$s form a linear pair.)
2. $m\angle 1 + m\angle 2 = m\angle 4 + m\angle 5$ (Substitution)
3. $m\angle 1 = m\angle 5$ (Given)
4. $\angle 2 \cong \angle 4$ (Subtraction Property)

17. Methods of proof may vary. An example is given. Since $\overline{AB} \parallel \overline{CD}$, $\angle ABC \cong \angle BCD$. (If two $\parallel$ lines are intersected by a transversal, then alternate interior $\angle$s are $\cong$.) Similarly, since $\overline{BC} \parallel \overline{DE}$, $\angle BCD \cong \angle CDE$. By the Transitive Property, $\angle ABC \cong \angle CDE$.   **21.** If a triangle is isosceles, then the triangle is equilateral; False.

Page 245   Checking Key Concepts

1. If two lines are intersected by a transversal and same-side interior angles are supplementary, then the lines are parallel.

Pages 245–248   Exercises and Applications

1. -32 **3.** 13   **5.** parallel; Answers may vary. An example is given. $m\angle YTV = 36°$, so $\angle XYW$ and $\angle YTV$ are congruent, and since corresponding angles are congruent, the lines are parallel.   **7.** not parallel; The labeled corresponding angles cannot be congruent.   **9.** $\overline{AD}$ and $\overline{BC}$
11. $\overline{AB}$ and $\overline{DC}$

21. Statements	Reasons
1. $\angle 1 \cong \angle 3$	1. Given
2. $\angle 2 \cong \angle 1$	2. Vertical $\angle$s are $\cong$.
3. $\angle 2 \cong \angle 3$	3. Transitive Property
4. $j \parallel k$	4. If two lines are intersected by a transversal and corresponding $\angle$s are $\cong$, then the lines are $\parallel$.
5. $\angle 4$ and $\angle 5$ are supplements.	5. If two $\parallel$ lines are intersected by a transversal, then same-side interior $\angle$s are supplementary.

25. Proofs may vary. An example is given.

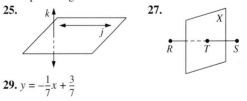

Reasons
1. Given
2. If two $\parallel$ lines are intersected by a transversal, then corresponding $\angle$s are $\cong$.
3. Given
4. Substitution Property (Steps 2 and 3)
5. The sum of the angle measures of a $\triangle$ is 180°.
6. Substitution Property (Steps 4 and 5)
7. Subtraction Property
8. Def. of right $\angle$
9. Def. of right $\triangle$

27. $y = \dfrac{1}{3}x - 6$

Page 251   Checking Key Concepts

1. $\overleftrightarrow{CE} \parallel \overleftrightarrow{FD}$ **3.** $\overleftrightarrow{CE} \parallel \overleftrightarrow{AB}$; If two lines are both parallel to a third line, then the lines are parallel.

Pages 252–255   Exercises and Applications

1. They are parallel; in a plane, if two lines are both perpendicular to a third line, then the two lines are parallel.
3. $x = 15$; $m\angle LPQ = 65°$ **11.** $4\sqrt{6}$ **17. b.** 4 units
23. (1) Given; (2) $\angle 3 \cong \angle 4$; (3) If two lines are intersected by a transversal and corresponding $\angle$s are $\cong$, then the lines are $\parallel$; (4) $p \parallel m$; (5) $n \parallel p$ (If two lines are both $\parallel$ to a third line, then the two lines are $\parallel$.)   **25, 27.** Answers may vary. Examples are given.

25.

27.

29. $y = -\dfrac{1}{7}x + \dfrac{3}{7}$

Page 255   Assess Your Progress

1. Since the sum of the angle measures of a triangle is 180°, $m\angle J + m\angle K + m\angle L = 180°$. It is given that $m\angle J + m\angle L = m\angle K$, so by the Substitution Property, $m\angle K + m\angle K = 180°$. $2 \cdot m\angle K = 180°$, and so $m\angle K = 90°$. $\angle K$ is a right angle by the definition of right angle and JKL is a right triangle by the definition of right triangle.

2. Statements	Reasons
1. $\angle 1 \cong \angle 2$	1. Given
2. $\overleftrightarrow{FE} \parallel \overleftrightarrow{DC}$	2. If two lines are intersected by a transversal and corresponding $\angle$s are $\cong$, then the lines are $\parallel$.
3. $\overleftrightarrow{DC} \parallel \overleftrightarrow{BA}$	3. Given
4. $\overleftrightarrow{FE} \parallel \overleftrightarrow{BA}$	4. If two lines are both $\parallel$ to a third line, then the two lines are $\parallel$.

3. $\overline{MN} \parallel \overline{PO}$; If two lines are intersected by a transversal and same-side interior angles are supplementary, then the lines are parallel. **4.** $\overline{RQ} \parallel \overline{ST}$; Answers may vary. An example is given. $\angle RQT$ and $\angle T$ are both right angles. In a plane, if two lines are both perpendicular to a third line, then the lines are parallel. **5.** 36 **6.** 42

Page 258 Checking Key Concepts

1. a. $\overline{AE}, \overline{AD}, \overline{BC}$, and $\overline{BF}$ **b.** Answers may vary. An example is given. Planes $ABCD$ and $EFGH$ are parallel and $\overleftrightarrow{AB}$ and $\overleftrightarrow{HG}$ are the lines of intersection when $ABCD$ and $EFGH$ are intersected by plane $ABGH$. By the Intersecting Planes Theorem, $\overleftrightarrow{AB} \parallel \overleftrightarrow{HG}$.

Pages 259–262 Exercises and Applications

1, 3. Answers may vary. Examples are given.
1. 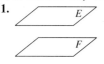 **3.**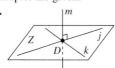

5. a. They are parallel; if two parallel planes are intersected by a third plane, the lines of intersection are parallel.
b. a trapezoid **7.** $\angle FDE$ and $\angle BDC$
11. $m \angle QNR = 45°$; $m \angle MPN = 30°$
13. Answers may vary. An example is given.

Given: plane $P \perp$ line j;
plane $R \perp$ line j
Prove: plane $P \parallel$ plane R

19. a. 60° **b.** No; since $m \angle AED \neq 90°$, $\overleftrightarrow{AE}$ is not perpendicular to $\overleftrightarrow{ED}$, so $\overleftrightarrow{AE}$ cannot be perpendicular to plane CDE.
25. D **27.** $\sqrt{97}$ **33.** Yes.

Page 265 Checking Key Concepts

1, 3. Use the indicated method. **1.** the Congruent Angles Construction **3.** the Perpendicular Bisector Construction

Pages 266–269 Exercises and Applications

1, 3. Use the indicated method. **1.** the Congruent Angles Construction **3.** the Perpendicular Bisector Construction

5. Constructions may vary. In the example, $m \angle PQR = x°$, $m \angle SQR = y°$, and $m \angle PQS = (x - y)°$.

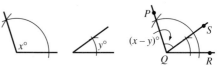

7. a. The same radius was used to draw the arcs used to locate points E, D, and H. **13. a.** Using M and N as centers and a radius greater than $\frac{1}{2} MN$, draw arcs that intersect below line g at a point R. Draw $\overleftrightarrow{QR}$. $\overleftrightarrow{QR} \perp g$ **23. a.** Use the Congruent Angles Construction and the Perpendicular Bisector Construction. **25.**

Page 269 Assess Your Progress

1. plane $P \parallel$ plane Q **2.** They are parallel; if two parallel planes are intersected by a third plane, the lines of intersection are parallel. **3.** 122°; 58° **4–6.** Use the indicated method. **4.** the Congruent Angles Construction **5.** the Perpendicular Bisector Construction **6.** the Parallel Lines Construction

Pages 274–275 Chapter 5 Assessment

1. trapezoid; bases **2.** transversal **3.** flow proof **4.** $\angle 1$ and $\angle 3$ **5.** $\angle 2$ and $\angle 4$ **6.** $\angle 2$ and $\angle 3$ **8.** $z = 59$; $y = 17$; $x = 112$ **9.** $p = r = 95°$ **10.** $s = 55$; $q = 35$
11. a. $\angle X$ and $\angle Y$ are supplementary since $\overline{WX} \parallel \overline{YZ}$; if two parallel lines are intersected by a transversal, then same-side interior angles are supplementary. **b.** Yes; $\angle Y$ and $\angle Z$ are not supplementary. If they were, $\overline{WZ}$ and $\overline{XY}$ would have to be parallel and $WXYZ$ would not be a trapezoid. (They may, however, be congruent.)
12. Given: $\overleftrightarrow{AY} \parallel \overleftrightarrow{CZ}$; $\overrightarrow{AB}$ bisects $\angle XAY$; $\overrightarrow{CD}$ bisects $\angle ACZ$.
Prove: $\overleftrightarrow{AB} \parallel \overleftrightarrow{CD}$
13.

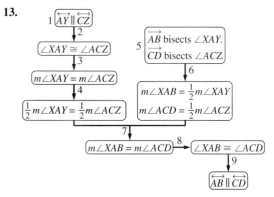

Reasons

1. Given
2. If two ∥ lines are intersected by a transversal, then corresponding ⊿ are ≅.
3. Def. of ≅
4. Multiplication Property
5. Given
6. Def. of bisector
7. Substitution Property
8. Def. of ≅
9. If two lines are intersected by a transversal and corresponding ⊿ are ≅, then the lines are ∥.

14. If two lines are intersected by a transversal and alternate interior angles are congruent, then the lines are parallel. **15.** In a plane, two lines perpendicular to the same line are parallel. **16.** If two lines are intersected by a transversal and corresponding angles are congruent, then the lines are parallel. **17.** If two lines are intersected by a transversal and same-side interior angles are supplementary, then the lines are parallel. **18.** If two lines are both parallel to a third line, then they are parallel to each other. **19.** j is parallel to l; thus, l must be parallel to k.

CHAPTER 6

Page 281   Checking Key Concepts

1. Yes. **3.** Yes. **5.** It is between 6 ft long and 32 ft long. **7.** It is less than 16 m long, and like every length, greater than 0 m long.

Pages 282–284   Exercises and Applications

1. Yes. **3.** Yes. **5.** No. **7.** It is between 1 ft long and 7 ft long. **9.** It is between 3 cm long and 21 cm long. **11.** It is between $\frac{1}{2}$ in. long and 14 in. long. **13.** $5 < AB < 27$ **15.** $4 < AB < 6$ **17.** $AB > 4$ **19.** B **21, 23.** Answers may vary. Examples are given. **21.** Yes; the three points are not collinear. **23.** No; all three points lie on the line $y = x$. **27.** Answers may vary. Examples are given.

Student	Lengths	Triangle?
1	7, 5, 2	No
1	5, 4, 2	Yes
1	2, 2, 6	No
2	3, 11, 6	No
2	4, 10, 3	No
2	10, 8, 1	No
3	9, 2, 1	No
3	11, 2, 1	No
3	6, 1, 10	No
4	7, 2, 11	No
4	7, 8, 9	Yes
4	9, 11, 10	Yes

41.

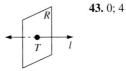

43. 0; 4

Page 288   Checking Key Concepts

1. △CBA **3. a.** 18 **b.** 5

Pages 289–291   Exercises and Applications

1. Yes; △FME ≅ △ELF. **3.** Yes; △WXY ≅ △WVU. **5.** $\overline{BC}$ **7.** $WC = 12$ **9.** 21 **13.** △CEF **15. a.** No. **b.** Yes. **25.** Reflexive Property

Page 291   Assess Your Progress

1. No. **2.** No. **3.** Yes. **4.** It is greater than 3 in. and less than 5 in. **5.** It is greater than 2 ft and less than 6 ft. **6.** It is greater than 6 cm and less than 30 cm. **7.** △AJC ≅ △BJG and △ACG ≅ △BGC **8. a.** 15 **b.** 3.5

Page 295   Checking Key Concepts

1. Yes; SAS. **3.** Yes; SSS. **5.** $\overline{RQ}$ ≅ $\overline{RS}$

7.

Statements	Reasons
1. $\overline{XY}$ ≅ $\overline{ZW}$; $\overline{XY}$ ∥ $\overline{ZW}$	1. Given
2. ∠ WZX ≅ ∠ YXZ	2. If two ∥ lines are intersected by a transversal, then alternate interior ⊿ are ≅.
3. XZ ≅ ZX	3. Reflexive Property
4. △XYZ ≅ △ZWX	4. SAS Postulate

Pages 295–298   Exercises and Applications

1. ∠ACB **5.** Yes; △ABD ≅ △CBD; SAS Postulate. **7.** No. **9.** Yes; △VWT is congruent to △SWR; SAS Postulate

11.

Statements	Reasons
1. $\overline{AB}$ ≅ $\overline{CB}$; $\overline{BE}$ ≅ $\overline{BD}$	1. Given
2. ∠ ABE ≅ ∠ CBD	2. Vertical ⊿ are ≅.
3. △ABE ≅ △CBD	3. SAS Postulate

19. No. **21.** Answers may vary. An example is given. It is given that $\overline{PQ}$ ≅ $\overline{RQ}$ and that $\overline{PS}$ ≅ $\overline{RS}$. By the Reflexive Property, $\overline{QS}$ ≅ $\overline{QS}$. Then, by the SSS Postulate, △PQS ≅ △RQS.

23. a.

The coordinates of the vertices of the image of △ABC are (1, −7), (6, −3), and (3, −1).

b.

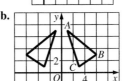

The coordinates of the vertices of the image of △ABC are (−1, 7), (−6, 3), and (−3, 1).

c. Yes; the lengths of the sides of each image of $\triangle ABC$ are the same as the lengths of the sides of $\triangle ABC$, so by the SSS Postulate, each image is congruent to $\triangle ABC$.
25. $\triangle MON \cong \triangle QOP$ and $\triangle MQN \cong \triangle QMP$ **27.** $1:120$
29. $x° = 38°$

Page 302   Checking Key Concepts

1. In any $\triangle ABC$, $\overline{AB}$ is the side included between $\angle A$ and $\angle B$. The non-included sides for $\angle A$ and $\angle B$ are $\overline{AC}$ and $\overline{BC}$.   **3.** SAS Postulate   **5.** ASA Postulate, AAS Theorem
7. none

Pages 303–305   Exercises and Applications

1. AAS Theorem, ASA Postulate   **3.** SAS Postulate
5. SSS Postulate   **7.** $\triangle PRU \cong \triangle PTQ$; AAS Theorem, ASA Postulate   **9.** $\triangle XWZ \cong \triangle YZW$; AAS Theorem, ASA Postulate   **13.** Answers may vary. An example is given. Since $JKLM$ is a rectangle, $\angle M$ and $\angle L$ are right angles. Therefore, $\triangle JMP$ and $\triangle KLP$ are right triangles. A rectangle is a parallelogram, and the opposite sides of a parallelogram are congruent, so $\overline{JM} \cong \overline{KL}$. $\overline{JP} \cong \overline{KP}$ (given), so $\triangle JMP \cong \triangle KLP$ by the HL Theorem.   **15.** Since $\overline{PQ} \parallel \overline{VS}$, $\angle P \cong \angle SVT$. (If two $\parallel$ lines are intersected by a transversal, then corresponding $\angle$s are $\cong$.) Since $\overline{QU} \parallel \overline{ST}$, $\angle QUP \cong \angle T$. (If two $\parallel$ lines are intersected by a transversal, then corresponding $\angle$s are $\cong$.) $\overline{PQ} \cong \overline{VS}$ (Given), so $\triangle PQU \cong \triangle VST$ by the AAS Theorem.   **17. a.** Answers may vary. There may be one or two ways to do Step 4. When there are two ways, only one of them results in a triangle that is congruent to $\triangle ABC$.   **b.** There is just one way to do Step 4, and that way gives a triangle that is congruent to $\triangle ABC$.   **21.** $\overline{JL}$
23. $\overline{CA}$ **25.** $128\frac{4}{7}°$; $51\frac{3}{7}°$

Page 309   Checking Key Concepts

1. $\triangle DGH$ and $\triangle EGF$ **3.** $\triangle DEF$ and $\triangle EDH$ **5.** $\triangle XVY$ and $\triangle ZVW$ or $\triangle WVX$ and $\triangle YVZ$ **7.** P, R **9.** isosceles

Pages 309–312   Exercises and Applications

1. $\triangle ABE$ and $\triangle DCE$ **3.** $\triangle ABC$ and $\triangle DCB$ or $\triangle ABD$ and $\triangle DCA$

5.

Statements	Reasons
1. $\overline{PS} \parallel \overline{QR}$	1. Given
2. $\angle PSQ \cong \angle RQS$	2. If two $\parallel$ lines are intersected by a transversal, then alternate interior $\angle$s are $\cong$.
3. $\angle P \cong \angle R$	3. Given
4. $\overline{QS} \cong \overline{QS}$	4. Reflexive Property
5. $\triangle PSQ \cong \triangle RQS$	5. AAS Theorem
6. $\overline{PQ} \cong \overline{RS}$	6. Def. of $\cong \triangle$s

7. (1) $\overline{PB}$; exactly one; (2) perpendicular bisector; (3) SAS Postulate; (4) $PA = PB$; Definition of congruent triangles
9. 5

17.

Statements	Reasons
1. $\angle WQX \cong \angle YQZ$	1. Vertical $\angle$s are $\cong$.
2. $\overline{WQ} \cong \overline{YQ}$ and $\overline{XQ} \cong \overline{ZQ}$	2. Given
3. $\triangle WQX \cong \triangle YQZ$	3. SAS Postulate
4. $\angle W \cong \angle Y$	4. Def. of $\cong \triangle$s
5. $\overline{WX} \parallel \overline{YZ}$	5. If two lines are intersected by a transversal and alternate interior $\angle$s are $\cong$, the lines are $\parallel$.

19. D   **21.** 1.8   **23.** 2.7   **25.** 86°

Page 312   Assess Your Progress

1. No.   **2.** Yes; SSS Postulate or SAS Postulate.   **3.** Yes; SSS Postulate.   **4.** Yes; ASA Postulate or AAS Theorem.
5. Yes; HL Theorem or SSS Postulate.   **6.** Yes; ASA Postulate or AAS Theorem.   **7.** $\triangle YVZ$ and $\triangle XVW$
8. $\triangle XWZ$ and $\triangle YZW$ **9.** Answers may vary. An example is given.

Page 315   Checking Key Concepts

1. 9   **3.** 42°

Pages 316–318   Exercises and Applications

3. $\angle PKD$ and $\angle PDK$ **5.** the roofs in Lithuania, Japan, and Germany   **7.** (1) $\overline{AB} \cong \overline{AC}$; (2) Reflexive Property; (3) SAS Postulate; (4) $\angle B \cong \angle C$ and Def. of $\cong \triangle$s   **9.** 2
11. 35°   **13.** 5.5   **15.** 12; 120°   **19.** As shown in Ex. 7 on page 316, if you could pick up $\triangle PQR$ and flip it over so that $\overline{PQ}$ lands on $\overline{PR}$, the resulting triangle would be congruent to $\triangle PQR$. "Flipping the triangle over" is equivalent to reflecting $\triangle PQR$ over the perpendicular bisector of $\overline{QR}$, which is the line of symmetry. (This line also bisects $\angle P$.)
21. Draw the bisector of $\angle XZY$. (Angle Bisector Construction) Let P be the intersection of the bisector and $\overline{XY}$. $\angle XZP \cong \angle YZP$ by the definition of angle bisector. $\angle X \cong \angle Y$ (given) and $\overline{ZP} \cong \overline{ZP}$ (Reflexive Property). $\triangle XPZ \cong \triangle YPZ$ (AAS Theorem), and so $\overline{XZ} \cong \overline{YZ}$ by the definition of congruent triangles.   **25.** If a point is on the perpendicular bisector of a segment, then the point is equidistant from the endpoints of the segment.
27. $\overline{DC}$ and $\overline{DE}$

Page 321   Checking Key Concepts

1. $\overline{EQ}$ **3.** Yes; if $\overline{ED} \cong \overline{EF}$, then $\overline{EQ}$ is both an altitude from E and the bisector of $\angle DEF$.   **5.** 11 cm

1. a. 90°   **b.** 14   **3. a.** In an isosceles triangle, the altitude to the base of the triangle is also the bisector of the vertex angle.   **b.** Answers may vary. An example is given. Plan of Proof: Use the diagram on page 320. Use the HL Theorem to prove that $\triangle DAB \cong \triangle DAC$. Then $\angle ADB \cong \angle ADC$, so $\overrightarrow{DA}$ bisects $\angle BDC$.   **7.** $y = -x + 6$ **11.** 8   **13. a.** No; each angle bisector lies inside the angle it bisects, that is, inside the triangle in which it is drawn.   **b.** The circle touches each side of the triangle in just one point; the incenter of a triangle is equidistant from its sides.   **15.** Yes; if a triangle is obtuse, then the lines containing the altitudes intersect outside the triangle.   **19. a.** 130°   **b.** 22   **21. a.** 12   **b.** 8

Page 325   Assess Your Progress

1. 6   **2.** 62°   **3.** line n **4.** $\overline{JS}$ **5.** 47°   **6.** 94

Pages 330–331   Chapter 6 Assessment

1. A   **2.** C, E   **3.** C, D, F   **4.** E   **5.** B   **6.** No.
7. It is between 14 cm long and 2 cm long.   **8.** $\overline{AB}$; $\overline{AC}$
9. $\triangle DCG$, $\triangle DFG$; $\triangle BCD$, $\triangle BFD$; $\triangle BCG$, $\triangle BFG$
10. polygon $KVMR$ **11.** $x = 15$; $y = \dfrac{25}{7}$; $z = 22$
12. $\triangle XTY \cong \triangle ZTW$; SAS Postulate
13. $\triangle TUV \cong \triangle TWV$; SSS Postulate   **14.** none
15. $\triangle SPR \cong \triangle QRP$; ASA Postulate or AAS Theorem
16. none   **17.** $\triangle EKF \cong \triangle GHF$; ASA Postulate or AAS Theorem

19.

Statements	Reasons
1. Line m is the perpendicular bisector of $\overline{CD}$.	1. Given
2. $\overline{FC} \cong \overline{FD}$	2. If a point is on the perpendicular bisector of a segement, then the point is equidistant from the endpoints of the segment.
3. $\angle C \cong \angle D$	3. If two sides of a triangle are $\cong$, then the $\angle$ opposite the sides are $\cong$.

20. 120°   **21. a.** $\overline{KN}$ **b.** $\overline{JM}$ **22.** $\overline{KN}$ is the perpendicular bisector of $\overline{JL}$.

Pages 332–333   Algebra Review/Preview

1. 37.68   **3.** 1.84   **5.** 6   **7.** 400 ft   **9.** $l = \dfrac{A}{w}$ **11.** $w = \dfrac{1}{2}P - l$

13. $r = \sqrt[3]{\dfrac{3V}{4\pi}}$ **15.** $\dfrac{1}{8}$ **17.** $\dfrac{25}{2}$ **19.** $\dfrac{10}{21}$ **21.** $3\dfrac{1}{3}$ **23.** 6

25. $(-3, 4)$ **27.** no solution   **29.** $\left(-1\dfrac{1}{18}, \dfrac{5}{12}\right)$

CHAPTERS 4–6

Pages 334–335   Cumulative Assessment

1. $5\sqrt{5} \approx 11.2$; $\left(\dfrac{1}{2}, 0\right)$ **3.** $y = \dfrac{1}{2}x - 4$
5. $(x - 3)^2 + (y + 5)^2 = 125$

7.

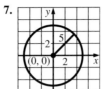

11. 0; $RA = \sqrt{(-a - 0)^2 + (0 - b)^2} = \sqrt{a^2 + b^2}$; $RB = \sqrt{(a - 0)^2 + (0 - b)^2} = \sqrt{a^2 + b^2}$; $RA = RB$ **13.** 25

15.

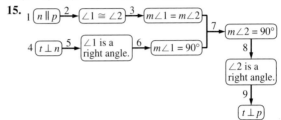

Reasons

1. Given
2. If two ∥ lines are intersected by a transversal, then corresponding $\angle$s are $\cong$.
3. Def. of congruent angles
4. Given
5. Def. of perpendicular lines
6. Def. of right angle
7. Substitution Property
8. Def. of right angle
9. Def. of perpendicular lines

17. $\angle 1 \cong \angle 2$ (given), $\angle 2 \cong \angle 3$ (vertical angles), and $\angle 3 \cong \angle 4$ (given). Thus, $\angle 1 \cong \angle 4$ (Transitive Property), and so $\overleftrightarrow{AB} \parallel \overleftrightarrow{DE}$ (alternate interior angles are congruent).
19. a. $\overline{MR}$ **b.** $\angle Z$ **c.** $\triangle MAR$ **21.** SSS Postulate
23. Answers may vary. An example is given. Since $\angle 1 \cong \angle 2$, $\overline{PT} \cong \overline{ST}$ (converse of the Isosceles Triangle Theorem). $\angle 3 \cong \angle 4$ (given). $\triangle PTR \cong \triangle STQ$ (AAS Theorem), so $\overline{PR} \cong \overline{SQ}$ (def. of congruent triangles). Therefore, $\overline{PQ} \cong \overline{RS}$ (Segment Addition Postulate and Subtraction Property).
25. never

CHAPTER 7

Page 342   Checking Key Concepts

1. Answers may vary. An example is given.

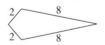

3. a. 7.5   **b.** 15

Pages 343–345   Exercises and Applications

1. $AD = 4$; $CD = 5$ **3. a.** $\overline{WX} \cong \overline{WZ}$ because they are radii of the same circle. Similarly, $\overline{YX} \cong \overline{YZ}$.   **b.** Draw two circles with centers A and C and different radii that intersect in two points, B and D. Draw $\overline{AB}$, $\overline{BC}$, $\overline{CD}$, and $\overline{AD}$. $ABCD$ has two pairs of congruent sides, but opposite sides are not

congruent. Two adjacent sides are radii of circle A and two are radii of circle C and A and C have different radii. $ABCD$ is a kite. **5.** If two rectangles are congruent, then their diagonals are congruent. (Given rectangle $PQRS \cong$ rectangle $TUVW$, if you draw diagonals $\overline{PR}$ and $\overline{TV}$, $\triangle RQP \cong \triangle VUT$ by the SAS Postulate.) Then since the diagonals of any rectangle are congruent, $\overline{BD}$, $\overline{BF}$, $\overline{HD}$, and $\overline{HF}$ are congruent and $DBFH$ is equilateral and, therefore, a rhombus.
7. a. $\sqrt{89} \approx 9.4$ **b.** 5 **9. a.** 3 **b.** 6
11. Answers may vary. An example is given.
Given: $ABCD$ is a rhombus.
Prove: $\overline{AC} \perp \overline{BD}$

$ABCD$ is a rhombus, so $ABCD$ is a parallelogram. Then $\overline{EA} \cong \overline{EC}$ since the diagonals of a parallelogram bisect each other. $\triangle ABE \cong \triangle CBE$ by the SSS Postulate. $\angle AEB \cong \angle CEB$, so $\overline{AC}$ and $\overline{BD}$ form congruent adjacent angles; therefore, $\overline{AC} \perp \overline{BD}$.
13. Construct a line through A parallel to $\overline{BC}$ and a line through C parallel to $\overline{AB}$. (Through a point not on a line, there is exactly one line parallel to the given line.) Label E, the point where the constructed lines intersect. $ABCE$ is a rectangle. Since $ABCE$ is a parallelogram, its diagonals bisect each other. Since $ABCE$ is a rectangle, its diagonals are congruent. Then $AD = BD = CD$. **15.** 8 **17.** 5
19. $\sqrt{73} \approx 8.5$ **31. a.** 9 **b.** 75° **33. a.** 90° **b.** 28°

Page 349 Checking Key Concepts

1. $VWXY$ is a parallelogram because $\overline{XY}$ and $\overline{WV}$ are both parallel and congruent. **a.** 14 **b.** 6 **c.** 10.3 **3.** Answers may vary. An example is given. Let $PQRS$ be a parallelogram and point T be the intersection of the diagonals. Since the diagonals bisect each other and vertical angles are congruent, $\triangle PTQ \cong \triangle RTS$ and $\triangle PTS \cong \triangle RTQ$ (SAS Postulate). Then $\overline{PQ} \cong \overline{SR}$ and $\overline{PS} \cong \overline{QR}$, so $PQRS$ is a parallelogram.

Pages 349–352 Exercises and Applications

1. $ABCD$ is a parallogram because both pairs of opposite angles are congruent. **a.** 130° **b.** 3 **c.** 5
3. $ABCD$ must be a parallelogram because both pairs of opposite sides are congruent.
5. $FGHI$ may or may not be a parallelogram. In the example, $FGHI$ is not a parallelogram.
17, 19. Answers may vary. Examples are given.
17. (4, 2) is the midpoint of both $\overline{AC}$ and $\overline{DB}$. $ABCD$ is a parallelogram because its diagonals bisect each other.

19. 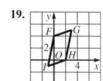 slope of $\overline{FG}$ = slope of $\overline{IH} = \frac{1}{3}$; $FG = IH = \sqrt{10}$; $FGHI$ is a parallelogram because one pair of sides, $\overline{FG}$ and $\overline{IH}$, are both congruent and parallel.
21.  Given: $JKLM$; $\angle J \cong \angle L$, $\angle K \cong \angle M$
Prove: $JKLM$ is a parallelogram.

Statements	Reasons
1. $\angle J \cong \angle L$ and $\angle K \cong \angle M$	1. Given
2. $m\angle J = m\angle L$ and $m\angle K = m\angle M$	2. Def. of $\cong$
3. $m\angle J + m\angle L + m\angle K + m\angle M = 360°$	3. The sum of the angle measures of a quad. is 360°.
4. $2m\angle K + 2m\angle L = 360°$ $2m\angle L + 2m\angle M = 360°$	4. Substitution Property
5. $m\angle K + m\angle L = 180°$ $m\angle L + m\angle M = 180°$	5. Algebra
6. $\overline{JK} \parallel \overline{ML}$ and $\overline{KL} \parallel \overline{JM}$	6. If two lines are intersected by a transversal and the same-side interior $\angle$ are supplementary, then the lines are $\parallel$.
7. $JKLM$ is a parallelogram.	7. Def. of parallelogram

23. $JKLM$ is a kite with $\overline{JK} \cong \overline{JM}$. Suppose that $\overline{JK}$ is congruent to $\overline{KL}$ or to $\overline{ML}$ as well. Both possibilities present a contradiction. If $\overline{JK} \cong \overline{ML}$, then two opposite sides of $JKLM$ are congruent and $JKLM$ is not a kite. Similarly, if $\overline{JK} \cong \overline{KL}$, then $\overline{JM} \cong \overline{KL}$ and $JKLM$ is not a kite. Therefore, $\overline{JK}$ is congruent to exactly one other side of the kite.
27. a. $(c, 0)$ **b.** b **c.** $a + c$ **d.** If one pair of sides of a quadrilateral is both parallel and congruent, the quadrilateral is a parallelogram. $\overline{MN}$ and $\overline{OP}$ are both parallel and congruent. **31.** $12\sqrt{2} \approx 17.0$ **33.** 6; $(-1, 4)$
35. $\sqrt{13} \approx 3.6$; $\left(3, 2\frac{1}{2}\right)$ **37.** $\sqrt{73} \approx 8.5$; $\left(0, -\frac{1}{2}\right)$

Page 355 Checking Key Concepts

1. parallelogram **3.** square

Pages 356–359 Exercises and Applications

1. The quadrilaterals are not necessarily parallelograms.
3. square **5.** rhombus **7.** rectangle **9.** rectangle
15, 17, 19. Answers may vary. Numerical examples should follow the given form, where each variable represents a positive number. **15.** $(0, a)$, $(0, -a)$, $(-b, 0)$, $(b, 0)$ $(a \neq b)$
17. If the diagonals of a rectangle are perpendicular, the rectangle must be a square. **19.** If the diagonals of a parallelogram are perpendicular, the parallelogram must be a rhombus. See Exs. 15 and 16. (The rhombus may or may not be a square.)

21. Given: quadrilateral $ABCD$; $\overline{AC}$ is the perpendicular bisector of $\overline{DB}$; $\overline{DB}$ is not the perpendicular bisector of $\overline{AC}$.

Prove: $ABCD$ is a kite.

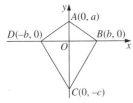

$AD = \sqrt{(-b-0)^2 + (0-a)^2} = \sqrt{b^2 + a^2}$;
$AB = \sqrt{(b-0)^2 + (0-a)^2} = \sqrt{b^2 + a^2}$;
$CD = \sqrt{(-b-0)^2 + (0-(-c))^2} = \sqrt{b^2 + c^2}$;
$CB = \sqrt{(b-0)^2 + (0-(-c))^2} = \sqrt{b^2 + c^2}$;
Since $\overline{DB}$ is not the perpendicular bisector of $\overline{AC}$, $|a| \neq |c|$, so $a^2 \neq c^2$ and $\sqrt{b^2 + a^2} \neq \sqrt{b^2 + c^2}$. Then $ABCD$ has two pairs of congruent sides but opposite sides are not congruent. $ABCD$ is a kite. **27.** $\overline{AC} \cong \overline{DB}$ (They are both diameters of the circle.)

29. a.

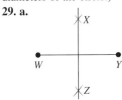

Draw a segment $\overline{WY}$. When you construct its perpendicular bisector, let X and Y be the points where the two pairs of arcs intersect. $\overline{WY}$ and $\overline{XZ}$ are perpendicular bisectors of each other. (You may choose different points X and Y by drawing any two arcs with the same radius and center at the midpoint of $\overline{WY}$.)
b. rhombus; The diagonals bisect each other, so $WXYZ$ is a parallelogram. The diagonals are perpendicular, so $WXYZ$ is a rhombus. **35.** $50°$ **37.** $y = -3x + 20$

Page 359 Assess Your Progress

1. a. 5 **b.** 10 **c.** $2\sqrt{21} \approx 9.2$ **2. a.** 15 **b.** 9 **c.** 18
3. a. 8 **b.** 16 **c.** $8\sqrt{2} \approx 11.3$ **4.** Suppose that $\angle W \cong \angle Y$. Since it is given that $\overline{WZ} \cong \overline{WX}$ and $\overline{YZ} \cong \overline{YX}$, and $\overline{WY} \cong \overline{WY}$ by the Reflexive Property, $\triangle WZY \cong \triangle WXY$ by the SSS Postulate. Then $\angle Z \cong \angle X$. However, if $\angle W \cong \angle Y$, then $WXYZ$ is a parallelogram since both pairs of opposite angles are congruent. This is a contradiction, since opposite sides of a kite are not parallel. The assumption that $\angle W \cong \angle Y$ must not be true. So $\angle W \not\cong \angle Y$. **5–7.** Answers may vary. Examples are given. **5.** rhombus; Both pairs of opposite sides are parallel, so the quadrilateral is a parallelogram by definition; it is a rhombus because the diagonals are perpendicular. **6.** rectangle; The diagonals bisect each other, so the quadrilateral is a parallelogram; it is a rectangle because the diagonals are congruent. **7.** square; It is given that the quadrilateral is equiangular. Then both pairs of opposite angles are congruent and the quadrilateral is an equiangular parallelogram, that is, a rectangle. Since two consecutive sides are congruent, it is a rhombus. A rectangular rhombus is a square.

Page 363 Checking Key Concepts

1. 12 **3.** 64 **5.** 234 ft^2 **7.** $56\sqrt{11} \approx 185.7$ in.^{2}

Pages 364–366 Exercises and Applications

1. 180.5 **3.** 16 **5.** 36 **11.** 18 **13.** 60 in.^{2}
15. $m = \frac{1}{2}(b_1 + b_2)$ **17.** rectangle; 21 **19.** triangle; 28
21. parallelogram; 15 **23.** If a rectangle and a parallelogram that is not a rectangle have the same base and height, they have the same area. To see why, picture cutting the parallelogram along the perpendicular segment and positioning the resulting triangle as shown.

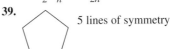

25.

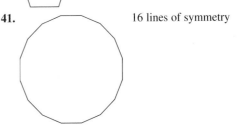

$ABCD$ is a rhombus; it is equilateral and $\overline{AC}$ and $\overline{BD}$ are perpendicular and bisect each other. Then $\overline{AC}$ and $\overline{BD}$ divide the rhombus into four congruent right triangles, each with area $\frac{1}{2}ab$. The area of the rhombus is $2ab$.
$AC = \sqrt{(0-0)^2 + (a-(-a))^2} = \sqrt{4a^2} = 2a$,
$BD = \sqrt{(b-(-b))^2 + (0-0)^2} = \sqrt{4b^2} = 2b$,
so $\frac{1}{2}AC \cdot BD = \frac{1}{2}(2a)(2b) = 2ab$. Then the area of the rhombus is half the product of the lengths of the diagonals.

27. a. 48; 48 **b.** Let n be any positive integer. To divide a triangle into n smaller triangles whose areas are equal, divide the base of the triangle into n congruent segments. The triangles all have the same height, h, as the original triangle. The base of each is $\frac{1}{n}$ times that of the original triangle, so each has area $\frac{1}{2} \cdot \frac{b}{n} \cdot h = \frac{bh}{2n}$. **37.** rhombus

39.

5 lines of symmetry

41.

16 lines of symmetry

Page 370 Checking Key Concepts

1. 36 **3.** $65\sqrt{90.21} \approx 617$ **5.** $\frac{169\pi}{4} \approx 133$

Pages 371–373 Exercises and Applications

1. 480 **3.** $98\sqrt{3} \approx 170$ **5.** $121\pi \approx 380$
7. $\frac{25\pi}{2} - 25 \approx 14$ **9.** $7\pi \approx 22$

15. Each side was constructed using the same compass radius. **17.** Connect only vertices A, C, and E or only vertices B, D, and F. To show that $\triangle ACE$ is equilateral, draw $\overline{AC}$, $\overline{AE}$, and $\overline{EC}$. Since $ABCDEF$ is a regular hexagon, $\triangle FEA$, $\triangle BAC$, and $\triangle DCE$ are congruent by the SAS Postulate, and $\overline{AC} \cong \overline{AE} \cong \overline{EC}$. **21.** about 13.1 **23.** 359.0 **29.** 49 **31, 33.** Answers may vary. Examples are given.

31. **33.**

Page 378 Checking Key Concepts

1. 147; 182 **3.** $576\sqrt{3} \approx 997.7$; $192\sqrt{3} + 288 \approx 620.6$

5. $96\pi \approx 301.6$; $80\pi \approx 251.3$

Pages 378–381 Exercises and Applications

1. $\frac{891\pi}{4} \approx 699.8$; $\frac{279\pi}{2} \approx 438.3$ **3.** 307.2; 281.6

5. 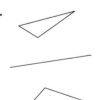 $81\sqrt{3} \approx 140.3$; $\frac{81\sqrt{3} + 216}{2} \approx 178.1$

9. $\frac{45\pi}{2} \approx 70.7 \text{ m}^3$ **11.** 60 in.^{3}

21. $V = 24$ **23.** $V = 192$

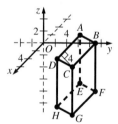 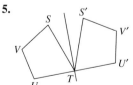

25. **27.**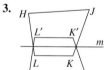

Page 381 Assess Your Progress

1. $2\sqrt{21} \approx 9.2$ **2.** 32.5 **3.** $18\sqrt{10} \approx 56.9$
4. $54\sqrt{3} \approx 93.5$ **5.** $\frac{65\sqrt{78.75}}{2} \approx 288.4$ **6.** $64\pi \approx 201.1$
7. $128\pi \approx 402.1$; $96\pi \approx 301.6$ **8.** 1750; 990
9. $240\sqrt{5} \approx 536.7$; $92\sqrt{5} + 300 \approx 505.7$

Pages 386–387 Chapter 7 Assessment

2. indirect **3.** base **5. a.** 2 **b.** 100° **6. a.** 10 **b.** 8
7. a. 6 **b.** $3\sqrt{2}$
8. 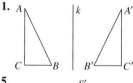 If both pairs of opposite angles of a quadrilateral are congruent, the quadrilateral is a parallelogram.
9. not necessarily
10. If both pairs of opposite sides of a quadrilateral are congruent, the quadrilateral is a parallelogram.
11. Answers may vary. An example is given. Suppose that a kite is a parallelogram. Then both pairs of opposite sides are congruent. But opposite sides of a kite are not congruent, so a kite is not a parallelogram. **12.** rectangle **13.** square
14. rhombus

15.

Statements	Reasons
1. $\overline{AB} \parallel \overline{DC}$; $\overline{AB} \cong \overline{DC}$	1. Given
2. $ABCD$ is a parallelogram.	2. If one pair of opposite sides of a quadrilateral is both parallel and congruent, the quadrilateral is a parallelogram.
3. $\overline{AC}$ and $\overline{BD}$ bisect each other.	3. The diagonals of a parallelogram bisect each other.
4. $\overline{EA} \cong \overline{EC}$; $\overline{EB} \cong \overline{ED}$	4. Def. of bisector
5. $\angle AED \cong \angle CEB$	5. Vertical $\angle\!s$ are $\cong$.
6. $\triangle AED \cong \triangle CEB$	6. SAS Postulate

16. 28 **17.** 16 **18.** 36 **19–24.** Answers are given to the nearest tenth. **19.** 158.7 **20.** 8.8 **21.** 160.2
22. a. 53.4 ft^2 **b.** 1084.5 ft^3 **23. a.** 21 cm^3 **b.** 42.3 cm^2
24. a. 332.6 in.^{3} **b.** 192 in.^{2}

CHAPTER 8

Page 394 Checking Key Concepts

1. $\overline{P'Q'}$; $\overline{R'S'}$ **3.** $\triangle S'P'Q'$ **5.** quadrilateral $PQRS$

Pages 394–397 Exercises and Applications

1. 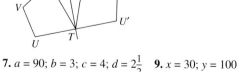 **3.**

5.

7. $a = 90$; $b = 3$; $c = 4$; $d = 2\frac{1}{2}$ **9.** $x = 30$; $y = 100$

15. H, O, T, U, V, W, X, Y; Answers may vary. Examples are given.

```
Y   W   M   M   Y   W
O   A   A   Y   A   H
U   X   T   T   M   I
T       H   H       M
H
```

17. F, G, J, L, N, P, Q, R, S, Z   **27.** $18\pi \approx 56.5$
29. $\triangle DCE$ **31.** $\triangle GLF$ **33.** $y = 2x - 2$ **35.** $y = 7$

Page 401   Checking Key Concepts

1.
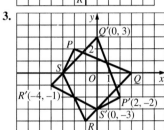

3.
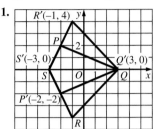

Pages 401–404   Exercises and Applications

1. $N'(-2, 3)$; $O'(0, 0)$; $P'(-4, 0)$
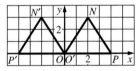

3. $H'(3, 4)$; $J'(-1, 2)$; $K'(1, -2)$

5. $Q'(-3, 0)$; $R'(1, -2)$; $S'(1, 2)$
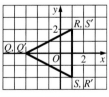

7. $M'(4, 0)$; $N'(2, 3)$; $O'(0, 0)$

9. $D'(2, 0)$; $E'(4, 2)$; $F'(4, 4)$; $G'(-1, 1)$
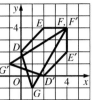

17. $(8, 0)$
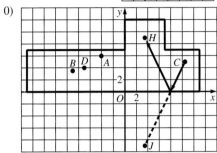

23.
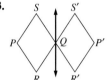

25.
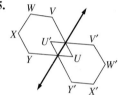

29, 31. Answers may vary. Examples are given.
29.

31.

Page 407   Checking Key Concepts

1.
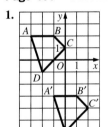

3.
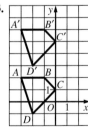

5. $(a, b) \rightarrow (a - 3, b + 4)$

Pages 408–411   Exercises and Applications

1.
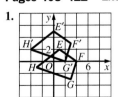

3.
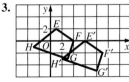

5. $(a, b) \rightarrow (a - 7, b + 7)$ **7.** $(a, b) \rightarrow (a, b + 4)$ **9.** Every point moves to the right 3 units; $(a, b) \rightarrow (a + 3, b)$.
11. Every point moves up 5 units; $(a, b) \rightarrow (a, b + 5)$.
13. D   **17.** The top flight can be translated down by a distance equal to the distance between the floors.

19. Answers may vary. An example is given. lines q and s
25. a, b. Answers may vary. An example is given. $\triangle ABC$ has vertices $A(0, 1)$, $B(2, 4)$, and $C(5, 5)$. **a.** $A'(3, 6)$; $B'(5, 9)$; $C'(8, 10)$ **b.** $A''(-1, 8)$; $B''(1, 11)$; $C''(4, 12)$ **c.** $(a, b) \rightarrow (a - 1, b + 7)$ **27.** acute **29.** Answers may vary. An example is given.

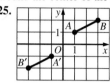

Page 411 Assess Your Progress

1. $L'M'$; $K'L'$ **2.** $\triangle K'M'N'$
3.

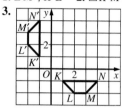

4.

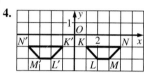

5. $(a, b) \rightarrow (a + 5, b - 7)$ **6.** $(a, b) \rightarrow (a - 3, b + 4)$

Page 414 Checking Key Concepts

1. $\triangle DEF$ **3.** $\triangle GHI$ **5.** $D(-2, -1)$; $E(-4, 1)$; $F(-1, 4)$
7. $J(2, 1)$; $K(4, -1)$; $L(1, -4)$

Pages 415–418 Exercises and Applications

1, 3. Estimates may vary. **1.** about $90°$ **3.** about $90°$
5.

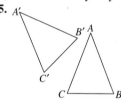

7.

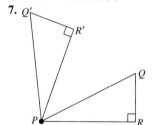

9.

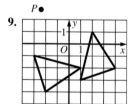

11. Yes; $180°$. **13.** No.
17.

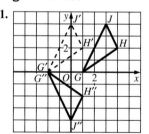

19, 21. The center of each rotation is the center of the square. **19.** $90°$ rotation **21.** no rotation
23. a–c. Answers may vary. Students may choose any two lines that intersect at an angle of $45°$. Reflecting Triangle 1 over one of the lines and reflecting its image over the other has the same effect as rotating Triangle 1 either $90°$ or $-90°$ around the center of the square.
25.

27.

29. a $180°$ rotation around the origin **31.** No; a $90°$ rotation of quadrant D does not have the same effect as a reflection of quadrant D over the x-axis.
35. $(a, b) \rightarrow (a - 3, b + 6)$ **37.** $\triangle DFG$ and $\triangle EFH$
39. $\triangle ABH$ and $\triangle CBG$
41. $(5, 0)$ **43.** $\left(0, 1\frac{1}{2}\right)$

Page 422 Checking Key Concepts

1. Yes. **3.**

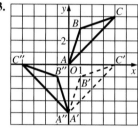

Pages 423–425 Exercises and Applications

1.

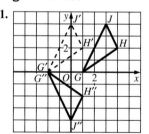

3.

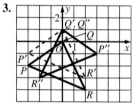

7, 9. Answers may vary. Examples are given.
7.

9.

11. Yes.

15. D **17.** generation 2; generation 3
21.

0   1   2

A translation left 1 unit followed by a reflection over a horizontal line moves generation 0 to generation 2.

23. A glide reflection is a translation followed by a reflection over a line parallel to the translation. Then, if j and k are parallel lines and line m is perpendicular to both j and k, reflection over j, then k, then m has the same effect as a glide reflection.

25. translation length: 0.6 cm (6 mm)

27. translation length: 11 in.

31.

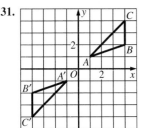

35. The third side is horizontal and is 4 units long. The other endpoint of the segment is (2, 2), so its length is 2, which is half the length of the third side. The segment is horizontal, so it is parallel to the third side.

Page 429 Checking Key Concepts

1.

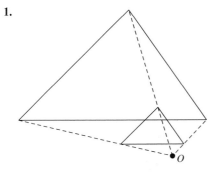

3. B; $\frac{1}{2}$

Pages 430–433 Exercises and Applications

1.

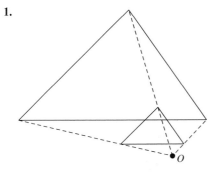

3.

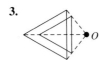

5.

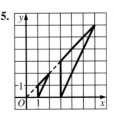

7. a.

Figure	Scale	Image
(2, 0)	3	(6, 0)
(3, 5)	3	(9, 15)
(−2, 4)	3	(−6, 12)

b. (ka, kb)

c.

Figure	Scale (Ex. 5)	Image	Figure	Scale (Ex. 6)	Image
(3, 0)	$\frac{1}{3}$	(1, 0)	(−1, −1)	4	(−4, −4)
(3, 3)	$\frac{1}{3}$	(1, 1)	(−2, 3)	4	(−8, 12)
(6, 6)	$\frac{1}{3}$	(2, 2)	(−4, −3)	4	(−16, −12)

11. O; 4 **13.**

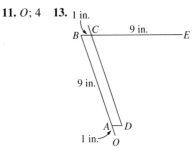

15. center: P; scale factor: $\frac{8}{5}$

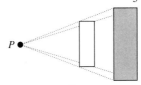

25. $XY = 2\sqrt{2} \approx 2.8$; $XZ = YZ = \sqrt{10} \approx 3.2$

Page 433 Assess Your Progress

1–4. The image of a point P after a 90° rotation around the origin is labeled P'. The image after a 180° rotation around the origin is labeled P''.

1.

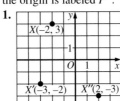

2.

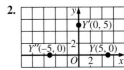

3.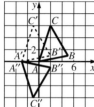

4.

9. a. **b.** (7, 0)

10. $(a, b) \rightarrow (a + 3, b - 1)$ **11.** $(a, b) \rightarrow (a - 1, b - 1)$

5.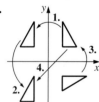

6.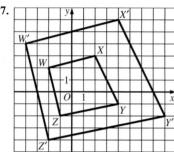

12.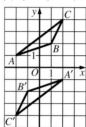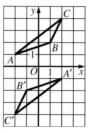

14. 6.2 cm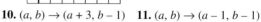

15. 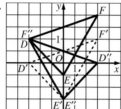 $D''(-1, 4)$; $E''(0, -1)$; $F''(3, -5)$

7.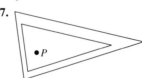

16.

8.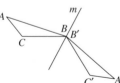

17.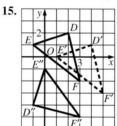

Pages 438–439   Chapter 8 Assessment

1–4.
1. reflection
2. translation
3. rotation
4. glide reflection

5.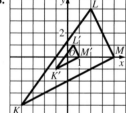

6.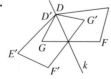

Pages 440–441   Algebra Review/Preview

1. cannot be simplified   **3.** cannot be simplified

5. $7n^2 - 3n$ **7.** $\frac{13}{12}x^2$ **9.** $-1.4x^2$ **11.** $7\sqrt{6}$ **13.** 37

15. $\frac{3\sqrt{5}}{4}$ **17.** 11 **19.** 105 **21.** $36\sqrt{2}$ **23.** ± 10

25. $\pm 3\sqrt{6}$ **27.** $\pm\sqrt{13}$ **29.** $\pm\sqrt{21}$ **31.** $\pm 3\sqrt{2}$ **33.** ± 4
35. 12 **37.** 20.5 **39.** 15 **41.** $\pm\sqrt{15}$ **43.** ± 21 **45.** ± 6

CHAPTER 9

Page 449   Checking Key Concepts

1. Yes; corresponding angles are congruent.
(The measure of the third angle of the larger triangle is $180° - (112° + 30°) = 38°$.) Lengths of corresponding sides are in proportion. $\frac{12}{8} = \frac{15}{10} = \frac{22.5}{15} = \frac{3}{2}$ **3.** $\triangle TUV \sim \triangle CAB$;
$\frac{TU}{CA} = \frac{UV}{AB} = \frac{TV}{CB}$ **5.** $\frac{1}{4}$

7.

8. 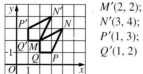 $M'(2, 2)$;
$N'(3, 4)$;
$P'(1, 3)$;
$Q'(1, 2)$

Pages 449–452   Exercises and Applications

1. 8 **3.** 2 **5.** 4 **7.** No; since *PQRS* is a parallelogram, $m \angle P = 120°$ and $m \angle S = m \angle Q = 60°$. Similarly, since *TUVW* is a parallelogram, $m \angle U = 45°$ and $m \angle T = m \angle V = 135°$. Then no two angles of the figures are congruent. **9.** Yes; $\angle AEB \cong \angle CED$ since vertical angles are congruent. Since $\overleftrightarrow{AB} \parallel \overleftrightarrow{DC}$, $\angle B \cong \angle D$ and $\angle A \cong \angle C$. So corresponding angles are congruent. Also, $\frac{AE}{CE} = \frac{EB}{ED} = \frac{AB}{CD} = \frac{3}{4}$. **11.** $x = 10.4$; $y = 90$

13. always

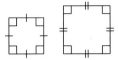

15. sometimes

17. sometimes

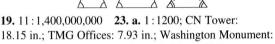

19. $11 : 1,400,000,000$ **23. a.** $1 : 1200$; CN Tower: 18.15 in.; TMG Offices: 7.93 in.; Washington Monument: 5.55 in.; Empire State Building: 12.5 in. **27.** D
31. $1 : 35\frac{3}{16}$; Workers can multiply any length on the model by $35\frac{3}{16}$ to find the appropriate length on the monument.
33. none **35.** none

Page 456   Checking Key Concepts

1. by the SAS Similarity Theorem; Vertical angles are congruent and $\frac{7}{10} = \frac{14}{20}$. **3.** Yes; since the ratio of the lengths of any two corresponding sides is $\frac{5}{8}$, the triangles are similar by the SSS Similarity Theorem. **5.** They are similar. **7.** They are similar by the SSS Similarity Theorem.

Pages 456–457   Exercises and Applications

1. Yes; the triangles are similar by the SAS Similarity Theorem since the two vertical angles are congruent and the sides including the angles are in proportion. **3.** Yes; it can be shown the triangles are similar by the AA Similarity Postulate. Because the two lines are parallel, it can be shown that two angles of one triangle are congruent to two angles of the other triangle. **5.** by the SAS Similarity Theorem; The sides including the congruent angles are in proportion; $\frac{20}{32} = \frac{15}{24} = \frac{5}{8}$. **7.** Since $\overline{BC} \parallel \overline{ED}$ and $\overline{AB} \parallel \overline{DC}$, $\angle BCA \cong \angle DEC$ and $\angle A \cong \angle ECD$. (If two $\parallel$ lines are intersected by a transversal, then alternate interior $\triangle$ are $\cong$.) Then $\triangle ABC \sim \triangle CDE$ by the AA Similarity Postulate.
15. a. Since the right angles are congruent and the sides including the right angles are in proportion $\left(\frac{3}{6} = \frac{4}{8} = \frac{1}{2}\right)$, the triangles are similar. The ratio of the lengths of corresponding sides is $\frac{1}{2}$, so $\frac{5}{c} = \frac{1}{2}$ and $c = 10$. **b.** $d = 50$; $e = 20$; $f = 24$

---

17. Answers may vary. An example is given.

Statements	Reasons
1. $\overline{AB} \perp \overline{AE}$ and $\overline{ED} \perp \overline{AE}$.	1. Given
2. $\overline{AB} \parallel \overline{ED}$	2. In a plane, two lines $\perp$ to the same line are $\parallel$.
3. $\angle A \cong \angle E$ and $\angle B \cong \angle D$.	3. If two $\parallel$ lines are intersected by a transversal, alternate interior $\triangle$ are $\cong$.
4. $\triangle ABC \sim \triangle EDC$	4. AA Similarity Postulate

25. Sketches may vary. Examples are given.
a. not necessarily

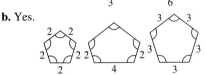

b. Yes.

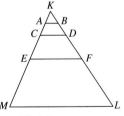

29. Yes; corresponding angles are congruent and the lengths of corresponding sides are in proportion. **31.** No; $\frac{4}{6} \neq \frac{6}{10}$.

Page 464   Checking Key Concepts

1. True. **3.** True. **5. a.** 1 **b.** 3 **c.** 4

Pages 464–467   Exercises and Applications

1. $x = 3\frac{3}{4}$ **3.** $z = 2\frac{2}{9}$ **5.** $\frac{3}{x}$ **7.** 2 cm; 2.25 cm; 1.5 cm
9. a. Answers may vary. An example is given.

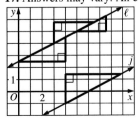

b. $CD = 8$ cm; $EF = 16$ cm; $ML = 32$ cm
c. $EM = 12$ cm; $FL = 24$ cm; $KF = 24$ cm

15. $\frac{AB}{FE}$; $\frac{AC}{FD}$

17. Answers may vary. An example is given.

a. In each pair of triangles, each leg of one triangle is parallel to a leg of the other. The parallel lines can be used to show that two angles of one triangle are congruent to two angles of the other. Then the triangles are similar by the AA Similarity Postulate. (Alternatively, the parallel lines can be used to show that one angle of one triangle is congruent to one angle of the other. Then, since one angle of each triangle is a right angle, the triangles are similar by the AA Similarity Postulate.) **b.** Yes; the reasoning is similar to that described in part (a). However, additional parallel lines (horizontals and verticals) may be needed to show the triangles are similar. **19.** Answers may vary. An example is given.

If $\frac{a}{b} = \frac{c}{d}$, then $1 \div \frac{a}{b} = 1 \div \frac{c}{d}$ or $\frac{b}{a} = \frac{d}{c}$.   **21.** C   **25.** 104 ft^2

27. 60 in.^{2} **29.** 0.001   **31.** 0.04

Page 467   Assess Your Progress

1. 10   **2.** 4.2   **3.** 8   **4.** No.   **5.** Yes.   **6.** $x = 29$; $y = 18\frac{1}{3}$

7. $x = 2\frac{2}{3}$; $y = 5\frac{7}{9}$ **8.** $\frac{5}{2}$ **9.** $\frac{5}{y}$ **10.** $\frac{2+x}{x}$

Page 471   Checking Key Concepts

1. $\frac{4}{3}$ **3.** $\frac{4}{3}$ **5.** 250

Pages 471–474   Exercises and Applications

1. small triangle: $A = 25.2$; large triangle: $P = 37.8$; $A \approx 56.7$ **3.** $EF = 6$; $A = 63$ **5.** 9   **7. a.** $9:25$

b. $3\sqrt{15}$; $5\sqrt{15}$ **9.** not similar   **11.** $\frac{13}{6}$; $\frac{13}{6}$; $\frac{169}{36}$

13. $2:5$; $4:25$ **15.** doubled; doubled; quadrupled
21. a. No; the ratios of the lengths of corresponding sides are not proportional.   **b.** The front faces of the boxes are what you see when you look at the shelf. If they are similar, you might make assumptions about the volumes of the boxes that are not necessarily true.   **25.** The length of a side of the image square is kx. Its area is $(kx)^2 = k^2x^2$.
29.

Statements	Reasons
1. $\overline{AB} \parallel \overline{CD}$	1. Given
2. $\angle ECD \cong \angle EAB$; $\angle ABE \cong \angle EDC$	2. If two $\parallel$ lines are intersected by a transversal, then alternate interior $\angle$ are $\cong$.
3. $\triangle ABE \sim \triangle CDE$	3. AA Similarity Postulate

31. $\frac{32}{43} \approx 74\%$ **33.** $114\frac{2}{7}\% \approx 114.3\%$ **35.** 375%   **37.** $\frac{1}{200}$

39. $\frac{19}{20}$

Page 478   Checking Key Concepts

1. $\frac{16}{49}$ **3.** $\frac{w}{z}$ **5.** $\frac{1}{2}$

Pages 478–481   Exercises and Applications

1. $\frac{6}{25}$ **3.** $\frac{4-\pi}{4} \approx 21.5\%$ **5.** $\frac{1}{2}$ **9.** $\frac{4}{23}$ **11. a.** $\frac{25\pi}{1296} \approx 6\%$

b. $\frac{125\pi}{1296} \approx 30\%$ **19.** B   **23.** $\frac{x+y}{y}$ **25.** right   **27.** obtuse

Page 481   Assess Your Progress

1. large trapezoid: $A = 102$, $P = 49$; small trapezoid: $A = 18\frac{36}{49}$ **2.** $P = 324$; $A = 12{,}285$ **3.** $\frac{3}{5}$ **4.** $\frac{2}{3}$ **5.** $\frac{1}{3}$ **6.** $\frac{4}{9}$

Pages 486–487   Chapter 9 Assessment

3. Yes; AA Postulate or SAS Similarity Theorem.   **4.** No.
5. Yes; SSS Similarity Theorem.   **6.** Yes; SAS Similarity Theorem.   **7.** $r = 14$, $s = 18$ **8.** $a = 4.5$, $b = 10$
9. $n = \frac{40}{9}$; $m = \frac{98}{9}$ **10.** $x = 5.2$; $y = 4.8$ **11.** 6.48
12. $A = 51.2$ m^2, $f = 3.6$ m   **13.** $V \approx 202.2$ cm^3, $h = 12$ cm
14. a. 24 cm^2 **b.** 243 cm^3; $\frac{2^3}{3^3} = \frac{72}{x}$ **15.** $\frac{1}{4}$ **16.** $\frac{4}{49}$

CHAPTERS 7–9

Pages 488–489   Cumulative Assessment

1. Since $\overline{WX} \parallel \overline{YZ}$ and $\overline{WX} \cong \overline{YZ}$, $WXYZ$ must be a parallelogram.   **3.** $60°$; $120°$ **5.** $2\pi \approx 6.28$ **7.** 128
9. Answer is given to the nearest whole number.
$2880\pi \approx 9048$; $768\pi \approx 2413$ **13. a.** $D'(2, 4)$, $E'(-1, -3)$, $F'(-2, 0)$ **b.** $D'(4, -2)$, $E'(-3, 1)$, $F'(0, 2)$ **15.** $D(-6, -2)$, $E(1, 1)$, $F(-2, 2)$ **21.** Yes; SAS Similarity Theorem.
23. $3:5$ **25.** 37.5 square units   **27.** about 0.59

CHAPTER 10

Page 496   Checking Key Concepts

1. $\angle HGF$ **3.** FH **5.** 15

Pages 497–499   Exercises and Applications

1. $\triangle JKL$, $\triangle JMK$, $\triangle KML$ **3.** 15   **5.** $6\sqrt{5}$ **7.** 1   **11.** 18
13. $x = 5\sqrt{5}$; $y = 10\sqrt{5}$ **15.** $u = 5.4$; $v = 9.6$; $w = 7.2$
17. $\triangle ACD \sim \triangle CBD$. (If the altitude is drawn to the hypotenuse of a right triangle, then the two triangles formed are similar to the original triangle and to each other.) Then, since corresponding sides of similar triangles are in proportion, $\frac{BD}{CD} = \frac{CD}{AD}$, or $\frac{e}{d} = \frac{d}{f}$.   **19.** (2) If an altitude is drawn to the hypotenuse of a right triangle, then the length of each leg is the geometric mean of the lengths of the hypotenuse and the segment of the hypotenuse adjacent to that leg. (3) a^2; b^2; a property of proportions (4) $a^2 + b^2$; Addition Property (5) $a^2 + b^2$; Distributive Property (6) c; Segment Addition Postulate (7) Substitution Property
27. $(3, -1)$ **29.** $\left(\frac{7}{2}, -\frac{1}{2}\right)$ **31.** $b = 4\sqrt{6} \approx 9.80$

Page 502   Checking Key Concepts

1. 3; $3\sqrt{2}$ **3.** 5; $5\sqrt{3}$ **5.** $4\sqrt{2}$; $4\sqrt{2}$

Pages 503–506   Exercises and Applications

1. 10   **3.** 4   **5.** $n = \frac{5}{\sqrt{3}}$; $m = \frac{10}{\sqrt{3}}$

11. about 13,317.16 mm^2 **13.** 11,032 mm^2 to the nearest square millimeter   **15.** $6\sqrt{3} \approx 10.4$ in.; The height is the length of an altitude of an equilateral triangle. The altitude determines two 30-60-90 triangles with shorter leg 6 in. long. The altitude is the longer leg of each triangle.
17. about 17 ft 5 in.   **19.** $x = 45$; $y = 12$ **21.** $h = 2\sqrt{2}$; $f = \sqrt{6}$; $g = \sqrt{3}$ **23.** C

25. a. $AB = \frac{1}{2}x$ because $\overline{AC}$ was folded in half; $AD = x$ because D was chosen so that $AD = AC$. **b.** 30-60-90; $AD = x$ and $AB = \frac{1}{2}x$, so by the Pythagorean theorem, $DB = \frac{\sqrt{3}}{2}x$. Then $\triangle ADB$ is a 30-60-90 triangle because it is similar to any 30-60-90 triangle by the SSS Similarity Theorem. **c.** $\overrightarrow{AE}$ bisects $\angle DAC$ as $\triangle EAD$ and $\triangle EAC$ are congruent. **d.** Given: $\triangle ADB$ obtained by paperfolding as described.

Prove: $m \angle EAC = 30°$

$\overline{AD} \cong \overline{AC}$ and $\overline{DB}$ bisects $\overline{AC}$, so $AB = \frac{1}{2}AD$. By the Pythagorean theorem, $DB = \frac{AB\sqrt{3}}{2}$, so $\triangle ADB$ is a 30-60-90 triangle with $m \angle DAC = 60°$. From the way the paper was folded, $\triangle EAD \cong \triangle EAC$. Thus, by the definition of congruent triangles, $\angle EAD \cong \angle EAC$. But $m \angle EAD + m \angle EAC = m \angle DAC$. So $m \angle EAC = \frac{1}{2}m \angle DAC = \frac{1}{2}(60°) = 30°$.

27. Given: Equilateral $\triangle ABC$;
$\overrightarrow{BD}$ bisects $\angle ABC$.

Prove: $AD = \frac{x}{2}$ and $BD = \frac{\sqrt{3}}{2}x$.

$m \angle ABD = 30°$ (Def. of angle bisector); $\overline{BD} \perp \overline{AC}$ and $AD = DC$ (the bisector of the vertex angle of an isosceles triangle is the perpendicular bisector of the base). Therefore, $\triangle BAD$ is a 30-60-90 triangle. Since $AB = AC$, $AD = \frac{1}{2}x$. By the Pythagorean theorem,

$x^2 = \left(\frac{1}{2}x\right)^2 + (BD)^2$. So $(BD)^2 = x^2 - \frac{1}{4}x^2 = \frac{3}{4}x^2$. Thus,

$BD = \frac{\sqrt{3}}{2}x$.   **31.** AA Similarity Postulate   **33.** $\frac{7}{4}$ **35.** $-\frac{4}{3}$

Page 506   Assess Your Progress
1. 9   **2.** $2\sqrt{10}$ **3.** $7\sqrt{15}$ **4.** $c = 7$; $d = 7\sqrt{5}$
5. $t = \frac{11}{\sqrt{2}}$; $s = \frac{22}{\sqrt{2}}$ **6.** 36   **7.** $n = 30$; $q = 5$; $p = 5\sqrt{3}$
8. $j = 9$; $k = \frac{9}{\sqrt{2}}$ **9.** $a = 2\sqrt{3}$; $b = 2\sqrt{3} - 2$; $c = 4$

Page 510   Checking Key Concepts
1. $\frac{1}{2\sqrt{2}}$; $2\sqrt{2}$ **3. a.** about 31,569 ft   **b.** about 7.4°

Pages 511–513   Exercises and Applications
1. $\frac{3}{4}$; $\frac{4}{3}$ **3.** $\frac{5}{4}$; $\frac{4}{5}$ **5.** 2.4751   **7.** 572.9572   **9.** 15.9°
11. 51.7°   **13.** 7.3   **15.** $s = 5.1$; $r = 40.4°$
19. $m \angle P = 18.9°$; $m \angle Q = 71.1°$ **21.** $m \angle Y = m \angle Z = 63.4°$; $m \angle X = 53.2°$ **25.** about 51.3 ft

27. In $\triangle ABC$, $0 < m \angle A < 45°$. Since $m \angle A + m \angle B = 90°$, $45° < m \angle B < 90°$. This means that $b > a$ because the side opposite the larger angle is longer than the side opposite the smaller angle. Since $\tan A = \frac{a}{b}$ and $\frac{a}{b} < 1$, $\tan A < 1$.
31. Yes.   **33.** No.

Page 517   Checking Key Concepts
1. $\frac{3}{\sqrt{13}} \approx 0.8321$; $\frac{2}{\sqrt{13}} \approx 0.5547$; 56°   **3.** about 234 ft

Pages 517–520   Exercises and Applications
1. $\sin A = \cos B = \frac{5}{13} \approx 0.3846$; $\cos A = \sin B = \frac{12}{13} \approx 0.9231$
3. $\sin A = \frac{5}{7} \approx 0.7143$; $\cos A = \frac{2\sqrt{6}}{7} \approx 0.6999$;
$\sin B = \frac{5}{\sqrt{34}} \approx 0.8575$; $\cos B = \frac{3}{\sqrt{34}} \approx 0.5145$
5. 0.0628   **7.** 0.3387   **9.** 51.9°   **11.** 16.3°   **13.** Answers are given to two decimal places. $r = 5.25$; $s = 7.85$
15. $m \angle A = 36.9°$; $m \angle B = 53.1°$ **17.** $m \angle X = m \angle Z = 50.0°$; $m \angle Y = 80°$ **21. a.** about 24.7 in.   **b.** $x \approx 35.2$ in.; $y \approx 30.8$ in.   **c.** about 39.5 in.   **23.** $\frac{1}{\sqrt{2}}$ **25.** $\frac{1}{2}$ **27.** $\frac{1}{2}$
29. a. $d = 300t$ **b.** $h = d \cos 15° = 300t \cos 15° \approx 289.78t$; $v = d \sin 15° = 300t \sin 15° \approx 77.65t$ **c.** about 2897.8 ft; about 776.5 ft   **31.** B   **33.** $(\sin A)^2 + (\cos A)^2 = \left(\frac{a}{c}\right)^2 + \left(\frac{b}{c}\right)^2 = \frac{a^2 + b^2}{c^2} = 1$ (By the Pythagorean theorem, $a^2 + b^2 = c^2$.)
37. 71.6°   **39.** 55.2°   **41.** $\left(\frac{3}{2}, 2\right)$ **43.** $(2, -2)$

Page 525   Checking Key Concepts
1–3. Answers may vary.
1. $|\overrightarrow{AB}| = 5$

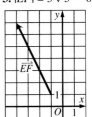

3. $|\overrightarrow{EF}| = 3\sqrt{5} \approx 6.7$

5. Answer is given to the nearest tenth. 56.3   **7.** $(8, -77)$

Pages 525–528   Exercises and Applications
1, 3. Values of the variables are given to the nearest tenth.
1. $(-4, -2)$; 26.6   **3.** $(250, 100)$; 21.8
5. $|\overrightarrow{KL}| = \sqrt{73} \approx 8.5$

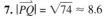

7. $|\overrightarrow{PQ}| = \sqrt{74} \approx 8.6$

9. $(2, 7)$ **11.** $(3, 10)$ **13.** $(7, 11)$ **15. a.** $(3600, -1100)$ **b.** $(3600, -1100)$ **c.** Both represent paths from checkpoint P to S; any path from checkpoint P to S can be represented by a vector with initial checkpoint P and terminal checkpoint S.
17. 91 ft

19. a. $\vec{AB} + \vec{CD} = (9, 0)$

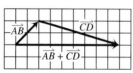

b. $\vec{AB} + \vec{CD} = (9, 0)$

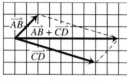

c. In the parallelogram method, the length of the side opposite $\vec{CD}$ equals $|\vec{CD}|$. Since $\vec{AB}$ is the same in both diagrams, the diagonal of the parallelogram has the same length as the third side of the triangle in the second diagram.
23. $a = 60\sqrt{3}$; $b = 60$; $(60\sqrt{3}, 60)$
25. $e = f = 2\sqrt{2}$; $(-2\sqrt{2}, -2\sqrt{2})$ **27. a.** about 219.2 mi/h **b.** about 50.6 mi/h **31.** 33.7° **33.** 83.7° **35.** 45.0°
37. $V = Bh$, where B is the area of a base and h is the height of the prism.

Page 528 Assess Your Progress

1. 85.8° **2.** 11.5° **3.** 41.4° **4.** 49.5 **5.** 8.6 **6.** 9.1
7. $|\vec{PQ}| = \sqrt{58} \approx 7.6$ **8.** $|\vec{RS}| = 5\sqrt{89} \approx 47.2$

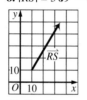

9. $(-8, -10)$ **10.** $(3, -7)$

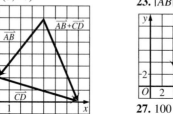

11. $(-2, -5)$

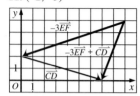

Page 531 Checking Key Concepts

1, 3, 5. Answers are given to the nearest tenth. **1.** 43.2
3. 404.8 **5.** 232.0 in.^{2}

Pages 532–534 Exercises and Applications

Answers are given to the nearest tenth. **1.** 33.5 **3.** 31.2
5. 110.0 **7.** 522 cm^2 **9.** 210 m^2 **11. a.** 6.4; 7.7 **b.** 13.7
c. 68.5

13. a. Answers may vary. An example is given.
b. about 0.025 in.^{2}

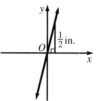

15. 138.9 m^3 **17.** 1211.2 cm^3
23. The area of $\triangle XYZ$ is $A = \frac{1}{2}(XZ)(YZ)$.
Since $\tan X = \frac{YZ}{XZ}$, $YZ = XZ \cdot \tan X$.
By substitution, $A = \frac{1}{2}(XZ)^2 \tan X$.

25. E **29.** $\sqrt{2} \approx 1.4$ **31.** $432\pi \approx 1357.2$ cm^2

Page 538 Checking Key Concepts

1. 5.2; 14.0 **3.** 51.4 cm^3; 98.8 cm^2

Pages 539–541 Exercises and Applications

1. 37.7; 75.4 **3.** 314.2; 282.7 **5.** 593.8; 497.1
7, 9. Answers may vary. Examples are given.
7.   **9.**

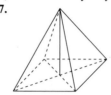

15. $r = 6$; $h = 12$; $s = 6\sqrt{5} \approx 13.4$ **21. a.** 8.0 **b.** 49.0
23. $|\vec{AB}| = 7$ **25.** $|\vec{EF}| = \sqrt{462,500} \approx 680.0$

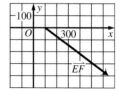

27. 100

Page 541 Assess Your Progress

1. 36.9 **2.** 186.2 **3.** 344.4 **4.** 615.8; 459.6 **5.** 326.7; 320.0 **6.** 5641.5; 2299.0

Pages 546–547 Chapter 10 Assessment

1. Answers may vary. Examples are given.

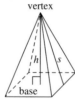

2. $\sin R = \frac{TS}{RS}$; $\cos R = \frac{RT}{RS}$; $\tan R = \frac{TS}{RT}$; $\sin S = \frac{RT}{RS}$; $\cos S = \frac{ST}{RS}$;
$\tan S = \frac{RT}{TS}$ **3.** Find the square root of their product.

4. No; a vector has both magnitude and direction.   **5.** An angle of elevation is formed with the horizontal and the line of sight to a point above the viewing point. An angle of depression is formed with the horizontal and the line of sight to a point below the viewing point.   **6.** $x = 2\sqrt{10}$; $y = 2\sqrt{14}$; $z = 2\sqrt{35}$ **7.** $m = \frac{121}{21}$

8. $f = 4\sqrt{7}$; $g = 12$ **9.** $s = \frac{95}{7}$; $t = \sqrt{95}$ **10.** $a = 4$; $b = 4\sqrt{3}$; $c = 4\sqrt{3}$; $d = 4\sqrt{6}$ **11.** $x = 60$; $y = 22$

12. a. $\sqrt{\frac{5}{12}}$ **b.** $10\sqrt{10}$ **c.** 8   **13.** 128   **14.** $\frac{25}{4}\sqrt{3}$

15. $\frac{49}{8}\sqrt{3}$ **16.** 0.6101   **17.** 0.7923   **18.** 0.7701

19. about 1.4 ft   **20.** about 32.5 ft tall   **21.** $m\angle B \approx 97.2°$, $m\angle A = m\angle C \approx 41.4°$

22. 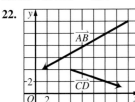 $|\overrightarrow{AB}| = 17$; $|\overrightarrow{CD}| = 3\sqrt{10}$

23. $(-6, -11)$ **25–29.** Answers may vary slightly due to rounding. Answers are given to the nearest tenth.
25. 296.1   **26.** 223.7 cm²   **27.** 2520; 1279.5   **28.** 8796.5; 3078.8   **29.** 1299.0; 779.4

Pages 548–549   Algebra Review/Preview

1. $2y^3$ **3.** $5\pi r^2$ **5.** $\frac{5}{2}s^2$ **7.** cannot be simplified

9. $-x^3 + x^2 - 2x$ **11.** 7×3 **13.** 23 losses   **15. a.** not possible   **b.** $\begin{bmatrix} 5 \\ -9 \\ 8 \end{bmatrix}$ **c.** not possible   **d.** $\begin{bmatrix} 6 & -2 \\ 3 & -4 \\ 0 & 4 \end{bmatrix}$

19. $\pm\sqrt{x^2 - r^2}$ **21.** 2   **23.** ± 6 **25.** 5   **27.** $\frac{3V}{\pi r^2}$

CHAPTER 11

Page 556   Checking Key Concepts

1. $\overline{AD}$ or $\overline{DC}$ **3.** $\angle ADC$ **5.** $\overarc{AC}$ and $\overarc{CDA}$, or $\overarc{AD}$ and $\overarc{ACD}$
7. 60°

Pages 557–559   Exercises and Applications

1. 50°   **3.** 25°   **5.** 310°   **7.** 136°; 104°   **9.** 62°; 127°
13. $x = 15$; $z = 16$; $(4x + 20)° = 80°$; $(10z + 120)° = 280°$
15. $x = 22$; $(x + 100)° = 122°$; $(5x + 10)° = 120°$; $(4x + 30)° = 118°$ **17.** 28.5   **19.** 50   **29.** Answers are given to the nearest tenth. 183.3; 213.7   **31.** $\frac{23}{36} \approx 63.9\%$

Page 562   Checking Key Concepts

1. $x = 40$; $y = 60$ **3.** $x = 55$; $y = 80$

Pages 563–566   Exercises and Applications

1, 3, 5. Answers may vary. Examples are given.   **1.** $\overline{DA}$
3. $\angle BFE$ **5.** 35   **7.** $y = 90$; $z = 20$ **9.** $x = 30$; $z = 160$

11. Statements

1. Secants $\overline{AP}$ and $\overline{CP}$ intersect as shown.
2. Draw chord $\overline{BC}$.
3. $m\angle 3 = \frac{1}{2}m\overarc{AC}$; $m\angle 2 = \frac{1}{2}m\overarc{BD}$
4. $m\angle 3 = m\angle 2 + m\angle 1$
5. $m\angle 1 = m\angle 3 - m\angle 2$
6. $m\angle 1 = \frac{1}{2}m\overarc{AC} - \frac{1}{2}m\overarc{BD}$
7. $m\angle 1 = \frac{1}{2}(m\overarc{AC} - m\overarc{BD})$
8. $m\angle 1 = \frac{m\overarc{AC} - m\overarc{BD}}{2}$

Reasons

1. Given
2. For any two points, there is exactly one line through the two points.
3. The measure of an inscribed angle is half the measure of the intercepted arc.
4. The measure of an exterior angle of a triangle is equal to the sum of the measures of the two interior angles that are not adjacent to it.
5. Subtraction Property
6. Substitution Property (Steps 3 and 5)
7. Distributive Property
8. Algebra

13. 43°

15. Statements

1. Draw chord $\overline{AB}$.
2. $m\angle PAB = \frac{1}{2}m\overarc{AB}$
3. $m\angle ABC = \frac{1}{2}m\overarc{AC}$
4. $m\angle ABC = m\angle PAB + m\angle P$
5. $m\angle P = m\angle ABC - m\angle PAB$
6. $m\angle P = \frac{1}{2}m\overarc{AC} - \frac{1}{2}m\overarc{AB}$
7. $m\angle P = \frac{m\overarc{AC} - m\overarc{AB}}{2}$

Reasons

1. For any two points, there is exactly one line through the two points.
2. The measure of an angle formed by a tangent and a chord is half the measure of the intercepted arc.
3. The measure of an inscribed angle is half the measure of the intercepted arc.
4. The measure of an exterior angle of a triangle is equal to the sum of the measures of the two interior angles that are not adjacent to it.
5. Subtraction Property
6. Substitution Property (Steps 2, 3, and 5)
7. Distributive Property

25. $(a, b) \rightarrow (a - 5, b + 2)$

27. Answers may vary. An example is given.

Page 566   Assess Your Progress

1. 228°   **2.** 43°   **3.** 73°   **4.** Each angle of the triangle is an inscribed angle with measure 60°. Then each of the arcs has measure 120°. (The measure of an inscribed angle is equal to half the measure of the intercepted arc.)   **5.** $3x° = 75°$; $(2x + 5)° = 55°$ **6.** $t° = 40°$ **7.** $(6b − 10)° = 158°$; $(2b + 14)° = 70°$

Page 570   Checking Key Concepts

1. $49°$ **3.** $262°$ **5.** $x = 20$ **7.** $x = y = 15$

Pages 570–572   Exercises and Applications

1. $t = 18$ **3.** $z = 75$; $y = 15$ **5.** $t = 21$; $x = 42$
9. $\overarc{AB} \cong \overarc{CD}$, so $m\,\overarc{AB} = m\,\overarc{CD}$. The measure of a central angle is equal to the measure of its intercepted arc, so $m\angle APB = m\,\overarc{AB}$ and $m\angle DPC = m\,\overarc{CD}$. Then $m\angle APB = m\angle DPC$ or $\angle APB \cong \angle DPC$. $\overline{PA}, \overline{PB}, \overline{PD}$, and $\overline{PC}$ are all congruent since they are radii of the same circle. By the SAS Postulate, $\triangle PAB \cong \triangle PDC$, so corresponding sides $\overline{AB}$ and $\overline{CD}$ are congruent.   **17.** (1) Given; (2) Def. of perpendicular lines; (3) Reflexive Property; (4) For any two points, there is exactly one line through the points; (5) Def. of radius; (6) HL Theorem; (7) Def. of congruent triangles
21. $\angle ABE$; $\angle BAE$ **23.** $\overrightarrow{CD}$ **25.** 0.3907

Page 575   Checking Key Concepts

1. $BC \cdot BE = BA^2$; $EG \cdot GC = FG \cdot GD$; $BA = BD$; $BC \cdot BE = BD^2$ **3.** $2z = 30$

Pages 576–578   Exercises and Applications

1. True.   **3.** False.   **5.** False.   **7.** $CD \cdot CG = CE \cdot CF$, so $\dfrac{CD \cdot CG}{CE \cdot CG} = \dfrac{CE \cdot CF}{CE \cdot CG}$ and $\dfrac{CD}{CE} = \dfrac{CF}{CG}$.   **9.** $CB^2 = CE \cdot CF$, so $\dfrac{CB^2}{CB} = \dfrac{CE \cdot CF}{CB}$, or $CB = \dfrac{CE \cdot CF}{CB}$.   **11.** $x = 6\sqrt{2} \approx 8.5$
13. $y = 4$; $y + 3 = 7$ **15.** $x = 12$ **17.** (1) For any two points, there is exactly one line through the points; (2) $\angle AEB$; (3) $\angle C$; (5) Def. of similar figures; (6) $CA \cdot CD$; a property of proportions   **19.** about 26,100 mi

21. Statements	Reasons
1. Draw $\overline{AB}$ and $\overline{BD}$.	1. For any two points, there is exactly one line through the points.
2. $\angle C \cong \angle C$	2. Reflexive Property
3. $m\angle CBA = \frac{1}{2}m\,\overarc{BA}$	3. The measure of an angle formed by a tangent and a chord is half the measure of the intercepted arc.
4. $m\angle D = \frac{1}{2}m\,\overarc{BA}$	4. The measure of an inscribed angle is equal to half the measure of the intercepted arc.
5. $m\angle CBA = m\angle D$ or $\angle CBA \cong \angle D$.	5. Substitution Property (Steps 3 and 4)
6. $\triangle ABC \sim \triangle BDC$	6. AA Similarity Theorem
7. $\dfrac{AC}{CB} = \dfrac{CB}{CD}$	7. Def. of similar figures
8. $AC \cdot CD = CB^2$	8. A property of proportions

29. $\dfrac{81}{\pi}$ cm$^2 \approx 25.8$ cm^2 **31.** $70°$ **33.** $220°$

Page 581   Checking Key Concepts

1. 74 in.^{2}; 15 in.   **3.** 236 m^2; 63 m   **5.** 64 in.

Pages 582–585   Exercises and Applications

1. 24 in.^{2} **3.** 56 in.^{2} **5.** 19 ft^2 **9.** 8 cm   **11.** 19 in.
13. 23 cm   **15.** 5.6 in.   **23.** about 20.9 ft
31. about 416 ft^2; 500 ft^3

Page 585   Assess Your Progress

1. 15   **2.** 4   **3.** 1.2   **4.** 101 m^2; 59 m

Page 588   Checking Key Concepts

1. $\overline{CT}$ **3.** $\overleftrightarrow{BD}$ **5.** $\overline{VT}, \overline{VC}$, or $\overline{VH}$ **7.** about 113 in.^{3}
9. about 1 in.

Pages 588–591   Exercises and Applications

1. B   **3.** C   **5.** A   **7.** 314 in.^{2}; 524 in.^{3} **9.** 129 cm^2; 137 cm^3 **13, 15.** Answers are given to the nearest tenth.
13. 1.7 ft   **15.** 3.0 ft   **17. a.** about 201,062,000 mi^2
b. about 58,308,000 mi^2 **21.** about 4.25 in.
23. about 449 in.^{3} **29. a.** 8:1   **b.** 4:1   **c.** The thickness is divided by 4.   **33.** about 5 in.^{2} **35.** about 94 ft^2 **37.** $\dfrac{7}{5}$

Page 594   Checking Key Concepts

1. 300,000 ft^2 **3.** 4:9   **5.** 16:81

Pages 594–596   Exercises and Applications

1. a. 3.7:1   **b.** 13.4:1   **c.** 49:1   **3.** about 8 oz
5. about 37 oz   **7. a.** $a:b$; $a^3:b^3$ **b.** smaller solid: $\frac{2}{3}\pi a^3$;
larger solid: $\frac{2}{3}\pi b^3$ **c.** $\dfrac{\frac{2}{3}\pi a^3}{\frac{2}{3}\pi b^3} = \dfrac{a^3}{b^3}$; Yes.   **9. a.** 2:1

b, c. Answers may vary.   **b.** about 1.26:1   **c.** about 1.59:1
d. about 1870 mm^2 **11.** As the size increases, the ratio of surface area to volume decreases.   **17.** the Parallel Postulate (Through a point not on a line, there is exactly one line parallel to the given line.)

Page 599   Checking Key Concepts

1.

3. perpendicular lines m and l

5. No; the figure for Ex. 4 provides a counterexample. In fact, through a point not on a line, there are infinitely many lines perpendicular to the given line.

Pages 600–603   Exercises and Applications

3. No; if you tried to do so, you would end up drawing a line. On a sphere, a ray could not extend in only one direction.

5. The figure shows a quadrilateral that has four congruent angles. Yes; a square or rectangle is a quadrilateral with four congruent angles.

7.

9. The triangle in Ex. 7 is a counterexample. Considering two sides of the triangle as two lines and the third side as the transversal, corresponding right angles are formed, but the lines are not parallel. **11. a.** 120°; 108° **b.** No; the sum of the measures of the angles at each vertex would be 348°, which would leave a gap at each vertex. **13.** about 111.4°; Since the sum of the measures of the angles at each vertex on a soccer ball is 360°, the measure of each interior angle of a pentagon is about 360° − 2(124.3°) = 111.4°. **15.** 100°20′ **17. a.** 4200′ **b.** about 4200 nautical miles **c.** about 4830 mi **25.** 100 : 49 **27.** A′(−2, 1); B′(0, −3); C′(5, 3)

Page 603 Assess Your Progress

1. Answers may vary. An example is given.

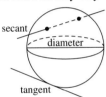

2. about 113 in.^{2}; about 113 in.^{3} **3.** about 105.7 cm^2; about 102.2 cm^3 **4. a.** 25 : 144 **b.** 125 : 1728 **5.** No; if you draw a diagonal, you divide the quadrilateral into two triangles. The sum of the measures of the interior angles of the each triangle is greater than 180°. Then the sum of the measures of the interior angles of the quadrilateral is greater than 360°.

Pages 608–609 Chapter 11 Assessment

1. arcs in congruent circles whose measures are equal

2. two points on a circle and all points of the circle not on the minor arc between the two given points

3. an angle whose vertex is on a circle and whose sides contain chords of the circle

4. a line, ray, or segment that contains a chord

5. a line in the same plane as a circle that intersects the circle in only one point

6. the intersection of a sphere with a plane containing the center of the sphere

7. 59° **8.** 52° **9.** 116°

10.

Statements	Reasons
1. $\overline{AB} \cong \overline{CD}$	1. Given
2. $\angle ADB \cong \angle CBD$	2. In a circle or congruent circles, angles that intercept the same arc or congruent arcs are congruent.
3. $\overline{AD} \parallel \overline{BC}$	3. If two lines are intersected by a transversal and alternate interior angles are congruent, then the lines are parallel.

11. $\sqrt{119} \approx 10.9$ **12.** 24 **13.** $x + 1 = 8$; $x − 1 = 6$

14.

Statements	Reasons
1. $\angle MNP \cong \angle RSP$; $\angle NMP \cong \angle SRP$	1. In a circle or congruent circles, angles that intercept the same arc or congruent arcs are congruent.
2. $\overarc{MN} \cong \overarc{RS}$	2. Given
3. $\overline{MN} \cong \overline{RS}$	3. Congruent arcs have congruent corresponding chords.
4. $\triangle MNP \cong \triangle RSP$	4. ASA Postulate

15–19. Answers are given to the nearest tenth. **15.** 40.8 ft^2; 13.6 ft **16.** 351.9 in.^{2}; 58.6 in. **17.** 47.0 m^2; 13.4 m **18.** 113.1 m^2; 113.1 m^3 **19.** 132.7 cm^2; 143.8 cm^3 **20.** 28 cm **21. a.** 3 : 5 **b.** 9 : 25 **c.** 27 : 125 **22.** In spherical geometry, the sum of the measures of the interior angles of a triangle is greater than 180°. In Euclidean geometry, the sum of the measures of the interior angles of a triangle is equal to 180°.

CHAPTER 12

Page 616 Checking Key Concepts

1. 12; a hexagon

3. $\begin{bmatrix} 4.5 & 3 & −6 & −4.5 \\ 6 & 1.5 & 1.5 & 6 \end{bmatrix}$

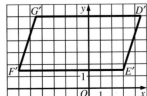

Pages 616–618 Exercises and Applications

1. $\begin{bmatrix} -3 \\ 2 \end{bmatrix}$ **3.** $\begin{bmatrix} -2 & 0 \\ -2 & 4 \end{bmatrix}$ **5.** $\begin{bmatrix} 0 & 3 & -2 \\ 4 & 0 & -2 \end{bmatrix}$ **7.** $\begin{bmatrix} 15 \\ -5 \end{bmatrix}$

9. $\begin{bmatrix} 0 & 9 & -6 \\ 0 & -3 & -12 \end{bmatrix}$ **11.** $\begin{bmatrix} 10 & 5 & 15 \\ 5 & 5 & 10 \end{bmatrix}$

13. (7) dilation of a point with center (0, 0) and scale factor 5; (8) dilation of a line segment, with center (0, 0) and scale factor 4; (9) dilation of a triangle, with center (0, 0) and scale factor 3; (10) dilation of a quadrilateral, with center (0, 0) and scale factor 2.5; (11) no transformation (the same figure before and after); (12) dilation of a line segment, with center (0, 0) and scale factor $\frac{2}{3}$.

15. $4\begin{bmatrix} 2 & 3 & 0 \\ -1 & 4 & \frac{1}{4} \end{bmatrix} = \begin{bmatrix} 8 & 12 & 0 \\ -4 & 16 & 1 \end{bmatrix}$

17. $\frac{1}{2}\begin{bmatrix} -4 & 0 & 4 \\ 0 & 6 & 0 \end{bmatrix} = \begin{bmatrix} -2 & 0 & 2 \\ 0 & 3 & 0 \end{bmatrix}$

19. $2\begin{bmatrix} -2 & 0 & 3 & 0 \\ 0 & 2 & 0 & -2 \end{bmatrix}$

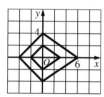

29. a. 2 × 8 **b.** 2 × 8 **33.** $\sqrt[3]{\frac{9}{4}}$ **35.** $x = 12$; $y = 15$; $z = 20$

37. 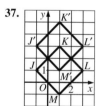 **39.**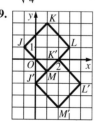

Page 621 Checking Key Concepts

1. $\begin{bmatrix} 4 \\ 4 \end{bmatrix}$ **3.** $\begin{bmatrix} 4 & 7 \\ 3 & 0 \end{bmatrix}$

5. (1) The point (0, 3) is translated to the right 4 units and up 1 unit, to image point (4, 4). (2) The triangle with vertices (0, 3), (3, −4), and (7, −2) is translated to the left 1 unit and up 5 units. The image triangle has vertices (−1, 8), (2, 1), and (6, 3). (3) The line segment connecting (1, 5) and (4, 2) is translated to the right 3 units and down 2 units. The image segment has endpoints (4, 3) and (7, 0). (4) A quadrilateral with vertices (−2, 0), (3, −1), (2, 6), and (−5, 5) is translated up 4 units. The image quadrilateral has vertices (−2, 4), (3, 3), (2, 10), and (−5, 9).

Pages 622–625 Exercises and Applications

1. $\begin{bmatrix} 2 \\ 4 \end{bmatrix}$ **3.** $\begin{bmatrix} 0 & 4 \\ 3 & -1 \end{bmatrix}$

5. (1) The point (5, −1) is translated to the left 3 units and up 5 units. (2) A triangle with vertices (1, −3), (2, 4), and (−1, 6) is the image after a translation to the right 3 units and up 2 units. (3) A line segment with endpoints (−1, 3) and (3, −1) is translated to the right 1 unit. (4) A quadrilateral with vertices (0, −2), (3, 2), (0, 3), and (−5, 1) is translated to the right 2 units and down 4 units.

7. $\begin{bmatrix} -2 & -2 \\ 0 & 0 \end{bmatrix}$ **9.** $\begin{bmatrix} 6 & 6 & 6 & 6 \\ -1 & -1 & -1 & -1 \end{bmatrix}$ **11. a.** $\begin{bmatrix} -4 & -4 & -4 \\ 4 & 4 & 4 \end{bmatrix}$

b. $\begin{bmatrix} -8 & -12 & -4 \\ 1 & 6 & 7 \end{bmatrix}$

c.

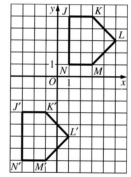

13. a. $\begin{bmatrix} 1 & 3 & 5 & 3 & 1 \\ 5 & 5 & 3 & 1 & 1 \end{bmatrix}$ **b.** $\begin{bmatrix} -4 & -4 & -4 & -4 & -4 \\ -8 & -8 & -8 & -8 & -8 \end{bmatrix}$

c. $\begin{bmatrix} -3 & -1 & 1 & -1 & -3 \\ -3 & -3 & -5 & -7 & -7 \end{bmatrix}$

15. $\begin{bmatrix} 4 & 4 & 4 \\ -3 & -3 & -3 \end{bmatrix} + \begin{bmatrix} -1 & 1 & 0 \\ 4 & 4 & 1 \end{bmatrix} = \begin{bmatrix} 3 & 5 & 4 \\ 1 & 1 & -2 \end{bmatrix}$

17. $\begin{bmatrix} 0 & 0 & 0 & 0 \\ -4 & -4 & -4 & -4 \end{bmatrix} + \begin{bmatrix} -5 & -4 & -1 & -2 \\ -1 & 1 & 1 & -1 \end{bmatrix} =$

$\begin{bmatrix} -5 & -4 & -1 & -2 \\ -5 & -3 & -3 & -5 \end{bmatrix}$ **19.** $\begin{bmatrix} -2 & 0 & 6 & 4 \\ 6 & 3 & 6 & 9 \end{bmatrix}$ **31.** about 2.93

Page 625 Assess Your Progress

1. $\begin{bmatrix} 0 & 20 \\ 25 & 15 \end{bmatrix}$ **2.** $\begin{bmatrix} 1.5 & 4.5 & 6 \\ 0 & 3 & 4.5 \end{bmatrix}$ **3.** $\begin{bmatrix} 1 & 2 & 1.5 & 0.5 \\ 0 & 2 & 3 & -1 \end{bmatrix}$

4. $\begin{bmatrix} 0 & 50 & 10 \\ 0 & -20 & -30 \end{bmatrix}$ **5.** $\begin{bmatrix} -1 & -1 & -1 & -1 & -1 \\ 6 & 6 & 6 & 6 & 6 \end{bmatrix}$ **6.** $\begin{bmatrix} 5 & 5 & 5 & 5 \\ 0 & 0 & 0 & 0 \end{bmatrix}$

Page 629 Checking Key Concepts

1. $\begin{bmatrix} 16 & -24 & 65 \\ 26 & -16 & 5 \end{bmatrix}$

3. The transformation matrix should be first, so the matrices cannot be multiplied in the order given.

Pages 630–633 Exercises and Applications

1. a. $\begin{bmatrix} 11 \\ 1 \end{bmatrix}$

3. a. $\begin{bmatrix} 1 & 3 & -1 & -5 & -4 \\ 7 & -1 & -3 & 3 & 10 \end{bmatrix}$

b.

b.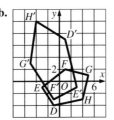

5.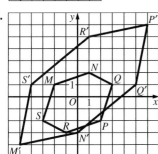

7.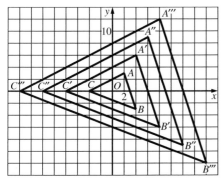

9. a. $\begin{bmatrix} 4 & 8 & -8 \\ 6 & -6 & 0 \end{bmatrix}$; $\begin{bmatrix} 6 & 12 & -12 \\ 9 & -9 & 0 \end{bmatrix}$; $\begin{bmatrix} 8 & 16 & -16 \\ 12 & -12 & 0 \end{bmatrix}$

b. a dilation about center $(0, 0)$ with scale factor n; $\begin{bmatrix} n & 0 \\ 0 & n \end{bmatrix} = n\begin{bmatrix} 1 & 0 \\ 0 & 1 \end{bmatrix}$, so $\begin{bmatrix} n & 0 \\ 0 & n \end{bmatrix}A = nA$.

13. a. $\begin{bmatrix} 181 & 103 & 146 \\ 485 & 443 & 362 \end{bmatrix}$ **b.** $\begin{bmatrix} 290 & 212 & 255 \\ 328 & 286 & 205 \end{bmatrix}$

c. a translation 109 units right and 157 units down **17.** B

21. $\begin{bmatrix} 2 & 6 & 8 \\ 0 & 4 & 6 \end{bmatrix}$ **23. a.** $\begin{bmatrix} -2 & 4 & 4 & -2 \\ -1 & -1 & -3 & -3 \end{bmatrix}$ **b.** $\begin{bmatrix} 2 & -4 & -4 & 2 \\ 1 & 1 & 3 & 3 \end{bmatrix}$

c. $\begin{bmatrix} 1 & 1 & 3 & 3 \\ -2 & 4 & 4 & -2 \end{bmatrix}$

Page 636 Checking Key Concepts

1. $\begin{bmatrix} 1 & 0 \\ 0 & -1 \end{bmatrix}$ **3.** $\begin{bmatrix} 3 & 1 & 1 & 3 \\ 0 & -1 & -3 & -3 \end{bmatrix}$ **5.** $\begin{bmatrix} 0 & 1 & 3 & 3 \\ 3 & 1 & 1 & 3 \end{bmatrix}$

Pages 637–640 Exercises and Applications

1. $\begin{bmatrix} 7 \\ 2 \end{bmatrix}$; the x-axis; **3.** $\begin{bmatrix} -2 \\ 7 \end{bmatrix}$; line $y = x$ **5.** $\begin{bmatrix} -1 & 0 \\ 0 & 1 \end{bmatrix}$

7. $\begin{bmatrix} -2 & -2 & 1 & 0 \\ 0 & -2 & -2 & 0 \end{bmatrix}$ **9.** $\begin{bmatrix} 0 & 2 & 2 & 0 \\ -2 & -2 & 1 & 0 \end{bmatrix}$

11. $A'(24, -10)$; $A'(-24, 10)$ **13.** $C'(-15, 27)$; $C'(15, -27)$

17. $\begin{bmatrix} -1 & 0 & 3 \\ -2 & 3 & 0 \end{bmatrix}$; $\begin{bmatrix} 1 & 0 & -3 \\ -2 & 3 & 0 \end{bmatrix}$ **19.** $\begin{bmatrix} -3 & -1 & -3 & -5 \\ 3 & 3 & -1 & -1 \end{bmatrix}$;

$\begin{bmatrix} -3 & -3 & 1 & 1 \\ 3 & 1 & 3 & 5 \end{bmatrix}$ **21. a.** (a) $\begin{bmatrix} 2 & 2 & 5 & 5 \\ 0 & 2 & 3 & 0 \end{bmatrix}$ (b) $\begin{bmatrix} -2 & -2 & -5 & -5 \\ 0 & 2 & 3 & 0 \end{bmatrix}$

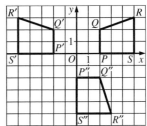

(c) $\begin{bmatrix} 0 & 2 & 3 & 0 \\ -2 & -2 & -5 & -5 \end{bmatrix}$

(d) $P''Q''R''S''$ is the rotation of $PQRS$ about $(0, 0)$ by $270°$.

25. $A'(-237, 572)$ **27.** $C'(365, 374)$

29. $\begin{bmatrix} -85 & -69 & -27 & 27 & 69 & 85 \\ 0 & 50 & 81 & 81 & 50 & 0 \end{bmatrix}$

33. $\begin{bmatrix} -1 & 0 \\ 0 & 1 \end{bmatrix}\begin{bmatrix} 3 & 3 & 0 \\ -3 & 4 & -1 \end{bmatrix} = \begin{bmatrix} -3 & -3 & 0 \\ -3 & 4 & -1 \end{bmatrix}$

37. $\begin{bmatrix} 2 & -6 & -6 & 2 \\ -6 & -6 & -2 & -2 \end{bmatrix}$

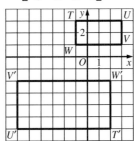

39. $t = 6$ **41.**

Page 644 Checking Key Concepts

1. a. $\begin{bmatrix} -1 & 2 & 2 & -3 \\ 5 & 5 & 3 & 3 \end{bmatrix}$

b.

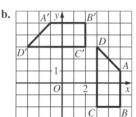

90° rotation

Pages 644–647 Exercises and Applications

1. $\begin{bmatrix} 1 & 0 & -3 \\ 4 & 2 & 5 \end{bmatrix}$

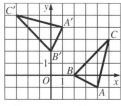

3. $\begin{bmatrix} -1 & -3 & -5 & -2 & 0 \\ 0 & 0 & 1 & 3 & 3 \end{bmatrix}$

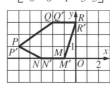

5. $\begin{bmatrix} 0 & 10 & -20 \\ -15 & 0 & 0 \end{bmatrix}$

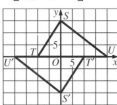

17. 0.9659 **19.** $x = 7\sqrt{2}$ **21.** $x = 20$

Page 647 Assess Your Progress

1. $x = 2; y = 12$ **2.** $g = -2; h = -4$

3. $\begin{bmatrix} 2 & 1 & -2 & 1 \\ 1 & 3 & 1 & -2 \end{bmatrix}$

4. $\begin{bmatrix} 1 & 2 & 4 \\ 1 & -3 & 1 \end{bmatrix}$

5. $\begin{bmatrix} 1 & 2 & 4 \\ -1 & 3 & -1 \end{bmatrix}$

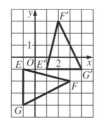

6. $\begin{bmatrix} -1 & -2 & -4 \\ 1 & -3 & 1 \end{bmatrix}$

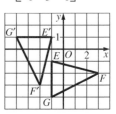

Pages 654–655 Chapter 12 Assessment

1. $\begin{bmatrix} 3 & 2 & 1 \\ 0 & 5 & 1 \end{bmatrix}; 2 \times 3$ **2.** $3 \cdot \begin{bmatrix} 1 & 6 \\ 5 & 2 \end{bmatrix} = \begin{bmatrix} 3 & 18 \\ 15 & 6 \end{bmatrix}$

3. $\begin{bmatrix} 1 & 1 \\ 2 & 3 \end{bmatrix} + \begin{bmatrix} 2 & 1 \\ 0 & 1 \end{bmatrix} = \begin{bmatrix} 3 & 2 \\ 2 & 4 \end{bmatrix}$ **4.** $\begin{bmatrix} 2 \\ 1 \end{bmatrix}$ **5.** $\begin{bmatrix} -3 \\ 0 \end{bmatrix}$

6. $\begin{bmatrix} 2 & 3 \\ 1 & -2 \end{bmatrix}$ **7.** $\begin{bmatrix} -2 & -3 \\ -2 & 0 \end{bmatrix}$ **8.** $\begin{bmatrix} -3 & 2 & 3 \\ 2 & 1 & -2 \end{bmatrix}$

9. $\begin{bmatrix} -3 & 2 & 3 & -2 & -3 \\ 2 & 1 & -2 & -2 & 0 \end{bmatrix}$ **10.** $\begin{bmatrix} -10 & 5 & 25 \\ 0 & 15 & -5 \end{bmatrix}$ **11.** $\begin{bmatrix} 40 & -16 \end{bmatrix}$

12. $\begin{bmatrix} 0 & 0 & 0 \\ 0 & 0 & 0 \end{bmatrix}$ **13.** $\begin{bmatrix} -15 \\ -1 \end{bmatrix}$ **14.** $\begin{bmatrix} -3 & -2 & -1 \\ 1 & -2 & 2 \end{bmatrix}$

15. $\begin{bmatrix} -7 & -4 & -2 & -4 \\ 2 & 4 & 2 & 0 \end{bmatrix}$ **16.** $\begin{bmatrix} 2 & 2 & 2 & 2 \\ 0 & 0 & 0 & 0 \end{bmatrix} + \begin{bmatrix} -1 & 0 & -2 & -3 \\ 3 & 1 & -2 & 2 \end{bmatrix}$

17. $\begin{bmatrix} -3 & -3 & -3 \\ 2 & 2 & 2 \end{bmatrix} + \begin{bmatrix} 3 & 2 & 1 \\ 1 & -3 & 0 \end{bmatrix}$

18. $\begin{bmatrix} -1 & -1 & -1 & -1 & -1 \\ -3 & -3 & -3 & -3 & -3 \end{bmatrix} + \begin{bmatrix} 0 & -1 & -2 & -1 & 2 \\ 0 & 0 & 1 & 3 & 3 \end{bmatrix}$

19. $\begin{bmatrix} 3 & 18 \\ -9 & -38 \end{bmatrix}$ **20.** $\begin{bmatrix} 5 & 0 & -5 & -1 & 3 \\ -20 & 8 & 12 & -20 & -20 \end{bmatrix}$

21. $\begin{bmatrix} -3 & -2 & 2 & 1 \\ -8 & 0 & 7 & -2 \end{bmatrix}$

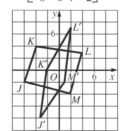

22. a. $\begin{bmatrix} -1 & 4 \\ -2 & 1 \end{bmatrix}$ **b.** $\begin{bmatrix} -2 & 1 \\ -1 & 4 \end{bmatrix}$ **c.** a reflection over $y = x$

23. $\begin{bmatrix} 0 & -3 & 1 & 2 \\ 4 & -1 & -3 & 0 \end{bmatrix}$

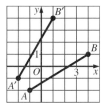

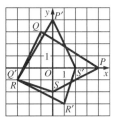

24. $\begin{bmatrix} 0 & 0 & 3 & 5 \\ -1 & -3 & -2 & 1 \end{bmatrix}$

25. $\begin{bmatrix} -1 & -3 & -2 & 1 \\ 0 & 0 & -3 & -5 \end{bmatrix}$

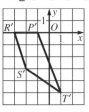

reflection over the
x-axis

reflection over $y = -x$

26. $\begin{bmatrix} 0 & 0 & -3 & -5 \\ -1 & -3 & -2 & 1 \end{bmatrix}$

27. $\begin{bmatrix} 0 & 0 & 3 & 5 \\ 1 & 3 & 2 & -1 \end{bmatrix}$

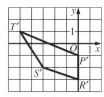

rotation of 180°
about the origin

no transformation

28. $\begin{bmatrix} -1 & -3 & -2 & 1 \\ 0 & 0 & 3 & 5 \end{bmatrix}$

29. $\begin{bmatrix} 1 & 3 & 2 & -1 \\ 0 & 0 & -3 & -5 \end{bmatrix}$

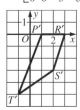

rotation of 90°
about the origin

rotation of 270°
about the origin

CHAPTERS 10–12

Pages 656–657 Cumulative Assessment

1. *JMK*, *KML* **3.** 31°; 59° **5. a.** 41.0°, 139.0°, 41.0°,
139.0° **b.** about 262 cm² **9.** about 610; about 784
11. 13,824; 82,944 **13.** $x = 90$; $y = 90$ **15.** $t = 7\sqrt{2} \approx$
9.90 **17.** $16\pi \approx 50.3$; $20\pi \approx 62.8$; $144\pi \approx 452$

21. $\begin{bmatrix} 4 & -1 & -3 & 0 \\ 0 & 3 & -1 & -2 \end{bmatrix}$

EXTRA PRACTICE

Pages 659–661 Chapter 1

1. $-14, -17$ **3.** 16, 25 **5.** 112, -224 **7.**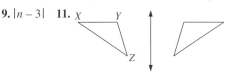

9. $|n - 3|$ **11.**

13. Answers may vary. An example is given.

15. reflection symmetry **17.** translation symmetry
19. If I forget to fill the tank, then my car runs out of gas.
21. hypothesis: $x = 10$; conclusion: $2x - 5 = 15$
23. True. **25.** False; Carmine is in Huma. **27.** F
29. $\overleftrightarrow{DH}$ **31.** *AEH* and *AEF* **33.** 4 **35.** 2 **37.** 5 **39.** 10
41. 6 **43.** 70° **45.** 110° **47.** 90° **49.** True. **51.** True.
53. 6 **55.** 14 **57.** $\left(-\frac{1}{2}, 3\right)$ **59.** $\left(\frac{1}{2}, -1\right)$ **61.** $(-3, -2)$

Pages 661–663 Chapter 2

1. $\angle AFH$ **3.** $\angle JGD$ **5.** 70° **7.** 30° **9.** 60° **11.** $x = 54$
13. $w = 48$, $z = 44$, $2z = 88$ **15.** $b = 36$, $c = 50$ **17.** $x = 60$
19. $a = 40$, $b = 35$ **21.** equilateral quadrilateral
23. equilateral pentagon **25.** regular quadrilateral or square
27. 1260° **29.** $x° = 65°$ **31.** $z° = 135°$; $(z - 15)° = 120°$
33. 144°; 36° **35. a.** 8 cm **b.** 8 cm **c.** 125° **37. a.** 40°
b. 55° **c.** 125° **d.** 85° **39.** 9.8 m **41.** 90° **43.** 40°
45. pentagonal prism **47.** 3 **49.** 10

Pages 663–665 Chapter 3

1. inductive **3.** deductive **5.** 17; Each term is 3 more than
the previous term. **7.** $\frac{1}{27}$; Each term is $\frac{1}{3}$ of the previous
term. **9.** Reflexive Property **11.** Definition of congruent
segments **13.** Definition of congruent segments
15. Segment Addition Postulate; Addition Property
17. False. **19.** 145° **21.** 71° **23.** Two angles that form
a linear pair are supplementary. **25.** If a wildflower is
yellow, then it is goldenrod. False. **27.** If I take my
umbrella, it will rain. False. **29. a.** If two angles are
congruent, then their complements are congruent.

b. If the complements of two angles are congruent, then the angles are congruent. **31.** 4.47; $2\sqrt{5}$ **33.** ±5.66; $\pm4\sqrt{2}$ **35.** -12.25; $-5\sqrt{6}$ **37.** Yes. **39.** No. **41.** No. **43.** $b = 5$ **45.** $a = 16$ **47.** $c = 169$ **49.** obtuse **51.** right **53.** right **55.** Inverse: If Jo does not study hard, then she will not do well on the test. Contrapositive: If Jo does not do well on the test, then she did not study hard. **57.** Inverse: If Bill was not born in France, then he cannot speak French. Contrapositive: If Bill cannot speak French, then he was not born in France.

Pages 665–667   Chapter 4

1. 5 **3.** $\sqrt{85} \approx 9.22$ **5.** $3\sqrt{5} \approx 6.71$ **7.** (4, 4) **9.** (–4, 5) **11.** (5, –1) **13.** $AB = 5$, $BC = 5$, $AC = 8$; isosceles triangle **15.** $AB = 2\sqrt{5} \approx 4.47$, $BC = 5$, $CD = 2\sqrt{5} \approx 4.47$, $DA = 5$; parallelogram **17.** –2 **19.** –2 **21.** 3 **23.** 1 **25.** $y = 2x + 1$ **27.** $y = \frac{1}{4}x - \frac{11}{4}$ **29.** $x = 5$ **31.** $y = \frac{7}{2}x - 20$ **33.** perpendicular **35.** parallel **37.** 0 **39.** undefined **41.** 3 **43.** Yes. **45.** No. **47.** $(x - 2)^2 + (y - 5)^2 = 16$ **49.** $(x + 2)^2 + (y + 5)^2 = 20.25$

51. 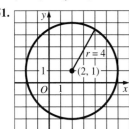 **53.**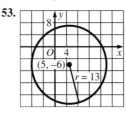

55. $H(c - a, 0)$ **57.** $J(a, 0)$ **59.** $M(\sqrt{3}, 1)$; $MA = 2$, $MB = 2$, $MO = 2$ **61.** (–3, 0, 6) **63.** 5 **65.** $2\sqrt{42} \approx 12.96$

Pages 667–670   Chapter 5

1. alternate interior angles **3.** same-side interior angles **5.** alternate interior angles **7.** same-side interior angles **9.** $x = 60$, $y = 22$, $z = 15$ **11.** $m\angle 1 = 45°$, $m\angle 2 = 45°$ **13.** $m\angle 5 = 105°$, $m\angle 6 = 75°$ **15.** $\overline{LM}$ and $\overline{KN}$ **17.** 105° **19.** (1) corresponding angles; (2) $\angle 2 \cong \angle 3$; (3) Transitive **21.** alternate interior; Substitution **23.** $t = 40$ **25.** parallel; $78° + 102° = 180°$ **27.** not parallel; $x + 10 \neq x + 6$ **29.** $\overline{PM}$ **31.** 63° **33.** 3 **35.** 104°

37. 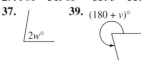 **39.** $(180 + v)°$

Pages 670–672   Chapter 6

1. Yes. **3.** Yes. **5.** No. **7.** Yes. **9.** No. **11.** It is between $1\frac{1}{4}$ in. long and $7\frac{3}{4}$ in. long. **13.** It is between 1 ft long and 11 ft long. **15.** It is between 6 cm long and 22 cm long. **17.** It is between 9 cm long and 31 cm long. **19.** $6 < AB < 14$ **21.** $AB = CB$ and $AB < 9$ **23.** $AB < 12$ **25.** $\triangle FCE$ **27.** $\angle FEC$ **29.** $\overline{FE}$ **31.** $\triangle YXW$

33. $\angle Y$ **35.** $x = 57$ **37.** Yes; SSS Postulate **39.** No. **41.** Yes; SAS Postulate **43.** (1) Given; (2) midpoint; (3) Given; (4) $\overline{AE} \cong \overline{ED}$; Def. of midpoint; (5) Vertical angles are $\cong$; (6) SAS Postulate **45.** none **47.** ASA Postulate **49.** SSS Postulate **51.** $\triangle BAD \cong \triangle ABC$ **53.** $\triangle ACB \cong \triangle BDA$ or $\triangle ACE \cong \triangle BDE$ **55.** $\overline{PT} \cong \overline{QT}$ **57.** $\angle 2 \cong \angle 3$ **59.** 90° **61.** 8

Pages 672–675   Chapter 7

1. a. 5 **b.** 5 **c.** $5\sqrt{3}$ **3. a.** 6 **b.** 12 **c.** 10 **5. a.** 12 **b.** 13 **c.** 20 **d.** 11 **e.** 20 **7.** Both pairs of opposite sides are parallel. (Def. of parallelogram) **a.** 67° **b.** 10 **c.** 20 **9.** Both pairs of opposite sides are congruent. **a.** 90° **b.** 45° **c.** 4 **11.** rectangle **13.** square **15.** rectangle **17.** 20 **19.** 216 **21.** 36 **23.** 242 cm^2 **25.** $16\sqrt{3} \approx$ 27.7 in.^{2} **27.** 16 **29.** $9\pi \approx 28.27$ **31.** $16.81\pi \approx 52.81$ **33.** $36\pi - 72 \approx 41.10$ **35.** $V = 170$; S.A. = 193 **37.** $V = 250\sqrt{3} \approx 433.01$; S.A. = $300 + 50\sqrt{3} \approx 386.60$ **39.** 96 m^3

Pages 675–677   Chapter 8

1.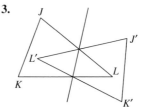

3.

5.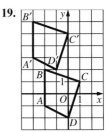

7. $A'(-2, 3)$, $B'(-3, -1)$, $C'(-1, 0)$ **9.** $G'(1, 1)$, $H'(-1, 3)$, $J'(-3, 1)$, $K'(-2, -1)$ **11.** $D'(-1, 0)$, $E'(-2, -2)$, $F'(1, -3)$, $G'(2, 0)$ **13.** $P'(2, 0)$, $Q'(3, 2)$, $R'(0, 1)$ **15.** $W'(-1, -1)$, $X'(2, -1)$, $Y'(3, 1)$, $Z'(2, 2)$

17. 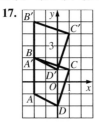 **19.**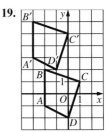

21. $(a, b) \rightarrow (a + 2, b + 1)$

23.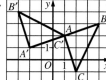

25.

27.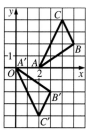

29.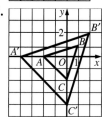

31.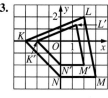

33.

59. 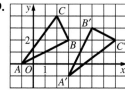 $|\overrightarrow{HJ}| = \sqrt{97} \approx 9.8$

61. 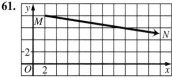 $|\overrightarrow{MN}| = \sqrt{409} \approx 20.2$

63. $(-1, 3)$ **65.** $(-4, 10)$ **67.** about 105   **69.** about 346

71. 1728   **73.** $V = \dfrac{500\sqrt{2}}{3} \approx 235.7$;
$S.A. = 100\sqrt{3} + 100 \approx 273.2$

Pages 683–685   Chapter 11

1. $270°$ **3.** $120°$ **5.** $310°$ **7.** $305°$ **9.** $194°$ **11.** $n = 35$
13. $x = 25$ **15.** $z = 70$ **17.** $x = 20$ **19.** $x = 105$
21. $x = 3, y = 4$ **23.** $v = 14, z = \sqrt{85}$ **25.** $w = 15$
27. $y \approx 2.34$ **29.** $m = 4$ **31.** $w = 7$ **33.** $s = 3.6$
35. $\frac{2}{3}\pi$ ft^2 ≈ 2.1 ft^2 **37.** 18π cm$^2 \approx 56.5$ cm^2
39. $54\frac{4}{9}$ cm$^2 \approx 54.4$ cm^2 **41.** $3\frac{1}{9}\pi$ in. ≈ 9.8 in.
43. $S.A. = 576\pi$ in.$^2 \approx 1809.6$ in.^{2}; $V = 2304\pi$ in.$^3 \approx 7238.2$ in.^{3} **45.** $S.A. = 256\pi$ m$^2 \approx 804.2$ m^2;
$V = 682\frac{2}{3}\pi$ m$^3 \approx 2144.7$ m^3 **47.** $\frac{128\pi}{3}$ ft^3 **49.** False.

Pages 685–686   Chapter 12

1. $\begin{bmatrix} -2 \\ 1 \end{bmatrix}$ **3.** $\begin{bmatrix} -2 & 1 \\ 1 & 0 \end{bmatrix}$ **5.** $\begin{bmatrix} 8 \\ -6 \end{bmatrix}$ **7.** $\begin{bmatrix} 4 & 3 & -3 \\ -2 & 1 & 5 \end{bmatrix}$

9. $\begin{bmatrix} 20 & -40 \\ 15 & 30 \end{bmatrix}$ **11.** $\begin{bmatrix} -3 \\ 9 \end{bmatrix}$ **13.** $\begin{bmatrix} 2 & 4 & 5 \\ -4 & 0 & -1 \end{bmatrix}$ **15.** $\begin{bmatrix} -3 & -3 \\ 0 & 0 \end{bmatrix}$

17. $\begin{bmatrix} 0 & 0 & 0 & 0 \\ -4 & -4 & -4 & -4 \end{bmatrix}$ **19.** $\begin{bmatrix} -4 & -1 \\ 10 & 3 \end{bmatrix}$

21. $\begin{bmatrix} 0 & -0.5 & 1.5 \\ 2.4 & 3.8 & 12.6 \end{bmatrix}$

23.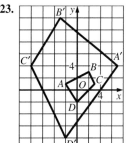
25. $(-1, 2)$; line $y = x$
27. $(1, -2)$; line $y = -x$

Pages 677–679   Chapter 9

1. $x = 12$ **3.** $z = 3\sqrt{5}$ **5.** $x = 9$ **7.** Yes; corresponding sides are proportional.   **9.** No; corresponding sides are not proportional.   **11.** $x = 8.25, y = 2.75$ **13.** $x = 3\frac{2}{3}$ **15.** No; only one pair of corresponding angles are congruent.
17. No; only two pairs of corresponding sides are in proportion.   **19.** SAS Similarity Theorem   **21.** SAS Similarity Theorem or SSS Similarity Theorem   **23.** AA Similarity Postulate   **25.** $y = 6$ **27.** $w = 12.5$ **29.** $y = 9.6$
31. $\frac{y}{5}$ **33.** $2\frac{2}{9}$ **35.** $12\pi \approx 37.70$ **37.** 22.5 **39.** $\frac{1}{2} = 0.5$
41. $\frac{16}{49} \approx 0.33$ **43.** $\frac{5}{14} \approx 0.36$ **45.** $\frac{3}{15} \approx 0.2$

Pages 680–682   Chapter 10

1. SQR **3.** SQ **5.** 9   **7.** $1\frac{1}{2}$ **9.** 1   **11.** $x = 2\sqrt{2} \approx 2.8$
13. $u = 8\frac{1}{3}$ **15.** $x = 8$ **17.** $x = 10$ **19.** $z = 5\sqrt{2} \approx 7.1$
21. $x = 6\sqrt{2} \approx 8.5$ **23.** $p = 10.5$ **25.** $x = 45$,
$y = 6\sqrt{2} \approx 8.5$ **27.** $\tan A = \frac{4}{3}, \tan B = \frac{3}{4}$ **29.** 0.5317
31. 11.4301 **33.** $y \approx 5.5$
35. $\sin A = \frac{2\sqrt{5}}{5}, \cos A = \frac{\sqrt{5}}{5}, \sin B = \frac{\sqrt{5}}{5}, \cos B = \frac{2\sqrt{5}}{5}$
37. $\sin A = \frac{3}{5}, \cos A = \frac{4}{5}, \sin B = \frac{4}{5}, \cos B = \frac{3}{5}$
39. 0.1478 **41.** 0.9781 **43.** $75.2°$ **45.** $11.6°$ **47.** $50.3°$
49. $56.3°$ **51.** $m \approx 3.53, n \approx 7.52$ **53.** $x \approx 53.1$
55. $z \approx 41.4$ **57.** $(2, 3); y \approx 56.3$

TOOLBOX

Page 687 Using a Protractor

1. $45°$ **2.** $105°$ **3.** $90°$ **4.** $25°$ **5.** $70°$ **6.** $135°$ **7.** $60°$
8. $115°$

Page 688 Using a Compass

1–8. Answers may vary.

Page 689 Transformations

1. 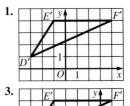 **2.**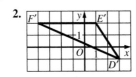

3. 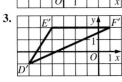 **4.**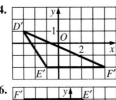

5.   **6.**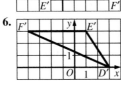

Page 690 Recognizing Symmetry

1. no symmetry **2.** reflection and rotational symmetry
3. reflection symmetry **4.** rotational symmetry
5–7. Answers may vary. Examples are given.
5. **6.** **7.**

Page 692 Simplifying and Evaluating Expressions

1. -8 **2.** 49 **3.** -168 **4.** 14 **5.** -9 **6.** -18 **7.** -1
8. 53 **9.** 3 **10.** 2 **11.** $-\frac{5}{9}$ **12.** 13 **13.** $3x-1$
14. $4x^2-4x^3$ **15.** $4x^3-10x$ **16.** $11x+21$
17. $t^2+9t+36$ **18.** $9x^2+15x$ **19.** x **20.** $-23m$
21. $5x^3+7x^2+7x+7$ **22.** $x+1$ **23.** $-7x+12$
24. $7x^2-17x$ **25.** -23 **26.** 239 **27.** -56 **28.** 32 **29.** 10
30. 25 **31.** -45 **32.** 45 **33.** 2 **34.** 8 **35.** -13 **36.** 45
37. 23 **38.** 12 **39.** 5 **40.** 23

Page 693 Translating Phrases into Variable Expressions

1. A child y years old must pick up $3y$ toys. **2.** The cost of
m main dishes, s side dishes, and d drinks is
$4.25m+0.75s+d$. **3.** If there is d amount of orange
juice, each glass has $\frac{d}{8}$ amount. **4.** Monica will charge $20h$
for working h hours. **5.** The hurricane will move $45h$ miles
in h hours.

Page 693 (continued)

6. m mosaics will require $44m$ yellow tiles, $125m$ white tiles,
$90m$ blue tiles, $78m$ red tiles, and $13m$ green tiles.

Page 693 Algebraic Properties

1. a. commutative **b.** distributive **2. a.** associative
b. commutative **c.** associative **3. a.** distributive
b. commutative

Page 694 Evaluating Equations for Given Values

1. $y=-4$ **2.** $y=1$ **3.** $y=0$ **4.** $y=12$ **5.** $y=11$
6. $y=14$ **7.** $y=-\frac{1}{4}$ **8.** $y=24$

Page 695 Translating Sentences into Equations

1. The cost c of soda in dollars equals 0.80 times the number of bottles b. $c=0.80b$ **2.** The total number of pieces p
is 17 times the number of weeks w. $p=17w$ **3.** Elaine's
speed s in feet per second is the distance run d in feet
divided by 15 s. $s=\frac{d}{15}$ **4.** The cost of using the program c
in dollars is 1 plus 0.05 times the number of minutes the
program is used, m. $c=1+0.05m$ **5.** The total weight w
of a container of food from the deli is the weight of the food
f plus 0.01 lb. $w=f+0.01$ **6.** The number n of oxygen
atoms required is half the number of hydrogen atoms, h.
$n=\frac{h}{2}$

Page 696 Solving One-Step Equations

1. $y=5$ **2.** $x=13$ **3.** $y=9$ **4.** $t=-4$ **5.** $x=\frac{8}{3}$
6. $z=45$ **7.** $t=32$ **8.** $x=-1$ **9.** $y=5$ **10.** $w=\frac{9}{8}$
11. $y=4$ **12.** $x=-3$

Page 696 Solving Two-Step Equations

1. $x=\frac{1}{2}$ **2.** $x=1$ **3.** $t=4$ **4.** $x=-\frac{7}{6}$ **5.** $y=2$ **6.** $x=9$
7. $x=9$ **8.** $y=-4$ **9.** $x=-\frac{77}{12}$ **10.** $x=\frac{23}{3}$
11. $t=\frac{131}{24}$ **12.** $y=0.44$

Page 697 Graphing Points

1–8.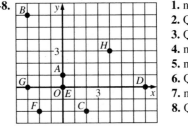

1. no quadrant
2. Quadrant II
3. Quadrant IV
4. no quadrant
5. no quadrant
6. Quadrant III
7. no quadrant
8. Quadrant I

9, 10. Answers may vary. Examples are given. **9.** $(2, -1)$
10. $(-2, 0)$

Pages 699–700 Graphing Linear Equations

1. slope 3; y-intercept 0 **2.** slope -2; y-intercept 0
3. slope -1; y-intercept 2 **4.** slope 0; y-intercept -4
5. slope -5; y-intercept -1 **6.** slope 8; y-intercept -3
7. slope 1; y-intercept 2 **8.** slope 13.8; y-intercept 122

9. slope 3; y-intercept -3

10. slope $-\frac{5}{4}$; y-intercept $-\frac{3}{2}$

11. slope $\frac{1}{2}$; y-intercept $\frac{1}{2}$

12. slope $-\frac{2}{3}$; y-intercept 3

13.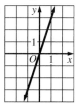

14.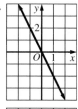

15.

16.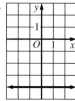

17.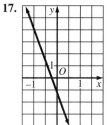

18.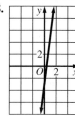

19.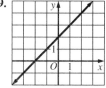

20.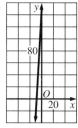

21.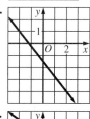

22.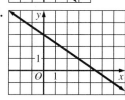

23.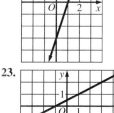

24.

25. slope 0; y-intercept 2

26. slope 1; y-intercept 12

27. slope $-\frac{1}{2}$; y-intercept -2

28. slope 2; y-intercept -2

29. slope $\frac{1}{3}$; y-intercept -5

30. slope -5; y-intercept -12

31. slope $\frac{4}{3}$; y-intercept -8

32. slope $\frac{1}{3}$; y-intercept 3

33. slope $-\frac{1}{3}$; y-intercept 6

Page 701 Simplifying Radicals

1. 4 **2.** $5\sqrt{3}$ **3.** $\sqrt{2}$ **4.** 6 **5.** $\frac{2\sqrt{2}}{3}$ **6.** $\frac{1}{8}$ **7.** $\frac{\sqrt{15}}{2}$

8. $\frac{2\sqrt{3}}{5}$ **9.** $5\sqrt{2}$ **10.** $5\sqrt{5}$ **11.** $-5\sqrt{5}$ **12.** -6 **13.** 49

14. 9 **15.** $\frac{1}{2}$ **16.** $\frac{\sqrt{6}}{2}$

Page 702 Solving Simple Quadratic Equations

1. $x = 5$ or $x = -5$ **2.** $x = 5\sqrt{2}$ or $x = -5\sqrt{2}$ **3.** $x = \sqrt{6}$ or $x = -\sqrt{6}$ **4.** $x = 6$ or $x = -6$ **5.** $x = 1$ or $x = -1$ **6.** $x = 4$ or $x = -4$ **7.** $x = \sqrt{10}$ or $x = -\sqrt{10}$ **8.** $x = 0$ **9.** $x = 2$ or $x = -2$ **10.** $x = 3$ or $x = -3$ **11.** $x = \sqrt{29}$ or $x = -\sqrt{29}$ **12.** $x = 3$ or $x = -3$

Page 703 Working with Formulas

1. $t = \frac{1}{3}$ h **2.** $d = 25$ mi **3.** $d = 35$ mi **4.** $s = 66\frac{2}{3}$ mi/h

5. $A = 0.16\pi$ **6.** $r = 3$ **7. a.** $c = \frac{ad}{4b}$ **b.** $a = 16$ in.

Page 703 Finding Perimeter

1. 19 ft **2.** 26 **3.** 36 **4.** 16 in. **5.** 14 ft **6.** 21 units

Page 704 Finding Circumference

1. 16π in. **2.** 4π cm **3.** 2π in. **4.** 24π in. **5.** 4π

6. 28π m **7.** 60π **8.** 9π **9.** 9 in. **10.** 17 **11.** $\frac{1}{2}$

12. $\frac{1}{\pi}$ mi

Page 705 Finding Area and Volume

1. 12 **2.** 2 **3.** 168 **4.** 3.5 **5.** 2.25 **6.** 60 **7.** 4 **8.** 20

9. 10 **10.** 27 **11.** 1400 **12.** 6 **13.** 96 **14.** 0.125

15. $l = 2$ **16.** $w = 3.75$ **17.** $w = \frac{1}{3}$ **18.** $A = 50$

Page 705 Solving Inequalities

1. $y < 16$

2. $n > \frac{9}{7}$

3. $n \le 2$

4. $n > 0$

5. $x \le 1$

6. $x < \frac{15}{14}$

7. $y < -8$

8. $n \le -1$

Pages 706–707 Graphing Systems of Equations and Inequalities

1.
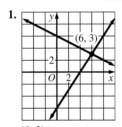
(6, 3)

2.
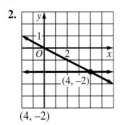
(4, −2)

3.
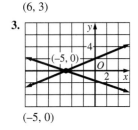
(−5, 0)

4.
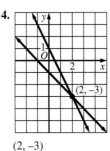
(2, −3)

5.
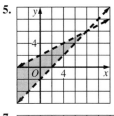

6.
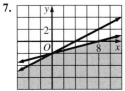

7.
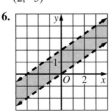

8.
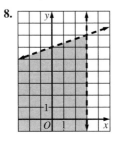

Page 707 Creating a Ratio

1. $\frac{4}{1}$ **2.** $\frac{9}{20}$ **3.** $\frac{1}{5}$ **4.** $\frac{3}{7}$ **5.** $\frac{1}{3}$ **6.** $\frac{1}{8}$ **7.** $\frac{12}{5}$ **8.** $\frac{1}{100}$ **9.** $\frac{1}{6}$
10. about 3.3 miles per hour **11.** $40 per hour **12.** $7500 per year **13.** about 6.4 hours per day **14.** 3.25 cats per hour **15.** 24 miles per day **16.** 120 cycles per second
17. 0.25 inches per minute **18.** 30 apples per basket

Page 708 Solving Proportions

1. $x = 2$ **2.** $x = 7$ **3.** $n = \frac{1}{3}$ **4.** $y = 26$ **5.** $x = 27$

6. $x = 20$ **7.** $m = \frac{4}{5}$ **8.** $x = 35$

Page 709 Finding Averages

1. mean: 7.25; median: 7 **2.** mean: 19.4; median: 20
3. mean: 5.125; median: 6 **4.** mean: 11.75; median: 11.5
5. mean: 14.16; median: 16 **6.** mean: 1.857142; median: 2
7. mean: −1; median: −3 **8.** mean: 19; median: 18

Page 709 Scientific Notation

1. 0.000018 **2.** 0.0000000000455 **3.** 5,200,000,000
4. 381.2 **5.** 98.4 **6.** 0.000000057 **7.** 0.39 **8.** 42,000
9. 8.8×10^{-4} **10.** 1.2×10^{-1} **11.** 1.245×10^{6}
12. 1.31×10^{-2} **13.** 6×10^{-11} **14.** 4.5×10^{4}
15. 1.58×10^{8} **16.** 2.3×10^{5} **17.** 6.2×10^{3}
18. 5×10^{-3} **19.** 1.1×10^{-12} **20.** 3.48651×10^{6}

Page 710 Probability

1. $\frac{3}{4}$, or 0.75 **2.** $\frac{10}{29} \approx 0.34$ **3.** $\frac{2}{5}$, or 0.4 **4.** $\frac{4}{7} \approx 0.57$

5. $\frac{3}{5}$, or 0.6 **6.** $\frac{1}{2}$, or 0.5 **7.** $\frac{46}{833} \approx 0.06$ **8.** $\frac{1}{2}$, or 0.5

9. 0 **10.** $\frac{3}{20}$, or 0.15

Page 711 Matrices

1. $\begin{bmatrix} 6 & 5 \\ 2 & 13 \end{bmatrix}$ **2.** $\begin{bmatrix} 7 \\ 10 \end{bmatrix}$ **3.** $\begin{bmatrix} 2 & -4 \\ 2 & 1 \end{bmatrix}$ **4.** $\begin{bmatrix} 1 \\ -5 \\ -5 \end{bmatrix}$ **5.** $\begin{bmatrix} 10 & 7 \\ 8 & 16 \end{bmatrix}$

6. $\begin{bmatrix} 1 & 11 & -6 \\ 2 & -2 & -4 \end{bmatrix}$ **7.** $\begin{bmatrix} 2 & 0 \\ 0 & 2 \end{bmatrix}$ **8.** $\begin{bmatrix} -4 & \frac{3}{2} \\ 3 & -2 \end{bmatrix}$ **9.** $\begin{bmatrix} 72 \\ -40 \end{bmatrix}$

10. $\begin{bmatrix} 3 & 12 & -6 \\ -12 & -3 & 3 \end{bmatrix}$ **11.** $\begin{bmatrix} -3 & 0 \\ 0 & -3 \end{bmatrix}$ **12.** $\begin{bmatrix} -6 & 0 & 10 \\ 16 & -4 & -2 \\ 18 & 14 & 0 \end{bmatrix}$

13. $\begin{bmatrix} -\frac{25}{3} & -5 \\ -5 & 5 \\ \frac{5}{3} & -10 \end{bmatrix}$ **14.** $\begin{bmatrix} \frac{9}{2} \\ 3 \\ -12 \end{bmatrix}$

TECHNOLOGY HANDBOOK

Pages 722–725 Calculator Practice

1. −12 **2.** 31.85 **3.** 3.4 **4.** about 10.3 **5.** 4096
6. about 1436.8 **7.** 1.804 **8.** 51.5° **9.** The shape of the scatter plot is similar to half of a parabola.

10. $\begin{bmatrix} -3 & 0 & 5 \\ 4 & 8 & 3 \end{bmatrix}$ **11.** $\begin{bmatrix} 5 & 20 & 45 \\ 15 & 35 & 10 \end{bmatrix}$ **12.** $\begin{bmatrix} -3 & -7 & -2 \\ -1 & -4 & -9 \end{bmatrix}$

13. $\begin{bmatrix} -3 & -7 & -2 \\ 1 & 4 & 9 \end{bmatrix}$

Page 725 Spreadsheet Practice

14, 15.

Type of ball	r(cm)	V(cm^3)		
Golf ball	2.1	38.79238609		
Tennis ball	3.3	150.5325536		
Softball	4.8	463.2466863		
Volleyball	10.5	4849.048261		
Basketball	11.9	7058.777513		
		Pythagorean	Triples	
n		n^2−1	2n	n^2+1
2		3	4	5
3		8	6	10
4		15	8	17
5		24	10	26
6		35	12	37
7		48	14	50
8		63	16	65
9		80	18	82
10		99	20	101

Pages 728–729 Software Practice

16.

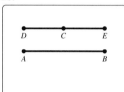

17.

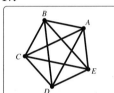

18.

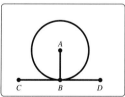

19.

$m\angle A = 115°$
$m\,\overline{AB} = 8$ cm

20. a, b.

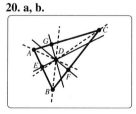

c. The point at which the three angle bisectors of a triangle intersect is equidistant from the sides of the triangle.

21.

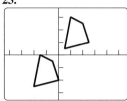

22.

23.

24.

25.

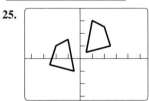

APPENDIX 2

Pages 740–741 Exercises and Applications

1. I win and you win. **3.** I do not win and you do not win. **5.** If you win or they win, then I do not win. **7.** I win or both you and they win. **9.** $q \rightarrow r$ **11.** $\sim p \rightarrow \sim(s \lor r)$

13.

p	$\sim p$	$p \land \sim p$
T	F	F
F	T	F

contradiction

15.

p	$\sim p$	$\sim(\sim p)$
T	F	T
F	T	F

17.

p	q	$\sim p$	$\sim p \lor q$
T	T	F	T
T	F	F	F
F	T	T	T
F	F	T	T

19.

p	q	$\sim p$	$\sim p \land q$
T	T	F	F
T	F	F	F
F	T	T	T
F	F	T	F

21. (1) "It is not the case that p or q is true" has the same meaning as "It is not the case that p is true, and it is not the case that q is true." (2) "It is not the case that both p and q are true" has the same meaning as "It is not the case that p is true or it is not the case that q is true."

23.

p	q	$p \rightarrow q$	$q \rightarrow p$	$(p \rightarrow q) \land (q \rightarrow p)$
T	T	T	T	T
T	F	F	T	F
F	T	T	F	F
F	F	T	T	T

25. a contradiction, since there are all F's in the final column of the truth table:

p	q	$p \land q$	$\sim p$	$\sim q$	$\sim p \lor \sim q$	$(p \land q) \land (\sim p \lor \sim q)$
T	T	T	F	F	F	F
T	F	F	F	T	T	F
F	T	F	T	F	T	F
F	F	F	T	T	T	F

Selected Answers **SA39**

27. a tautology, since there are all T's in the final column of the truth table:

p	q	$\sim p$	$p \vee q$	$(p \vee q) \wedge \sim p$	$[(p \vee q) \wedge \sim p] \to q$
T	T	F	T	F	T
T	F	F	T	F	T
F	T	T	T	T	T
F	F	T	F	F	T

29.

p	q	$p \to q$	$\sim q$	$(p \to q) \wedge \sim q$	$\sim p$	$[(p \to q) \wedge \sim q] \to \sim p$
T	T	T	F	F	F	T
T	F	F	T	F	F	T
F	T	T	F	F	T	T
F	F	T	T	T	T	T

31.

p	q	r	s	$p \to q$	$q \to r$	$r \to s$	$(p \to q) \wedge (q \to r) \wedge (r \to s)$
T	T	T	T	T	T	T	T
T	T	T	F	T	T	F	F
T	T	F	T	T	F	T	F
T	T	F	F	T	F	T	F
T	F	T	T	F	T	T	F
T	F	T	F	F	T	F	F
T	F	F	T	F	T	T	F
T	F	F	F	F	T	T	F
F	T	T	T	T	T	T	T
F	T	T	F	T	T	F	F
F	T	F	T	T	F	T	F
F	T	F	F	T	F	T	F
F	F	T	T	T	T	T	T
F	F	T	F	T	T	F	F
F	F	F	T	T	T	T	T
F	F	F	F	T	T	T	T

$p \to s$	$[(p \to q) \wedge (q \to r) \wedge (r \to s)] \to (p \to s)$
T	T
F	T
T	T
F	T
T	T
F	T
T	T
F	T
T	T
T	T
T	T
T	T
T	T
T	T
T	T
T	T

16 rows